Amide Bond Activation

Amide Bond Activation

Concepts and Reactions

Edited by Michal Szostak

WILEY-VCH

Editor

Prof. Michal Szostak
Rutgers University
Department of Chemistry
73 Warren Street
Newark, NJ 07102
USA

Library of Congress Card No.: applied for

British Library Cataloguing-in-Publication Data
A catalogue record for this book is available from the British Library.

Bibliographic information published by the Deutsche Nationalbibliothek
The Deutsche Nationalbibliothek lists this publication in the Deutsche Nationalbibliografie; detailed bibliographic data are available on the Internet at <http://dnb.d-nb.de>.

Print ISBN: 978-3-527-34831-2
ePDF ISBN: 978-3-527-83024-4
ePub ISBN: 978-3-527-83026-8
oBook ISBN: 978-3-527-83025-1

Typesetting Straive, Chennai, India
Printing and Binding CPI Group (UK) Ltd, Croydon, CR0 4YY

C108960_280922

Contents

Preface

Every student of chemical science knows that the amide bond represents a privileged motif in chemistry, biochemistry, and structural biology underpinning the structure of proteins. The amide bond resonance is the key to the vibrant utility of amides as the most common moieties in medicinal chemistry, widely used connectors in synthetic polymers, and, most crucially, as pivotal linkages of life in peptides.

Since the seminal studies by Pauling in 1930s outlining approximately 40% N–C double bond character of the amide linkage, the stability of amides has been considered vital and fascinating in organic chemistry. This delocalization of the nitrogen lone pair, $n_N \rightarrow \pi^*_{C=O}$, renders the amide bond geometrically planar, while the N–C(O) moiety is inherently extremely robust to synthetic transformations. However, defying this conventional knowledge on stability of amides, the last decades have witnessed tremendous advances in the development of new paradigms in the chemistry or amides.

The early work by Lukeš, Pracejus, and Brown and then the seminal studies by Stoltz, Greenberg, Kirby, and Aubé on hyper-reactive bridged lactams opened the door to the most recent avenues in chemoselective activation of the N–C(O) amide bond by metal insertion. The structural characterization of a number of these non-planar lactams together with well-defined electronic modification of the altered amidic resonance enabled the design of acyclic amides as resonance tunable, highly reactive acyl electrophiles in organic synthesis. These reactions are closely tied to the emergence of new concepts for transformations of amides by decarbonylative, radical, and nucleophilic reaction pathways. Simultaneously, the significance of new methods for transformations of amides by electrophilic and Lewis acid mediated activation manifolds cannot be overstated. This X–C(O) bond activation paradigm goes far beyond amides and is related to the recently discovered transformations of carboxylic acid derivatives, as exemplified by chemoselective O–C(O) activation of esters. Altogether, these thriving discoveries have laid the framework for re-defining the classical amide bond as a collection of different amide functional groups characterized by a gradual continuum of electrophilic, nucleophilic, and metal-catalyzed reactivity. From the synthetic standpoint, these powerful methods provide novel means for functional group interconversion in organic synthesis and grant invaluable insight into the fundamental properties of the classical amide bond.

The book is organized into 15 chapters with Chapters 1–3 providing an overview of the structural properties of different classes of non-planar and electronically activated amides. Chapters 4 and 5 cover reactions of amides by tetrahedral intermediates, including recent advances by electrophilic functionalization, with a focus on carbon nucleophiles. Chapters 6 and 7 detail the reactions of amides with heteroatom nucleophiles, amines, and alcohols, enabled by ligand design, amide tuning, and catalyst identification. Chapters 8–10 describe powerful cross-coupling methods initiated by N–C(O) oxidative addition to transition metals with a focus on nickel and palladium catalysis as well as decarbonylative pathways. Chapters 11 and 12 address prominent reactivity of amides by radical mechanisms, including notable advances in photoredox catalysis, and C–H functionalization enabled by amide bond chelation. A critical aspect of the underlying mechanistic processes of amide bond activation is addressed in Chapter 13. This chapter ties the mechanisms of amide bond activation with ester bond activation as a common theme of acyl bond activation of carboxylic acid derivatives. The book is concluded with Chapters 14 and 15 that are focused on O–C(O) activation of esters by acyl and decarbonylative pathways, a manifold that finds many similarities with amide bond activation.

I am indebted to all the authors and coauthors for their participation in the project. Each chapter has been carefully assembled by a cast of international leaders in this field who have generously contributed their perspective and excitement on the amide bond chemistry. We hope that the work described in this book will encourage researchers to investigate amide bonds and spark an even greater interest in this essential field.

Newark
September 2022

Michal Szostak

1

Bridged Lactams as Model Systems for Amidic Distortion

Tyler J. Fulton, Yun E. Du, and Brian M. Stoltz

California Institute of Technology, Division of Chemistry and Chemical Engineering, Pasadena, California, 91125, United States

1.1 Introduction and Scope

Amide linkages are pervasive functional groups present in peptides, natural products, and bioactive materials [1]. The high degree of resonance contribution of the amide nitrogen n_N to $\pi^*_{C=O}$ (15–20 kcal/mol stabilization from resonance form **2**, Scheme 1.1) imparts approximately 40% double bond character and renders typical amides planar and less reactive toward nucleophilic attack compared with other carbonyls and carboxylic acid derivatives (Scheme 1.1) [2]. Disruption of the n_N to $\pi^*_{C=O}$ resonance via distortion of the amide bond is a mode of amide activation with a rich history dating back to the 1938 when Lukeš first proposed that incorporation of an amide nitrogen into the bridgehead position of a bicyclic system would be "sterically impossible" and violate Bredt's rule [3]. Lukeš further hypothesized that if amides bearing a bridgehead nitrogen were successfully synthesized, they would bear properties more akin to those of ketones. Bridged lactams have since served as a remarkable tool for the understanding of the amide bond and its properties. This chapter provides a review of bridged lactams up to 2020. A comprehensive discussion of bridged lactams is outside of the scope of this entry; therefore, the present chapter is focused on the reactivity and properties of bridged lactams as models for amidic distortion. The most recent comprehensive review of bridged lactams was published by Szostak and Aubé in 2013 [4], with an update of the field covering the period of 2014–2018 from Szostak [5]. For a full accounting of the synthesis of bridged lactams, we direct the reader to these recent reviews [4, 5].

The nomenclature and classifications of twisted amides were originally delineated by Yamada [6] and have become standard for discussion within the field [7]. Destabilization of amide resonance **2** can be affected by several general means: intramolecular steric repulsion, intramolecular steric restriction, intermolecular steric restriction, electronic delocalization (e.g. classical anomeric amides), and conformational effects (Figure 1.1) [4]. It is important to note that *destabilized amide* is a more general term, whereas *nonplanar amide* refers specifically to an amide that has been geometrically destabilized via steric and/or electronic factors, i.e. the

Amide Bond Activation: Concepts and Reactions, First Edition. Edited by Michal Szostak.

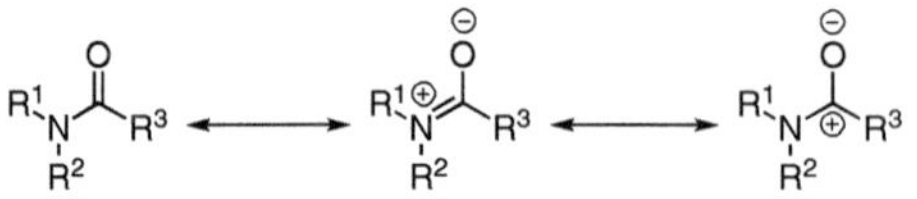

1 2 3

Scheme 1.1 Resonance structures of the amide bond.

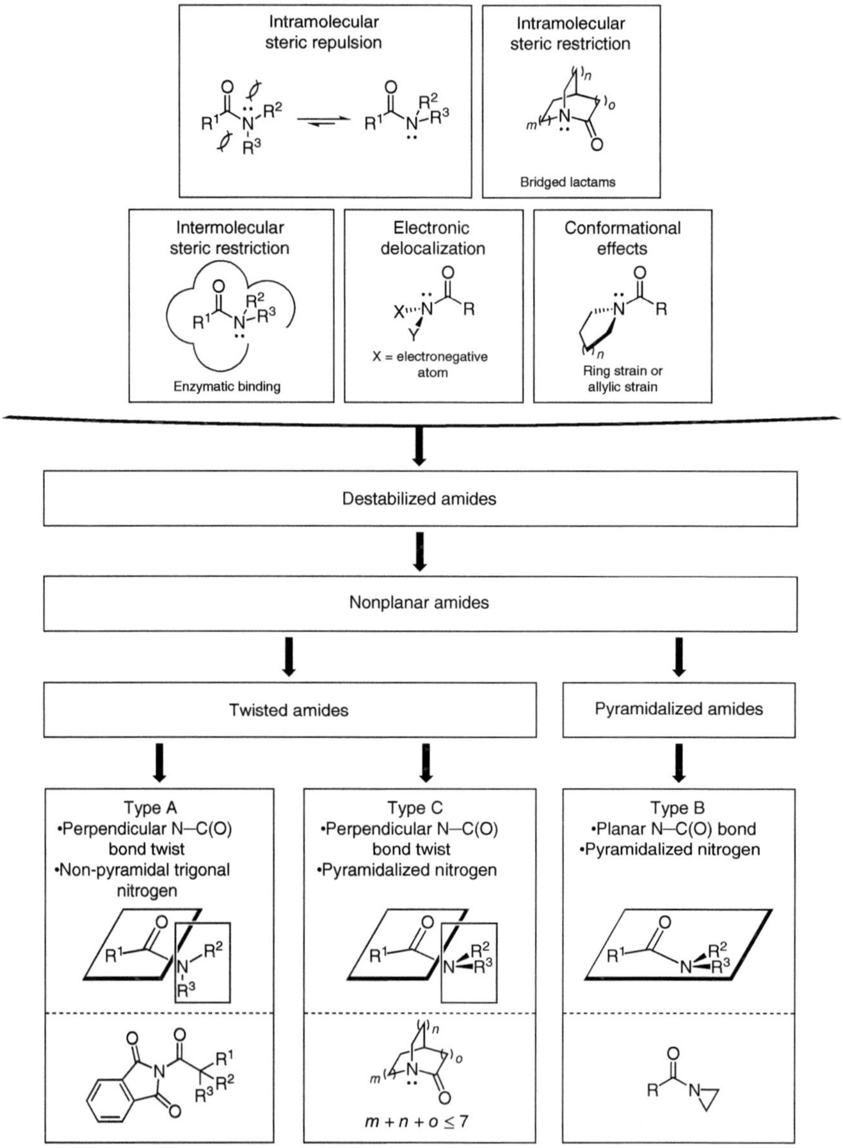

Figure 1.1 Amidic distortion and types of nonplanar amides.

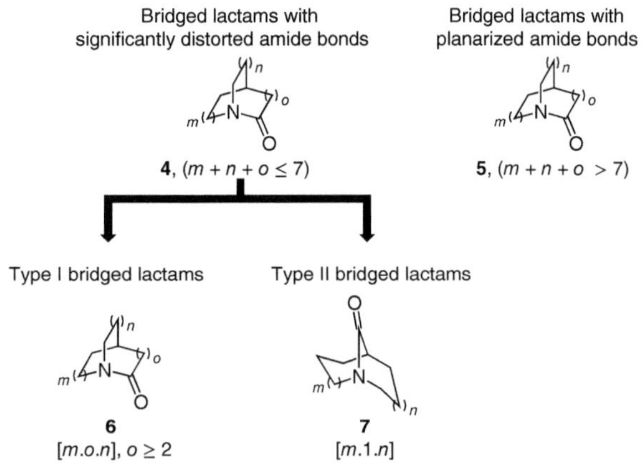

Figure 1.2 Types of bridged lactams.

contribution of resonance form **2** has been diminished. Qualitatively, nonplanar amides can be divided into Types A, B, and C as described by Yamada [6]. Type A amides contain a perpendicularly twisted N–C(O) bond with a nonpyramidalized trigonal geometry at nitrogen. Type B amides contain a significant pyramidalization of the nitrogen atom with a planar N–C(O) bond geometry. Type C amides are a combination of Type A and Type B wherein the N–C(O) bond is twisted with a pyramidalized nitrogen. Both Type A and Type C amides are known as *twisted amides*, while Type B amides are known as *pyramidalized amides*.

Bridged lactams represent one specific class of nonplanar twisted amides wherein the amide nitrogen is embedded in the ring fusion of a bridged bicyclic scaffold, consequently resonance form **2** constitutes an anti-Bredt double bond in a bridged lactam. Steric restriction of the amide bond in such a way remains the most successful strategy by far for distortion of the amide bond. Bridged lactams therefore constitute privileged chemical motifs for understanding the structure and reactivity of distorted amide bonds in general [8]. In addition to the insights into the nature of the amide bond, amidic distortion is critical for understanding cis/trans-isomerization in peptides for protein folding [9]. Bridged lactams are divided into Type I scaffolds where the N–C(O) bond is on a bridge consisting of two or more carbon atoms or Type II where the N–C(O) bond is on a single carbon bridge (Figure 1.2). Type II bridged lactams are generally more strained than the corresponding Type I bridged lactams; however, they tend to be more resistant to hydrolysis due to medium ring scaffolding effects [10]. Bridged lactams wherein the 1-azabicyclo core contains more than 10 atoms will not be covered as these structures display physical properties and reactivity akin to that of planar amides.

1.2 General Properties of Bridged Lactams

1.2.1 Parameters of Amide Bond Distortion

The Winkler–Dunitz distortion parameters provide a quantitative assessment of amide bond distortion based on the twist angle (τ), *N*-pyramidalization (χ_N),

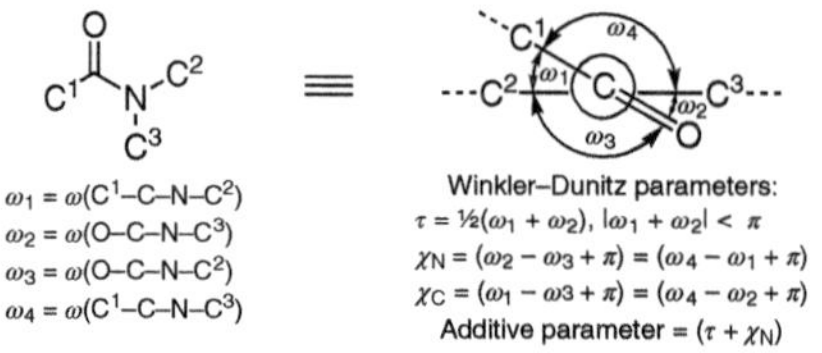

Figure 1.3 Winkler–Dunitz parameters for quantitative assessment of amide bond distortion.

and *C*-pyramidalization (χ_C) (Figure 1.3) [11]. The twist angle τ describes the magnitude of rotation about the N–C(O) bond. Quantitatively, τ is 0° for a planar amide bond and 90° for a fully orthogonal nonplanar amide bond. Both χ_N and χ_C describe the tetrahedral character on the nitrogen and carbon, respectively, and range from 0° for a fully planar amide to 60° for fully pyramidalized amide bonds. In 2015, Szostak identified the additive Winkler–Dunitz distortion parameter ($\tau + \chi_N$) as a means to more accurately predict structural and energetic properties [12]. The additive parameter is an extremely powerful tool for predicting the properties of distorted amides, corresponding linearly to N–C(O) bond length, *N*- vs. *O*-protonation affinity, infrared frequencies, resonance energies, atomic charges, and frontier molecular orbital energies [12]. Nearly the entire Winkler–Dunitz scale is represented by Type II bridged lactams prepared via synthetic efforts [5, 8]. Type I bridged lactams tend to be more flexible and thus do not span the entire range of the Winkler–Dunitz scale, limiting their usefulness in comprehensive understanding of distorted amides (Figure 1.3).

It should be noted that χ_C values for bridged lactams are generally close to 0° (even for relatively undistorted bridged lactams) due to the heightened contribution of amino-ketone resonance form **1** (Scheme 1.1, *vide supra*). At this stage, it is important to note polarized resonance contributor **3** [13], which corresponds to a nitrogen-to-carbon π-electron transfer with little oxygen resonance overlap. Polarized resonance contributor **3** describes the considerably larger N–C(O) bond lengthening compared with a considerably smaller C=O bond lengthening observed upon amide bond distortion [1, 14, 15]. According to the Wiberg–Bader theory of atoms in molecules (AIM), amide resonance stabilization is attributed to the hybridization change of nitrogen from sp^3 to sp^2 in planarizing an orthogonal amide bond. The validity of this polarized resonance model **3** has been a topic of considerable ongoing debate, with recent extensive computational modeling of Type II bridged lactam amide bond rotation demonstrating structural and energetic changes consistent with classical resonance form **1** [12].

An important and underappreciated aspect of amidic distortion is the bending of the amide carbonyl oxygen toward nitrogen. Notably, C=O bending is indicative of impending N–C(O) bond cleavage [16, 17]. In 2016, Stoltz and coworkers proposed the bending angle parameter ξ to describe the deviation from the imaginary CCN angle bisector (Figure 1.4) [16]. The ξ value is calculated by the equation shown in Figure 1.4, with a positive ξ corresponding to bending toward nitrogen and a negative ξ corresponding to bending toward carbon. The amide carbonyl bending effect was first discussed by Bürgi and Schmidt in 1985 in the context of crystallographic and molecular orbital calculations of lactones and lactams [17]. In these studies, an anomeric effect was proposed to reduce a destabilizing mixing of an oxygen *p*-type

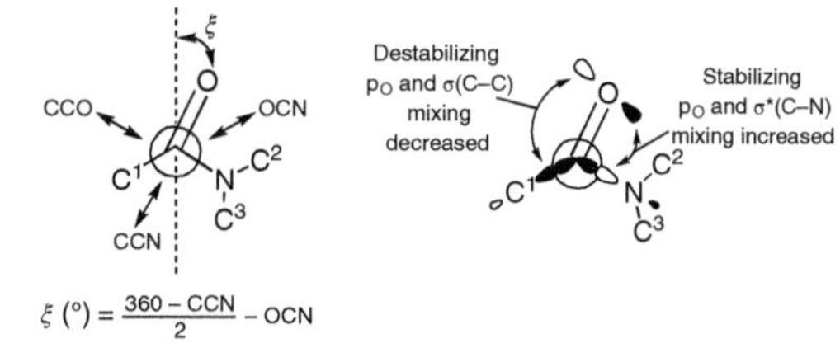

Figure 1.4 Definition of Bürgi C=O bending parameter ξ proposed by Stoltz and MO proposal for stabilizing effect of C=O bending.

lone pair orbital with the C–C(O) bonding orbital with simultaneous stabilizing mixing of the oxygen *p*-type orbital with the C–N antibonding orbital. Reported values of ξ deviate significantly from computational models [12, 16, 17]; therefore only ξ values derived from X-ray crystal structures will be discussed. Bridged lactam crystal structures published prior to the formalization of the ξ parameter will be presented with ξ values determined from crystal structures available via the Cambridge Crystallographic Data Centre when possible, thus some of the values shown here are not reported in the primary literature.

In addition to the Bürgi–Winkler–Dunitz parameters, amidic distortion has been quantified by the sum of three bond angles at nitrogen, θ, and the hinge angle of the carbonyl carbon with respect to the plane defined by the nitrogen atom and its two substituents, α (Figure 1.5) [18]. For an idealized sp^2-hybridized nitrogen atom $\theta = 360°$ and $\alpha = 180°$, whereas for a sp^3-hybridized nitrogen atom $\theta = 328.4°$ and $\alpha = 125°$.

1.2.2 Bond Lengths, Bond Angles, and Spectroscopic Properties of Bridged Lactams

Throughout the later sections of this chapter, examples of bridged lactams will be presented with their respective Winkler–Dunitz parameters and other relevant spectroscopic properties. Kirby's twisted amide (**8**) is presented in the following text in comparison to computationally predicted values for *N*-methyl-2-piperidone (**9**) as an instructive example of the amino-ketone nature of bridged lactams engendering electrophilicity akin to an acid chloride (Figure 1.6) [15, 19]. Kirby's amide (**8**) contains a nearly perpendicular amide bond ($\tau = 90.5°$) with a 0.150 Å longer N–C(O) bond but only a 0.037 Å longer N–C(O) bond when compared with lactam **9**.The θ value of 325.7° indicates that the nitrogen atom is sp^3 hybridized compared to the nearly ideal sp^2 hybridization of lactam **9** ($\theta = 358.9°$). Notably, the θ value of 325.7° is identical to the sum of bond angles at nitrogen in 1,4-diazabicyclo[2.2.2]octane (DABCO) and reminiscent of a typical θ value for trialkylamines [20]. In addition to these physical parameters, both the ^{13}C NMR and IR spectroscopic values are indicative of a aldehyde-like C=O bond in Kirby's amide **8**, whereas **9** displays spectroscopic properties typical of a lactam C=O bond [21]. The aldehydic nature of the

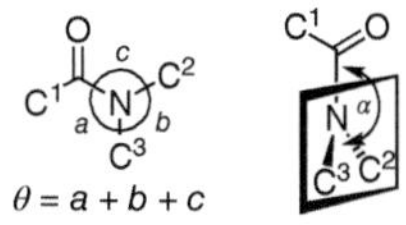

Figure 1.5 Definition of amide angle parameters θ and α.

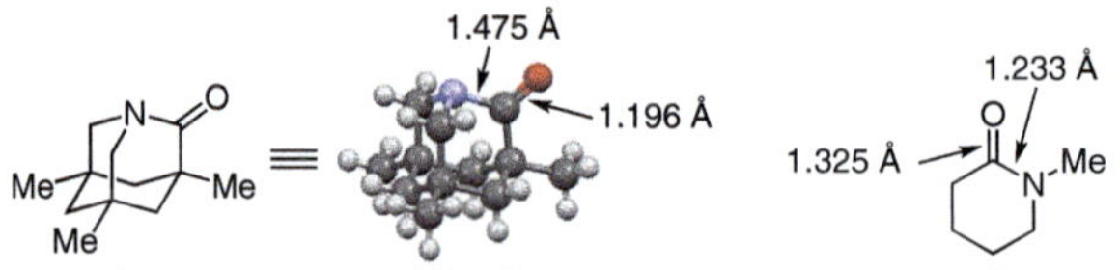

8 [X-ray]	9
τ = 90.5°, ξ = 5.7°	τ = 2.5°
θ = 325.7°	θ = 358.9°
Sum of bond angles at C=O = 359.9°	Sum of bond angles at C=O = 359.9
^{13}C C=O δ = 200 ppm, IR $\nu_{C=O}$ = 1732 cm^{-1}	^{13}C C=O δ = 165 ppm, IR $\nu_{C=O}$ = 1653 cm^{-1}

Figure 1.6 Comparison of Kirby's amide **8** and *N*-methyl-2-piperidone (**9**).

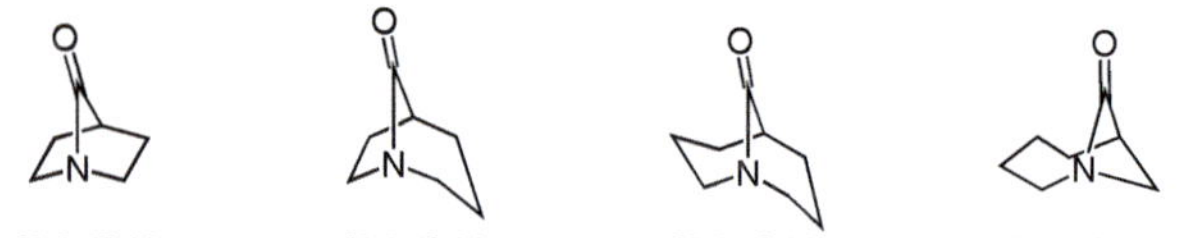

[2.2.1] **10**	[3.2.1] **11**	[3.3.1] **12**	[4.1.1] **13**
N–C(O): 1.474 Å	N–C(O): 1.447 Å	N–C(O): 1.446 Å	N–C(O): 1.422 Å
C=O: 1.201 Å	C=O: 1.209 Å	C=O: 1.212 Å	C=O: 1.207 Å
τ = 90.00°	τ = 65.40°	τ = 90.00°	τ = 38.23°
χ_N = 71.44°	χ_N = 63.10°	χ_N = 53.20°	χ_N = 57.48°
χ_C = 0.00°	χ_C = 7.66°	χ_C = 0.00°	χ_C = 6.33°

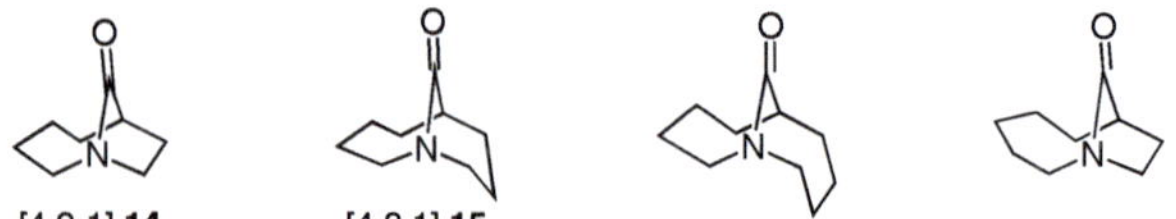

[4.2.1] **14**	[4.3.1] **15**	[4.4.1] **16**	[5.2.1] **17**
N–C(O): 1.401 Å	N–C(O): 1.391 Å	N–C(O): 1.409 Å	N–C(O): 1.381 Å
C=O: 1.218 Å	C=O: 1.225 Å	C=O: 1.223 Å	C=O: 1.223 Å
τ = 36.90°	τ = 41.78°	τ = 47.52°	τ = 21.97°
χ_N = 47.69°	χ_N = 36.01°	χ_N = 41.78°	χ_N = 32.11°
χ_C = 11.17°	χ_C = 15.42°	χ_C = 11.53°	χ_C = 7.62°

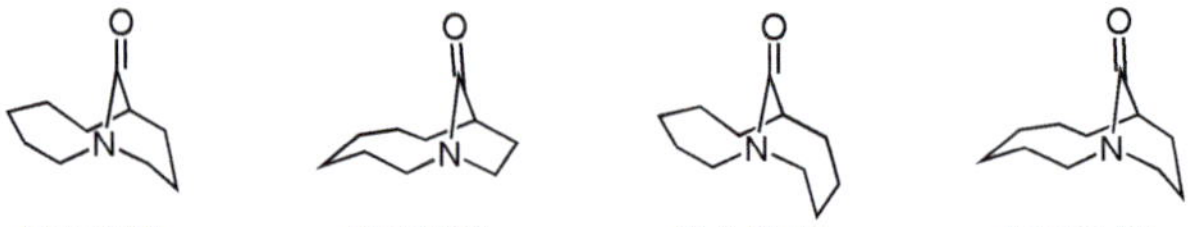

[5.3.1] **18**	[6.2.1] **19**	[5.4.1] **20**	[6.3.1] **21**
N–C(O): 1.372 Å	N–C(O): 1.365 Å	N–C(O): 1.378 Å	N–C(O): 1.367 Å
C=O: 1.230 Å	C=O: 1.228 Å	C=O: 1.231 Å	C=O: 1.233 Å
τ = 25.53°	τ = 11.32°	τ = 37.94°	τ = 18.84°
χ_N = 17.39°	χ_N = 5.99°	χ_N = 14.10°	χ_N = 12.19°
χ_C = 10.67°	χ_C = 4.94°	χ_C = 13.87°	χ_C = 7.97°

Figure 1.7 Geometric parameters for optimized structures of Type II bridged lactams calculated at the MP2/6-311++G(d,p) level. Source: Modified from Szostak et al. [12b].

N–C(O) bond is further characterized by the loss of CO from Kirby's amide upon electron-impact ionization.

Szostak utilized *ab initio* molecular orbital calculations to study the relationships between the structures of synthetically accessible Type II bridged lactams spanning the Winkler–Dunitz scale and their properties (Figure 1.7) [12]. In this series, increasing additive parameter ($\tau + \chi_N$) had excellent linear correlation with increasing N–C(O) bond length (Chart 1.1). Importantly, plotting of τ vs. N–C(O) bond length or χ_N vs. N–C(O) bond length resulted in more scattered correlations

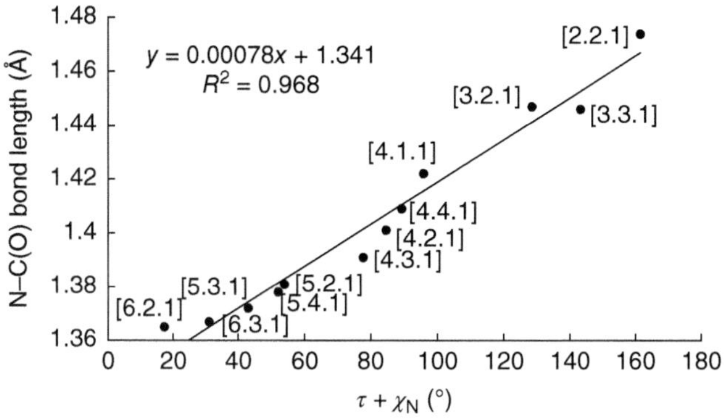

Chart 1.1 Plot of N–C(O) bond length vs. additive parameter for Type II bridged lactams in Figure 1.7. Source: Modified from Szostak et al. [12b].

(R^2 = 0.87 and 0.88, respectively, not shown). While the N–C(O) bond increased in length with increasing distortion, the C=O bond length did not appreciably change, consistent with amino-ketone resonance (**1** and **2**, Figure 1.1). This model should serve as a predictive tool of amide bond N–C(O) length upon distortion from planarity.

1.2.3 *N*- vs. *O*-protonation and Methylation and Structural Effects of *N*-coordination

A planarized amide bond favors *O*-protonation by approximately 10–15 kcal/mol (Scheme 1.2) [1]. The resulting *O*-protonated amide **22** benefits from approx. 40 kcal/mol of resonance stabilization [22] with resonance forms **22** and **23**, whereas *N*-protonated amide **24** cannot benefit from this resonance stabilization [9c]. Notably, *O*-protonation results in significant shortening of the N–C(O) with increased double bond character. In planarized amide bonds, the nitrogen lone pair contributes significantly to the HOMO of the amide system with its contribution denoted as π_N [1, 23]. As the amide bond is distorted from planarity and the nitrogen becomes pyramidalized, the lone pair contribution to the HOMO shifts toward lone pair character (n_N), resulting in a shift in *N*/*O*-protonation

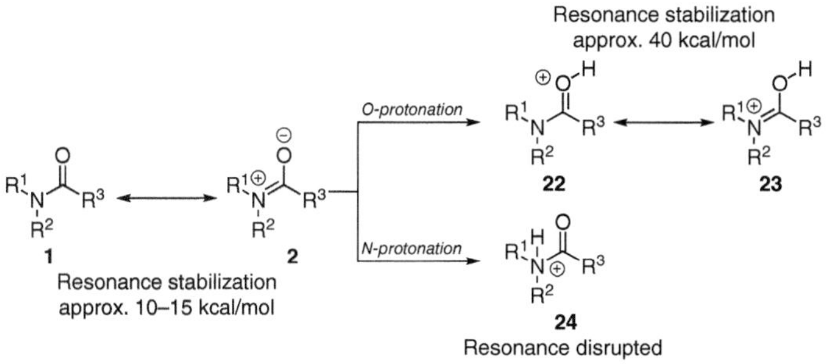

Scheme 1.2 *N*- vs. *O*-protonation of amides. Source: Greenberg et al. [1].

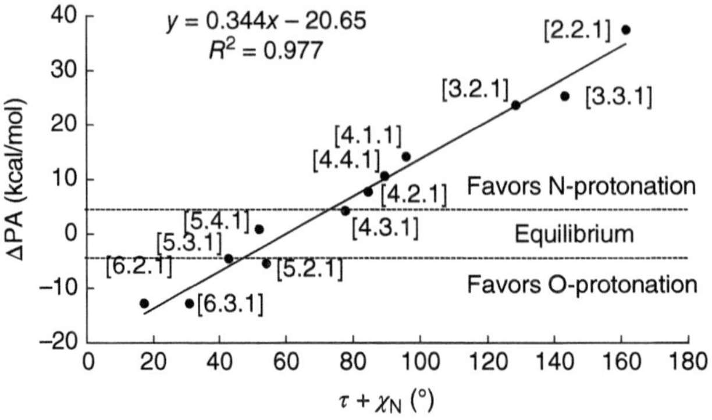

Chart 1.2 Plot of ΔPA (defined as $PA_N - PA_O$) for Type II bridged lactams in Figure 1.7.

aptitude [1, 24]. The *N*-protonation of amides is an important mode of reactivity for understanding processes such as σ C–N bond cleavage and *N*-acyl transfer reaction [25].

Before discussing specific examples of *N*- and *O*-protonation/methylation, it is informative to discuss the relationship between the additive Winkler–Dunitz parameter and *N*- vs. *O*-protonation affinity. In their 2015 studies, Szostak and coworkers found excellent linear correlation between the additive Winkler–Dunitz parameter and the difference in *N*- vs. *O*-proton affinity (PA), ΔPA, defined as $PA_N - PA_O$ for Type II bridged lactams shown in Figure 1.6 (*vide supra*) (Chart 1.2) [12]. This model divides the bridged lactams studied into three regimes: (i) lactams favoring *O*-protonation (ΔPA = −4.5 to −12.8 kcal/mol): [5.2.1], [5.3.1], [6.2.1], and [6.3.1], (ii) the [5.4.1] lactam that is expected to be in equilibrium between *O*- and *N*-protonation (ΔPA = 0.9 kcal/mol), and (iii) lactams favoring *N*-protonation (ΔPA = 4.3–37.5 kcal/mol): [2.2.1], [3.2.1], [3.3.1], [4.1.1], [4.2.1], and [4.3.1]. This provides a powerful method for prediction of the site of amide protonation and is in excellent agreement with examples comprehensive examinations of protonation affinity by Greenberg [22, 24, 26]. Notably, the *O*-protonation models of amides **14–21** had significantly longer C=O bonds than N–C(O) bonds, further supporting the classical amide resonance model (resonance contributors **1** and **2**). Szostak later modeled ΔPA in a series of Tröger's base twisted amides; however, these extremely twisted cases do not provide an ideal setting for studying *N*- vs. *O*-protonation [27].

The first experimental observations of *N*-protonation and *N*-methylation in bridged lactams were disclosed by Pracejus [28] and Yahkontov [29]. Here, some key examples of *N*- vs. *O*-protonation and methylation site selectivity are presented, along with relative parameters to demonstrate the relationship between amidic distortion and reactivity. Kirby's bridged lactam **8** undergoes quantitative *N*-methylation with excess Meerwein's salt as a desiccant to afford moisture-sensitive bridged lactam salt **25** [15b] (Scheme 1.3).

Treatment of lactam **8** with HCl did not allow isolation because of an inseparable hydrolysis by-product, thus characterization of the cation **26** was enabled treatment with anhydrous *p*-TsOH in $CDCl_3$. Spectroscopically, a significant shift of the

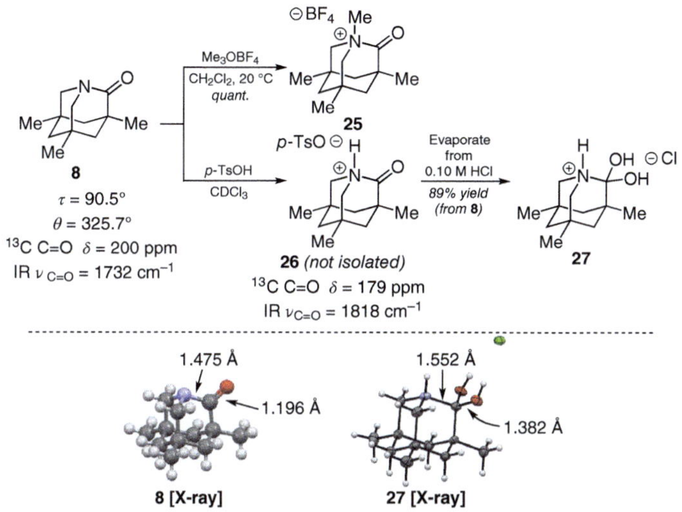

Scheme 1.3 *N*-methylation and *N*-protonation of Kirby's twisted amide **8**.

carbonyl ^{13}C signal from 200 to 179 ppm and IR signal frequency from 1732 to 1818 cm^{-1} was observed, although the IR frequency shift was extremely sensitive to trace moisture. Evaporation of tosylate salt **26** from 0.10 M HCl provided crystalline hydrate **27**. Notably, the 1.552 Å N–C bond is one of the longest N–C bonds recorded to date and the C–O(H) bonds are unusually short at 1.382 Å. This hydrate may be important for understanding intermediates in *N*-acyl transfer reactions in distorted amides [25].

Brown [30] demonstrated that [3.2.2] lactam **28** undergoes *N*-methylation selectively, whereas a more flexible (and therefore less distorted) [3.3.2] analog (**31**) is selectively methylated at oxygen (Scheme 1.4). Although specific parameters of

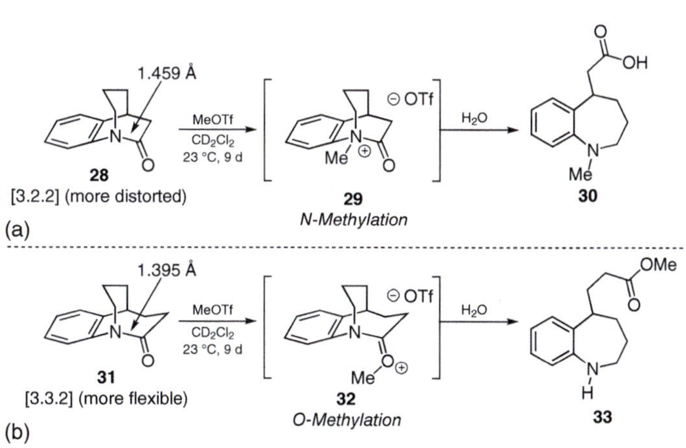

Scheme 1.4 *N*-methylation vs. *O*-methylation in [3.2.2] vs. [3.3.2] systems.

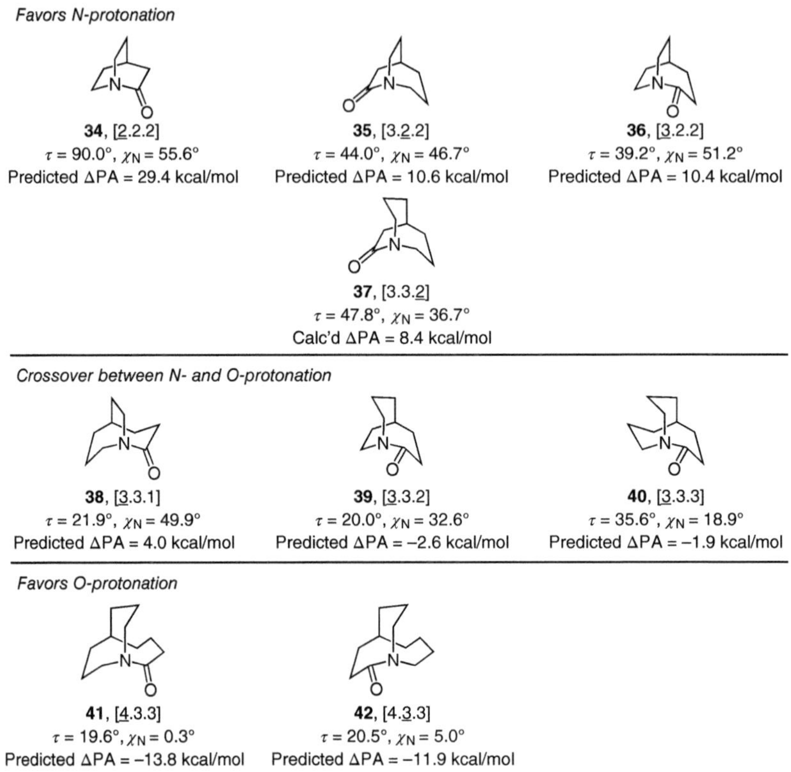

Figure 1.8 *Ab initio* MO calculated structures of Type I bridged lactams and distortion parameters by Greenberg with predicted ΔPA (defined as $PA_N - PA_O$) calculated using the trendline from Chart 1.2.

τ and χ_N are not directly reported, the calculated N–C(O) bond lengths of 1.459 and 1.395 Å for lactams **28** and **31**, respectively, demonstrate that **28** contains a much more distorted amide bond than **31**. It should be mentioned that Brown is careful to indicate that it is unclear if these reactions are the result of kinetic or thermodynamic selectivity. Greenberg later predicted *N*- vs. *O*-protonation with *ab initio* molecular orbital (MO) calculations for [2.2.2] to [4.3.3] bridged lactam scaffolds related to the Brown systems (Figure 1.8) [22, 26]. The "crossover" point between *N*- and *O*-site selection was found in [3.3.1], [3.3.2], and [3.3.3] systems. We present predicted ΔPA values in Figure 1.7 calculated using the trendline of Chart 1.2 [12] for this discussion, noting these values agree with Greenberg's predictions for *N*- vs. *O*-protonation selectivity.

Greenberg demonstrated an experimental example of crossover between *N*-protonation and *O*-protonation with [3.3.1] lactam **43** ($\tau = 19.8°$ and $\chi_N = 49.7°$) (Scheme 1.5) [31]. When lactam **43** is treated with H_2SO_4, a 4 : 1 ratio of *N*-protonation (**44**) and *O*-protonation (**45**) is measured at equilibrium. Amidic distortion is accompanied by a large degree of nitrogen pyramidalization, with significantly smaller pyramidalization of the carbonyl carbon (*vide supra*). On this basis, Greenberg proposed that the degree of distortion for complexation of other

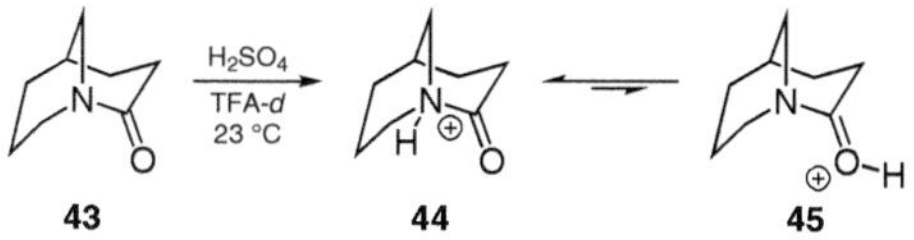

43 **44** **45**

τ = 19.8°, χ_N = 49.7°, χ_C = 7.2°
θ = 340.5°
Predicted ΔPA = 3.3 kcal/mol

4 : 1 ***44*:*45*** *observed*

Scheme 1.5 *N*- and *O*-protonation crossover in a [3.3.1] lactam system. Source: Modified from Sliter et al. [31].

species should vary significantly. For example, complexation of "hard" Lewis acids such as Na^+ should intrinsically favor *O*-binding and thus significant distortion of the amide bond would be required to achieve preferential *N*-binding, whereas "soft" Lewis acids such as Cu^{2+} do not require a large degree of nitrogen pyramidalization for preferential *N*-binding [1, 31]. Importantly, amide bond *N*-binding of Cu^{2+} has been demonstrated as a mode for lowering the barrier to *cis/trans* peptidyl-prolyl isomerization in β-2-microglobulin [32].

In general, the *N*-coordination of amides causes significant distortion of the amide bond. Of the bridged lactams that have been synthesized to date, the *N*-coordinated bridged lactams have proven to be the most distorted and reactive of the series. The first direct crystallographic comparison between bridged lactams and their *N*-protonated form was reported by Aubé in 2010 (Scheme 1.6) [33]. *N*-protonation of lactam **46** with *p*-TsOH enabled isolation and crystallization of salt **47**, which displays extensive pyramidalization of the amide nitrogen compared with parent lactam **46**. The increase of χ_N from 31.6° to 52.1° demonstrates that nitrogen atom has significantly more sp^3 character (χ_N = 60° for an idealized sp^3 nitrogen atom). Comparison of the crystal structures of **46** and **47** also reveals a flattening (χ_C = 12.8° to χ_C = 1.4°) and slight shortening of the C=O bond (1.218 to 1.192 Å), a significant twisting of the C–N bond (τ = 51.5° to τ = 81.9°), and a large increase in C=O bending toward nitrogen (ξ = 0.2° to ξ = 5.1°).

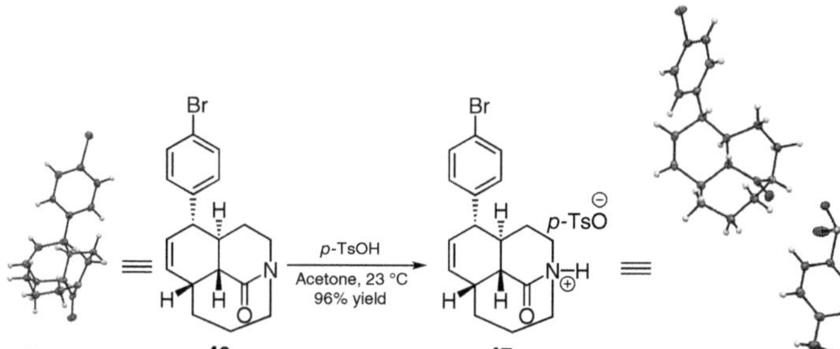

[X-ray] **46** **47** **[X-ray]**

Amide	N–C(O) (Å)	C=O (Å)	χ_N (°)	χ_C (°)	τ (°)	ξ (°)
46	1.387	1.218	36.1	12.8	51.5	0.2
47	1.502	1.192	52.1	1.4	81.9	5.1

Scheme 1.6 First direct crystallographic comparison of a bridged lactam and its *N*-protonated form. Source: Modified from Szostak et al. [33].

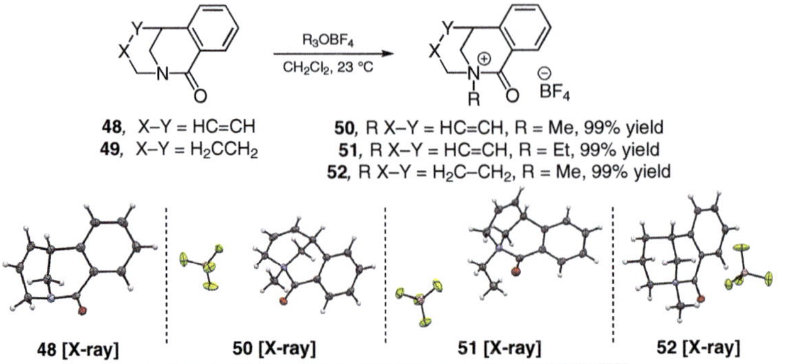

Amide	N–C(O) (Å)	C=O (Å)	χ_N (°)	χ_C (°)	τ (°)	ξ (°)
48	1.386	1.226	49.7	7.1	30.7	0.4
50	1.554	1.192	58.3	1.0	44.0	6.0
51	1.546	1.192	60.4	0.2	41.7	5.6
52	1.536	1.196	59.3	3.6	55.2	5.3

Scheme 1.7 Crystallographic comparison of a bridged lactam and *N*-alkylated bridged lactams. Source: Modified from Hu et al. [34].

Similarly, the first crystallographic structures of *N*-methylated bridged lactams and a parent bridged lactam were demonstrated in 2016 by Szostak (Scheme 1.7) [34]. As observed in the *N*-protonation of lactam **46**, *N*-alkylation of lactam **48** results in a dramatic lengthening and twisting of the N–C(O) bond, a slight shortening of the C=O bond, an increased pyramidalization of nitrogen, and a significant bending of the carbonyl toward nitrogen. Szostak also found that while lactam **48** protonates exclusively on nitrogen, *O*-coordination could be achieved nearly quantitatively via treatment with $AgSbF_6$, resulting in a rare example of an *O*-coordinated amide crystal structure (Scheme 1.8). Directly comparing the crystallographic data demonstrates the incredible structural impact of *N*- and *O*-coordination, with *O*-coordination reinforcing double bonding character of the N–C(O) bond. Interestingly, complex **53** is a rare example of a molecular wire where

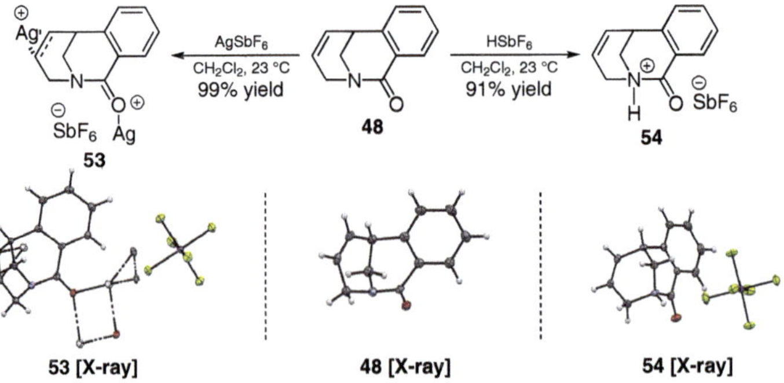

Amide	N–C(O) (Å)	C=O (Å)	χ_N (°)	χ_C (°)	τ (°)	ξ (°)
48	1.386	1.226	49.7	7.1	30.7	0.4
53	1.355	1.262	47.7	5.8	33.3	1.9
54	1.539	1.196	55.0	0.7	42.0	6.9

Scheme 1.8 Divergent *N*- vs. *O*-coordination and crystallographic data.

each silver cation is coordinated to the olefin and an amide oxygen in an infinite chain structure [35].

N-coordination of bridged lactams has enabled isolation and characterization of some of the most twisted amides prepared to date with τ values c. 90°, indicating complete orthogonality of the amide bond. In closing this discussion, we highlight the structural features of some of these lactams and their parent lactams. It is important to note that in many cases, the parent lactams have not been successfully isolated and characterized, thus data presented is computationally derived. The reactivity of these systems will be further discussed in Section 1.3 (*vide infra*).

In 2006, Stoltz completed the synthesis of Type I [2.2.2] lactam **34·HBF$_4$**, which bears an extremely twisted amide bond (Scheme 1.9) [36, 37]. Isolation as the tetrafluoroborate salt via the Aubé–Schmidt reaction from azide **55** under rigorously anhydrous conditions was necessary to avoid nucleophilic attack from a more coordinating counter-anion [38]. Similarly, in 2016 Stoltz reported the synthesis of the [2.2.1] Type II bridged amides **10·HBF$_4$** and **10·BF$_3$**, utilizing the Aubé–Schmidt strategy [16]. Compound **10·HBF$_4$** was stable enough for rapid characterization by spectroscopic techniques, but readily decomposed to the more stable trifluoroborate **10·BF$_3$**, which provided X-ray quality crystals. The crystallographic data of **34·HBF$_4$** and **10·BF$_3$** reveal two of the most twisted amide bonds known to date. In both molecules, the amide bond achieves complete orthogonality, the nitrogen atoms are fully pyramidalized, and a large bend of the C=O bond toward

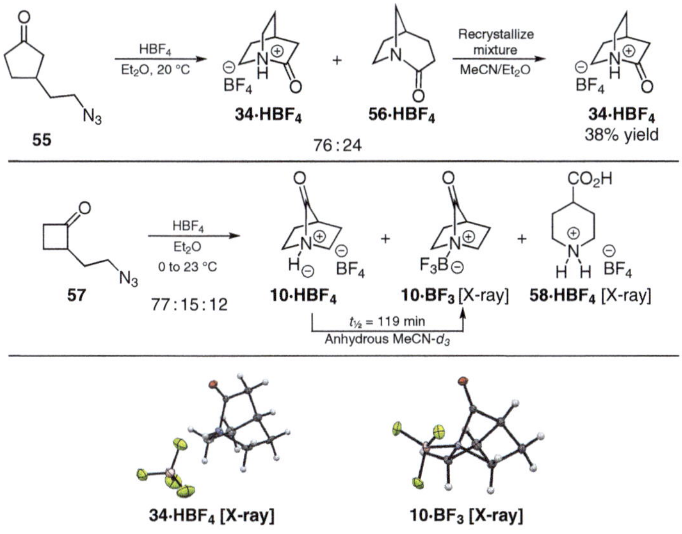

Amide	N–C(O) (Å)	C=O (Å)	χ_N (°)	χ_C (°)	τ (°)	ξ (°)
34 [calculated]	1.504	1.167	57.6	0.0	89.9	—
34·BF$_4$ [X-ray]	1.526	1.192	59.5	0.2	90.9	5.8
10 [calculated]	1.474	1.201	71.4	0	90.0	—
10·BF$_3$ [X-ray]	1.526	1.186	69.8	0	90.0	5.8

Scheme 1.9 Synthesis of [2.2.2] and [2.2.1] *N*-coordinated bridged lactams via the Aubé–Schmidt reaction with computational and crystallographic structural data. Source: Adapted from Yakhontov et al. [36] and Tani et al. [37].

nitrogen of 5.8° is measured. Aside from Stoltz's *N*-coordinated lactams, Kirby's 1-adamantan-2-one-based bridged lactams are the only bridged lactams that have a fully orthogonal amide bond with an additive Winkler–Dunitz parameter exceeding 100° in their neutral form [5, 15, 19, 39].

1.2.4 Twisted Amide Basicity and pK_a Measurements

The preferential *N*-protonation of bridged lactams is concomitant with a substantial increase in amide basicity. Pracejus and Yakhontov measured the pK_a of the conjugate acids of methylated 2-quinuclidoniums **59–61** and **62**, respectively, which were sterically shielded from hydrolysis (Figure 1.9) [28b, 29a, c]. The *N*-protonated 2-quinuclidinones [**59**·**H**]$^+$–[**62**·**H**]$^+$ (pK_a = 5.33–6.37) are situated between *O*-protonated *N*,*N*-dimethylformamide ([**63**·**H**]$^+$, pK_a = 0.0) and quinuclidinium (**64**, pK_a = 10.65) on a pK_a scale, demonstrating their significantly enhanced basicity relative to planar amides. Kirby subsequently validated the Pracejus and Yakhontov measurements utilizing a stopped-flow indicator method [40] to estimate the pK_a value of 5.2 for lactam **8** [15b]. Few pK_a values have been reported for bridged lactams, ostensibly due to their lability in nucleophilic solvents.

Stoltz and Julian examined kinetic proton affinity of 2-quinuclidonium [**34**·**H**]$^+$ in the gas phase via competitive fragmentation of proton bound dimers relative to reference bases [41]. An experimental PA value of 964.2 kJ/mol was determined, well in agreement with the computationally predicted PA of 944.3 kJ/mol. In contrast, a typical amide has a much lower PA within the range of 880–900 kJ/mol [42]. 2-Quinuclidonium ([**34**·**H**]$^+$) is hydrolyzed instantaneously in water, and the reverse reaction through either the amino acid or acid chloride has not been achieved in solution phase; however, collisional excitation of hydrolyzed derivative **59** in the gas phase leads to formation of [**34**·**H**]$^+$ (Scheme 1.10) [37]. Stoltz and Julian also calculated a significantly higher PA of 982.0 kJ/mol for Yakhontov's more hydrolytically stable lactam **62**, demonstrating that bridged lactam nucleophilicity does not correlate with increased hydrolytic lability [41, 43].

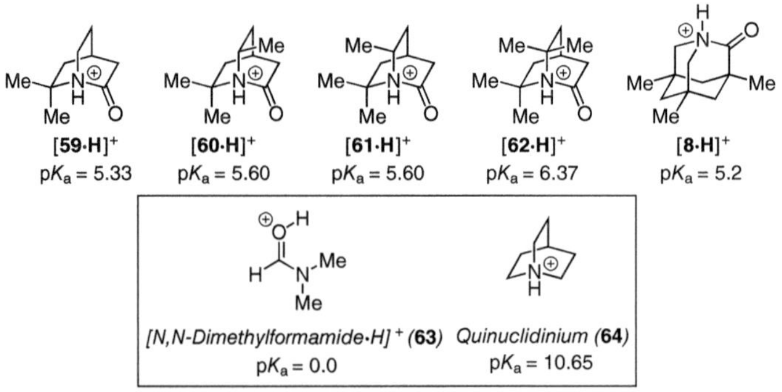

Figure 1.9 pK_a measurements (H_2O) of 2-quinuclidone derivatives [**59**·**H**]$^+$–[**62**·**H**]$^+$ and Kirby's lactam [**8**·**H**]$^+$.

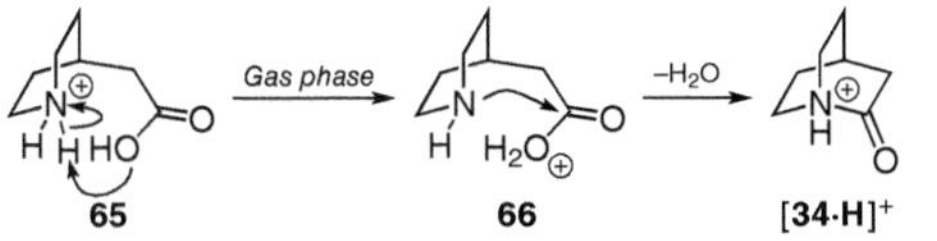

Scheme 1.10 Gas phase synthesis of 2-quiniclidone [**34**·**H**]$^+$. Source: Modified from Tani et al. [37].

1.3 Reactivity of Bridged Lactams

1.3.1 Reactivity of the Lactam Nitrogen

1.3.1.1 Hydrolysis of the N–C(O) Bond

The amide bond is well known for being recalcitrant toward hydrolysis, with half-lives ($t_{1/2}$) in neutral aqueous solution of hundreds of years [44]. Distortion of the amide linkage has a remarkable effect on this half-life [1], with some of the most strained bridged lactams undergoing instantaneous hydrolysis in water. Consequently, the increased sensitivity of bridged lactams often precludes the utilization of nucleophilic solvents (e.g. methanol, DMSO) in their synthesis and adds to the challenge of isolation and spectroscopic characterization of bridged lactams [4]. The study of bridged lactams hydrolysis has direct implications for the biological cleavage of peptide bonds that occur as a result of amidic distortion [1]. Here, we provide a general overview of bridged lactam hydrolysis in relation to amidic distortion. For a more comprehensive review, we direct the reader to a previously published reviews [4, 5]. Additionally, several theoretical studies have been performed to address the mechanism of hydrolysis in bridged lactams [45].

Early studies by Blackburn on the alkaline hydrolysis rate of strained *N*-aryl lactams demonstrated a dramatic rate increase correlated with increased twisting of the amide bond [46]. These studies demonstrated that distortion of the amide bond accelerates both the rate of nucleophilic attack at the amide carbonyl and the rate of nitrogen expulsion via collapse of the tetrahedral intermediate. Subsequently, Blackburn [47] and Brown [48] published a series of studies on the rate of hydrolysis of benzo-fused 2-quinuclidones (Scheme 1.11). The series of benzo-fused bridged lactams undergo hydrolysis in both acidic and basic water with rates several orders of magnitude larger compared with a planarized acetamide model (**73**). The best correlation between rate of hydrolysis and the structure of the bridged lactam is obtained when considering the additive Winkler–Dunitz parameter [12], which accounts for both amide bond twist and nitrogen pyramidalization.

Highly distorted amides typically undergo irreversible hydrolysis to the corresponding amino acid, [16, 39] with the notable exception of Kirby's 1-adamant-2-one **8** [15, 19]. While highly twisted bridged lactams provide insight into some of the more extreme cases of amide bond distortion, their lability preclude studies in nucleophilic solvents. In 2009, Aubé and coworkers demonstrated that medium-bridged lactams, while still significantly distorted (see Scheme 1.6 for distortion parameters), are thermodynamically and kinetically stable at acidic, neutral, and basic

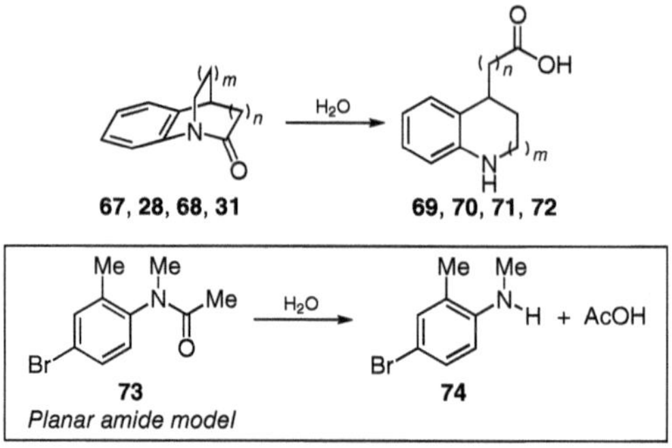

Amide	N–C(O) (Å)	C=O (Å)	χ_N (°)	χ_C (°)	τ (°)	ξ (°)	*Base hydrolysis* k_3 (M/s)	*Acid hydrolysis* k_3/K_3 (M/s)
67 ($n = m = 1$) [calculated]	—	—	63.4	0.0	90.0	—	2.6×10^2	2.3×10^4
28 $n = 1$, $m = 2$ [X-ray]	1.401	1.216	57.2	9.0	71.2	2.1	6.0×10^1	5.6×10^1
68 $n = 2$, $m = 1$ [X-ray]	1.413	1.225	52.8	11.0	75.6	2.4	1.7×10^1	3.0×10^1
31 $n = m = 2$ [X-ray]	1.370	1.233	38.6	4.3	34.4	0.0	5.1×10^{-4}	1.2×10^{-4}
73 [X-ray]	1.338	1.235	3.7	−1.5	1.3	0.4	2.2×10^{-5}	2.2×10^{-7}

Scheme 1.11 Rates of acidic and basic hydrolysis of benzo-fused 2-quinuclidones.

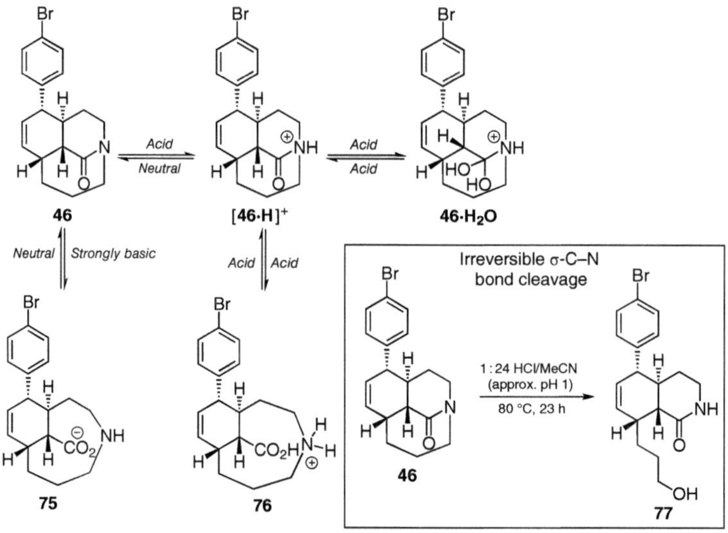

Scheme 1.12 pH-dependent speciation of medium-bridged lactam **46**. Source: Modified from Szostak et al [10].

pH in water (Scheme 1.12) [10]. Aubé and coworkers discovered that amide bond hydrolysis of amide **46** was reversible by performing a series of extraction and NMR experiments. Under basic conditions, the amide bond is hydrolyzed and conjugate base **75** is observed. In acid, speciation is observed between the *N*-protonated species [**46**·**H**]$^+$, hydrate **46**·$\mathbf{H_2O}$, and the amide hydrolysis conjugate acid product **76**. Only after prolonged heating in pH 1 MeCN does medium-bridged lactam **46** undergo irreversible decomposition via σ-C–N bond cleavage. Importantly, these medium-bridged lactams represent an underexplored region of the

Winkler–Dunitz scale and could provide a potential avenue for study of the amide bond in nucleophilic solvents [49].

1.3.1.2 Cleavage of the σ C–N Bond

Cleavage of the σ C–N is a common pathway of reactivity for bridged lactams upon *N*-coordination. As discussed in Section 1.2.3, coordination of a bridged lactam nitrogen leads to increased bending of the C=O bond toward nitrogen, activating the σ C–N bond toward nucleophilic cleavage. The first examples of such reactivity were reported by Yakhontov; however, these examples were limited to tetramethyl-substituted 2-quinuclidone (**62**, Figure 1.9) for its ability to form tertiary carbocations [29a, c]. In efforts toward the total synthesis of stenine [50], Aubé observed hydrogenolysis of the σ C–N bond in a medium bridged lactam. A series of medium-bridged lactam analogs were subsequently investigated, demonstrating facile hydrogenolysis (with debromination) to lactam **78** [51], oxidative cleavage with DDQ to aldehyde **79**, and iodide cleavage of the σ C–N bond via *N*-methylation to iodide **80** (Scheme 1.13) [52]. Other examples demonstrated that the olefin was not required for the hydrogenolysis to proceed. Interestingly, utilization of deuterium gas in EtOD led to incorporation of a single deuterium atom on the propyl chain, implicating that cleavage may occur via direct insertion of Pd or displacement via a Pd hydride.

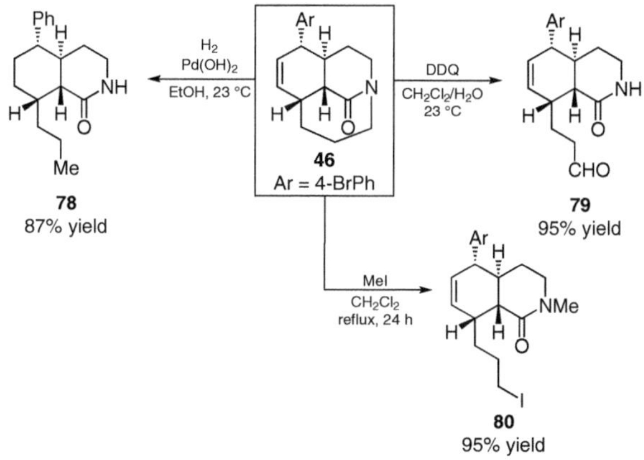

Scheme 1.13 σ C–N bond cleavage of a medium-bridged lactam. Source: Adapted from Lei et al. [52].

Szostak reported a "sew-and-cut" strategy for the two-step functionalization of isoquinoline-2-one ring systems (Scheme 1.14) [53]. Intramolecular Pd-catalyzed Heck reaction of aryl iodides **81a–s** generated bridged lactams with an additive Winkler–Dunitz parameter, $\tau + \chi_N$, in the range of 50°–60°. Cleavage of the σ C–N was then affected by a range of alkyl halides affording *N*-alkylated isoquinolin-2-ones **83a–s** in generally excellent yield. Control experiments suggest that the activation of the amide bond via *N*-alkylation is rate limiting. Interestingly, the lactams Szostak studied are more pyramidalized than twisted, possibly

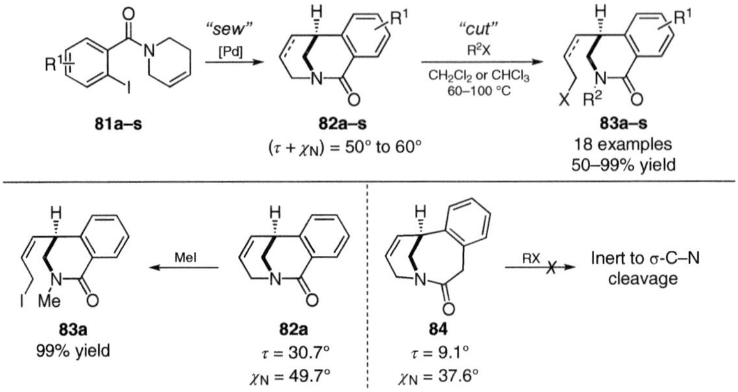

Scheme 1.14 "Sew-and-cut" approach to isoquinoline-2-one ring systems. Source: Adapted from Hu et al. [53].

indicating that the pyramidalization of the amide bond is more activating toward σ C–N bond cleavage. Ring-expanded bridged lactam **84** with a less distorted amide bond ($\tau + \chi_N = 46.7°$) was inert to cleavage of the σ C–N bond with alkyl halides.

1.3.2 Reactivity of the Carbonyl Group

The heightened electrophilicity of the twisted amide carbonyl coupled with the higher *O*-basicity of planarized amides allows an opportunity for divergent chemoselectivity in complex systems [4, 5]. Steric restriction of the amide bond in bridged lactams often enables isolation of stable tetrahedral intermediates resulting from 1,2-addition to the carbonyl [25]. A comprehensive review of reactions of the lactam carbonyl is not possible here; instead we provide a broad overview of general reactivity to highlight the amino-ketone resonance contribution in bridged lactams.

1.3.2.1 Heteroatom Nucleophiles

Yakhontov reported first nucleophilic cleavage of bridged lactam amide bonds with tetramethyl 2-quinuclidinone **62** with nucleophiles including alcohols, amines, hydrazines, and hydroxylamines (Scheme 1.15) [29a, c]. This reactivity is quite general for bridged lactam carbonyls despite the steric shielding in the present example (**62**) [4, 5]. In 2019, Szostak compared and modeled the reactivity of

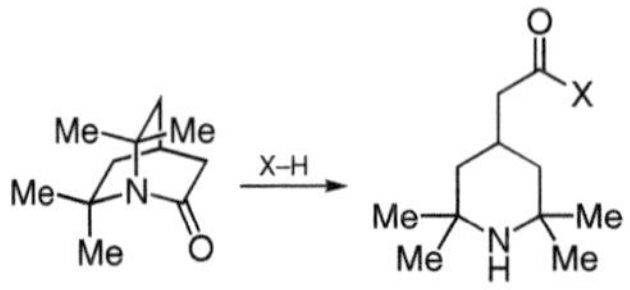

62 85a–f

Entry	X–H	Conditions	Yield (%)
1	H_2O	Reflux, 30 min	94
2	MeOH	Reflux, 5 h	96
3	Morpholine	Reflux, 12 h	97
4	NH_2OH	EtOH, reflux, 2 h	62
5	NH_2NH_2	Xylenes, reflux, Dean–Stark, 5 h	63
6	NH_2NHPh	Xylenes, reflux, Dean–Stark, 12 h	64

Scheme 1.15 Bridged lactam bond cleavage with heteroatom nucleophiles. Source: Adapted from Levkoeva et al. [29a, c].

bridgehead lactam **82** against less reactive acyclic 3° anilides in intermolecular transamination with non-nucleophilic amines (not shown) [54, 55]. The reactivity of bridgehead lactams toward heteroatom nucleophiles has also been exploited in polymerization processes by Hall [56].

Condensation reactions with heteroatom nucleophiles including diols [15b], hydrazines [15b, 57], and amines [15b] are generally limited to adamantane or medium-bridged lactams as the scaffolding effects of these systems are required for product stability [4]. Ma and coworkers reported an elegant total synthesis of indole alkaloid communesin F wherein a bridged lactam facilitated the rapid synthesis of the assembly of the natural product core via a one-pot Staudinger/intramolecular condensation sequence (Scheme 1.16) [58].

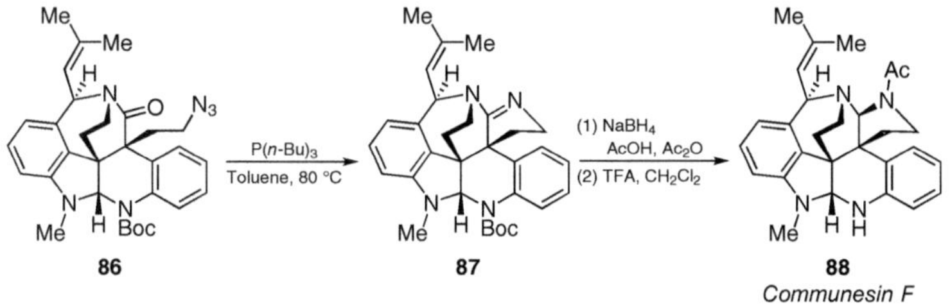

Scheme 1.16 Staudinger/intramolecular condensation in the total synthesis of communesin F by Ma. Source: Adapted from Zuo et al. [58].

1.3.2.2 Organometallics

Aubé reported the formation of ketones upon treatment of medium-sized bridged lactam **89** with an excess of organometallic reagents (Scheme 1.17, top) [57]. Despite addition of excess nucleophile, no addition to the ketone was observed. When tricyclic bridged lactams were treated with organolithium reagents, both the hemiaminal **92** and ketone **93** could be observed (Scheme 1.17, bottom). The steric bulk

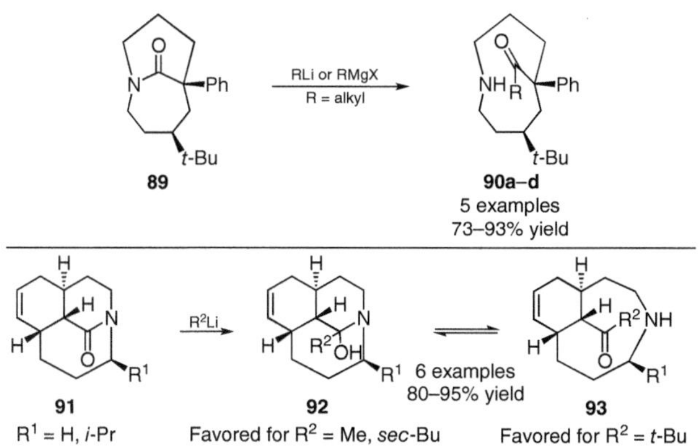

Scheme 1.17 Organometallic addition to medium-bridged lactams. Source: Adapted from Szostak et al. [57].

of the R^2 group determined the major product in the equilibrium, with hemiaminal **92** favored for R^2 = Me and *sec*-Bu and amino-ketone favored **93** for R^2 = *t*-Bu.

1.3.2.3 Reduction of the Carbonyl

While amides are typically inert to reduction by sodium borohydride, the ketone-like nature of the carbonyl in bridged lactams enables reactivity. Subsequent to the first reports of adamantane-based bridged lactam reduction by borohydride from Denzer and Ott [59] and Kirby [15b], Aubé demonstrated that moderately distorted bridged lactams **94** and **46** can be reduced (Scheme 1.18) [57]. Placement of an electron-withdrawing group α such as SO_2Me (**96**) to the bridgehead lactam results in an interesting C–C bond scission leading to formamide **98** in excellent yield. The stability of the hemiaminal products **95** and **99** are likely due to poor overlap between the nitrogen lone pair and the σ*(C–O).

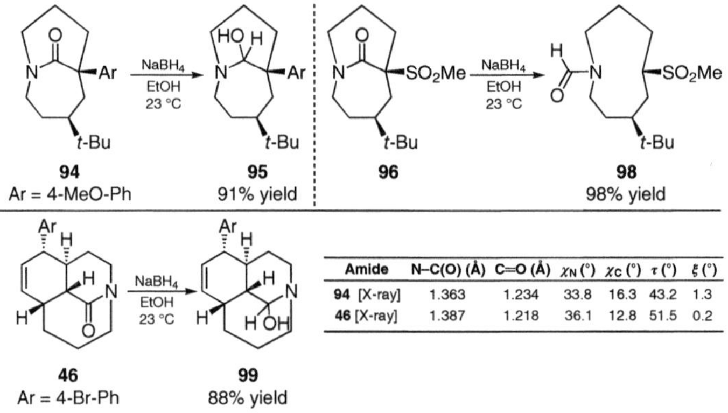

Amide	N–C(O) (Å)	C═O (Å)	χ_N (°)	χ_C (°)	τ (°)	ξ (°)
94 [X-ray]	1.363	1.234	33.8	16.3	43.2	1.3
46 [X-ray]	1.387	1.218	36.1	12.8	51.5	0.2

Scheme 1.18 Sodium borohydride reduction of bridged lactams.

In general, the more flexible a bridged lactam system is the more facile the full reduction of the amide bond is. This is best highlighted by examination of Thomas' efforts toward the total synthesis of stemofoline wherein reduction of rigid tropanone **100** with $LiAlH_4$ does not proceed past hemiaminal **101**. A subsequent transformation of the hemiaminal to the corresponding amino chloride is required for full reduction to amine **102** (Scheme 1.19, top) [60]. In contrast, Dolby achieved full reduction of the amide functionality in the highly flexible lactam **103** with a [6.6.2] ring system, which is more similar to a planarized amide (*vide supra*) [61].

1.3.2.4 Olefination and Epoxidation Reactions

The amino-ketone behavior of bridged lactams has also been demonstrated by their reactivity toward olefination and epoxidation conditions. In planar amides, olefination is typically impeded by the delocalization of nitrogen lone pair into the π*(CO) orbital, decreasing the sp^2 characteristic of the bond, thus rendering the planar amide bond inactive toward olefination. Epoxidation is similarly disrupted by resonance, and ring-opened products are obtained rather than epoxides. However, both types of reactions may be illustrated with sufficiently twisted

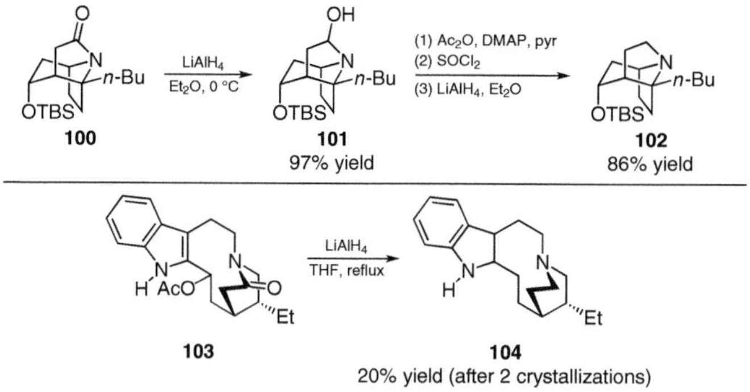

Scheme 1.19 Comparison of rigid and flexible bridged lactam reduction. Source: Adapted from Beddoes et al. [60a].

amide bonds as demonstrated by the examples in the following text from Aubé (Scheme 1.20) [57, 62]. Petasis olefination of bicyclic bridged lactam **89** generated olefin **105**, while Corey–Chaykovsky epoxidation formed stable amino epoxide **106**. Corey–Chaykovsky epoxidation was also effective in tricyclic bridged lactams such as **46**. Wittig olefinations are also effective in bridged lactam systems as first demonstrated by Kirby [15b].

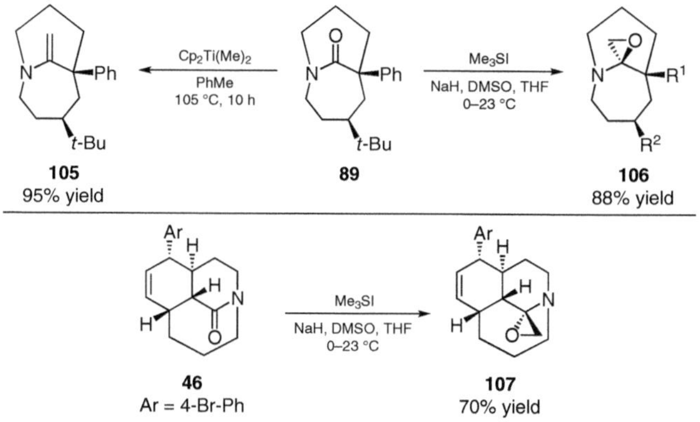

Scheme 1.20 Olefination and epoxidation of bridged lactams. Source: Adapted from Szostak et al. [57, 62].

1.3.2.5 Enolate and Conjugate Addition Chemistry

In limited examples, flexible bridged lactams have been utilized as nucleophiles in enolate functionalization [63] and in Friedel–Crafts reactions in indole-fused systems [64]. In [6.3.1] bridged lactam **108**, Ban demonstrated that several electrophiles could be utilized to accomplish C-alkylation and C-acylation (Scheme 1.21). With methyl iodoacetate and methyl bromoacetate, moderate yields of α-halogenation products **109c** and **109d** were obtained. In addition to α-functionalization of bridged lactams, Judd and coworkers demonstrated Rh-catalyzed conjugate addition of an aryl boronic acid with α,β-unsaturated bridged lactam **110** (Scheme 1.22) [65].

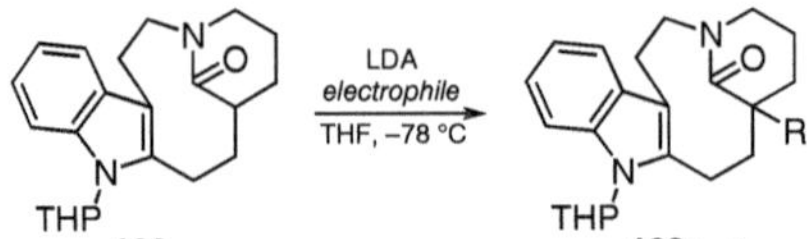

108 → 109a–g

Entry	Electrophile	Product	R	Yield (%)
1	Et–I	**109a**	Et	84
2	Allyl bromide	**109b**	Allyl	71
3	I–CH_2CO_2Me	**109c**	I	42
4	Br–CH_2CO_2Me	**109d**	Br	34
5	Cl–CH_2CO_2Me	**109e**	$C(O)CH_2Cl$	66
6	Ac_2O	**109f**	CO_2Me	90
7	$CH_2C(SOMe)(SMe)$	**109g**	$CH_2CH(SOMe)(SMe)$	71

Scheme 1.21 Enolate reactivity of [6.3.1] lactam **108**.

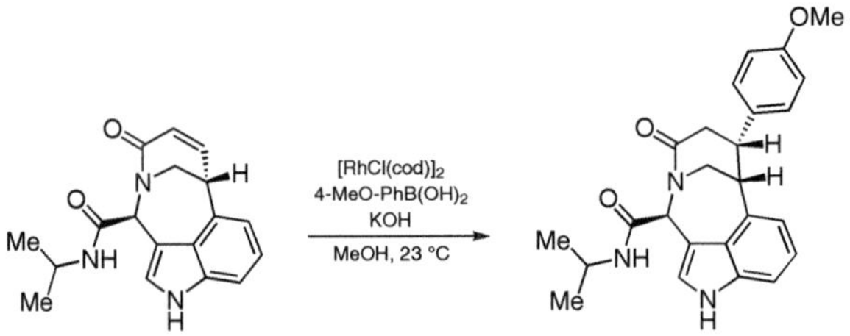

110 → **111**
87% yield, >98 : 2 dr

Scheme 1.22 Rh-catalyzed conjugate addition with bridged lactam **110**. Source: Adapted from Ribelin et al. [65].

1.3.3 Polymerization Reactions

The strain of bridged lactams has been harnessed to drive polymerization reactions [56]. Inspired by reports from Stoltz [37] that the twisted amide 2-quinuclidone (**34**) formed polymeric material in attempts to neutralize the compound Gutekunst designed the first living polymerization of twisted amides with simple primary alkyl iodides employed as initiators (Scheme 1.23) [66]. Bridged lactam monomer

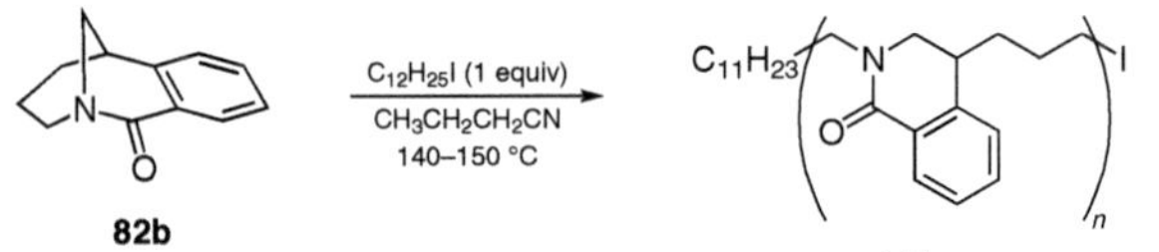

82b → **112**

Entry	Target DP (*n*)	Temp (°C)	Time (h)	Conversion (%)	$M_{n,theo}$ (kg/mol)	$M_{n,SEC}$ (kg/mol)	$Đ$
1	10	140	12	98	2.1	2.1	1.21
2	25	140	24	96	4.8	5.1	1.20
3	50	140	30	94	9.4	9.4	1.20
4	75	140	48	94	13.5	12.1	1.26
5	100	150	52	>95	18.1	17.3	1.26
6	200	150	64	91	34.4	24.3	1.46

Scheme 1.23 Halide-rebound polymerization of twisted amide **82b**. Source: Adapted from Fu et al. [66].

82b was observed to form polymers of linearly increasing molecular weights via halide-rebound polymerization. Gutekunst also reported the living ring-opening metathesis polymerization (ROMP) of an unsaturated bridged lactam with Grubbs III (Scheme 1.24) [67]. The calculated strain energy of 11.3 kcal/mol served as the primary driving force toward ROMP. Upon ring-opening, the amide could reintegrate into the conjugated planar system, promoting propagation of the polymer. To understand the role of amide distortion in the polymerization, another ROMP experiment was performed with a methine carbon in the place of the bridgehead nitrogen, which resulted in no conversion.

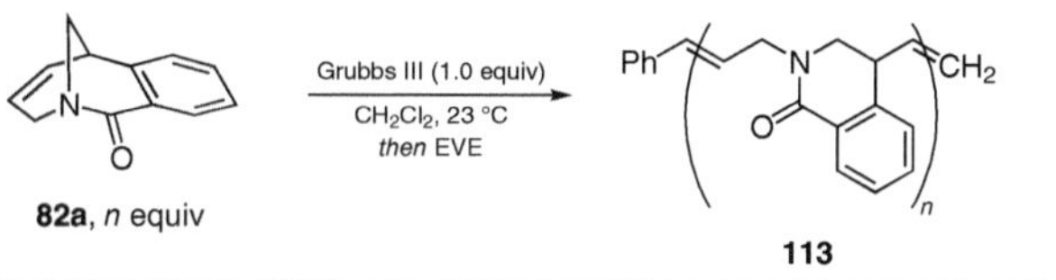

Entry	Target DP (*n*)	Time (h)	Conversion (%)	$M_{n,theo}$ (kg/mol)	$M_{n,SEC}$ (kg/mol)	Đ
1	10	2	95	1.9	1.9	1.21
2	25	2	93	4.4	4.6	1.20
3	50	2	91	8.6	9.0	1.20
4	100	2	91	16.9	15.2	1.20
5	200	2	89	27.6	33.1	1.26
6	500	3	91	84.7	63.4	1.46

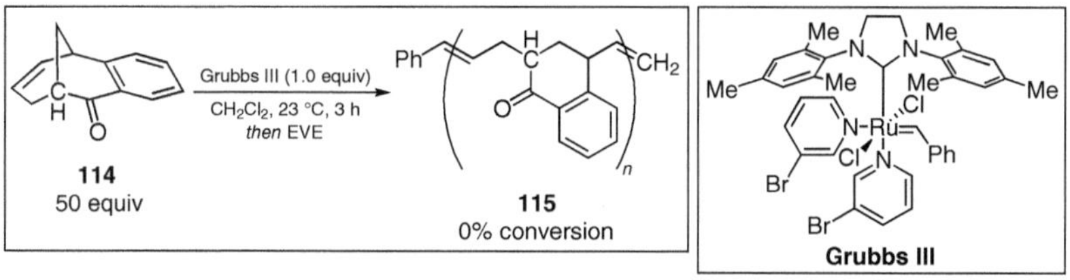

Scheme 1.24 Living ROMP of unsaturated twisted amide **82a**. Source: Modified from Xu et al. [67].

1.3.4 Miscellaneous Reactions

1.3.4.1 Ring Opening via Olefin Metathesis

Szostak reported the selective C=C bond cleavage of twisted lactam **82a** containing alkene moieties by ring-opening metathesis with Grubbs II catalyst without any products resulting from N–C(O) or σ N–C bond cleavage (Scheme 1.25) [68]. Cinnamyl, aliphatic, and electron-deficient olefins could be utilized to produce the ring-opened products **116** and **117** with moderate to excellent selectivity for ring-opened product **116**. Computational modeling predicts that the ring strain of the bridged lactam **82a** (14.2 kcal/mol) is significantly higher than that of analogous amine (0.8 kcal/mol), ketone (0.0 kcal/mol), and hydrocarbon (1.0 kcal/mol). The analogous planar anilide resulted in >95% recovery of the starting material, highlighting the essential role of amidic strain energy of in the ring-opening metathesis.

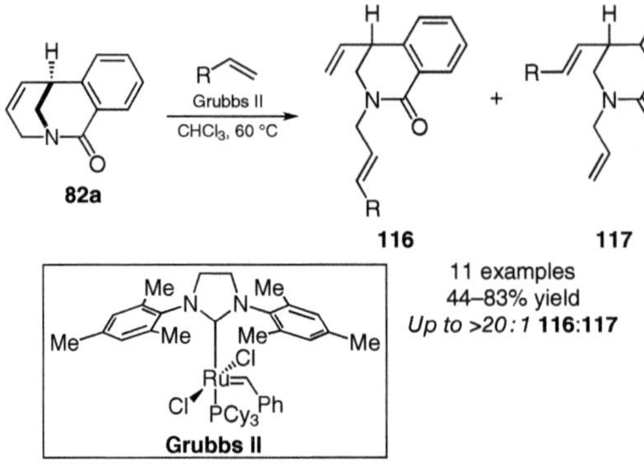

Scheme 1.25 Ring-opening metathesis of bridged lactam **82a**. Source: Adapted from Zhao et al. [68].

1.4 Conclusions and Outlook

Bridged lactams are the most successful models for studying the full range of amide bond distortion. The structure and reactivity of distorted amide bonds has wide-ranging applications, from amide bond rotation and cleavage in biological systems to applications in synthetic chemistry as reactive intermediates. The wealth of crystallographic and structural data now available provide valuable quantification of amide bond distortion with powerful predictive capabilities. While many valuable contributions to bridged lactam chemistry have been made since Lukeš first conceived of a bridged lactam in 1938 [3], there are many challenges that remain. First, no neutral bridged lactam with the exception of Kirby's adamantane-based bridged lactams has been prepared with an additive parameter $\tau + \chi_N$ exceeding 100° or 60% of maximum amide bond distortion [5]. Additionally, although nearly the entire Bürgi–Winkler–Dunitz scale is represented by bridged lactams synthesized to date, examples spanning the higher range of the distortion scale lack the structural diversity required for ample understanding of their stability [4, 5]. Finally, synthetic utilization of bridged lactams has mainly focused on hydrolysis and reactions akin to those of activated carbonyls. Given these limitations and the ubiquity of the amide bond, continued development of bridged lactams toward a comprehensive understanding of amide bond distortion will be invaluable.

References

1 Greenberg, A., Breneman, C.M., and Liebman, J.F. (ed.) (2000). *The Amide Linkage: Structural Significance in Chemistry, Biochemistry, and Materials Science*. New York: Wiley.

2 (a) Pauling, L. (1940). *The Nature of the Chemical Bond*. London: Oxford University Press. (b) Pauling, L. (1931). *J. Am. Chem. Soc.* 53: 1367–1400. (c) Liebman, J.F. and Greenberg, A. (2019). *Struct. Chem.* 30: 1631–1634.

3 Lukeš, R. (1938). *Collect. Czech. Chem. Commun.* 10: 148–152.
4 Szostak, M. and Aubé, J. (2013). *Chem. Rev.* 113: 5701–5765.
5 Szostak, R. and Szostak, M. (2019). *Molecules* 24: 274–290.
6 Yamada, S. (1999). *Rev. Heteroatom Chem.* 19: 203–236.
7 Liu, C. and Szostak, M. (2017). *Chem. Eur. J.* 23: 7157–7173.
8 Li, G., Ma, S., and Szostak, M. (2020). *Trends Chem.* 2: 914–928.
9 (a) Liu, J., Albers, M.W., Chen, C.M. et al. (1990). *Proc. Natl. Acad. Sci. U. S. A.* 87: 2304–2308. (b) Fisher, G. and Schmid, F.X. (1999). Peptidyl-prolyl cis/trans isomerases. In: *Molecular Chaperones and Folding Catalysts: Regulation, Cellular Functions and Mechanisms* (ed. B. Bakau). Boca Raton, FL: CRC Press. (c) Cox, C. and Lectka, T. (2000). *Acc. Chem. Res.* 33: 849–858. (d) Fisher, G. (2000). *Chem. Soc. Rev.* 29: 119–127.
10 Szostak, M., Yao, L., and Aubé, J. (2009). *J. Org. Chem.* 74: 1869–1875.
11 (a) Winkler, F.K. and Dunitz, J.D. (1971). *J. Mol. Biol.* 59: 169–182. (b) Dunitz, J.D. and Winkler, F.K. (1975). *Acta Crystallogr., Sect. B: Struct. Crystallogr. Cryst. Chem.* 31: 251–263.
12 (a) Szostak, R., Aubé, J., and Szostak, M. (2015). *Chem. Commun.* 51: 6395–6398. (b) Szostak, R., Aubé, J., and Szostak, M. (2015). *J. Org. Chem.* 80: 7905–7927.
13 (a) Wieberg, K.B. (1999). *Acc. Chem. Res.* 32: 922–929. (b) Wiberg, K.A. (2000). Origin of the amide rotational barrier. In: *The Amide Linkage: Structural Significance in Chemistry, Biochemistry, and Materials Science* (ed. A. Greenberg, C.M. Breneman and J.F. Liebman). New York: Wiley. (c) Kemnitz, C.R. and Loewen, M.J. (2007). *J. Am. Chem. Soc.* 129: 2521–2528.
14 Yamada, S. (1993). *Angew. Chem. Int. Ed.* 32: 1083–1085.
15 (a) Kirby, A.J., Komarov, I.V., Kowski, K., and Rademacher, P. (1999). *J. Chem. Soc., Perkin Trans.* 2: 1313–1316. (b) Kirby, A.J., Komarov, I.V., and Feeder, N. (2001). *J. Chem. Soc., Perkin Trans.* 2: 522–529.
16 Liniger, M., Vander Velde, D.G., Takase, M.K. et al. (2016). *J. Am. Chem. Soc.* 138: 969–974.
17 Nørskov-Lauritsen, L., Bürgi, H.-B., Hofmann, P., and Schmit, H.R. (1985). *Helv. Chim. Acta* 68: 76–82.
18 Otani, Y., Nagae, O., Naruse, Y. et al. (2003). *J. Am. Chem. Soc.* 125: 15191–15199.
19 (a) Kirby, A.J., Komarov, I.V., and Feeder, N. (1998). *J. Am. Chem. Soc.* 120: 7101–7102. (b) Kirby, A.J., Komarov, I.V., Wothers, P.D., and Feeder, N. (1998). *Angew. Chem. Int. Ed.* 37: 785–786.
20 Sorriso, S. (1982). *Chemistry of Functional Groups, Supplement F, Part 1* (ed. S. Patai), 1. Chichester: Wiley-Interscience.
21 Silverstein, R.M., Bassler, G.C., and Morrill, T.C. (1991). *Spectroscopic Identification of Organic Compounds*. New York: Wiley.
22 Greenberg, A. and Veranzi, C.A. (1993). *J. Am. Chem. Soc.* 115: 6951–6957.
23 (a) Treschanke, L. and Rademacher, P. (1985). *THEOCHEM* 122: 35–45. (b) Treschanke, L. and Rademacher, P. (1985). *THEOCHEM* 122: 47–56.
24 Morgan, J. and Greenberg, A. (2014). *J. Chem. Thermodyn.* 73: 206–212.
25 Adler, M., Alder, S., and Boche, G.J. (2005). *Phys. Org. Chem.* 18: 193–209.

26 Greenberg, A., Moore, D.T., and Dubois, T.D. (1996). *J. Am. Chem. Soc.* 118: 8658–8668.
27 Szostak, R. and Szostak, M. (2019). *J. Org. Chem.* 84: 1510–1516.
28 (a) Pracejus, H. (1959). *Chem. Ber.* 92: 988–998. (b) Pracejus, H., Kehlen, M., Kehlen, H., and Matschiner, H. (1965). *Tetrahedron* 21: 2257–2270. (c) Pracejus, H. (1965). *Chem. Ber.* 98: 2897–2905.
29 (a) Levkoeva, E.I., Nikitskaya, E.S., and Yakhontov, L.N. (1970). *Dokl. Akad. Nauk* 192: 342. (b) Kostyanovsky, R.G., Mikhlina, E.E., Levkoeva, E.I., and Yakhontov, L.N. (1970). *Org. Mass Spectrom.* 3: 1023–1029. (c) Levkoeva, E.I., Nikitskaya, E.S., and Yakhontov, L.N. (1971). *Khim. Geterotsikl. Soedin.* 378–384.
30 Werstiuk, N.H., Brown, R.S., and Wang, Q. (1996). *Can. J. Chem.* 74: 524–532.
31 Sliter, B., Morgan, J., and Greenberg, A. (2011). *J. Org. Chem.* 76: 2770–2781.
32 Eakin, C.M., Berman, A.J., and Miranker, A.D. (2006). *Nat. Struct. Mol. Biol.* 13: 202–208.
33 Szostak, M., Yao, L., Day, V.W. et al. (2010). *J. Am. Chem. Soc.* 132: 8836–8837.
34 Hu, F., Lalancette, R., and Szostak, M. (2016). *Angew. Chem. Int. Ed.* 55: 5062–5066.
35 Schwab, P.F.H., Levin, M.D., and Michl, J. (1999). *Chem. Rev.* 99: 1863–1934.
36 The original report of the synthesis of 2-quinuclidone reported by Yakhontov includes only nitrogen elemental analysis and has been called into question by Pracejus [30] and Stoltz [39]. For the original report by Yakhontov, see:Yakhontov, L.N. and Rubsitov, M.V. (1957). *J. Gen. Chem. USSR* 27: 83–87.
37 Tani, K. and Stoltz, B.M. (2006). *Nature* 441: 731–734.
38 Aubé, J. and Milligan, G.L. (1991). *J. Am. Chem. Soc.* 113: 8965–8966.
39 Komarov, I.V., Yanik, S., Ishchenko, A.Y. et al. (2015). *J. Am. Chem. Soc.* 137: 926–930.
40 Hine, J., Craig, J.C. Jr., Underwood, J.G. II, and Via, F.A. (1970). *J. Am. Chem. Soc.* 92: 5194–5199.
41 Ly, T., Krout, M., Pham, D.K. et al. (2007). *J. Am. Chem. Soc.* 129: 1864–1865.
42 Hunter, E.P. and Lias, S.G. (2005). Proton Affinity Evaluation. In: *NIST Chemistry Web Book*, NIST Standard Reference Database Number 69 (ed. P.J. Linstrom and W.G. Mallard), 20899. National Institute of Standards and Technology: Gaithersburg, MD.
43 Greenberg, A., Wu, G., Tsai, J.-C., and Chiu, Y.-Y. (1993). *Struct. Chem.* 4: 127–129.
44 (a) Radzicka, A. and Wolfenden, R. (1996). *J. Am. Chem. Soc.* 118: 6105–6109. (b) Smith, R.M. and Hansen, D.E. (1998). *J. Am. Chem. Soc.* 120: 8910–8913.
45 (a) Lopez, X., Mujika, J.I., Blackburn, G.M., and Karplus, M. (2003). *J. Phys. Chem.* 107: 2304–2315. (b) Mujika, J.I., Mercero, J.M., and Lopez, X. (2003). *J. Phys. Chem. A* 31: 6099–6107. (c) Mujika, J.I., Mercero, J.M., and Lopez, X. (2005). *J. Am. Chem. Soc.* 127: 4445–4453. (d) Mujika, J.I., Formoso, E., Mercero, J.M., and Lopez, X. (2006). *J. Phys. Chem. B* 110: 15000–15011. (e) Wang, B. and Cao, Z. (2011). *Chem. Eur. J.* 17: 11919–11929.
46 Blackburn, G.M. and Plackett, J.D. (1972). *J. Chem. Soc., Perkin Trans.* 2: 1366–1371.

47 Blackburn, G.M., Skaife, C.J., and Kay, I.T. (1980). *J. Chem. Res., Synop* . 294–295.

48 (a) Somayaji, V. and Brown, R.S. (1986). *J. Org. Chem.* 51: 2676–2686. (b) Wang, Q.P., Bennet, A.J., Brown, R.S., and Santarsiero, B.D. (1990). *Can. J. Chem.* 68: 1732–1739. (c) Bennet, A.J., Wang, Q.-P., Slebocka-Tilk, H. et al. (1990). *J. Am. Chem. Soc.* 112: 6383–6385. (d) Wang, Q.P., Bennet, A.J., Brown, R.S., and Santarsiero, B.D. (1991). *J. Am. Chem. Soc.* 113: 5757–5765.

49 Szostak, M. and Aubé, J. (2011). *Org. Biomol. Chem.* 9: 27–35.

50 Golden, J.E. and Aubé, J. (2002). *Angew. Chem. Int. Ed.* 41: 4316–4318.

51 For additional examples of hydrogenolysis of σ C–N bonds, see:Szostak, M. and Aubé, J. (2009). *Org. Lett.* 11: 3878–3881.

52 Lei, Y., Wrobleski, A.D., Golden, J.E. et al. (2005). *J. Am. Chem. Soc.* 127: 4552–4553.

53 Hu, F., Nareddy, P., Lalancette, R. et al. (2017). *Org. Lett.* 19: 2386–2389.

54 Li, G., Ji, C.-L., Hong, X., and Szostak, M. (2019). *J. Am. Chem. Soc.* 141: 11161–11172.

55 For a recent review on transamidation reactions, see:Li, G. and Szostak, M. (2020). *Synthesis* 52: 2579–2599.

56 Hall, H.K. Jr., and El-Shekeil, A. (1983). *Chem. Rev.* 83: 549–555.

57 Szostak, M., Yao, L., and Aubé, J. (2010). *J. Am. Chem. Soc.* 132: 2078–2084.

58 Zuo, Z., Xie, W., and Ma, D. (2010). *J. Am. Chem. Soc.* 132: 13226–13228.

59 Denzer, M. and Ott, H. (1969). *J. Org. Chem.* 34: 183–187.

60 (a) Beddoes, R.L., Davies, M.P.H., and Thomas, E.J. (1992). *J. Chem. Soc., Chem. Commun.* 538–540. (b) Baylis, A.M., Davies, M.P.H., and Thomas, E.J. (2007). *Org. Biomol. Chem.* 5: 3139–3155.

61 (a) Dolby, L.J. and Sakai, S. (1964). *J. Am. Chem. Soc.* 86: 1890–1891. (b) Dolby, L.J. and Sakai, S. (1967). *Tetrahedron* 23: 1–9.

62 Szostak, M. and Aubé, J. (2009). *J. Am. Chem. Soc.* 131: 13246–13247.

63 Yoshida, K., Sakuma, Y., and Ban, Y. (1987). *Heterocycles* 25: 47–50.

64 Schill, G., Löwer, H., Priester, C.U. et al. (1987). *Tetrahedron* 43: 3729–3745.

65 Ribelin, T.P., Judd, A.S., Akritopopulou-Zanze, I. et al. (2007). *Org. Lett.* 9: 5119–5122.

66 Fu, L., Xu, M., Yu, J., and Gutekunst, W.R. (2019). *J. Am. Chem. Soc.* 141: 2906–2910.

67 Xu, M., Bullard, K.K., Nicely, A.M., and Gutekunst, W.R. (2020). *Chem. Sci.* 10: 9729–9734.

68 Zhao, Q., Lalancette, R., Szostak, R., and Szostak, M. (2020). *ACS Catal.* 10: 737–742.

2

Modification of Amidic Resonance Through Heteroatom Substitution at Nitrogen: Anomeric Amides

Stephen A. Glover and Adam A. Rosser

University of New England, School of Science and Technology, Department of Chemistry, Madgwick Drive, Armidale, NSW 2351, Australia

2.1 Introduction

Amide linkages are structurally and chemically integral to the functionality of a range of molecules such as peptides, proteins, lactams, and many synthetic polymers [1]. While, in valence bond terms, they are composed of both a carbonyl and an amino functional group, joined by a single bond between the carbon and nitrogen, there is unanimity in the contemporary understanding of the resonance interaction between the nitrogen and the carbonyl in amides. Rather than the simplistic resonance hybrid of structures I and III (Figure 2.1), the electronic interaction is best seen as that between the LUMO of the carbonyl, $\pi^*_{C=O}$, and the HOMO of the amide nitrogen, the nitrogen atom lone pair, n_N, which resides in a $N2p_z$ orbital (Figure 2.2). The small contribution of the carbonyl oxygen to the LUMO dictates limited charge transfer to oxygen and transfer of lone pair density mostly to carbon, in line with the full resonance model presented in Figure 2.1. The major factor in the geometry of amides, and the restricted rotation about the amide bond, is this strong π-overlap between the nitrogen lone pair and the $C2p_z$ component of the LUMO, which dominates the $\pi^*_{C=O}$ orbital on account of the well-understood polarization toward oxygen in the $\pi_{C=O}$ [3, 4]. In addition, the strength of the partial nitrogen–carbon double bond character, so generated, imparts pseudo vinylic character to nitrogen. Accordingly, amide nitrogens are unlike those of amines, which, through their lone pair, are inherently basic in nature. Amide carbonyls are also resistant to nucleophilic addition reactions on account of the reduced positive charge at carbon.

1 | 2 | 3 | 4

a: X = Cl, Y = OR
b: X = OAc, Y = OR
c: X = Y = OR
d: X = OR, Y = NR₂
e: X = OR, Y = SR
f: X = Cl, Y = NR₂
g: X = OR, Y = O–
h: X = OR, Y = N
i: X = NR₂, Y = SR
j: X = Cl, Y = SR
k: X = OR, Y = N(OR)COR

3a: R = H
3b: R = OMe

4a: R = Me
4b: R = OMe

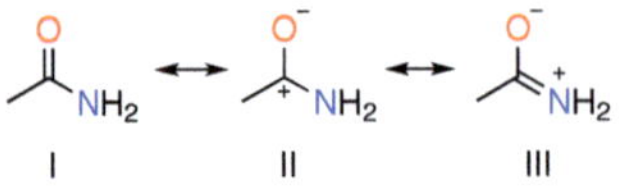

Figure 2.1 Resonance hybrid contributions in amides showing that the majority of charge transfer occurs between nitrogen and carbon. Source: Glover et al. [2]/MDPI/CC BY 4.0.

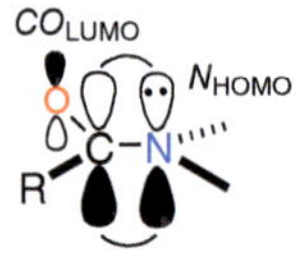

Figure 2.2 The interaction between the nitrogen HOMO and the carbonyl LUMO in amides. Source: Glover et al. [2]/MDPI/CC BY 4.0.

Amide resonance can be diminished by limiting the overlap between the nitrogen lone pair, n_N, and $\pi^*_{C=O}$ orbitals. In most known instances, this can occur by enforced twisting of the amino group about the N–C(O) bond and/or pyramidalizing the nitrogen, which amounts to introducing "s" character into the $N2p_z$ orbital. Other chapters in this book deal with the chemical consequences of these deformations; however, it is instructive to analyze these processes in some detail.

Computational modelling of N–C(O) rotation and nitrogen pyramidalization in *N,N*-dimethylacetamide **1**, at the B3LYP/6-31G(d) level, illustrates the energetic changes therefrom (Figure 2.3) [2, 5]. Distortion of the amide linkage can be characterized by Winkler–Dunitz parameters χ and τ, where $\chi = 60°$ for a fully pyramidal nitrogen and $\tau = 90°$ for a completely twisted amide where the lone pair orbital on nitrogen is orthogonal to the carbonyl $C2p_z$ orbital [6, 7]. Deformation from the non-twisted sp^2 planar ground state, where $\chi = \tau = 0°$ (Figure 2.3a), through rehybridization of nitrogen to sp^3 (Figure 2.3b), is accompanied by an increase in energy (~27 kJ/mol), but the majority of amide resonance remains intact. A much larger increase in energy (~100 kJ/mol) results from twisting the N–C(O) bond through 90° while maintaining sp^2 planarity at nitrogen (Figure 2.3d), and as the nitrogen is allowed to relax to sp^3 hybridization (Figure 2.3c), a fully twisted amide devoid of resonance is obtained. The release of ~31 kJ/mol in this final step is indicative of the concomitant twisting and pyramidalization observed in a variety of twisted amides. Thus, at this level of theory, the difference in energy between this relaxed, twisted form and the ground state, ~70 kJ/mol, can be regarded as an estimate of the barrier to rotation in *N,N*-dimethylacetamide and is usually equated with its resonance stabilization energy.

Typically, amides exhibit restricted rotation about the N–C(O) bond with measurable barriers in the order of 67–84 kJ/mol. From Figure 2.3, the energetic pathway is sigmoid with little change in energy upon moderate pyramidalization ($\chi = 0°–40°$) and minor twisting ($\tau = 0°–20°$). Clearly, rotation without pyramidalization is energetically unfavourable and many examples of twisted amides are testament to this. The shift to sp^3 hybridization at the amidic nitrogen is clearly demonstrated when twisting of the N–C(O) bond is geometrically enforced by tricyclic and bicyclic bridged lactams [8–15]. Kirby's "most twisted amide"

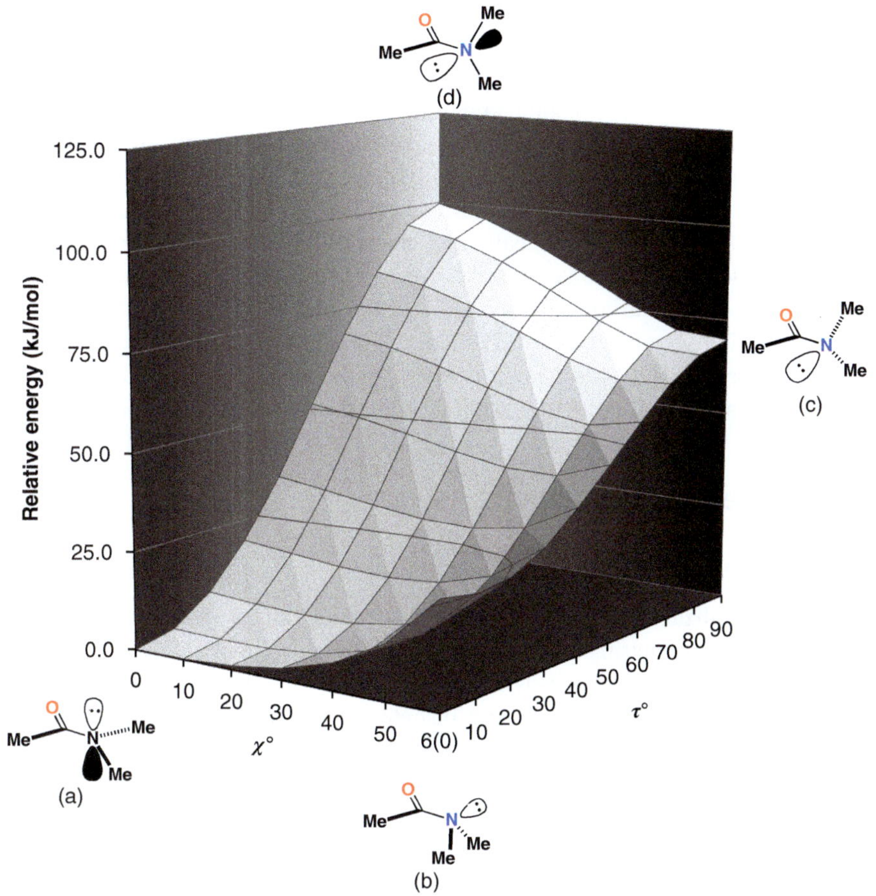

Figure 2.3 A B3LYP/6-31G(d) generated energy surface for the deformation of the amide moiety in *N,N*-dimethylacetamide **1** [5]. χ and τ are Winkler–Dunitz parameters for pyramidalization and twist, respectively [6, 7]. (a) The ground state planar structure for *N,N*-dimethylacetamide; (b) untwisted system with sp^3 hybridized nitrogen; (c) 90° rotation about *N–C(O)*, while maintaining sp^3 hybridization at nitrogen; and (d) the fully twisted moiety obtained through 90° rotation about *N–C(O)* while maintaining sp^2 hybridized nitrogen. Source: Glover et al. [2]/MDPI/CC BY 4.0.

3,5,7-trimethyl-1-azaadamantan-2-one, synthesized in 1998 [12, 14], and Tani and Stoltz's 2-quinuclidone, synthesized in 2006 [16], exemplify the fully twisted amide, geometrically and chemically. Intramolecular steric hindrance is another source of nonplanar twisted amides [17, 18], as exemplified by the thioglycolurils and other systems [19, 20]. Ring strain in non-twisted amides, such as 1-acylaziridines [21, 22] and *N*-acyl-7-azabicyclo[2.2.1]heptanes [23–26] can result in pyramidality at the amide nitrogen, despite retaining a noticeable n_N–$\pi^*_{C=O}$ interaction. In contrast, Szostak recently reported that a series of *N,N*-diBoc-4-substituted benzamides constitute a class of highly twisted amides with minimal nitrogen pyramidalization [27].

The structural changes accompanying the loss of amide resonance include the lengthening of the N–C(O) bond and minor shortening of the (N)C=O bond. Comparing the fully twisted trimethyl-1-azaadamantan-2-one to an analogous, unstrained tertiary δ-lactam, the N–C(O) bond lengthens by 0.123 Å from 1.352 to 1.475 Å, and the (N)C=O bond shortens minimally by 0.037 Å from 1.233 to 1.196 Å [14, 28, 29]. Spectroscopically and chemically, the amide carbonyl trends toward ketonic behavior at large twist angles. In tricyclic 3,5,7-trimethyl-1-aza-2-adamantanone, the carbonyl carbon ^{13}C NMR resonance is at 200.0 ppm, and the IR carbonyl vibrational frequency (1732 cm^{-1}) is significantly higher than regular amides (1680 cm^{-1}) [12, 14, 28]. Disconnection of the nitrogen lone pair, n_N, results in the known hydrolytic instability of highly twisted amides [8, 11, 12, 14, 16, 28, 30].

2.2 Properties of Anomeric Amides

2.2.1 Structural Properties

While pyramidalization, which is intrinsic to twisting, mixes "N2s" character into the N2p_z orbital, increasing "p" character in sp^2 hybrid orbitals achieves the same result. Consequently, another way in which amides may be dispossessed of their resonance is through heteroatom and bisheteroatom substitution by electronegative heteroatoms at the amide nitrogen in **2**. Electronegative atoms demand an electron density redistribution, which is facilitated by a shift in the hybridization at nitrogen from sp^2 toward sp^3 in accordance with Bent's rule [31, 32], and the effect is increased with increasing electronegativity of the connected atoms and, importantly, with bisheteroatomic rather than monoheteroatomic substitution. Reduced n_N–$\pi^*_{C=O}$ overlap due to pyramidalization (Figure 2.4) together with the increased "2s" character, and therefore a lowering in energy of the lone pair on nitrogen, results in structural and chemical differences in comparison to traditional amides. We named the class of bisheteroatom-substituted amides, **2**, "anomeric amides" on account of the pronounced anomeric effects that can and do occur between the heteroatoms (see Section 2.3.1) [33–42]. However, their physical, theoretical, spectroscopic, and chemical properties differ radically from those of conventional primary or secondary and tertiary amides, and most notably with respect to their propensity for reactivity involving both the amide nitrogen and carbonyl carbon atoms. Their unique properties have been reviewed in recent years [2, 38, 39].

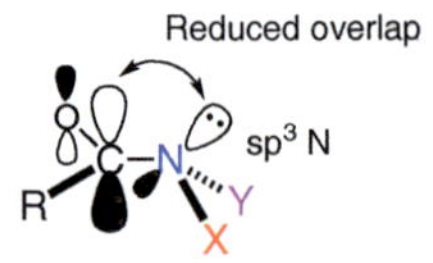

Figure 2.4 Reduced n_N–$\pi^*_{C=O}$ overlap due to pyramidalization. Source: Glover et al. [2]/MDPI/CC BY 4.0.

In the first detailed theoretical study on these amides in 1996, we reported on the B3LYP/6-31G(d) theoretical properties of the corresponding formamides, *N*-methoxyformamide **3a**, and *N,N*-dimethoxyformamide **3b** [33]. The substitution of a hydrogen at nitrogen in formamide by methoxyl in **3a** led to an increase in N–C(O) bond length from 1.362 to 1.380 Å, and a second substitution at nitrogen in **3b** resulted in a similar increase to 1.396 Å; the carbonyl bond contracted marginally by 0.006 Å. Nitrogen in *N*-methoxyformamide and *N,N*-dimethoxyformamide becomes distinctly pyramidal with average angles at nitrogen of ~115°. With similar degrees of pyramidality (χ), the increased *N*–C(O) bond length in *N,N*-dimethoxyformamide could not solely be attributed to deformation at nitrogen. Energetic lowering of the lone pair electrons is also responsible for reduced overlap with the adjacent carbonyl C2p_z orbital. This is dramatically exemplified by the rotational barriers for formamide, *N*-methoxyformamide **3a**, and *N,N*-dimethoxyformamide **3b**, which were computed to be of the order of 75, 67, and 29 kJ/mol, respectively [33].

More recently, the impact of heteroatom substitution at amide nitrogen has been clearly demonstrated by comparing deformation surfaces of *N,N*-dimethylacetamide **1** (Figure 2.3) to those of *N*-methoxy-*N*-methylacetamide **4a** (Figure 2.5) and *N,N*-dimethoxyacetamide **4b** (Figure 2.6) [2]. In contrast to *N,N*-dimethylacetamide (Figure 2.3), the lowest energy forms of **4a** and **4b** clearly deviate from planarity at nitrogen with χ_0 in the region of 40° and 50°, respectively. In both structures, the highest point corresponding to $\chi = 0°$, $\tau = 90°$ is metastable and the fully twisted planar, and therefore nonconjugated forms relax to fully pyramidal conformations ($\tau = 90°$, $\chi = 60°$), the energy of which approximates the B3LYP/6-31G(d) barriers to amide isomerization in each case, which are 67 and 44 kJ/mol, respectively. While the reduction in the energy barrier for **4a**, relative to that for *N,N*-dimethylacetamide, is negligible, the attachment of two electronegative oxygens to the amide nitrogen in **4b** radically lowers the isomerization barrier.

Inversion barriers at nitrogen are low on account of the gain in resonance stabilization in the planar transition states for inversion, which are represented by structures (a) in Figures 2.5 and 2.6. Planarization in the hydroxamic ester is less costly than in *N,N*-dimethoxyacetamide (~6.3 vs. 14.6 kJ/mol, respectively). The barrier is higher for *N,N*-dimethoxyacetamide where resonance can only be achieved by loss of N2s character as the lone pair assumes pure 2p_z character.

2.2.2 Natural Bond Order Analysis

Natural bond order (NBO) analysis [31, 43–45] of ground state structures of *N,N*-dimethylacetamide **1** (at χ, $\tau = 0°$ in Figure 2.3a), *N*-methoxy-*N*-methylacetamide **4a** (at $\chi = 40°$, $\tau = 0°$, Figure 2.5), and *N,N*-dimethoxyacetamide **4b** (at $\chi = 50°$, $\tau = 0°$, Figure 2.6), computed at the B3LYP/6-31G(d) level, gives a measure of the change in hybridization upon heteroatom substitution. The amide nitrogens contribute a total of 2s (100%) and 2p (300%) to the bonding and NBO analysis

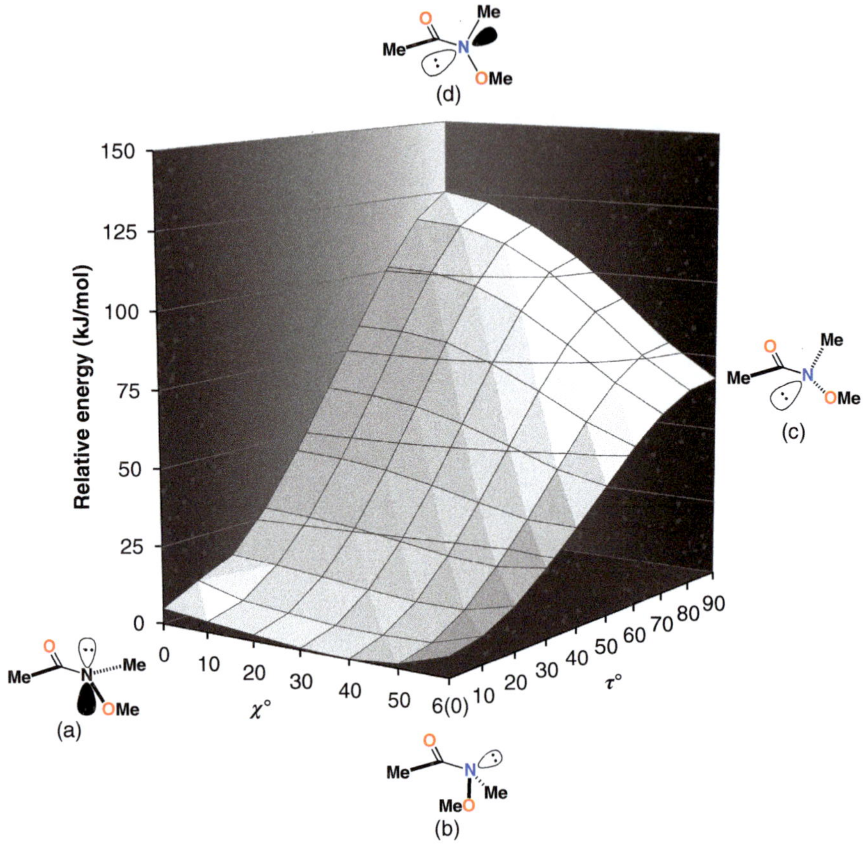

Figure 2.5 B3LYP/6-31G(d) energy surfaces for deformation of *N*-methoxy-*N*-methylacetamide **4a**. (a) Planar untwisted structure; (b) untwisted system with sp^3 hybridized nitrogen; (c) 90° rotated structure with sp^3 hybridization at nitrogen; and (d) the fully twisted moiety with sp^2 hybridized nitrogen. Source: Glover et al. [2]/MDPI/CC BY 4.0.

gives the percentage contribution of 2s and 2p orbital density to the nitrogen lone pair, and the hybrid orbital contributions from nitrogen to the N–O and N–C bonds attached to nitrogen (Table 2.1).

In *N*,*N*-dimethylacetamide, which is fully planar at nitrogen, the lone pair is pure 2p, while the hybrid orbitals bonding to the carbonyl carbon and *N*-methyl carbons are formally sp_2 in character. While the composition of the nitrogen hybrid orbital contribution to the N–C(O) bond is nearly the same in all three structures, in **4a** some 2p character is moved from the nitrogen lone pair to the hybrid orbital bonded to oxygen, and, accordingly, the lone pair gains some 9% 2s character. This shift is compounded in **4b** where the nitrogen lone pair gains some 19% 2s character and nitrogen "p" character is increased in both N–O bonds. It is clear that the reduction in $N2p_z$ character and concomitant increase in 2s character in the nitrogen lone pair in **4b** must reduce the capacity for orbital overlap with the carbonyl $C2p_z$ orbital without twisting about the N–C(O) bond, with a consequent reduction in formal resonance.

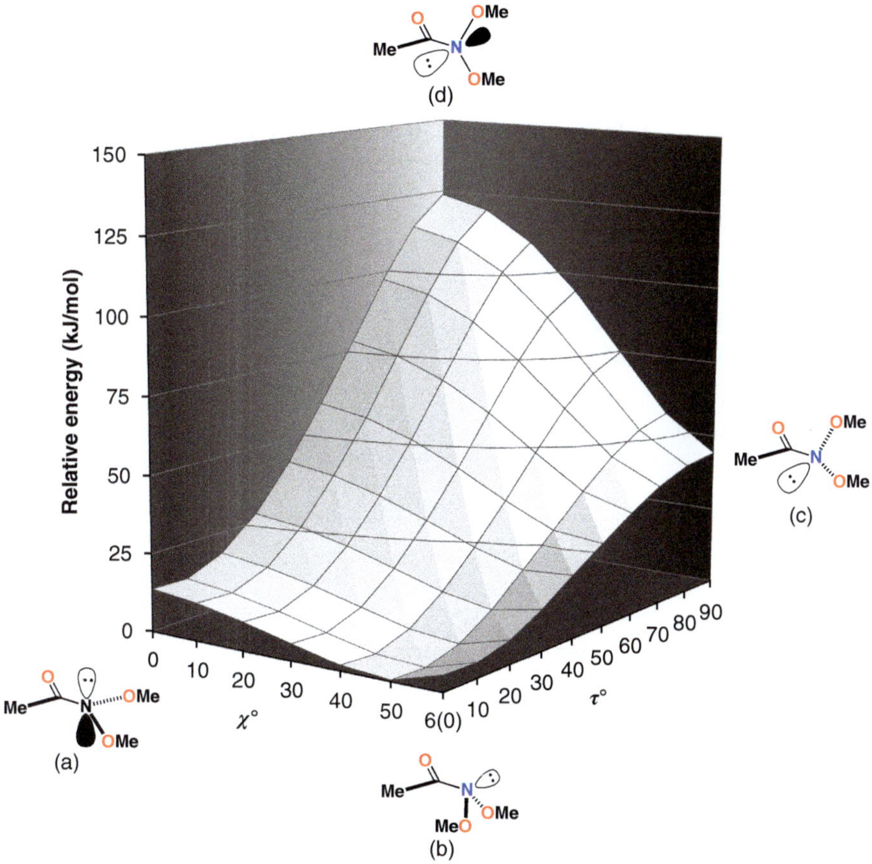

Figure 2.6 B3LYP/6-31G(d) energy surfaces for deformation of *N*,*N*-dimethoxyacetamide **4b**. (a) Planar untwisted structure; (b) untwisted system with sp^3 hybridized nitrogen; (c) 90° rotated structure with sp^3 hybridization at nitrogen; and (d) the fully twisted moiety with sp^2 hybridized nitrogen. Source: Glover et al. [2]/MDPI/CC BY 4.0.

2.2.3 Theoretical Determination of Amide Bond Resonance

While rotational barriers are a measure of resonance in amides, they comprise of not only a resonance stabilization that must be overcome, but they may also be influenced by steric compression in the ground state. Computational chemistry provides an ideal means of measuring the extent and modifications of amidic resonance at the ground state using suitably constituted models at an acceptable level of MO theory. It has become the norm to measure and quantify the extent of an amide resonance interaction relative to that of *N*,*N*-dimethylacetamide **1**, which is arbitrarily assigned as 100% *amidicity* [5, 46]. Several methods have been devised to calculate amidicities. Mucsi first used a method based on relative enthalpies of hydrogenation of the amide carbonyls, which works adequately for simple, acyclic amides. However, their results for cyclic amides, in particular β-lactam [46], and penicillin [47] have been refuted [48].

Table 2.1 Natural bond order analysis for nitrogen, computed resonance energies, and amidicities for *N,N*-dimethylacetamide **1**, *N*-methoxy-*N*-methylacetamide **4a**, and *N,N*-dimethoxyacetamide **4b**.

Structure	N LP	N–C(O)	N–C	N–O	RE (kJ/mol)	Amidicity (%)
1	s (0%) p (100%)	s (36.7%) p (63.2%)	s (32.2%) p (67.7%); s (31.0%) p (69.0%)	—	−76.0[a)]/−77.5[b)]	100
4a	s (9.2%) p (90.8%)	s (37.2%) p (62.7%)	s (33.8%) p (66.2%)	s (19.8%) p (80.1%)	−62.1[a)]/−61.3[b)]	80[a)]/81[b)]
4b	s (18.7%) p (81.2%)	s (37.7%) p (62.2%)	—	s (20.7%) p (79.2%); s (22.9%) p (77.0%)	−35.9[a),b)]	47[a)]/46[b)]

a) COSNAR method.
b) TA method.

In a 2012 publication, we outlined two concurring isodesmic methods for estimating the amidicity of amides and lactams [5, 48]. The "carbonyl *s*ubstitution, *n*itrogen *a*tom *r*eplacement" (COSNAR) method, developed by Greenberg, can be used to evaluate the energy stabilization when the target amide is generated from corresponding ketone and amine according to isodesmic Eq. (2.1) [49–51]. Steric or substituent effects are conserved throughout the reaction, so steric corrections are not required. For reference *N,N*-dimethylacetamide itself, RE = $\Delta E_{\mathrm{COSNAR}} = -77.5\,\mathrm{kJ/mol}$.

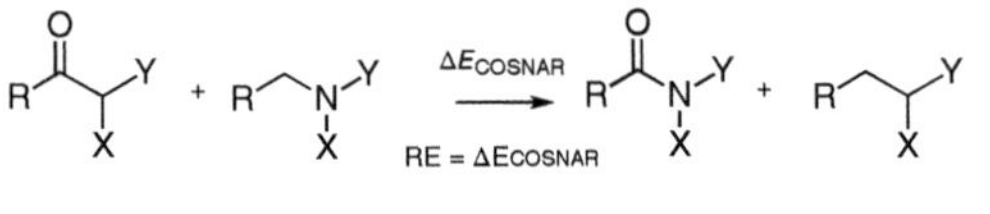

(2.1)

In our second approach, which we called the "transamidation method" (TA), the energy change is determined when *N,N*-dimethylacetamide **1** transfers the carbonyl oxygen to a target amine according to Eq. (2.2). The limiting energy increase for formation of fully twisted and unstrained 1-azaadamantan-2-one by the corresponding reaction with 1-azaadamantane constitutes complete loss of resonance stabilization ($\Delta E_{\mathrm{TA}} = 76.0\,\mathrm{kJ/mol}$), and the $\mathrm{RE_{TA}}$ for *N,N*-dimethylacetamide is therefore −76.0 kJ/mol. For simple primary amides and secondary or tertiary alkylamides, the intrinsic resonance energy $\mathrm{RE_{TA}}$ is given by Eq. (2.3). However, where heteroatom substituents are present at nitrogen, ΔE_{TA} must be corrected for any additional inductive destabilization of the carbonyl (ΔE_{ind}) in the absence of resonance, which we obtain isodesmically from reactions such as Eq. (2.4) [40]:

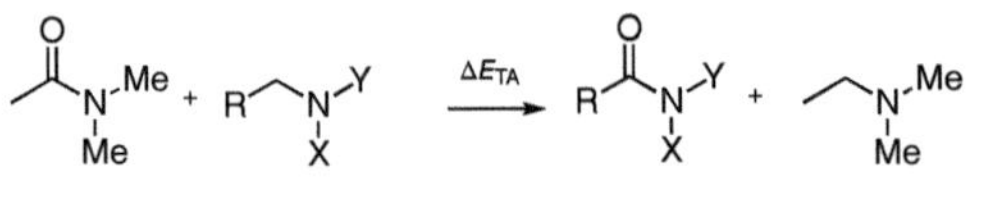

(2.2)

$$\mathrm{RE_{TA}} = [-76.0 + \Delta E_{\mathrm{TA}}] \quad \mathrm{kJ/mol} \tag{2.3}$$

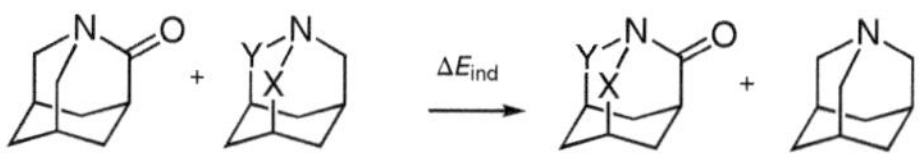

(2.4)

In the cases of anomeric amides, the residual resonance by the TA method, $\mathrm{RE_{TA}}$, is given by Eq. (2.5):

$$\mathrm{RE_{TA}} = [-76.0 + (\Delta E_{\mathrm{TA}} - \Delta E_{\mathrm{ind}})] \quad \mathrm{kJ/mol} \tag{2.5}$$

In both the TA and COSNAR approaches, zero-point energies largely cancel, and meaningful results are conveniently obtained without the need for full frequency calculations [50, 52]. RE by both methods, the negative of the traditional representation of resonance stabilization energy, should correlate [5, 40–42, 53, 54], and RE as a percentage of −76.0 kJ/mol (or −77.5 kJ/mol in the case of COSNAR) yields the amidicity relative to *N*,*N*-dimethylacetamide **1**.

The resonance energy and amidicity of *N*-methoxy-*N*-methylacetamide **4a** has been determined by COSNAR (−62.1 kJ/mol, 80% amidicity) and TA (−61.3 kJ/mol, 81% amidicity) and accords nicely with the lowest rotational barrier from Figure 2.5 of 67.5 kJ/mol. In contrast, for the bisoxyl-substituted acetamide **4b**, $\mathrm{RE_{COSNAR}}$ and $\mathrm{RE_{TA}}$ were determined at −35.9 kJ/mol, respectively, just 47% or 46% that of *N*,*N*-dimethylacetamide [40]. From Figure 2.6 the rotational barrier was 44.4 kJ/mol, and therefore most of the barrier can be accounted for by loss of resonance.

2.2.4 Sources of Anomeric Amides

Since the early 1990s, we have encountered a range of *N*,*N*-bisheteroatom-substituted amides, primarily as a result of our investigations of the mutagenic activity, structure, and reactivity of *N*-acyloxy-*N*-alkoxyamides (NAAs) **2b**, which are a class of anomeric amides [38, 55–68]. Originally, we synthesized these as possible sources of resonance-stabilized *N*-acyl-*N*-alkoxynitrenium ions [69] by analogy with carcinogenic arylamine metabolites, *N*-acetoxy-*N*-acetylarylamines, which are known to generate long-lived, resonance-stabilized *N*-acetyl-*N*-arylnitrenium ions, the ultimate DNA-damaging agents (Scheme 2.1i) [70–75]. Since *N*-acyl-*N*-alkoxynitrenium ions are similarly resonance-stabilized [69], our original premise was that, as a consequence, NAAs would be similarly active (Scheme 2.1ii). *N*-Acyloxy-*N*-alkoxyamides were found to be uniformly mutagenic; however, we found that they are direct-acting mutagens, which react intact with nucleophilic centers in DNA by S_N2 displacement of carboxylate (Scheme 2.1iii) [38, 55, 57, 58, 60–65, 68].

As consequence of our extensive quantitative structure-activity relationship (QSAR) investigations into their mutagenic activity, we have synthesized well over a hundred structures by treatment of a range of *N*-alkoxy-*N*-chloroamides **2a**, which are ONCl forms of anomeric amides, with silver or sodium carboxylate

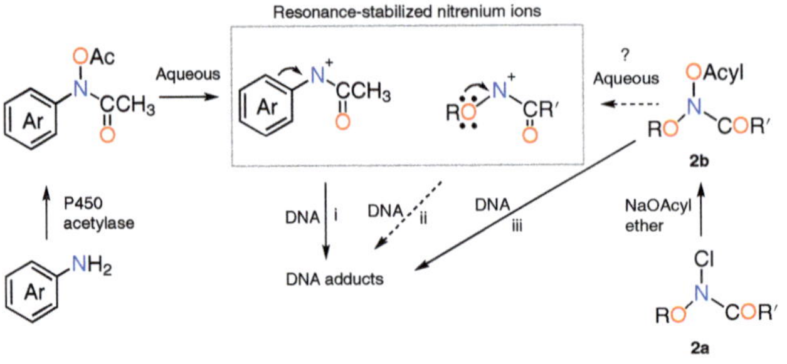

Scheme 2.1 Relationship between *N*-acyloxy-*N*-alkoxyamides and aromatic amine metabolites.

salts in anhydrous solvents [64, 65, 67, 68]. Additionally, NAAs react with a variety of nucleophiles to produce other anomeric amide congeners. Reaction with aromatic amines generates reactive intermediates *N*-alkoxy-*N*-aminoamides **2d**, NNO anomeric amides [63, 66, 76]. We encountered a relatively stable NNO system in studies of the structure and thermal decomposition of *N,N′*-diacyl-*N,N′*-dialkoxyhydrazines **2k**, a dimeric form of **2d** [36, 77]. Thiols generate reactive *N*-alkoxy-*N*-alkylthiylamides **2e**, an SNO system [78]. In related studies *N,N*-dialkoxyamides **2c**, an ONO system, were synthesized by solvolysis of *N*-alkoxy-*N*-chloroamides **2a** in aqueous alcohols and, more recently, through the reaction of hydroxamic esters with hypervalent iodine reagents in appropriate alcohols [41, 77, 79]. We have also encountered several anomeric reactive intermediates through reactions of NAAs with hydroxide ions, which generate *N*-alkoxy-*N*-hydroxamate anions **2g** [64], and reaction with azide, which ultimately generates 1-acyl-1-alkoxydiazenes and aminonitrenes **2h** [80, 81]. Other anomeric amides of theoretical interest to us are *N*-amino-*N*-chloroamides **2f**, an NNCl congener [42], as well as *N*-alkylthiyl-*N*-aminoamides **2i** and *N*-alkylthiyl-*N*-chloroamides **2j** (NNS and SNCl systems, respectively). A number of carbamate and urea analogs of **2a–c** and **2k** are also known [82–86].

2.2.5 Experimental Evidence for Reduced Resonance in Anomeric Amides

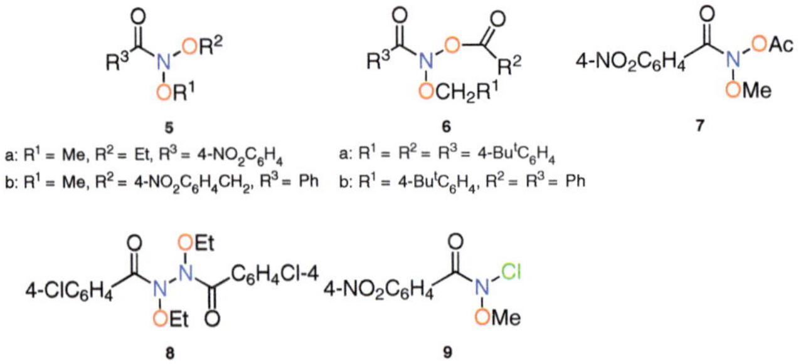

The unusual structure of anomeric amides is borne out by X-ray crystallography and infrared and NMR spectroscopy, each of which is a reporter of their reduced

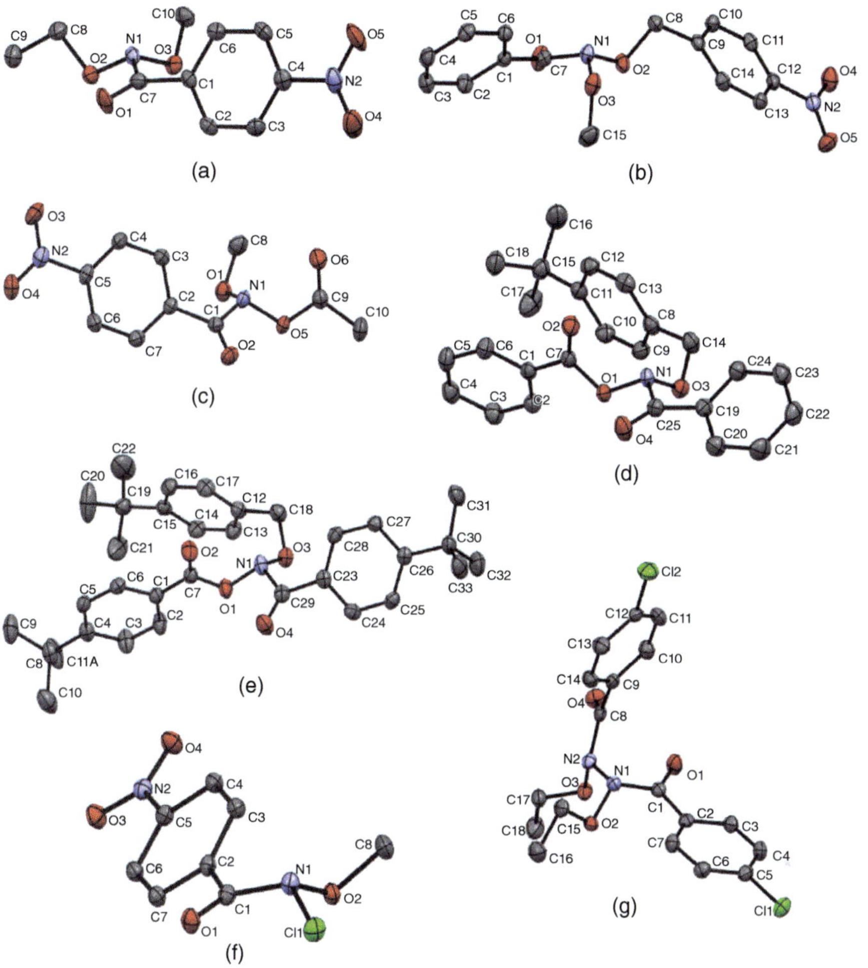

Figure 2.7 X-ray structures for (a) *N*-ethoxy-*N*-methoxy-4-nitrobenzamide **5a**, (b) *N*-methoxy-*N*-(4-nitrobenzyloxy)benzamide **5b**. Source: (a, b) Adapted from Glover et al. [40], (c) *N*-acetoxy-*N*-methoxy-4-nitrobenzamide **7**. Source: Adapted from Shtamburg et al. [87], (d) *N*-benzoyloxy-*N*-(4-*tert*-butylbenzyloxy)benzamide **6b**, (e) *N*-(4-*tert*-butylbenzoyloxy)-*N*-(4-*tert*-butylbenzyloxy)-4-*tert*-butylbenzamide **6a**. Source: (d, e) Adapted from Gillson et al. [37], (f) *N*-chloro-*N*-methoxy-4-nitrobenzamide **9.** Source: Adapted from Shtamburg et al. [87], and (g) *N*,*N*′-4-chlorobenzoyl-*N*,*N*′-diethoxyhydrazine **8.** Source: Adapted from Glover et al. [36].

resonance. X-ray structures can detect both conformation at nitrogen and variations in bond lengths, particularly the N–C(O) bond lengths. Spectroscopically, reduced resonance would be expected to impact upon the amide carbonyl stretch frequencies in their infrared spectra and on carbonyl ^{13}C shifts, as well as line widths in their ^{1}H NMR spectra, which are influenced by barriers to *E*/*Z* isomerization.

The solid-state structures of a number of stable anomeric amides (**5–9**) have been determined by X-ray crystallography (Figure 2.7) and relevant structural parameters from these are given in Table 2.2.

Table 2.2 Selected structural data for anomeric amides (**5–9**) from X-ray structures (Figure 2.7).

Structure[a]	N–C(O) (Å)	(N)C=O (Å)	$\theta^{b)}$ (°)	χ_N (°)	τ (°)	Anomeric twist[c] (°)
5a ONO [40]	1.409	1.206	331.2	58.3	6.7	**C10–O3–N1–O2: 95.9** C8–O2–N1–O3: −114.1
5b ONO [40]	1.421	1.211	334.5	55.6	13.9	**C8–O2–N1–O3: −101.6** C15–O3–N1–O2: −63.8
6a ONOAc [37]	1.439	1.205	323.5	65.33	15.51	**C18–O3–N1–O1: 96.21** C7–O1–N1–O3: −141.60
6b ONOAc [37]	1.441	1.207	324.1	65.62	13.93	**C14–O3–N1-O1: 96.66** C7–O1–N1–O3: −137.6
7 ONOAc [87]	1.411	1.203	330.3	59.7	−14.0	**C8–O1–N1–O5: −91.83** C9–O5–N1–O1: 116.15
8 (N1) NNO [36]	1.412	1.213	343.2	47.33	−8.49	LP(N2)–N2 –N1–O2: 47.3
(N2) NNO	1.410	1.207	341.1	48.90	−11.35	**LP (N1)–N1–N2–O3: 178.6**
9 ONCl [87]	1.408	1.204	337.5	52.5	−13.3	**C8–O2–N1–Cl1: −84.2**

a) XNY designation.
b) Sum of angles at anomeric nitrogen.
c) Anomeric alignments in boldface.
Source: Glover et al. [2]/MDPI/CC BY 4.0.

Our structures for **5**, **6**, and **8** and those of Shtamburg for **7** and **9** provide clear evidence of reduced amide resonance. The reported average N–C(O) bond length in acyclic amides, generated from Cambridge Structural Database (CSD), is 1.359 Å (median 1.353 Å) [37, 88], which is significantly shorter than the average of 1.418 Å from these seven X-ray structures. The data in Table 2.2 shows that as the combined electron demands of X and Y increase, both the N–C(O) bond length and pyramidalization at nitrogen increase. The N–C(O) bond lengths in **6a** and **6b**, which are highly pyramidal at nitrogen ($\chi > 65°$), are particularly long (1.44 Å). The electron demand of the acyloxyl group, together with the alkoxyl group (estimated total electronegativity >6.8), must apportion more "s" character to the nitrogen lone pair orbital. The *N*,*N*-dialkoxybenzamide **5b** has a shorter N–C(O) bond length (1.42 Å) and is considerably less pyramidal at nitrogen ($\chi = 56°$). The single NNO representative, hydrazine **8**, has both shorter N–C(O) bonds (1.41 Å) and much less pyramidal nitrogens ($\chi = 47°$ and 49°), which is reflective of the reduced combined electron demand of nitrogen and oxygen (total electronegativity 6.5). Electron demand in the ONCl amide **9** is similar (total electronegativity 6.6) as are the N–C(O) bond length (1.41 Å) and pyramidality at nitrogen ($\chi = 53°$). Substituent effects are evident; the N–C(O) bonds of 4-nitrobenzamides **5a** (1.41 Å), **7** (1.41 Å), and **9** (1.41 Å) are shorter than those of respective benzamides **5b**, **6a**, and **6b**, which can be accounted for by the electron demand of the conjugated 4-nitro group, which imposes greater positive charge at the amide carbonyl carbon, thereby facilitating better amide nitrogen lone pair overlap. However, their N–C(O) bond lengths are still 0.09 Å longer than the N–C(O) bond in their precursor hydroxamic ester, *N*-methoxy-4-nitrobenzamide (1.32 Å) [87], which is nearly the same as that in a *N,N*-dialkyl-4-nitrobenzamide (1.34 Å) [89]. It is informative that while the (N)C=O bond (average 1.21 Å) contracts slightly compared to simple acyclic amides (1.23 Å), there is little to no correlation between change in (N)C=O bond length and degree of lone pair dislocation. Again, this is attributed to bias toward carbon in the LUMO of carbonyls, which results in little transfer of nitrogen lone pair electron density to oxygen, making the (N)C=O bond insensitive to the degree of amide resonance (Figure 2.2) [3, 4, 34, 39].

Deviation from the usual sp^2 hybridization at the amide nitrogen ($\chi_N = 0°$) is significant and also in line with the electron demands of substituents. *N*-Acyloxy-*N*-alkoxyamides **6a**, **6b**, and **7** ($\chi_N = 65.33°$, 65.62°, and 59.7°, respectively) are pyramidalized at nitrogen to the extent of, and beyond, what is expected for a pure sp^3 hybridization. However, though similarly substituted at nitrogen, the 4-nitrobenzamide **7** is less pyramidal than the benzamides **6a** and **6b**, presumably, once again, on account of greater positive charge at the carbonyl carbon and attendant increase in nitrogen lone pair attraction. Both *N,N*-dialkoxyamides **5a** and **5b** ($\chi_N = 58.3°$ and 55.6°, respectively) are also strongly pyramidalized. For Shtamburg's *N*-alkoxy-*N*-chloroamide **9**, X-ray data reveals a χ_N of 52.5°, similar to that of *N,N*-dialkoxyamides **5a** and **5b**. Both amide nitrogens in hydrazine **8**, a stable form of an *N*-alkoxy-*N*-aminoamide **2d**, are the least pyramidal with χ_{N1} and χ_{N2} of 47° and 49°, respectively.

Despite high degrees of pyramidalization in this set of anomeric amides, there is minimal twist about the N–C(O) bond ($\tau = 6.7°–15.5°$, Table 2.2). This indicates that lone pair orbital overlaps with the carbonyl $C2p_z$ orbitals, though clearly less effective on electronic and geometric grounds, remains a stabilizing influence. As can be seen in the B3LYP/6-31G(d) deformation surface for *N,N*-dimethoxyacetamide **4b** (Figure 2.6), the strongly pyramidal structure requires twist angles (τ) beyond 20° before there is significant loss of stabilization.

Pyramidal nitrogens and corresponding anomeric properties have also been observed in the structures of a number of urea and carbamate analogs of anomeric amides **10–14**, studied by Shtamburg and coworkers, and selected structural data are presented in Table 2.3 [82–85]. In all cases, the anomeric amide nitrogens are distinctly pyramidal with χ_N values close to 60°. In the cases of ureas **10**, **11**, and **12**, the α nitrogens are planar and N^α–C(O) bonds are much shorter than bonds to the anomeric nitrogens. Ureas are an interesting case where the N^α lone pair resonance is in competition with that of the anomeric nitrogen. Consequent to this, the N^α–C(O) bond is shorter than in urea; diminished resonance from the anomeric nitrogen permits stronger resonance with N^α. This is borne out computationally: at the B3LYP/6-311+G(d,p) level, the N^α–C(O) and N–C(O) bonds in **12a** are 1.361 and 1.442 Å, respectively, while the N–C(O) bonds in urea are computed to be 1.376 Å. Where substituent electronic effects at the anomeric nitrogen are largely similar as in **11b** and **13**, it can be deduced that the stronger conjugative effect of the α-nitrogen lone pair in the urea **11b**,

Table 2.3 Selected structural data for ureas (**10–12**) and carbamates (**13–14**) from X-ray structures.

Structure[a]	N–C(O)[b] (Å)	(N)C=O (Å)	θ[c] (°)	χ_N[b] (°)	τ[b] (°)	Anomeric twist[d] (°)
10a ONCl [82]	1.443(1.321)	1.226	329.1	59.9	8.2	**C–O–N–Cl: −90.9**
10b ONCl [82]	1.472(1.135)	1.210	325.8	61.9	−13.4	**C–O–N–Cl: −100.1**
11a ONOAc [83]	1.426(1.330)	1.222	333.6	−57.1	−6.8	**C–O–N–Oacyl: −104.0** C(O)–O–N–O: −69.0
11b ONOAc [84]	1.441(1.321)	1.233	323.7	64.6	11.8	**C–O–N–Oacyl: −98.2** C(O)–O–N–O : 119.2
12a ONO [84]	1.438(1.320)	1.220	331.8	57.4	0.8	**C–O2–N1–O1: −89.3** C–O1–N1–O2: 55.2
12b ONO [90]	1.441(1.357)	1.204	324.0	64.5	3.8	**C–O2–N1–O1: −88.2** C–O1–N1–O2: 127.8
13 ONOAc [83]	1.424	1.198	334.2	−56.3	2.9	**C–O–N–Oacyl: −95.5** C(O)–O–N–O : 67.0
14 NNO [85] (N1)/(N2)[e]	1.408	1.194	340.0	59.8	−9.4	**LP** **(N1)–N1–N2–O : 189.2**

a) XNY designation.
b) Anomeric nitrogen; N^{α}–C(O) bond lengths in parentheses.
c) Sum of angles at anomeric nitrogen.
d) Anomeric alignments in boldface.
e) Equivalent nitrogens.
Source: Modified from Glover et al. [2].

relative to the α-oxygen in the carbamate **13**, results in significantly less anomeric amide resonance interaction and, hence, a longer N–C(O) bond and significant increase in pyramidality at the amide nitrogen. Comparison of the ONCl structures of ureas **10a** and **10b** with amide **9** (Table 2.2) again indicates greater pyramidality at the anomeric nitrogen and longer amide bonds in the ureas where competing acyl nitrogen resonance reduces the anomeric amide resonance interaction.

10

a: R^1 = H, R^2 = Me
b: R^1 = 4-$NO_2C_6H_4$, R^2 = $C(Me)_2CO_2Me$

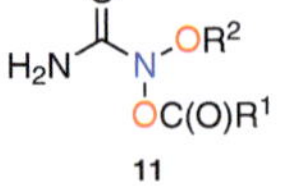

11

a: R^1 = Me, R^2 = Et
b: R^1 = 4-ClC_6H_4, R^2 = Bu

12

a: R = H
b: R = 4-$NO_2C_6H_4$

13

a: R = 4-ClC_6H_4

14

2.2.6 Spectroscopic Properties of Anomeric Amides

Spectroscopic properties of anomeric amides are strongly influenced by reduced resonance due to electronegativity of substituents at nitrogen. Infrared and ^{13}C NMR data for a diverse range of ONCl, ONOAcyl [34, 38, 39], and ONO systems [39, 79] have been reported as well as for a number of stable NNO systems, the *N,N′*-dialkoxy-*N,N′*-diacylhydrazines [36, 91, 92] and typical frequencies and chemical shifts are presented in Table 2.4. Representative infrared carbonyl stretch frequencies for stable anomeric amides in solution are significantly higher (1700–1750 cm^{-1}) [36, 66, 79, 93, 94] (A. Taherpour and S.A. Glover, unpublished work) than those of their precursor hydroxamic esters (1650–1700 cm^{-1}) and typical primary (1690 cm^{-1}), secondary (1665–1700 cm^{-1}), and tertiary (1630–1670 cm^{-1}) amides [34, 39, 95]. As usual, arylamide frequencies were lower than those of alkylamides. From structural data there is a slight contraction of the (N)C=O bond, which is consistent with the limited charge transfer to oxygen in accord with Wiberg's amide model (Figure 2.2) [3, 4]. The increase in $\nu_{C=O}$ can be attributed primarily to electronic destabilization of the single-bond resonance form of the carbonyl (Figure 2.1, II) as electron density is pulled toward the electronegative heteroatoms on nitrogen, resulting in a more ketonic carbonyl bond. Likewise, relative to their hydroxamic ester precursors (with one oxygen at nitrogen), the carbonyl carbons of anomeric amides resonate downfield by approximately 8 ppm. Deshielding of the carbonyl carbon is an expected consequence of the diminished resonance due to the electron-withdrawing substituents. However, like acid chlorides and anhydrides, the carbonyls resonate upfield of ketones, relative to which

Table 2.4 Typical spectroscopic data for stable anomeric alkanamides and arylamides (RC(O)NXY) and their hydroxamic ester precursors.

System[a)]	X	Y	R	Amide[b)] ν (cm^{-1}) ($\delta^{13}C$)	Hydroxamic ester ν (cm^{-1}) ($\delta^{13}C$)
2a ONCl [93]	OBu	Cl	Me	1740 (175.3)	1678 (167.9)
2a ONCl [39]	OBu	Cl	Ph	1719 (174.2)	1654 (165.7)
2b ONOAc [93]	OBu	OAc	Me	1746 (176.2)	1678 (167.9)
2b ONOAc [94]	OBu	OAc	Ph	1732 (173.9)	1654 (165.7)
2c ONO (A. Taherpour and S.A. Glover, unpublished work)	OBu	OBu	Me	1707[c)] (174.1)	1678 (167.9)
2c ONO [79]	OMe	OMe	Ph	1711 (174.3)	1683 (166.4)
2k NNO [36]	N(Ac)OBn	OBn	Me	1735/1711 (172.5)	1682 (168.1)
2k NNO [36]	N(OEt)Bz	OEt	Ph	1708 (170.0)	1685 (166.5)

a) XNY designation.
b) IR in $CHCl_3$; NMR in $CDCl_3$.
c) Neat.
Source: Glover et al. [2]/MDPI/CC BY 4.0.

the electron density at the carbonyl is higher on account of the residual N–C(O) double bond character.

Common primary, secondary, and tertiary amides have significant *cis-trans* isomerization barriers to rotation about the N–C(O) bond, due to the stabilizing effect of amide resonance in their ground state [3]. Restricted rotation through lone pair overlap with the adjacent carbonyl C2p_z orbital, as illustrated in Figure 2.2, often results in observation of different chemical shifts of *cis* and *trans* conformers in their ^{1}H NMR spectra, from which barriers to isomerization can be deduced [96, 97]. The computed surface for *N*,*N*-dimethylacetamide **1** in Figure 2.3 indicates that the barrier (the difference between the completely planar (a) and the fully rotated-pyramidal forms (c)) is of the order of 70 kJ/mol. The rotational barrier for *N*-methoxy-*N*-methylacetamide **4a** from Figure 2.5 is approximately 67 kJ/mol and, if this is representative, hydroxamic esters should exhibit only a slight reduction in resonance. In accordance with this, many hydroxamic esters have broadened ^{1}H NMR signals at ambient temperatures. Anomeric amides with lower resonance should have lower *cis–trans* isomerization barriers as exemplified in the model *N,N*-dimethoxyacetamide in Figure 2.6. Accordingly, the isomerization barrier in bisoxyl-substituted amides **2b** and **2c** are too low to be measured by usual dynamic ^{1}H NMR methods. All proton signals of *N,N*-dimethoxy-4-toluamide remained sharp down to 180 K (in d_4-methanol) and the ^{1}H NMR signals for *N*-acetoxy-*N*-benzyloxybenzamide remained isochronous down to 190 K (in d_8-toluene) [34]. The barrier for *N*-benzyloxy-*N*-chlorobenzamide could likewise not be determined by dynamic NMR [39].

Figure 2.8 illustrates the dramatic difference between ^{1}H NMR spectra of *N*-butoxyacetamide and its *N*-chloro- or *N*-acetoxyl derivatives as a consequence of reduced resonance in the anomeric structures, which clearly have much lower isomerization barriers. At room temperature, isomerizations of the anomeric

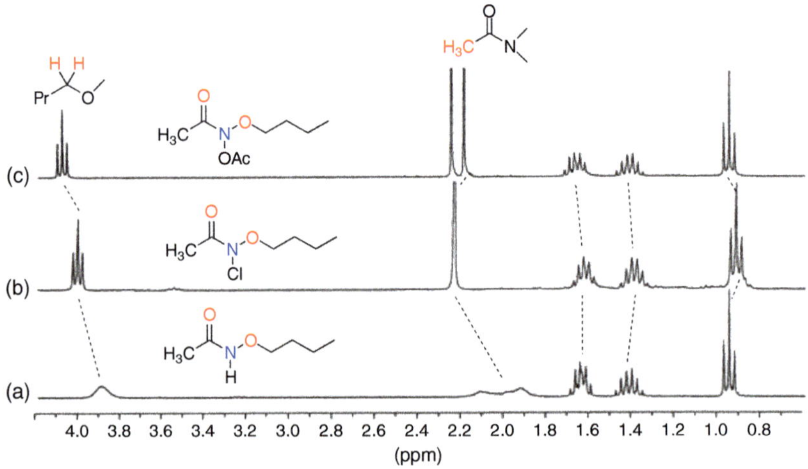

Figure 2.8 ^{1}H NMR spectra (300 K, $CDCl_3$) of (a) *N*-butoxyacetamide and anomeric amides, (b) *N*-butoxy-*N*-chloroacetamide, and (c) *N*-acetoxy-*N*-butoxyacetamide. Source: Modified from Glover et al. [2].

amides are in the fast exchange region as opposed to the slow exchange region in the case of the hydroxamic ester. Earlier theoretical studies by Glover and Rauk support this assertion [33, 35]. For example, a comparison of formamide **15a** and *N*-methoxyformamide **15b** with *N*-chloro-*N*-methoxyformamide **15d**, calculated at the B3LYP/6-31G(d) level, showed a small reduction in the *cis–trans* isomerization barrier of 73 kJ/mol in formamide, to 66 kJ/mol in *N*-methoxyformamide. However, the introduction of a second electronegative heteroatom, Cl, reduced the barrier to only 32 kJ/mol. The barrier in *N,N*-dimethoxyformamide **15c** was similar (30 kJ/mol), indicating that monosubstitution alone is insufficient to impact upon the isomerization barrier [33]. A theoretical isomerization barrier for *N*-methoxy-*N*-(dimethylamino)formamide, which, by analogy with typical NNO systems, should retain about 70% the resonance in *N,N*-dimethylacetamide, was computed to be higher at 53 kJ/mol [35]. Such an amide isomerization barrier in the hydrazine, *N,N′*-diacetyl-*N,N′*-di(4-chlorobenzyloxy)hydrazine (**20d**) was measurable at $\Delta G^{\ddagger}{}_{278} = 54.0$ kJ/mol, in line with this theoretical barrier [36].

Chirality at nitrogen has not been observed in the ^{1}H NMR of acyclic anomeric amides. The intrinsic barrier to inversion at nitrogen in bisheteroatom-substituted amides is expected to be much lower than those of analogous amines. The transition state for inversion in anomeric amides is expected to be planar, where the nitrogen lone pair would reside in the $N2p_z$ orbital and can interact with the carbonyl. The unfavourable redistribution of “p” character is offset by a stabilizing interaction with the carbonyl $C2p_z$ orbital. Several theoretical estimates put these barriers at about 10 kJ/mol [33, 35, 39]. Inspection of the deformation surface of *N,N*-dimethoxyacetamide **4b** (Figure 2.6) provides insight into this difference. The inversion barrier is represented by the energy of structure (a) and is ~14 kJ/mol. The difference in energy between the planar (d) and pyramidal (c) twisted structures, at 80 kJ/mol, provides an estimate for the inversion barrier for *N,N*-dimethoxyamine. Rudchenko has measured the inversion barriers for *N,N*-dialkoxyureas and *N,N*-dialkoxyamines at $\Delta G^{\ddagger} = 38–46$ and 84–92 kJ/mol, respectively [98]. The value for *N,N*-dialkoxyamines is in accord with the estimate from analysis of the energy surface in Figure 2.6. The barrier for ureas would be higher than that for *N,N*-dialkoxyamides amides on account of competing resonance by the N^{α} at the transition state.

Interestingly, while we have not found evidence for chirality at nitrogen in *N,N*-dialkoxyamides in solution, dialkoxyamides **5a** and **5b** actually crystallize with two enantiomeric pairs in their unit cell [40].

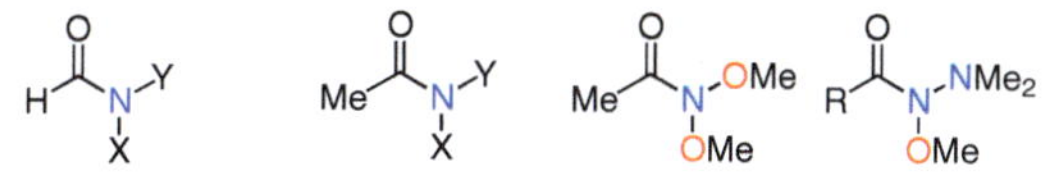

15 **16** **4b** **17**

15
a: X = Y = H
b: X = H, Y = OMe
c: X = Y = OMe
d: X = Cl, Y = OMe
e: X = OMe, Y = NMe_2

16
a: X = Cl, Y = OMe
b: X = OAc, Y = OMe
c: X = OMe, Y = NMe_2
d: X = OMe, Y = SMe
e: X = Cl, Y = NMe_2
f: X = OMe, Y = O^-
g: X = OMe, Y = N

17
a: R = MeO
b: R = H_2N

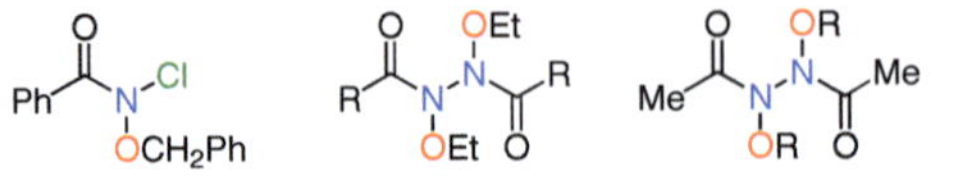

2.2.7 Theoretical Structures and Amidicities of Model Anomeric Amides

Structures of anomeric amides have been modelled at B3LYP/6-31G(d) level in the simplified acetamide, carbamate, and urea systems, and ground state models of **16a–e**, **4b**, and **17a,b** display high degrees of pyramidality, long N–C(O) bonds, and slightly shortened (N)C=O bonds, in line with the electron demands of nitrogen substituents (Figure 2.9, Table 2.5). Where applicable, the minimum energy structures are reasonable approximations of their respective X-ray structure counterparts. All structures exhibit minimal twist about the N–C(O) bond. SNO and NNCl analogs, **16d** and **16e**, have only been generated as intermediates in reactions but their theoretical structures are in line with those of anomeric systems **4b** and **16a–c**. N–C(O) bond lengths and pyramidality at nitrogen (χ) are broadly in line with the gross electronegativity of substituents at nitrogen, and in the case of acetamides **4b** and **16a–d**, change is in concert with the percentage 2s character in the anomeric nitrogen lone pairs. The chlorine-bearing systems **16a** and particularly **16e** need a higher level of theory that corrects for dispersion but at B3LYP/6-31G(d) and M06/6-311++G(d,p) levels, geometries are similar. **16e** is completely planar (and untwisted) for steric reasons [42]. Nonetheless, its N–C(O) bond (1.396 Å) is comparatively long when compared to *N*,*N*-dimethylacetamide (1.378 Å). The carbamate and the urea **17a** (χ = 46.3°) and **17b** (χ = 49.6°) are both more pyramidal than the corresponding acetamide **16c** (χ = 41.8°), presumably as a result of competing resonance from the α-oxygen and α-nitrogen lone pairs. This is more evident with urea **17b**, resulting in a much longer N–C(O) bond than found in **17a**. In the NNO urea **17b**, the α nitrogen is essentially planar and the N^{α}–C(O) is short (1.371 Å) when compared with that computed for urea itself (1.378 Å). There is synergistic resonance in ureas; diminished resonance to the anomeric nitrogen facilitates better interaction with the α nitrogen lone pair.

The resonance energy and the amidicity of anomeric amides **4b, 16a–e**, and **17a** have been calculated at the B3LYP/6-31G(d) level by both the TA and COSNAR methods and by the COSNAR method using dispersion corrected M06/6-311++G(d,p). ΔE_{COSNAR} (Eq. (2.1)), reaction energies (ΔE_{TA}) (Eq. (2.2)), inductive stabilization corrections (ΔE_{ind}) (Eq. (2.4)), and resultant RE_{TA} together with COSNAR and TA amidicities are presented in Table 2.6. Ureas such as model **17b** are not amenable to either computational methods. Dispersion corrected methods were deemed necessary with the ONCl and particularly NNCl systems on account of evidence of charge transfer from O or N to Cl (see Section 2.3.1).

Table 2.5 Selected structural data for B3LYP/6-31G(d) lowest energy conformers of model acetamides **4b**, **16a–e**, and **17a,b**.

Structure[a)]	N–C(O) (Å)	(N)C=O (Å)	(N-X,N-Y) (Å)	θ[b)] (°)	χ (°) (%2s)[c)]	τ (°)	Anomeric twist angles[d)] (°)
16a ONCl [42]	1.432	1.207	Cl: 1.787 O2: 1.389	337.6	52.3 (26%)	−5.3	**C2–O2–N1–Cl:88.8**
16b ONOAc [53]	1.429	1.209	O2: 1.423 O3: 1.395	332.1	58.0 (25%)	2.3	**C4–O3–N1–O2: 101.1** C2–O2–N1–O3: −111.0
4b ONO [40]	1.417	1.212	O2: 1.387 O3: 1.412	342.9	48.1 (19%)	8.5	C2–O2–N1–O3: 66.6 **C3–O3–N1–O2: 83.8**
16c NNO [54]	1.404	1.217	O1: 1.430 N2: 1.387	346.5	41.8 (12%)	5.4	**LP (N2)–N2–N1–O1: 190°**
16d SNO [53]	1.408	1.215	S1: 1.717 O2: 1.420	352.4	31.7 (8%)	−4.6	C4–S1–N1–O2: −79° **C3–O2–N1–S1: −86.6°**
16e NNCl [42]	1.396	1.212	Cl: 1.820 N2: 1.351	360.0	0	0	**LP(N2)N2–N1–Cl: 180°**
17a NNO [53]	1.406	1.212	O1: 1.440 N2: 1.383	342.0	46.3	0	**LP (N2)–N2–N1–O1: 169.4**
17b NNO	1.428 1.371[e)]	1.217	O1: 1.434 N2: 1.397	340.0 352.8[e)]	49.6 28.6[e)]	−1.7 −4.0[e)]	**LP (N2)–N2–N1–O1: 167.5**

a) XNY designation.
b) Sum of angles at anomeric nitrogen.
c) Percentage 2s character in the anomeric nitrogen lone pair from NBO analysis.
d) Anomeric alignments in bold face.
e) N^{α} data.
Source: Modified from Glover et al. [2].

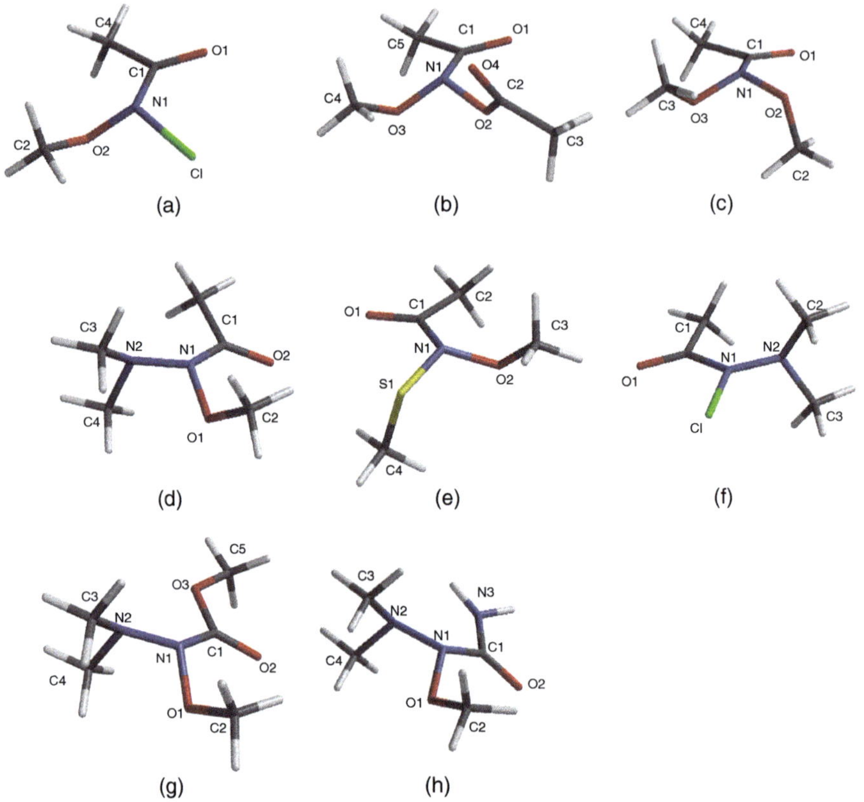

Figure 2.9 B3LYP/6-31G(d) lowest energy ground state conformers of model anomeric acetamides: (a) *N*-chloro-*N*-methoxyacetamide **16a**, (b) *N*-acetoxy-*N*-methoxyacetamide **16b**, (c) *N,N*-dimethoxyacetamide **4b**, (d) *N*-methoxy-*N*-(dimethylamino)acetamide **16c**, (e) *N*-methoxy-*N*-methylthiylacetamide **16d**, (f) *N*-chloro-*N*-(dimethylamino)acetamide **16e**, (g) *O*-methyl *N*-methoxy-*N*-(dimethylamino)carbamate **17a**, and (h) *N*-methoxy-*N*-dimethyl-aminourea **17b**. Source: Glover et al. [2]/MDPI/CC BY 4.0.

The TA and COSNAR methodologies give almost identical resonance energies for all the anomeric amides **(4b, 16a–e)** at the B3LYP/6-31G(d) level. ΔE_{COSNAR} determined at M06 with the expanded basis set yields very similar results to B3LYP/6-31G(d) for *N,N*-dimethylacetamide **1** and the anomeric amides with the exception of **16e**, where resonance is computed to be just over 60% that of *N,N*-dimethylacetamide and substantially higher than that predicted from B3LYP/6-311++G(d,p), which was identical to the B3LYP/6-31G(d) value (29 kJ/mol) [42]. B3LYP predicts RE parity for **16a** and **16e**, but this is no longer observed with M06, which yields a result in line with the lower overall combined electronegativity of nitrogen and chlorine as opposed to oxygen and chlorine at the amide nitrogen. These results impute the necessity for inclusion of dispersion corrections in treatments of molecules where anomeric interactions are likely to have a pronounced influence [42].

Table 2.6 B3LYP/6-31G(d), B3LYP/6-311++G(d,p), and M06/6-311++G(d,p)-derived resonance energies and amidicities of **1** and model anomeric acetamides **4b, 16a–e**, and carbamate **17a**.

Structure[a)]	ΔE_{COSNAR}[b)] (kJ/mol)	ΔE_{TA} (kJ/mol)	ΔE_{ind} (kJ/mol)	RE_{TA}[b)] (kJ/mol)
16a ONCl [42]	−29.6(38%)[c)]	69.9	23.4	−29.5(39%)
	−27.2(36%)[d)]	—	—	—
	−34.4(45%)[e)]	—	—	—
16b ONOAc [53]	−39.7(52%)[c)]	65.3	29.7	−40.5(53%)
	−39.5 (52%)[e)]			
4b ONO [40]	−36.0(47%)[c)]	58.2	18.0	−36.0(47%)
	−39.5(53%)[e)]			
16c NNO [54]	−52.3(69%)[c)]	35.1	10.0	−51.0(67%)
	−55.7(73%)[e)]			
16d SNO [53]	−48.6(64%)[c)]	26.8	5.0	−48.6(64%)
	−47.3(62%)[e)]			
16e NNCl [42]	−28.7(38%)[c)]	64.5	17.6	−29.0(37%)
	−28.6(38%)[d)]	—	—	—
	−48.7(64%)[e)]	—	—	—
17a NNO [53]	−47.7 (63%)[c)]	31.8	14.6	−50.6(67%)
1 [42]	−75.9 (100%)[c)e)]			

a) XNY designation.
b) Amidicity in parentheses.
c) At B3LYP/6-31G(d).
d) At B3LYP/6-311++G(d,p).
e) At M06/6-311++G(d,p).
Source: Modified from Glover et al. [2].

It is clear that resonance in anomeric amides is impaired, broadly in line with the gross electronegativity or negative inductive effect of the atoms/groups bonded to nitrogen. Interestingly, the NNO and SNO acetamides, **16c**, and **16d** have very similar resonance energies despite the lower electronegativity of sulfur. From Bent's rule, the opposite effect would be expected, since s-character in the lone pair orbital should decrease and resonance should increase with decreasing electronegativity of X. This is likely to be a manifestation of the role of orbital size, since the influence of second period elements is significant; nitrogen increases p-character in the bond to sulfur to effect better overlap with the larger 3p orbital of sulfur. Consequently, the nitrogen lone pair gains more "s" character resulting in reduced amide resonance similar to that of the NNO system **16c** [31, 32, 99]. Comparing all the M06 values, ONCl and ONO systems have about half the resonance of *N*,*N*-dimethylacetamide, while NNO, NNCl, and SNO are computed to preserve some 60–70% of the resonance of *N*,*N*-dimethylacetamide.

Clearly, this reduction in resonance in anomeric amides must render them more reactive than simple amides or hydroxamic esters. However, as will be seen, the reactivity patterns are unlike those of highly twisted amides, which are vulnerable to facile hydrolysis and amide bond cleavage [8, 11, 12, 14, 16, 28, 30]. In all anomeric amide structures, crystal, or computational data, the extent of twist is low, signifying appreciable interaction between the nitrogen lone pair and the carbonyl. Even in the ONO systems, with ~50% amidicity, the degree of twist is low. For *N,N*-dimethoxyacetamide **4b**, which has less than 50% of the amide character of *N,N*-dimethylacetamide **1**, proton affinity calculations at B3LYP/6-31G(d) level predict that, like *N,N*-dimethylacetamide **1** ($PA_O - PA_N = 50.3$ kJ/mol), protonation is still more favourable on the carbonyl oxygen as opposed to nitrogen ($PA_O - PA_N = 41.5$ kJ/mol) [40].

2.3 Reactivity of Anomeric Amides

The configuration, the reduced resonance, and atypical geometries at the amide nitrogen impact greatly upon the chemical behavior of anomeric amides. While the reduction in N–C(O) double bond character would be expected to slant reactivity of these amides toward those of esters, anhydrides, or carboxylic acid chlorides, which undergo addition reactions at the carbonyl leading to acyl substitution, the predominant reactivity of anomeric amides takes place at the amide nitrogen, where reactivity profiles tend toward that of saturated carbon and, in particular, S_N1, S_N2, and elimination reactions. Acyl bond cleavage is found where these processes are disfavoured by poor leaving groups. However, scission occurs in a rearrangement process known as the *He*teroatom *R*earrangement *O*n *N*itrogen (HERON) reaction, which is an intramolecular acyl substitution process leading to heteroatom-stabilized nitrenes [36, 77, 100–104]. This and substitution/elimination at nitrogen are all driven by the bisheteroatom substitution pattern at nitrogen, which leads to very pronounced anomeric effects between substituent atoms at the nitrogen, all of which are electronegative atoms of the p-block and possess lone pairs of electrons. To understand the reactivity of XNY anomeric amides, it is essential, first, to explore the nature and extent of the *anomeric* interactions that operate between donor Y and acceptor X through the amide nitrogens.

2.3.1 The Anomeric Effect

The synthesis, reactivity, and anomeric properties of acyclic and cyclic dialkoxyamines have been documented in an earlier review by Rudchenko [98]. They formulated an understanding of structure and reactivity of such amines based on strong n_O–σ^*_{NO} anomeric interactions. Anomeric amides are exemplars of a broader class of XNY systems featuring an anomeric effect [34, 39]. With different heteroatoms at nitrogen, there are two possible anomeric interactions designated as n_X–σ^*_{NY} and n_Y–σ^*_{NX}, and, where X and Y are different electronegative atoms, one of these interactions will be favoured over the other (Figure 2.10). When X and Y are the same,

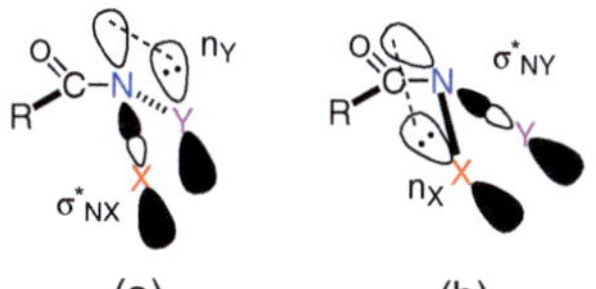

Figure 2.10 Two possible anomeric interactions in anomeric amides: (a) $n_Y-\sigma^*_{NX}$ and (b) $n_X-\sigma^*_{NY}$. Source: Modified from Glover et al. [2].

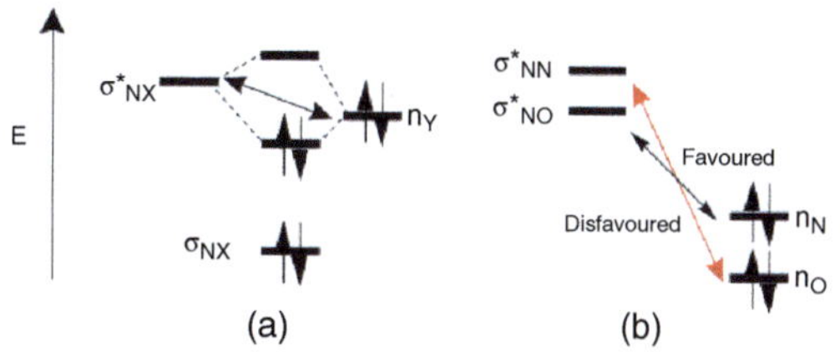

Figure 2.11 (a) Stabilization of a lone pair through a $n_Y-\sigma^*_{NX}$ anomeric interaction; (b) the energetics of the anomeric effect in NNO systems **16c**. Source: Modified from Glover et al. [2].

other factors such as conformation may impose a preferential anomeric interaction. By analogy with anomeric carbon centers [34, 39, 45, 105, 106], the relative electronegativities of heteroatoms X and Y at nitrogen and the relative sizes of interacting orbitals contribute to the strength of an anomeric interaction. Heteroatoms Y and X directly influence the relative energies of n_Y and σ^*_{NX}, which, in turn, affect the net stability gain for the lone pair electrons (Figure 2.11a).

As the electronegativity of X and Y increases by going across the p-block row on the periodic table, σ^*_{NX} (and σ_{NX}) decreases in energy, as does n_Y. Additionally, as X decreases in electronegativity by going down a p-block group, σ^*_{NX} decreases in energy due to reduced orbital overlap [31, 34, 39, 45, 105]. An optimal anomeric effect can be achieved when Y is an early p-block element and X is a p-block element to the right of Y on the periodic table; more specifically an anomeric stabilization will be greater when the energy gap between n_Y and σ^*_{NX} is lower (Figure 2.11b) [107, 108]. In the unusual case of anomeric amides, where the nitrogen may range between sp^2 (planar) and sp^3 hybridized (pyramidal), the geometry of the central nitrogen atom in XNY systems also plays a role, as pyramidal nitrogen is more conducive to edge on n_Y and σ^*_{NX} overlap than is planar, sp^2 hybridized nitrogen (Figure 2.12a) [34, 39].

Like XCY configurations [45, 107], in an XNY system, a stabilizing $n_Y-\sigma^*_{NX}$ anomeric effect, for example, can cause the amide to adopt a gauche conformation in which the lone pair of Y is coplanar with the vicinal N–X bond (Figure 2.12b). In anomeric amides, when n_Y is divalent oxygen or sulfur, one lone pair is in an

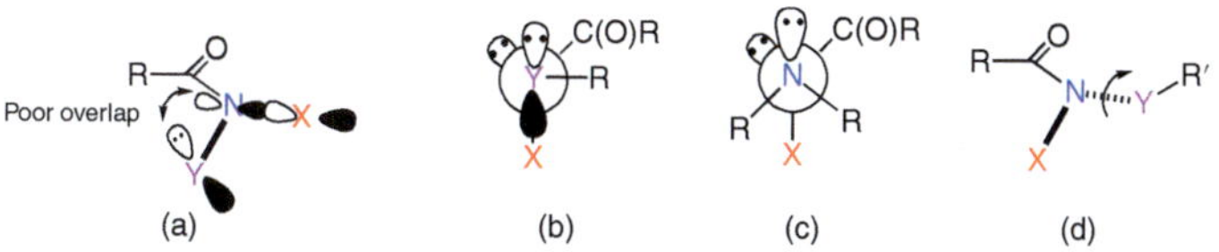

Figure 2.12 (a) sp^2-hybridized nitrogen hinders overlap with heteroatom Y in an $n_Y-\sigma^*_{NX}$ system; (b) optimum conformation in an $n_Y-\sigma^*_{NX}$ anomeric interaction; (c) optimum conformation in an $n_N-\sigma^*_{NX}$ anomeric interaction; (d) restricted rotation about the N–Y bond in an $n_Y-\sigma^*_{NX}$ anomeric interaction. Source: Modified from Glover et al. [2].

s-type orbital, while the other is in a p-type orbital and an R-Y-N-X dihedral angle close to |90°| aligns the p-type lone pair on those atoms with the vicinal σ^*_{NX} (Figure 2.12b). Where the donor is n_N, an optimum anomeric effect would need the lone pair on nitrogen to be *antiperiplanar* to X, LP(N)-N-N-X = 180° (Figure 2.12c). Consequences of the n_Y–σ^*_{NX} interaction include an increased barrier to rotation about the N–Y bond (Figure 2.12d), a contraction of the N–Y bond, and an extension of the N–X bond as electrons from Y populate the σ^*_{NX} orbital.

In each of the X-ray structures in Figure 2.7, there is clear evidence of these anomeric interactions (Table 2.2, boldface torsion angles). On the basis of energetics, the expected anomeric interactions are n_O–σ^*_{NOAc} in **6a**, **6b**, and **7**, n_N–σ^*_{NO} in **8**, and n_O–σ^*_{NCl} in **9**. In **5a** and **5b** one n_O–σ^*_{NO} would be expected to prevail. Dihedral angles about the N–O bonds in **5–7** and **9** show a preferred p-type lone pair alignment with the adjacent σ^*_{NO}, σ^*_{NOAc}, or σ^*_{NCl} bond. In hydrazine **8** (Figure 2.7g), the lone pair on N1 is almost perfectly aligned with the N2–O3 bond, while the N2 lone pair makes an angle of only 47° to the N1–O2 bond (1.403 Å), which is significantly shorter than the anomerically destabilized bond N2–O3 (1.411 Å). The structure is asymmetrical with a donor N1 and recipient N2, and, in fact, hydrazines **19** in the condensed phase and **20** in solution present two amide carbonyls in their infrared spectra [36]. Similar, preferred anomeric interactions are observable in ureas **10–12** and carbamates **13** and **14** (Table 2.3, boldface torsion angles).

All computed structures for **16**, **4b**, and **17** exhibit a preferred anomeric interaction (Table 2.5, boldface torsion angles), and, besides the ONO system where one n_O–σ^*_{NO} is clearly favoured, an anomeric n_O–σ^*_{NX} or n_N–σ^*_{NO} prevails in ONCl, ONOAc, and NNO systems in line with expectations based on electronegativity. In NNO and NNCl systems, where a donor nitrogen is present, the n_N–σ^*_{NO} and n_N–σ^*_{NCl} interactions are clearly in evidence. Torsion angles are similar in the SNO system but the size of the sulfur p-orbital and lower energy of the σ^*_{NS} renders an n_S–σ^*_{NO} anomeric interaction less likely than a n_O–σ^*_{NS} stabilization.

In anomeric systems, strong n_Y–σ^*_{NX} interactions should increase the bond order of the N–Y bond, which should impose barriers to rotation about those bonds. For *N*-chloro-*N*-methoxyformamide **15d**, the N–O and N–C(O) rotation barriers have been estimated at the B3LYP/6-31G(d) level at 44.7 and 29.2 kJ/mol, respectively [33], while the theoretical barrier to rotation about the N–N bond in *N*-methoxy-*N*-(dimethylamino)formamide **15e** was computed at ~60 kJ/mol [35]. Experimental measurements for both anomerically induced barriers have been made employing dynamic ^{1}H NMR. An N–O anomeric rotational barrier of $\Delta G^\ddagger$ = 43 kJ/mol has been determined for *N*-chloro-*N*-benzloxybenzamide **18**, where the benzylic methylenes become diastereotopic at 217 K in d_8-toluene. However, as no amide isomerization could be detected, that barrier must be significantly lower [34]. An anomerically induced rotational barrier for N–N′ bond in a number of *N*,*N*-diacyl-*N*,*N*-dialkoxyhydrazines **19a–e** and **20a–e** has been measured in dynamic ^{1}H NMR studies [36]. Methylene signals of *N*,*N*′-diethoxy (in **19a–e**) and *N*,*N*′-dibenzyloxy groups (in **20a–e**), which were diastereotopic at room temperature, as a consequence of restricted rotation about the N–N′ bond, coalesced at higher temperatures (T_c = 316–346 K) from which rotational free energy barriers of the order of 60–70 kJ/mol could be determined [36]. These compare favourably with the theoretically calculated N–N′ rotational barrier of 60 kJ/mol for **15e** [35].

The anomeric effects not only dictate stereochemistry at nitrogen, but, as will be seen, combined with reduced amide resonance, they also have a profound impact upon the reactivity of anomeric amides.

2.3.2 Reactivity at the Anomeric Amide Nitrogen

Due to the characteristic, diminished amide resonance in all XNY amides, along with anomeric destabilization of NX bonds by the donor Y, this class of amide has been shown to react at nitrogen, leading to displacement of the X group in S_N2 reactions and elimination of *N*-substituents leading to S_N1-type processes.

2.3.2.1 S_N2 Reactions

In XNY systems, a moderate n_Y–σ^*_{NX} negative hyperconjugation leads, through neighboring group participation, to weakening of the N–X bond, which encourages S_N2 reactions at nitrogen [34, 38, 39]. Nucleophilic attack at the amide carbonyl is not competitive with this reaction process.

The increased electrophilicity at nitrogen in *N*-acyloxy-*N*-alkoxyamides **21** leaves them vulnerable to attack by arylamines such as *N*-methylaniline (Scheme 2.2i) [61, 63, 66, 76], azide (Scheme 2.2ii) [81], hydroxide (Scheme 2.2iii) [64], and thiols (Scheme 2.2iv) [78]; the outcomes from which are anomerically substituted intermediates **22–24** and **26** that undergo secondary reactions including HERON reactions

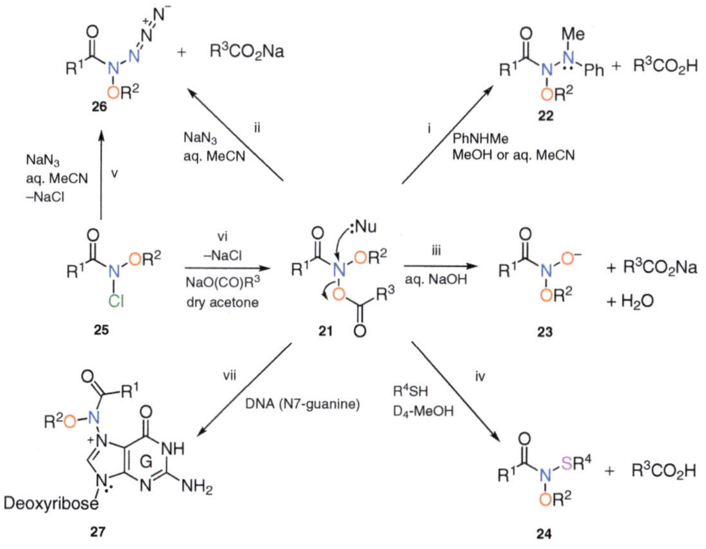

Scheme 2.2 S_N2 reactions of *N*-acyloxy-*N*-alkoxyamides **21** and *N*-alkoxy-*N*-chloroamides **25**. Source: Modified from Glover et al. [2].

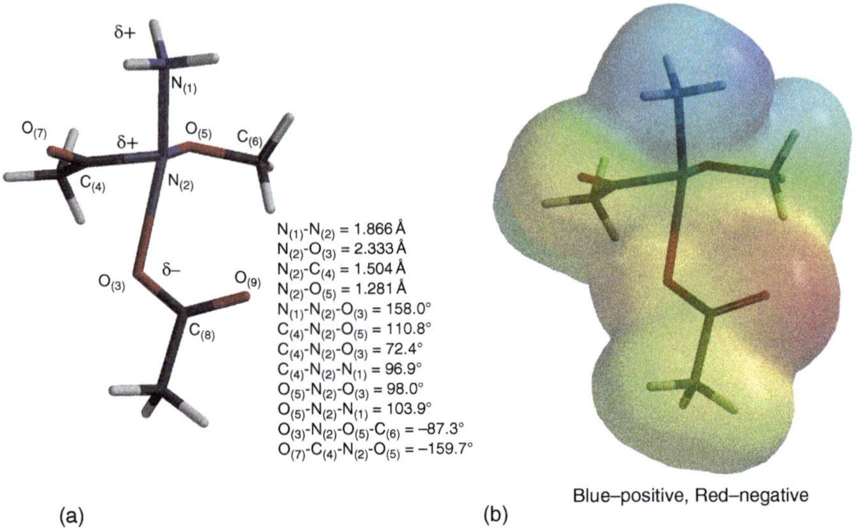

Figure 2.13 (a) HF/6-31G* predicted charge separation in transition states for reactions of *N*-acetoxy-*N*-methoxyacetamide with ammonia and (b) associated electrostatic potential energy surface.

(Section 2.3.3). All four reactions have been studied kinetically and are bimolecular in nature. Furthermore, *N*-acyloxy-*N*-alkoxyamides **21** themselves may be synthesized in S_N2 reactions between *N*-alkoxy-*N*-chloroamides **25** and sodium carboxylates (Scheme 2.2vi) [62, 64, 65, 67, 76, 87, 94, 109]. *N*-Alkoxy-*N*-chloroamides **25** also react bimolecularly with azide generating the reactive *N*-alkoxy-*N*-azidoamides **26** (Scheme 2.2v) [81].

Reactions of **21** with amines and thiols have been modelled at the AM1 and HF/6-31G(d) levels, which revealed significant charge separation in the transition states and partial alkoxynitrenium ion character as exemplified in Figure 2.13 [76, 110]. Relative to the ground state, at the transition state ammonia gains positive charge ($\Delta q = +0.4$), while the acetoxyl group becomes more negatively charged ($\Delta q = -0.7$). The remaining positive charge ($\Delta q = +0.3$) resides on the central nitrogen and the methoxyl group. These reactions should be favoured by electron-donor substituents on the nucleophile and electron-acceptor substituents on the acyloxyl group.

The S_N2 reaction of *N*-methylaniline with a wide range of NAAs **21** according to Scheme 2.3 has been studied in detail, and relative rate constants, Arrhenius activation energies, and entropies of activation are in accord with a transition state with significant charge separation [38, 63, 76, 111]. Activation energies are of the order of 40–60 kJ/mol, while entropies of activation (−90 to −160 J/K/mol) are strongly negative owing to a greater degree of solvation in the charge separated transition state [112]. As expected, for a series of alkanoyloxyl and benzoyloxyl acetamides, the rate constants correlated negatively with pK_A and molar refractivity of leaving groups according to Eq. (2.6) [76]. In addition, the rates for reactions of *N*-methylaniline with a series of *N*-*p*-substituted benzoyloxyl leaving groups (Scheme 2.3, R^1 = Bn, R^2 = aryl, R^3 = Ph) correlated with Hammett σ constants and $\rho = 1.7$ [111]. The ease of transition state formation is dictated by the stabilization of developing negative charge on the carboxyl leaving group.

$$\ln k^{303} = -4.9(\pm 0.2)\mathrm{p}K_\mathrm{A} - 0.017(\pm 0.005)\mathrm{MR} \qquad (2.6)$$

$$(r = 0.995, s = 0.14)$$

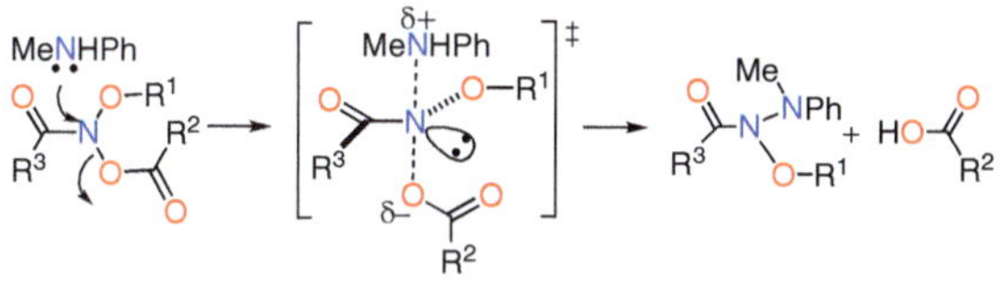

Scheme 2.3 S_N2 reaction of *N*-acyloxy-*N*-alkoxyamides with *N*-methylaniline.

Hydroxide and thiol, in reactions iii and iv in Scheme 2.2, correlate similarly with Hammett σ constants with $\rho = 0.6$ and $\rho = 1.1$, respectively [64, 78, 111]. In addition, a series of anilines reacted bimolecularly, according to Scheme 2.3, and rate constants correlated with Hammett σ^+ constants ($\rho = -0.9$) [63, 111]. In most respects, the S_N2 reactions are electronically and geometrically reminiscent of those at carbon centers and are accelerated by electron-donor groups on the nucleophile and electron-withdrawing groups on the leaving group. Not only do nucleophiles react preferentially at nitrogen rather than the carbonyl carbon, the amide carbonyl possibly facilitates S_N2 reactivity in line with enhanced S_N2 reactivity in phenacyl bromides [113]. In addition, bimolecular reaction rates are strongly impeded with branching α to the carbonyl [59], which is analogous to the resistance to S_N2 reactions of α-halo ketones bearing substituents and branching at the α′ position [114, 115].

Anomeric substitution at nitrogen in NAAs renders this class of amides as direct-acting mutagens. Mutagenicity toward *Salmonella typhimurium* in the Ames reverse mutation assay does not require pre-metabolic activation [116, 117]. Our DNA damage studies on plasmid DNA at physiological pH, as well as extensive structure–activity relationships [38, 56–60, 62, 64, 65, 67, 68, 93, 94], point to binding of NAAs **21** intact into the major groove of DNA, where an S_N2 reaction occurs at the most nucleophilic center, the electron-rich N7 in guanine (Scheme 2.2vii) [118–121]. All three side chains (R^1, R^2, and R^3) of **21** have an impact upon both DNA damage profiles and mutagenicity levels. An S_N1 mechanism, yielding electrophilic *N*-acyl-*N*-alkoxynitrenium ions, was ruled out since only R^1 and R^2 would influence binding and reactivity. Moreover, mutagenic activity is also radically reduced when there is branching α to the carbonyl in parallel with the impaired S_N2 reactivity with anilines [58, 59, 76]. We have recently used the mutagenic activity of *N*-acyloxy-*N*-alkoxyamides **21** to show how hydrophobicity and intercalating side chains impact upon DNA binding [55–57].

2.3.2.2 Elimination Reactions (S_N1-Type Processes)

In an anomeric amide with a strong n_Y–σ^*_{NX} interaction, where X has a high electron affinity and Y is a strong electron donor, polarization can lead to elimination of X^-, leaving a Y-stabilized nitrenium ion **28** (Scheme 2.4) [69]. The stronger the anomeric effect, the more readily the elimination is expected to occur. In the case of *N*-alkoxy-*N*-chloroamides **25**, elimination can be facilitated by Lewis acid complexation with X and by the use of polar solvents [39]. For example, treatment of *N*-chloro-*N*-(2-phenylethyloxy)- and *N*-chloro-*N*-(3-phenylpropyloxy)amides, **29a** and **29b**, with silver tetrafluoroborate in diethyl ether, initiates a ring closing reaction to form *N*-acyl-1*H*-3,4-dihydro-2,1-benzoxazines **30a**, and *N*-acyl-1,3,4,5-tetrahydro-2,1-benzoxazepines **30b**, respectively (Scheme 2.5). Chloride elimination generates *N*-acyl-*N*-alkoxynitrenium ions, which are strongly stabilized by delocalization of the positive charge onto oxygen [69, 122]. This methodology has been used widely since its discovery in 1984 by Glover [123, 124] and Kikugawa [125–131]. In addition, treatment of *N*-alkoxy-*N*-chloroamides **25** with silver carboxylates in diethyl ether allows the nitrenium ion to be scavenged by carboxylate in a versatile reaction, which has been used to synthesize a range of *N*-acyloxy-*N*-alkoxyamides **21** [67, 94].

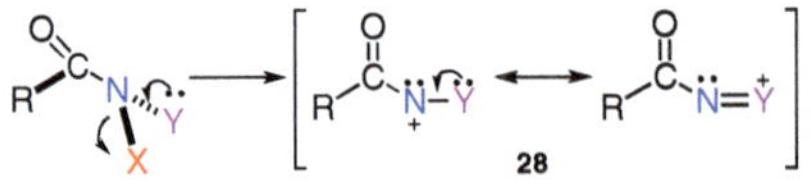

Scheme 2.4 Nitrenium ion formation by elimination of X due to a strong n_Y–σ^*_{NX} interaction. Source: Glover et al. [2]/MDPI/CC BY 4.0.

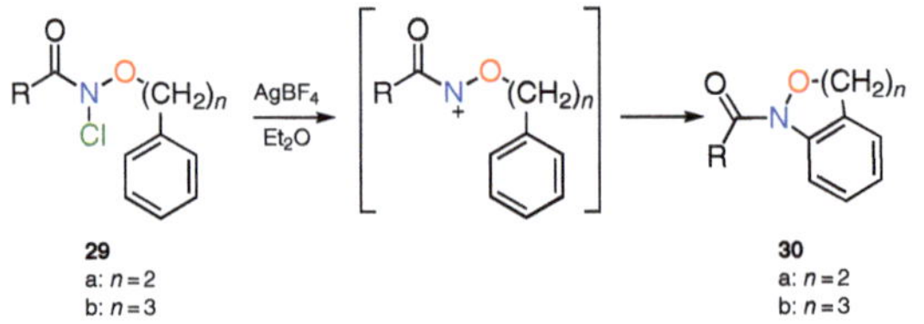

Scheme 2.5 Cyclization by silver ion promoted elimination reactions. Source: Glover et al. [2]/MDPI/CC BY 4.0.

Elimination of chloride in the alcoholysis of **25** to give nitrenium ion provided a synthetic pathway to *N*,*N*-dialkoxyamides **5** (Scheme 2.6) [77, 79]. We recently reported a more versatile synthesis effected by PIFA oxidation of hydroxamic esters **31** in the presence of an appropriate alcohol, which proceeds through a reactive *N*-(trifluoroacetoxyiodobenzene) derivative **32**, itself a form of anomeric amide [41, 79]. Similar hypervalent iodine oxidations have been used in nitrenium ion cyclizations onto aromatic rings [128, 132, 133].

N-Alkoxy-*N*-benzoylnitrenium ions **34** are generated through $A_{Al}1$ acid-catalyzed solvolysis of *N*-acetoxy-*N*-alkoxybenzamides **33** (Scheme 2.7) [65, 67, 68]. Acetoxyl,

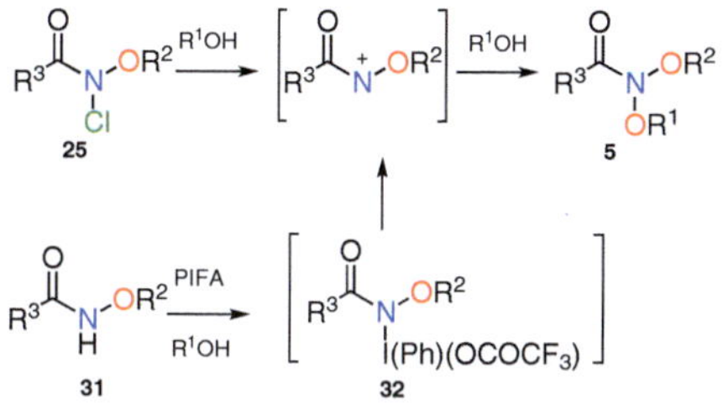

Scheme 2.6 *N*-Acyl-*N*-alkoxynitrenium ion-mediated syntheses of *N*,*N*-dialkoxyamides **5**. Source: Modified from Glover et al. [2].

upon protonation with a catalytic amount of mineral acid, is eliminated from **33** as acetic acid, and the nitrenium ions are trapped by water to form anomeric *N*-alkoxyhydroxamic acids **35**. The anomeric **35** undergoes secondary reactions to form a range of products.

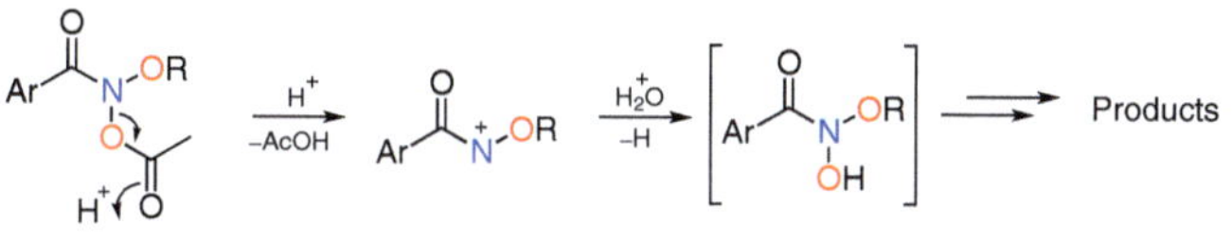

Scheme 2.7 Acid-catalyzed hydrolysis of *N*-acetoxy-*N*-alkoxybenzamides. Source: Modified from Glover et al. [2].

The elimination reactions of *N*-alkoxy-*N*-chloroamides **25** and the acid-catalyzed solvolysis reactions of **33**, both of which proceed through intermediacy of *N*-acyl-*N*-alkoxynitrenium ions, can better be reevaluated in terms of anomeric destabilization in combination with their reduced amidicities.

Shtamburg has reported analogous S_N2 and S_N1 reactivity at anomerically substituted nitrogen in both ureas and carbamates. Treatment of *N*-alkoxy-*N*-chloroureas and -carbamates with sodium salts of carboxylic acids in acetonitrile converted them to *N*-acyloxy-*N*-alkoxyureas and -carbamates [82, 86]. *N,N*-Dialkoxyureas and -carbamates can be synthesized by substitution of carboxylate, in these, by suitable alcohols. *N,N*-Dialkoxyureas have been synthesized by methanolysis of *N*-alkoxy-*N*-chloroureas with methanol in the presence of sodium acetate [82, 86]. Shtamburg's group has also reported interchange of carboxylate substituents of *N*-acyloxy-*N*-alkoxyureas, –carbamates, and -benzamides in the presence of sodium carboxylates in acetonitrile [134]. Tertiary amines and pyridine generate stable *N*-alkoxy-*N*-hydrazinium or -pyridinium chlorides by S_N2 reaction with *N*-alkoxy-*N*-chloroureas and -carbamates [84, 135].

2.3.3 Amide Bond Scission Reactions: The HERON Reaction

Despite the reduced resonance characteristics of anomeric amides, nucleophilic addition to the amide carbonyl, which would lead to acyl substitution reactions, does not compete with substitution reactions at nitrogen. However, in the absence of nucleophiles, or when elimination reactions are disfavoured by poor leaving groups at nitrogen, the anomeric interaction promotes an unusual thermal rearrangement, which results in N–C(O) bond scission.

Unique to suitably constituted anomeric amides, the HERON reaction is facilitated by weakened amide resonance but is driven by a strong n_Y–σ^*_{NX} anomeric destabilization of the N–X bond [2, 77, 100–104]. In such amides, when the X moiety of an anomerically destabilized N–X bond is a poor leaving group, the amide can undergo a concerted rearrangement involving the migration of X to the carbonyl carbon and the ejection of a Y-stabilized nitrene (Scheme 2.8). In essence, the reaction involves an internal S_N2 reaction at the amide carbonyl carbon. Where the HERON reaction is disfavoured by weak n_Y–σ^*_{NX} interactions, competitive homolysis of the N–X bond under thermolytic conditions generates Y-stabilized amidyl radicals.

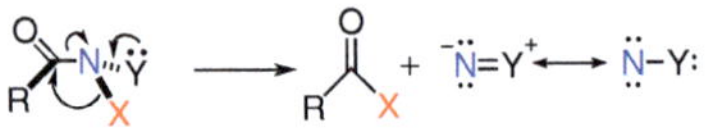

Scheme 2.8 The HERON reaction of an anomeric amide. Source: Modified from Glover et al. [2].

2.3.3.1 HERON Reactions of *N*-Alkoxy-*N*-Aminoamides

We discovered the HERON reaction in the early 1990s during research into S_N2 reactivity of *N*-acyloxy-*N*-alkoxyamides **21**, specifically the reaction between *N*-acetoxy-*N*-butoxybenzamide and *N*-methylaniline according to Scheme 2.3 (R^1 = Bu, R^2 = Me, R^3 = Ph) [66, 102]. In a polar solvent such as methanol, *N*-methylaniline attacks the amide nitrogen, replacing the acetoxyl side chain to form, in this case, an unstable intermediate, *N*-butoxy-*N*-(*N*′-methylanilino) benzamide **36**, which undergoes the HERON reaction to form butyl benzoate **37**, and 1-methyl-1-phenyldiazene **38**, an aminonitrene (Scheme 2.9i). Aminonitrenes are reactive intermediates with a singlet ground state, which persist long enough under reaction conditions to dimerize to tetrazenes [136–139], in this case, **39** (Scheme 2.9ii) [63, 66].

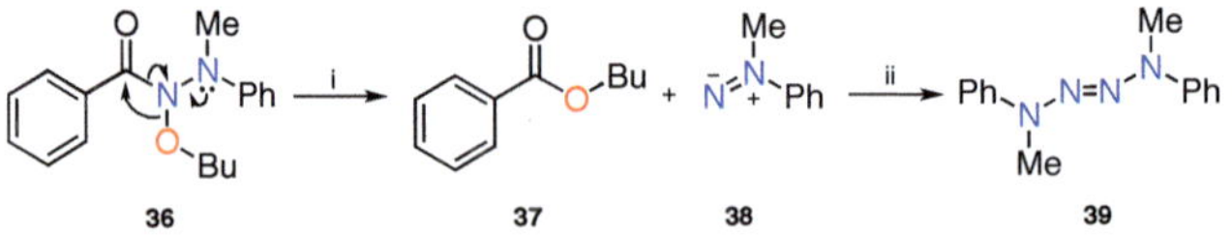

Scheme 2.9 The first HERON reaction of *N*-butoxy-*N*-(*N*′-methylanilino)benzamide **36**. Source: Glover et al. [2]/MDPI/CC BY 4.0.

N,N′-Diacyl-*N,N*′-dialkoxyhydrazines **40**, which can be isolated from oxidation of hydroxamic esters, were known to decompose thermally to generate esters, purportedly via consecutive four-centered mechanisms [92, 140]. We and Barton and coworkers studied the decomposition at about the same time, and, by using asymmetric hydrazines **40**, both our groups established the operation of consecutive three-center mechanisms rather than two four-center processes [77, 91]. **40**, a form of *N*-amino-*N*-alkoxyamides **2d**, undergo tandem HERON reactions to form two equivalents of ester and a molecule of nitrogen; the *N*-acyl-*N*-alkoxyaminonitrenes **42**, formed with ester **41** in the initial HERON reaction (Scheme 2.10, HERON i), rapidly undergo a second rearrangement to form a molecule of nitrogen and ester before dimerization of the *N*-acyl-*N*-alkoxyaminonitrene can occur (Scheme 2.10, HERON ii) [36, 77]. Step ii can also be regarded as a HERON process, driven by a high energy electron pair on 1,1-diazene, a charge separated form of aminonitrene **42**. Barton found that the concerted nature of these decompositions facilitated formation of a range of highly hindered esters, and, recently, Zhang has utilized the reaction to generate hindered esters from *N,N′*-diacyl-*N,N′*-dialkoxyhydrazines **40**, which they synthesized by *N*-bromosuccinimide oxidation of hydroxamic esters [141].

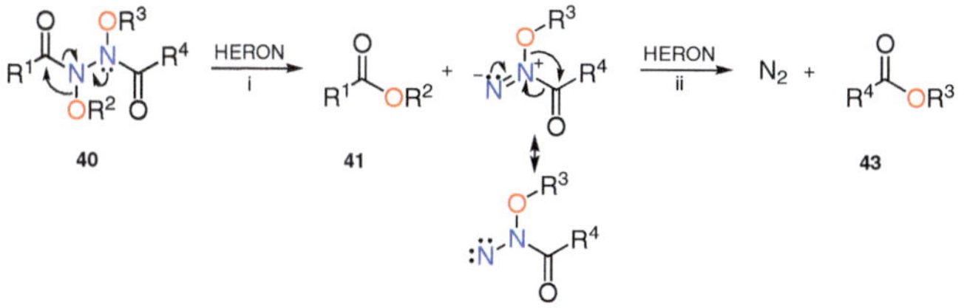

Scheme 2.10 Tandem HERON reactions of *N,N′*-diacyl-*N,N′*-dialkoxyhydrazines **40**. Source: Glover et al. [2]/MDPI/CC BY 4.0.

The HERON reactions of **22** and **40** have been modelled and validated theoretically [36, 77, 91, 102, 142]. At the B3LYP/6-31G(d) level, the HERON reaction of *N*-methoxy-*N*-(dimethylamino)formamide **15e**, a model representative of *N*-butoxy-*N*-(*N′*-methylanilino)benzamide **36**, and *N,N′*-diacyl-*N,N′*-dialkoxyhydrazines **40** [36], confirmed that the $n_N–\sigma^*_{NO}$ anomeric destabilization resulted in migration of the methoxyl group with an activation barrier of only 90 kJ/mol and the overall reaction was exothermic by 23 kJ/mol. The transition state accessible from the lowest energy (*syn*) conformer of **15e** was found to be that depicted in Figure 2.14a. The transition state for HERON reaction of *N*-methoxy-*N*-(dimethylamino)acetamide **16c** is similar and occurs with a barrier of 95 kJ/mol [54]. These barriers are in accord with experimental activation energies determined from Arrhenius studies of the unimolecular decomposition of hydrazines **19** and **20** in mesitylene, which demonstrated E_A's of between 99 and 126 kJ/mol for the initial, rate determining step (Scheme 2.10, HERON i).

In the model NNO HERON reaction (Figure 2.14a), methoxyl migrates in a plane largely orthogonal to the *N1-C1-O1* plane. The transition state is accompanied by development of partial positive charge of +0.5 on the dimethylamino group, partial negative charge of −0.3 on the migrating methoxyl group, while charge on the formyl group is unchanged. The theoretical charge separation in the transition

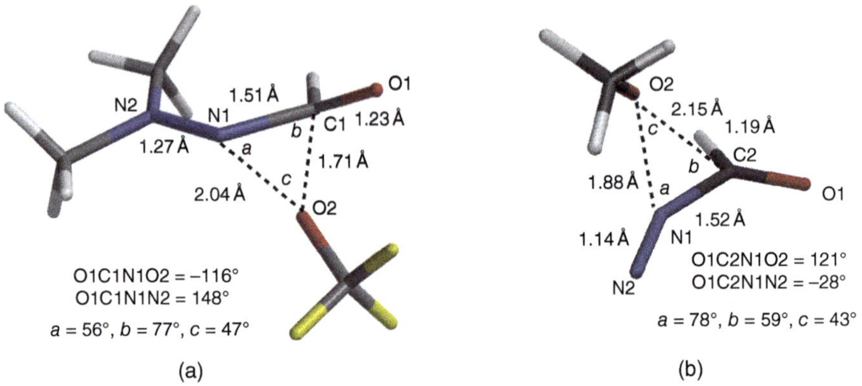

Figure 2.14 Twisted HERON transition states of (a) *N*-methoxy-*N*-(dimethylamino)formamide **15e** and (b) 1-formyl-1-methoxydiazene **42** (R^3 = Me, R^4 = H) at B3LYP/6-31G(d) level [36, 80]. Source: Modified from Glover et al. [2].

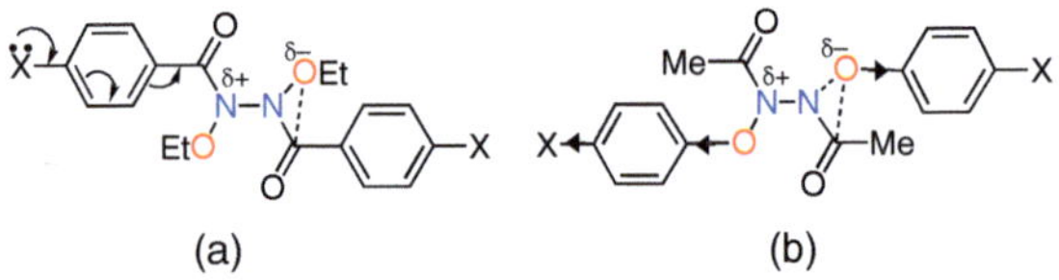

Figure 2.15 Influence on the HERON transition state of (a) electron-rich benzoyl groups in **19** and (b) electron-deficient benzyloxy groups in **20.** Source: Modified from Glover et al. [2].

state was confirmed by a negative Hammett σ^+ ($\rho = -0.35$) correlation for **19** and a positive Hammett σ ($\rho = 1.02$) for **20.** Donor groups at X (Figure 2.15a) facilitate rearrangement of **19** by stabilizing developing positive charge at the donor nitrogen atom (since charge is unaltered at the formyl group in model transition state (Figure 2.14a), the electronic effect of X at the recipient carbon would be minimal). Electron-withdrawing groups at X (Figure 2.15b) stabilize negative charge at the migrating oxygen [36]. However, since withdrawing substituents in **20** also lower the energy of the donor nitrogen lone pair, this series reacts about 2 orders of magnitude slower than **19.**

2.3.3.2 Other HERON Reactions

The second step in the tandem HERON reaction of *N,N′*-diacyl-*N,N′*-dialkoxyhydrazines **40** (Scheme 2.10, HERON ii) was modelled by rearrangement of 1-formyl-1-methoxydiazene **42** (R^3 = Me, R^4 = H) to methyl formate and N_2. At both B3LYP/6-31G(d) and CCSD(T)//B3P86 levels, the HERON reaction was found to have an extremely small E_A of between 5 and 12 kJ/mol and to be highly exothermic ($\Delta E = -400$ kJ/mol) [80, 142]. We encountered this process upon the reaction of *N*-acyloxy-*N*-alkoxyamides **21** and *N*-alkoxy-*N*-chloroamides **25** with azide (Scheme 2.2ii,v), which generates ester and two molecules of nitrogen [81]. Intermediate *N*-alkoxy-*N*-azidoamides **26** are highly unstable, losing nitrogen to generate 1-acyl-1-alkoxydiazenes **44,** which react further by a HERON process to give nitrogen and ester (Scheme 2.11). The transition state for the rearrangement of *N*-formyl-*N*-methoxydiazene, modelled at the B3LYP/6-31G(d) level, is depicted in Figure 2.14b. Once again, the methoxyl group migrates in a plane approximately orthogonal to the N1-C2-O1 plane in an early transition state with little N–C(O) bond cleavage. Overall, the decomposition of *N*-azido-*N*-methoxyformamide to two molecules of nitrogen and methyl formate was computed to be exothermic by some 575 kJ/mol [80].

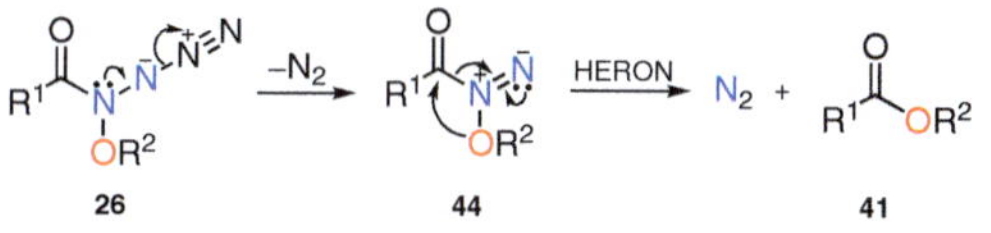

Scheme 2.11 Decomposition and HERON reaction of *N*-alkoxy-*N*-azidoamides **26**. Source: Glover et al. [2]/MDPI/CC BY 4.0.

Importantly, inspection of both HERON transition states in Figure 2.14 showed that amide resonance in the ground states is largely forfeited, as migration occurs in a plane more or less perpendicular to the carbonyl, twisting the nitrogen lone pair away from alignment with the $\pi^*_{C=O}$ orbital and into the N–C–O plane. However, anomeric destabilization of the N–O bond was preserved along the reaction coordinate, driving the reaction forward. The N–C(O) bond is largely intact at the transition state but breaks as the O2–C bond forms in an internal, S_N2-like reaction at the carbonyl carbon. Significantly, a tetrahedral intermediate is avoided by this process.

Apart from NNO anomeric amides **2d** such as **36** and **40**, as well as 1-acyl-1-alkoxydiazenes **2h**, we have observed thermal HERON reactivity of *N*-acyloxy-*N*-alkoxyamides **2b** themselves, which give anhydrides in competition with homolysis. **47** (Scheme 2.12) generates anhydrides **48** and alkoxyamidyls **46,** each of which undergoes secondary reactions leading to esters **49** and **50** and dioxazoles **45** [109].

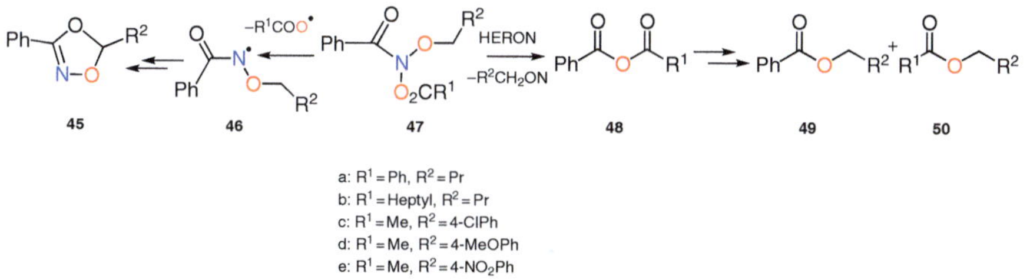

Scheme 2.12 HERON and radical decomposition pathways for *N*-acyloxy-*N*-alkoxyamides in toluene at 90 °C.

Acyclic *N,N*-dialkoxyamides **2c** are stable at room temperature, but at elevated temperatures, in mesitylene, they have been found to undergo homolysis to alkoxyl and alkoxyamidyl radicals [79]. However, we have observed HERON reactions of several cyclic forms of **2c**. PIFA oxidation of β- and γ-hydroxyhydroxamic esters **51** and **55** gives γ- and δ-oxazinolactams, **52** and **56**. In the NMR at r.t., **52** was observed to convert to the hyponitrite **54**, while **56** underwent ring contraction to γ-lactone **57** (with formation of **58**) (Scheme 2.13) [41].

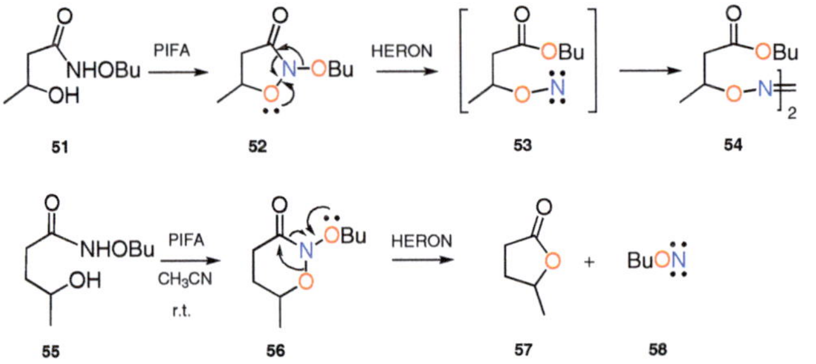

Scheme 2.13 HERON reactions of γ-oxazinolactam **52** and δ-oxazinolactam **56**. Source: Adapted from Glover et al. [41].

N-Alkoxyhydroxamates **23**, which are generated by $B_{Al}2$ reaction of hydroxide with *N*-acyloxy-*N*-alkoxyamides (Scheme 2.2iii), are also strongly activated to HERON reactivity and rearrange intramolecularly to form esters. Reaction of the hydroxide substitution product from **21**, a hydroxamic acid, with excess base generates the anion **23**, which rearranges to ester **41** (Scheme 2.14) [64]. Shtamburg also reported that methoxide reaction at the acyloxyl group in *N*-acetoxy-*N*-ethoxybenzamide (**21**, R^1 = Ph, R^2 = Et, R^3 = Me) generated ethyl benzoate (**41**, R^1 = Ph, R^2 = Et), presumably via the same intermediate anion [86].

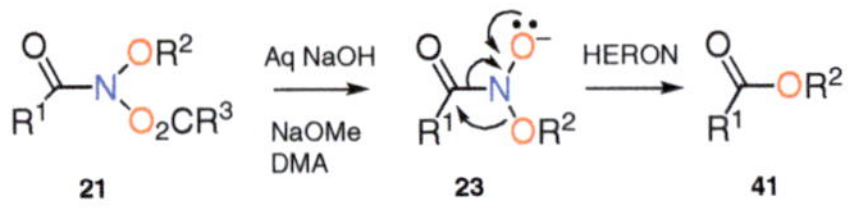

Scheme 2.14 HERON reaction of hydroxamate anion **23**. Source: Adapted from Glover et al. [64].

O-Alkyl-*N*-benzyloxy-*N*-(*N'*-methylanilino)carbamates **60**, formed by *N*-methylaniline reaction with *O*-alkyl-*N*-acetoxy-*N*-benzyloxycarbamates **59**, rearrange to dialkylcarbonates **61** and tetrazene **62** (Scheme 2.15) (K.L. Cavanagh and S.A. Glover, unpublished data).

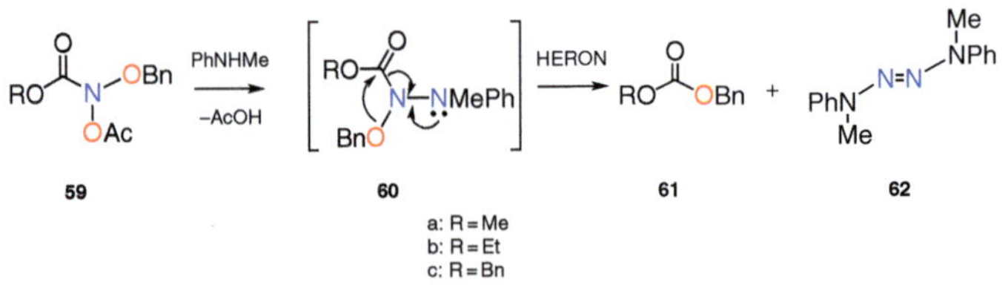

Scheme 2.15 HERON reactions of *O*-alkyl-*N*-benzyloxy-*N*-(*N'*-methylanilino)carbamates.

2.3.3.3 The Role of the $n_Y-\sigma^*_{NX}$ Anomeric Effect and Resonance in HERON Reactions

All these HERON processes have been modelled theoretically and the role of substituents and different anomeric combinations has been evaluated [53, 54]. The spatial arrangement of critical atoms at the transition states, which are exemplified in Figure 2.14, is common to transition state complexes for all these reactions (Figure 2.16) [2]. Notably, the anomeric configuration influences both the ground state resonance in these amides as well as impacting upon the ease of the rearrangement process. Since migration always occurs in a plane largely orthogonal to the N–C(O) plane, the resonance energy, which is lost at the HERON transition state, is a component of the overall E_A. The activation energy component for the actual rearrangement process, E_{rearr}, can therefore be estimated as (E_A + RE) [53, 54]. B3LYP/6-31G(d) calculations of ground state resonance energies and E_A's for all these processes, together with analysis of the transition state geometries of model reactions, indicate that the overriding determinant of HERON reactivity is the strength of the anomeric effect, rather than the reduced resonance in these systems [53, 54].

In Table 2.7, the computed E_A's for a range of model HERON processes are deconvoluted into their transamidation resonance, RE_{TA} (from Table 2.6), and rearrangement E_{rearr} components. By way of example for the NNO model system **16c**, the resonance energy (RE_{TA}) of −51 kJ/mol (67% amidicity), which is lost in the reaction is a component of the overall E_A, computed to be 95 kJ/mol, so the actual rearrangement process E_{rearr} can be estimated at 44 kJ/mol [53, 54]. The B3LYP/6-31G(d) resonance energy of *N,N*-dimethoxyacetamide **4b** is only −36 kJ/mol (47% amidicity), but the E_A for HERON of **4b** is 156 kJ/mol, yielding a E_{rearr} of 120 kJ/mol. When compared with an amino nitrogen lone pair, the low energy p-type oxygen lone pair is a weak anomeric donor resulting in a very high rearrangement component. For this reason, rather than a HERON reaction, *N,N*-dialkoxyamides undergo thermal homolysis to alkoxyamidyl and alkoxyl free radicals a process that clearly has a lower E_A [79].

Based on the deconvolution in Table 2.7, overall, the decreasing order of E_{rearr} for migration of methoxyl in XNY systems, which can be regarded as the order of increasing effectiveness of an $n_Y-\sigma^*_{NX}$ interaction and destabilization of the N–OMe bond, is ONOAc > ONS ~ ONO > ONN > ONO$^-$ > ONNitrene. The order of RE_{TA} is ONN > ONS > ONOAc > ONO > ONO$^-$ > ONNitrene and is broadly inverse of the total combined electronegativity of atoms attached to nitrogen. However, the overall activation energies decrease in the order ONOAc > ONS > ONO > ONN > ONO$^-$ > ONNitrene, similar to the decrease in E_{rearr}. Clearly, the dominant influence in HERON reactivity is the strength of the anomeric effect rather than the decrease in amidicity.

Several other salient deductions from this analysis can be made. Nitrogen lone pair donation lowers E_{rearr} for **16c** and **16g**. 1-Acetyl-1-methoxydiazene **16g**, a model for intermediates **42** and **44** in the decomposition of hydrazine dimers (Scheme 2.10) and *N*-azido-*N*-alkoxyamides (Scheme 2.11), respectively, has particularly low resonance energy of only −14.8 kJ/mol to overcome and negligible E_{rearr}. The nitrene is a

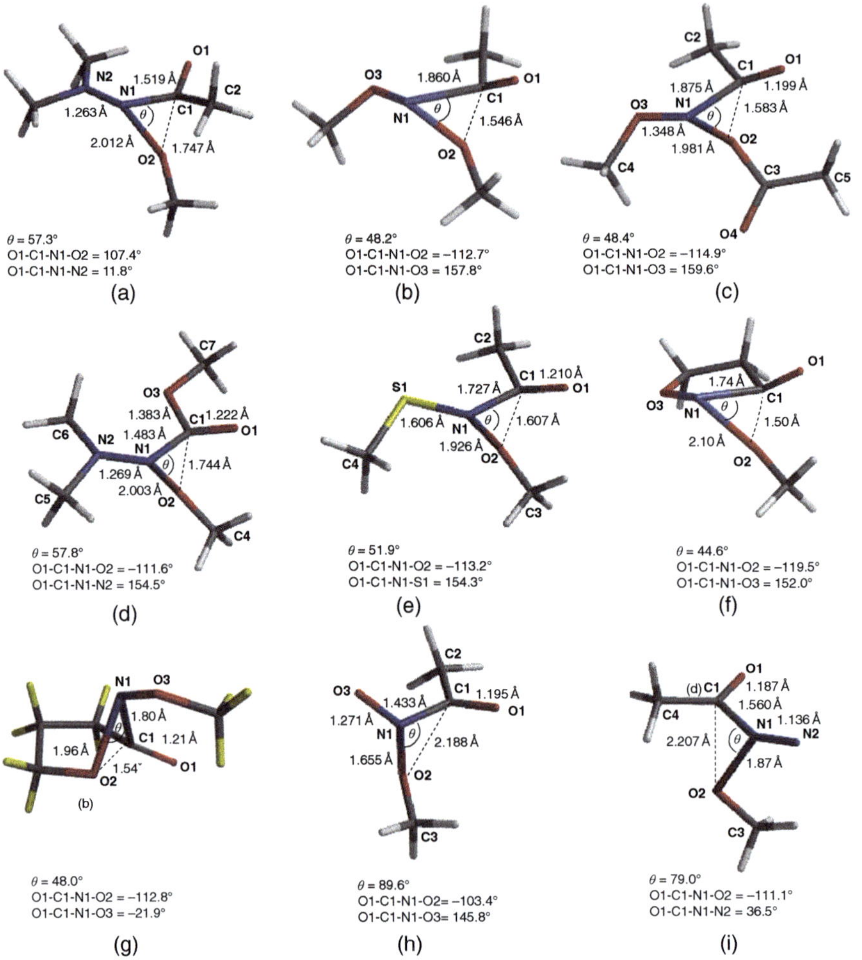

Figure 2.16 B3LYP/6-31G(d) optimized transition states of favoured HERON reactions of (a) *N*-methoxy-*N*-(dimethylamino)acetamide **16c**, (b) *N,N*-dimethoxyacetamide **4b**, (c) *N*-acetoxy-*N*-methoxyacetamide **16b**, (d) *O*-methyl-*N*-methoxy-*N*-(dimethylamino) carbamate **17a**, (e) *N*-methoxy-*N*-methylthiylacetamide **16d**, (f) *N*-methoxy-γ-oxazinolactam (ring opening), (g) *N*-methoxy-δ-oxazinolactam (ring contraction), (h) *N*-methoxyacetohydroxamate **16f**, and (i) 1-acetyl-1-methoxydiazene **16g**.

strong anomeric donor, and this, combined with the fact that a tetrahedral intermediate is not formed along the reaction pathway, a limiting stereochemical requirement in Fischer esterification, has facilitated the synthesis, in good yields, of highly hindered esters from both the decomposition of hydrazines [91, 141] and from the reaction of azide with *N*-alkoxy-*N*-chloroamides **25** (Scheme 2.2v), an alternative source of *N*-alkoxy-*N*-azidoamides **26** [81].

The data in Table 2.7 indicates that *N*-methoxy-γ- and δ-oxazinolactams **52** and **56** (Scheme 2.13) have a lower E_{rearr} than the acyclic form **4b**, which we have attributed to conformationally enhanced n_O–σ^*_{NO} anomeric driving forces [41].

Table 2.7 B3LYP/6-31G(d) activation energies (E_A), resonance energies (RE_{TA}), and rearrangement energies (E_{rearr}) for HERON reactions of model anomeric amides; all values in kJ/mol.

Migratory mode	E_A	XNY[a)]	RE_{TA} [b)]	E_{rearr}
MeO from *anti* **16c**	95.0	ONN	−51.5	43.5
MeO from **4b**	156.5	ONO	−35.9	120.5
MeO from *syn* **17a**	92.4	ONN	−50.7	41.8
AcO from *syn* **16b**	181.0	AcONO	−40.5	140.5
MeO from *syn* **16b**	207.9	ONOAc	−40.5	167.4
AcO from *anti* **16b**	181.5	AcONO	−40.5	141.0
MeO from *anti* **16b**	182.8	ONOAc	−40.5	142.3
MeO from *anti* **16d**	174.1	ONS	−48.6	125.4
MeO from **16f**	48.0	ONO^-	−32.4[c)]	15.6
MeO from **16g**	8.8	ONNitrene	−14.8[c)]	−6.0
Ring opening **52**	113.0	ONO_{endo}	−19.2[d)]	93.8
Ring contraction **56**	145.2	ONO_{exo}	−36.7[d)]	108.4
Ring opening of **56**	136.4	ONO_{endo}	−36.7[d)]	99.7

a) n_Y donor to σ^*_{NX}.
b) From Table 2.6.
c) Data from Reference [53].
d) Data from Reference [41].
Source: Adapted from Glover et al. [53].

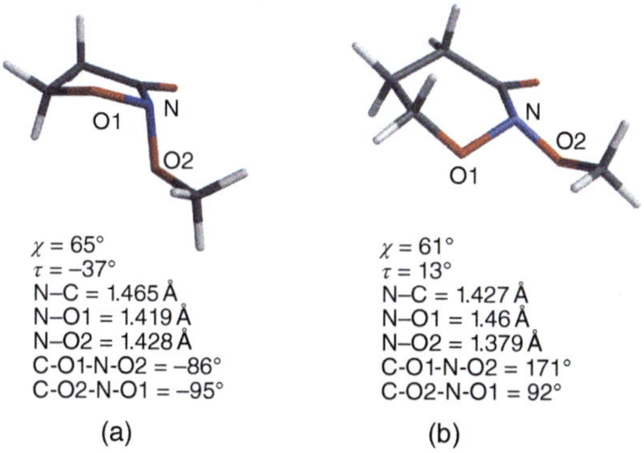

Figure 2.17 Ground state structures of models (a) *N*-methoxy-γ-oxazinolactam **52** and (b) chair *N*-methoxy-δ-oxazinolactam **56**.

Analysis of B3LYP/6-31G(d) optimized ground states (Figure 2.17) predict excellent n_{Oendo}–σ^*_{NOexo} alignment in **52**, facilitating a HERON ring opening and a strong n_{Oexo}–σ^*_{NOendo} anomeric effect in **56** favouring ring contraction (N–O2 bond is 0.08 Å shorter than the N–O1 bond). In the chair conformation of **56**, the n_{Oendo}–σ^*_{NOexo} is switched off disfavouring ring opening.

Finally, several other anomeric amides cannot be observed directly, but are implicated in reactions or from reaction products. While thiols react bimolecularly with *N*-acyloxy-*N*-alkoxyamides **21** to generate the SNO systems, *N*-alkoxy-*N*-alkylthiylamides **24** (Scheme 2.16), these intermediates have been found to react rapidly with reactant thiols at sulfur-generating hydroxamic esters **63** and disulfides **64** [78]. HERON reactivity of model **16d** through methoxyl migration is predicted to have a high E_{rearr}, making it uncompetitive with bimolecular reaction with thiol.

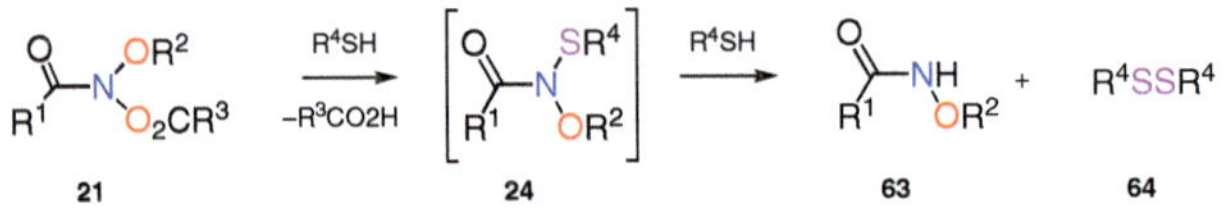

Scheme 2.16 Reaction of *N*-acyloxy-*N*-alkoxyamides with alkylthiols.

N-Amino-*N*-chloroamides **2f**, with strong anomeric destabilization of the N–Cl bond, would be expected to eliminate chloride readily. Chlorination of *N′*,*N′*-dimethylbenzhydrazide **65** with *tert*-butylhypochlorite generates *N*-benzoyl derivative **69**, presumably by reaction of reagent *N′*,*N′*-dimethylbenzhydrazide with either hydrazinium ion **67** or benzoyl chloride **68**, which would be formed by a HERON reaction of anomeric amide **66** (Scheme 2.17). However, B3LYP/6-31G(d) calculations suggest that elimination is the favoured pathway for these anomeric intermediates [42].

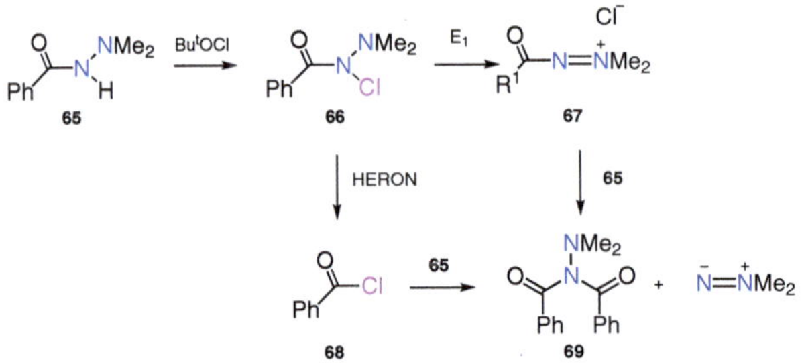

Scheme 2.17 Reactions of an anomeric *N*-amino-*N*-chloroamide.

Finally, the historical background to the HERON reaction and extensive reviews of this unusual process have appeared recently [2, 103, 104].

2.4 Concluding Remarks

The impact of introducing two substituents bonded through electronegative heteroatoms of the p-block is as marked as it is unusual. The combination of specifically two heteroatoms, through their electronegativity, impacts the amide resonance as well as the conformation at nitrogen. Both facets are borne out by ample experimental evidence. Pyramidality at nitrogen together with reduced lone pair overlap, in the traditional amide sense, is evident in structural, spectroscopic, and theoretical analysis of congeners of the class of anomeric amides, as well as the related anomeric carbamates and ureas. Analysis of computed energetic changes to resonance as a result of bisheteroatom substitution sheds light on isomerization processes and barriers to rotation and inversion. While resonance is reduced, it is still present through hybrid lone pair overlap with the carbonyl carbon. However, this alone does not account for the reduction in amide characteristics. Adoption of 2s character by the lone pair renders that interaction less favourable than in conventional amide bonds.

The reduction in resonance does not translate to expected susceptibility of the amide bond to acyl substitution reactions. These reactions are uncompetitive with anomerically driven reactions at the nitrogen. With the partial sp^3 character, the anomerically substituted nitrogens behave like saturated carbon centers and, with suitable leaving groups at nitrogen, both S_N1 and S_N2 processes occur. Where these are disfavoured by poor leaving groups, anomeric donor destabilization results in the HERON rearrangement that involves scission of the amide bond. This reaction is unique both in the unusual migration of an *N*-substituent to the carbonyl and in the resultant generation of stabilized nitrenes.

Though the reduction in resonance is integral to the activation of the amide bond, overall reactivity trends are dictated by the attendant anomeric destabilization of groups at nitrogen.

In the course of publication of this work, a remarkable application of anomeric amides in skeletal editing by nitrogen deletion was reported in Nature [143, 144]. Based on our work, Levin and coworkers used a designer NNOAc anomeric amide, *N*-benzyloxy-*N*-pivaloyloxy-4-trifluoromethylbenzamide **1′** as a 1,1-diazene source. S_N2 displacement of pivalate by aliphatic amines generated NNO anomeric amides **2′**, which through the HERON reaction produced ester **3′** and 1,1-diazenes **4′**. Unlike the 1-methyl-1-phenyldiazene from the authors' reactions, which dimerize to tetrazene (Section 2.3.3.1), the dialkyl diazenes undergo thermally and entropically favourable nitrogen extrusion to generate a carbon–carbon bond between nitrogen substituents **5′**, most likely by solvent cage radical combination. The reaction is extremely versatile and has been elaborated to include nitrogen atom deletions from a wide range of secondary amines and more recently primary amines [145]. However, it is important to point out that almost all *N*-acyloxy-*N*-alkoxyamides we have made are direct-acting mutagens, and our QSAR studies with *Salmonella typhimurium* TA100 predict that the reagent in question is likely to be significantly mutagenic than most [56–58]. It would be wise to advocate the synthesis and wide employment of the reagent with appropriate precautions.

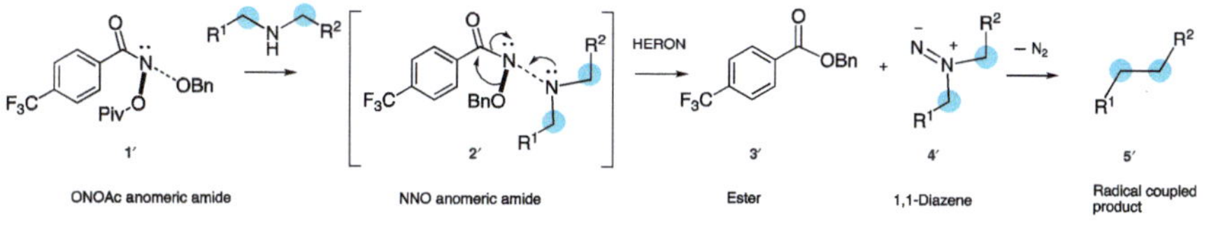

References

1 Greenberg, A., Breneman, C.M., and Liebman, J.F. (ed.) (2003). *The Amide Linkage: Structural Significance in Chemistry, Biochemistry and Materials Science*. Hoboken, NJ: Wiley.

2 Glover, S.A. and Rosser, A.A. (2018). Heteroatom substitution at amide nitrogen—resonance reduction and HERON reactions of anomeric amides. *Molecules* 23: 2834.

3 Wiberg, K.B. (2003). Origin of the amide rotational barrier. In: *The Amide Linkage. Structural Significance in Chemistry, Biochemistry and Materials Science* (ed. A. Greenberg, C.M. Breneman and J.F. Liebman), 33. New York: Wiley.

4 Wiberg, K.B. and Breneman, C.M. (1992). Resonance interactions in acyclic systems. 3. Formamide internal rotation revisited. charge and energy redistribution along the C–N bond rotational pathway. *J. Am. Chem. Soc.* 114: 831–840.

5 Glover, S.A. and Rosser, A.A. (2012). Reliable determination of amidicity in acyclic amides and lactams. *J. Org. Chem.* 77: 5492–5502.

6 Dunitz, J.D. (1979). *X-Ray Analysis and Structure of Organic Molecules*. London: Cornell University Press.

7 Winkler, F.K. and Dunitz, J.D. (1971). The non-planar amide group. *J. Mol. Biol.* 59: 169–182.

8 Szostak, M. and Aube, J. (2013). Chemistry of bridged lactams and related heterocycles. *Chem. Rev.* 113: 5701–5765.

9 Szostak, M. and Aube, J. (2011). Medium-bridged lactams: a new class of non-planar amides. *Org. Biomol. Chem.* 9: 27–35.

10 Szostak, R. and Szostak, M. (2019). Chemistry of bridged lactams: recent developments. *Molecules* 24: 274.

11 Kirby, A.J., Komarov, I.V., and Feeder, N. (1998). Spontaneous, millisecond formation of a twisted amide from the amino acid, and the crystal structure of a tetrahedral intermediate. *J. Am. Chem. Soc.* 120: 7101–7102.

12 Kirby, A.J., Komarov, I.V., and Feeder, N. (2001). Synthesis, structure and reactions of the most twisted amide. *J. Chem. Soc., Perkin Trans.* 2: 522–529.

13 Kirby, A.J., Komarov, I.V., Kowski, K., and Rademacher, P. (1999). Distortion of the amide bond in amides and lactams. Photoelectron-spectrum and electronic structure of 3,5,7-trimethyl-1-azaadamantan-2-one, the most twisted amide. *J. Chem. Soc., Perkin Trans.* 2: 1313–1316.

14 Kirby, A.J., Komarov, I.V., Wothers, P.D., and Feeder, N. (1998). The most twisted amide: structure and reactions. *Angew. Chem. Int. Ed.* 37: 785–786.

15 Morgan, J. and Greenberg, A. (2014). Novel bridgehead bicyclic lactams: (a) molecules predicted to have *O*-protonated and *N*-protonated tautomers of comparable stability; (b) hyperstable lactams and their *O*-protonated tautomers. *J. Chem. Thermodyn.* 733: 206–212.

16 Tani, K. and Stoltz, B.M. (2006). Synthesis and structural analysis of 2-quinuclidonium tetrafluoroborate. *Nature* 441: 731–734.

17 Yamada, S. (1998). Chemistry of highly twisted amides. *Rev. Heteroatom. Chem.* 19: 203–236.

18 Yamada, S. (2000). Sterically hindered twisted amides. In: *The Amide Linkage: Structural Aspects in Chemistry, Biochemistry, and Material Science* (ed. A. Greenberg, C.M. Breneman and J.F. Liebman), 215–246. Hoboken, NJ, ch. 8: Wiley.

19 Yamada, S. (1993). Structure and reactivity of an extremely twisted amide. *Angew. Chem. Int. Ed.* 32: 1083–1085.

20 Cow, C.N., Britten, J.F., and Harrison, P.H.M. (1998). X-ray crystal structure of 1,6-diacetyl-3,4,7,8-tetramethyl-2,5-dithioglycoluril a highly twisted acetamide. *J. Chem. Soc., Chem. Commun.* 1147–1148.

21 Anet, F.A.L. and Osyani, J.M. (1967). Nuclear magnetic resonance spectra and nitrogen inversion in 1-acylaziridinies. *J. Am. Chem. Soc.* 89: 352–356.

22 Boggs, G.R. and Gerig, J.T. (1969). Nitrogen inversion in *N*-benzoylaziridines. *J. Org. Chem.* 34: 1484–1486.

23 Ohwada, T., Achiwa, T., Okamoto, I. et al. (1998). On the planarity of amide nitrogen. Intrinsic pyramidal nitrogen of *N*-acyl-7-azabicyclo[2.2.1]heptanes. *Tetrahedron Lett.* 39: 865–868.

24 Otani, Y., Nagae, O., Naruse, Y. et al. (2003). An evaluation of amide group planarity in 7-azabicyclo[2.2.1]heptan amides. Low amide bond rotation barrier in solution. *J. Am. Chem. Soc.* 125: 15191–15199.

25 Otani, Y., Futaki, S., Kiwada, T. et al. (2005). Synthesis of non-planar peptides bearing the 7-azabicyclo[2.2.2]heptane skeleton, and possible self-organized structures. *Pept. Sci.* 41: 173–174.

26 Ohwada, T., Ishikawa, S., Mine, Y. et al. (2011). 7-Azabicyclo[2.2.2]heptane as a structural motif to block mutagenicity of nitrosomines. *Biorg. Med. Chem.* 19: 2726–2741.

27 Meng, G., Shi, S., Lalancette, R. et al. (2017). Reversible twisting of primary amides via ground state *N—C(O)* destabilization: highly twisted rotationally inverted acyclic amides. *J. Am. Chem. Soc.* 140: 727–734.

28 Komarov, I.V., Yanik, S., Ishchenko, A.Y. et al. (2015). The most reactive amide as a transition-state mimic for cis-trans interconversion. *J. Am. Chem. Soc.* 137: 926–930.

29 Allen, F.H., Kennard, O., Watson, D.G. et al. (1987). Tables of bond lengths determined by X-ray and neutron diffraction. Part 1. Bond lengths in organic compounds. *J. Chem. Soc., Perkin Trans.* 2: S1–S19.

30 Kirby, A.J., Komarov, I.V., Wothers, P.D. et al. (1999). Stereoelectronic interactions between hetero-atoms. *Pure Appl. Chem.* 71: 385–391.

31 Alabugin, I.V., Bresch, S., and Manoharan, M. (2014). Hybridization trends for main group elements and expanding the Bent's rule beyond carbon: more than electronegativity. *J. Phys. Chem. A* 118: 3663–3677.

32 Bent, H.A. (1961). An appraisal of valence-bond structures and hybridization in compounds of the first-row elements. *Chem. Rev.* 61: 275–311.

33 Rauk, A. and Glover, S.A. (1996). A computational investigation of the stereoisomerism in bisheteroatom-substituted amides. *J. Org. Chem.* 61: 2337–2345.

34 Glover, S.A. (1998). Anomeric amides—structure, properties and reactivity. *Tetrahedron* 54: 7229–7272.

35 Glover, S. and Rauk, A. (1999). Conformational stereochemistry of the HERON amide, *N*-methoxy-*N*-dimethylaminoformamide: a theoretical study. *J. Org. Chem.* 64: 2340–2345.

36 Glover, S.A., Mo, G., and Rauk, A. (1999). HERON rearrangement of *N,N′*-diacyl-*N,N′*-dialkoxyhydrazines – a theoretical and experimental study. *Tetrahedron* 55: 3413–3426.

37 Gillson, A.-M.E., Glover, S.A., Tucker, D.J., and Turner, P. (2003). Crystal structures and properties of mutagenic *N*-acyloxy-*N*-alkoxyamides – "most pyramidal" acyclic amides. *Org. Biomol. Chem.* 1: 3430–3437.

38 Glover, S.A. (2008). *N*-Acyloxy-*N*-alkoxyamide – structure, properties, reactivity and biological activity. *Adv. Phys. Org. Chem.* 42: 35–123.

39 Glover, S.A. (2009). *N*-Heteroatom-substituted hydroxamic esters. In: *The Chemistry of Hydroxylamines, Oximes and Hydroxamic, Acids, Part 2* (ed. Z. Rappoport and J.F. Liebman), 839–923. Chichester: Wiley.

40 Glover, S.A., White, J.M., Rosser, A.A., and Digianantonio, K.M. (2011). Structures of *N,N*-dialkoxyamides – pyramidal anomeric amides with low amidicity. *J. Org. Chem.* 76: 9757–9763.

41 Glover, S.A., Rosser, A.A., Taherpour, A., and Greatrex, B.W. (2014). Formation and HERON Reactivity of Cyclic *N,N*-dialkoxyamides. *Aust. J. Chem.* 67: 507–520.

42 Glover, S.A., Rosser, A.A., and Spence, R.M. (2014). Studies of the structure, amidicity and reactivity of *N*-chlorohydroxamic esters and *N*-chloro-β,β-dialkylhydrazides: anomeric amides with low resonance energies. *Aust. J. Chem.* 67: 1344–1352.

43 Reed, A.E., Curtiss, L.A., and Weinhold, F. (1988). Intermolecular interactions from a natural bond orbital, donor-acceptor viewpoint. *Chem. Rev.* 88: 899–926.

44 Weinhold, F., Landis, C.R., and Glendening, E.D. (2016). What is NBO analysis and how is it useful? *Int. Rev. Phys. Chem.* 35: 399–440.

45 Alabugin, I.V. (2016). *Stereoelectronic Effects: A Bridge Between Structure and Reactivity*, 1e. Chichester/Hoboken, NJ: Wiley.

46 Mucsi, Z., Tsai, A., Szori, M. et al. (2007). A quantitative scale for the extent of conjugation of the amide bond. Amidity Percentage as a chemical driving force. *J. Phys. Chem. A* 111: 13245–13254.

47 Mucsi, Z., Chass, G.A., Abranyi-Balogh, P. et al. (2013). Penicillin's catalytic mechanism revealed by inelastic neutrons and quantum chemical theory. *J. Phys. Chem. Chem. Phys.* **15**: 20447–20455.

48 Glover, S.A. (2019). Comment on "Penicillin's catalytic mechanism revealed by inelastic neutrons and quantum chemical theory" by Z. Mucsi, G. A. Chass, P.Abranyi-Balogh, B. Jojart, D.-C. Fang, A. J. Ramirez-Cuesta, B. Viskolczc and

I. G. Csizmadia, Phys. Chem. Chem. Phys., 2013, 15, 20447. *J. Phys. Chem. Chem. Phys.* 21: 18012–18025.
49 Greenberg, A. (2003). The amide linkage as a ligand: its properties and the role of distortion. In: *The Amide Linkage. Structural Significance in Chemistry, Biochemistry and Materials Science* (ed. A. Greenberg, C.M. Breneman and J.F. Liebman), 47–83. New York: Wiley.
50 Greenberg, A., Moore, D.T., and DuBois, T.D. (1996). Small and medium-sized bridgehead lactams: a systematic ab initio molecular orbital study. *J. Am. Chem. Soc.* 118: 8658–8668.
51 Greenberg, A. and Venanzi, C.A. (1993). Structures and energetics of two bridgehead lactams and their *N*- and *O*-protonated froms: An ab initio molecular orbital study. *J. Am. Chem. Soc.* 115: 6951–6957.
52 Hehre, W.J., Radom, L., Schleyer, v.R., and Pople, J.A. (1986). *Ab Initio Molecular Orbital Theory*. New York: Wiley.
53 Glover, S.A. and Rosser, A.A. (2016). The Role of substituents in the HERON reaction of anomeric amides. *Can. J. Chem.* 94: 1169–1180.
54 Glover, S.A. and Rosser, A.A. (2015). HERON reactions of anomeric amides: understanding the driving force. *J. Phys. Org. Chem.* 28: 215–222.
55 Glover, S.A. and Schumacher, R.R. (2021). Mutagenicity of *N*-acyloxy-*N*-alkoxyamides as an indicator of DNA intercalation: the role of fluorene and fluorenone substituents as DNA intercalators. *Mutat. Res. Genet. Toxicol. Environ. Mutagen.* 863–864: 503299.
56 Glover, S.A. and Schumacher, R.R. (2016). The effect of hydrophobicity upon the direct mutagenicity of *N*-acyloxy-*N*-alkoxyamides – bilinear dependence upon LogP. *Mutat. Res. Genet. Toxicol. Environ. Mutagen.* 795: 41–50.
57 Banks, T.M., Clay, S.F., Glover, S.A., and Schumacher, R.R. (2016). Mutagenicity of *N*-Acyloxy-*N*-alkoxyamides as an indicator of DNA intercalation Part 1: evidence for naphthalene as a DNA intercalator. *Org. Biomol. Chem.* 14: 3699–3714.
58 Glover, S.A., Schumacher, R.R., Bonin, A.M., and Fransson, L.E. (2011). Steric effects upon the direct mutagenicity of *N*-acyloxy-*N*-alkoxyamides – probes for drug-DNA interactions. *Mutat. Res. Genet. Toxicol. Environ. Mutagen.* 722: 32–38.
59 Andrews, L.E., Bonin, A.M., Fransson, L.E. et al. (2006). The role of steric effects in the direct mutagenesis of *N*-acyloxy-*N*-alkoxyamides. *Mutat. Res.-Genet. Toxicol. Environ. Mutagen.* 605: 51–62.
60 Andrews, L.E., Banks, T.M., Bonin, A.M. et al. (2004). Mutagenic *N*-acyloxy-*N*-alkoxyamides – probes for drug – DNA interactions. *Aust. J. Chem.* 57: 377–381.
61 Banks, T.M., Bonin, A.M., Glover, S.A., and Prakash, A.S. (2003). Mutagenicity and DNA damage studies of *N*-acyloxy-*N*-alkoxyamides – the role of electrophilic nitrogen. *Org. Biomol. Chem.* 1: 2238–2246.
62 Bonin, A.M., Banks, T.M., Campbell, J.J. et al. (2001). Mutagenicity of electrophilic *N*-acyloxy-*N*-alkoxyamides. *Mutat. Res.-Genet. Toxicol. Environ. Mutagen.* 494: 115–134.

63 Campbell, J.J. and Glover, S.A. (1999). (Synopsis)Bimolecular reactions of mutagenic *N*-acyloxy-*N*-alkoxybenzamides with aromatic amines. *J. Chem. Res., Synop.* 8: 474–475.

64 Glover, S.A., Hammond, G.P., and Bonin, A.M. (1998). A comparison of the reactivity and mutagenicity of *N*-benzoyloxy-*N*-benzyloxybenzamides. *J. Org. Chem.* 63: 9684–9689.

65 Bonin, A.M., Glover, S.A., and Hammond, G.P. (1994). Reactive intermediates from the solvolysis of mutagenic *O*-alkyl *N*-acetoxybenzohydroxamates. *J. Chem. Soc., Perkin Trans.* 2: 1173–1180.

66 Campbell, J.J. and Glover, S.A. (1992). Bimolecular reactions of mutagenic *N*-acetoxy-*N*-alkoxybenzamides and *N*-methylaniline. *J. Chem. Soc., Perkin Trans.* 2: 1661–1663.

67 Campbell, J.J., Glover, S.A., Hammond, G.P., and Rowbottom, C.A. (1991). Evidence for the formation of nitrenium ions in the acid-catalysed solvolysis of mutagenic *N*-acetoxy-*N*-alkoxybenzamides. *J. Chem. Soc., Perkin Trans.* 2: 2067–2079.

68 Campbell, J.J., Glover, S.A., and Rowbottom, C.A. (1990). Solvolysis and mutagenesis of *N*-acetoxy-*N*-alkoxybenzamides – evidence for nitrenium ion formation. *Tetrahedron Lett.* 31: 5377–5380.

69 Glover, S.A. and Scott, A.P. (1989). MNDO properties of heteroatom and phenyl substituted nitrenium ions. *Tetrahedron* 45: 1763–1776.

70 Chakraborty, M., Jin, K.J., Glover, S.A., and Novak, M. (2010). Characterisation of 4-(benzothiazol-2-yl)phenylnitrenium ion from a putative metabolite of a model anti-tumor drug. *J. Org. Chem.* 75: 5296–5304.

71 Novak, M. and Rajagopal, S. (2002). Correlations of nitrenium ion selectivities with quantitative mutagenicity and carcinogenicity of the corresponding amines. *Chem. Res. Toxicol.* 15: 1495–1503.

72 Cheng, B. and McClelland, R.A. (2001). Reaction of arylnitrenium ions with guanine derivatives: *N*1-methylguanosine and *N*2,*N*2-dimethylguanosine. *Can. J. Chem.* 79: 1881–1886.

73 Kennedy, S.A., Novak, M., and Kolb, B.A. (1997). Reactions of ester derivatives of carcinogenic *N*-(4-biphenylyl)hydroxylamine and the corresponding hydroxamic acid with purine nucleosides. *J. Am. Chem. Soc.* 119: 7654–7664.

74 McClelland, R.A., Kahley, M.J., and Davidse, P.A. (1996). Reactivity of the 4-biphenylyl and 2-fluorenylnitrenium ions with heterocyclic and carbon nucleophiles. *J. Phys. Org. Chem.* 9: 355–360.

75 Novak, M. and Kennedy, S.A. (1995). Selective trapping of *N*-acetyl-*N*-(4-biphenylyl)nitrenium and *N*-acetyl-*N*-(2-fluorenyl)nitrenium ions by 2′-deoxyguanosine in aqueous solution. *J. Am. Chem. Soc.* 117: 574–575.

76 Cavanagh, K.L., Glover, S.A., Price, H.L., and Schumacher, R.R. (2009). S_N2 substitution reactions at the amide nitrogen in the anomeric mutagens, *N*-acyloxy-*N*-alkoxyamides. *Aust. J. Chem.* 62: 700–710.

77 Buccigross, J.M., Glover, S.A., and Hammond, G.P. (1995). Decomposition of *N*,*N*′-diacyl-*N*,*N*′-dialkoxyhydrazines revisited. *Aust. J. Chem.* 48: 353–361.

78 Glover, S.A. and Adams, M. (2011). Reaction of *N*-acyloxy-*N*-alkoxyamides with biological thiols. *Aust. J. Chem.* 64: 443–453.

79 Digianantonio, K.M., Glover, S.A., Johns, J.P., and Rosser, A.A. (2011). Synthesis and thermal decomposition of *N,N*-dialkoxyamides. *Org. Biomol. Chem.* 9: 4116–4126.

80 Glover, S.A. and Rauk, A. (2002). A computational investigation of the structure of the novel anomeric amide *N*-azido-*N*-methoxyformamide and its concerted decomposition to methyl formate and nitrogen. *J. Chem. Soc., Perkin Trans.* 2: 1740–1746.

81 Glover, S.A. and Mo, G. (2002). Hindered ester formation by S_N2 azidation of *N*-acetoxy-*N*-alkoxyamides and *N*-Alkoxy-*N*-chloroamides – novel application of HERON rearrangements. *J. Chem. Soc., Perkin Trans.* 2: 1728–1739.

82 Shtamburg, V.G., Shishkin, O.V., Zubatyuk, R.I. et al. (2006). *N*-Chloro-*N*-alkoxyureas: synthesis, structure and properties. *Mendeleev Commun.* 16: 323–325.

83 Shishkin, O.V., Zubatyuk, R.I., Shtamburg, V.G. et al. (2006). Pyramidal amide nitrogen in *N*-acyloxy-*N*-alkoxyureas and *N*-acyloxy-*N*-alkoxycarbamates. *Mendeleev Commun.* 16: 222–223.

84 Shtamburg, V.G., Shishkin, O.V., Zubatyuk, R.I. et al. (2007). Synthesis, structure and properties of *N*-alkoxy-*N*-(1-pyridinium)urea salts, *N*-alkoxy-*N*-acyloxyureas and *N,N*-dialkoxyureas. *Mendeleev Commun.* 17: 178–180.

85 Shtamburg, V.G., Shishkin, O.V., Zubatyuk, R.I. et al. (2013). Synthesis and structure of *N*-alkoxyhydrazines and *N*-alkoxy-*N′*,*N′*,*N′*-trialkylhydrazinium salts. *Mendeleev Commun.* 23: 289–291.

86 Shtamburg, V.G., Klots, E.A., Pleshkova, A.P. et al. (2003). Geminal systems. 50. Synthesis and alcoholysis of *N*-acyloxy-*N*-alkoxy derivatives of ureas, carbamates, and benzamides. *Russ. Chem. Bull.* 52: 2251–2260.

87 Shtamburg, V.G., Tsygankov, A.T., Shishkin, O.V. et al. (2012). The properties and structure of *N*-chloro-*N*-methoxy-4-nitrobenzamide. *Mendeleev Commun.* 22: 164–166.

88 Allen, F.H. (2002). The Cambridge Structural Database: a quarter of a million crystal structures and rising. *Acta Crystallogr., Sect. B: Struct. Sci* B58: 380–388.

89 Nishii, Y., Hirai, T., and Mashima, K. (2017). CCDC 1536567: Experimental Crystal Structure Determination. https://doi.org/10.5517/ccdc.csd.cc1nkxpt

90 Shtamburg, V.G., Tsygankov, A.V., Gerasimenko, M.V. et al. (2011). New approach to *N,N*-dialkoxy-*N′*-arylureas and *N,N*-dialkoxycarbamates. *Mendeleev Commun.* 21: 50–52.

91 De Almeida, M.V., Barton, D.H.R., Bytheway, I. et al. (1995). Preparation and thermal decomposition of *N,N′*-diacyl-*N,N′*-dialkoxyhydrazines: synthetic applications and mechanistic insights. *J. Am. Chem. Soc.* 117: 4870–4874.

92 Cooley, J.H., Mosher, M.W., and Khan, M.A. (1968). Preparation and reactions of *N,N′*-diacyl-*N,N′*-dialkoxyhydrazines. *J. Am. Chem. Soc.* 90: 1867–1871.

93 Banks, T.M. (2003). Reactivity, Mutagenicity and DNA Damage of *N*-Acyloxy-*N*-alkoxyamides, Ph.D., University of New England.

94 Gerdes, R.G., Glover, S.A., Ten Have, J.F., and Rowbottom, C.A. (1989). *N*-Acetoxy-*N*-alkoxyamides – a new class of nitrenium ion precursors which are mutagenic. *Tetrahedron Lett.* 30: 2649–2652.

95 Williams, D.H. and Fleming, I. (1980). *Spectroscopic Methods in Organic Chemistry*, 3e. Maidenhead: McGraw Hill Ltd.

96 Jackman, L.M. (1975). Rotation about partial double bonds in organic molecules. In: *Dynamic Nuclear Magnetic Resonance Spectroscopy* (ed. L.M. Jackman and F.A. Cotton). Academic Press 203–252.

97 Günther, H. (1995). *NMR Spectroscopy- Basic Principles, Concepts, and Applications in Chemistry*, 2e. Chichester: Wiley.

98 Rudchenko, V.F. (1993). Synthesis, reactions, and properties of ONO systems. *Chem. Rev.* 93: 725–739.

99 Alabugin, I.V., Bresch, S., and dos Passos Gomez, G. (2015). Orbital hybridization: a key electronic factor in control of structure and reactivity. *J. Phys. Org. Chem.* 28: 147–162.

100 The Heron Rearrangement (2013). *Merck Index – Named Reaction Index*, 15e. Royal Society of Chemistry https://www.rsc.org/Merck-Index/reaction/r197/.

101 The HERON Rearrangement (2006). *The Merck Index*, 14e (ed. M.J. O'Neil), ONR-43. Whitehouse Station: Merck & Co., Inc.

102 Buccigross, J.M. and Glover, S.A. (1995). Molecular orbital studies of *N* to *C* migrations in *N*,*N*-bisheteroatom-substituted amides – HERON rearrangements. *J. Chem. Soc., Perkin Trans.* 2: 595–603.

103 Glover, S.A., Rauk, A., Buccigross, J.M. et al. (2005). Review: The HERON reaction: origin, theoretical background and prevalence. *Can. J. Chem.* 83: 1492–1509.

104 Glover, S.A. (2017). Development of the HERON reaction: a historical account. *Aust. J. Chem.* 70: 344–361.

105 Alabugin, I.V. and Zeidan, T.A. (2002). Stereoelectronic effects and general trends in hyperconjugative acceptor ability of σ bonds. *J. Am. Chem. Soc.* 124: 3175–3185.

106 Rauk, A. (1994). *Orbital Interaction Theory of Organic Chemistry*. New York: Wiley.

107 Eliel, E.L., Wilen, S.H., and Mander, L.N. (1994). *Stereochemistry of Organic Compounds*. New York: Wiley.

108 Epiotis, N.D., Cherry, W.R., Shaik, S. et al. (1977). *Topics in Current Chemistry*, Structural Theory of Organic Chemistry, vol. 70. Berlin: Springer-Verlag.

109 Johns, J.P., van Losenoord, A., Mary, C. et al. (2010). Thermal decomposition of *N*-acyloxy-*N*-alkoxyamides – a New HERON reaction. *Aust. J. Chem.* 63: 1717–1729.

110 Glover, S.A. (2001). S_N2 reactions at amide nitrogen – theoretical models for reactions of mutagenic *N*-acyloxy-*N*-alkoxyamides with bionucleophiles (OT-308CP). *Arkivoc*, Part xii 143–160.

111 Campbell, J.J. and Glover, S.A. (1999). (Microfiche) Bimolecular reactions of mutagenic *N*-(acyloxy)-*N*-alkoxybenzamides with aromatic amines. *J. Chem. Res. (M)* 8: 2075–2096.

112 Isaacs, N.S. (1995). *Physical Organic Chemistry*, 2e. New York: Longman Scientific and Technical.

113 Forster, W. and Laird, R.M. (1982). The mechanism of alkylation reactions. Part 1. The effect of substituents on the reaction of phenacyl bromide with pyridine in methanol. *J. Chem. Soc., Perkin Trans.* 2: 135–138.

114 De Kimpe, N. and Verhé, R. (1988). The chemistry of α-haloketones, α-haloaldehydes and α-haloimines. In: *Chemistry of Functional Groups* (ed. S. Patai and Z. Rappoport), 38. Chichester: Wiley.

115 Thorpe, J.W. and Warkentin, J. (1973). Stereochemical and steric effects in nucleophilic substitution of α-halo ketones. *Can. J. Chem.* 51: 927–935.

116 Maron, D.M. and Ames, B.N. (1983). Revised methods for *Salmonella* mutagenicity tests. *Mutat. Res.* 113: 173–215.

117 Mortelmans, K. and Zeiger, E. (2000). The Ames Salmonella/microsome mutagenicity assay. *Mutat. Res. Fundam. Mol. Mech. Mutagen.* 455: 29–60.

118 Pullman, A. and Pullman, B. (1980). Electrostatic effect of macromolecular structure on the biochemical reactivity of the nucleic acids. Significance for chemical carcinogenesis. *Int. J. Quantum Chem.* 7: 245–259.

119 Kohn, K.W., Hartley, J.A., and Mattes, W.B. (1987). Mechanism of sequence selective alkylation of guanine-N7 positions by nitrogen mustards. *Nucleic Acids Res.* 15: 10531–10544.

120 Warpehoski, M.A. and Hurley, L.H. (1988). Sequence selectivity of DNA covalent modification. *Chem. Res. Toxicol.* 1: 315–333.

121 Prakash, A.S., Denny, W.A., Gourdie, T.A. et al. (1990). DNA-directed alkylating ligands as potential antitumor agents: sequence specificity of alkylation by aniline mustards. *Biochemistry* 29: 9799–9807.

122 Schroeder, D., Grandinetti, F., Hrusak, J., and Schwarz, H. (1992). Experimental and *ab initio* MO studies on [H2,N,O]+ ions in the gas phase: characterization of the isomers H2NO+, HNOH+ and NOH2+ and the mechanism of unimolecular dehydrogenation of [H2,N,O]+. *J. Phys. Chem.* 96: 4841–4845.

123 Glover, S.A., Goosen, A., McCleland, C.W., and Schoonraad, J.L. (1984). *N*-Alkoxy-*N*-acylnitrenium ions as possible intermediates in intramolecular aromatic substitution: novel formation of *N*-acyl-3,4-dihydro-1*H*-2,1-benzoxazines and *N*-Acyl-4,5-dihydro-1*H*,3*H*-2,1-benzoxazepine. *J. Chem. Soc., Perkin Trans.* 1: 2255–2260.

124 Glover, S.A., Rowbottom, C.A., Scott, A.P., and Schoonraad, J.L. (1990). Alkoxynitrenium ion cyclisations: evidence for different mechanisms in the formation of benzoxazines and benzoxazepines. *Tetrahedron* 46: 7247–7262.

125 Kikugawa, Y. and Kawase, M. (1984). Electrophilic aromatic substitution with a nitrenium ion generated from *N*-chloro-*N*-methoxyamides. Application to the synthesis of 1-methoxy-2-oxindoles. *J. Am. Chem. Soc.* 106: 5728–5729.

126 Kawase, M., Kitamura, T., and Kikugawa, Y. (1989). Electrophilic aromatic substitution with *N*-methoxy-*N*-acylnitrenium ions generated from *N*-chloro-*N*-methoxy amides: syntheses of nitrogen heterocyclic compounds bearing a *N*-methoxy amide group. *J. Org. Chem.* 54: 3394–3403.

127 Kikugawa, Y., Shimada, M., Kato, M., and Sakamoto, T. (1993). A new synthesis of *N*-alkoxy-2-ethoxyarylacetamides from *N*-alkoxy-*N*-chloroarylacetamides with triethylamine in ethanol. *Chem. Pharm. Bull.* 41: 2192–2194.

128 Miyazawa, E., Sakamoto, T., and Kikugawa, Y. (2003). Synthesis of spirodienones by intramolecular ipso-cyclization of *N*-methoxy-(4-halegenophenyl)amides using [hydroxy(tosyloxy)iodo]benzene in trifluoroethanol. *J. Org. Chem.* 68: 5429–5432.

129 Kikugawa, Y., Kawase, M., Miyake, Y. et al. (1988). A convenient synthesis of eupolauramine. *Tetrahedron Lett.* 29: 4297–4298.

130 Kikugawa, Y., Shimado, M., and Matsumoto, K. (1994). Cyclization with nitrenium ions generated from *N*-methoxy- or *N*-allyloxy-*N*-chloroamides with anhydrous zinc acetate. Synthesis of *N*-hydroxy- and *N*-methoxynitrogen heterocyclic compounds. *Heterocycles* 37: 293–301.

131 Kikugawa, Y. (1996). Uses of hydroxamic acids and *N*-alkoxyimidoyl halides in organic synthesis. *Rev. Heteroatom Chem.* 15: 263–299.

132 Correa, A., Tellitu, I., Dominguez, E. et al. (2004). An efficient, PIFA mediated approach to benzo-, naphtho-, and heterocycle-fused pyrrolo[2,1-c][1,4]diazepines. An advantageous access to the antitumor antibiotic DC-81. *J. Org. Chem.* 70: 2256–2264.

133 Li, X., Yang, L., Zhang, X. et al. (2014). Construction of 1,4-benzodiazepine skeleton from 2-(arylamino)benzamides through $PhI(OAc)_2$-mediated oxidative C–N bond formation. *J. Org. Chem.* 79: 955–962.

134 Shtamburg, V.G., Tsygankov, A.V., Klots, E.A., and Kostyanovsky, R.G. (2004). Geminal systems, part 51. Acyloxy group exchange in *N*-acyloxy-*N*-alkoxyamides. *Mendeleev Commun.* 14: 208–210.

135 Shtamburg, V.G., Shtamburg, V.V., Klots, E.A. et al. (2020). Nuclophilic substitution in *N*-alkoxy-*N*-chlorocarbamates as a way to *N*-alkoxy-*N′*,*N′*,*N′*-trimethylhydrazinium chlorides. *Eur. Chem. Bull.* 9: 28–32.

136 Hinsberg, W.D. III, and Dervan, P.B. (1978). Synthesis and direct spectroscopic observation of a 1,1-dialkyldiazene. Infrared and electronic spectrum of *N*-(2,2,6,6-tetramethylpiperidyl)nitrene. *J. Am. Chem. Soc.* 100 (5): 1608–1610.

137 Hinsberg, W.D. III, and Dervan, P.B. (1979). Kinetics of the thermal decomposition of a 1,1-dialkyldiazene, *N*-(2,2,6,6-tetramethylpiperidyl)nitrene. *J. Am. Chem. Soc.* 101 (20): 6142–6144.

138 Hinsberg, W.D. III, Schultz, P.G., and Dervan, P.B. (1982). Direct studies of 1,1-diazenes. Syntheses, infrared and electronic spectra, and kinetics of the thermal decomposition of *N*-(2,2,6,6-tetramethylpiperidyl)nitrene and *N*-(2,2,5,5-tetramethylpyrrolidyl)nitrene. *J. Am. Chem. Soc.* 104: 766–773.

139 Schultz, P.G. and Dervan, P.B. (1980). Synthesis and direct spectroscopic observation of *N*-(2,2,5,5-tetramethylpyrrolidyl)nitrene. Comparison of five- and six-membered cyclic 1,1-dialkyldiazenes. *J. Am. Chem. Soc.* 102 (2): 878–880.

140 Cooley, J.H., Stone, D.H., and Oguri, H. (1977). Electronic effects in multicenter rearrangements of compounds with nitrogen-nitrogen bonds. *J. Org. Chem.* 42: 3096–3097.

141 Zhang, N., Yang, R., Zhang-Negrerie, D. et al. (2014). Direct conversion of *N*-alkoxyamides to carboxylic esters through tandem NBS-mediated oxidative homocoupling and thermal denitrogenation. *J. Org. Chem.* 78: 8705–8711.

142 Thomson, L.M. and Hall, M.B. (2000). Theoretical study of the thermal decomposition of *N,N′*-diacyl-*N,N′*-dialkoxyhydrazines: a comparison of HF, MP2, and DFT. *J. Phys. Chem. A* 104: 6247–6252.

143 Kennedy, S.H., Dherange, B.D., Berger, K.J., and Levin, M.D. (2021). Skeletal editing through direct nitrogen deletion of secondary amines. *Nature* 593: 223–227.

144 Zippel, C., Seibert, J., and Brase, S. (2021). Skeletal editing – nitrogen deletion of secondary amines by anomeric amide reagents. *Angew. Chem. Int. Ed.* 60: 19522–19524.

145 Berger, K.J., Driscoll, J.L., Yuan, M. et al. (2021). Direct deamination of primary amines via isodiazene intermediates. *J. Am. Chem. Soc.* 143: 17366–17373.

3

Amide Bond Activation by Twisting and Nitrogen Pyramidalization

Yuko Otani and Tomohiko Ohwada

Department of Organic and Medicinal Chemistry, Graduate School of Pharmaceutical Science, The University of Tokyo, 7-3-1 Hongo, Bunkyo-ku, Tokyo 113-0033, Japan

3.1 Introduction

Amide bonds are the fundamental bonds in proteins and many other biologically active molecules. The planarity of amide bonds is a common feature of natural and synthetic molecules containing amide moieties, which often determine the overall three-dimensional structure of the molecule. While the amide cis–trans isomerization$^+$ occurs via a transitional nonplanar structure [1], there are amides that take nonplanar structure in the ground state. ($^+$About amide geometry, while Z-E notation follows IUPAC nomenclature, we use cis-trans notation in most cases, which matches the definition of peptide along the main chain.)

Amide nonplanarity is associated with a reduced rotational barrier and a weakened amide N–C(O) bond, i.e. susceptibility to hydrolysis or high ability to transfer an acyl group [2–5]. As most compounds with a nonplanar amide show high reactivity toward water, they have rarely been used in structural units of functional molecules.

The structural unit to induce moderate nonplanarity with high chemical stability is introduced in Section 3.2. Amides of bicyclic 7-azabicyclo[2.2.1]heptane are focused in this section (Figure 3.1). The boat-shaped structure of monocyclic cyclohexane among the equilibrium between the chair-shape and the boat-shape is bridged by a nitrogen to form the 7-azabicyclo[2.2.1]heptane. Therefore, it has robust three-dimensional structure, and the nitrogen in the skeleton is strained, which are the causes of the unique structural properties. The rotational kinetics of these nonplanar amides and exceptional stability toward alkaline hydrolysis are also described.

Section 3.3 describes studies of biasing the cis–trans equilibrium of nonplanar tertiary amides to one side by introducing a single substituent at the bridgehead position. Application of these findings to generate nonnatural helical structures composed of bicyclic amino acids is also described. While the homooligomer of bicyclic amino acids adopts a helical structure, the heterooligomer incorporating α-amino acid takes a β-strand structure at the α-amino acid site.

Amide Bond Activation: Concepts and Reactions, First Edition. Edited by Michal Szostak.

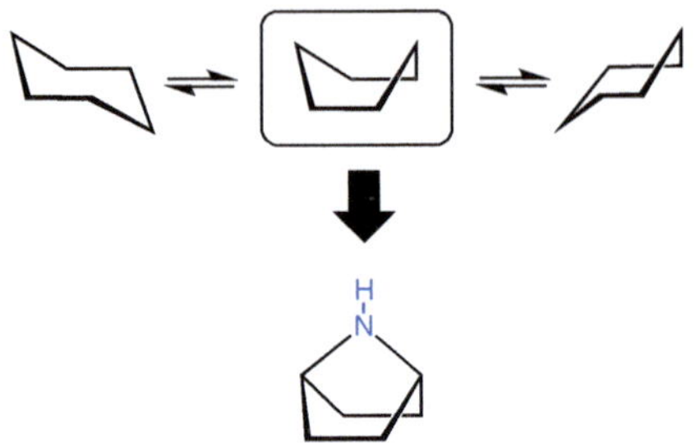

Figure 3.1 7-Azabicyclo[2.2.1]heptane.

The utility of these unnatural helix peptides as inhibitors of protein–protein interactions (PPIs) is discussed in Section 3.4. PPIs are often mediated by recognition of helical, stranded, or coiled secondary structures in the interface region. It was shown that a stable helical trimer of a bicyclic β-amino acid with all-*trans* amide bonds (**tAbh**-AA) inhibits the p53-MDM2/MDMX interaction, which plays a role in regulating p53 function.

Section 3.5 deals with the amide isomerism in bicyclic amino acid dipeptide derivatives with stapled side chains. We found that the *cis–trans* ratio and the rotational barrier of the amide changed dramatically depending on the length of the stapled carbon chain.

3.2 Nonplanar Amides that Are Sufficiently Stable for Chemical Modification

3.2.1 Nonplanar Amides

Amide bonds are fundamental components of proteins and many other biologically active molecules. They are characteristically planar, and this property often determines the overall three-dimensional structure of the molecule. *Cis–trans* isomerization of an amide bond occurs via transitional nonplanar structures [1]. These are (i) a twisted structure resulting from rotation of the N–C(O) bond and (ii) a nitrogen pyramidal structure resulting from nitrogen inversion (Figure 3.2). These transitional structural changes can occur simultaneously. Notably, several molecules that adopt such nonplanar amide structures in the ground state have been reported. However, most nonplanar amides are extremely reactive due to the weakening of the N–C(O) bond, and consequently they are easily hydrolyzed, and/or they readily transfer the acyl group to other molecules [2–5].

We have searched for a simple molecular unit containing a nonplanar amide that can tolerate various synthetic reaction conditions. Finally, we focused on acyl derivatives of 7-azabicyclo[2.2.1]heptane, which contains a conformationally constrained fused bicyclic ring. This skeleton is also found in epibatidine, a natural alkaloid.

We discovered that amides of 7-azabicyclo[2.2.1]heptane take nonplanar structures with a tilted and twisted amide in the solid state and in solution [6, 7]. The planarity of amide nitrogen can be expressed in terms of three angular parameters:

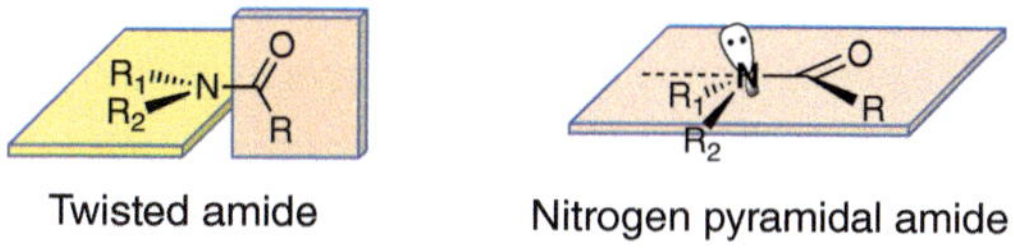

Figure 3.2 Nonplanar amides.

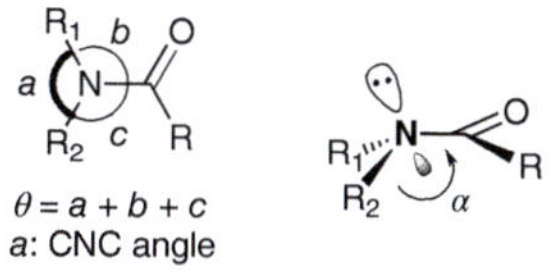

Figure 3.3 Definition of angle parameters.

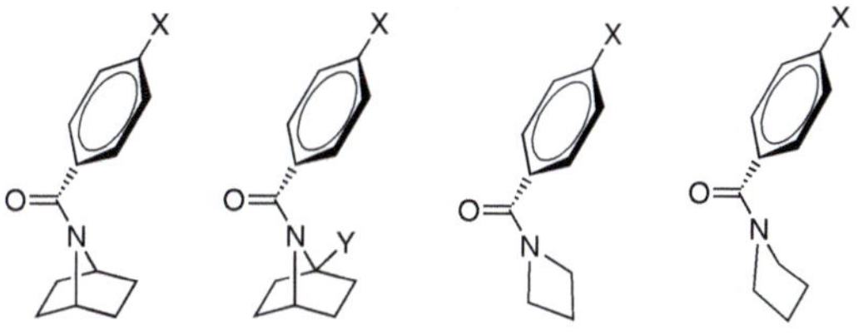

Figure 3.4 Bicyclic and monocyclic amides.

the sum of the three bond angles around the nitrogen atom (θ), the hinge angle of the *N*-acyl group relative to the plane defined by the nitrogen atom and the two adjacent carbon atoms (α), and the torsion angle (τ) [8], which is the average of the two torsion angles $\omega_1(RCNR_2)$ and $\omega_2(OCNR_1)$ (Figure 3.3). The angle θ of an ideal triangular planar nitrogen atom is 360°, while the hinge angle α and the torsion angle τ of an ideal planar amide are 180° and 0°, respectively.

The amide structures of various bicyclic and monocyclic amides were analyzed by X-ray crystallography (Table 3.1). The crystal structure of *N*-benzoyl-7-azabicyclo[2.2.1]heptane has a nitrogen pyramidal structure with the angle α in the range of 149.0°–153.8° and the twist angle $|\tau|$ in the range of 6.9°–15.1° (series **1** in Figure 3.4, Table 3.1). Since θ ranges from 346.2° to 350.1°, the bicyclic amide has an intermediate character of the nitrogen atom between sp^3 (ideally, $\theta = 328.4°$) and sp^2 (ideally, $\theta = 360°$). The bridgehead-substituted bicyclic amide (series **2**) also takes a nonplanar structure irrespective of the kind of substituent at the bridgehead position. Azetidine amides with smaller rings (series **3**) showed some degree of nonplanarity with $\theta = 354.2°$, $\alpha = 161.0°$, and $\tau = 4.9°$, but the bicyclic amides showed higher degrees of nonplanarity. The amide of pyrrolidine, a monocyclic five-membered amine (series **4**), has θ about 359°, α about 172°, and τ 3° and 7°, and the nitrogen atom character is close to sp^2. This suggests that the nonplanarity in the bicyclic amide is due not only to the reduction of the CNC angle (a) by the ring skeleton, which causes the nitrogen to become sp^3-like, but also to 1,3-arylic strain with respect to the pseudo C–N double bond.

Nonplanar amides are expected to have a reduced amide rotational barrier due to the reduced double-bond character of the amide N–C bond (Scheme 3.1). Among the resonance structures of benzamide, the higher the contribution of resonance structure B, the greater the double-bond character of the N–C bond and the higher the

Table 3.1 Crystal structural data of N-benzoyl amides of bicyclic and monocyclic amines. The number in the first column represents structure number shown in Figure 3.4.[a)]

	X	θ (°)	α (°)	$\|\tau\|$ (°)	N–C bond (Å)	CNC angle (°)
1	H[b)]	349.5(2)	153.2	15.1	1.356(3)	97.2(2)
	Me[c)]	346.2(2)	149.0	9.8	1.361(2)	97.1(1)
	Cl[c)]	349.0(2)	152.4	8.5	1.352(4)	97.6(2)
	NO_2[b),d)]	347.1(3)	150.2	13	1.354(4)	96.9(2)
		347.5(3)	153.8	6.9	1.350(4)	97.8(2)
2	H (Y = Me)[d),e)]	349.2(1)	152.6	1.2, 12.8	1.342(2)	97.5(1)
	H (Y = COOMe)	339.3(1)	141.5	12.3	1.367(1)	96.6(1)
	H (Y = C≡CH)	339.7(1)	141.9	22.8	1.376(1)	96.6(1)
		341.5(1)	143.8	22.9	1.371(1)	96.6(1)
3	Me[c)]	354.2(3)	161.0	4.9(6)	1.339(4)	94.6(3)
4	OMe[c)]	359.2(2)	171.3	3.0(3)	1.345(3)	111.5(2)
	NO_2[b)]	359.3(2)	172.5	7.0(4)	1.337(2)	111.6(2)

a) Standard deviations are shown in parentheses.
b) Ohwada et al. [6].
c) Otani et al. [7].
d) Two kinds of molecules are present in a unit cell.
e) Hosoya et al. [9].

rotational barrier. Since resonance structure B requires the nitrogen to adopt a planar structure, it is expected that the contribution of B will be lower in bicyclic amides. Electron-donating substituents on the benzene ring reduce the double-bond character of the N–C bond (stabilizing D), while electron-withdrawing groups increase the double-bond character.

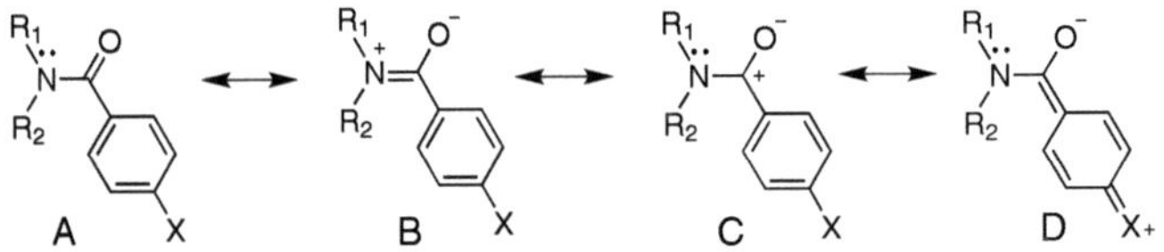

Scheme 3.1 Resonance structures of benzamide.

To show that the amide of 7-azabicyclo[2.2.1]heptane is nonplanar in solution, the rotational barriers were measured by means of variable temperature ^1H NMR spectroscopy in deuterated tetrachloroethane. The free energy of rotation ($\Delta G^{\ddagger}$) was estimated by the coalescence temperature method and line-shape analysis, and the results of the two methods were in good agreement (Table 3.2). The rotational barriers of the bicyclic amides (**1**) were lower than those of the monocyclic amides (**3** and **4**). The decrease in the rotational barrier of these amides of 7-azabicyclo[2.2.1]heptane indicates that the double-bond character of the

Table 3.2 Experimental evaluation of rotational barriers by the coalescence temperature method and by line-shape analysis in $CDCl_2CDCl_2$. The number in the first column represents structure number shown in Figure 3.4.

		$\Delta G^{\ddagger}$ (kcal/mol)	
	X	Coalescence method[a]	Line-shape analysis[b]
1	OMe	14.6	14.5
	Me	15.0	15.0
	F	14.9	
	H	**15.3**	**15.0**
	Cl	15.1	14.9
	CN	15.5	
	NO_2	15.6	15.4
3	**H**	**16.0**	
4	**H**	**16.9**	**17.1**

a) Rotational barriers estimated at the coalescence temperature.
b) Rotational barriers estimated at 25 °C.

N–C(O) bond is reduced, i.e. the planarity of the amide is distorted in solution. The four-membered azetidine amide (**3**) has a lower rotational barrier than the five-membered pyrrolidine amide (**4**). These differences were consistent with the trends in the degree of amide nonplanarity in the crystal structures.

The rotational barrier is also affected by the substituent at the para-position of the benzene ring (Figure 3.4). It was found that the rotational barriers were well correlated with the Hammett's $\sigma_p{}^+$ constants. The greater the electronegativity of the substituent is, the higher the rotational barrier is. This trend was observed not only for bicyclic amides but also for monocyclic azetidine and pyrrolidine amides [7], indicating the involvement of similar electronic effects in the rotation process of these amides (Scheme 3.1).

3.2.2 Thioamides

Thioamide is known to have greater tendency to adopt planar structure than amide. Its rotational barrier is higher than that of amides by 2–3 kcal/mol, corresponding to a higher degree of N–C double-bond character. The solid and solution structures of bicyclic thioamide were investigated [10]. It was found that the thioamide derivatives of bicyclic amine also show a similar tendency.

3.2.3 Chemical Stability of Nitrogen Pyramidal Amides

In order to estimate the strength of the amide bond, the reactivities of planar and nonplanar amides were compared. Kinetic studies of the base-catalyzed hydrolysis of the benzamides of 7-azabicyclo[2.2.1]heptane (**1** and **2**), nonplanar cyclic amine

Table 3.3 Base-catalyzed hydrolysis rates of benzamides in two solvent systems. The number in the first column represents structure number shown in Figure 3.4. Relative reaction rates are shown in parentheses (referenced to **1** (X = Cl)).

		$-k_{obs}$ (s^{-1})	
	X	**Methanol-d_4 – D_2O (1 : 1), NaOD (10 equiv), 70 °C**	**1,4-Dioxane-d_8 – D_2O (1 : 1), NaOD (10 equiv), 70 °C**
1	Cl	1.1×10^{-5} (1)	2.8×10^{-6} (1)
2	Cl (Y = CH_2OMe)	3.0×10^{-6} (0.3)	1.2×10^{-6} (0.4)
3	Cl	9.7×10^{-4} (88)	2.3×10^{-4} (82)
4	Cl	7.0×10^{-5} (6)	1.8×10^{-5} (6)

(**3**), and planar cyclic amine (**4**) were conducted (Figure 3.4) [11]. Alkaline hydrolysis was monitored by NMR. The starting amides (0.05 mmol) and NaOD (0.5 mmol) were dissolved in a mixture of organic solvent (1,4-dioxane-d_8 or methanol-d_4) and water (1/1 [v/v], 500 μl in total), and reactions were carried out at 70 °C. The consumption of amide was estimated from the ^{1}H NMR integration values. The use of excess deuteroxide anion guarantees pseudo first-order kinetics, so rate constants were calculated using the following first-order equation:

$$\frac{d[\text{Amide}]}{dt} = -k[\text{Amide}]$$

that is,

$$\ln[\text{Amide}] = -kt + \ln\,[\text{Amide}]_0$$

where k represents the reaction constant.

Hydrolysis rates under these conditions ($-k_{obs}$) are shown in Table 3.3. Azetidine amide (**3** [X = Cl]) reacted the most rapidly, followed by pyrrolidine amide (**4** [X = Cl]), which were hydrolyzed faster than unsubstituted bicyclic amides (**1**). The bridgehead-substituted bicyclic amides (**2**) were least reactive. Thus, the order of reaction rate is as follows: **3** > **4** > **1** > **2**. This order is very different from the order of the degree of nitrogen pyramidalization of the amides (**2** ~> **1** > **3** > **4**).

In order to understand the low reactivity of the bicyclic amides to hydrolysis, the reaction mechanism was analyzed by means of DFT calculations. Transition structures for the hydroxide ion ($^-$OH)-catalyzed hydrolysis of the amides, including five explicit water molecules [12], were calculated (Figure 3.5). It was suggested that in the bicyclic amides, the water molecules in the bridgehead-unsubstituted **1** were placed further from the amide carbonyl carbon as compared with those in monocyclic **3**. The water molecules in the bridgehead-substituted **2** were located a little further from the amide carbonyl carbon as compared with those in bridgehead-unsubstituted **1**. Therefore, a possible reason for the low reactivity of **1** and **2** is inefficient hydration of the hydroxide ion due to steric congestion.

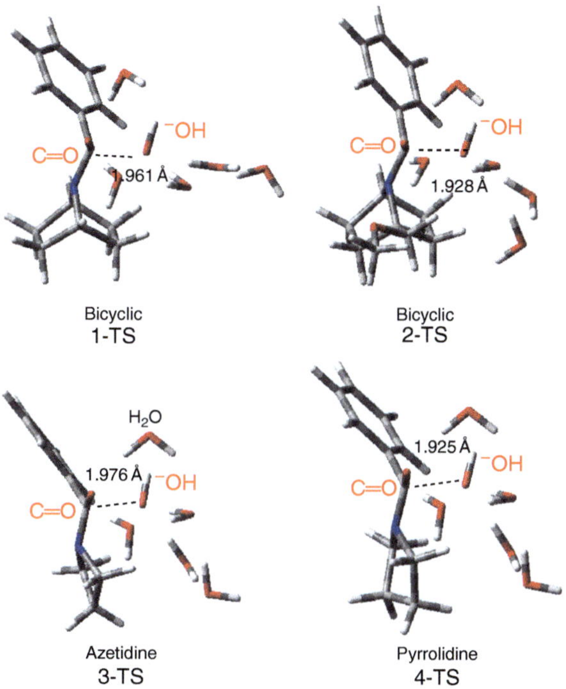

Figure 3.5 Transition structures of hydroxide anion addition to the amides **1–4** optimized at the M06-2X/6-31+G(d) level in the presence of real water molecules with SMD as a solvent model (solvent = water). Source: Adapted from Ocampo et al. [11].

3.3 Application to Amino Acids: Artificial Helices Composed of Bicyclic Amino Acids

3.3.1 Conformational Preference of Bicyclic β-Amino Acids

The bicyclic scaffold with high chemical stability was then applied to conformationally constrained β-amino acid, 7-azabicyclo[2.2.1]heptane-2-carboxylic acid, and Abh amino acid (Abh-AA, Figure 3.6a).

We systematically investigated the intrinsic conformational preference of dipeptides of bridged bicyclic β-amino acids by means of accelerated molecular dynamics simulation and DFT calculations (Figure 3.6b) [13]. Dipeptides (CH_3CO-X-NHMe, X = bicyclic amino acid) were simulated as the minimal structures that express conformational properties in peptides. It was expected that the Abh-AA derivatives have the smallest tilt angle, i.e. the highest degree of nitrogen pyramidalization. Bicyclic amino acid with the amino nitrogen incorporated in the ring system forms a tertiary amide bond, and the amide *cis–trans* preference was estimated. Also the conformational restriction along the main chain was examined in terms of the torsion angles, ϕ, θ, and ψ. While the dihedral angles ϕ and θ are fixed by the bicyclic ring skeleton, the rotation of the C-terminal carbonyl group (ψ) is also restricted, which is in contrast with monocyclic β-proline (Figure 3.7).

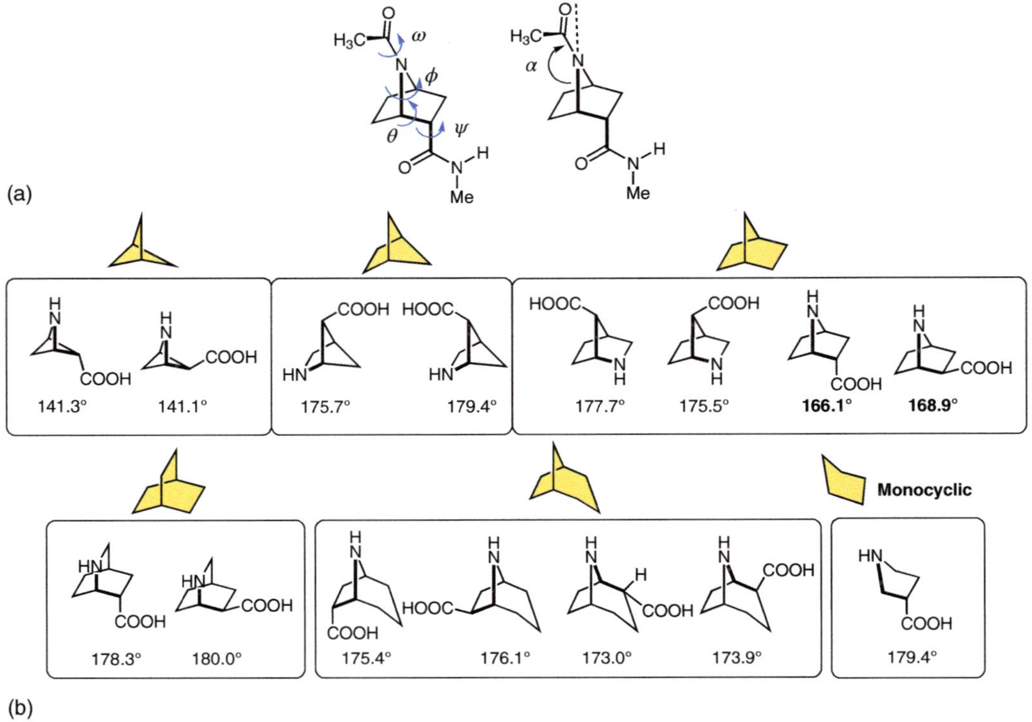

Figure 3.6 (a) Main chain dihedral angles (ω, ϕ, θ, and ψ) and tilt angle (α) of Ac-Abh-AA-*exo*-NHMe. (b) Bicyclic and monocyclic β-amino acids (**X**). Calculated tilt angles (α) of the acetyl amide of CH_3CO-**X**-NHMe are shown below the amino acid X. (B3LYP/6-311++G(d), SCRF = IEFPCM, solvent = methanol). Source: Modified from Otani et al. [13].

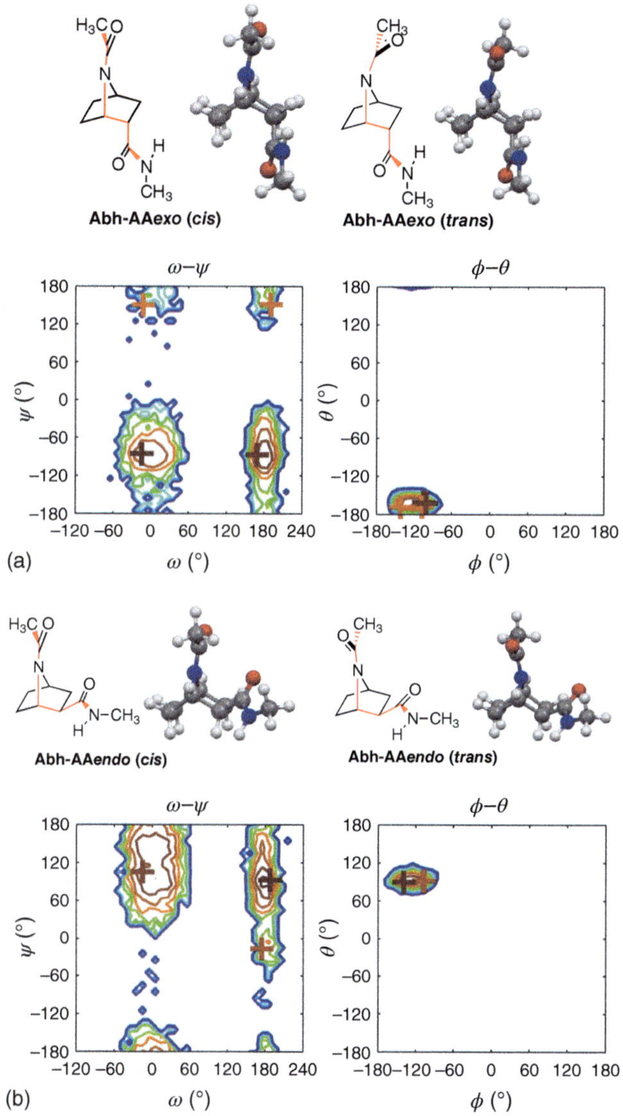

Figure 3.7 Distribution of ω–ψ and ϕ–θ in bicyclic (a, b) and monocyclic (c) β-amino acids obtained by replica exchange with solute tempering (REST) simulations in methanol. DFT-optimized structures are indicated by + symbols, whose color reflects the free energy: 0–1.5 kcal/mol from brown to orange; 1.5–2.5 kcal/mol: from orange to green; 2.5–3.3 kcal/mol: from green to light blue; 3.3–3.5 kcal/mol: from light blue to blue. Source: Modified from Otani et al. [13].

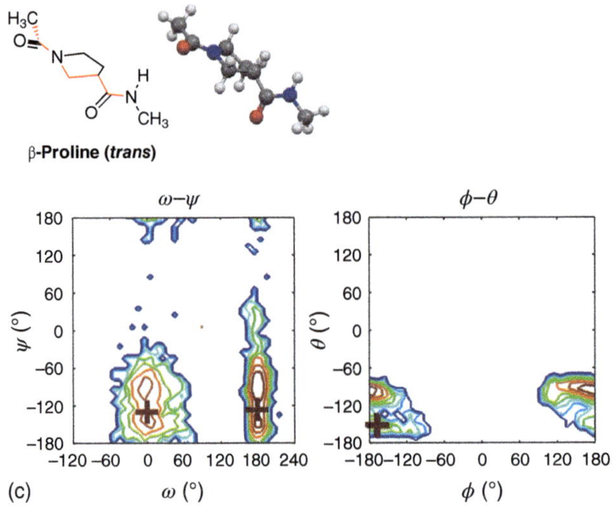

Figure 3.7 (*Continued*)

3.3.2 Bridgehead-Substituted Bicyclic Amino Acids

Controlling the *cis–trans* isomerization of the amide bond is very important for stabilizing the ordered structure. While secondary amides strongly favor *trans* structures along the carbon main chain, biasing the *cis–trans* equilibrium of tertiary amides to either side is especially difficult because the steric difference between *cis* and *trans* is small.

Proline, a natural α-amino acid with a tertiary amide bond, also exists in an equilibrium mixture of *trans* and *cis* forms (Figure 3.8) [14, 15]. Many synthetic proline mimics in which this equilibrium is controlled by steric and stereoelectronic effects have been reported [16–18], but complete control of the *cis–trans* preference is still difficult. There are even fewer examples of controlling the *cis–trans* preference of tertiary amide bonds in β-amino acids, i.e. where the amino and carboxyl groups are separated by two carbons.

It was found that substitution of the bridgehead of bicyclic amino acids completely biases the *E-Z* equilibrium of the amide to the Z form, both in the solid and in solution. X-ray crystallographic analysis showed that *N*-benzoyl amides of bicyclic amine substituted with a methyl, a methoxymethyl (–CH_2OMe), an ester (–COOMe), or an ethynyl (–C≡CH) group take Z-structure, that is, the carbonyl oxygen and the bridgehead substituent are on the same side with respect to the amide bond (Figure 3.9, Table 3.1). This is mainly due to the steric repulsion

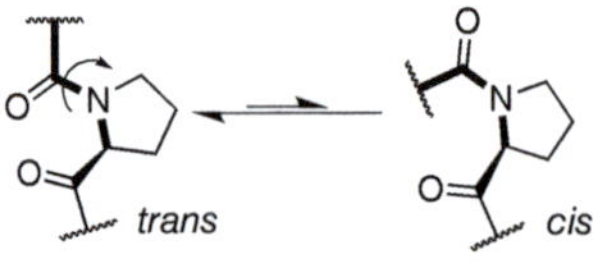

Figure 3.8 The *cis–trans* equilibrium of the tertiary amide in proline.

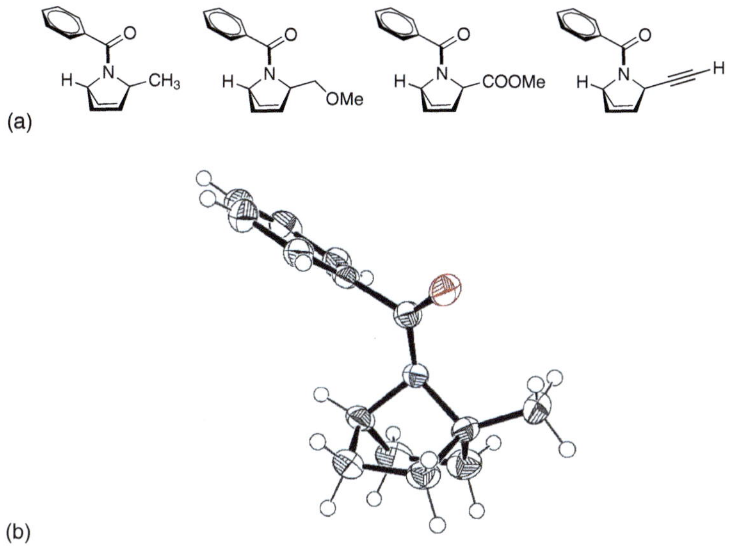

Figure 3.9 (a) Preference for the Z-form of bridgehead-substituted bicyclic benzoyl amides. (b) Crystal structure of *N*-benzoyl-7-azabicyclo[2.2.1]heptane substituted with a methyl group at the bridgehead position. Source: Hosoya et al. [9].

between the phenyl group and the substituent protruding slightly upward from the bridgehead position. All of them showed a single conformer in ^{1}H NMR, and NOESY measurements showed NOE between the phenyl group and the hydrogen at the bridgehead position on the other side.

Even in molecules where a natural amino acid is attached to the nitrogen of a bridgehead-substituted bicyclic amine, only the Z-amide form was observed (Figure 3.10) [19]. First, a bicyclic amine with a benzyloxymethyl group at the bridgehead position (Abh(OBn)) was used as a model compound, and various α-amino acids were linked to its N-terminal side. In this case, the bridgehead substituent is oriented to the nitrogen side, and this steric effect endows the amine with a different reactivity from that of proline. As a result of various investigations, the yield was improved by using 2-chloro-4,6-dimethoxy-1,3,5-triazine (CDMT) and *N*-methylmorpholine (NMM) as a base, and Fmoc-Trp-Abh(OBn) was obtained in 79% yield. The use of CDMT also gave improved yields for other amino acids. We examined the protected forms of 15 amino acids with polar and less-polar side chains (Gly, Ala, Val, Leu, Ile, β-proline, Phe, Trp, Tyr, Asp, Lys, Asn, His, Ser, and Arg) and succeeded in coupling 14 amino acids, except for Arg. ^{1}H NMR and NOESY spectroscopy showed that all the α-amino acid-Abh dimers took only the *Z*-form, and the *E*-form was not observed.

The conformation distribution of α-amino acids was further investigated by metadynamics simulation (Figure 3.10), which is a kind of molecular dynamics method that enables efficient conformation sampling [20]. By adding multiple bias potentials to the low potential energy region, the conformation distribution of the α-amino acids can be determined. In this study, we investigated the effect of multiple bias

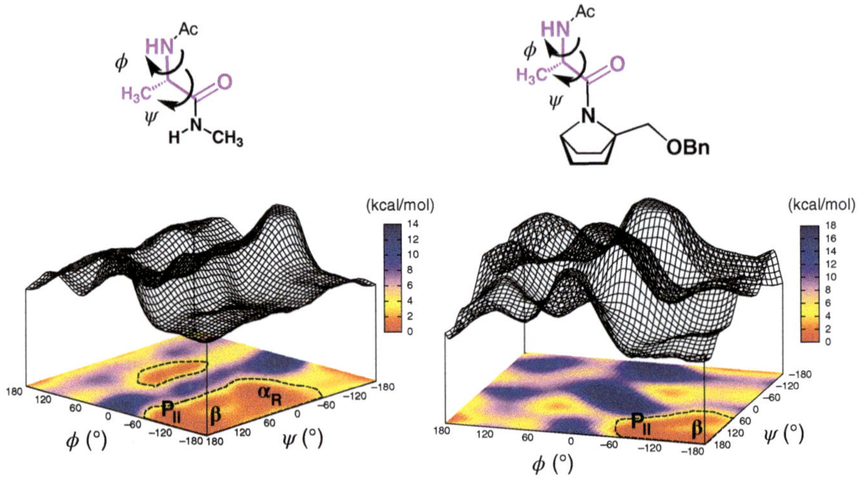

Figure 3.10 Schematic free energy landscapes of Ac-Ala-NHMe (**5**) and Ac-Ala-Abh(CH_2OBn) (**6**).

potentials on the free energy topography. While alanine dipeptide (Ac-Ala-NHMe, **5**) is stable for all conformations of β, PII, and α_R, the stable conformation range of Ac-Ala-Abh(OBn) (**6**) is smaller (Figure 3.10). This was confirmed by measurement of the $^3J(H_N, H_\alpha)$ coupling constants by ^{1}H-NMR. The $^3J(H_N, H_\alpha)$ value of alanine dipeptide **5** is 7.68 Hz in DMSO-d_6 or 6.03 Hz in H_2O, whereas the ^3J value of **6** is larger, 7.97 Hz in DMSO-d_6 or 6.89 Hz in H_2O, indicating that **6** is more likely to adopt an elongated conformation.

3.3.3 Application to Artificial Helices and Strand Mimics

Because the amide of 7-azabicyclo[2.2.1]heptane is chemically stable and conformationally constraining, it was employed as amino acid skeleton (Figure 3.11). Syntheses of **tAbh-AA** and **cAbh-AA** were accomplished, and chiral separations were performed by introducing a chiral auxiliary followed by recrystallization and column chromatography.

We also showed that **cAbh** with a bridgehead substituent (–CH_2OMe) introduced at the 4-position of β-amino acids has a *cis*-amide form, while **tAbh** with a bridgehead substituent (-CH_2OMe) introduced at the 1-position has a *trans*-amide form, and homooligomers linking the same amino acids take all-*cis* and all-*trans* amide helices, respectively (Figure 3.11) [9, 21].

3.3.3.1 Heterooligomers

Short hybrid peptides containing one or more bicyclic β-amino acids can form extended helix or β-strand-like structure without interpeptide hydrogen bonding, i.e. sheet formation (Figure 3.12) [22, 23]. This is related to compound **6** shown in Figure 3.10.

The active sites of various proteases have an elongated β-strand structure, which can recognize peptides and non-peptide ligands. Many studies have aimed to mimic

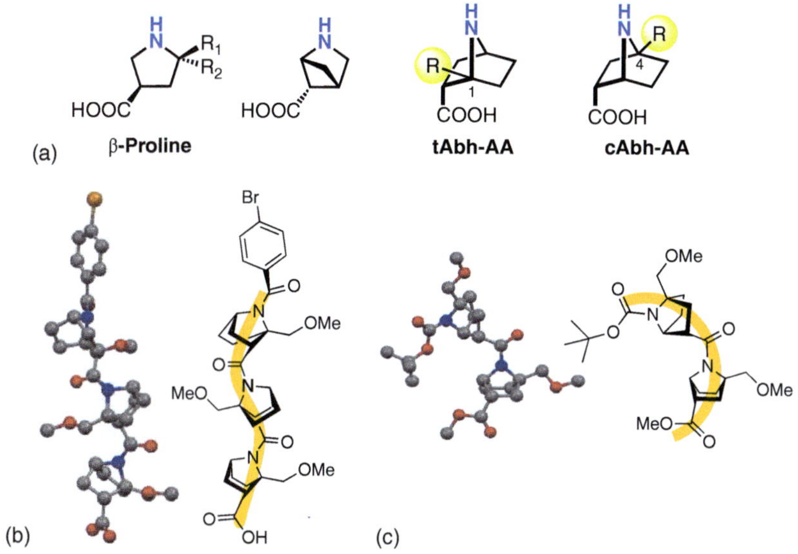

Figure 3.11 (a) Conformationally constrained bicyclic β-amino acid and bridgehead-substituted bicyclic β-amino acids. (b) Crystal structure of *p*-bromobenzoyl-**tAbh-AA** trimer. (c) Crystal structure of Boc-**cAbh-AA** dimer. Gray: carbon; blue: nitrogen; red: oxygen; orange: bromine. Hydrogens are omitted for clarity. Source: Adapted from Hosoya et al. [9] and Wang et al. [21].

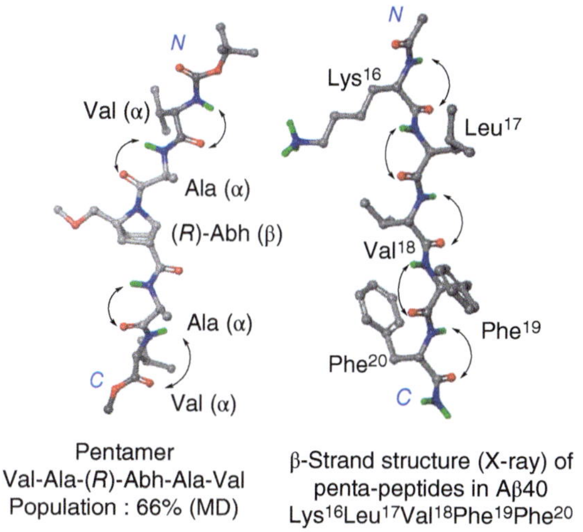

Figure 3.12 Representative structure of Abh-heterooligomer (left), together with a typical β-strand structure of part of an Aβ peptide fibril (right). Left: Val-Ala-(*R*)-Abh-Ala-Val. A 60% populated structure in MD calculations in $CHCl_3$. Right: peptapeptide in Ab40, Lys-Leu-Val-Phe-Phe, taken from the crystal structure (PDB code: 2LMP). Oxygen is shown in red, nitrogen in blue, carbon in gray, and polar hydrogen in green (only polar hydrogen atoms are shown). Double-headed curved arrows indicate hydrogen bonding characteristic of β -strand structures. The lengths of the amino acid residues are similar: left: 16.46 Å; right: 16.51 Å. Source: Adapted from Zhai et al. [19, 22].

the hydrogen-bonded dimeric β-sheet structure and designed molecules to stabilize hydrogen bonds with artificial hydrogen bond-locking structures [24–26]. However, although short peptide segments (3–10 amino acids) mediate many biomolecular recognition events, short peptides themselves rarely assume distinct structures in solution, and, in particular, single β-strand structures of short peptides are not stabilized in water (Figure 3.12 [right]: a typical β-strand structure of the pentapeptide unit in Aβ oligomer). The homooligomer of Abh-AA adopts a helical structure, whereas the heterooligomer incorporating α-amino acid takes a β-strand structure at the α-amino acid site. Bicyclic β-proline amino acids (Abh-AA) are highly conformationally constrained due to severe restrictions on almost all of the main chain dihedral angles (φ, θ, and ψ; see Figure 3.6). These conformational constraints promote and propagate the β-chain formation of β-α dipeptides, α-β dipeptides, α-β-α tripeptides, α-α-β-α tetrapeptides, and α-α-β-α-α pentapeptides in protic and non-protic solvents (Figure 3.12). Surprisingly, these β-chain mimics do not form intermolecular hydrogen bonds or aggregates and are stable in isolation; the noninteracting β-chain sequence induced by the single substitution of Abh-AA may weaken intermolecular β-chain interactions, thus providing a tool to explore the importance of β-chain/β-sheet structure in PPIs and aggregation.

3.4 Applications of Helical Peptides as Inhibitors of p53-MDM2/MDMX Interaction

In PPI, the recognition site often contains secondary structures, such as helix, strand, and coil. Among compounds that inhibit PPIs, peptidomimics containing artificial amino acids have attracted much attention. One potential target is the p53-MDM2/MDMX interaction. p53 functions as a tumor suppressor, and PPI with MDM2 suppresses the p53 activity by creating a negative feedback loop of MDM2-p53 [27]. Inhibitors of the p53-MDM2 interaction, i.e. antagonists of MDM2/MDMX, may be candidates for therapeutic agents that prevent p53 degradation and restore p53 cellular activity [28, 29]. Crystallographic analysis revealed a deep binding pocket between MDM2 and MDMX, to which the N-terminal domain of p53 binds in an α-helical conformation [30, 31]. Three amino acid residues of p53 (Phe19, Trp23, and Leu26) are essential for the interaction with MDM2 and MDMX.

The helical structure of the bicyclic β-proline homoligomers shows clear differences in shape, such as pitch and diameter per turn, compared with the typical α-helix structure of α-amino acids. The *cis*-amide helix is estimated to have about 4 residues/turn and the *trans*-amide helix, about 2.7 residues/turn (Figure 3.11) [9, 21]. Although the natural α-helix and helix of **tAbh**-AA have different numbers of residues per turn and different pitch, we investigated whether our helical peptides can bind to MDM2/MDMX analogously to p53.

Initial screening of a library of diverse Abh-AA oligomers revealed that homooligomers as short as the trimer with an N-terminal *p*-bromobenzoyl group (**pBrBz-(tAbh-AA)$_3$-OMe**) exhibit some MDM2 and MDMX-inhibitory activity in

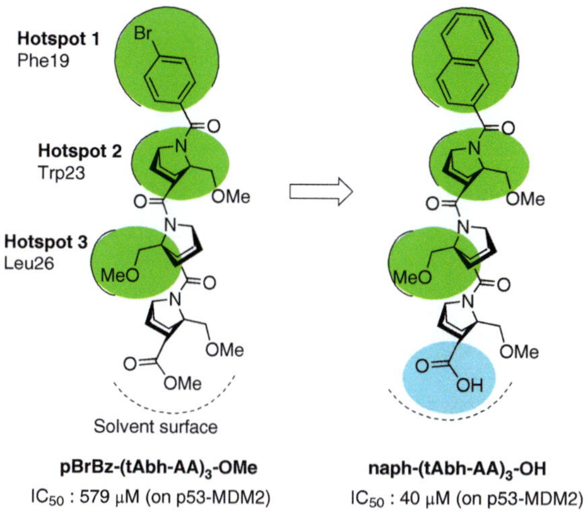

Figure 3.13 Helical tripeptides as inhibitors of p53-MDM2/MDMX interaction. Source: Adapted from Su et al. [32].

a protein-level ELISA assay (Figure 3.13) [32]. It is surprising that a short peptide consisting of only 3 residues can apparently maintain a stable helical structure and cover the helical region corresponding to the 15-residue peptide of p53.

The binding poses obtained from docking simulation followed by molecular dynamics simulation of the peptide–protein complex indicate that the hydrophobic surface of the tAbh-AA helix fits well into the pocket of MDM2 and inhibits its interaction with p53. The *p*-bromobenzoyl group occupies hotspot 1, mimicking the Phe19 residue of p53 and having a π–π interaction with Tyr67 of MDM2. The first and second bicyclic residues bind to hotspots 2 and 3, mimicking the binding of Trp23 and Leu 26, respectively (Figure 3.13). The C-terminal part is exposed to the solvent surface.

Next, we modified the structure of the three-residue peptide based on the putative binding mode. The IC_{50} was improved to 40 μM by modifying the N-terminal to a naphthyl group and the C-terminal to a carboxy group (Figure 3.13, **naph-(tAbh-AA)$_3$-OH**). Hydrolysis of the C-terminus was found to improve both the inhibitory activity and solubility. Furthermore, Differential scanning fluorometry (DSF) and Isothermal titration calorimetry (ITC) analysis confirmed that the peptide binds directly to MDM2 via an entropy-driven process [33]. The activity of the peptide in cells was also examined, and it was found that the developed tripeptide was membrane permeable, as it activated p53 and induced p53 target genes in SJSA-1 cells [33].

3.5 Nonplanar Lactam Amide Spinning

Although the *cis–trans* isomerization of amides in bridgehead-substituted bicyclic amines is completely skewed, we wondered whether the balance of *cis–trans* isomerization or nitrogen pyramidalization in amides with this backbone could be modified [34].

3.5.1 Lactam Amide Rotation

Shi et al. analyzed the isomerization process of α-proline amide located in a disulfide-cyclized peptide containing two cysteine (Cys) residues [35]. They reported that the *cis–trans* isomer ratio and the *cis–trans* isomerization rate depend on the size of the ring. Nitrogen-pyramidalized tertiary amides of 7-azabicyclo[2.2.1]heptane are sufficiently stable to permit further synthetic transformations. Furthermore, bicyclic amino acids with carboxylic acid functionality (Abh-AA) can be regarded as conformationally constrained derivatives of β-proline. In 7-azabicyclo[2.2.1]heptane-2-carboxylic acid with a bridgehead substituent, the *cis–trans* equilibrium of the amide is completely biased to one side (*cis* or *trans* amide), regardless of the solvent, due to the steric effect of the bridgehead substituent. This led us to speculate that stapling the side chains of this dimer of linked amino acids would have a distinctive effect on the thermodynamics and kinetics of rotation of the nonplanar amide lactam.

Ring-closing metathesis (RCM) reactions were used to staple bicyclic dimers with alkenyl side chains of various lengths at the bridgehead position. The resulting mixture of *E/Z* isomers of the olefin was hydrogenated to give stapled amides with linker lengths (Cn) varying from 8 to 12 carbons (**7** (C8), **8** (C9), **9** (C10), and **10** (C11)). NMR spectroscopic study showed that the *cis/trans* ratio changed dramatically depending on the linker length (Figure 3.14): the percentage of *cis*-amides decreased from 92% with the short linker **7** (**C8**) to 14% with the long linker **10** (**C11**) (Table 3.4), compared with nearly 0% for the open-chain amide.

The rotational rates of amide bonds in bicyclic dimers with different linker lengths were measured using exchange spectroscopy (EXSY). The free energies of rotation ($\Delta G_{t\rightarrow c}^{\ddagger}$ and $\Delta G_{c\rightarrow t}^{\ddagger}$) were derived from the enthalpy of activation ($\Delta H^{\ddagger}$) and entropy of activation ($\Delta S^{\ddagger}$), which were obtained from Eyring plots of rotational rates vs. temperature.

The rotational rate from *trans* amide to *cis* amide (k_{tc}) consistently decreased as the linker length was increased, while the rotational rate from *cis* to *trans* (k_{ct}) showed a slight increase as the linker length was increased (Table 3.4). These changes in rotational rate (k) are expected to be linked to changes in the free energy

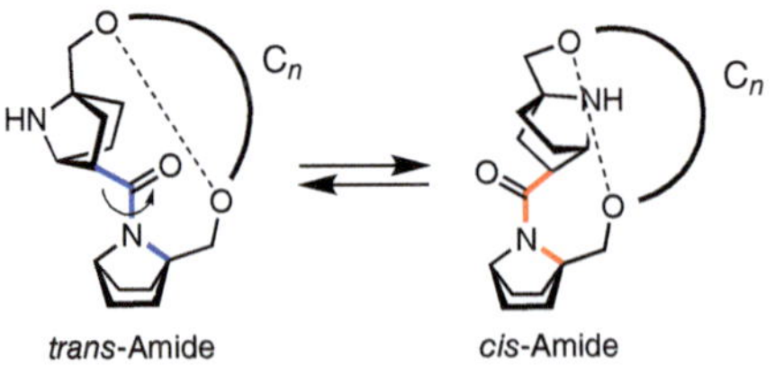

Figure 3.14 Amide isomerization of stapled Abh dimer lactams. The distance between two oxygen atoms is shown in a dotted line.

Table 3.4 The ratio of *cis*-amide and the rotational barrier in CD_2Cl_2 at −50 °C.a)

	Cis (%)	$\Delta G^{\ddagger}_{t\rightarrow c}$ (kcal/mol)	$\Delta G^{\ddagger}_{c\rightarrow t}$ (kcal/mol)
7 (C8)	92	(12.1)	(13.1)
8 (C9)	86	11.4 (12.3)	12.4 (13.0)
9 (C10)	27	12.3 (13.0)	11.8 (12.5)
10 (C11)	14	12.9	11.9

a) The rotational barriers measured in CD_3OD are shown in parentheses.

of rotation. Indeed, $\Delta G_{t\rightarrow c}^{\ddagger}$ decreased and $\Delta G_{c\rightarrow t}^{\ddagger}$ slightly increased with decreasing linker length. The rotational barriers $\Delta G_{t\rightarrow c}^{\ddagger}$ of **8** (C9)–**10** (C11) (at −50 °C) ranged from 11.4, 12.3, and 12.9 kcal/mol in CD_2Cl_2, while $\Delta G_{t\rightarrow c}^{\ddagger}$ of **8** (C9)–**10** (C11) ranged from 12.4, 11.8, and 11.9 kcal/mol (Table 3.4).

To examine this structural factor, the distance between the two oxygen atoms located at the ends of the linker was measured for the DFT-optimized energy-minimum structures. It was found that the oxygen–oxygen distances were 6.0 Å for the *cis*-amide in **7** (C8) and 7.0 Å for the *trans*-amide in **7** (C8) and 6.3 Å for the *cis*-amide of **10** (C11) and 8.2 Å for the *trans*-amide of **10** (C11) (Figure 3.14). It was expected that the smaller the lactam ring, the greater the ring strain and, therefore, the more destabilized the *trans* amide would be.

It turns out from metadynamics simulation that the lactam amide can rotate 360° owing to nitrogen pyramidalization (Figure 3.15) [36], which allows the entire molecule to escape the strain. That is, in state a, the amide group is tilted in the opposite direction to the carbon at position 2 ($\phi = 150°$, inward tilting) (Figure 3.15). From there, the amide rotates 180° to state c. After reaching state c, the amide group changes the direction of nitrogen pyramidalization and is tilted in the direction of the carbon at position 2 ($\phi = 100°$, outward tilting) and rotates 360°. The direction of the tilting of the pyramidal nitrogen atom is synchronized with the direction of the semicircular rotation of the amide.

3.6 Conclusion and Prospects

Amides have a planar structure due to π resonance. However, in transitional structures, the amide bond is twisted, and the sp^2 nitrogen atom pyramidalizes to sp^3. In very limited cases, such nonplanar amides can exist stably as well-defined structures. As described in this chapter, amides of some bicyclic amines exhibit distinctive helical structures. The pyramidalization of the amide nitrogen and the twisting of the amide bond are synchronized, and the isomerization of the amide is accelerated due to the weakening of the amide π-resonance. Thus, the structural properties also activate the dynamic properties. We speculate that such phenomena can occur even

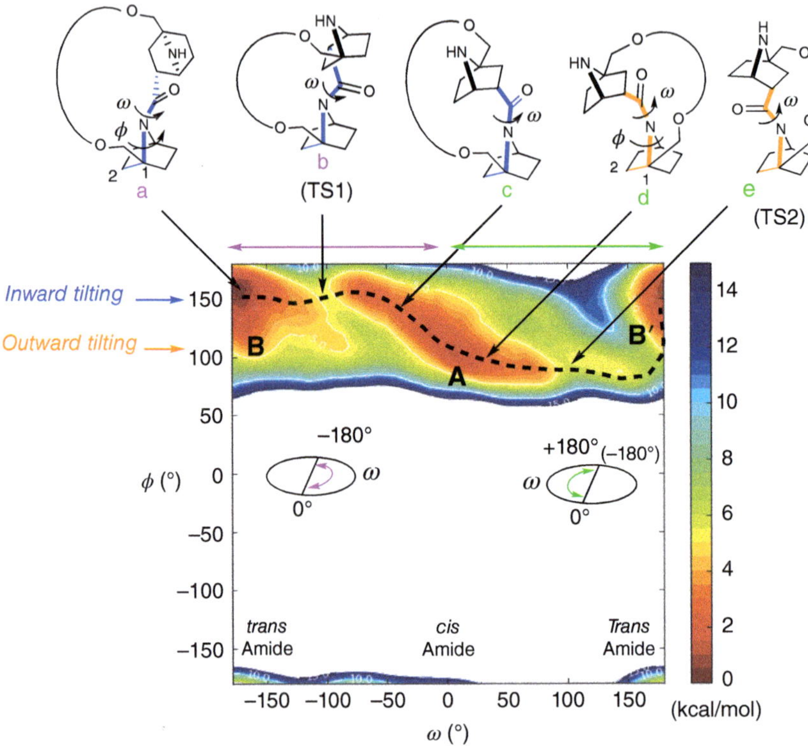

Figure 3.15 Energy landscape of bicyclic lactam amide rotation **8** (C9) obtained by metadynamics simulation in chloroform at 300 K. The low-energy path through a to e is shown by a dotted line. Source: Otani et al. [34]/Springer Nature/CC BY 4.0.

in the absence of bicyclic structures and might be fairly general. Further studies of nonplanar amides might be fruitful.

References

1 Fischer, G. (2000). Chemical aspects of peptide bond isomerisation. *Chem. Soc. Rev.* 29 (2): 119–127.

2 Kirby, A.J., Komarov, I.V., and Feeder, N. (1998). Spontaneous, millisecond formation of a twisted amide from the amino acid, and the crystal structure of a tetrahedral intermediate. *J. Am. Chem. Soc.* 120 (28): 7101–7102.

3 Komarov, I.V., Yanik, S., Ishchenko, A.Y. et al. (2015). The most reactive amide as a transition-state mimic for *cis–trans* interconversion. *J. Am. Chem. Soc.* 137 (2): 926–930.

4 Wang, Q.P., Bennet, A.J., Brown, R.S., and Santarsiero, B.D. (1991). Distorted amides as models for activated peptide N-C(O) units. 3. Synthesis, hydrolytic profile, and molecular structure of 2,3,4,5-tetrahydro-2-oxo-1,5-propanobenzazepine. *J. Am. Chem. Soc.* 1: 5757–5765.

5 Yamada, S. (1993). Structure and reactivity of a highly twisted amide. *Angew. Chem. Int. Ed.* 32 (7): 1083–1085.
6 Ohwada, T., Achiwa, T., Okamoto, I. et al. (1998). On the planarity of amide nitrogen. Intrinsic pyramidal nitrogen of *N*-acyl-7-azabicyclo[2.2.1]heptanes. *Tetrahedron Lett.* 39 (8): 865–868.
7 Otani, Y., Nagae, O., Naruse, Y. et al. (2003). An evaluation of amide group planarity in 7-azabicyclo[2.2.1]heptane amides. Low amide bond rotation barrier in solution. *J. Am. Chem. Soc.* 125 (49): 15191–15199.
8 Winkler, F.K. and Dunitz, J.D. (1971). Nonplanar amide group. *J. Mol. Biol.* 59 (1): 169–182.
9 Hosoya, M., Otani, Y., Kawahata, M. et al. (2010). Water-stable helical structure of tertiary amides of bicyclic β-amino acid bearing 7-azabicyclo[2.2.1]heptane. Full control of amide cis-trans equilibrium by bridgehead substitution. *J. Am. Chem. Soc.* 132 (42): 14780–14789.
10 Hori, T., Otani, Y., Kawahata, M. et al. (2008). Nonplanar structures of thioamides derived from 7-azabicyclo[2.2.1]heptane. Electronically tunable planarity of thioamides. *J. Org. Chem.* 73 (22): 9102–9108.
11 Gutiérrez, O., de Velasco, D., Su, A. et al. (2018). Unexpected resistance to base-catalyzed hydrolysis of nitrogen pyramidal amides based on the 7-azabicyclic[2.2.1]heptane scaffold. *Molecules* 23 (9): 2363.
12 Xiong, Y. and Zhan, C.G. (2006). Theoretical studies of the transition-state structures and free energy barriers for base-catalyzed hydrolysis of amides. *J. Phys. Chem. A* 110 (46): 12644–12652.
13 Otani, Y., Park, S., and Ohwada, T. (2020). Conformational preference of bicyclic β-amino acid dipeptides. *Chirality* 32 (6): 790–807.
14 Dugave, C. and Demange, L. (2003). Cis-trans isomerization of organic molecules and biomolecules: implications and applications. *Chem. Rev.* 103 (7): 2475–2532.
15 Lu, K.P., Finn, G., Lee, T.H., and Nicholson, L.K. (2007). Prolyl cis-trans isomerization as a molecular timer. *Nat. Chem. Biol.* 3 (10): 619–629.
16 Panasik, N., Eberhardt, E.S., Edison, A.S. et al. (1994). Inductive effects on the structure of proline residues. *Int. J. Pept. Protein Res.* 44 (3): 262–269.
17 Sonntag, L.-S., Schweizer, S., Ochsenfeld, C., and Wennemers, H. (2006). The "azido gauche effect"-implications for the conformation of azidoprolines. *J. Am. Chem. Soc.* 128 (45): 14697–14703.
18 Pandey, A.K., Naduthambi, D., Thomas, K.M., and Zondlo, N.J. (2013). Proline editing: a general and practical approach to the synthesis of functionally and structurally diverse peptides. Analysis of steric versus stereoelectronic effects of 4-substituted prolines on conformation within peptides. *J. Am. Chem. Soc.* 135 (11): 4333–4363.
19 Zhai, L., Wang, S., Nara, M. et al. (2018). Application of C-terminal 7-azabicyclo[2.2.1]heptane to stabilize β-strand-like extended conformation of a neighboring α-amino acid. *J. Org. Chem.* 83 (21): 13063–13079.

20 Laio, A. and Parrinello, M. (2002). Escaping free-energy minima. *Proc. Natl. Acad. Sci. U. S. A.* 99 (20): 12562–12566.

21 Wang, S., Otani, Y., Liu, X. et al. (2014). Robust trans-amide helical structure of oligomers of bicyclic mimics of β-proline: impact of positional switching of bridgehead substituent on amide cis-trans equilibrium. *J. Org. Chem.* 79 (11): 5287–5300.

22 Zhai, L., Otani, Y., Hori, Y. et al. (2020). Peptide-based short single β-strand mimics without hydrogen bonding or aggregation. *Chem. Commun.* 56 (10): 1573–1576.

23 Wang, S., Otani, Y., Zhai, L. et al. (2019). Overall shape constraint of alternating α/β-hybrid peptides containing bicyclic β-proline. *Org. Lett.* 21 (19): 7813–7817.

24 Nowick, J.S., Holmes, D.L., Mackin, G. et al. (1996). An artificial beta-sheet comprising a molecular scaffold, a beta-strand mimic, and a peptide strand. *J. Am. Chem. Soc.* 118 (11): 2764–2765.

25 Nowick, J.S. (2008). Exploring β-sheet structure and interactions with chemical model systems. *Acc. Chem. Res.* 41 (10): 1319–1330.

26 Phillips, S.T., Rezac, M., Abel, U. et al. (2002). "@-Tides": the 1,2-dihydro-3(6H)-pyridinone unit as a beta-strand mimic. *J. Am. Chem. Soc.* 124 (1): 58–66.

27 Lane, D.P. (1992). The guardian of the genome. *Nature* 358 (2): 15–16.

28 Neochoritis, C., Estrada-Ortiz, N., Khoury, K., and Dömling, A. (2014). p53-MDM2 and MDMX antagonists. *Annu. Rep. Med. Chem.* 49: 167–187.

29 Azzarito, V., Long, K., Murphy, N.S., and Wilson, A.J. (2013). Inhibition of α-helix-mediated protein-protein interactions using designed molecules. *Nat. Chem.* 5 (3): 161–173.

30 Kussie, P.H., Gorina, S., Marechal, V. et al. (1996). Structure of the MDM2 oncoprotein bound to the p53 tumor suppressor transactivation domain. *Science (80-.)* 274 (5289): 948–953.

31 Popowicz, G.M., Czarna, A., and Holak, T.A. (2008). Structure of the human Mdmx protein bound to the p53 tumor suppressor transactivation domain. *Cell Cycle* 7 (15): 2441–2443.

32 Su, A., Wang, S., Sada, A. et al. (2019). Non-naturally occurring helical molecules can interfere with p53–MDM2 and p53–MDMX protein–protein interactions. *Chem. Pharm. Bull.* 67 (10): 1139–1143.

33 Su, A., Tanabe, Y., Aoki, K. et al. (2021). Elaboration of non-naturally occurring helical tripeptides as p53-MDM2/MDMX interaction inhibitor. *Chem. Pharm. Bull.* 69 (7): 681–692.

34 Otani, Y., Liu, X., Ohno, H. et al. (2019). Amide nitrogen pyramidalization changes lactam amide spinning. *Nat. Commun.* 10 (461): 1–9.

35 Shi, T., Rabenstein, D.L., Shi, T., and Rabenstein, D.L. (2004). Unexpectedly fast cis/trans isomerization of Xaa-Pro peptide bonds in disulfide-constrained cyclic peptides unexpectedly fast cis/trans isomerization of Xaa-Pro peptide bonds in disulfide-constrained cyclic peptides. *J. Am. Chem. Soc.* 126 (3): 790–796.

36 Shah, N.H., Butterfoss, G.L., Nguyen, K. et al. (2008). Oligo(*N*-aryl glycines): a new twist on structured peptoids oligo(*N*-aryl glycines): a new twist on structured peptoids. *J. Am. Chem. Soc.* 130 (49): 16622–16632.

4

Transition-Metal-Free Reactions of Amides by Tetrahedral Intermediates

Marco Blangetti[1], Karen de la Vega-Hernández[2,3], Margherita Miele[1,2], and Vittorio Pace[1,2]

[1] *University of Torino, Department of Chemistry, Via Pietro Giuria 7, 10125 Torino, Italy*
[2] *University of Vienna, Department of Pharmaceutical Chemistry, Althanstrasse 14, 1090 Vienna, Austria*
[3] *Institute of Chemical Research of Catalonia (ICIQ), The Barcelona Institute of Science and Technology, Avinguda dels Països Catalans 16, 43007 Tarragona, Spain*

4.1 Introduction

The ubiquitous abundance of organic compounds featuring the amidic linkage makes them privileged materials for realizing synthetic operations levered on conceptually distinct rationales [1–10]. Being carboxylic derivatives, it is conceivable to assume for amides a reactivity profile analogous to that of other members of the class, consisting in a nucleophilic addition/elimination sequence (i.e. the acyl nucleophilic substitution) [11–13]. The constitutive essence of the amide bond justifies the reluctance of this functional group to take part in productive nucleophilic addition processes which would ultimately increase the synthetic potential of this functional group [14]. As a consequence of the greater degree of delocalization of the lone pair on the nitrogen atom accounting for the increased stabilization of the electrophilic carbonyl carbon, amides exhibit significant tamed electrophilic reactivity toward classical nucleophiles in comparison to other members of the carboxylic class (Scheme 4.1) [15–17]. An additional aspect makes amides rather inert substrates: because of the antiperiplanar spatial orientation between one of the oxygen lone pairs and the C–N bond, a secondary electronic delocalization is established through overlapping this oxygen pair orbital and the antibonding C–N sigma bond orbital (σ^*) [18, 19]. Collectively, these phenomena result in a planar array and thus govern the overall chemistry of a general amide functionality [20, 21].

Considering a model amide **1**, it is possible to assume that it may accept up to two nucleophilic elements on the sp^2-hybridized carbon, thus conducting to a fully substituted (tertiary) amine **2** [22]. Such a trivial concept is unfortunately not easy to be validated experimentally because of the aforementioned high stability of the amide bond (resonance of 15–20 kcal/mol) [1]: in particular, the first nucleophilic addition to the amide carbonyl group, furnishing the tetrahedral intermediate **1a**,

Amide Bond Activation: Concepts and Reactions, First Edition. Edited by Michal Szostak.

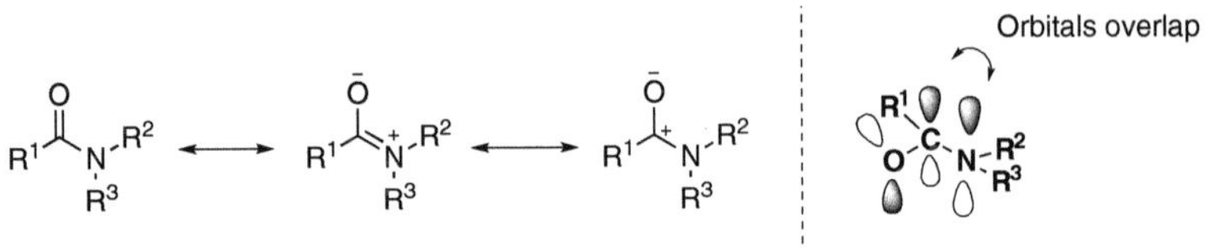

Scheme 4.1 Basics of the amide bond.

requires harsh conditions whenever carried out in the absence of activation factors (Scheme 4.2) [23]. However, once the tetrahedral intermediate is formed, full evidence of the synthetic potential of the transformation is manifested, as recognized in the seminal work by Bouveault in 1904 [24]: (i) carbonyls (ketone and aldehydes, **3**) are formed after hydrolysis or (ii) amines (**2**) could be generated via the judicious conversion of the tetrahedral intermediate **1a** into an iminium ion **1b** which because of its exceptional electrophilicity [25, 26] could easily react with a second equivalent of the nucleophile. In theory, the addition of a different nucleophile might produce the fully substituted amine **2** (R^3, R^4 different); however, because of the presence of the first nucleophile in the reaction environment, the successful installation of the second element is elusive.

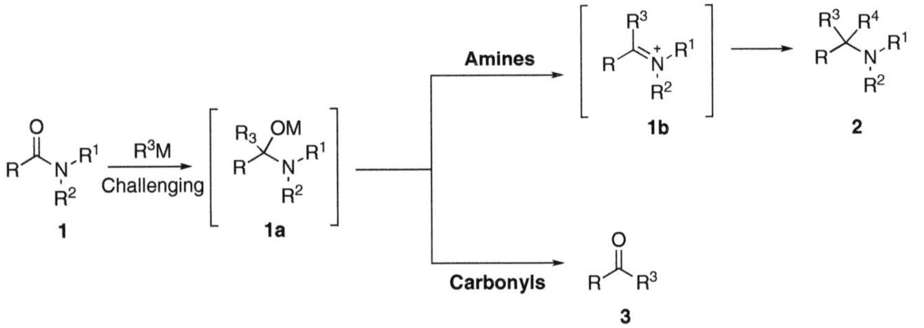

Scheme 4.2 General context of the addition of organometallic reagents to amides.

To enable these transformations to be productive, it becomes essential ensuring the effectiveness of the initial nucleophilic addition event to amides [27]. The first part of the chapter will present protocols levered on the direct addition of organometallic elements to amides – leading to tetrahedral intermediates – in the absence of external activating elements. Subsequently, the treatment of starting amides with electrophilic agents will be commented. In order to rationalize the chapter's organization, in each part, we will first abord the synthesis of carbonyl derivatives and then the preparation of amines.

4.2 Synthesis of Carbonyls from Amides

4.2.1 Addition to Canonical Amides

E. A. Evans [28], Scilly [29], Olah and coworker [30, 31], Carreira and Du Bois [32], among others [33–37] successfully realized the addition of organolithiums and organomagnesiums to classical tertiary amides for the preparation of carbonyls. In this context, the synthesis of enantioenriched ketones from *N*-acylpseudoephedrines (**4**) and various RLi reagents developed by Myers et al. in 1997 is highly remarkable (Scheme 4.3) [25]. To avoid the common side phenomena associated with the employment of highly polar carbanions, organolanthanums and organoceriums introduced, respectively, by Collins et al. [38, 39] and Imamoto et al. [40] could be considered as valuable alternatives.

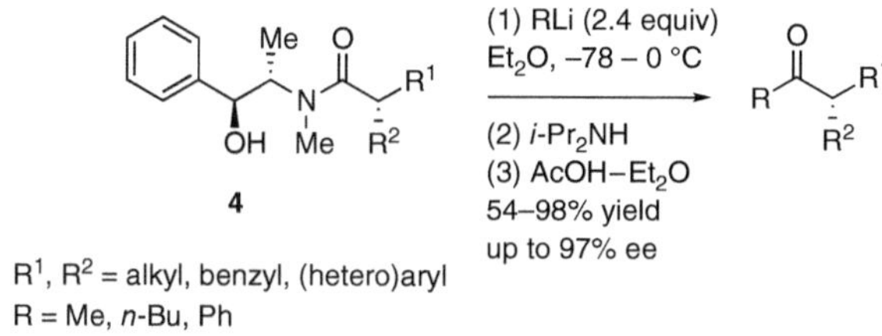

Scheme 4.3 Myers' synthesis of enantioenriched ketones *via* the addition of RLi reagents to *N*-acylpseudoephedrines.

In 2016, Feringa and coworkers have described a one-pot approach to functionalized ketones **6** starting from simple *N,N*-dimethylbenzamides **5** (Scheme 4.4) [41]. This method entails the initial addition of organolithium reagents to benzamides, followed by a palladium-catalyzed α-arylation of the resulting ketones using the commercially available $Pd[P(t\text{-}Bu)_3]_2$ as a catalyst. Noteworthily, the lithium *N,N*-dimethylamide generated from the collapse of the tetrahedral intermediate **5a** avoids the addition of an external base. The methodology tolerates a range of functional groups with complete selectivity toward the monoarylated products.

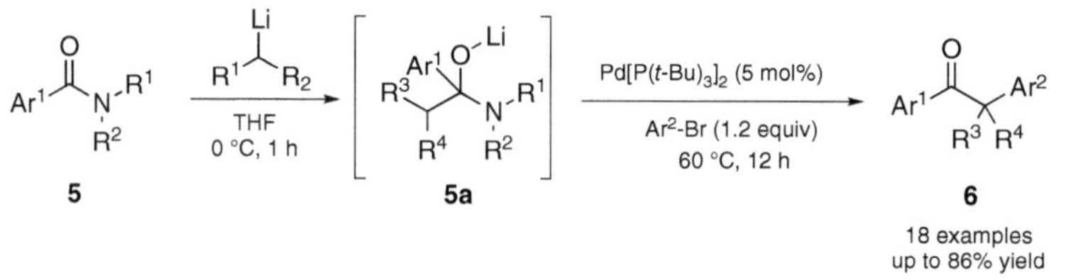

Scheme 4.4 Feringa's combined nucleophilic addition – α-arylation of amides *en route* to ketones.

Simple *N,N*-dialkyl-halobenzamides **7** can be efficiently converted into functionalized ketones **8** by a tandem nucleophilic addition/Buchwald–Hartwig cross-coupling procedure, as described in 2019 by Feringa and coworkers (Scheme 4.5) [42]. Starting from these easily accessible amides, the strategy entails the preliminary addition of organolithium reagents to form a stable tetrahedral intermediate **7a**, which slowly releases the lithium amide coupling partner upon the addition of the $Pd_2(dba)_3$/SPhos catalytic system.

Miyoshi and coworkers have recently shown that simple *N,N*-dimethylamides **9** can be efficiently monoalkylated by organostrontium reagents generated *in situ*

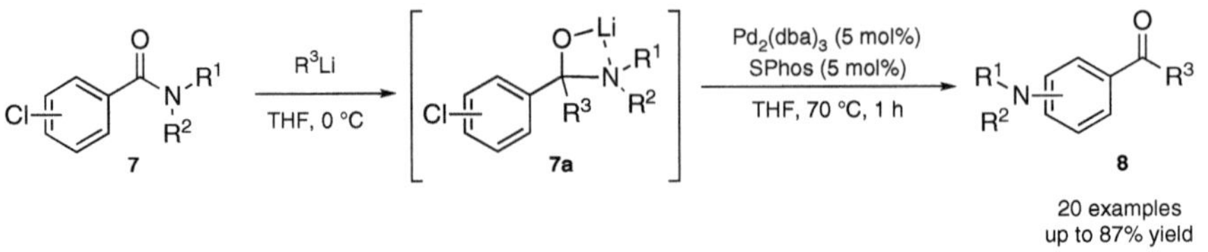

Scheme 4.5 Feringa's combined nucleophilic addition – Buchwald–Hartwig amination *en route* to ketones.

from strontium metal and alkyl (or aryl) iodides (Scheme 4.6) [43]. The corresponding ketones **10** were prepared with high chemoselectivity in high yields and with complete suppression of the undesired overaddition by-product (carbinols **10a**). The reactions proceed smoothly under mild conditions, allowing also the use of enolizable aliphatic amides, as well as poor electrophilic substrates such as tetramethylurea (TMU).

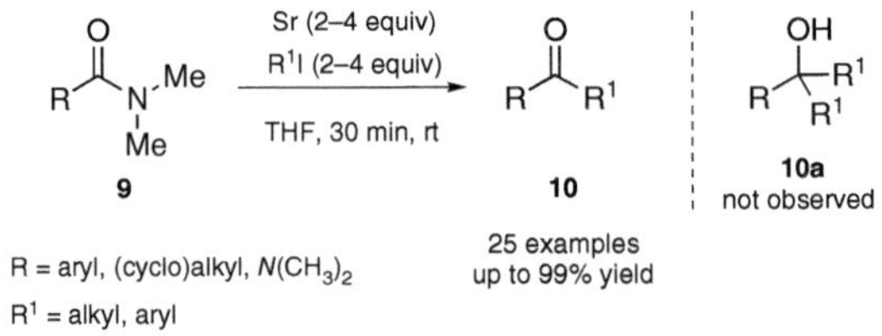

Scheme 4.6 Chemoselective acylation of organostrontium reagents.

Recent investigations by Szostak pointed out the effectiveness of *N*-acyl azetidines **11** as valuable precursors of ketones **12** in the reaction with Grignard and RLi reagents, as a consequence of the nitrogen pyramidalization imparted by the four-membered ring (Scheme 4.7) [44]. The peculiar geometrical properties of these *N*-acyl cyclic amines can be summarized as follows [45, 46]: (i) the half pyramidalization of the amide bonds increases the N–C(O) reactivity because of diminished amide bond delocalization [47] and (ii) the significant ring strain in four-membered rings and the high barrier for nitrogen inversion (90° angle) decrease the aptitude for collapse of the tetrahedral intermediate **11a** [48]. The method is flexible for a variety of starting materials, thus enabling a high chemoselective synthesis of ketones using alkyl, alkynyl, and (hetero)aryllithiums. The excellent stability of the intermediate **11a** was deducted under forcing thermal and stoichiometric conditions: no trace of decomposition phenomena was detected. An intermolecular competitive experiment with a Weinreb amide [49] demonstrated the higher reactivity of *N*-acylazetidines **11**, thus showing that the nucleophilic addition is kinetically favored to the *N*-pyramidalized amide bond.

Subsequently, Prandi–Capriati–Blangetti demonstrated that also *N*-acylpyrrolidines **13** smoothly underwent the nucleophilic addition of organolithiums under aerobic conditions in CPME (Scheme 4.8) [50]. Both aliphatic and (hetero)aromatic *N*-acylpyrrolidines have been found to successfully react with commercial solutions of aliphatic and (hetero)aromatic organolithiums giving the desired ketones

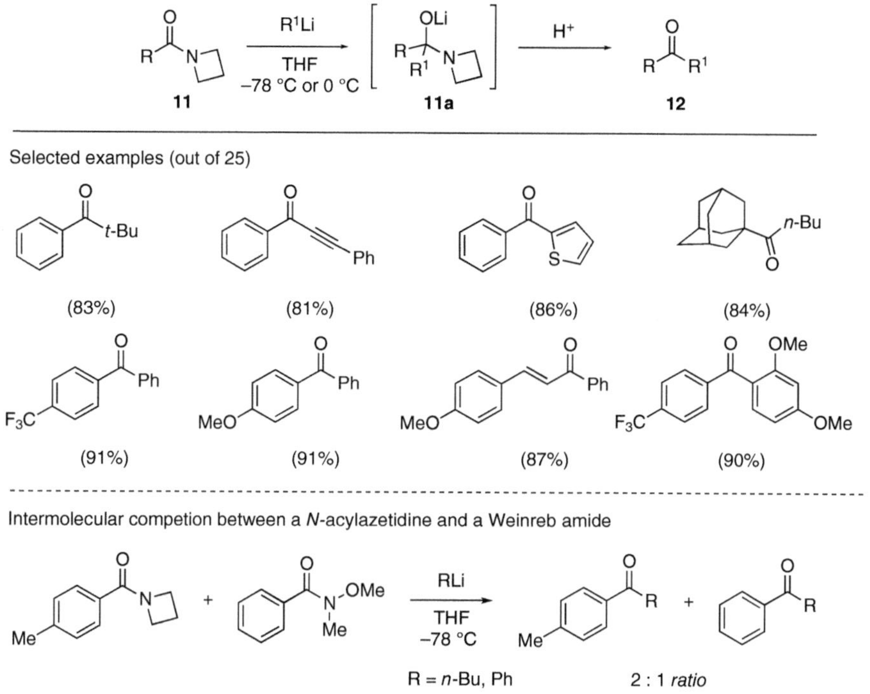

Scheme 4.7 Szostak's chemoselective synthesis of ketones with *N*-acylazetidines.

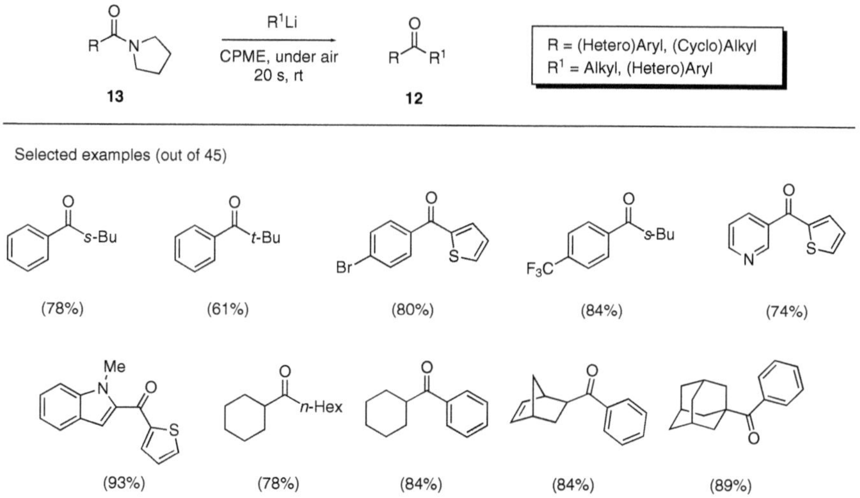

Scheme 4.8 Use of *N*-acylpyrrolidines as acylating agents of RLi reagents.

obtained in up to 93% yield without noticing the overaddition phenomena. Some practical aspects of this methodology make it interesting from a sustainable perspective: reactions are run in open air and at room temperature and reach to completion within short times (20 seconds). High tolerance to several electron-donating, electron-deficient, and sterically hindered functional groups is manifested, as well as no interference is displayed by groups featuring the potential direct *ortho*-metalating effect. Detailed DFT calculations and NMR spectroscopic investigations support the stability of the involved tetrahedral intermediate.

Liu and coworkers designed a chemodivergent protocol enabling the addition of 1,1-diborylalkanes **14** as pronucleophiles to primary, secondary, tertiary amides, and lactams (**15**, Scheme 4.9) [51]. Upon forming the α-boryl carbanion **16** – via deprotonation or Lewis base coordination – a tetrahedral intermediate (**15a**) was generated. Depending on the fate – imparted by the nature of the starting amide – the intermediate could evolve through two possible divergent outcomes: (i) the B–N elimination conducting to an enolate **17** (tertiary amides and lactams) or (ii) the B–O elimination leading to an enamine **18** (primary and secondary amides). These nucleophilic species could be further functionalized under electrophilic conditions, thus providing convenient entries to (α-functionalized) ketones **19a,b** or enamine derivatives **20**. *Ceteris paribus*, primary amides could be transformed into

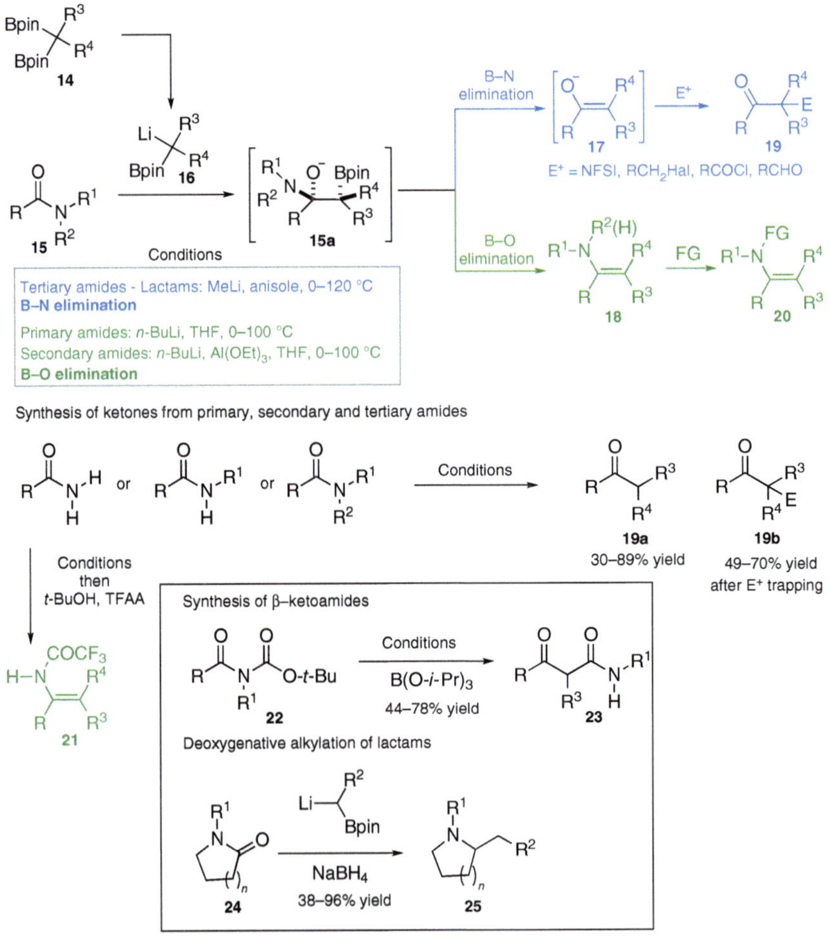

Scheme 4.9 Acylation of α-borylcarbanions with amide derivatives.

N-trifluoroacetylenamines (**21**) after treating with TFAA, whereas *N*-Boc amides (**22**) afforded β-ketoamides (**23**) after treatment with a boronic ester. Notably, upon subsequent deoxygenative conditions, lactams (**24**) could be transformed into α-functionalized cyclic amines (**25**).

4.2.2 Variation of the Amide Structure

Since long time, the installation on the nitrogen atom of amides of electron-withdrawing groups enabling the taming of the lone-pair delocalization has been considered an effective solution to increase the inherent low electrophilicity of these carboxylic analogues [52–54]. Recently, Lee and coworkers adopted this strategy for the preparation of 1,3-diketones (**28**, Scheme 4.10) [55]: *N*-tosyl-, *N*-triflyl-, *N*-mesyl-, and *N*-Boc-substituted tertiary amides (**26**), including *N*-benzoyl saccharin and *N*-benzoyl succinimide, have been reacted with enolizable ketones (**27**) in the presence of lithium hexamethyldisilazide (LiHMDS) as a base at room temperature. Moreover, primary and secondary amides also reacted efficiently with acetophenone derivatives in a one-pot reaction to give the corresponding products.

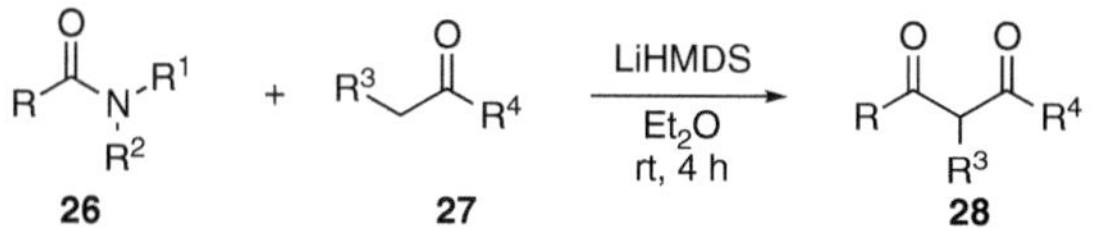

R = (hetero)aryl, (cyclo)alkyl
R^1 = Ts, Boc, Tf, Ms, benzoyl, H
R^2 = Ar, Me, H
R^3 = alkyl R^4 = aryl, alkyl

74 examples
up to 96% yield

Scheme 4.10 Use of EWG-substituted amide derivatives for the synthesis of 1,3-diketones.

The modification of the amide mainly at the level of the nitrogen atom opened in the late 1970s new directions for the productive nucleophilic addition to amides *en route* to carbonyls. Comins and Meyer acylated Grignard reagents with pyridyl *N*-methyl amides (**29**, forming stable six-membered chelates, Scheme 4.11a) [34–36]; whenever an excess of Grignards was used under THF refluxing conditions, the intermediate (**29a**) decomposed giving the corresponding carbinol **31** instead of the desired ketone **30**. Olah and Arvanaghi used *N*-formylpiperidine (**32**) for accessing aldehydes (**33**) with both organolithium and organomagnesium (Scheme 4.11b) [30, 31]. Probably, it was the introduction of *N*-methyl-*N*-methoxy amides (**34**) by Weinreb and Nahm in 1981 the most significant advancement achieved through the structural modification of the amide backbone (Scheme 4.11c) [49, 56–59]. The introduction of the –OMe group in this sense positively modulates the electrophilicity of the carbonyl and more importantly imparts an outstanding stability to the resulting tetrahedral intermediate (**34a**): because of the Lewis base nature, it coordinates the oxygen-bonded metal, thus forming a highly stable five-membered chelate. Accordingly, the further addition of the nucleophile is prevented, and only upon acidic (aqueous) treatment, the C=O is restored, thus furnishing carbonyls (**30**). The impact of Weinreb amides in synthesis is undoubtedly one of the major breakthroughs realized for enabling the controlled addition of nucleophilic polar organometallics to carboxylic derivatives (e.g. acid halides, esters, and anhydrides), which are known to easily undergo polyaddition processes leading to carbinols rather than (targeted) carbonyls.

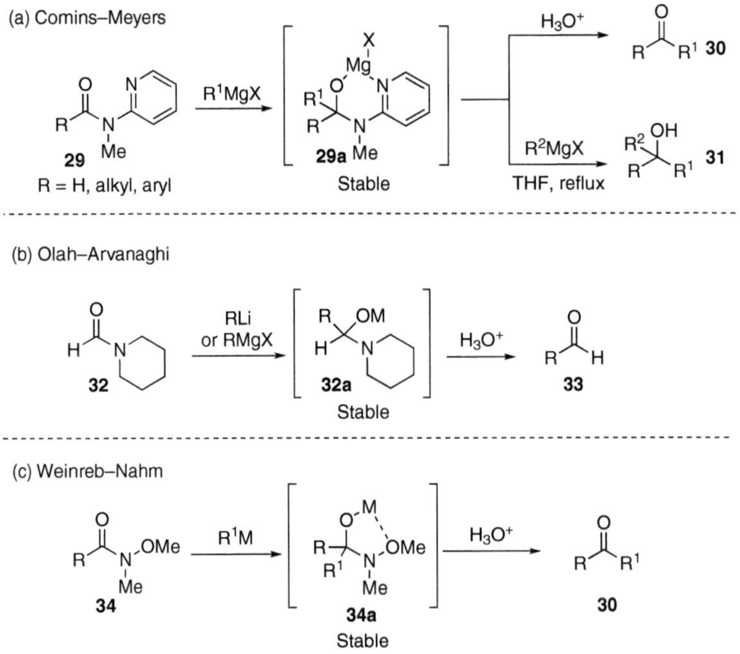

Scheme 4.11 Classical modifications of amides for enabling the addition of organometallics.

In recent years, Pace and coworkers extensively used Weinreb amides (**34**) for preparing α-(poly)-substituted methylketones (**35**) via addition–elimination sequences run with functionalized methyllithiums (LiCHXY) [60, 61] or activated halomethyl-silanes (Scheme 4.12) [59, 62, 63]. Accordingly, once the conditions for generating these rather unstable carbanion-like species are defined, the protocol is flexible for obtaining with remarkable chemocontrol ketones also in the presence of sensitive functional groups such as halides [64–67], nitriles [68], and ethers (oxy- [69], thio- [70], and seleno-). The assembling of α-functionalized ketones via this strategy is further validated by the preparation of optically active ketones starting from chiral Weinreb amides: complete integrity of the stereocenters is noticed, thus representing an interesting approach toward chiral substituted ketones.

The induction of stereochemistry within the same reaction would be highly desirable and, in this context, Pace and coworkers translated the well-developed Hoppe–Beak chemistry [71–74] – involving the enantioselective generation of organolithium compounds in the presence of sparteine – to the acylation of Weinreb amides to access chiral ketones (**37a** and **37b**) [75]. The control over both enantiomers of a given ketone was achieved by simply exchanging between

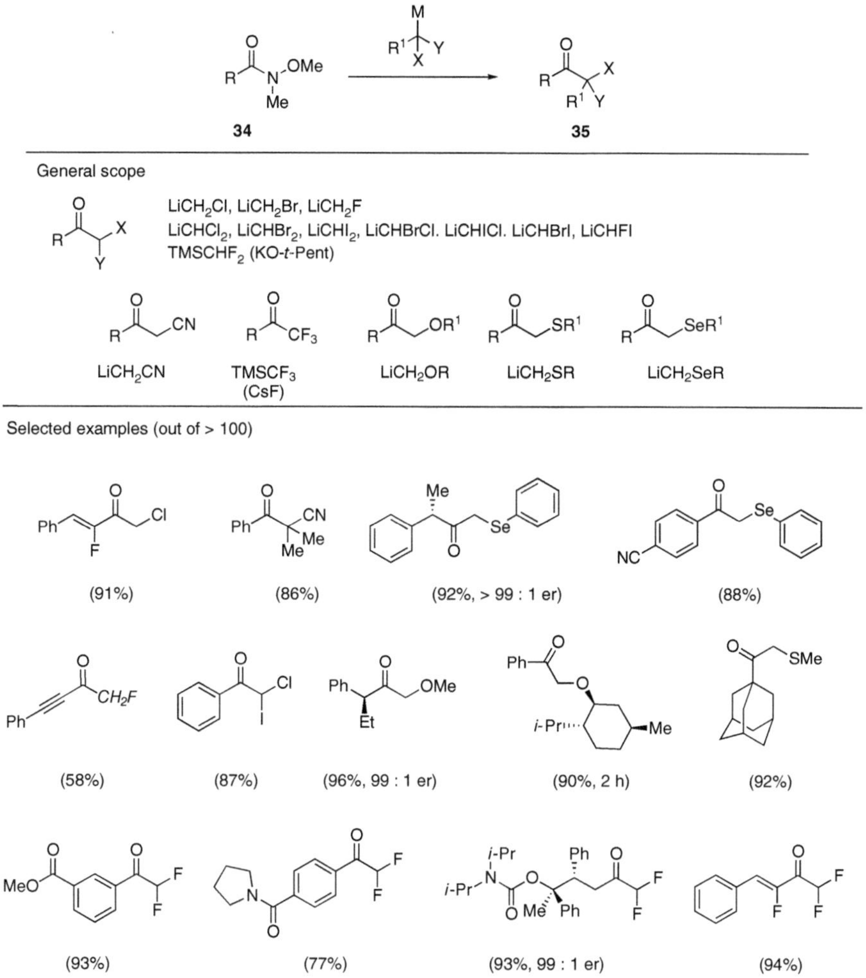

Scheme 4.12 Use of Weinreb amides for acylating substituted carbanions.

the two enantioforms of sparteine (Scheme 4.13). The reaction is wide in scope: (hetero)aromatic, conjugated, and aliphatic Weinreb amides were well tolerated. The acylation was equally efficient with enantiopure organolithium reagents (Hoppe's carbamates (**36a**) and *N*-Boc pyrrolidine (**36b**)), thus providing a modular access to α-oxyketones (**37a**) and cyclic α-aminoketones (**37b**) in uniformly high yields and optical purities. Similarly, enantioenriched α-oxy-α-methylbenzyllithium reagents (**38**) reacted with Weinreb amides to provide α-quaternary ketones (**39**, without requiring sparteine). Using CPME as the reaction solvent [76] was not only beneficial for maintaining the chiral integrity of the nucleophile but also for recycling up to c. 80% the chiral inductor sparteine.

Szostak developed a valuable protocol for preparing biaryl ketones (**41**) by arylation of Weinreb amides (**40**) with functionalized Grignard reagents generated via Mg/halide exchange promoted by *i*-PrMgCl-LiCl on easily available haloarenes

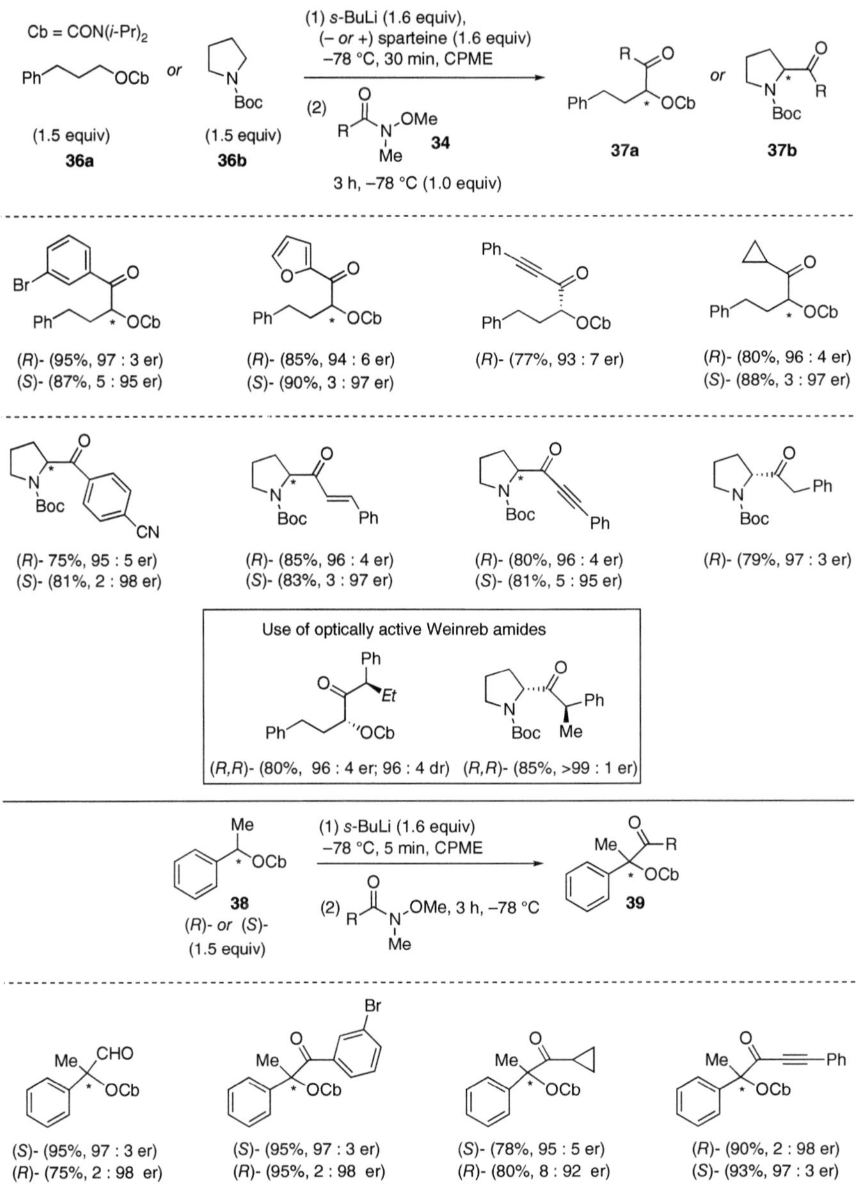

Scheme 4.13 Synthesis of chiral α-oxyketones, α-aminoketones, and α-quaternary ketones from Weinreb amides via asymmetric organolithium chemistry.

(Scheme 4.14) [77]. The study is not only remarkable for the preparative robustness (41 examples, up to 85% yield) but also for its complementarity with the acylation of *N,N*-Boc$_2$ amides (**42**) with aryl Grignard reagents previously published by the same research group [77, 78]. With the latter amides – more reactive than Weinreb congeners – the reaction proceeds under kinetic control of the (transient) tetrahedral intermediate (**42a**). On the other hand, using the less electrophilic

Weinreb amides, a more thermodynamically stable metal-chelated intermediate (**40a**) is formed prior to its collapse into the desired ketones, thus suppressing undesired overaddition phenomena (carbinol formation), somehow observed when excess of Grignards were added to *N,N*-Boc_2 amides [79]. For instance, the collapse of transient intermediates is thermodynamically favored in the case of *N,N*-Boc_2 amides as a consequence of the tamed nucleophilicity of the leaving group *N*Boc_2 compared to *N*Me(OMe) of Weinreb analogues.

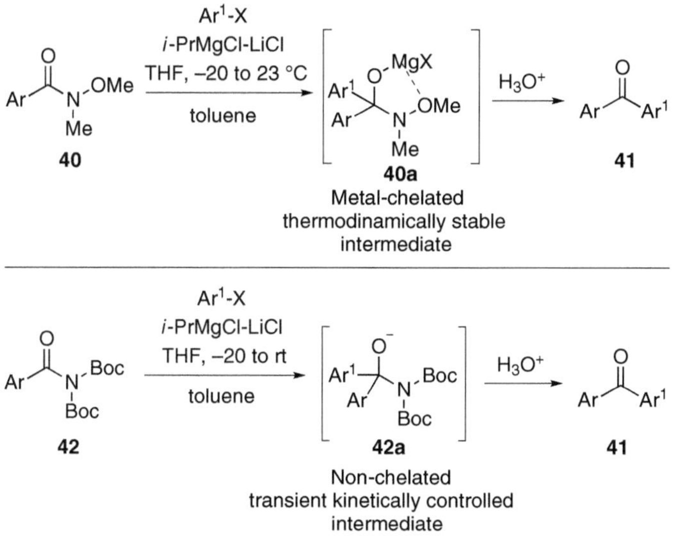

Scheme 4.14 Synthesis of biaryl ketones through the addition of Grignard reagents to Weinreb amides and *N,N*-Boc_2 amides.

In 2015, Sarpong introduced an unprecedented carbonyl dication synthetic equivalent – *N*-methoxy-*N*-methyl-1*H*-pyrrole-1-carboxamide (CLAmP, **43**) – enabling the attack of two nucleophilic elements for the preparation of unsymmetrical ketones (**44**, Scheme 4.15) [80]. This unique reagent features two nucleofugal groups, each of which forms a differently stable tetrahedral intermediate upon the sequential monoaddition of the nucleophiles. In particular, after the first installation, the corresponding intermediate (**43a**) – embodying both the pyrrole and the Weinreb-type fragments – is stable enough (from some hours to 40 hours at low temperatures) at such an extent that its collapse can be finely modulated by varying the experimental conditions. Thus, upon this first controlled intermediate breaking, a new carbonyl Weinreb amide-type electrophile (**43b**) is generated and is subjected to the second nucleophilic addition, furnishing a new intermediate (**43c**) stable enough to hamper undesired overaddition phenomena. *In situ* IR analysis is diagnostic for this mechanistic scenario. Some practical aspect is necessary to maximize the efficiency: (i) the use of ArLi as the first nucleophiles is preferable; (ii) in general, the second addition is not influenced by the nature of the organolithiums; and (iii) the addition of Grignards is well tolerated, except for sp^2-hybridized analogues (when used first nucleophiles) or when are consecutively added RMgX and RLi. For this latter circumstance, drawbacks are conveniently skipped installing first the organolithium.

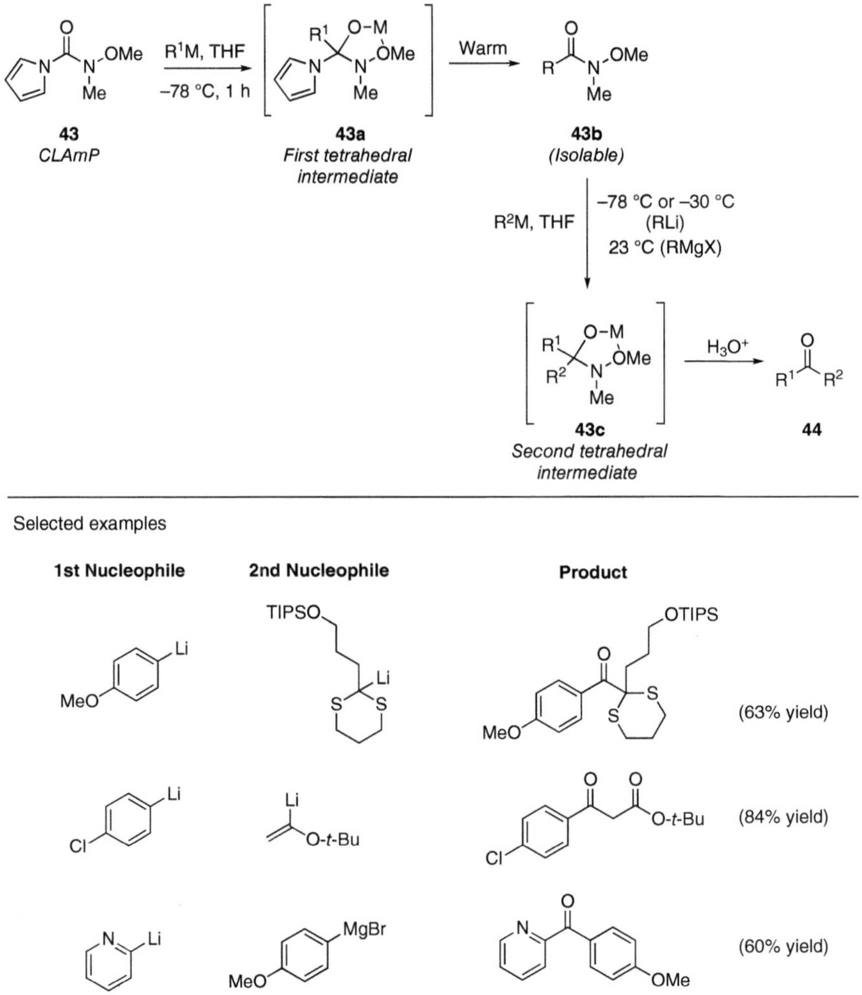

Scheme 4.15 Synthesis of unsymmetrical ketones through the addition of two nucleophiles to Sarpong's CLAmP.

The nucleophilic addition of polar organometallics to *N*-alkoxylactams featuring a α-halo substituent (**45**) has been advantageously exploited by Ueda in 2020 for a straightforward ring-contractive rearrangement (Scheme 4.16) [81]. Accordingly, the Weinreb-type tetrahedral intermediate generated (**45a**) collapses, furnishing the lower cyclic homolog with an external ketone moiety (**46**), as a consequence of the extrusion of the halogen. A variety of Grignard and organolithium reagents including (hetero)aryl and alkenyl groups were suitable for the nucleophilic addition. It is interesting to note the key role played by the α-leaving group as only α-bromo was effective for the sequential ring contraction; neither α-iodo nor α-chloro furnished the desired products. Notably, the methodology showed promising potential for the development of a diastereoselective version: an excellent diastereoselectivity while upholding the stereofidelity of the chiral (−)-*cis*-*N*-benzyloxylactam was observed in a preliminary experiment.

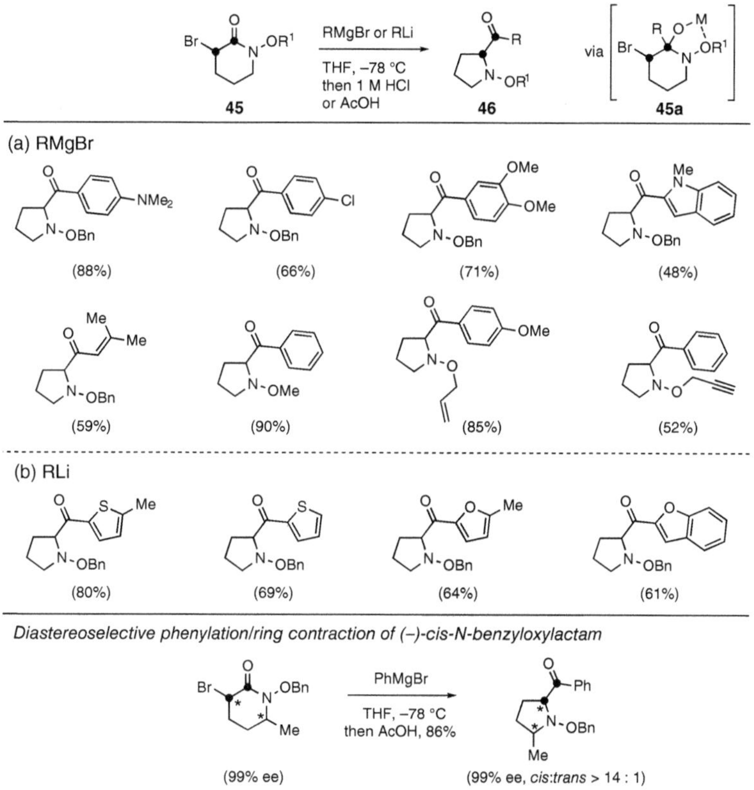

Scheme 4.16 Preparation of α-acylpyrrolidines from α-bromo *N*-alkoxylactams and organometallic reagents through nucleophilic addition/ring contractive rearrangement.

In 2016, Schwartz disclosed a new and highly reactive Weinreb amide (*N*-methoxy-*N*-methylcyanoformamide (**48**), obtained from the imidazolyl-containing Weinreb amide (**47**)) enabling the direct synthesis of α-carbonyl Weinreb amides (**49**) from lithium enolates (Scheme 4.17) [82]. Addition reactions of these nucleophiles – obtained from enones, alkyl and aryl ketones, lactones, and saturated cyclic ketones – to **48** afforded the corresponding α-keto Weinreb amides (**49**) in excellent yields within only 15–30 minutes at −78 °C (Scheme 4.17a). Additionally, the performance of **48** – in reactions with diverse organometallic reagents – rendered this reactive species a suitable precursor for the installation of the Weinreb amide

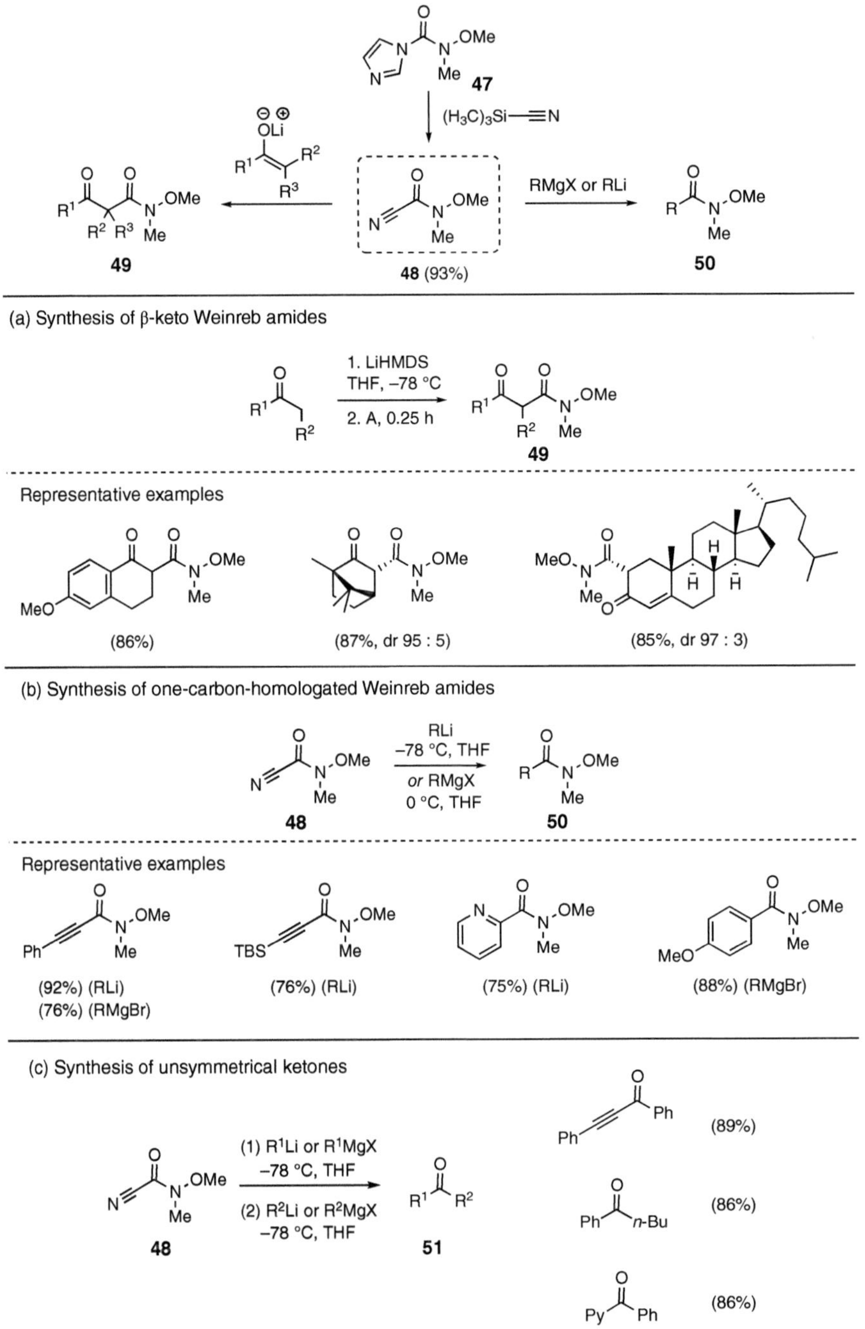

Scheme 4.17 Synthesis of β-keto Weinreb amides (a), one-carbon-homologated Weinreb amides (b) and unsymmetrical ketones (c) from *N*-methoxy-*N*-methylcyanoformamide.

functionality, thus affording the one-carbon-homologated Weinreb amides (**50**, Scheme 4.17b). Moreover, the reaction of **48** with two sequential nucleophiles furnished unsymmetrical ketones (**51**, Scheme 4.17c), likely because of the action of **48** as a carbonyl dication synthon, in analogy to the above discussed Sarpong's CLAmP [80].

4.2.3 Isolation of Tetrahedral Intermediates

Tetrahedral intermediates formed upon the addition of nucleophilic elements to carboxylic derivatives constitute short-lived species during acyl-transfer processes whose synthetic use remains still eclipsed, as a consequence of their lability, which makes the isolation highly challenging [28, 83–87]. Chan and coworkers [88], followed by Brandange, observed that the addition of organolithiums to *N*-acylpyrroles resulted in the formation of unusually stable pyrrolyl carbinols [89, 90]. A similar observation was done by Fukuyama when indolylamides were reduced with DIBAL-H, thus furnishing stable hemiaminals [91], as also shown later by Hou through the $LiAlH_4$-mediated reduction of carbamates [92]. D. A. Evans confirmed the existence of a pyrrolyl carbinol (**53**) upon $LiBH_4$-mediated reduction of a *N*-acylpyrrole derivative (**52**, Scheme 4.18a) [93]. Despite the substitution pattern of the acyl fragment (R^1), the addition of hydride-type and alkyl- or aryl-type nucleophiles (RLi or RMgX) to *N*-acylpyrroles (**54**) resulted in the formation of stable adducts (**55**) whose chemical integrity was preserved even during chromatographic purification and, for instance, the X-ray analysis confirmed the structure (Scheme 4.18b). This unique feature of *N*-acylpyrroles in providing tetrahedral intermediates is attributed to the dual effect of the heteroaromatic ring in activating the nucleophilic addition and difficulting the pyrrole extrusion. In this sense, the reactivity of *N*-acylpyrroles shows a significant degree of analogy to that of ketones because of the nitrogen lone pair delocalization in the heteroaromatic system rather than into the amide carbonyl function [94]. Also, the pyrrole aromaticity prevents nitrogen protonation, thus impeding the collapse of the intermediate. Some points merit mention: (i) the analogous addition of organometallics to 1,1′-carbonyldipyrrole gave less stable intermediates and (ii) pyrrolyl carbinols did not decompose under mildly acidic (PPTS) or basic conditions (pyridine, NEt_3) at rt. Conversely, upon treatment with catalytic DBU, the intermediates smoothly provided the corresponding carbonyl compounds following kinetics depending on the steric congestion. The study also highlighted the synthetic importance of these intermediates (Scheme 4.18c): they act as valuable masked carbonyl equivalents whose C=O functionalities were revealed under basic conditions. For example, upon treating a pyrrolyl carbinol (**56**) with MeLi, the subsequent pyrrole elimination gave an intermediate ketone (**56a**) amenable of addition of a second equivalent of the same MeLi, thus yielding carbinol (**57**). Furthermore, an OH group of intermediate **56** is susceptible of silylation (**58** and **59**), and after treatment with tetrabutylammonium fluoride (TBAF), the carbonyl moiety (**60**) is restored.

In the course of investigations toward the application of a lithiated Hoppe's carbamate (**36a**) in synthesis, Pace et al. isolated diastereopure pyrrolyl carbinols (**62–63**)

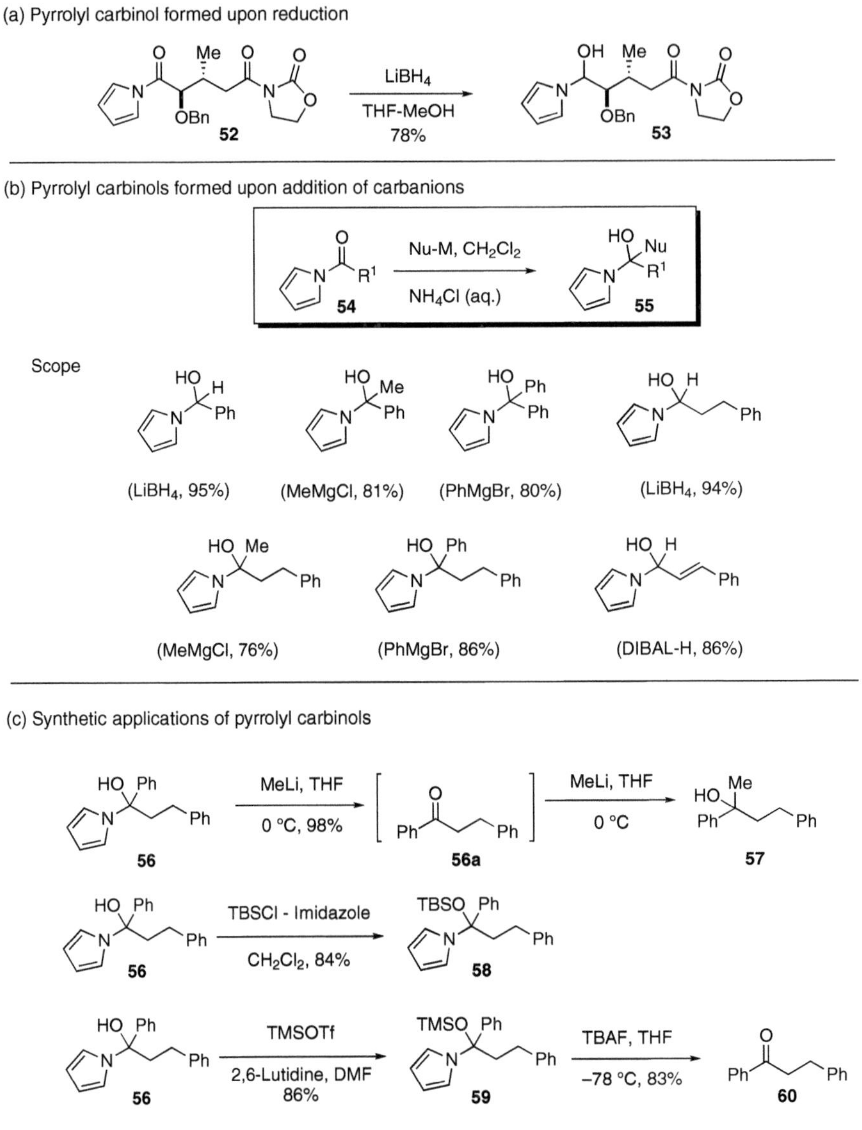

Scheme 4.18 Isolation and synthetic use of pyrrolyl carbinols.

from the corresponding *N*-acylpyrrole (**61**) in excellent dr and er (Scheme 4.19) [75]. As mentioned above, the use of both enantioforms of sparteine enabled the preparation of these enantiopure stable tetrahedral intermediates which showed a remarkable chemical integrity during the subsequent chromatographic purification and HPLC analysis.

Aubé and Szostak documented that a significant bond distortion of amides imparts a reactivity profile similar to ketones, thus qualitatively acting as Evans' *N*-acylpyrroles [93] in reactions involving nucleophilic organometallics

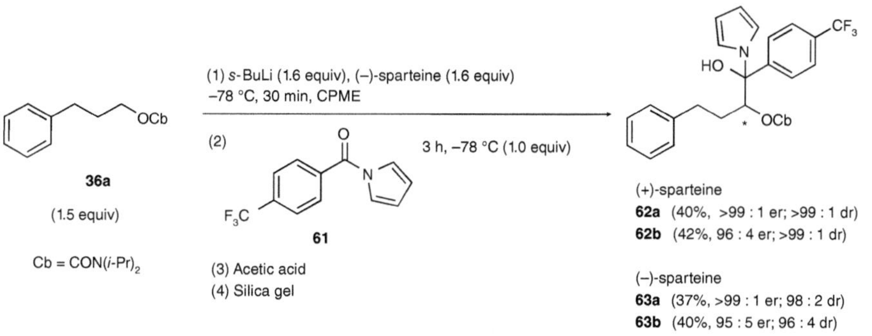

Scheme 4.19 Isolation of optically active pyrrolyl carbinols using lithiated Hoppe's carbamates.

(Scheme 4.20) [95]. In this context, they isolated sterically hindered hemiaminals (**65**) upon the addition of primary and secondary organolithiums to tricyclic bridged lactams (**64**) [48, 96–98]; however, the use of *t*-BuLi resulted in the formation of an aminoketone adduct (**66**).

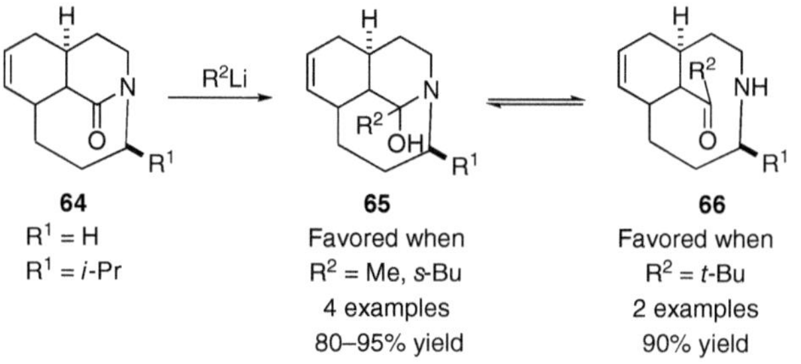

Scheme 4.20 Stable hemiaminals from tricyclic bridged lactams.

Pace and coworkers isolated and characterized the tetrahedral intermediates generated via the addition of lithium carbenoids to Weinreb amides (**34**), followed by the trapping as *O*-TMS protected hemiaminals (**67**, Scheme 4.21) [99]. Remarkably, the use of the commercially available *N*-trimethylsilyl imidazole (ImTMS) as the ideal trapping agent and performing the chromatographic purification on deactivated neutral alumina (Brockmann grade 3, Alox-BG3) were crucial for the isolation of such labile intermediates. Likewise, both mono- (Scheme 4.21a) or dihalo-methyllithium and related [63] (Scheme 4.21b) reagents were successful as carbenoid partners for aliphatic and (hetero)aromatic Weinreb amides. The employment of a chiral organolithium – generated as discussed before via Hoppe's chemistry (**36a**) – provided the corresponding Weinreb-type *O*-TMS protected hemiaminal (**69–70**, Scheme 4.21c) [75]; however, the high chemical instability of this species impeded an efficient separation of the diastereomers (40 : 60 dr mixture) via flash chromatography or preparative HPLC.

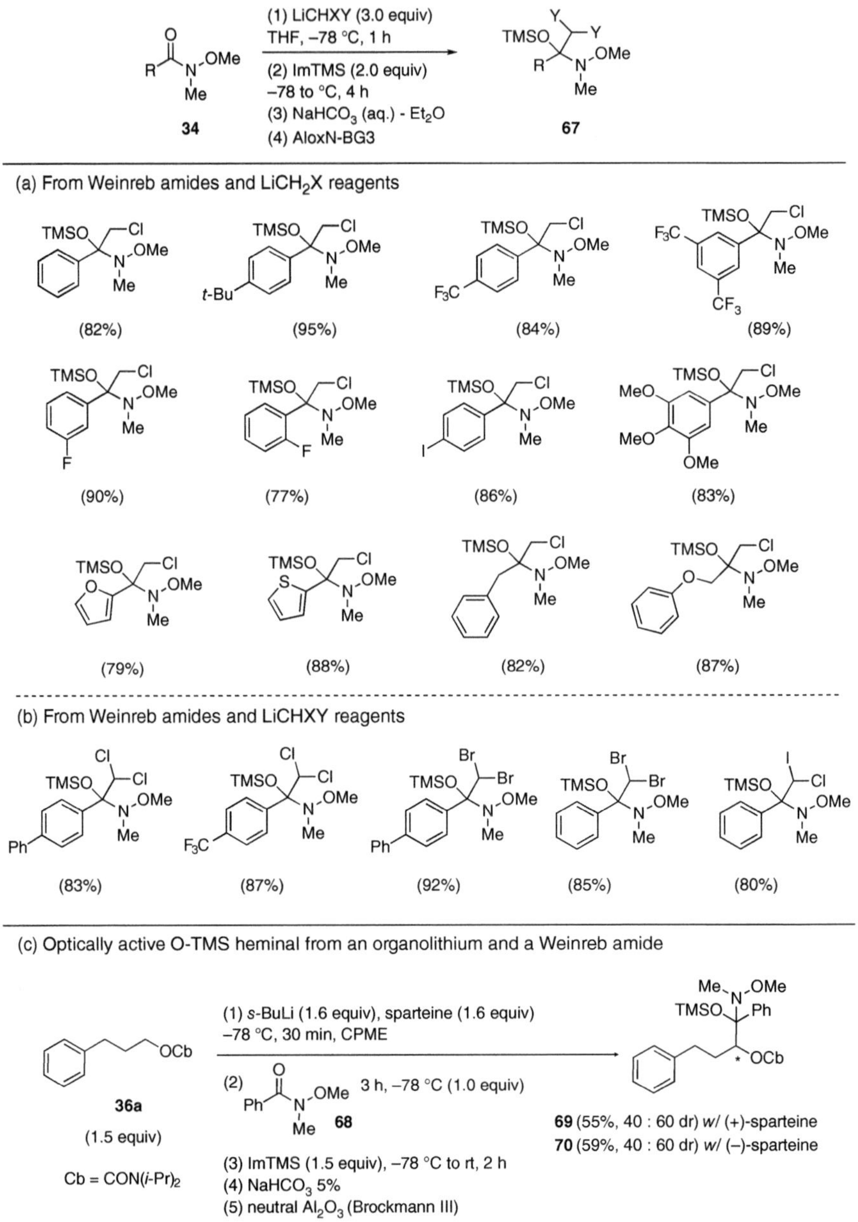

Scheme 4.21 *O*-TMS hemiaminals derived from Weinreb amides and lithium monohalo-, dihalo-carbenoids or an enantiopure lithiated Hoppe's carbamate.

Analogous *O*-TMS-protected hemiaminals (**71**) derived from *N*-acylpyrroles (**54**) could also be prepared under the same Pace's reaction conditions (Scheme 4.22). Collectively, both Weinreb- and pyrrole-type *O*-TMS hemiaminals could be used in the synthesis in a series of transformations, including the Feringa-organolithium coupling or the preparation of functionalized carbenoids [99].

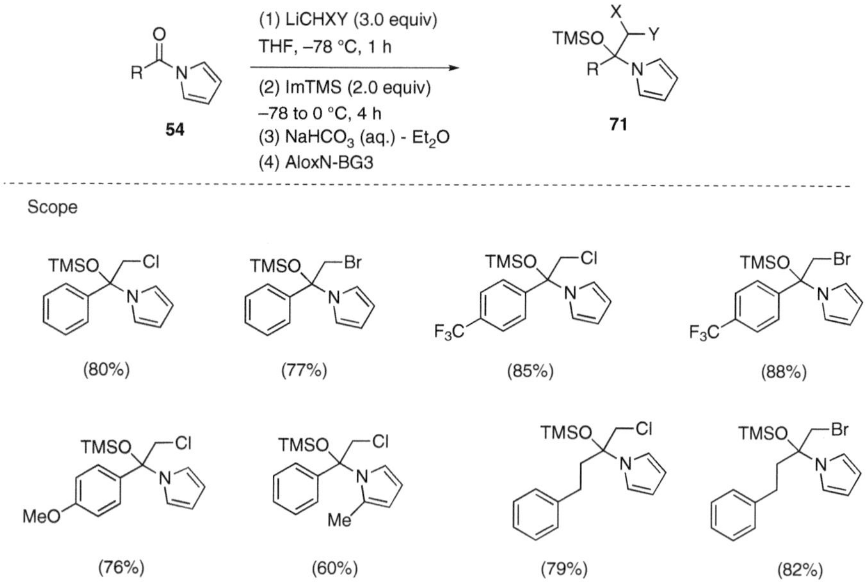

Scheme 4.22 *O*-TMS hemiaminals derived from *N*-acylpyrroles and lithium carbenoids.

4.3 Recent Uses of Amides and *N*-Alkoxyamides for the Synthesis of Amines

Amides are important building blocks not only for the synthesis of carbonyls but can also be advantageously employed for installing two nucleophilic elements on the carbonyl carbon for forming amines [22]. Considering that this chemistry has been thoroughly reviewed in recent years [11, 27, 100], the discussion will be mainly focused on the most recent advancements in the field.

Historically, hydride nucleophiles have been frequently engaged in the reduction of amides [101]. It is well documented that sodium borohydride ($NaBH_4$), lithium aluminum hydride ($LiAlH_4$), lithium triethoxyaluminum hydride [$LiAlH(OEt)_3$], and diisobutylaluminum hydride (DIBAL) are among the prototypical amide reducing agents, although selectivity issues could affect the use of these species [102]. To circumvent these shortcomings, Singaram reported a controlled reduction of tertiary amides (**72**) to amines (**73**), aldehydes (**74**), or alcohols (**75**) mediated by dialkylboranes and aminoborohydrides as mild and selective reducing agents (Scheme 4.23) [103]. Focusing on the complete reduction to amines, tertiary amides were effectively converted to the corresponding tertiary amines using 2 equiv of 9-borabicycle[3.1.1]nonane (9-BBN). Acyclic and cyclic tertiary amides (**72**) along with Weinreb amides cleanly underwent the reaction with excellent yields (Scheme 4.23a). Moreover, hindered lithium diisopropylaminoborohydride (*i*PrLAB) performed as a chemoselective reducing agent, allowing the selective C–O bond cleavage, thereby providing the corresponding amine (**74a**, Scheme 4.23b); conversely, less sterically crowded LAB reagents afforded the alcohol product (**75a**).

Formerly, the same group reported that lactams (**72b**) could be easily reduced to the corresponding cyclic amines (**74b**) by running the reaction with the non-hindered lithium dimethylaminoborohydride (lithium dimethylaminoborohydride (MeLAB), Scheme 4.23c) [104].

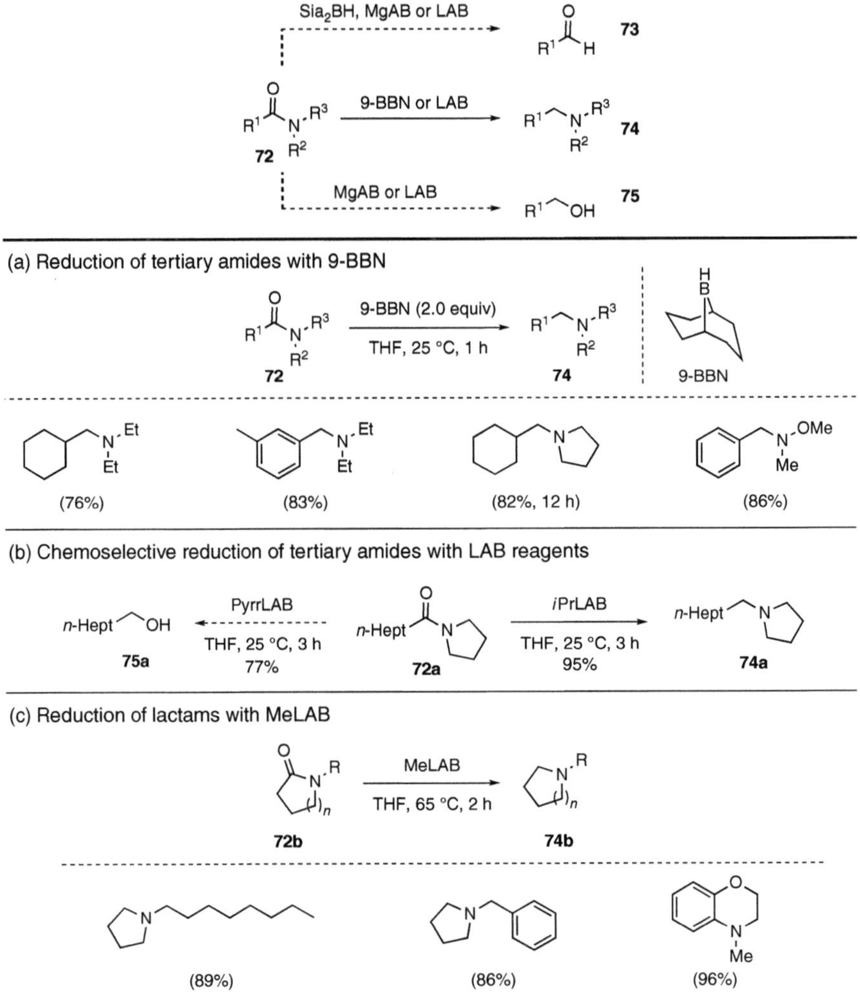

Scheme 4.23 Reduction of tertiary amides to amines with dialkylboranes and aminoborohydrides.

As illustrated in an inspiring review by Seebach in 2011, using the amide carbonyl as a placeholder for two *C*-nucleophilic residues continues to appear a big challenge for the generation of chemists [22, 105]. Initially developed strategies relied on increasing the amide electrophilicity through the installation of electron-withdrawing groups (e.g. Boc, Ts *inter alia*) and, in many instances, found some applicability to specific amide congeners (3-, 5-, and 6-membered lactams) [11]. Despite some remarkable solutions provided by organotitanium reagents for amides *gem*-dimethylation [106–109] or Kulinkovich-De Meijere-type cyclopropanations [110–112], effective and general tactics took long time to emerge.

In the course of the total synthesis of gephyrotoxin 223AB, Kibayashi successfully replaced the C=O group, respectively, with an alkyl (from a Grignard reagent) and a hydrogen (from $NaBH_3CN$), thus transforming with remarkable stereocontrol – guaranteed by the structure of the transient iminium intermediate – the amide into an amine [113]. More recent investigations by Chida and Sato demonstrated the versatility of *N*-alkoxy amides (**76**) as precursors of tetrahedral intermediates (**76a**), which upon trapping with Lewis acids generated electrophilic *N*-oxyiminium ions (**76b**) amenable for the second nucleophilic addition, yielding *N*-alkoxyamines (Scheme 4.24, **77**) [23, 27, 100, 114, 115]. The procedure entails the addition of a hydride as the first nucleophile (DIBAL-H) followed by the competent carbon nucleophile (stannane and trimethylsilyl cyanide (TMSCN)); when the tactic is applied to cyclic systems, diastereocontrol is usually observed. In the case of six-membered *N*-alkoxy lactams, two distinct nucleophiles (organolithium and stannane or TMSCN) can be sequentially added with the overall formation of a fully substituted *N*-alkoxyamine. The employment of the Schwartz reagent (Cp_2ZrHCl) as a convenient hydride source [116–118] provides the stable five-membered zyrconacycle, precursor of an *N*-oxyiminium cation amenable for the second addition of various *C*-nucleophiles.

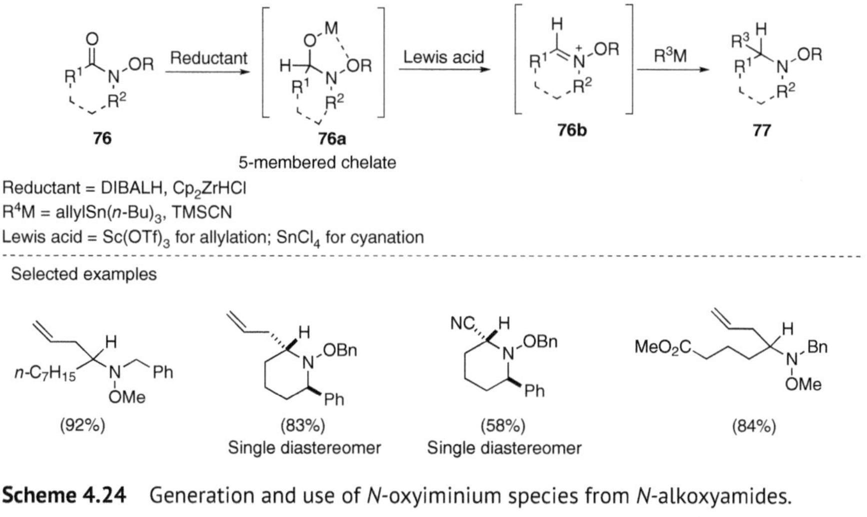

Scheme 4.24 Generation and use of *N*-oxyiminium species from *N*-alkoxyamides.

This tunable, consecutive double nucleophilic addition tactic illustrated has been adopted for solving key steps in stereoselective total syntheses, as documented by Chida–Sato (gephyrotoxin) [119], Kouklovsky–Vincent (isooxazololactams and porantheridine) [120–122], and Helmchen (prosophylline) [123]. The judicious selection of the firstly added nucleophile (hydride or carbanion) constitutes a key aspect for conferring stereocontrol.

More recently, Chida and Sato developed an efficient synthetic strategy for the conversion of *O*-protected *N*-hydroxy lactones **78** to fully substituted cyclic nitrones

79 by means of organolithium reagents (Scheme 4.25) [124]. Protection with the easily removable SEM (2-(trimethylsilyl)ethoxymethyl) group was essential to prevent the preliminary deprotonation of the *N*-hydroxy group and did not inhibit the nucleophilic addition of the organolithium reagent. Formation of the five-membered Weinreb-type chelated intermediate **78a** followed by the elimination/deprotection accomplished with *p*-toluenesulfonic acid (PTSA) afforded the corresponding cyclic nitrones **79** in good yields. The methodology was successfully applied to the total synthesis of the complex alkaloid cylindricine C, starting from the *O*-SEM-protected five-membered *N*-hydroxy lactam **80**, which upon addition of an organolithium provided the requested nitrone-containing building block **81**.

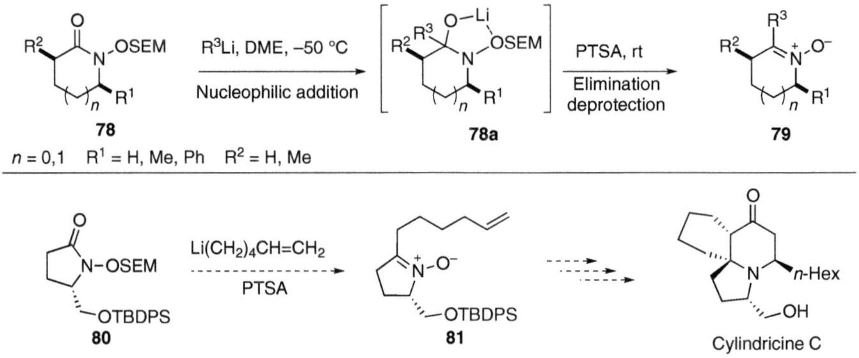

Scheme 4.25 Preparation of nitrones from *N*-alkoxylactams and organolithiums.

Chida and Sato reported a chemoselective reductive nucleophilic addition to tertiary and secondary amides (**72**) extensible also to *N*-methoxyamides, grounded on the use of the Schwartz reagent [Cp_2ZrHCl] as a competent hydride donor (Scheme 4.26) [125–127]. The strategy was focused on the initial formation of a four-membered chelated intermediate (zirconacycle **83**), which in the presence of an acid delivered an iminium ion **83a**. The generated iminium species then underwent the addition of mild nucleophiles to afford substituted amines (**84**) in a one-pot procedure. The combination of the high oxophilicity of the Schwartz reagent (enabling the direct reduction of the amide carbonyl group) with the excellent electrophilicity of the iminium ion (thus allowing the addition of mild nucleophiles) imparted a high chemoselective profile, as documented by the lack of interference of sensitive functional groups such as methyl esters, nitro, nitrile, sulfonamides groups or in the application to challenging total synthesis. The reductive allylation of tertiary amides and Weinreb amides gave good-to-excellent results in terms of both yield and chemoselectivity, while secondary amides could manifest some minor limitations.

Simple and useful reductive functionalization protocols entailing the use of the Schwartz reagent for the conversion of sugar-derived lactams to the corresponding cyclic amines [128–130] and fluoroacetamides to fluorinated imines [131] have been developed by Furman and coworkers. These methods allow the facile synthesis of polyhydroxylated cyclic amines and fluorinated amine moieties of valuable biological and pharmacological interest.

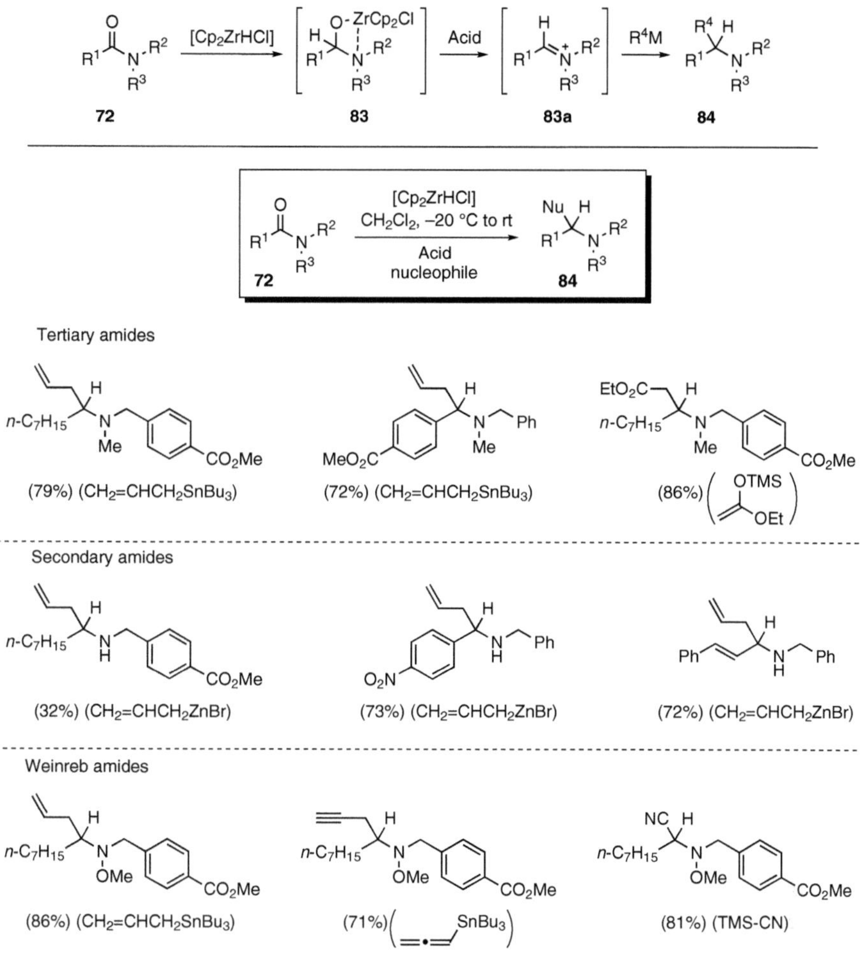

Scheme 4.26 Reductive nucleophilic addition to amides with the Schwartz reagent.

In order to avoid the stoichiometric use of the expensive Schwartz reagent, Chida and Sato developed a novel approach based on an iridium-catalyzed nucleophilic addition to *N*-methoxyamides applying the Nagashima's [132] catalytic method (Scheme 4.27). In this scenario, the addition of $(Me_2HSi)_2O$ (tetramethyldisiloxane [TMDS]) and Vaska's complex $[IrCl(CO)(PPh_3)_2]$ to *N*-methoxyamides (**85**) gave *N,O*-acetals (**85a**), which delivered *N*-oxyiminium ions (**85b**) after treatment with an acid [126]. Likewise, the subsequent nucleophilic addition enabled the formation of the corresponding *N*-methoxyamines (**86**). The reaction occurred in high yields with an outstanding level of chemoselectivity, as also documented during the total synthesis of stemoamide-type alkaloids in the presence of a more electrophilic γ-lactone moiety [133].

Chiral derivatives of traditional Weinreb amides have been successfully exploited by Sato and coworkers in the asymmetric total synthesis of the strained tricyclic alkaloid fasicularin (Scheme 4.28) [134]. The chiral auxiliary firstly allowed the

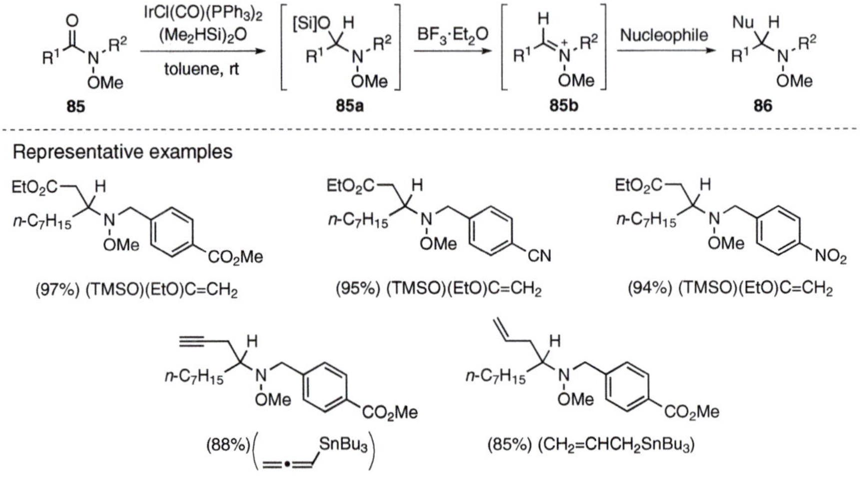

Scheme 4.27 Iridium-catalyzed reductive nucleophilic addition to *N*-methoxyamides.

stereoselective aza-spirocyclization reaction for the construction of the intermediate *N*-alkoxylactam (**87**), whose reduction has been accomplished by treatment with Schwartz reagent or Vaska complex. The subsequent Strecker reaction on the tetrahedral intermediate (**88**) provided the aminonitrile (**89**) with excellent diastereoselectivity. Further synthetic steps finally led to the target fasicularin in a 7.9% total yield over 11 steps.

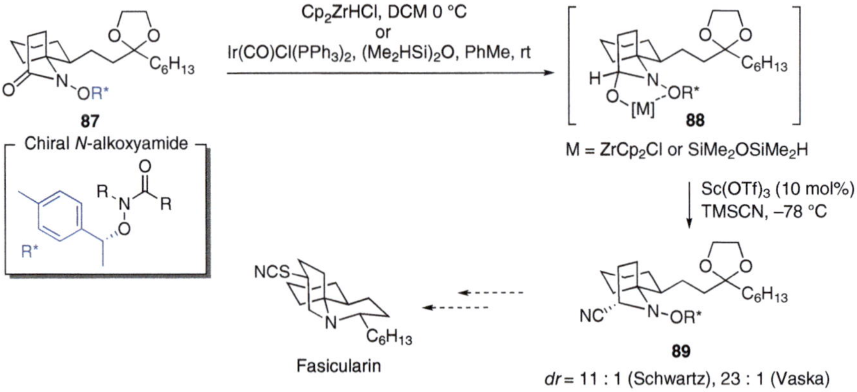

Scheme 4.28 Use of chiral *N*-alkoxyamides in the synthesis of fasicularin.

In recent years, a series of protocols levered on the transition-metal-catalyzed hydrosilylation – as the initial step – prior to the nucleophilic functionalization has been developed [135]. Dixon reported the reductive coupling of an extensive variety of tertiary amides and Grignard reagents using Vaska's complex and TMDS (Scheme 4.29) [136]. The reaction is wide in scope for both the tertiary amides (aliphatic amides, formamides, and (hetero)aromatic amides, **90**) and the Grignard reagents (alkyl, vinyl, alkynyl, and (hetero)aryl Grignard reagents), thus offering a straightforward access to structurally diverse tertiary amines (**91**) in good to excellent yields. Regardless of the use of more reactive nucleophilic partners, this protocol exhibited a remarkable functional group tolerance: ester, nitrile, carbamate, pinacol boronate ester, and aryl ketone were well tolerated under the reaction conditions. Notably, the methodology was applied to the synthesis of several bioactive compounds, as well as, for the late-stage functionalization of drug molecules.

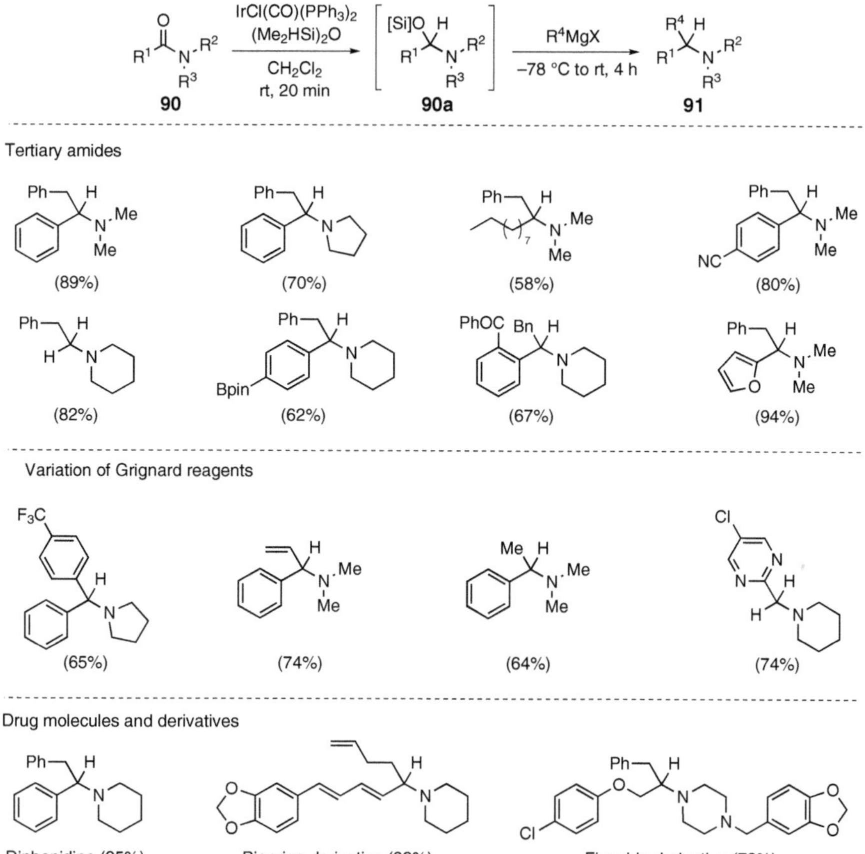

Scheme 4.29 Iridium-catalyzed reductive nucleophilic addition to tertiary amides with Grignard reagents.

Conceptually analogous is the reductive functionalization of tertiary amides (**90**) through sequential Ir-catalyzed hydrosilylation/Cu(I)-catalyzed alkynylation to access propargylic amines reported by Huang et al. (Scheme 4.30) [137]. The methodology displayed a broad substrate scope; both *tert*-amides with or without α-hydrogen were easily converted to the corresponding propargylic amines (**92**). In addition, the method showed an excellent chemoselectivity: either the amide or the alkyne coupling partner bearing sensitive groups such as aldehyde, nitro, ester, and cyano were compatible with the reaction conditions.

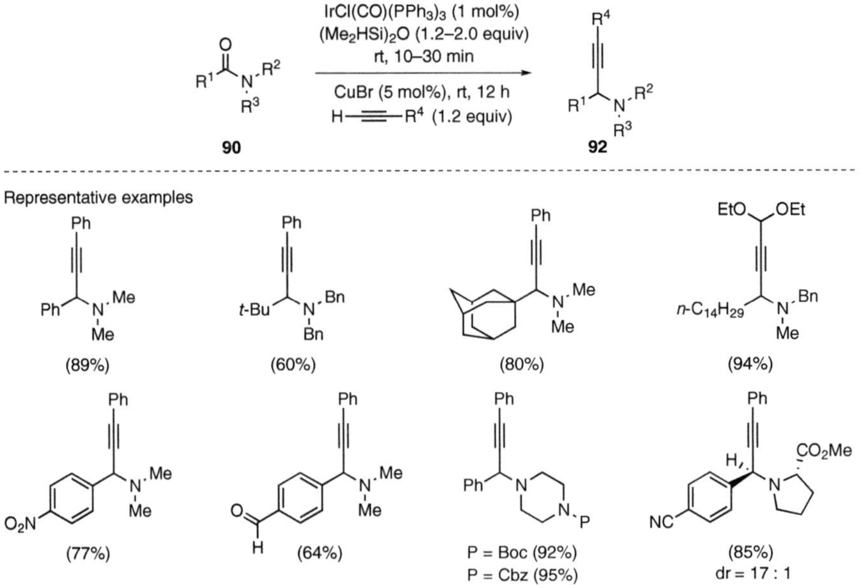

Scheme 4.30 Bis-metallic Ir- and Cu(I)-catalyzed reductive alkynylation of tertiary amides.

In 2018, Tokuyama and coworkers developed a reductive allylation of secondary amides (**93**) by using titanium hydride and an allylzinc reagent in a one-pot procedure (Scheme 4.31) [138]. Benzylic amides and six- to eight-membered lactams were smoothly transformed into the corresponding amines (**94**) in good to excellent yields. The protocol was highly chemoselective even in the presence of sensitive functional groups such as nitro and methyl ester. The synthetic potential of this methodology was validated in the total synthesis of (−)-castoramine.

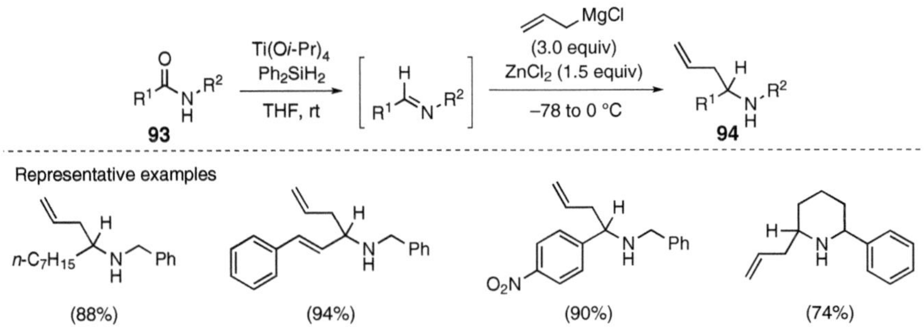

Scheme 4.31 Reductive allylation of secondary amides with titanium hydride and an allylzinc reagent.

Recently, Dixon and Chiba disclosed a transition-metal-free reductive nucleophilic addition to tertiary amides and lactams (**90**), benefiting from the controlled reduction step of the amide carbonyl group by NaH/NaI [139]. The resulting anionic hemiaminal (**90a**) was then treated with trimethylsilyl chloride and a nucleophile (Grignard reagents or tetrabutylammonium cyanide), enabling the synthesis of a wide series of tertiary amines (**95**) (Scheme 4.32).

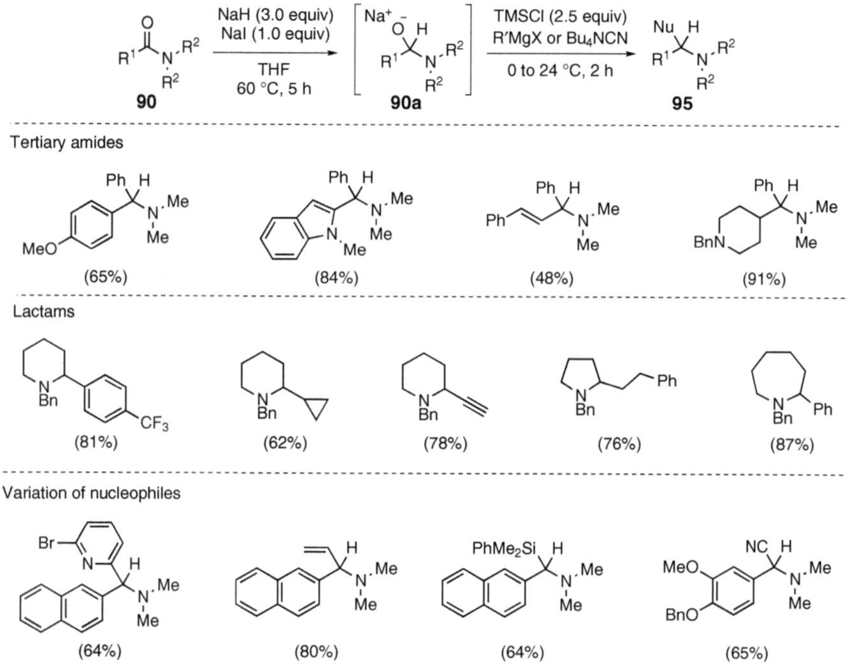

Scheme 4.32 Transition-metal-free reductive nucleophilic addition to tertiary amides and lactams using NaH/NaI.

Chiba and coworkers have developed an elegant transition-metal-free protocol for the synthesis of tertiary amines (**97**) by the iterative addition of organometallic reagents to simple non-activated *N*,*N*-dialkyl carboxamides (**96**, Scheme 4.33) [140]. The methodology entails the preliminary 1,2-addition of organolithium reagents to carboxamides to form the tetrahedral hemiaminal intermediate (**96a**). *O*-silylation with trimethylsilyl bromide furnished the corresponding silylated intermediate (**96b**), which evolved to the intermediate iminium ion (**96c**). The latter underwent the nucleophilic addition of Grignard reagents to give tertiary amines **97** in excellent yields after workup. The reaction is wide in scope and, remarkably, the nucleophilic trapping of **96c** using (trimethylsilyl)methylmagnesium bromide followed by acidic workup gives access to a series of terminal olefins (**98**) through an aza-Peterson process.

In addition to the abovementioned strategies, thioformamides (**99**) have been used to access 2-sulfanyl tertiary amines (**100**) via the sequential addition of α-thioorganolithiums and Grignard reagents (Scheme 4.34) [141–143]. The targeted products were obtained in moderate to good yields (18–78%) with good to high diastereoselectivities (dr = 70 : 30 to >95 : 5), engaging primary alkyl, vinyl, aryl, and alkynyl Grignard reagents.

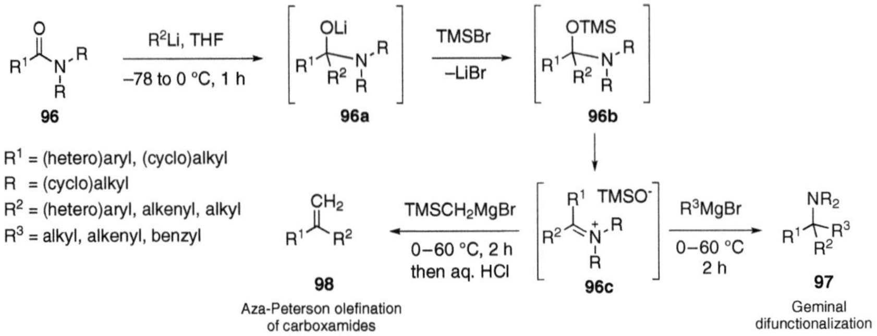

Scheme 4.33 Iterative addition of RLi and RMgX reagents for converting tertiary amides into *gem*-disubstituted amines.

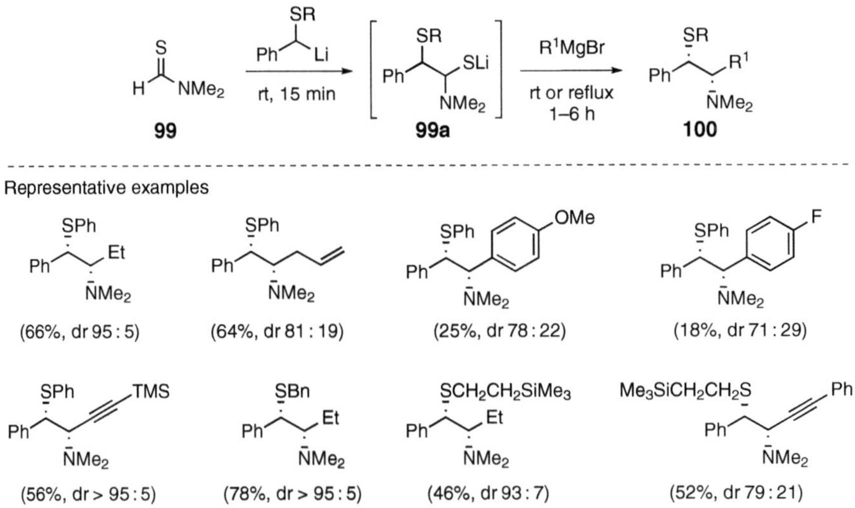

Scheme 4.34 Preparation of 2-phenyl-2-sulfanyl tertiary amines.

4.4 Electrophilic Amide Linkage Activation

4.4.1 General Concept

Firstly noticed by Ghosez and coworkers in 1981 [144], the treatment of an amide with a strong electrophilic agent (e.g. triflic anhydride) dramatically enhances the electrophilicity of the carbonyl group as a consequence of the formation of an iminium triflate ion [11, 145–149]. Mechanistic studies by Charette and Grenon revealed the nature of the active triflating agent (**101**), generated from pyridine and triflic anhydride, which upon reaction with secondary or tertiary amides (**102**) furnishes pyridinium triflates (**102a–102b**) susceptible of the nucleophilic attack (Scheme 4.35a) [150]. The intermediacy of these pyridinium triflates has also been confirmed more recently by Maulide through combined IR and NMR analysis [151]. The presence of the pyridine base has been the objective of debate: Movassaghi showed that although a sterically hindered pyridine (2,6-di-*tert*-butyl) did not react with Tf_2O, the latter could still activate the amide at some extent [152–155]. In this sense, Huang proposed the direct reaction of the amide (**102**) with triflic anhydride for the formation of the pivotal iminium triflate (**102a**) susceptible of nucleophilic attack, thus yielding a *O*-triflyl aminal (**103**) from which a nitrogen-assisted triflate anion elimination gave the iminium triflate (**104**, Scheme 4.35b) [156]. Finally, the subsequent attack of the second nucleophile gave the amine (**105**).

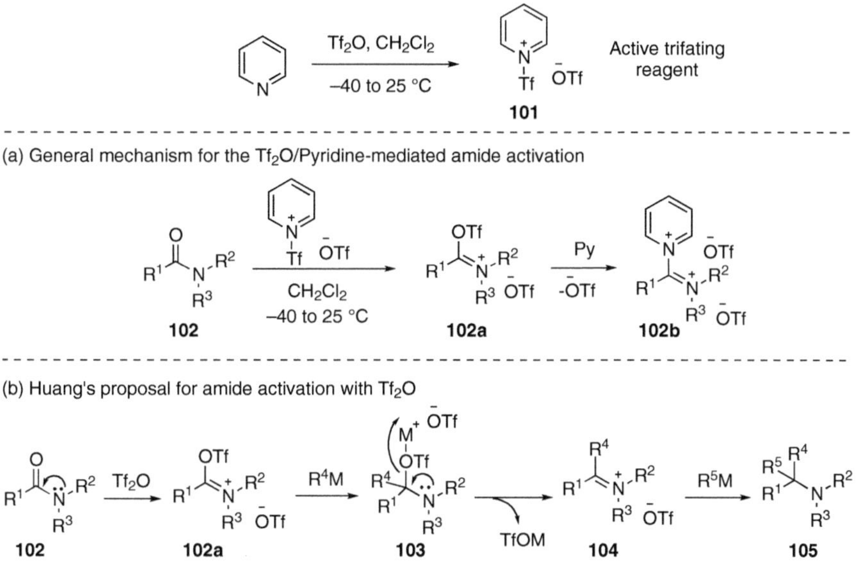

Scheme 4.35 Fundamentals of amide bond activation with electrophilic reagents.

4.4.2 Synthesis of Carbonyl-Like Compounds

The reactivity of iminium triflate intermediates toward nucleophiles has been deeply exploited in the past decade for the synthesis of carbonyl compounds and their derivatives. The pioneering works of Charette and coworkers [157] and Huang and coworkers [158] in 2012 have opened the way to the realization of fully chemoselective strategies for the synthesis of ketones from secondary amides and organometallic reagents such as organolithiums, organocuprates, Grignards, and organozinc (Scheme 4.36). In particular, Charette and coworkers have demonstrated that the iminium triflate intermediate (**106a**), generated *in situ* from secondary amides (**106**) by treatment with Tf_2O and 2-fluoropyridine, allows the fast chemoselective addition of organomagnesium and organozinc reagents in high yields (Scheme 4.36) [157].

This methodology shows a high substrate and organometallic reagent scope, and other electrophilic groups such as esters, nitriles, azides, azo group ketones, aryl, and alkyl halides are well tolerated. Remarkably, outstanding levels of chemoselectivity have been achieved even in the presence of a tertiary or a Weinreb amide. This approach overcomes the formation of overaddition products by providing stable

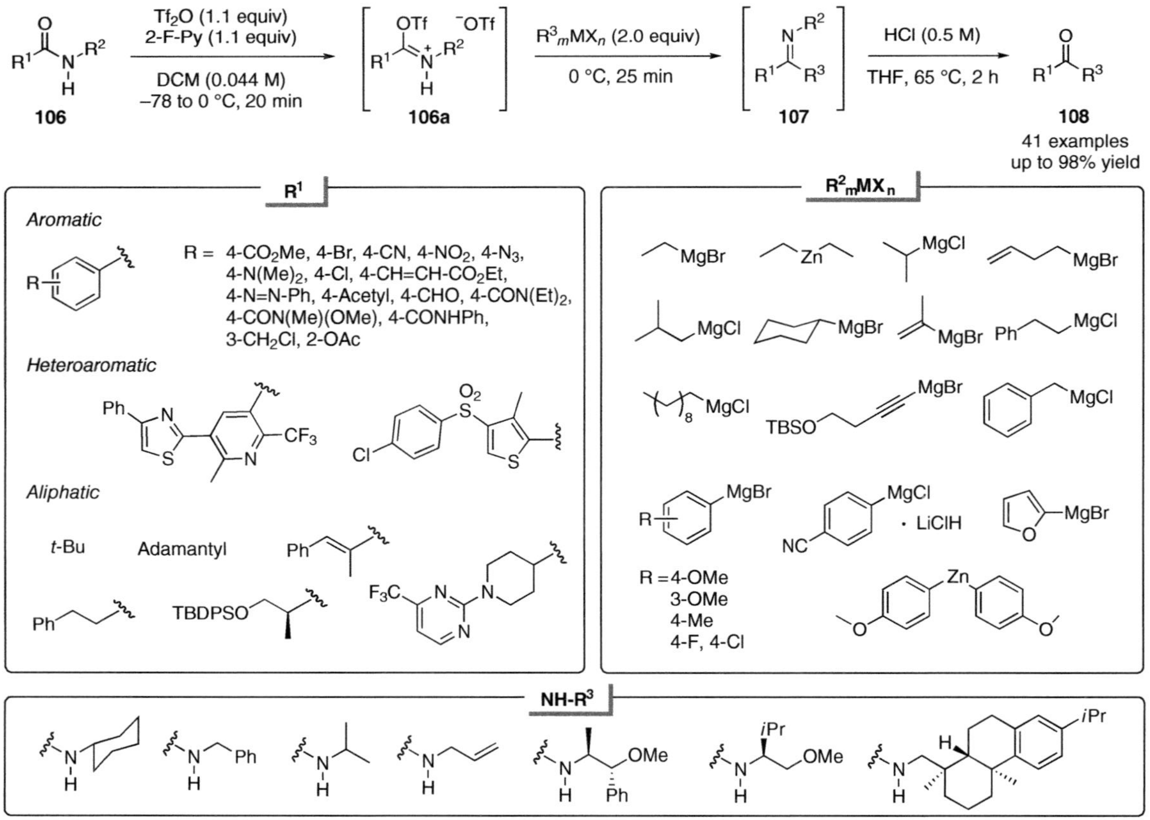

Scheme 4.36 Charette's addition of organometallic reagents to activated secondary amides.

ketimine intermediates (**107**), which can be converted into the desired ketones (**108**) after mild acidic hydrolysis.

Independent investigations by Huang and coworkers in the same year have disclosed an analogous protocol on secondary amides (**106**) involving the addition of organocerium reagents ($RCeCl_2$) to give iminium triflate intermediates (**106a**, Scheme 4.37) [158]. Primary, (cyclic)secondary, benzylic, and vinylic organoceriums generated *in situ* from the corresponding Grignard or organolithium reagents allow the preparation of ketones (**108**) in up to 90% isolated yields. The method has been applied to aliphatic and (hetero)aromatic secondary amides, including sterically hindered (R^1 = adamantyl) and unsaturated amide moieties.

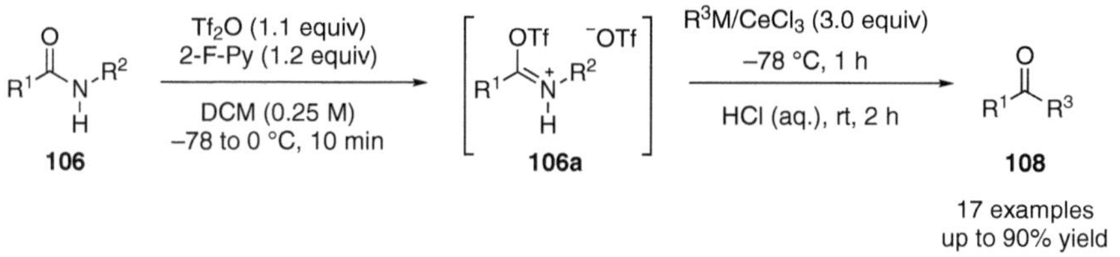

R^1 = Ph, 4-OMeC$_6$H$_4$, 4-*N*(Me)$_2$C$_6$H$_4$, 2-thienyl, cinnamyl, Me, Cy, *n*-C$_{11}$H$_{23}$, adamantyl

R^2M = *n*-BuMgBr, *i*PrMgBr, *c*-PentMgBr, PhMgBr, allylMgBr, EtMgBr, BnMgBr
n-BuLi, Ph—≡—Li

R^3 = *n*-Bu, Cy, Ph

Scheme 4.37 Huang's addition of organocerium reagents to activated secondary amides.

The direct transformation of activated amides into ketones by addition of organometallic reagents has been further developed in the following years. In 2015, Huang and coworkers disclosed a general method for the conversion of tertiary amides into ketones by addition of a variety of Grignard reagents (Scheme 4.38) [159]. Activation of tertiary amides (**102**) with Tf_2O in the presence of 2,6-di-*tert*-butyl-4-methylpyridine (DTBMP), followed by treatment of the intermediate iminium triflate (**102a**) with a variety of organomagnesiums (including Knochel's Grignard reagents), allows the synthesis of several functionalized aryl ketones inaccessible by classic Friedel–Crafts acylation methods, with remarkable yields up to 85%.

More recently, Wang and Hu have described the first general synthesis of α-alkyl ynones based on the strategy of electrophilic activation of amides (Scheme 4.39) [160]. The protocol involves the nucleophilic addition of air-stable 1-copper(I) alkynes (**112**) to preactivated morpholine amides (**111**), allowing the synthesis of a variety of ynones (**113**) without the formation of overaddition products with remarkable yields up to 87%. The reaction proceeds at high temperature (100 °C in DCE) through a highly reactive keteniminium triflate (**111b**), which afford the desired ketone (**112**) after acidic hydrolysis of the amino-enyne (**111c**).

In 2018, Gao and coworkers have elegantly shown that activated sterically hindered tertiary amides can be efficiently converted into aldehydes using a

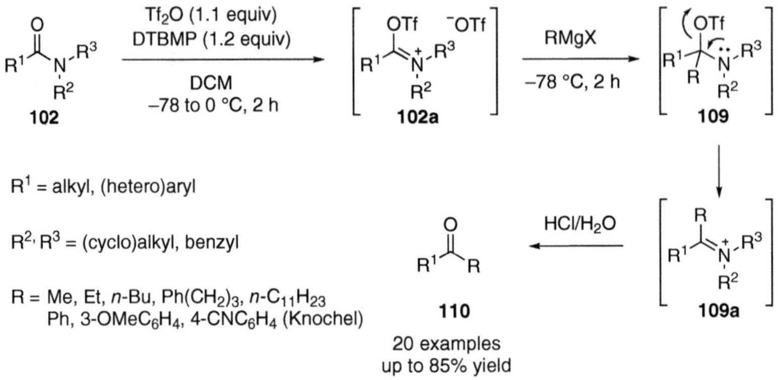

Scheme 4.38 Addition of Grignard reagents to activated tertiary amides.

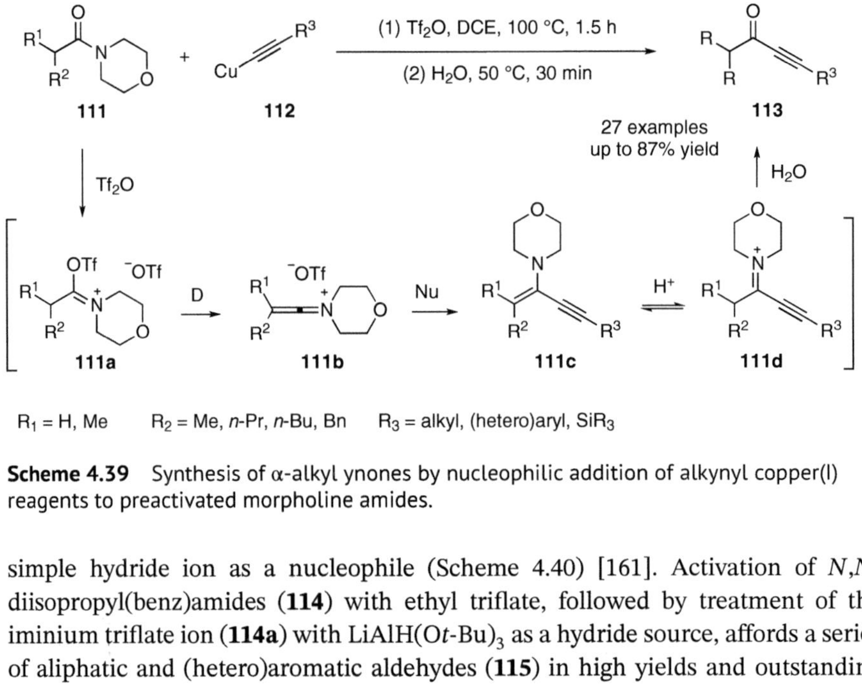

Scheme 4.39 Synthesis of α-alkyl ynones by nucleophilic addition of alkynyl copper(I) reagents to preactivated morpholine amides.

simple hydride ion as a nucleophile (Scheme 4.40) [161]. Activation of *N,N*-diisopropyl(benz)amides (**114**) with ethyl triflate, followed by treatment of the iminium triflate ion (**114a**) with $LiAlH(Ot\text{-}Bu)_3$ as a hydride source, affords a series of aliphatic and (hetero)aromatic aldehydes (**115**) in high yields and outstanding levels of chemoselectivity.

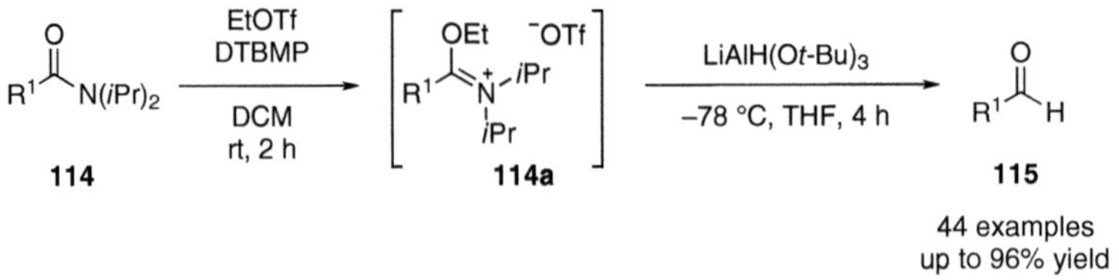

Scheme 4.40 Synthesis of aldehydes by nucleophilic addition of hydrides to preactivated sterically hindered amides.

The high electrophilic behavior of the intermediate iminium triflates (and of the corresponding nitrilium ions) generated from secondary amides has been largely exploited in recent years for the synthesis of ketones using a variety of non-metallic nucleophiles such as arenes, olefins, enamines, and amides. In 2016, the direct transformation of common secondary amides (**106**) into aromatic ketimines (**117**) and aromatic ketones (**118**) has been described by Huang and coworkers using arenes as mild nucleophiles for the C–C bond formation (Scheme 4.41) [162]. The protocol entails the preliminary activation of amides (**106**) with triflic anhydride *in situ*, followed by the intermolecular trapping of the resulting highly electrophilic nitrilium intermediate (**116**) with an arene. The reaction proceeds smoothly with several alkyl and (hetero)aryl amides showing high chemoselectivity and tolerability toward sensitive functional groups such as carbonyls, nitriles, nitro, and tertiary amido groups. However, only electron-rich and moderately electron-rich arenes can be used as nucleophiles.

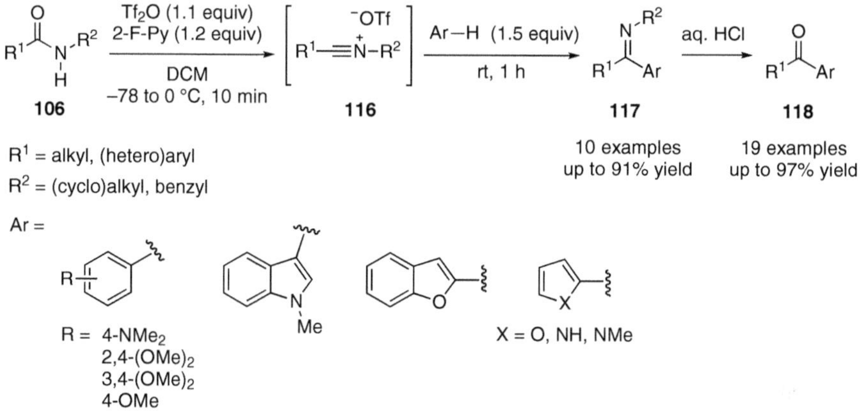

Scheme 4.41 Arylation of activated secondary amides.

Several efforts have been devoted to the intermolecular trapping of preactivated secondary amides using olefins as nucleophiles. In 2016, Huang and coworkers have reported the synthesis of α,β-unsaturated ketimines **120** and their corresponding enones **121** by nucleophilic addition of trisubstituted alkenes to nitrilium intermediates **119a** (Scheme 4.42a) [163]. Remarkably, evidence of the intermediate nitrilium ion **119a** has been assessed by ^{1}H and ^{13}C NMR and *in situ* IR monitoring of a model reaction. The reaction proceeds under mild reaction conditions, showing excellent chemoselectivity, and allows the metal- and protective group-free preparation of a variety of multi-functionalized compounds. The synthetic utility of this method was further demonstrated by the synthesis of okanin (Scheme 4.42b), a natural product used in traditional Chinese medicine, and by the preparation of an intermediate in the concise synthesis of a series of styryllactone natural products (Scheme 4.42c) [164].

An intramolecular approach has been reported by the same authors in 2019 for the efficient assembly of fused enimines and enones starting from ω-unsaturated

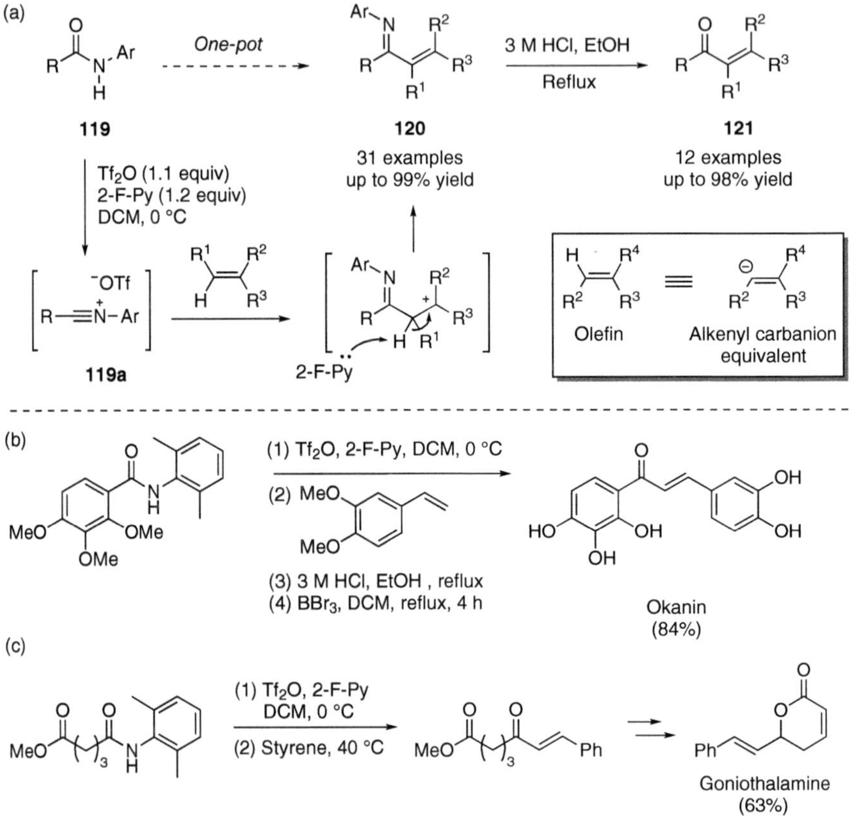

Scheme 4.42 (a) Synthesis of α,β-unsaturated ketimines and enones by nucleophilic addition of alkenes to activated secondary amides. (b) Application of the methodology to the synthesis of okanine and (c) (±)-goniothalamine.

amides (Scheme 4.43) [165]. Activation of the starting amides **122** with Tf_2O in the presence of 2-fluoropyridine, followed by intramolecular trapping of the resulting nitrilium ion **122a**, leads to the reductive alkenylation products **123**, which can be further converted into the corresponding enones **124** by mild acidic hydrolysis.

Olefins can be used as surrogates of highly reactive alkyl metallic reagents (as alkyl carbanion equivalents) for the direct transformation of unreactive secondary amides into saturated ketones **127**. Two independent investigations by Huang and Geng [166] and Maulide and coworkers [167] in 2019 have disclosed that secondary amides **125** can be exploited as powerful hydroacylating reagents of alkenes **126** upon activation with triflic anhydride. In particular, Huang and coworkers have developed a straightforward and sustainable method running at room temperature, which tolerates several sensitive functional groups on either alkenes or amides (Scheme 4.44), with a broad substrate scope and remarkable yields up to 90% [166].

Similar results have been reported by Maulide and coworkers (Scheme 4.45a) [167], and the best results have been achieved with *N*-allylated secondary

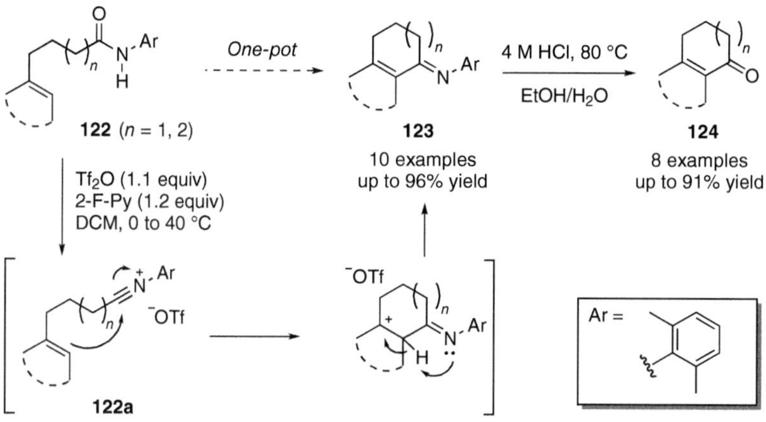

Scheme 4.43 Intramolecular alkenylation of activated ω-unsaturated amides.

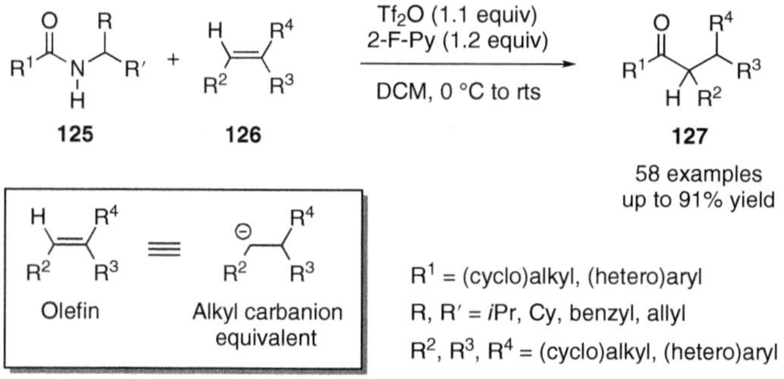

Scheme 4.44 Huang's hydroacylation of alkenes.

amides **128**. The protocol entails the preactivation of the starting amide **128** with triflic anhydride and 2-fluoropyridine and the subsequent intermolecular trapping of the nitrilium ion **128a** with a variety of terminal olefins **130**, which deliver the corresponding ketones **131** with high levels of chemo- and regioselectivity. The reaction mechanism has been elucidated by DFT calculations and isotope labeling experiments (Scheme 4.45b). Starting from a deuterated amide, selective D incorporation has been observed at the β carbon of the resulting ketone, while the use of a deuterated terminal olefin led to the selective deuterium incorporation in the α-position. These results suggests an intramolecular 1,5-hydride transfer occurring on intermediate **129a**, leading to the formation of an azoniaallene intermediate **129b** (Scheme 4.45a), analogously as reported by Huang and Geng in their preliminary mechanistic investigations [166]. Finally, quenching the reaction with $H_2{}^{18}O$ revealed full incorporation of ^{18}O in the ketone product, as a result of the hydrolysis process of intermediate **129b** (Scheme 4.45a).

Enamines represent another class of mild *C*-nucleophiles, which has been exploited for the conversion of activated secondary amides to ketones. In this context, Huang and coworkers have clearly demonstrated in 2019 that enamines derived from both aldehydes and ketones can efficiently act as powerful

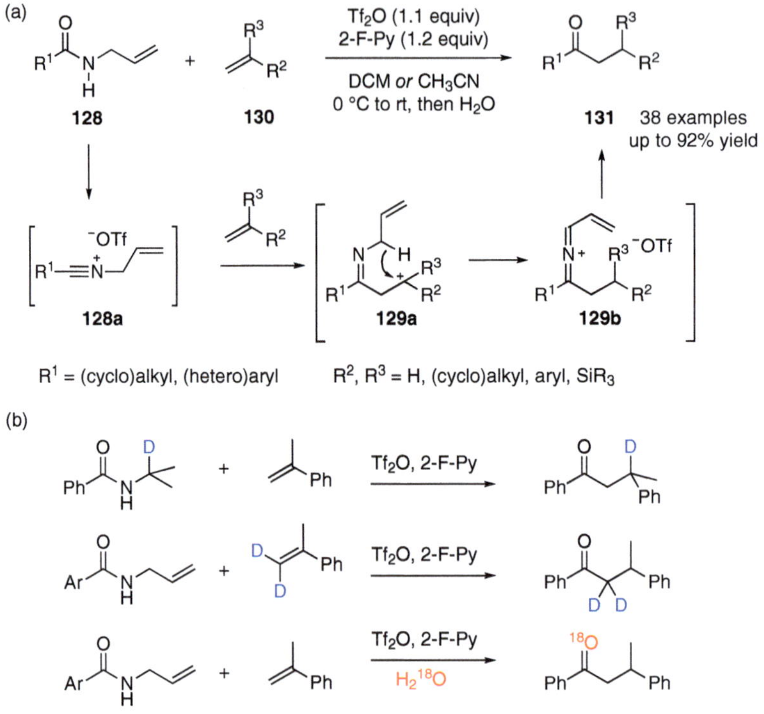

Scheme 4.45 (a) Maulide's hydroacylation of terminal olefins and (b) mechanistic studies by isotope labeling experiments.

secondary alkyl [168] or enolate anion [169] equivalents (Scheme 4.46). The nitrilium ion **132a** generated from the activation of secondary amides **132** with triflic anhydride can be smoothly trapped by enamines derived from ketones (Scheme 4.46, pathway a), and mild hydrolysis of the resulting intermediate **133a** led to the formation of the expected β-enaminone products **134** in high yields [169]. On the other hand, enamines derived from aldehydes behave as surrogates of secondary alkyl carbanions, and their intermolecular trapping of the nitrilium **132a** afforded the α,α-disubstituted ketones **135** after acidic hydrolysis and decarboxylation of the intermediate β-iminoiminium salt **133b** (Scheme 4.46, pathway b) [168].

Tertiary amides have also been exploited as pro-nucleophiles for the assembly of α,α-disubstituted ketones **138**. Huang and coworker have recently reported an elegant cross-coupling of Tf_2O-activated secondary amides with tertiary amides, giving access to ketones using tertiary amides as surrogates of alkyl carbanions [170]. The developed protocol entails the coupling of enamines **137a**, generated from tertiary amides **137** by catalytic partial reduction with Vaska's catalyst in the presence of poly(methylhydrosiloxane) (PMHS), with nitrilium ions **136a** followed by subsequent deformylation under acidic conditions (Scheme 4.47). The reaction showed excellent chemoselectivity and functional group tolerance both on the secondary amide and the tertiary amide coupling partner.

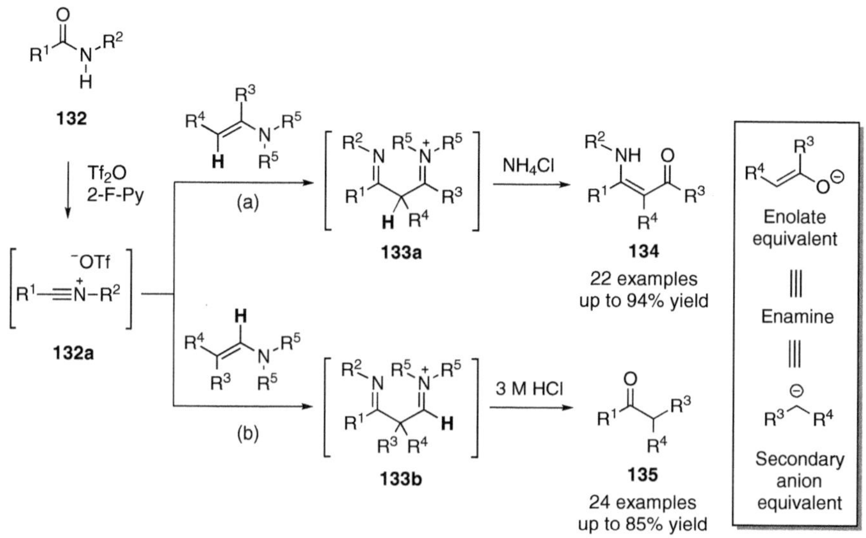

Scheme 4.46 Reactivity of enamines with activated secondary amides.

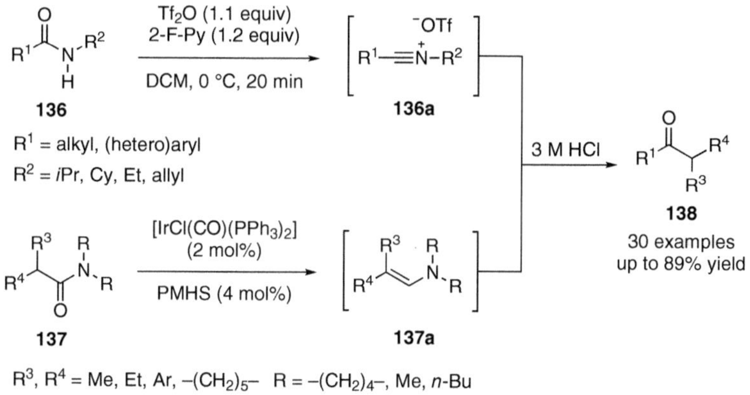

Scheme 4.47 Cross-coupling of tertiary amides with activated secondary amides.

4.4.3 Synthesis of Amine-Like Compounds

Imidoyl triflates show a high tendency to undergo nucleophilic attacks carried out with a plethora of reagents. Charette documented the direct transformation of tertiary amides **139** into the corresponding amines **140** via the treatment with the Hantzsch ester, notably proceeding with high chemocontrol in the presence of sensitive groups (nitriles, ketones, esters *inter alia*), as detailed in Scheme 4.48 [171].

In 2015, the direct reduction of secondary amides **141**, preactivated with Tf_2O in the presence of 2-fluoropyridine as base, has been reported by Huang and coworkers [172]. The nitrilium intermediate **141a** can be smoothly reduced by $NaBH_4$ in a single synthetic step, giving access to a variety of secondary amines

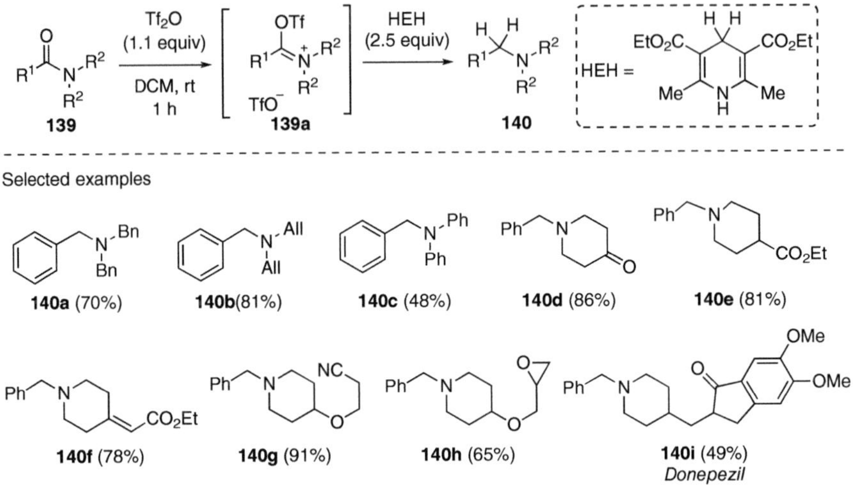

Scheme 4.48 Charette's synthesis of amines mediated by Hantzsch's ester.

142 (Scheme 4.49, top). The reaction works well with several types of secondary amides including aromatic, aliphatic, and α,β-unsaturated amides, showing a good functional group tolerance for a series of potentially reducible functional groups. Complementary to the Huang's protocol is the chemoselective reduction of secondary amides developed by Charette using the weaker hydride donor Et_3SiH: upon fine tuning of the work-up procedure, it was possible to isolate both the imine ($NaHCO_3$) and the aldehyde (citric acid – THF) [173]. Again, decorating substrates with interfering motifs was allowed and the chemoselective reduction of a secondary amide in the presence of a tertiary congener was feasible. Moreover, the intermediate imine could be fully reduced to the amine upon sequential treatment with Et_3SiH.

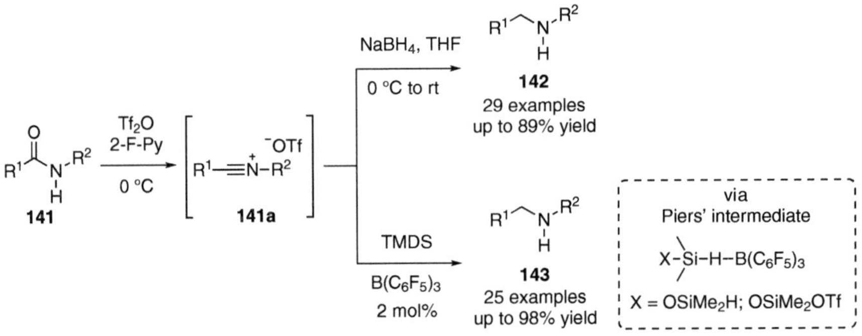

Scheme 4.49 Direct reduction of activated secondary amides.

Huang further implemented the methodology for the chemoselective one-pot reduction of secondary amides to amines [174]: upon the preactivation with Tf_2O, the generated nitrilium triflate intermediate **141a** could be reduced with

$NaBH_4$ or Cp_2ZrHCl-$NaBH_4$ for lactams [172, 175]. Alternatively, reduction can be accomplished by the $B(C_6F_5)_3$-catalyzed hydrosilylation of the intermediate with 1,1,3,3-TMDS (Scheme 4.49, bottom) [174]. The reaction proceeds via the Piers' intermediate [176] and tolerates a variety of functional groups, giving access to a broad set of functionalized secondary amides **143** in very good yields. Moreover, the method has been exploited for a concise synthesis of the alkaloid (−)-*epi*-pseudoconhydrine·HCl in four steps with an overall yield of 43%.

Extensive studies were conducted for designing methodologies enabling the one-pot transformations of secondary and tertiary amides into amines with concomitant C–C bond formation. In 2015, Huang et al. reported a straightforward protocol for the reductive alkylation of secondary amides (Scheme 4.50) [177]. This one-pot method entails the triflic anhydride-mediated activation of amide **144**, followed by the partial reduction of the nitrilium triflate intermediate **144a** with Et_3SiH and subsequent nucleophilic addition of several organometallic reagents (both hard and soft nucleophiles) to the activated intermediate imine **144c**. The reaction proceeds smoothly with high chemoselectivity for a wide set of substrates and nucleophiles, allowing the preparation of a bounce of alkylated secondary amides **145** in very good yields.

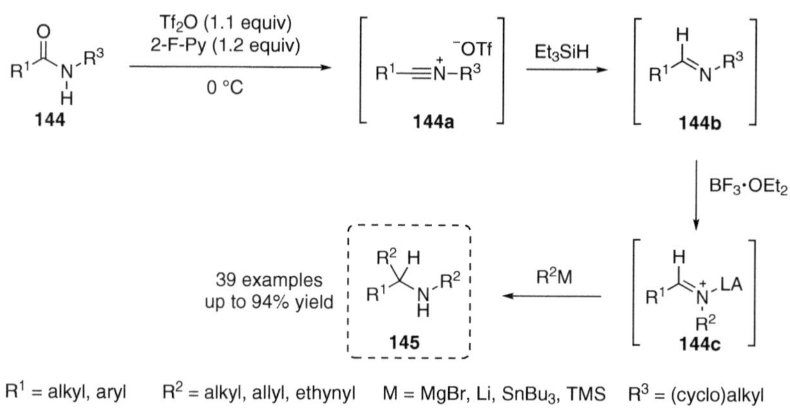

Scheme 4.50 Reductive alkylation of activated secondary amides.

A slightly modified procedure has been reported by the same research group few years later for the direct reductive dialkynylation of amides and lactams [178]. This one-pot method consists of *in situ* activation of amides **146** with Tf_2O in the presence of 2,6-di-*tert*-butyl-4-methylpyridine (DTBMP, which could be recovered during chromatographic isolation of products), followed by double addition of alkynyllithium reagents to the iminium triflate **146a** giving access to *gem*-alkynylation products **147** bearing an aza-quaternary carbon center in good yields (Scheme 4.51). The method is extremely wide in scope and easy to scale up to gram-scale preparations.

Secondary amides can be efficiently subjected to reductive alkenylation using enamines as surrogates of alkene carbanions [179]. In the method described by Huang and coworkers in 2018, secondary amides **148** are smoothly converted to

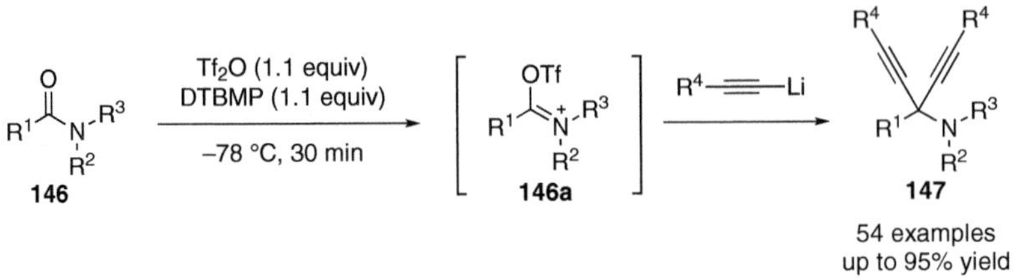

Scheme 4.51 Reductive bis-alkynylation of activated tertiary amides.

functionalized allylamines **152** using triflic anhydride/2-fluoropyridine as amide activator and enamines as unconventional alkenylation reagents (Scheme 4.52). In the presence of $NaBH_4$ as reducing agent and NaOH as base, the amino-borane intermediate **150**, arising from the reduction of iminium ion **149**, undergoes a Hoffmann-type elimination reaction to yield an α,β-unsaturated imine **151**, which is further reduced by $NaBH_4$ to give the final allylamine products **152** in high yields.

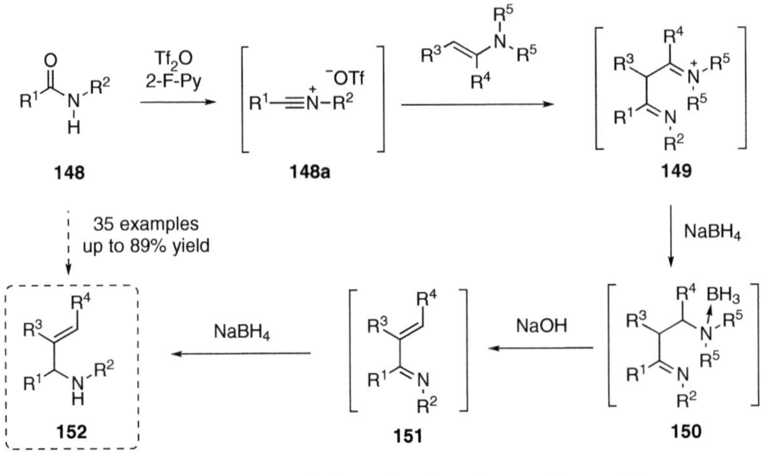

Scheme 4.52 Reductive alkenylation of activated secondary amides.

The logic of activating secondary amides has been extended by Huang for the first chemoselective direct reductive trifluoromethylation of amides (Scheme 4.53a) [180]. Thus, the nitrilium triflate **153a** is partially reduced by TMDS, and the resulting iminium ion **153b** is trifluomethylated in the presence of $TMSCF_3$ and a fluoride source (KHF_2). The α-CF_3-amines **154** are obtained in good yields starting from aromatic and aliphatic secondary amides, with good functional group tolerance.

In a similar manner, Anbarasan and coworker have described the reductive trifluoromethylation of Weinreb amides in the presence of indoles as nucleophiles (Scheme 4.53b) [181]. The reaction entails the trifluoromethylation of the starting Weinreb amides **155** in the presence of $TMSCF_3$ and CsF, followed by activation of the trifluomethylated hemiaminal **156** with triflic anhydride and nucleophilic trapping of the resulting iminium ion **156a** with substituted indoles. The intermediate

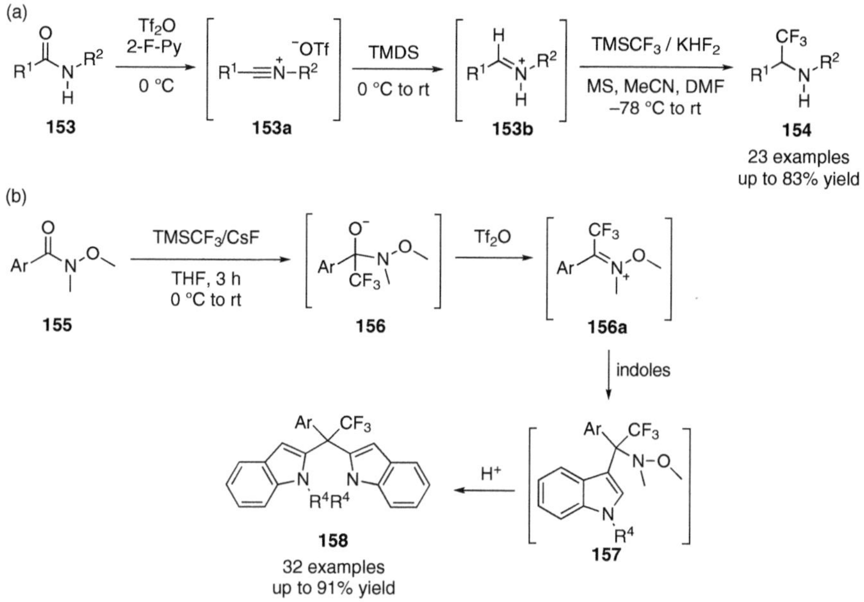

Scheme 4.53 Reductive trifluoromethylation of activated secondary amides.

α-trifluoromethylated amines **157** undergo rapid hydrolysis, giving access to a series of CF_3-functionalized bisindolylmethanes **158** in high yields.

The chemoselective reductive bis-phosphonylation of secondary and tertiary amides has also been described [182]. The nitrilium **159a** or iminium **159b** triflates, arising from the activation of secondary and tertiary amides **159**, respectively, can be efficiently trapped by diethyl phosphite, allowing the chemoselective synthesis of α-amino bisphosphonates **160**, an important class of biologically active compounds (Scheme 4.54, top). Moreover, activated secondary amides can be subjected to a subsequent geminal cyanation/phoshonylation sequence using TMSCN and dialkyl(aryl) phosphites, leading to α-amino-α-cyanophosphonic acid derivatives **161** in good overall yields (Scheme 4.54, bottom) [183].

An elegant transformation of activated secondary amides has been illustrated by Huang in 2016 with the reductive 1,3-dipolar cycloaddition of secondary aromatic *N*-(trimethylsilylmethyl)amides **162** with reactive dipolarophiles (Scheme 4.55) via the generation of non-stabilized NH azomethine ylides [184]. Amide activation with triflic anhydride, followed by partial reduction of the nitrilium triflate **162a** with TMDS, affords the intermediate iminium ion **162b**, and subsequent desilylation with CsF generates the ylides **163**, which is trapped *in situ* by a cycloaddition reaction in the presence of several dipolarophiles, leading to derivatives **164** in good yields, however with poor diastereoselectivities.

A conceptually analogous cycloaddition protocol for the assembly of structurally complex scaffolds based on the reductive functionalization of tertiary amides has been developed by De Mesmaeker and coworkers in the same year [185]. *N*-Allylamides **165** are efficiently activated by Tf_2O, and the intermediate keteniminium ion **165a** reacts spontaneously with an alkene to form the

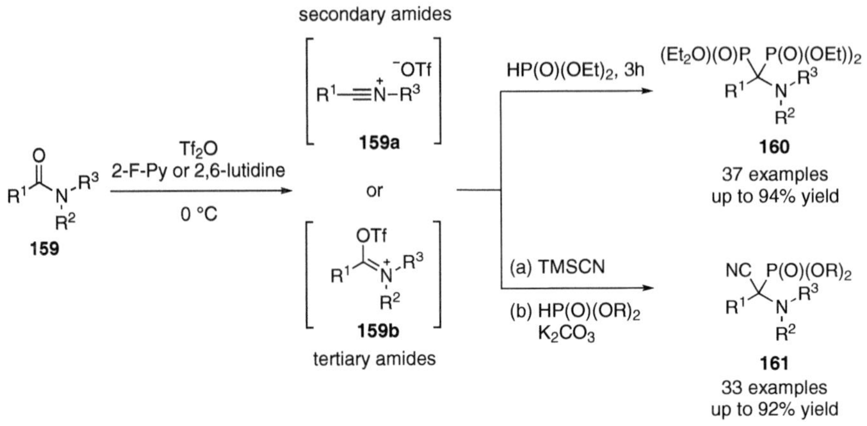

Scheme 4.54 Reductive bis-phosphonylation (top) and geminal cyanation/phoshonylation (bottom) of activated secondary and tertiary amides.

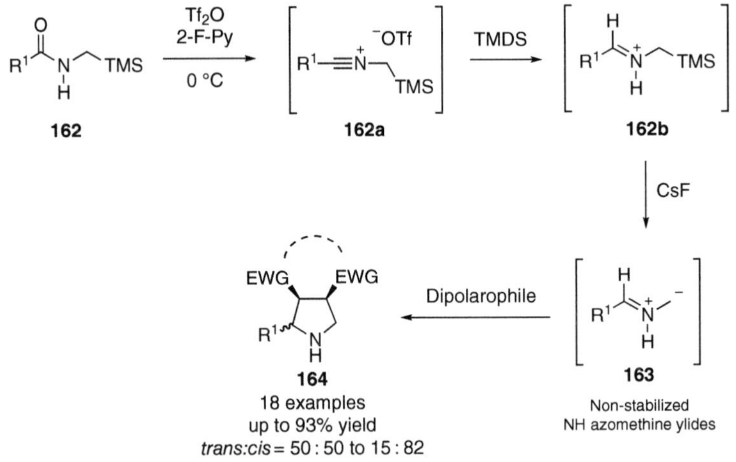

Scheme 4.55 Reductive 1,3-dipolar cycloaddition of secondary aromatic amides.

cyclobutaniminium salt **166**. Addition of several organometallic reagents (M = Li, Mg, Zn, Ce, and Si) to the resulting intermediate affords a series of functionalized cyclobutyl amines **167** with high levels of stereoselectivity (Scheme 4.56).

In 2019, the first enantioselective direct reductive bis-functionalization of secondary amides has been described (Scheme 4.57) [186]. Secondary *N*-arylamides **168**, activated with triflic anhydride, are converted into the 3-iminoindoles **169** after trapping of the nitrilium ion with isocyanides, and sequential organocatalytic enantioselective Mannich-type addition of a ketone in the presence of (L)-proline generates the corresponding 2,2-disubstituted 3-iminoindolines **170** with high yields and outstanding enantioselectivities.

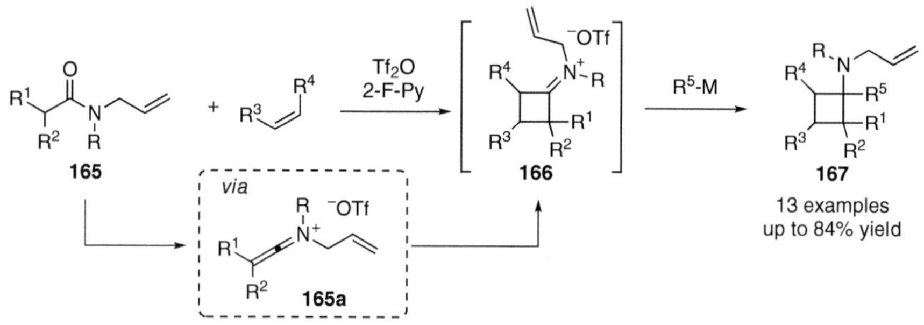

Scheme 4.56 Reductive [2+2] cycloaddition and nucleophilic functionalization of activated amides.

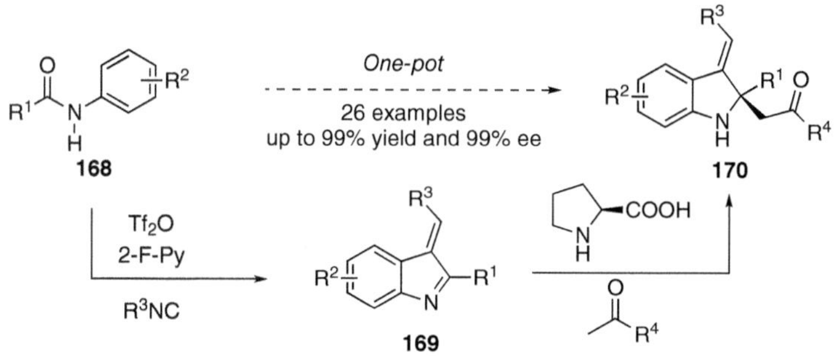

Scheme 4.57 Organocatalytic enantioselective double functionalization of secondary amides.

Other powerful direct transformations of activated amides have been developed for the synthesis of *N*-containing compounds. As an example, in 2015, Huang, Zheng, and coworkers have described for the first time an efficient reductive Ugi-type three-component reaction using common secondary amides as starting materials (Scheme 4.58) [187]. Activation of secondary amides **171** with triflic

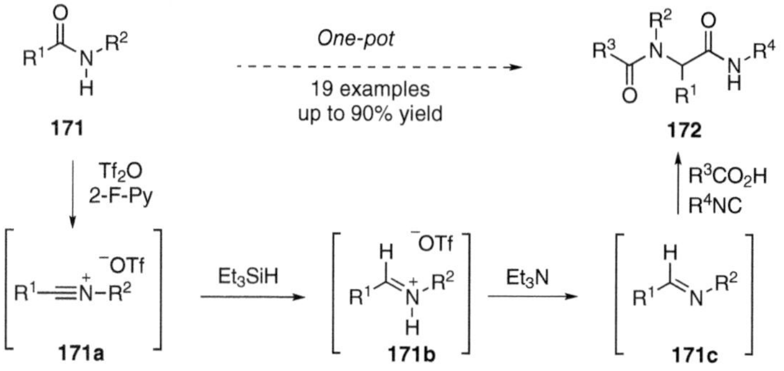

Scheme 4.58 Reductive Ugi-type reaction of activated secondary amides for the synthesis of α-acetamidoamides.

anhydride/2-F-pyridine, followed by partial reduction of the nitrilium triflate **171a** with Et_3SiH/NEt_3, generates the intermediate imine **171b**, which afford a series of functionalized α-acetamidoamides **172** upon treatment with substituted carboxylic acids and isocyanides in high yields. The reaction proceeds with high chemoselectivity and also tolerates the presence of carbonyls, usually incompatible with Ugi-type reactions.

In 2017, Huang et al. reported the transformation of lactams **173** into bicyclic vinylogous lactams **174** and **175** (Scheme 4.59) [188]. This one-pot approach relies on the *in situ* formation of terminal (**A**) and/or internal (**B**) silyl enol ethers with *tert*-butyldimethylsilyl triflate (TBDMSOTf), formation of the iminium triflate ion under the classical activation conditions, followed by a tandem cyclocondensation reaction.

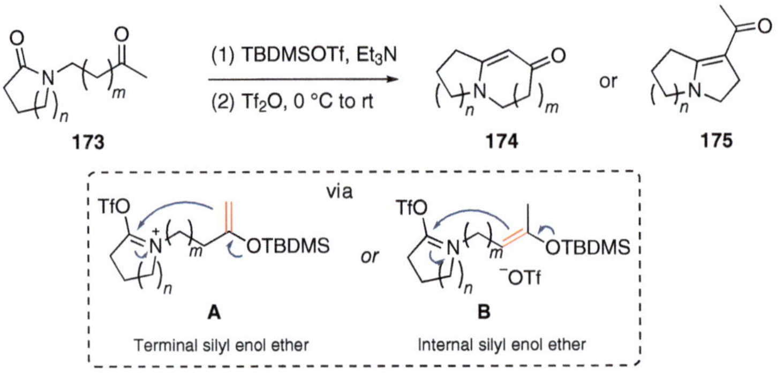

Scheme 4.59 Synthesis of vinylogous lactams from activated lactams.

4.4.4 Activation of Amides with Different Electrophilic Agents

In another approach, reduction of amides to amines has been realized via *O*-alkylation and catalytic hydrogenation in a one-pot procedure. Indeed, Kadyrov reported that amides **176** were converted into amide acetals **177** and imidates **178** by using several alkylating agents (triethyloxonium tertafluoroborate, methyl trifluoromethanesulfonate, dimethyl sulfate, and ethyl chloroformate) [189]. Hydrogenolysis of the amide acetals and imidates subsequently proceeded over standard hydrogenation catalysts (Pd and Pt on charcoal). A variety of secondary and tertiary amides were reduced to the corresponding amines **179** in good yields, as well as macrocyclic peptidomimetics and dipeptides without affecting the neighboring chiral center (Scheme 4.60).

Recently, transamidation of amides with amines has gained a renewed attention. In 2019, Song/Lee disclosed a metal-free transamidation of primary amides **180** with primary and secondary amines using TMSCl as an activator (Scheme 4.61) [190]. As a result, secondary (**181**) and tertiary (**182**) amides were synthesized in moderate to good yields. The precise role of the TMSCl in this transformation has not been clarified yet.

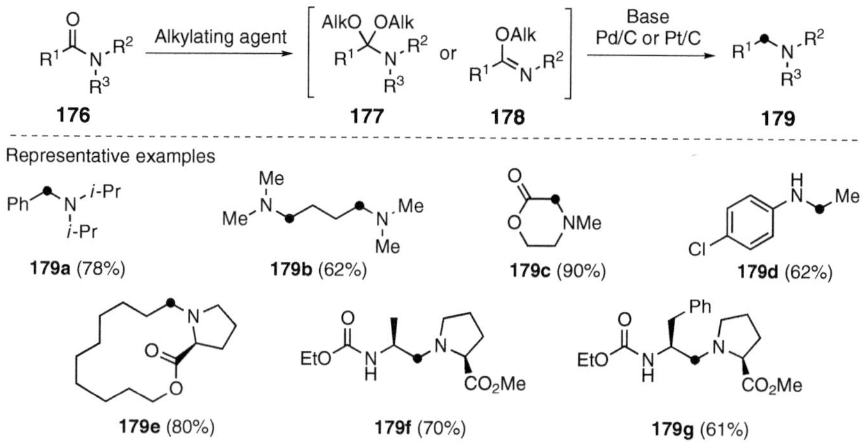

Scheme 4.60 Reduction of amides to amines via *O*-alkylation and catalytic hydrogenation.

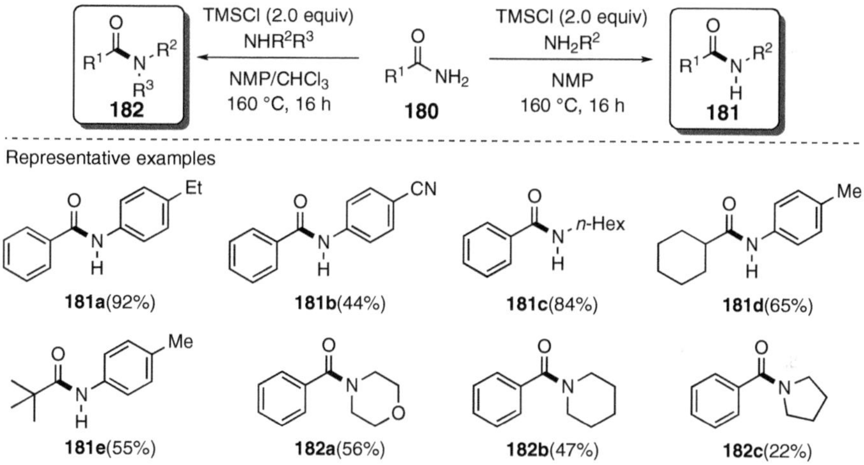

Scheme 4.61 Metal-free transamidation of primary amides with primary and secondary amines. NMP, *N*-Methyl-2-pyrrolidone.

4.5 Synthesis of Heterocycles

The chemistry of activated amides has played an important role in the panorama of synthetic methods suited for building *N*-containing heterocyclic frameworks [145]. Most of these protocols rely on the triflic anhydride-mediated electrophilic activation of secondary and tertiary amides, followed by nucleophilic intermolecular trapping of the intermediate iminium triflates/nitrilium ions and subsequent cascade cyclization reactions.

In this context, Charette and coworkers have reported in 2015 a general approach for the one-pot synthesis of 3,4,5-trisubstituted 1,2,4-triazoles starting from secondary amides **183** and hydrazides, followed by microwave-assisted cyclodehydration (Scheme 4.62) [191]. Hydrazides act as a nucleophilic trap of the iminium triflate **183a**, generated upon activation of amides **183** with Tf_2O in the presence of 2-fluoropyridine, and microwave irradiation promoted the cyclodehydration of intermediate **184**, leading to a variety of alkyl/aryl 3,4,5-trisubstituted 1,2,4-triazoles **185** in high yields.

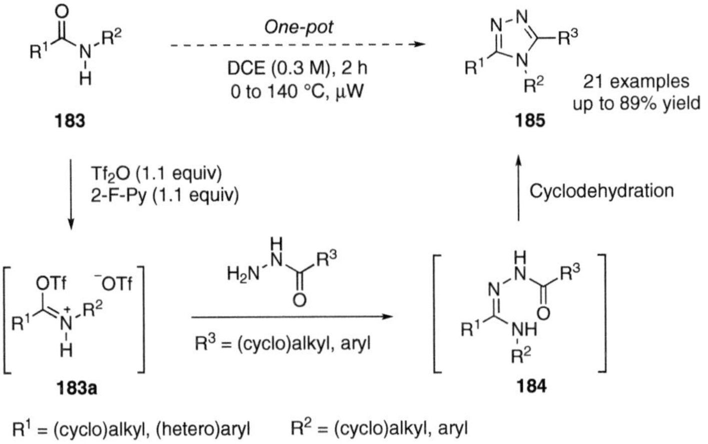

Scheme 4.62 Charette's synthesis of trisubstituted 1,2,4-triazoles.

In 2016, Mosey and coworkers have reported the synthesis of 1,2-disubstituted 2-imidazolines via a tandem reaction of activated *N*-(2-haloethyl) secondary amides **186** in the presence of primary amines (Scheme 4.63) [192]. Under electrophilic activation conditions, the iminium triflate **186a** can be efficiently trapped by primary alkyl amines and anilines, affording diverse 1,2-disubstituted 2-imidazolines **188** in high yields upon a 5-*exo*-tet cyclization of the intermediate amidine **187**.

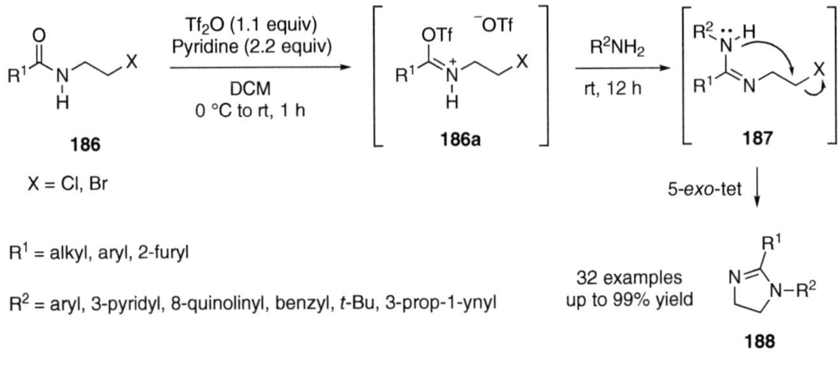

Scheme 4.63 Mosey's synthesis of 1,2-disubstituted 2-imidazolines. DCM, dichloromethane.

Huang and coworkers have described in 2017 an alkenylative cyclization methods for the synthesis of saturated azacycles starting from halogenated secondary amides **189** and olefins (Scheme 4.64) [193]. Triflic anhydride promoted amide activation and induced dehydrative coupling of the nitrilium ion **189a** with alkenes. *In situ* reduction of the olefinated intermediate **190** by $NaBH_4$ promotes a tandem cyclization reaction, giving access to α-vinylic substituted pyrrolidines ($n = 1$) and piperidines ($n = 2$) **191** in high yields (up to 98%). Moreover, the procedure led to indolizidine and quinolizidine ring systems starting from *N*-alkenyl secondary amides in remarkable yields. This protocol was successfully applied to the one-pot synthesis of the cytotoxic natural product caulophyllumine B and to the synthesis of δ-coniceine.

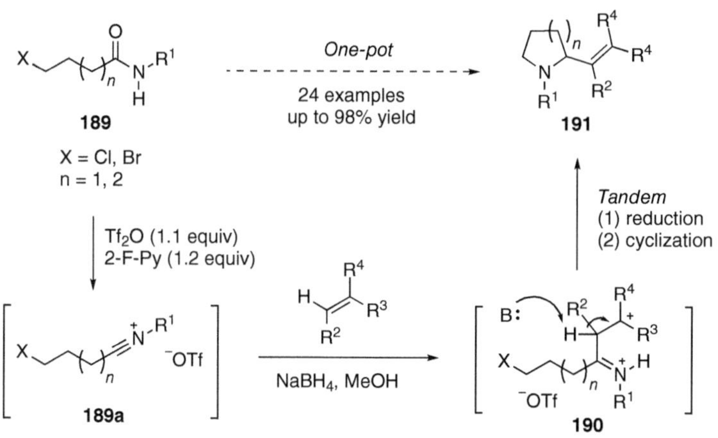

Scheme 4.64 Huang's synthesis of saturated azacycles.

The same authors have reported in 2018 the directed alkynylation of activated secondary *N*-aryl amides **192** as an efficient method for the synthesis of substituted quinolines (Scheme 4.65) [194]. Nucleophilic trapping of the nitrilium ion **192a** (observed by heteronuclear 2D NMR investigations) under electrophilic activation conditions with terminal alkynes, followed by an intramolecular Friedel–Crafts

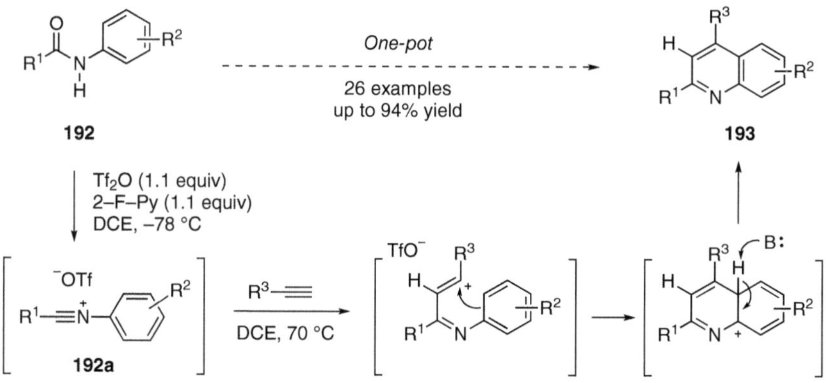

Scheme 4.65 Huang's synthesis of substituted quinolines.

process, leads to the formation of a variety of substituted quinolines **193** in high yields. This method tolerates several EWG and EDG substituents both on the benzamide and on the alkyne moieties.

More recently, Liu and coworkers have disclosed a straightforward protocol for the functionalization of benzoxazoles with activated *N*-acyl pyrrolidines **194** (Scheme 4.66) [195]. Several amides, preactivated with triflic anhydride in the presence of LiCl and Cs_2CO_3 as bases, have been efficiently converted into 2-alkylated/arylated benzoxazole skeletons **195** under mild conditions. This protocol offers a metal-free alternative to the common procedures for the functionalization of benzoxazoles, tolerates a variety of functional groups, and it is insensitive to moisture. The reaction mechanism proposed by the authors has been elucidated by ^{13}C-labeling NMR experiments tracking the ^{13}C signal of a labeled starting material. Briefly, the nitrilium ion **194a** is trapped by benzoxazole and hydrolyzed to afford the *N*-acylated benzoxazole intermediate **194b**. The latter undergoes a ring-opening/ring-closure pathway and regenerates the target compound **195a** after base-promoted dehydration of intermediate **194c**.

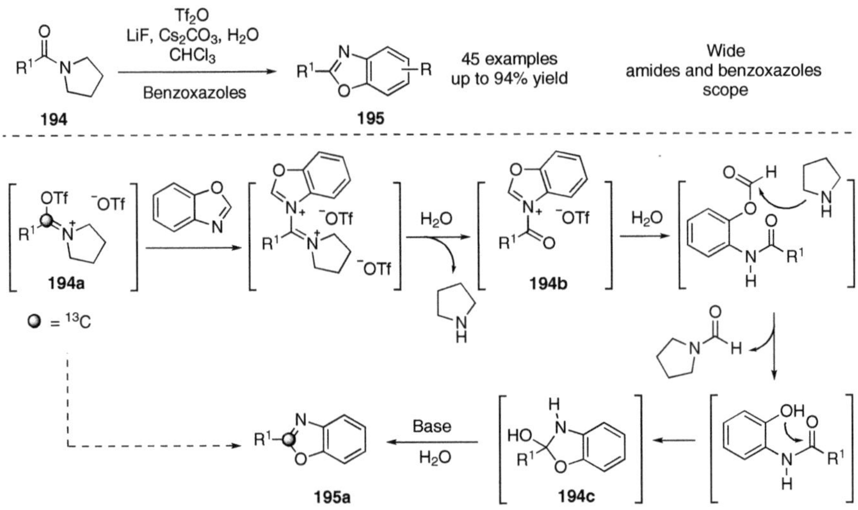

Scheme 4.66 Liu and Liang's functionalization of benzoxazoles with activated amides.

Maulide and coworkers have recently described the preparation of a novel class of seven-membered heterocyclic compounds starting from α-phthalimido-amides **196**, which are readily prepared from amino acids in an easy two-steps procedure (Scheme 4.67) [196]. The use of classical amide activation conditions in acetonitrile leads to the unexpected incorporation of the solvent into a fused seven-membered ring system. The mechanism involves (i) the formation of the iminium triflate **196a**, followed by (ii) nucleophilic trapping by one of the phthalimido carbonyl oxygens leading to the oxazolinium intermediate **196b**, and (iii) addition of the acetonitrile leading to the intermediate ketenimine **197**. This species undergoes a rapid ring enlargement to yield the final product **198** after hydrolysis of the iminium **197a**.

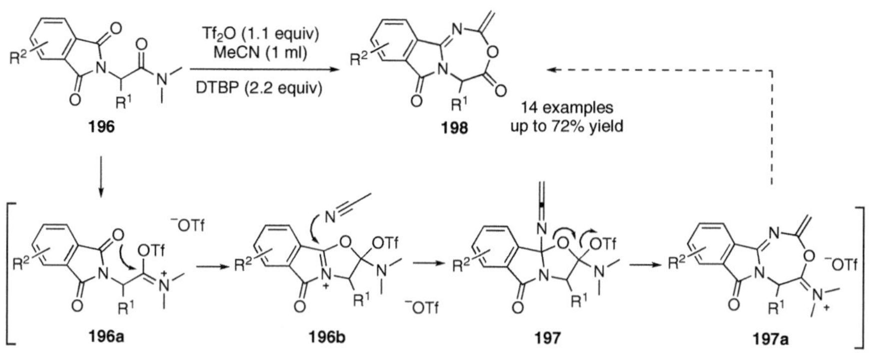

Scheme 4.67 Maulide's synthesis of new fused seven-membered heterocycles. DTBP, 2,6-*ditert*-butyl-pyridine.

A different method for the synthesis of heterocyclic rings starting from amides or lactams has been described by Huang and coworkers in 2017 (Scheme 4.68) [197]. This approach relies on the reductive activation of C–N bonds mediated by zirconocene Schwartz' reagent (Cp_2ZrHCl). Exploiting the reactivity of isocyanoacetates **204** as nucleophiles, the methodology developed by Huang efficiently converted activated amides or lactams to 5-methoxyoxazoles **202** or bicyclic

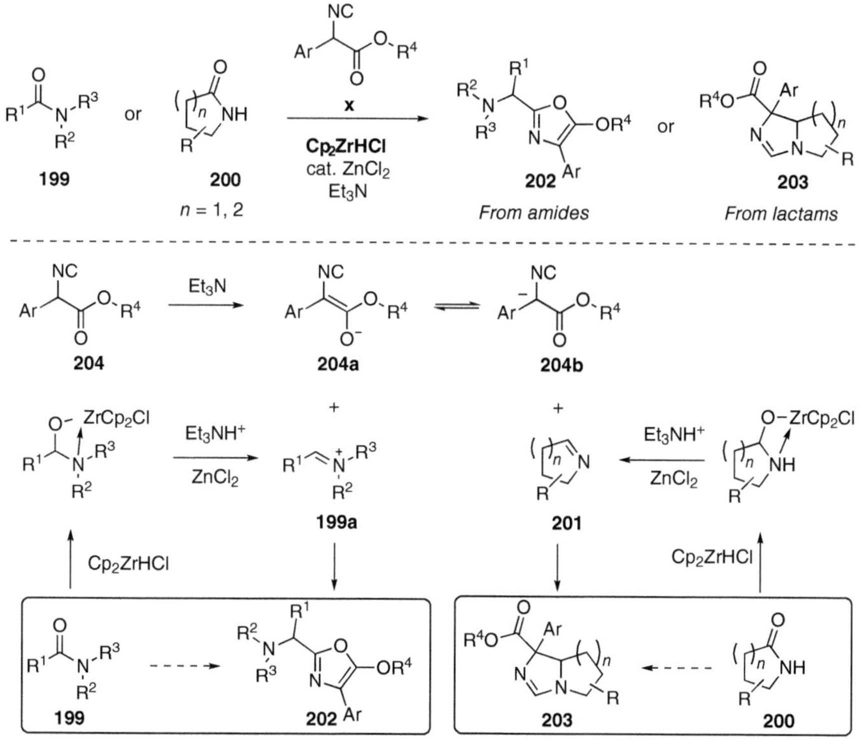

Scheme 4.68 Huang's synthesis of heterocycles by reductive activation of amides and lactams.

imidazolines **203**, respectively. Iminium ion **199a** or cyclic imine **201**, arising from the reductive activation of amides **199** or lactams **200**, are efficiently trapped by deprotonated aryl isocyanoacetates **204a** or **204b**, leading to the corresponding heterocycles in high yields.

4.6 Conclusions and Outlook

In summary, an overview of the recent transition-metal-free synthetic approaches for the transformation of the usually considered inert amide linkage, involving the formation of stable tetrahedral intermediates, has been presented. The challenging nucleophilic addition of organometallic reagents for the preparation of carbonyl-containing compounds and the reductive nucleophilic functionalization of amides have experienced a significant improvement in the past decade. The proposed strategies range from the use of simple non-activated amides to the structural modification of the amide backbone, mainly at the level of the nitrogen atom. Furthermore, the development of electrophilic activation strategies of the amide linkage enabled the development of fully chemoselective transformations *en route* to ketone and amine-like compounds, opening the way to intriguing perspectives in the panorama of the synthetic methods for building *N*-containing heterocyclic frameworks. There is no doubt that the current state-of-the-art experienced tremendous advancements, which ultimately revisited the paradigm of considering the amide functionality as the less reactive among carboxylic derivatives. Indeed, the fine-tuning of the activation and reaction conditions – supported by solid and well-understood logics – would rather shift amides into truly reactive elements in organic synthesis.

References

1 Greenberg, A., Breneman, C.M., and Liebman, J.F. (ed.) (2000). *The Amide Linkage. Selected Structural Aspects in Chemistry, Biochemistry and Materials Science*. New York: Wiley.
2 Pattabiraman, V.R. and Bode, J.W. (2011). *Nature* 480: 471–479.
3 Valeur, E. and Bradley, M. (2009). *Chem. Soc. Rev.* 38: 606–631.
4 de Figueiredo, R.M., Suppo, J.-S., and Campagne, J.-M. (2016). *Chem. Rev.* 116: 12029–12122.
5 Massolo, E., Pirola, M., and Benaglia, M. (2020). *Eur. J. Org. Chem.* 2020: 4641–4651.
6 Mahesh, S., Tang, K.-C., and Raj, M. (2018). *Molecules* 23: 2615.
7 Roughley, S.D. and Jordan, A.M. (2011). *J. Med. Chem.* 54: 3451–3479.
8 Hie, L., Fine Nathel, N.F., Shah, T.K. et al. (2015). *Nature* 524: 79–83.
9 Dander, J.E. and Garg, N.K. (2017). *ACS Catal.* 7: 1413–1423.
10 Hughes, A.B. (2011). *Amino Acids, Peptides, and Proteins in Organic Chemistry*. Weinheim: Wiley.

11 Pace, V., Holzer, W., and Olofsson, B. (2014). *Adv. Synth. Catal.* 356: 3697–3736.
12 Smith, M.B. and March, J. (2001). *March's Advanced Organic Chemistry: Reactions, Mechanisms, and Structure*, 5e. Weinheim: Wiley-VCH.
13 Dieter, R.K. (1999). *Tetrahedron* 55: 4177–4236.
14 Ruider, S.A. and Maulide, N. (2015). *Angew. Chem. Int. Ed.* 54: 13856–13858.
15 Li, G., Ma, S., and Szostak, M. (2020). *Trends Chem.* 2: 914–928.
16 Kemnitz, C.R. and Loewen, M.J. (2007). *J. Am. Chem. Soc.* 129: 2521–2528.
17 Meng, G., Zhang, J., and Szostak, M. (2021). *Chem. Rev.* 121: 12746–12783.
18 Bennet, A.J., Wang, Q.P., Slebocka-Tilk, H. et al. (1990). *J. Am. Chem. Soc.* 112: 6383–6385.
19 Slebocka-Tilk, H. and Brown, R.S. (1987). *J. Org. Chem.* 52: 805–808.
20 Deslongchamps, P. (1983). *Stereoelectronic Effects in Organic Chemistry*. Elmsford, NY: Pergamon Press.
21 Pauling, L. (1940). *The Nature of the Chemical Bond*. London: Oxford University Press.
22 Seebach, D. (2011). *Angew. Chem. Int. Ed.* 50: 96–101.
23 Yanagita, Y., Nakamura, H., Shirokane, K. et al. (2013). *Chem. Eur. J.* 19: 678–684.
24 Bouveault, L. (1904). *Bull. Chem. Soc. Fr.* 31: 1306–1322.
25 Myers, A.G., Yang, B.H., Chen, H. et al. (1997). *J. Am. Chem. Soc.* 119: 6496–6511.
26 Hwang, Y.C., Chu, M., and Fowler, F.W. (1985). *J. Org. Chem.* 50: 3885–3890.
27 Sato, T. and Chida, N. (2014). *Org. Biomol. Chem.* 12: 3147–3150.
28 Evans, E.A. (1956). *J. Chem. Soc.* 4691–4692.
29 Scilly, N.F. (1973). *Synthesis* 3: 160–161.
30 Olah, G.A. and Arvanaghi, M. (1981). *Angew. Chem. Int. Ed.* 20: 878–879.
31 Olah, G.A., Prakash, G.K.S., and Arvanaghi, M. (1984). *Synthesis* 3: 228–230.
32 Carreira, E.M. and Du Bois, J. (1995). *J. Am. Chem. Soc.* 117: 8106–8125.
33 Ding, S. and Jiao, N. (2012). *Angew. Chem. Int. Ed.* 51: 9226–9237.
34 Comins, D. and Meyers, A.I. (1978). *Synthesis* 5: 403–404.
35 Meyers, A.I. and Comins, D.L. (1978). *Tetrahedron Lett.* 19: 5179–5182.
36 Monticelli, S., Parisi, G., Rui, M. et al. (2016). *Nat. Prod. Commun.* 11: 1729–1732.
37 Izzo, P. and Safir, S. (1959). *J. Org. Chem.* 24: 701–703.
38 Collins, S., Hong, Y., Hoover, G.J. et al. (1990). *J. Org. Chem.* 55: 3565–3568.
39 Collins, S. and Hong, Y. (1987). *Tetrahedron Lett.* 28: 4391–4394.
40 Imamoto, T., Takiyama, N., Nakamura, K. et al. (1989). *J. Am. Chem. Soc.* 111: 4392–4398.
41 Wolters, A.T., Hornillos, V., Heijnen, D. et al. (2016). *ACS Catal.* 6: 2622–2625.
42 de Jong, J., Heijnen, D., Helbert, H. et al. (2019). *Chem. Commun.* 55: 2908–2911.
43 Miyoshi, N., Kimura, S., Kubo, S. et al. (2020). *Asian J. Org. Chem.* 9: 1660–1664.
44 Liu, C., Achtenhagen, M., and Szostak, M. (2016). *Org. Lett.* 18: 2375–2378.
45 Glover, S.A. and Rosser, A.A. (2012). *J. Org. Chem.* 77: 5492–5502.

46 Bennet, A.J., Somayaji, V., Brown, R.S. et al. (1991). *J. Am. Chem. Soc.* 113: 7563–7571.
47 Otani, Y., Nagae, O., Naruse, Y. et al. (2003). *J. Am. Chem. Soc.* 125: 15191–15199.
48 Szostak, M. and Aubé, J. (2013). *Chem. Rev.* 113: 5701–5765.
49 Nahm, S. and Weinreb, S.M. (1981). *Tetrahedron Lett.* 22: 3815–3818.
50 Ghinato, S., Territo, D., Maranzana, A. et al. (2021). *Chem. Eur. J.* 27: 2868–2874.
51 Sun, W., Wang, L., Hu, Y. et al. (2020). *Nat. Commun.* 11: 3113.
52 Giovannini, A., Savoia, D., and Umani-Ronchi, A. (1989). *J. Org. Chem.* 54: 228–234.
53 Kong, K., Moussa, Z., Lee, C. et al. (2011). *J. Am. Chem. Soc.* 133: 19844–19856.
54 Pepe, A., Pamment, M., Georg, G.I. et al. (2011). *J. Org. Chem.* 76: 3527–3530.
55 Chen, J., Xia, Y., and Lee, S. (2020). *Org. Chem. Front.* 7: 2931–2937.
56 Lipshutz, B.H., Pfeiffer, S.S., and Chrisman, W. (1999). *Tetrahedron Lett.* 40: 7889–7892.
57 Singh, J., Satyamurthi, N., and Aidhen, I.S. (2000). *J. Prakt. Chem.* 342: 340–347.
58 Balasubramaniam, S. and Aidhen, I.S. (2008). *Synthesis* 23: 3707–3738.
59 Senatore, R., Ielo, L., Monticelli, S. et al. (2019). *Synthesis* 51: 2792–2808.
60 Pace, V., Holzer, W., and De Kimpe, N. (2016). *Chem. Rec.* 16: 2061–2076.
61 Castoldi, L., Monticelli, S., Senatore, R. et al. (2018). *Chem. Commun.* 54: 6692–6704.
62 Citarella, A., Gentile, D., Rescifina, A. et al. (2021). *Int. J. Mol. Sci.* 22: 1398.
63 Miele, M., Citarella, A., Micale, N. et al. (2019). *Org. Lett.* 21: 8261–8265.
64 Pace, V., Castoldi, L., and Holzer, W. (2013). *J. Org. Chem.* 78: 7764–7770.
65 Pace, V., Holzer, W., Verniest, G. et al. (2013). *Adv. Synth. Catal.* 355: 919–926.
66 Pace, V. (2016). New perspectives in homologation processes for synthetic medicinal chemistry: lithium halocarbenoids at the helm. Habilitation Thesis. University of Vienna, Vienna.
67 Parisi, G., Colella, M., Monticelli, S. et al. (2017). *J. Am. Chem. Soc.* 139: 13648–13651.
68 Mamuye, A.D., Castoldi, L., Azzena, U. et al. (2015). *Org. Biomol. Chem.* 13: 1969–1973.
69 Pace, V., Murgia, I., Westermayer, S. et al. (2016). *Chem. Commun.* 52: 7584–7587.
70 Senatore, R., Ielo, L., Urban, E. et al. (2018). *Eur. J. Org. Chem.* 2018: 2466–2470.
71 Hoppe, D. and Hense, T. (1997). *Angew. Chem. Int. Ed.* 36: 2282–2316.
72 Hoppe, D. and Christoph, G. (2004). Asymmetric deprotonation with alkylithium-(−)-sparteine. In: *The Chemistry of Organolithium Compounds*, vol. 2 (ed. Z. Rappoport and I. Marek), 1055–1164. Chichester: Wiley.
73 Beak, P., Kerrick, S.T., Wu, S. et al. (1994). *J. Am. Chem. Soc.* 116: 3231–3239.
74 Kerrick, S.T. and Beak, P. (1991). *J. Am. Chem. Soc.* 113: 9708–9710.

75 Monticelli, S., Holzer, W., Langer, T. et al. (2019). *ChemSusChem* 12: 1147–1154.

76 Azzena, U., Carraro, M., Pisano, L. et al. (2019). *ChemSusChem* 12: 40–70.

77 Li, G. and Szostak, M. (2020). *Org. Biomol. Chem.* 18: 3827–3831.

78 Li, G. and Szostak, M. (2020). *Chem. Eur. J.* 26: 611–615.

79 Sureshbabu, P., Azeez, S., Muniyappan, N. et al. (2019). *J. Org. Chem.* 84: 11823–11838.

80 Heller, S.T., Newton, J.N., Fu, T. et al. (2015). *Angew. Chem. Int. Ed.* 54: 9839–9843.

81 Takeda, N., Kobori, Y., Okamura, K. et al. (2020). *Org. Lett.* 22: 9740–9744.

82 Nugent, J. and Schwartz, B.D. (2016). *Org. Lett.* 18: 3834–3837.

83 Adler, M., Adler, S., and Boche, G. (2005). *J. Phys. Org. Chem.* 18: 193–209.

84 Kirby, A.J., Komarov, I.V., and Feeder, N. (1998). *J. Am. Chem. Soc.* 120: 7101–7102.

85 Adler, M., Marsch, M., Nudelman, N.S. et al. (1999). *Angew. Chem. Int. Ed.* 38: 1261–1263.

86 Cox, C., Wack, H., and Lectka, T. (1999). *J. Am. Chem. Soc.* 121: 7963–7964.

87 Kirby, A.J., Komarov, I.V., Wothers, P.D. et al. (1998). *Angew. Chem. Int. Ed.* 37: 785–786.

88 Lee, S.D., Brook, M.A., and Chan, T.H. (1983). *Tetrahedron Lett.* 24: 1569–1572.

89 Brandänge, S., Holmgren, E., Leijonmarck, H. et al. (1995). *Acta Chem. Scand.* 49: 922–928.

90 Brandänge, S. and Rodriguez, B. (1987). *Acta Chem. Scand.* B41: 740–744.

91 Arai, E., Tokuyama, H., Linsell, M.S. et al. (1998). *Tetrahedron Lett.* 39: 71–74.

92 Hsu, H.-C. and Hou, D.-R. (2009). *Tetrahedron Lett.* 50: 7169–7171.

93 Evans, D.A., Borg, G., and Scheidt, K.A. (2002). *Angew. Chem. Int. Ed.* 41: 3188–3191.

94 Dahlqvist, K.I. and Forsen, S. (1969). *J. Phys. Chem.* 73: 4124–4129.

95 Szostak, M., Yao, L., and Aubé, J. (2010). *J. Am. Chem. Soc.* 132: 2078–2084.

96 Szostak, M. and Aubé, J. (2009). *J. Am. Chem. Soc.* 131: 13246–13247.

97 Szostak, M., Yao, L., and Aubé, J. (2009). *J. Org. Chem.* 74: 1869–1875.

98 Szostak, M. and Aube, J. (2011). *Org. Biomol. Chem.* 9: 27–35.

99 Castoldi, L., Holzer, W., Langer, T. et al. (2017). *Chem. Commun.* 53: 9498–9501.

100 Sato, T., Yoritate, M., Tajima, H. et al. (2018). *Org. Biomol. Chem.* 16: 3864–3875.

101 Volkov, A., Tinnis, F., Slagbrand, T. et al. (2016). *Chem. Soc. Rev.* 45: 6685–6697.

102 Ong, D.Y., Yen, Z., Yoshii, A. et al. (2019). *Angew. Chem. Int. Ed.* 58: 4992–4997.

103 Bailey, C.L., Joh, A.Y., Hurley, Z.Q. et al. (2016). *J. Org. Chem.* 81: 3619–3628.

104 Flaniken, J.M., Collins, C.J., Lanz, M. et al. (1999). *Org. Lett.* 1: 799–801.

105 Czerwiński, P.J. and Furman, B. (2021). *Front. Chem.* 9: 655849.

106 Reetz, M.T. (1986). *Organotitanium Reagents in Organic Synthesis*. Berlin: Springer.

107 Reetz, M.T., Westermann, J., Steinbach, R. et al. (1985). *Chem. Ber.* 118: 1421–1440.

108 Reetz, M.T., Westermann, J., and Steinbach, R. (1981). *J. Chem. Soc., Chem. Commun.* 237–239.

109 Denton, S.M. and Wood, A. (1999). *Synlett* 55–56.

110 Kulinkovich, O.G. (2003). *Chem. Rev.* 103: 2597–2632.

111 Kulinkovich, O.G. and de Meijere, A. (2000). *Chem. Rev.* 100: 2789–2834.

112 Tomashenko, O., Sokolov, V., Tomashevskiy, A. et al. (2008). *Eur. J. Org. Chem.* 2008: 5107–5111.

113 Iida, H., Watanabe, Y., and Kibayashi, C. (1985). *J. Am. Chem. Soc.* 107: 5534–5535.

114 Shirokane, K., Kurosaki, Y., Sato, T. et al. (2010). *Angew. Chem. Int. Ed.* 49: 6369–6372.

115 Yokoyama, T., Fukami, Y., Sato, T. et al. (2016). *Chem. Asian J.* 11: 470–473.

116 Schedler, D.J.A., Li, J., and Ganem, B. (1996). *J. Org. Chem.* 61: 4115–4119.

117 White, J.M., Tunoori, A.R., and Georg, G.I. (2000). *J. Am. Chem. Soc.* 122: 11995–11996.

118 Spletstoser, J.T., White, J.M., Tunoori, A.R. et al. (2007). *J. Am. Chem. Soc.* 129: 3408–3419.

119 Shirokane, K., Wada, T., Yoritate, M. et al. (2013). *Angew. Chem. Int. Ed.* 53: 512–516.

120 Vincent, G., Guillot, R., and Kouklovsky, C. (2011). *Angew. Chem. Int. Ed.* 50: 1350–1353.

121 Vincent, G., Karila, D., Khalil, G. et al. (2013). *Chem. Eur. J.* 19: 9358–9365.

122 Sancibrao, P., Karila, D., Kouklovsky, C. et al. (2010). *J. Org. Chem.* 75: 4333–4336.

123 Jäkel, M., Qu, J., Schnitzer, T. et al. (2013). *Chem. Eur. J.* 19: 16746–16755.

124 Hiraoka, S., Matsumoto, T., Matsuzaka, K. et al. (2019). *Angew. Chem. Int. Ed.* 58: 4381–4385.

125 Nakajima, M., Oda, Y., Wada, T. et al. (2014). *Chem. Eur. J.* 20: 17565–17571.

126 Nakajima, M., Sato, T., and Chida, N. (2015). *Org. Lett.* 17: 1696–1699.

127 Takahashi, Y., Sato, T., and Chida, N. (2019). *Chem. Lett.* 48: 1138–1141.

128 Szcześniak, P., Stecko, S., Staszewska-Krajewska, O. et al. (2014). *Tetrahedron* 70: 1880–1888.

129 Szcześniak, P. and Furman, B. (2021). *Org. Biomol. Chem.* 19: 6842–6846.

130 Więcław, M.M. and Furman, B. (2021). *Beilstein J. Org. Chem.* 17: 115–123.

131 Czerwiński, P.J. and Furman, B. (2019). *Chem. Commun.* 55: 9436–9439.

132 Motoyama, Y., Aoki, M., Takaoka, N. et al. (2009). *Chem. Commun.* 12: 1574–1576.

133 Yoritate, M., Takahashi, Y., Tajima, H. et al. (2017). *J. Am. Chem. Soc.* 139: 18386–18391.

134 Yamamoto, S., Komiya, Y., Kobayashi, A. et al. (2019). *Org. Lett.* 21: 1868–1871.

135 Matheau-Raven, D., Gabriel, P., Leitch, J.A. et al. (2020). *ACS Catal.* 10: 8880–8897.

136 Xie, L.-G. and Dixon, D.J. (2017). *Chem. Sci.* 8: 7492–7497.

137 Huang, P.-Q., Ou, W., and Han, F. (2016). *Chem. Commun.* 52: 11967–11970.

138 Itabashi, S., Shimomura, M., Sato, M. et al. (2018). *Synlett* 29: 1786–1790.

139 Ong, D.Y., Fan, D., Dixon, D.J. et al. (2020). *Angew. Chem. Int. Ed.* 59: 11903–11907.

140 Chen, J., Lim, J.W., Ong, D.Y. et al. (2022). *Chem. Sci.* 13: 99–104.

141 Murai, T. and Mutoh, N. (2016). *J. Org. Chem.* 81: 8131–8134.

142 Murai, T. and Mutoh, Y. (2012). *Chem. Lett.* 41: 2–8.

143 Murai, T. (ed.) (2019). *Chemistry of Thioamides*. Singapore: Springer Nature.

144 Falmagne, J.-B., Escudero, J., Taleb-Sahraoui, S. et al. (1981). *Angew. Chem. Int. Ed.* 20: 879–880.

145 Kaiser, D., Bauer, A., Lemmerer, M. et al. (2018). *Chem. Soc. Rev.* 47: 7899–7925.

146 Kaiser, D. and Maulide, N. (2016). *J. Org. Chem.* 81: 4421–4428.

147 Sisti, N.J., Zeller, E., Grierson, D.S. et al. (1997). *J. Org. Chem.* 62: 2093–2097.

148 Cravallee, C., Didier, E., and Pecquet, P. (1998). *Tetrahedron Lett.* 39: 4263–4266.

149 Sforza, S., Dossena, A., Corradini, R. et al. (1998). *Tetrahedron Lett.* 39: 711–714.

150 Charette, A.B. and Grenon, M. (2001). *Can. J. Chem.* 79: 1694–1703.

151 Peng, B., Geerdink, D., Farès, C. et al. (2014). *Angew. Chem. Int. Ed.* 53: 5462–5466.

152 Movassaghi, M., Hill, M.D., and Ahmad, O.K. (2007). *J. Am. Chem. Soc.* 129: 10096–10097.

153 Movassaghi, M. and Hill, M.D. (2006). *J. Am. Chem. Soc.* 128: 14254–14255.

154 Medley, J.W. and Movassaghi, M. (2009). *J. Org. Chem.* 74: 1341–1344.

155 Pelletier, G. and Charette, A.B. (2013). *Org. Lett.* 15: 2290–2293.

156 Xiao, K.-J., Luo, J.-M., Xia, X.-E. et al. (2013). *Chem. Eur. J.* 19: 13075–13086.

157 Bechara, W.S., Pelletier, G., and Charette, A.B. (2012). *Nat. Chem.* 4: 228–234.

158 Xiao, K.-J., Wang, A.-E., Huang, Y.-H. et al. (2012). *Asian J. Org. Chem.* 1: 130–132.

159 Huang, P.-Q., Wang, Y., Xiao, K.-J. et al. (2015). *Tetrahedron* 71: 4248–4254.

160 Weng, Y., Min, L., Zeng, X. et al. (2020). *Org. Lett.* 22: 8296–8301.

161 Xiao, P., Tang, Z., Wang, K. et al. (2018). *J. Org. Chem.* 83: 1687–1700.

162 Huang, P.-Q., Huang, Y.-H., and Xiao, K.-J. (2016). *J. Org. Chem.* 81: 9020–9027.

163 Huang, P.-Q., Huang, Y.-H., Geng, H. et al. (2016). *Sci. Rep.* 6: 1–10.

164 Huang, P.-Q. and Huang, Y.-H. (2017). *Chin. J. Chem* 35: 613–620.

165 Wu, D.-P., He, Q., Chen, D.-H. et al. (2019). *Chin. J. Chem.* 37: 315–322.

166 Geng, H. and Huang, P.-Q. (2019). *Chin. J. Chem.* 37: 811–816.

167 Li, J., Oost, R., Maryasin, B. et al. (2019). *Nat. Commun.* 10: 2327.

168 Liu, Y.P., Wang, S.R., Chen, T.T. et al. (2019). *Adv. Synth. Catal.* 361: 971–975.

169 Liu, Y.P., Zhu, C.J., Yu, C.C. et al. (2019). *Eur. J. Org. Chem.* 2019: 7169–7174.

170 Wang, S.R. and Huang, P.Q. (2019). *Chin. J. Chem.* 37: 887–891.

171 Barbe, G. and Charette, A.B. (2008). *J. Am. Chem. Soc.* 130: 18–19.

172 Huang, P.-Q. and Geng, H. (2015). *Org. Chem. Front.* 2: 150–158.

173 Pelletier, G., Bechara, W.S., and Charette, A.B. (2010). *J. Am. Chem. Soc.* 132: 12817–12819.

174 Huang, P.-Q., Lang, Q.-W., and Wang, Y.-R. (2016). *J. Org. Chem.* 81: 4235–4243.

175 Xiang, S.-H., Xu, J., Yuan, H.-Q. et al. (2010). *Synlett* 1829–1832.

176 Parks, D.J. and Piers, W.E. (1996). *J. Am. Chem. Soc.* 118: 9440–9441.

177 Huang, P.-Q., Huang, Y.-H., Xiao, K.-J. et al. (2015). *J. Org. Chem.* 80: 2861–2868.
178 Chen, H., Huang, Y.-H., Ye, J.-L. et al. (2019). *J. Org. Chem.* 84: 9270–9281.
179 Wang, A.-E., Yu, C.-C., Chen, T.-T. et al. (2018). *Org. Lett.* 20: 999–1002.
180 Chen, H., Ye, J.-L., and Huang, P.-Q. (2018). *Org. Chem. Front.* 5: 943–947.
181 Pandey, V.K. and Anbarasan, P. (2017). *J. Org. Chem.* 82: 12328–12336.
182 Wang, A.-E., Chang, Z., Sun, W.-T. et al. (2015). *Org. Lett.* 17: 732–735.
183 Chen, T.-T., Wang, A.-E., and Huang, P.-Q. (2019). *Org. Lett.* 21: 3808–3812.
184 Huang, P.-Q., Lang, Q.-W., and Hu, X.-N. (2016). *J. Org. Chem.* 81: 10227–10235.
185 Kolleth, A., Lumbroso, A., Tanriver, G. et al. (2016). *Tetrahedron Lett.* 57: 3510–3514.
186 Xu, Z., Wang, X.-G., Wei, Y.-H. et al. (2019). *Org. Lett.* 21: 7587–7591.
187 Zheng, J.-F., Qian, X.-Y., and Huang, P.-Q. (2015). *Org. Chem. Front.* 2: 927–935.
188 Huang, P.-Q. and Fan, T. (2017). *Eur. J. Org. Chem.* 2017: 6369–6374.
189 Kadyrov, R. (2019). *Adv. Synth. Catal.* 361: 185–191.
190 Yu, S., Ho, S.K., and Lee, S. (2019). *Asian J. Org. Chem.* 8: 1613–1616.
191 Bechara, W.S., Khazhieva, I.S., Rodriguez, E. et al. (2015). *Org. Lett.* 17: 1184–1187.
192 Ellsworth, A.A., Magyar, C.L., Hubbell, G.E. et al. (2016). *Tetrahedron* 72: 6380–6389.
193 Huang, P.-Q., Huang, Y.-H., and Wang, S.-R. (2017). *Org. Chem. Front.* 4: 431–444.
194 Ye, J.-L., Zhu, Y.-N., Geng, H. et al. (2018). *Sci. China Chem.* 61: 687–694.
195 Niu, Z.J., Li, L.H., Liu, X.Y. et al. (2019). *Adv. Synth. Catal.* 361: 5217–5222.
196 Bauer, A., Borsos, E., and Maulide, N. (2020). *Eur. J. Org. Chem.* 2020: 3971.
197 Zheng, J.-F., Hu, X.-N., Xu, Z. et al. (2017). *J. Org. Chem.* 82: 9693–9703.

5

Electrophilic Amide Bond Functionalization

Carlos R. Gonçalves and Daniel Kaiser

Institute of Organic Chemistry, Faculty of Chemistry, University of Vienna, Währinger Straße 38, 1090, Vienna, Austria

5.1 Introduction: Electrophilic Activation

When Charles-Adolphe Wurtz first delved into the reactivity of aldehydes in acidic media, he was not aware of the long-standing implications that his work would have on the future of organic chemistry. The aldol reaction, as it became known after its first disclosure in 1872, was named not after both acclaimed scientists that simultaneously discovered it, Wurtz and chemist-turned-composer Alexander Borodin, but instead after the eponymous first product, which was observed and characterized (Scheme 5.1) [1, 2].

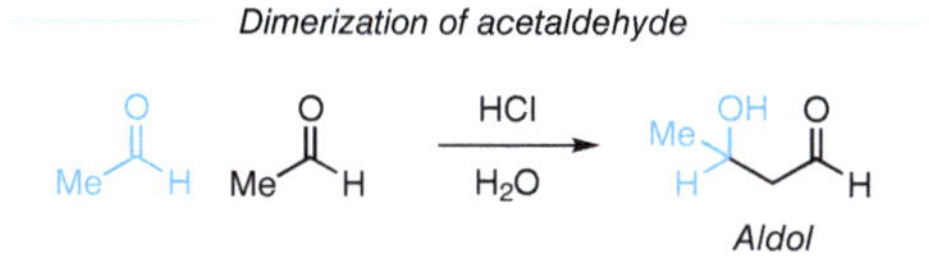

Scheme 5.1 Origin of the aldol reaction. Source: Adapted from Wurtz [1, 2].

The comparatively simple system of acetaldehyde combined with aqueous hydrochloric acid enabled initial elucidation of both the innate characteristics of aldehydes and this addition reaction as a whole. As a functional group, aldehydes were largely perceived as difficult to handle; hence, gaining an understanding of the process in which they react with themselves was of major importance. Nevertheless, it is crucial to point out that the unsung hero of this ordeal was the hydrochloric acid, acting as a vital component in this early example of an acid-catalyzed transformation, in which a simple proton is able to induce the formation of a high-energy state that enables a subsequent reaction.

This exact proposal was first rationalized decades later, in 1923, by Johannes Brønsted and Thomas Lowry [3, 4], who presented a generalization of the Arrhenius theory, including important definitions that helped shape the paradigm of acid–base theory.

Amide Bond Activation: Concepts and Reactions, First Edition. Edited by Michal Szostak.

One could break down the exact role of a Brønsted acid to that of a catalyst that increases the energy of a carbonyl electrophile by interaction with its nucleophilic domain, i.e. the oxygen lone pairs, making it more reactive (Scheme 5.2) [5]. This process can be succinctly named *electrophilic activation*, because the (unreactive) electrophile is being coerced into reacting with a nucleophilic species through activation with another, more reactive, electrophile. Apart from employing protons (by using Brønsted acids), more complex electrophilic reagents (generically termed "*E*") can also be used to promote this activation.

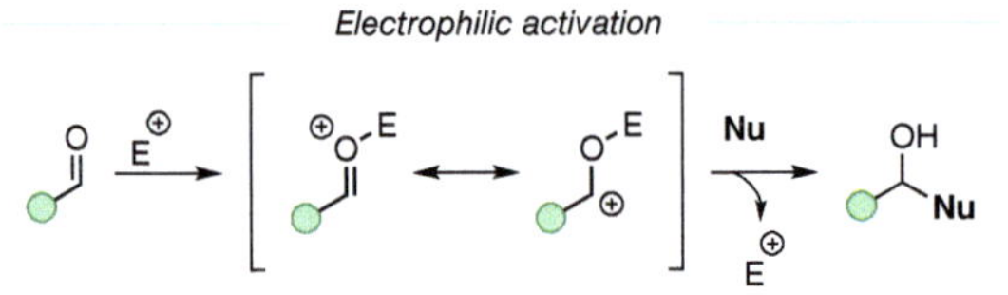

Scheme 5.2 Electrophilic activation of carbonyl derivatives using strong electrophiles. Source: Adapted from Fujita et al. [5].

5.2 Introduction: Electrophilic Activation of Amides

Developing new and efficient ways to guide a species toward doing something it is not naturally destined to do is, and will continue to be, one of the most important objectives in modern organic chemistry. In this sense, redirecting well-known concepts and established common knowledge has the potential to push the boundaries in terms of reactivity, leading to exciting novel discoveries.

An amicable target, and one that will be the topic of this chapter, is the carboxamide functionality. Amides are generally perceived as a symbol of chemical stability, due to their major role in the structural integrity of all life forms. The ubiquity of amides (and amines) in natural and biologically active molecules has made amide functionalization a long-standing topic in organic chemistry, as exemplified by textbook transformations such as the Bischler–Napieralski and Vilsmeier–Haack reactions (Scheme 5.3).

A possible misconception derived from the aforementioned (perceived) stability of amides would be to infer that the activation of an amide would inevitably lead to concomitant activation of similar functional groups, particularly carbonyls, thereby evoking side reactions and leading to limited applicability. It is the aim of this chapter to showcase that this, however, is far from the truth, and suitable activation can enable remarkably chemoselective reactions.

Analysis of the nature of the amide bond allows uncovering of the secrets behind its unique reactivity. As illustrated by the resonance structures shown in Scheme 5.4, there is a significant contribution of nitrogen lone pair conjugation, leading to a decrease in electrophilicity of the central carbon atom. Notably, this also increases the nucleophilicity of the oxygen atom, making it particularly susceptible to reaction with electrophilic activators, especially in comparison to other carbonyl analogs. This characteristic is the pith for the subsequent section, in which a

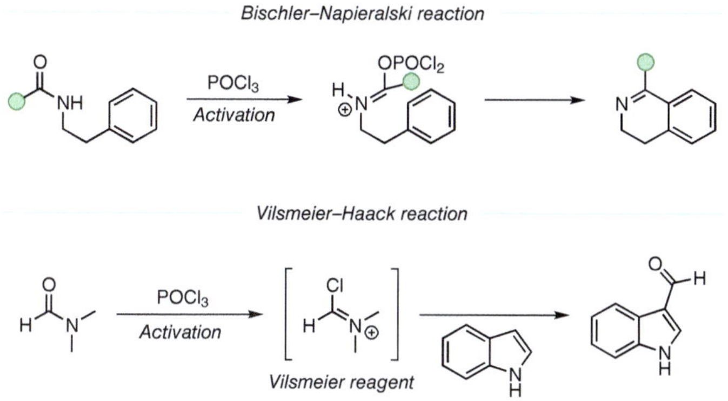

Scheme 5.3 Bischler–Napieralski and Vilsmeier–Haack reactions as examples of early electrophilic activation.

plethora of transformations exploiting the unique features of this exquisite moiety will be described [6–8].

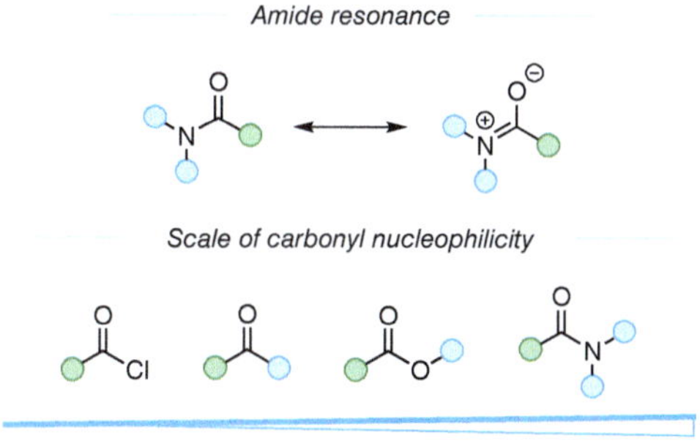

Scheme 5.4 Amide bond resonance properties and a comparison of the nucleophilicity with respect to other carbonyl derivatives.

5.3 Early Endeavors in Electrophilic Amide Activation

5.3.1 History of the Activation of Secondary Amides

As illustrated previously, the resonance of the amide bond and the resulting inherent stabilization renders it inaccessible to most nucleophiles. Conversely, the participation of the nitrogen lone pair concomitantly creates an opportunity for the exploitation of amides themselves as competent nucleophiles. In this sense, the end of the nineteenth century saw the development of electrophilic amide activation using common dehydrating agents as electrophiles (cf. also Scheme 5.3). While full dehydration of a primary amide afforded the expected nitrile [9], the reaction with secondary and tertiary amides initially failed to produce any identifiable products.

In 1877, Wallach reported that the treatment of secondary amides (**1**) with phosphorous pentachloride led to the formation of imidoyl chlorides (**4**) (Scheme 5.5) [9].

These species, formed from the corresponding imidoylium chlorides (**3**) by deprotonation, bear significance as the cornerstone of all electrophilic amide activation. Besides the observed instability of this class of compounds [10], readily reverting to the corresponding amides in the presence of moisture, it was also noted that different nucleophiles were able to replace the chloride by a facile addition–elimination mechanism. For example, the treatment of **2/3** with an aniline will inevitably give rise to an amidine (**4**) [11].

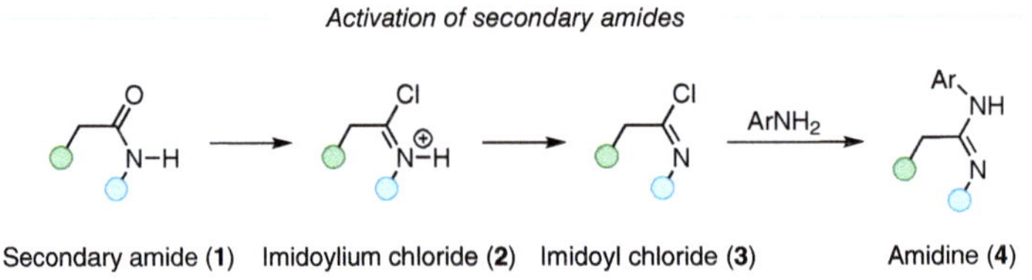

Scheme 5.5 Electrophilic activation of secondary amides for the synthesis of amidines. Source: Adapted from Kaiser and Maulide [8].

In a similar fashion, the prelude for modern electrophilic amide activation must also include the reaction of secondary amides with other activating agents, for example, oxalyl chloride [12].

Initially impaired by erroneous assignment of the product structure [13, 14], early studies failed to shed light on the mechanism of this reaction. Although the middle of the twentieth century offered comparatively few chemical tools to aid in mechanistic investigations, based on isolated by-products, Speziale and Smith were able to derive a mechanistic picture for this process [15] and proposed that, depending on which heteroatom of a secondary amide preferably acts as the nucleophile, treatment with oxalyl chloride can follow two possible pathways (Scheme 5.6). While nucleophilic attack of the nitrogen atom (Path A) leads to the initial formation of **7**, an O-nucleophilic reaction involving nitrogen lone-pair participation (Path B) affords intermediate **10**. While these initially formed species were virtually indistinguishable, premature quenching of the reaction with methanol offered a solution: aniline (**11**) was detected in the reaction mixture and could only have been formed through methanolysis of **10**, leading to the substantiated conclusion that the activation took place on the oxygen atom (Scheme 5.6, Path B).

5.3.2 History of the Activation of Tertiary Amides

Concomitantly, this mechanistic elucidation proved useful for the understanding of tertiaryamide activation, regarding the nature of which little evidence had been known. When other forms of electrophilic activation arose in the middle of the twentieth century, the ability to understand these increasingly complex reactions became exceedingly more important. One of the main protagonists of these endeavors was Léon Ghosez, whose contributions to and influence on the field of electrophilic amide activation resonate to this day [6, 10]. His group's early

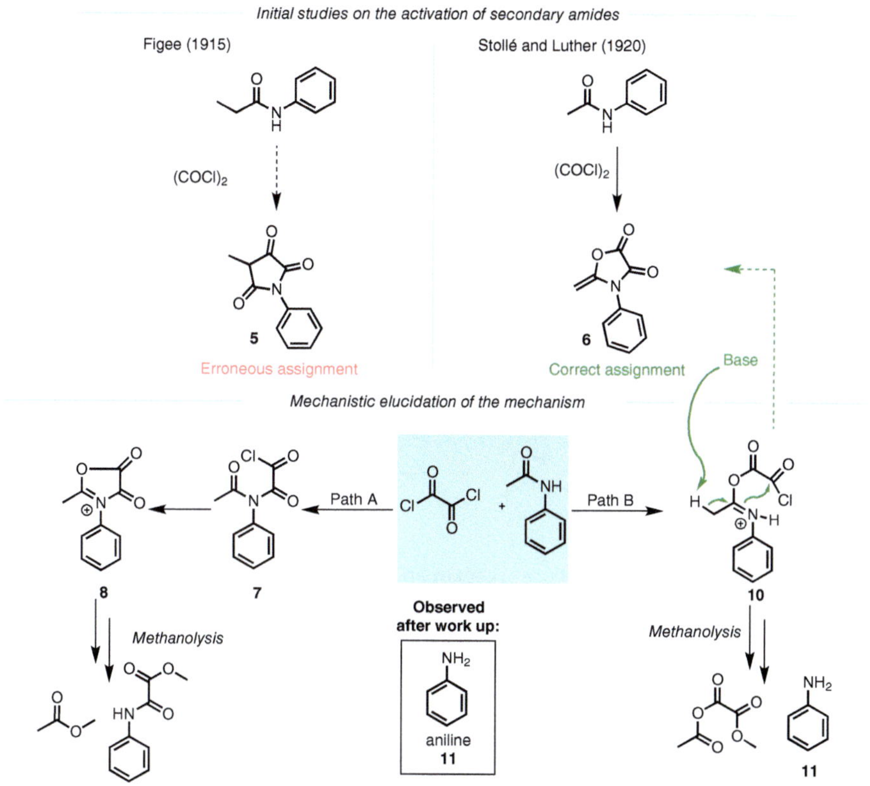

Scheme 5.6 Elucidation of the mechanism of the electrophilic activation of secondary amides with oxalyl chloride, capturing intermediates by methanolysis [13–15].

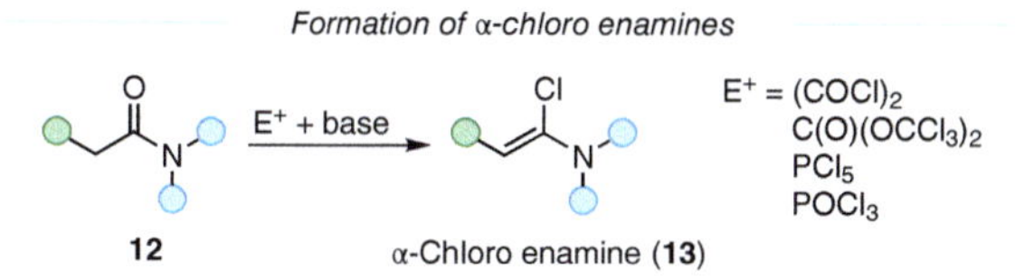

Scheme 5.7 Formation of α-chloro enamines by electrophilic activation of tertiary amides. Source: Adapted from Ott et al. [16], Wolf and Block [17], Speziale and Freeman [18], [19], Ficini and Barbara [20], Yakubovitch et al. [21] and Buyle and Viehe [22].

research focused mainly on the exploration of α-chloro enamines (**13**, Scheme 5.7) [16–22].

Among the different ways to generate these intermediates [10], one of the most prominent employed is triphosgene, a chemical that sees limited use in laboratory settings, despite having great importance in industrial processes [23]. The reaction of tertiary amides with triphosgene yields α-chloro enamines and was first explored using *N*,*N*-dimethylisobutyramide [24]. The tremendous impact that this specific α-chloro enamine (**14**) and its inventor have had on the field of electrophilic activation is reflected by its colloquial name: *Ghosez reagent* (Scheme 5.8).

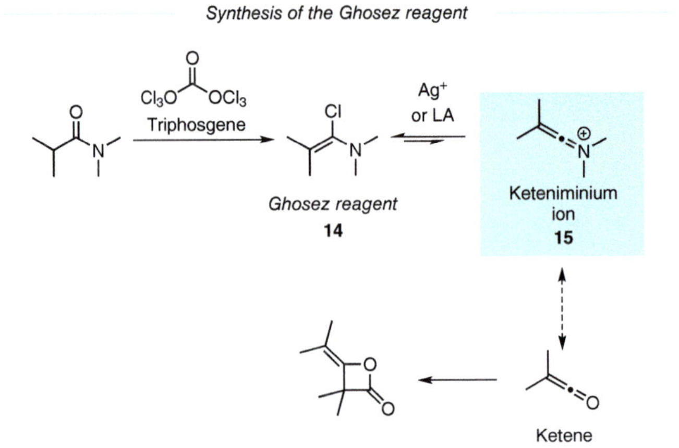

Scheme 5.8 Synthesis of the Ghosez reagent and a keteniminium ion.

An important feature of α-chloro enamines is that they are known precursors of the respective keteniminium ions (**15**), the cationic nitrogen analogs of ketenes, when subjected to silver salts or Lewis acids [25]. A significant number of new characteristics are introduced by this apparently innocuous atom replacement (Scheme 5.8). Alongside increased electrophilicity, the keteniminium ion (or keteneimmonium, as it was once referred to) is also easier to manipulate due to its increased accessibility and decreased propensity to form dimers. In contrast to ketenes, it is also important to note that keteniminium ions offer the possibility to induce stereoselectivity by placing chiral substituents on the nitrogen atom (*vide supra*, Schemes 5.13 and 5.17).

Groundbreaking work by Ghosez also showcased the versatility of the intermediate at hand, particularly in terms of applicability to different reaction classes. In addition to water, leading to regeneration of the starting material, a wide array of other nucleophiles, ranging from aromatics to alcohols, amines, and thiols, can be easily added to keteniminium ions [25]. However, nowadays keteniminium ions are most well known for a different type of transformation: [2+2]-cycloadditions [10]. The engagement of a keteniminium ion with an external or tethered π-system leads to the formation of a substituted cyclobutanone or cyclobutenone derivative (see Scheme 5.9 and Section 5.4.1). The iminium group, still present in the initial adduct (**16**), seems to be specifically tailored for the required antarafacial interaction (*vide infra*, Scheme 5.13), thus leading to high degrees of selectivity and performance.

Another approach that perfectly illustrates the synthetic prowess of keteniminium ions and α-chloro enamines is elimination to afford the corresponding ynamides (**17**) [26, 27]. While this transformation is brought about by action of an organic base, the

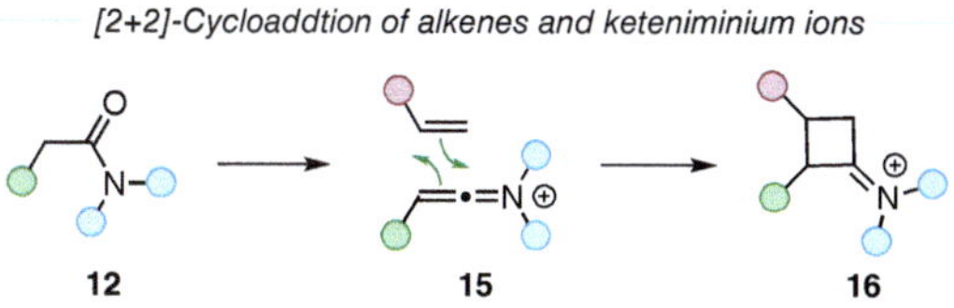

Scheme 5.9 Cycloaddition of alkenes and keteniminium ions.

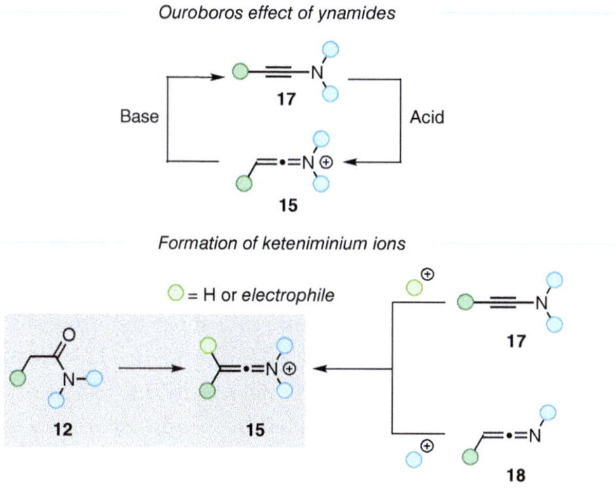

Scheme 5.10 Different ways to form keteniminium ions. Source: Adapted from Lambrecht et al. [33] and Deyrup and Kuta [34].

treatment of ynamides with a Brønsted acid leads to an *ouroboros*-like effect, and the ynamide is transformed back into a keteniminium ion (Scheme 5.10, top) [28].

For the sake of completeness, despite not being the aim of this chapter, it shall be mentioned that other methods for forming keteniminium ions also exist (Scheme 5.10, bottom). As previously stated, the activation of ynamides can also be achieved using a range of electrophiles, resulting in β,β-disubstituted keteniminium ions (Scheme 5.10, top path) [29–32], while the direct alkylation of the corresponding ketenimines (**18**) poses an alternative, which also circumvents amide starting materials (Scheme 5.10, bottom path) [33, 34].

By the 1970s, while the chemistry of keteniminium ions generated from imidoylium chlorides (**2**) had already been widely explored [10], there existed two important hurdles that prevented the broad application of this transformation. While the need for hazardous reagents has already been alluded to, the second obstacle limited synthetic utility, as α-chloro enamines themselves can act as nucleophiles to the keteniminium species, leading to aldol-like products (**19**) (Scheme 5.11, top) [35]. Fortuitously, triflic anhydride, used as an *in situ* activator in the presence of a pyridine base, provided the ideal solution to both problems.

The aforementioned changes, alongside the development of more powerful spectroscopic methods, enabled the detailed observation, driven by André Charette

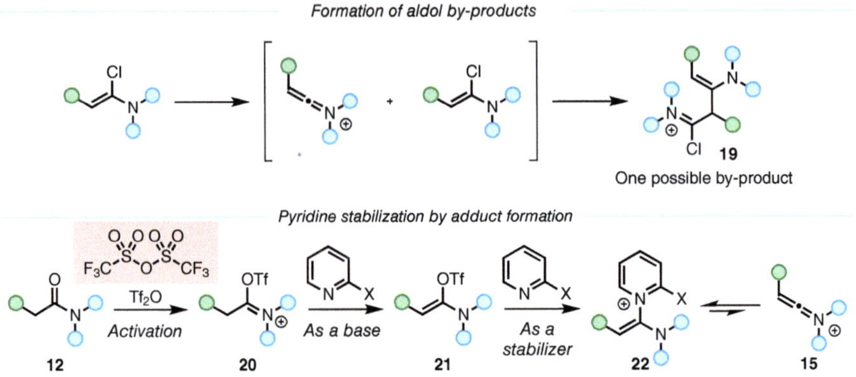

Scheme 5.11 Formation of aldol by-products with α-chloro enamines and pyridine stabilization of keteniminium ions formed with triflic anhydride. Source: Adapted from Falmagne et al. [35].

and coworker, and with it a more consolidated understanding, of the reaction mechanism [36, 37]. It soon became apparent that the role of the pyridine exceeded that of a simple base, as it seemed crucial to stabilization of the keteniminium intermediate (Scheme 5.11, bottom). In fact, such is the magnitude of stabilization that only the pyridinium adduct (**22**) can be detected by NMR (nuclear magnetic resonance) spectroscopy, highlighting the fleeting nature of the keteniminium species [36]. Experimental evidence further suggested that the reversibility of this adduct formation is vitally important. To ensure reversibility, the nucleophilicity of the pyridine must be attenuated through changes in steric or electronic properties (although the exact rationale is still unknown), thereby enabling the formation of the keteniminium ion and, as a consequence, productive reaction. This necessity is reflected in the use of bases such as 2,6-lutidine, 2,4,6-collidine, and the series of 2-halopyridines that have found widespread application in recent years [37].

5.4 Amide Bond Functionalization of Activated Tertiary Amides

5.4.1 [2+2]-Cycloadditions

First presented in Scheme 5.9, [2+2]-cycloadditions are among the most intriguing transformations enabled by the transient formation of keteniminium ions. While arguably encountered less ubiquitously than the [4+2]-cycloaddition, the initial description of [2+2]-cycloadditions actually predates the first account of the Diels–Alder reaction [38]. In 1908, Ciamician and Silber reported the observation that the exposure of carvone to "*Italian sunlight,*" [39] resulted in the formation of a cyclic compound of unknown structure. Following this revelation, 50 years should pass before the product was identified as carvonecamphor (**23**), the product of intramolecular [2+2]-cycloaddition (Scheme 5.12) [40, 41].

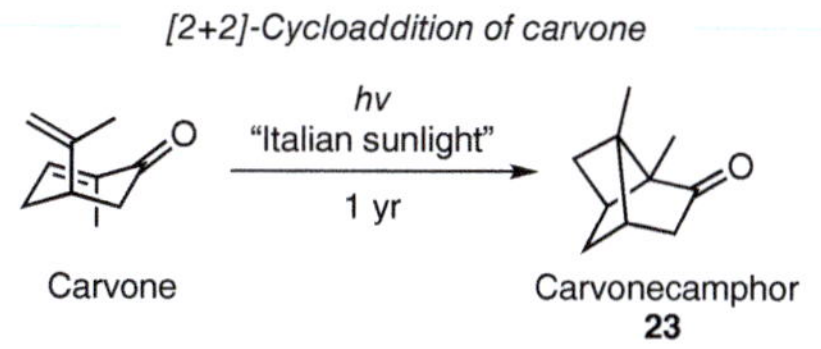

Scheme 5.12 Serendipitous discovery of the [2+2]-cycloaddition of carvone upon one-year exposure to sunlight. Source: Adapted from Buchi and Goldman [40] and Tsipi et al. [41].

When discussing [2+2]-cycloadditions, the role of ketenes, with respect to their discovery and mechanistic study, cannot be downplayed. The reaction of ketenes with simple olefins in the absence of light was first reported by Staudinger, who noted the unusual [2+2]-cycloaddition reaction at the beginning of the nineteenth century [42] – it was later found that this increased reactivity even extended to the [2+2]-cycloaddition with butadienes, forgoing the expected [4+2]-pathway [43]. Importantly, the cycloaddition of ketenes and olefins is also a prime example of a reaction that can be rationalized by and conversely used to exemplify the Woodward–Hoffmann rules [44, 45].

While [2+2]-cycloadditions between two simple olefins are disfavored, ketenes readily undergo these reactions, largely due to a lesser degree of steric constraint allowing for a suprafacial–antarafacial approach of the two fragments (Scheme 5.13, top) [46]. Indeed, even the smallest and simplest alkene, ethylene, struggles to undergo this fascinating transformation, due to crowding of the hydrogen atoms. While ketenes possess beneficial properties for [2+2]-cycloadditions, their reactivity comes as a double-edged sword, as they also tend to undergo rapid dimerization (Scheme 5.13, bottom and *cf*. Scheme 5.8), hampering their broad synthetic utility.

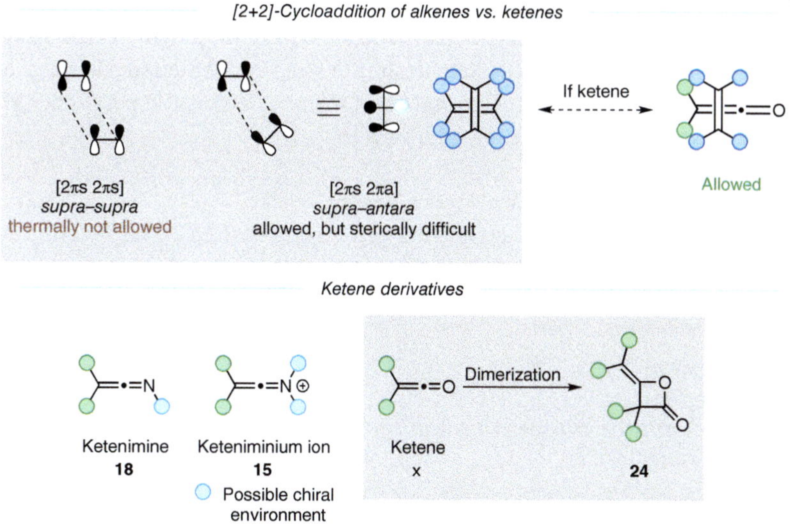

Scheme 5.13 Intricacies of [2+2]-cycloadditions. Ketene analogs avoid the problem of dimerization. Source: Adapted from Huisgen and Otto [48].

This fact is of the utmost importance because it contrasts beautifully with other more user-friendly ketene derivatives, as shown in the following text.

In contrast, nitrogen analogs of ketenes, such as ketenimines (**18**) [47], and keteniminium ions (**15**), offer significant advantages over the parent group, especially regarding increased stability resulting from a reduced tendency to dimerize (Scheme 5.13, **24**) [48]. Another marked advantage of keteniminium ions is their tremendous reactivity: even ethylene, commonly known to be unreactive in uncatalyzed cycloadditions [49–52], readily engages in [2+2]-cycloadditions with keteniminium ions at room temperature and atmospheric pressure [53] (Scheme 5.14). This initial observation has entailed numerous computational studies aimed at understanding and subsequently harnessing the full potential of this transformation [54]. Theoretical scrutinization has revealed that the concertedness of the reaction is strongly tied to the electronic features of the reactants, and [2+2]-cycloadditions of keteniminium ions tend to proceed via stepwise processes [55–57].

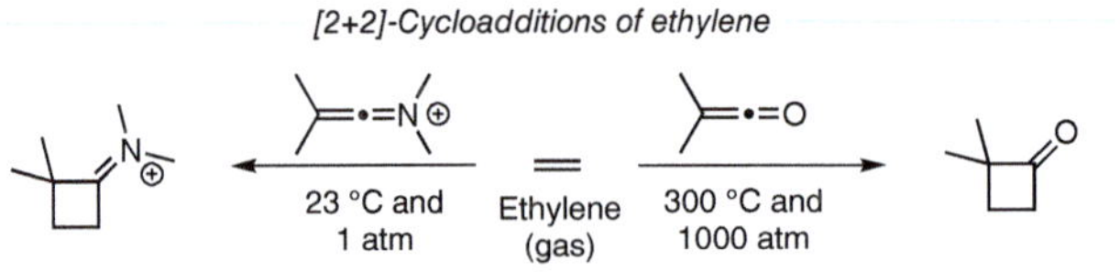

Scheme 5.14 Comparison between the cycloadditions of ethylene with ketene derivatives.

Realizing the potential of keteniminium ions as practical and reactive substitutes of ketenes [25], Ghosez' work was initially centered on the direct comparison of the two congeners [55, 58, 59]. The first natural steps into the exploration of additional chemical space involved the [2+2]-cycloadditions of keteniminium ions with the other major class of π-systems, alkynes (Scheme 5.15) [35]. Employing triflic anhydride for the *in situ* generation of the keteniminium ions, cyclobutenone derivatives (**25**) could be easily obtained, with more recent research showing the possibility for their direct engagement in [4+2]-cycloadditions [60].

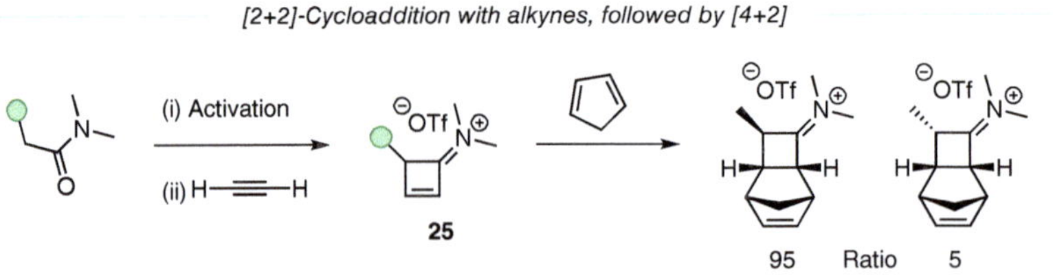

Scheme 5.15 Cycloaddition between keteniminium ions and alkynes. Source: Adapted from Falmagne et al. [35].

Rendering these initial approaches to small-ring formation intramolecular, Ghosez and coworkers were able to delve into natural product synthesis (Scheme 5.16). The first examples of intramolecular [2+2]-cycloadditions of keteniminium ions were reported in 1985 [59] and highlighted the prowess of amide activation with triflic anhydride. To date, [2+2]-cycloadditions of keteniminium ions are the go-to method for the formation of cyclobutanones in the synthesis of natural products [61–70].

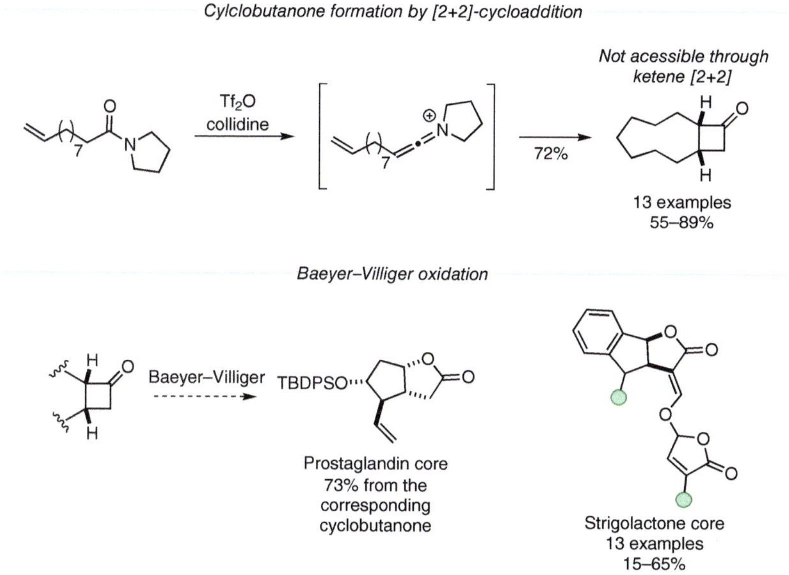

Scheme 5.16 Intramolecular [2+2]-cycloaddition and subsequent functionalization of the cyclobutanone to afford natural product scaffolds.

5.4.2 Stereoselective Cycloadditions

Following early applications of keteniminium ions, there was an enticing possibility yet to be explored. Structural comparison of keteniminium ions and ketenes reveals the possibility for the introduction of chiral information at the nitrogen substituents (*cf*. Scheme 5.13). Employing amides equipped with chiral substituents in [2+2]-cycloadditions endowed these processes with the possibility of diastereoselectivity, with hydrolysis of the resulting iminium ions yielding enantioenriched products. Ghosez' pioneering work on the use of chiral auxiliaries in the chemistry of keteniminium ions initially employed enantioenriched methoxymethyl pyrrolidine amide **26** [67], forming the target compound with only relatively low enantiomeric excess (Scheme 5.17). Despite this initial moderate success, this method still constitutes an important milestone for asymmetric electrophilic amide activation.

In pursuit of enantioenriched cyclobutanones, a wide range of chiral auxiliaries have been subject to investigation (Scheme 5.17, bottom) [68]. While poor stereoselectivity was observed for the aforementioned methoxymethyl pyrrolidine, even in an intramolecular reaction, more fruitful results were obtained when C2-symmetric pyrrolidine analogs were employed. While distal substitution of positions 3 and 4

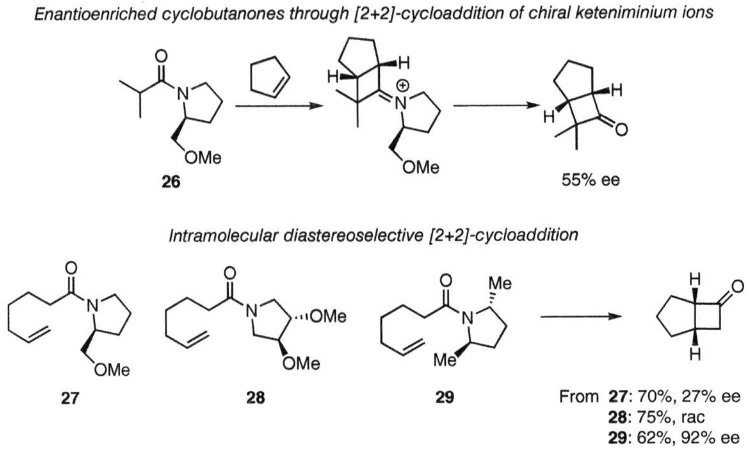

Scheme 5.17 Diastereoselective [2+2]-cycloaddition using chiral amides. Source: Adapted from Houge et al. [67]

of the pyrrolidine (**28**) failed to exert a positive influence on the enantioselectivity, α,α′-disubstitution (**29**) was shown to induce excellent levels of enantioselectivity, and this methodology was successfully applied to the synthesis of the four stereoisomers of 5-deoxystrigol (not shown) [25].

5.4.3 Nucleophile Addition

5.4.3.1 Carbon Nucleophiles

Following the development of [2+2]-cycloaddition reactions, the pursuit of highly functionalized molecular scaffolds led to the investigation of other ways to promote C–C bond formation. To this effect, the reactions of keteniminium ions with various nucleophilic carbon motifs, including electron-rich aromatics [69–73], enol ethers [74], and enamines [75], were extensively studied, resulting in numerous elegant syntheses of complex natural products [76–80].

The adaptation of the Vilsmeier–Haack reaction illustrates this point and once more highlights the power of electrophilic amide activation using triflic anhydride [81]. In contrast to classical Vilsmeier–Haack conditions, triflic anhydride enables the activation of virtually any tertiary amide, which opens the door to the formation of congested C–C bonds. This approach is particularly intriguing when intramolecular nucleophilic capture leads to (poly)cyclic products (Scheme 5.18), as electron-rich aromatics have proven particularly capable of adding to the keteniminium ion intermediate, with indoles constituting a privileged motif within this class [72, 73].

A further interesting example is found in Evans' synthesis of nakadomarin A, in which the iminium ion **30**, resulting from the initial nucleophilic addition of an indole to the keteniminium intermediate, is directly reduced to the desired α-secondary amine in a one-pot process (Scheme 5.19) [69, 82].

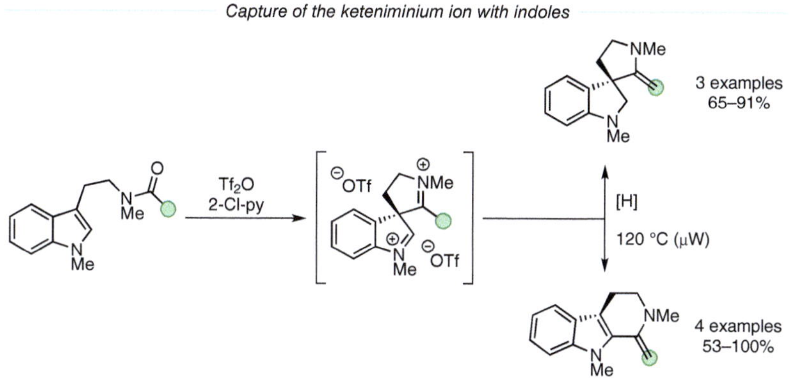

Scheme 5.18 Trapping of keteniminium ions with indole.

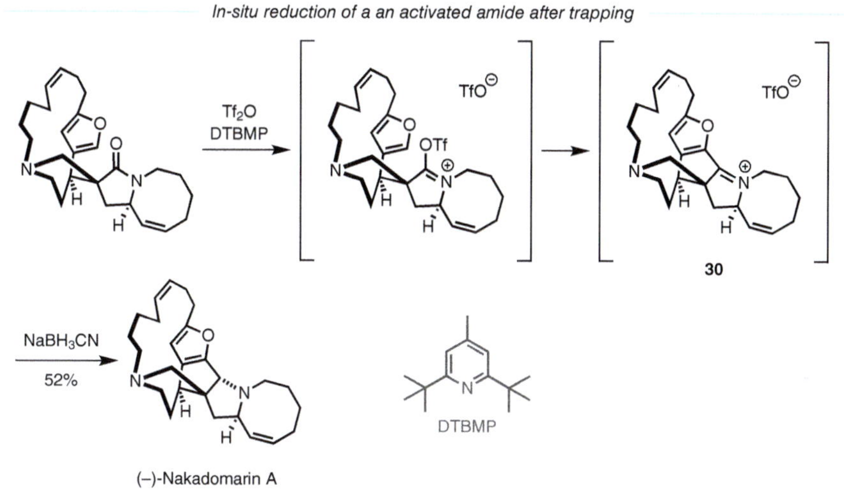

Scheme 5.19 Total synthesis of Nakadomarin A, making use of activation of congested tertiary amides. Source: Adapted from Bonazzi et al. [69] and Xiao et al. [85].

Apart from electron-rich aromatics, a wide range of other carbon nucleophiles can also be used for additions to keteniminium ions. Among these, silyl enol ethers [74] and simple alkenes, which do not participate in [2+2]-cycloadditions [83], stand out. A prominent example of such a reaction can be seen in Scheme 5.20. Overman and coworker were able to engage the 1,1-disubstituted alkene present in **31** in a sequence of nucleophilic addition (**32**) and pinacol-type rearrangement (activation), elegantly generating a bicyclic system (**34**) [83]. Notably, the steric inability for an antarafacial approach likely prevented any competing [2+2]-cycloaddition.

The high inherent reactivity of both keteniminium ions and organometallic reagents makes it nonobvious whether a productive and, moreover, selective reaction of these two partners is possible. The value of such a transformation, however, is undeniable, as the direct formation of α-tertiary amines from the

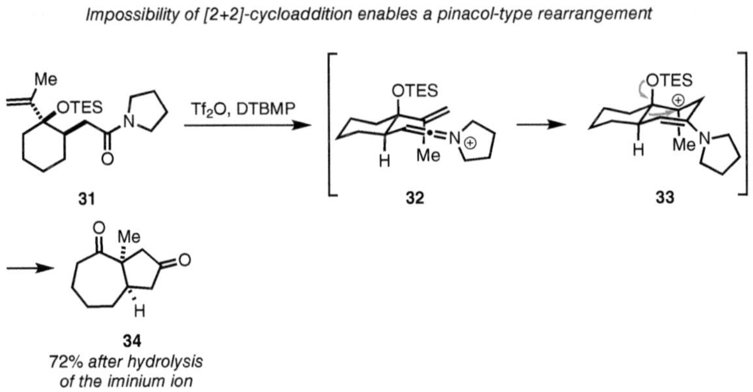

Scheme 5.20 Addition of an alkene to the keteniminium ion followed by pinacol-type rearrangement.

corresponding amides holds great promise for the synthesis of biologically active molecules – especially when considering that not one, but two C–C bonds, are formed in the process.

The selective addition of organometallic reagents to keteniminium ions was pioneered by the group of Pei-Qiang Huang [84]. Particularly, the reactions of keteniminium ions with Grignard reagents were shown to readily afford the desired amines in high yields [85, 86]. Notably, methods allowing for the sequential addition of different Grignard reagents were also developed, opening the doors for the construction of complex molecules [87]. This sequential alkylation boasts an additional beneficial feature, adding to its utility in natural product synthesis: diastereoselectivity. As shown in Huang's synthesis of (−)-FR901483, the prior installation of an enantioenriched chiral center enabled the highly diastereoselective addition of a vinyl Grignard reagent to the iminium intermediate **35**, laying the groundwork for the rapid construction of the target natural product (Scheme 5.21).

In the context of this synthesis (as well as that shown in Scheme 5.19), it is worth mentioning that the electrophilic activation of lactams does not follow the scenario shown in Scheme 5.11. Despite readily reacting with triflic anhydride, the impossibility of forming sp-hybridized – and, therefore, linear – keteniminium ions within the constraints of small- to medium-sized ring structures leads to a significant difference in the chemistry of lactams compared to their acyclic analogs.

The sequential addition of a Grignard reagent and a reductant offers a practical alternative to the aforementioned approach, delivering the α-secondary amines selectively (akin to Scheme 5.19). Several reducing agents have been used to this effect, ranging from hydrogen [86] to common hydride sources [88–90].

5.4.3.2 Hydridic Reduction

The synthesis of α-primary amines can also be easily achieved by employing the smallest possible nucleophile: the hydride. While the robust nature of amides makes them comparably unreactive to many reducing agents, thus lowering functional group compatibility to a minimum, keteniminium ions circumvent this

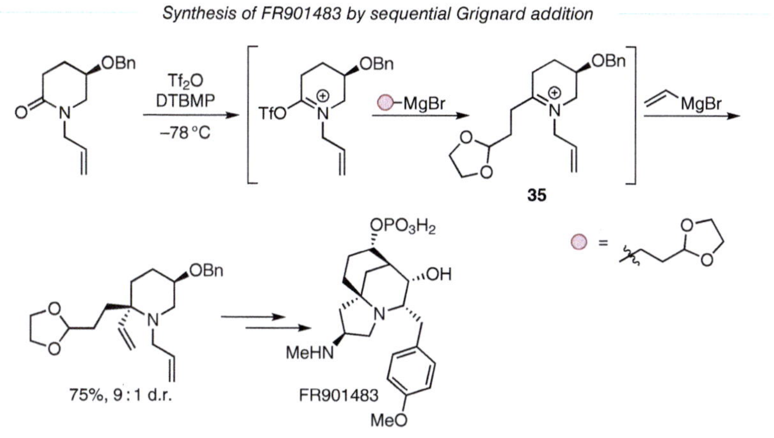

Scheme 5.21 Sequential addition of organometallics to an activated lactam, resulting in the total synthesis of FR901483.

limitation, enabling reduction with weak hydride sources. In this regard, Charette and coworker reported one of the first examples, using the classical Hantzsch ester as the hydride source (Scheme 5.22) [91]. The use of such a mild reductant enabled remarkable functional group tolerance, with all other tested carbonyl groups remaining untouched. Similarly, activated amides can also be reduced by a range of inorganic reductants, including sodium borohydride [92].

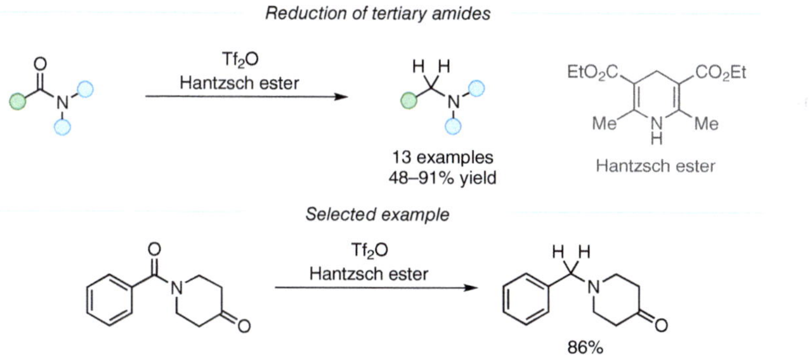

Scheme 5.22 Chemoselective reduction of tertiary amides using electrophilic activation and a mild reductant. Source: Adapted from Barbe and Charette [91].

5.4.3.3 Heteroatom Nucleophiles

The high electrophilicity of tertiary amide-derived keteniminium ions can also be exploited by intramolecular addition of heteroatom nucleophiles – a process that knows virtually no limitations. Oxygen nucleophiles have been most commonly employed, with the simplest of these being, of course, water itself. Despite simply restoring the starting material, this reaction has found application in the study of reaction mechanisms, as it has been used to form ^{18}O-labeled amides with exceptionally high levels of isotopic incorporation [93, 94]. Aside from water, a large number of other oxygen-based nucleophiles can be used in the same manner. The addition of alcohols to keteniminium ions, for example, is a simple method to transform amides into the corresponding esters (Scheme 5.23) [95]. This *de facto* esterification occurs under surprisingly mild conditions (ambient temperature and moderate acidity), with the key aspect being the formation of the corresponding ortho ester (**36**), which reverts to the desired ester during work up. The ortho ester itself can be isolated when tridentate alcohols are used [96].

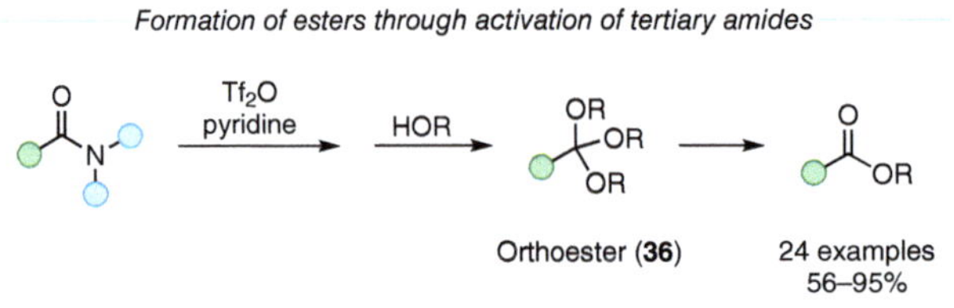

Scheme 5.23 Functional group interconversion using amide activation. Source: Adapted from Charette and Chua [95].

From an intramolecular point of view, the addition of heteroatom nucleophiles also offers intriguing possibilities, especially when ethers with tethered unsaturated groups are used (Scheme 5.24). In this regard, Maulide and coworkers were able to show that the treatment of compounds such as **37** with triflic anhydride does not deliver the initially expected bicyclic butanone **38** through [2+2]-cycloaddition, but rather evolves through an initial attack of the oxygen atom, forming **39** [7, 97]. The formation of **39** sets the stage for a [3,3]-sigmatropic rearrangement, forming **40**, which affords the corresponding α-allylated lactones (**41**) after hydrolysis. Notably, the configuration of the allyl moiety can be transposed to the final product, and it was shown that propargylic ethers also readily rearranged to the corresponding α-allenyl lactones.

In an alternative and complementary approach, Maulide and coworkers also reported the use of a chiral amide auxiliary, itself bearing the rearranging allyl moiety (Scheme 5.25) [98]. By employing O-allylated pseudoephedrine-derived amides (**42**), the authors were able to form diastereoenriched cyclic, cationic imidates (**43**), giving rise to the corresponding α-allylated aldehydes (**44**) or carboxylic acids (**45**) in enantioenriched form after hydrolysis or reduction, respectively.

The range of products that can be obtained through the reaction of activated amides with nitrogen nucleophiles goes far beyond the amidines mentioned in Section 5.3 [99] and can even involve the addition of a second heteroatom. In this context, the addition of 1,2-aminothiols to keteniminium ions provides thiazolines under remarkably mild conditions [100], with the cyclization occurring smoothly and allowing for the retention of peripheral stereocenters.

Owing to its mild conditions and exceptional chemoselectivity, the activation of tertiary amides is extremely valuable for the synthesis of complex scaffolds containing several heteroatoms. Apart from the aforementioned thiazolines, a vast number

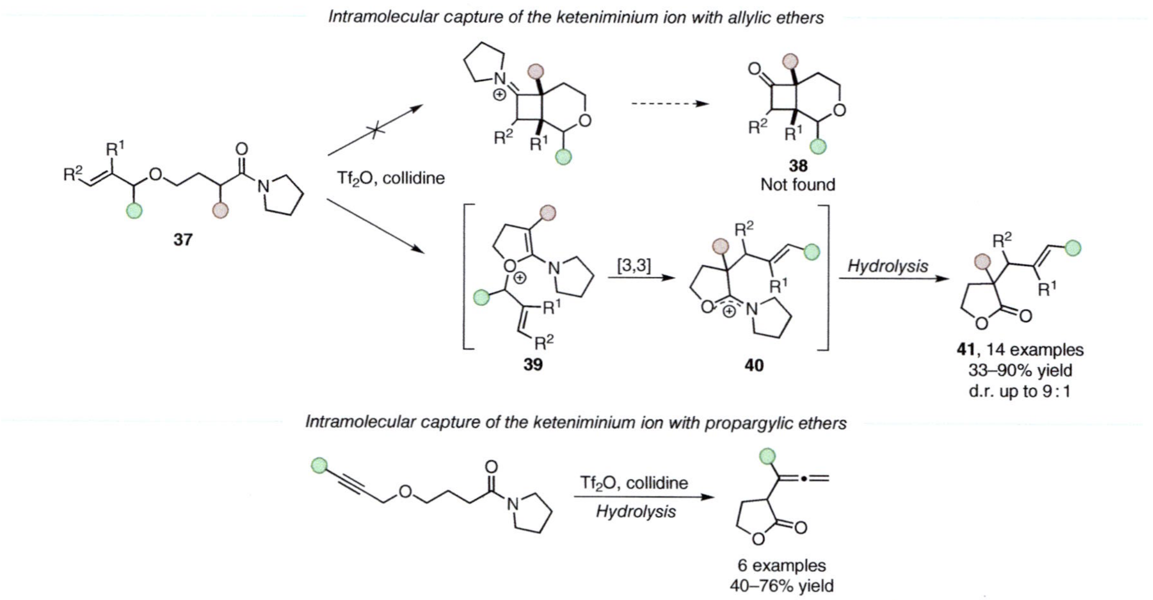

Scheme 5.24 Trapping of the keteniminium ion using tethered allylic and propargylic ethers, resulting in a [3,3]-sigmatropic rearrangement.

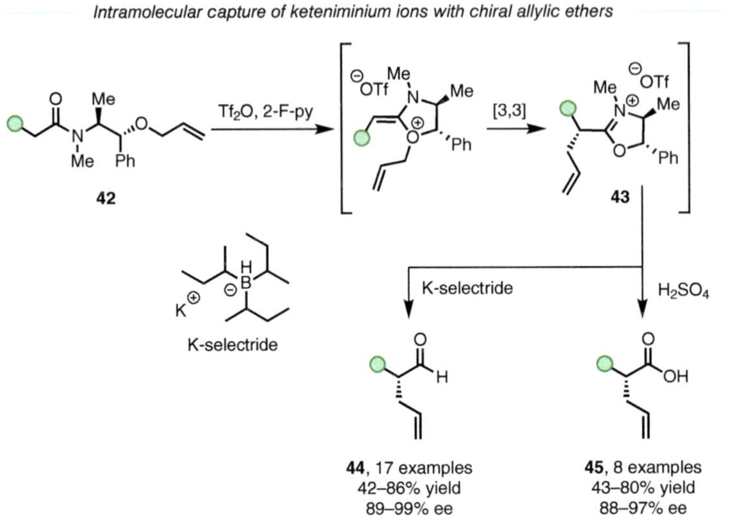

Scheme 5.25 Use of chiral allylic ethers as chiral auxiliaries on the allylation on the α-allylation of carbonyl derivatives. Source: Adapted from Peng et al. [98].

of other heterocycles can be synthesized by similar approaches, ranging from benzothiophenes [70] and benzofurans [101] to promising pharmaceutical compounds such as fluorazones [102]. Nevertheless, the creation of novel types of cyclic structures, elusive to other synthetic tools and strategies, is arguably one of the most enticing prospects of amide activation.

This can be exemplified by the synthesis of bicyclic alkoxyoxazolium salts (**46**) [103], which, despite their apparent simplicity, bear interesting proprieties and show intriguing reactivity, as they can readily undergo cycloadditions that benefit from their unique electronic characteristics (Scheme 5.26).

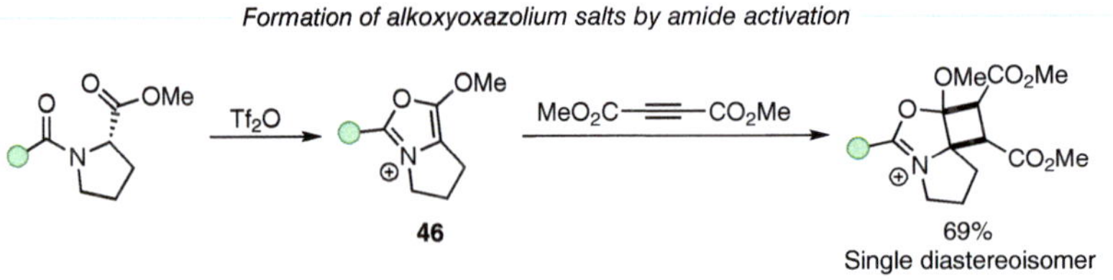

Scheme 5.26 Formation of alkoxyoxazolium salts (**46**) by electrophilic amide activation.

A class of compounds that share very particular similarities with these products are cyclic amidinium salts (**48**, Scheme 5.27), heterocycles that are traditionally challenging to synthesize due to their unusual disposition. An attractive strategy relies on amide activation and the use of azides, which are also known to undergo other types of transformations in the context of amide activation [104, 105]. Nucleophilic addition of an alkyl azide to a keteniminium ion leads to an umpolung event at the α-position of the amide, paving the way for the interception by the attached aryl moiety. The reaction shown in Scheme 5.27 is a prime example of a novel class of transformations, in which the activated amide is used as a handle to selectively functionalize the α-position [94, 106–110]. Replacing the pendant nucleophilic aryl group with a carbonyl, α-ketoazides were shown to deliver the corresponding oxazinones (**49**).

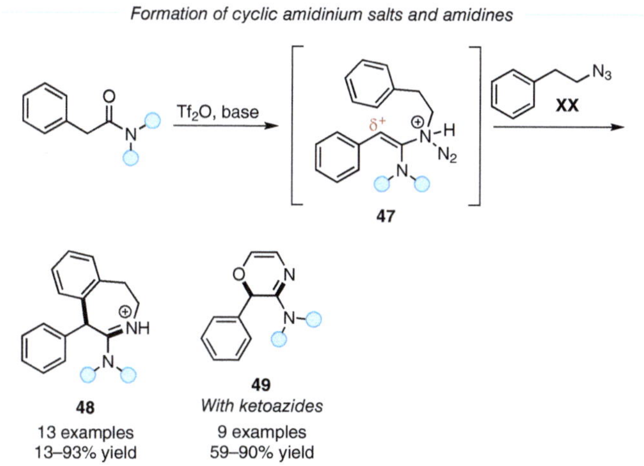

Scheme 5.27 Products formed through reaction of activated tertiary amides with alkyl azides.

As shown in this section, electrophilic activation offers a plethora of methods to easily functionalize tertiary amides with concomitant cleavage of one of the central amide bonds. In this way, complex molecular scaffolds can be readily accessed, and tertiary amides, commonly considered of low reactivity, can undergo rapid transformations to a wide range of useful functional groups. A similar effect of electrophilic activation on the reactivity, albeit under considerably changed circumstances, can be observed for secondary amides, which are the topic of the following section.

5.5 Amide Bond Functionalization of Activated Secondary Amides

When moving from tertiary to secondary amides, a significant shift in reactivity can be observed. In order to be able to discuss the reactivity profiles of activated secondary amides, it is important to highlight the intermediates that are intrinsically different from those derived from activation of tertiary amides (Scheme 5.28). While the initial step, nucleophilic attack of the amide carbonyl onto the activating agent, is comparable, deprotonation of the nitrogen swiftly forms an uncharged imidoyl triflate (**50**). At this point, and always in a crucial equilibrium, **50** can be stabilized by the pyridine base, resulting in the formation of a pyridinium amidine (**51**), which itself can be used for further transformations. Alternatively, **51** can undergo elimination of the triflate or pyridine, forming a nitrilium ion (**52**).

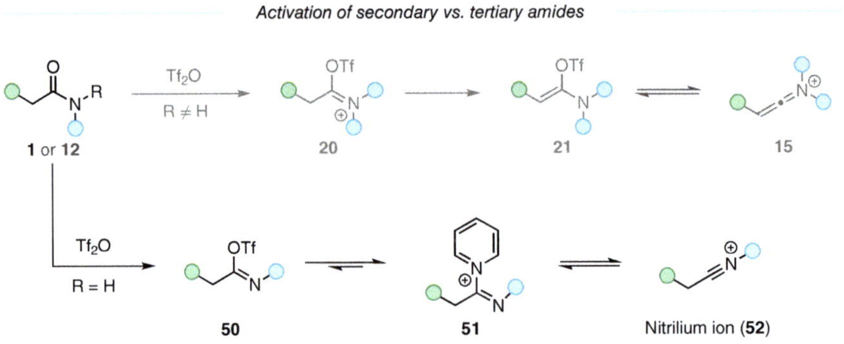

Scheme 5.28 Overview of the activation of secondary and tertiary amides.

5.5.1 Synthesis and Functionalization of Heterocycles

The best way to highlight the difference in reactivity between secondary and tertiary amides is by comparing their reaction with azides. While Section 5.4.3.3 showed that the reaction between activated tertiary amides and organic azides can give rise to amidine-containing heterocycles, activated secondary amides react with azides to form an entirely different class of heterocycles: tetrazoles (**53**, Scheme 5.29). While early reports showed that sodium azide, when added to an amide-derived nitrilium ion, leads to the formation of uncharged tetrazoles [111], the use of alkyl-substituted azides considerably expanded the scope of the reaction, forming charged tetrazolium salts (**54**) [105].

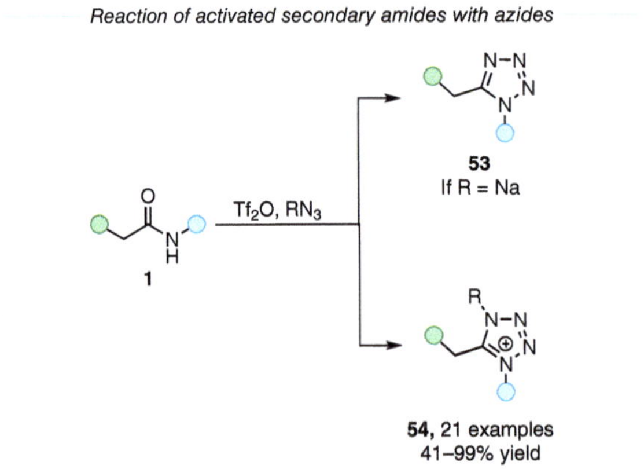

Scheme 5.29 Reactivity of azides in conjunction with nitrilium ions formed through activation of secondary amides.

While azides provided initial insight into the electrophilic reaction of nitrilium ions, other heteroatom nucleophiles have since proven more valuable, enabling widespread synthetic application of activated secondary amides. For example, using hydrazine derivatives, Charette and coworkers were able to synthesize interesting heterocyclic scaffolds, including aminoindazoles [112] and triazoles [113].

An additional appealing aspect of this type of transformation is that it can be applied to the synthesis of a wide range of biologically relevant motifs such as quinolines [114], isoquinolines [115], pyrimidines [116], and pyridines [117, 118], which are particularly privileged heterocycles due to their ubiquity in natural products and pharmaceuticals alike [119]. Interestingly, a clear evolution of pyridine synthesis through amide activation can be observed (Scheme 5.30). The first report by Movassaghi and coworkers used electrophilic amide activation to initially form azadienynes (**55**), which underwent cycloisomerization under ruthenium catalysis to afford the desired pyridines. Soon after, the same authors reported that an analogous transformation could also be initiated by the addition of π-nucleophiles, in particular alkynes and enol ethers, to **56**, even without any catalyst.

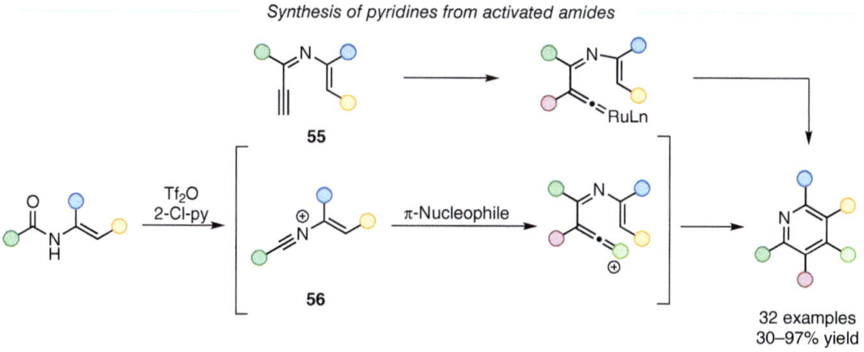

Scheme 5.30 Synthesis of pyridines using electrophilic activation of secondary amides.

Indoles, an equally important class of heterocycles [120], were successfully synthesized by exploiting the fleeting nature of the equilibrium between the pyridine adduct **58** and the nitrilium ion (*cf*. Scheme 5.28). In 2008, Wang and coworkers found that the domino reaction between *N*-arylamides and ethyl diazoacetate (EDA) to afford indoles did not proceed smoothly under standard amide activation conditions (Scheme 5.31) [121]. They postulated that the pyridine adduct **57**, formed by addition of the standard additive 2-chloropyridine, was not electrophilic enough to undergo substitution by the diazo ylide. However, by addition of a second, more electron-deficient base (2,6-dichloropyridine), a stabilized intermediate of increased electrophilicity was formed, drastically increasing the yield of the overall reaction. Notably, when only 2,6-dichloropyridine was used, no reaction occurred, highlighting the necessity of a carefully balanced system for the initial step of the amide activation.

As mentioned in Chapter 3, the use of unsubstituted pyridines is usually avoided due to the formation of a highly stable adduct, thwarting the generation of the usually desired reactive species (keteniminium ions with tertiary amides and nitrilium ions with secondary amides). However, the outset of this chapter hinted at the fact that the pyridinium adduct **51** (*cf*. Scheme 5.28) can also be employed for further transformations of the heteroaromatic moiety, affording functionalized pyridines.

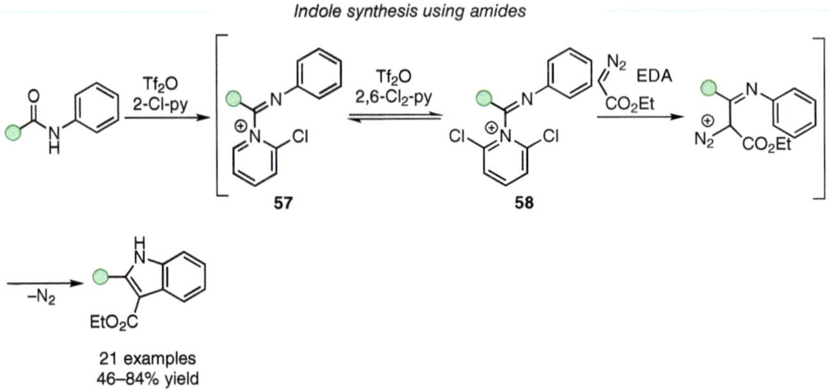

Scheme 5.31 Construction of indoles by addition of EDA to activated secondary amides. Source: Adapted from Wang and coworkers [121].

Utilizing this logic, Charette and coworkers were able to add a range of different nucleophiles, from enol ethers to organometallics, to different pyridinium ions generated by the addition of a pyridine to a nitrilium ion [122]. Highly reactive nucleophiles such as organometallic reagents add almost exclusively to the 2-position of the pyridine (Scheme 5.32) [123]. Employing an amide bearing a chiral auxiliary (**59**), the authors observed that the nucleophilic attack occurred with high diastereoselectivity, ensured by a bidentate interaction of the vicinal heteroatoms, which was crucial for the enantioselective synthesis of the natural product Julifloridine [124].

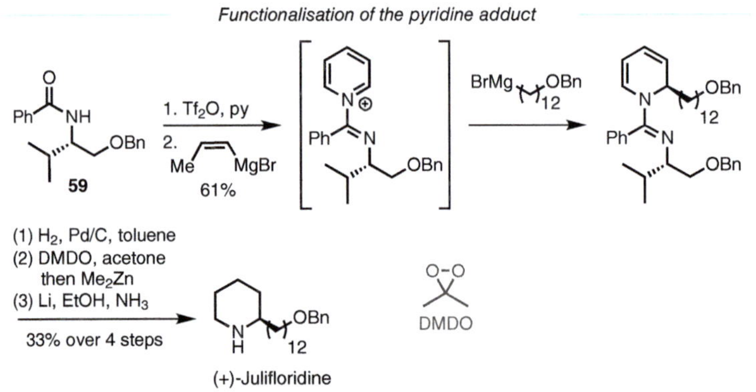

Scheme 5.32 Total synthesis of Julifloridine after functionalization of the pyridine adduct. Source: Adapted from Charette et al. [123].

An alternative and equally interesting strategy for the functionalization of pyridines had been explored many years before, using *N*-oxides bearing a halogen in the 2-position (Scheme 5.33). A report by Abramovitch showed that addition of a pyridine *N*-oxide derivative to an imidoyl chloride (**3**) led to the subsequent intramolecular substitution of the halogen by the newly formed amine, resulting in the formation of 2-pyridinamine derivatives (**60**) [125]. This process was subsequently further refined by Movassaghi, managing to achieve *N*-pyridinylation of aromatic secondary amides. While interesting products can be formed, this reaction suffers from the limitation that only benzamide derivatives are amenable, with competing processes being formal dehydrogenation (for branched amides) [126, 127] or nucleophilic addition to an intermittently formed α-electrophilic enolonium species (for linear aliphatic amides) [94, 106–110].

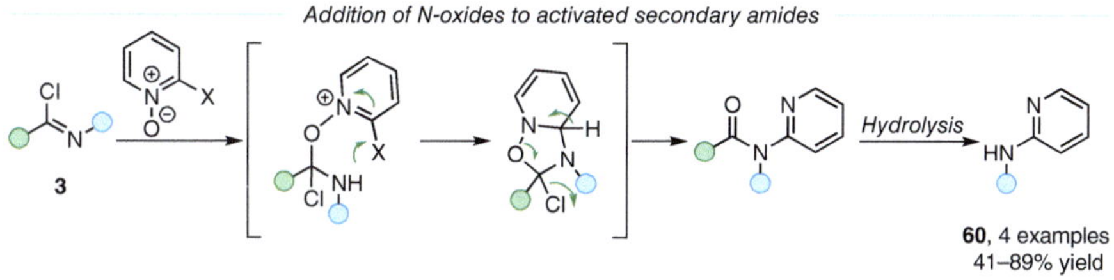

Scheme 5.33 Synthesis of 2-aminopyridines using N-oxides in combination with activated amides.

5.5.2 Ketone Synthesis

The direct comparison to intermediates stemming from activation of tertiary amides suggests that addition of an external nucleophile to an activated secondary amide (nitrilium ion) will generate an uncharged and, therefore, more stable species, thus less susceptible to a second addition reaction. With this knowledge, it can be inferred that ketones or imines as the final products are much more accessible. Following this logic, the Charette group accomplished the direct synthesis of various imines from the corresponding secondary amides, using a sequence of electrophilic activation and addition of a Grignard reagent (Scheme 5.34). Other carbon nucleophiles such as electron-rich aromatics can also be used for the same transformation [128].

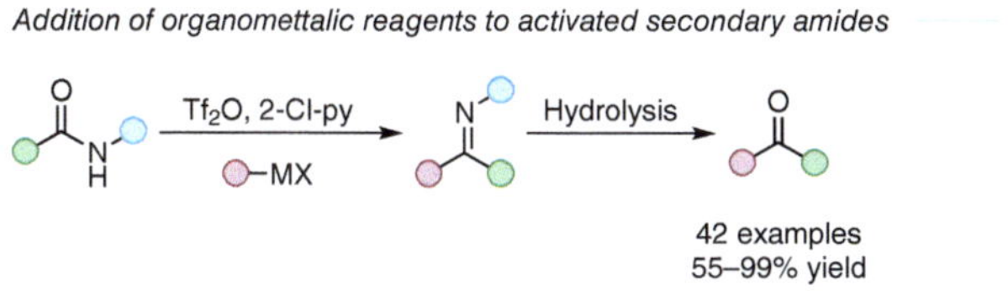

Scheme 5.34 Synthesis of ketones by direct addition of organometallic reagents.

Alternatively, similar ketone products can be formed by a more intricate reaction developed by the Maulide group (Scheme 5.35) [129]. Using activated amides in combination with electron-rich alkenes, the authors were able to develop a hydroacylation reaction in which initial attack of the olefin on the nitrilium ion (**61**) leads to the formation of a carbocationic intermediate (**62**). **62** is ideally arranged for a [1,5]-hydride transfer, giving rise to an azoniaallene (**63**), which in turn affords the desired ketone upon hydrolysis.

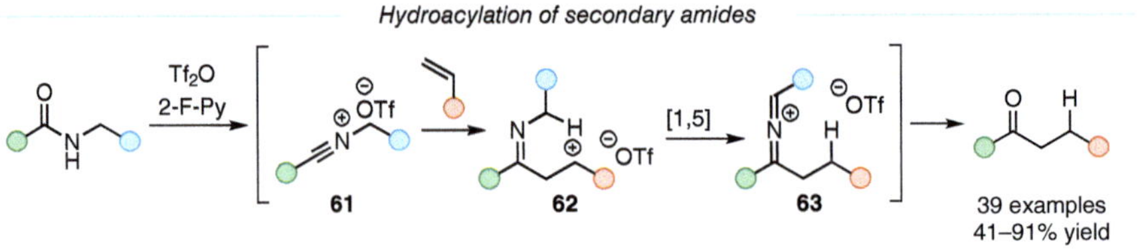

Scheme 5.35 Hydroacylation of secondary amides using alkenes. Source: Adapted from Li et al. [129].

In addition to the aforementioned nucleophiles, the use of enamines has led to intriguing results (Scheme 5.36). Huang and coworkers showed that the addition of enamines to activated secondary amides, followed by treatment with sodium borohydride, gave a diamine (**64**), which underwent direct elimination to the corresponding allylic amine (**65**) [130]. The compelling aspect of this transformation is the traceless nature of the enamine, specifically the nitrogen moiety, which ensures nucleophilic addition, yet makes no appearance in the final product.

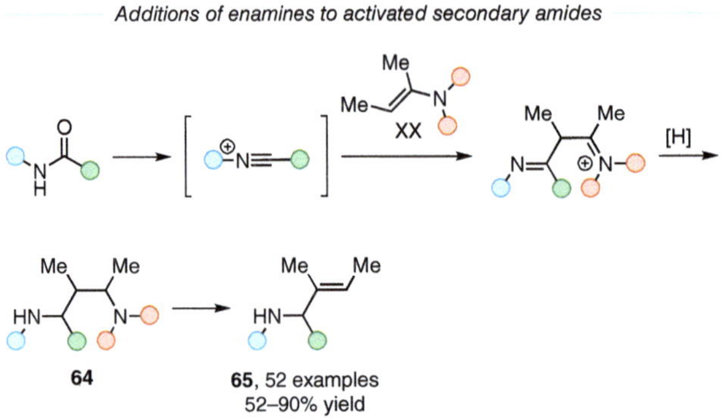

Scheme 5.36 The use of enamines as nucleophiles results in the formation of allyl amines.

In similarity to tertiary amides, the activation of secondary amides with electrophiles such as triflic anhydride enables the synthesis of a variety of different products. These transformations allow access to a wide range of nitrogen heterocycles, accessed through either elegant cyclization reactions or direct functionalization of pyridinium adducts, as well as carbonyl derivatives, formed through highly chemoselective addition reactions.

5.6 Conclusions

The low inherent electrophilicity of amides, while having historically hampered attempts at selective functionalization of the amide bond, enables remarkably chemoselective reactions with a range of electrophilic activators, the most prominent and widely used of which is triflic anhydride. This activation has been shown to drastically increase the electrophilic properties of the amide's central carbon, allowing addition and addition–elimination reactions with a plethora of, even weak, nucleophiles. Exploiting the unique properties of the amide functionality, advances in the area of amide activation have enabled applications that range from the (stereoselective) formation of small rings and α-functionalization of various carbonyl derivatives to the synthesis of amines, ketones, and heterocycles. Notably, many of these methods have even found application in the synthesis of complex natural products.

References

1 Wurtz, A. (1872). *Bull. Soc. Chim. Fr.* Part 2 17: 436.
2 Wurtz, A. (1872). *C. R. Hebd. Seances Acad. Sci.* 74: 1361.
3 Brönsted, J.N. (1923). *Recl. Trav. Chim. Pays-Bas* 42: 718–728.
4 Lowry, T.M. (1923). *J. Soc. Chem. Ind.* 42: 43–47.
5 Fujita, M., Fukuzumi, S., and Otera, J. (1993). *J. Mol. Catal.* 85: 143–148.
6 Kaiser, D., Bauer, A., Lemmerer, M., and Maulide, N. (2018). *Chem. Soc. Rev.* 47: 7899–7925.
7 Valerio, V., Madelaine, C., and Maulide, N. (2011). *Chem. Eur. J.* 17: 4742–4745.
8 Kaiser, D. and Maulide, N. (2016). *J. Org. Chem.* 81: 4421–4428.
9 Wallach, O. (1877). *Liebigs Ann. Chem.* 184: 1–127.
10 Ghosez, L., Haveaux, B., and Viehe, H.G. (1969). *Angew. Chem. Int. Ed.* 8: 454–455.
11 Wallach, O. (1887). *Liebigs Ann. Chem.* 238: 78–89.
12 Speziale, A.J. and Smith, L.R. (1962). *J. Org. Chem.* 27: 4361–4365.
13 Figee, T. (1915). *Recl. Trav. Chim. Pays-Bas* 34: 289–325.
14 Stollé, R. and Luther, M. (1920). *Ber. Dtsch. Chem. Ges.* 53: 314–317.
15 Speziale, A.J. and Smith, L.R. (1963). *J. Org. Chem.* 28: 1805–1811.
16 Ott, E., Dittus, G., and Weißenburger, H. (1943). *Ber. Dtsch. Chem. Ges.* 2: 84–88.
17 Wolf, V. and Block, W. (1960). *Liebigs Ann. Chem.* 1: 119–126.
18 Speziale, A.J. and Freeman, R.C. (1960). *J. Am. Chem. Soc.* 4: 903–909.
19 Speziale, A.J. and Freeman, R.C. (1962). *J. Am. Chem. Soc.* 10: 1868–1876.
20 Ficini, J. and Barbara, C. (1964). *Bull. Soc. Chim. Fr.* 871.
21 Yakubovitch, A.Y., Sergeev, A.P., and Befyaeva, T.N. (1965). *Dokl. Akad. Nauk SSSR* 161: 1362.
22 Buyle, R. and Viehe, H.G. (1969). *Tetrahedron* 25: 3447–3451.
23 Ryan, T.A., Ryan, C., Seddon, E.A., and Seddon, K.R. (1996). , Topics in Inorganic and General Chemistry, Elsevier Science, vol. 24, 3–932.
24 Haveaux, B., Dekoker, A., Rens, M. et al. (1979). *Org. Synth.* 59: 26.
25 Marchand-Brynaert, J. and Ghosez, L. (1972). *J. Am. Chem. Soc.* 94: 2869–2870.
26 Janousek, Z., Collard, J., and Viehe, H.G. (1972). *Angew. Chem.* 84: 993.
27 Wei, L.-L., Mulder, J.A., Xiong, H. et al. (2001). *Tetrahedron* 57: 459–466.
28 Evano, G., Coste, A., and Jouvin, K. (2010). *Angew. Chem.* 49: 2840–2859.

29 Zificsak, C.A., Mulder, J.A., Hsung, R.P. et al. (2001). *Tetrahedron* 57: 7575–7606.

30 Viehe, H.G., Buijle, R., Fuks, R. et al. (1967). *Angew. Chem.* 6: 77–78.

31 Theunissen, C., Métayer, B., Lecomte, M. et al. (2017). *Org. Biomol. Chem.* 15: 4399–4416.

32 Li, J.-L., Lin, E., Han, X.-L. et al. (2019). *Org. Lett.* 21: 4255–4258.

33 Lambrecht, J., Zsolnai, L., Huttner, G., and Jochims, J.C. (1982). *Chem. Ber.* 115: 172–184.

34 Deyrup, J.A. and Kuta, G.S. (1978). *J. Org. Chem.* 43: 501–505.

35 Falmagne, J.-B., Escudero, J., Taleb-Sahraoui, S., and Ghosez, L. (1981). *Angew. Chem., Int. Ed. Engl.* 20: 879–880.

36 Charrete, A.B. and Grenon, M. (2001). *Can. J. Chem.* 79: 1694–1703.

37 Medley, J.W. and Movassaghi, M. (2009). *J. Org. Chem.* 74: 1341–1344.

38 Diels, O. and Alder, K. (1928). *Liebigs Ann. Chem.* 460: 98–122.

39 Ciamician, G. and Silber, P. (1908). *Ber. Dtsch. Chem. Ges.* 41: 1071–1080.

40 Büchi, G. and Goldman, I.M. (1957). *J. Am. Chem. Soc.* 79: 4741–4748.

41 Tsipi, D., Gegiou, D., and Hadjoudis, E. (1987). *J. Photochem.* 37: 159–166.

42 Staudinger, H. (1912). *Die Ketene*. Stuttgart: Ferdinand Enke.

43 Huisgen, R. and Otto, P. (1968). *Tetrahedron Lett.* 9: 4491–4495.

44 Woodward, R. and Hoffman, R. (1965). *J. Am. Chem. Soc.* 87: 395.

45 Woodward, R. and Hoffman, R. (1969). *Angew. Chem., Int. Ed. Engl.* 8: 781.

46 Moore, H.W. and Wilbur, D.S. (1980). *J. Org. Chem.* 45: 4483.

47 Staudinger, H. (1907). *Liebigs Ann. Chem.* 356: 51–123.

48 Huisgen, R. and Otto, P. (1968). *J. Am. Chem. Soc.* 90: 5342–5343.

49 Patai, S. (1964). *The Chemistry of Alkenes*. Wiley.

50 Bartlett, P.D., Cohen, G.M., Elliott, S.P. et al. (1972). *J. Am. Chem. Soc.* 94: 2899–2902.

51 Cannell, L.G. (1972). *J. Am. Chem. Soc.* 94: 6867–6869.

52 Huisgen, R. and Feiler, L.A. (1969). *Ber. Dtsch. Chem. Ges.* 102: 3391–3404.

53 Hall, H.K. Jr., Smith, C.D., and Plorde, D.E. (1973). *J. Org. Chem.* 38: 2084–2087.

54 Ding, W.J. and Fang, D.C. (2001). *J. Org. Chem.* 66: 6673–6678.

55 Saimoto, H., Houge, C., Hesbain-Frisque, A. et al. (1983). *Tetrahedron Lett.* 24: 2251–2254.

56 Arrieta, A., Lecea, B., and Cossío, F.P. (1998). *J. Org. Chem.* 17: 5869–5876.

57 Díez, E., Fernández, R., Marqués-López, E. et al. (2004). *Org. Lett.* 6: 2749–2752.

58 Schmit, C., Sahraoui-Taleb, S., Differding, E. et al. (1984). *Tetrahedron Lett.* 52: 5043–5046.

59 Marko, I., Ronsmans, B., Hesbain-Frisque, A.M. et al. (1985). *J. Am. Chem. Soc.* 107: 2192–2194.

60 Lachia, M., Dakas, P.Y., and De Mesmaeker, A. (2014). *Tetrahedron Lett.* 55: 6577–6581.

61 Brady, W.T., Giang, Y.S.F., Weng, L., and Dad, M.M. (1986). *J. Org. Chem.* 52: 2216–2220.

62 Lachia, M., Wolf, H.C., and De Mesmaeker, A. (2014). *Bioorg. Med. Chem. Lett.* 24: 2123–2128.

63 Depré, D., Chen, L.Y., and Ghosez, L. (2003). *Tetrahedron* 59: 6797–6812.

64 Adam, J.M., Ghosez, L., and Houk, K.N. (1999). *Angew. Chem. Int. Ed.* 38: 2728–2730.

65 Chen, L.Y. and Ghosez, L. (1991). *Tetrahedron: Asymmetry* 2: 1181–1184.

66 Ghosez, L., Marko, I., and Hesbain-Frisque, A.M. (1986). *Tetrahedron Lett.* 27: 5211–5214.

67 Houge, C., Frisque-Hesbain, A.M., Mockel, A., and Ghosez, L. (1982). *J. Am. Chem. Soc.* 104: 2920–2921.

68 Chen, L.Y. and Ghosez, L. (1990). *Tetrahedron Lett.* 31: 4467–4470.

69 Bonazzi, S., Cheng, B., Wzorek, J.S., and Evans, D.A. (2013). *J. Am. Chem. Soc.* 135: 9338–9341.

70 Lumbroso, A., Behra, J., Kolleth, A. et al. (2015). *Tetrahedron Lett.* 56: 6541–6545.

71 Bélanger, G., Darsigny, V., Doré, M., and Lévesque, F. (2010). *Org. Lett.* 12: 1396–1399.

72 Medley, J.W. and Movassaghi, M. (2013). *Org. Lett.* 15: 3614–3617.

73 White, K.L., Mewald, M., and Movassaghi, M. (2015). *J. Org. Chem.* 80: 7403–7411.

74 Xiao, K.-J., Luo, J.-M., Ye, K.-Y. et al. (2010). *Angew. Chem. Int. Ed.* 49: 3037–3040.

75 Barabé, F., Nantel, M., Larouche-Gauthier, R., and Bélanger, G. (2005). *Org. Lett.* 7: 4431–4434.

76 White, K.L. and Movassaghi, M. (2016). *J. Am. Chem. Soc.* 138: 11383–11389.

77 Kang, T., White, K.L., Mann, T.J. et al. (2017). *Angew. Chem. Int. Ed.* 56: 13857–13860.

78 Mewald, M., Medley, J.W., and Movassaghi, M. (2014). *Angew. Chem. Int. Ed.* 53: 11634–11639.

79 Medley, J.W. and Movassaghi, M. (2012). *Angew. Chem. Int. Ed.* 51: 4572–4576.

80 Magnus, P., Gazzard, L., Hobson, L. et al. (2002). *Tetrahedron* 58: 3423–3443.

81 Shojia, T., Inoue, Y., and Ito, S. (2012). *Tetrahedron Lett.* 53: 1493–1496.

82 Jakubec, P., Kyle, A.F., Calleja, J., and Dixon, D.J. (2011). *Tetrahedron Lett.* 52: 6094–6097.

83 Overman, L.E. and Wolfe, J.P. (2002). *J. Org. Chem.* 67: 6421–6429.

84 Huang, P.-Q. (2010). *Chem. Eur. J.* 16: 12792–12796.

85 Xiao, K.-J., Luo, J.-M., Xia, X.-E. et al. (2013). *Chem. Eur. J.* 19: 13075–13086.

86 Deng, H.-Q., Qian, X.-Y., Li, Y.-X. et al. (2014). *Org. Chem. Front.* 1: 258–266.

87 Huo, H.-H., Xia, X.-E., Zhang, H.-K., and Huang, P.-Q. (2013). *J. Org. Chem.* 78: 455–465.

88 Xiao, K.-J., Wang, Y., Huang, Y.-H. et al. (2013). *J. Org. Chem.* 78: 8305–8311.

89 Liao, J.-C., Xiao, K.-J., Zheng, X., and Huang, P.-Q. (2012). *Tetrahedron* 68: 5297–5302.

90 Lindemann, C. and Schneider, C. (2016). *Synthesis* 48: 828–844.

91 Barbe, G. and Charette, A.B. (2008). *J. Am. Chem. Soc.* 130: 18–19.

92 Xiang, S.-H., Xu, J., Yuan, H.-Q., and Huang, P.-Q. (2010). *Synlett* 12: 1829–1832.
93 Porte, V., Di Mauro, G., Schupp, M. et al. (2020). *Chem. Eur. J.* 26: 15509–15512.
94 de la Torre, A., Kaiser, D., and Maulide, N. (2017). *J. Am. Chem. Soc.* 139: 6578–6581.
95 Charette, A.B. and Chua, P. (1998). *Synlett* 1998: 163–165.
96 Charette, A.B. and Chua, P. (1997). *Tetrahedron Lett.* 38: 8499–8502.
97 Valerio, V., Petkova, D., Madelaine, C., and Maulide, N. (2013). *Chem. Eur. J.* 19: 2606–2610.
98 Peng, B., Geerdink, D., and Maulide, N. (2013). *J. Am. Chem. Soc.* 135: 14968–14971.
99 Charette, A.B. and Grenon, M. (2000). *Tetrahedron Lett.* 41: 1677–1680.
100 Charette, A.B. and Chua, P. (1998). *J. Org. Chem.* 63: 908–909.
101 Dagoneau, D., Kolleth, A., Quinodoz, P. et al. (2019). *Helv. Chim. Acta* 102: e1900031.
102 Mátravölgyi, B., Hergert, T., Bálint, E. et al. (2018). *J. Org. Chem.* 83: 2282–2292.
103 Spinozzi, E., Bauer, A., and Maulide, N. (2019). *Eur. J. Org. Chem.* 31: 5230–5233.
104 Tona, V., de la Torre, A., Padmanaban, M. et al. (2016). *J. Am. Chem. Soc.* 138: 8348–8351.
105 Tona, V., Maryasin, B., de la Torre, A. et al. (2017). *Org. Lett.* 19: 2662–2665.
106 Kaiser, D., de la Torre, A., Shaaban, S., and Maulide, N. (2017). *Angew. Chem. Int. Ed.* 56: 5921–5925.
107 Gonçalves, C.R., Lemmerer, M., Teskey, C.J. et al. (2019). *J. Am. Chem. Soc.* 141: 18437–18443.
108 Kaiser, D., Teskey, C.J., Adler, P., and Maulide, N. (2017). *J. Am. Chem. Soc.* 139: 16040–16043.
109 Li, J., Berger, M., Zawodny, W. et al. (2019). *Chem* 5: 1883–1891.
110 Adler, P., Teskey, C.J., Kaiser, D. et al. (2019). *Nat. Chem.* 11: 329–334.
111 Thomas, E.W. (1993). *Synthesis* 6: 767–768.
112 Cyr, P., Régnier, S., Bechara, W.S., and Charette, A.B. (2015). *Org. Lett.* 17: 3386–3389.
113 Bechara, W.S., Khazhieva, I.S., Rodriguez, E., and Charette, A.B. (2015). *Org. Lett.* 17: 1184–1187.
114 Li, L.-H., Niu, Z.-J., and Liang, Y.-M. (2017). *Chem. Eur. J.* 61: 15300–15304.
115 Wezeman, T., Zhong, S., Nieger, M., and Bräse, S. (2016). *Angew. Chem. Int. Ed.* 55: 3823–3827.
116 Movassaghi, M. and Hill, M.D. (2008). *Org. Lett.* 10: 3485–3488.
117 Movassaghi, M. and Hill, M.D. (2006). *J. Am. Chem. Soc.* 128: 14254–14255.
118 Movassaghi, M., Hill, M.D., and Ahmad, O.K. (2007). *J. Am. Chem. Soc.* 129: 10096–10097.
119 Henry, G.D. (2004). *Tetrahedron* 60: 6043–6061.
120 Michael, J.P. (2005). *Nat. Prod. Rep.* 22: 627–646.

121 Cui, S.-L., Wang, J., and Wang, Y.-G. (2008). *J. Am. Chem. Soc.* 130: 13526–13527.
122 Charette, A.B., Mathieu, S., and Martel, J. (2005). *Org. Lett.* 7: 5401–5404.
123 Charette, A.B., Grenon, M., Lemire, A. et al. (2001). *J. Am. Chem. Soc.* 123: 11829–11830.
124 Lemire, A. and Charette, A.B. (2005). *Org. Lett.* 13: 2747–2750.
125 Abramovitch, R.A. and Singer, G.M. (1969). *J. Am. Chem. Soc.* 91: 5672–5673.
126 Da Costa, R., Gillard, M., Falmagne, J.-B., and Ghosez, L. (1979). *J. Am. Chem. Soc.* 101: 4381–4383.
127 Bauer, A. and Maulide, N. (2019). *Chem. Sci.* 10: 9836–9840.
128 Huang, P.-Q., Huang, Y.-H., and Xiao, K.-J. (2016). *J. Org. Chem.* 81: 9020–9027.
129 Li, J., Oost, R., Maryasin, B. et al. (2019). *Nat. Commun.* 10: 2327.
130 Wang, A.-E., Yu, C.-C., Chen, T.-T. et al. (2018). *Org. Lett.* 20: 999–1002.

6

Transamidation of Carboxamides and Amide Derivatives: Mechanistic Insights, Concepts, and Reactions

Paola Acosta-Guzmán, John Corredor-Barinas, and Diego Gamba-Sánchez

Universidad de los Andes, Laboratory of Organic Synthesis, Bio and Organocatalysis, Chemistry Department, Cra 1 No. 18A-12 Q:305, Bogotá 111711, Colombia

6.1 Introduction

Amides have been the focus of intensive research in recent years, not only because the amide bond is a fundamental structure in nature (peptides and proteins) but also because a large number of active compounds have at least one amide bond. In consequence, most of the research has been focused on the development of new strategies for the construction of amides; in other words, the amide function has been the target, and many amazing methodologies have been developed. The main reason for those developments is justified with some issues in classic reactions for amide bond formation, such as the production of waste materials and the stoichiometric use of coupling reagents, which are reputed by their extreme reactivity, dangerousness, or toxicity.

In recent years, amides have also been seen as synthetic intermediates, and the development of amide bond activation processes has afforded access to new reactions, and, most importantly, nowadays the amides are recognized not only by their stability and presence in some active compounds but also by their capacity to act as precursors of valuable and highly functionalized molecules. As part of those activation processes, the amides have also been used as sources of acyl groups; in that context, the transamidation reaction has gained enormous interest.

In principle, the transamidation reaction is the exchange of the amine moiety using an amide and an amine as the reagents; as usual, in exchange reactions, an equilibrium is established, and, consequently, traditional transamidations required an excess of one of the reagents (typically the amine). Recent developments have been concentrated in the activation of the amide bond to increase the electrophilicity of the carbonyl group. On the other hand, the development of new transformations in which the amine moiety is tacked off in a non-nucleophilic form is also appreciated, since those methods reduce the quantity of reagents needed to accomplish the desired transformation. It is noteworthy that the twist of the amide bond has also proved useful as the "activation" step.

Amide Bond Activation: Concepts and Reactions, First Edition. Edited by Michal Szostak.

This chapter aims to convince the reader of the power of the transamidation processes as an alternative in the synthesis of complex amides. In this context, the chapter starts with historical background, showing the pioneering developments, the initial problems in the equilibration of the transamidation process, and the first solutions. The next section will be focused on direct transamidation of carboxamides, which means reactions where the activation is made *in situ* by a catalyst, an acid, a base, or another reagent, which do not involve the formation of an intermediate derivative. We will divide the section into several subsections regarding the type of catalyst used or the activation process. The subsequent section will be focused on the activation of the amide by using an activating group or an analog activated function; notably, we will discuss the insertion of groups like Ts and Boc, among others, which induce a modification on the amide geometry increasing its electrophilicity. The section is divided into subsections according to the catalyst or promoter used. Finally, we will discuss the transamidation of atypical substrates, which means the use of reagents other than amines and amides or reactions that proceed by a different mechanism of activation; in this section we will also discuss some examples where the useful functions are generated *in situ*. At the end of the chapter, we will present the conclusions and some perspectives. It has to be highlighted that the aim of the chapter is not to show all the existent literature on this topic but to show and discuss some selected publications to illustrate the concepts.

6.2 Historical Background

The transamidation reaction, understood as the amine exchange between an amide and an amine (Scheme 6.1a), was discovered in 1889 by C. Hoffmann [1] when heating a mixture of acetamide and hydroxylammonium chloride (Scheme 6.1b); however, similar reactions had been previously described in the literature. Notably, between 1876 and 1895 several reports on thermal "transamidations" of urea and aromatic amines were observed and described without a rigorous study of the transformation [2].

It is noteworthy that even if – accurately speaking – the reaction of amines exchange between urea(s) and amines is not a real transamidation process, recent

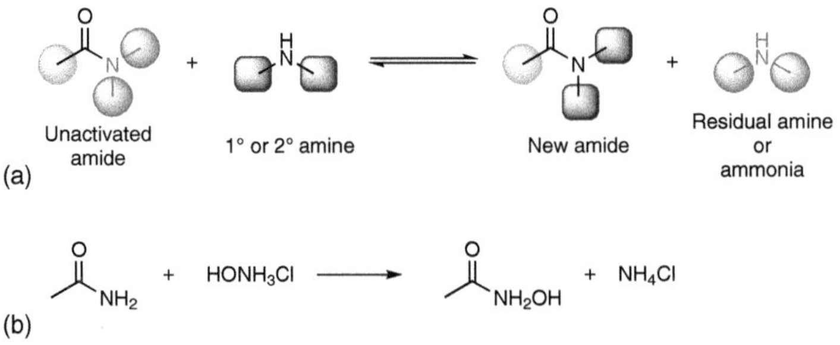

Scheme 6.1 (a) General transamidation reaction and (b) First transamidation.

literature uses to include urea(s) in the study of the transamidation scope and limitations; consequently during this chapter we will include urea as an amide analog.

After the discovery of Hoffman, the transamidation of urea and substituted ureas was studied by Tenney L. Davis [3], in a process called "urea dearrangement." It is well known that the nucleophilic addition of ammonia or amines to isocyanates will produce ureas, so Davis proposed a thermal decomposition of ureas to produce a mixture of products including ammonia, amines, isocyanates, and isocyanic acid, depending on the structure of the original urea (see Scheme 6.2).

The opposite process was known since the famous experiment of Wöhler, who synthesizes urea from a mineral product in 1828. Davis used Wöhler's observations and provided an explanation for the amine exchange when mixtures of ureas and amines were heated. This is not a transamidation process in the terms we know it now; however, it is certainly a plausible mechanism in modern transamidations involving ureas.

Despite all the studies on urea transformations, the transamidation of amides remained aside until 1895 when Hirst and Cohen published the reaction between formamide and some aromatic amines [2]; on their first attempts they used a mixture of formamide, phenylhydrazine, and acetic acid (Scheme 6.3a), then they realized that the hydrazine may be replaced by aromatic amines, and the reaction still proceeds (Scheme 6.3b). Some years later a study on the transamidation reaction was published, and the focus was the evaluation of the reactivity of several amides with aniline; unfortunately, only low yields were obtained using acetamide or benzamide, and extremely low conversion was observed with other amides.

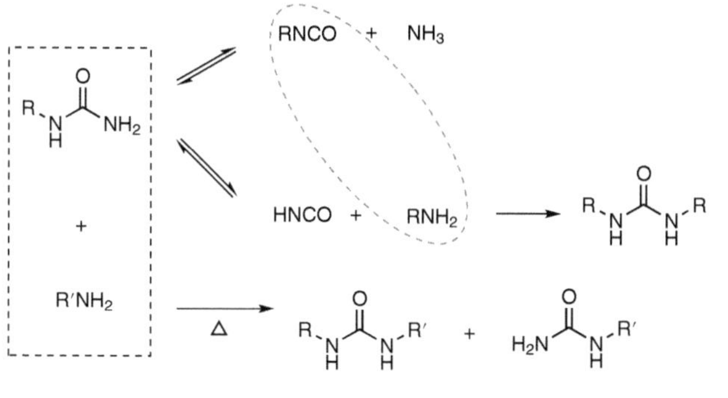

Scheme 6.2 Davis' urea dearrangement.

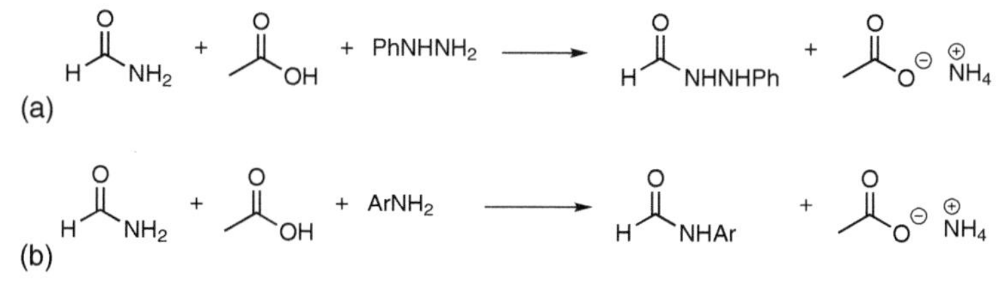

Scheme 6.3 Transamidation of formamide promoted by acetic acid. (a) With phenylhydrazine and (b) With aniline.

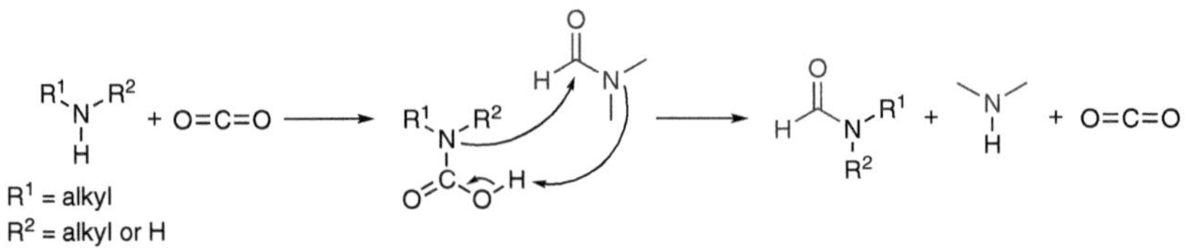

Scheme 6.4 Transamidation promoted by CO_2.

The first general procedure of transamidation was described by Elion and Galat [4]; they took the Hoffmann observations and demonstrate that the amines may be acylated with amides if their hydrochlorides react under the appropriate conditions. It has to be highlighted that all the successful examples of acylations of amines with amides have one of the following conditions: (i) they used an acidic medium provided by the amine hydrochloride (or acetic acid) or (ii) the temperature is very high. It is clear by now that the activation of the acyl group may be accomplished by the use of an acid partner or by increasing the temperature. Nevertheless, in all cases the reaction times were long and the yields were poor.

The first transamidation promoted by an external, weak electrophilic reagent (authors called it a catalyst) was described in 1968 using carbon dioxide [5]; unfortunately authors described that the reaction procedure was carried out under a stream of CO_2, making it impossible to determine the quantity of the promoter used for that process. They propose the formation of a carbamic acid *in situ* and a kind of rearrangement to afford the transamidation product as shown in Scheme 6.4.

The use of DMF as the formylating agent of amines was described a few years later, with the same issues showed by other thermal transamidations [6]. It was in 1994 that the first modern study on a transamidation reaction was described using aluminum chloride (a Lewis acid) as the promoter [7]; the reaction was described using primary and secondary aliphatic and aromatic amides and primary and secondary aliphatic and aromatic amines. The drawbacks of this method are the use of excess amine (2.5–3.5 equiv) and an excess of the promoter (1.3–2.3 equiv). Modern studies and the introduction of real catalysts were posteriorly described and will be discussed in the next section.

6.3 Direct Transamidation of Carboxamides

Throughout this chapter, we will discuss two different types of transamidations. First is the direct transamidation that refers to the direct substitution of the amine moiety using a promoter or catalyst to increase the electrophilicity of the carbonyl group. Second, we will discuss the activation of the amide bond by the insertion of electron-withdrawing groups on the nitrogen atom.

6.3.1 Mechanistic Insights

The activation of the carbonyl group by the interaction of the oxygen with a Lewis or a Brønsted acid is the classic activation way to promote transamidations. Nearly all authors working on these kinds of reactions have published a mechanistic proposal for their work; however, the differences between the metals are virtually negligible, and all of them are based on the studies described by Stahl and coworkers [8]. Having this in mind we will discuss Stahl's mechanistic proposals for transamidation of secondary amides and tertiary amides independently. It is noteworthy that the proposal for secondary amides is also applicable to primary amides. On the other hand, the transamidation of ureas should follow an entirely different mechanism. Finally, the few examples concerning transamidations of imides do not provide any insight for the mechanism; however, we can presume that most of the imides follow the same mechanism as tertiary amides; the differences will be discussed in the next section.

Stahl's proposal is summarized in Scheme 6.5; only one coordinated amide is shown for clarity; however, depending on the metal, two or three amides will be coordinated to the metallic center. The mechanism starts with the generation of the amidate complex **A**; the formation of this complex needs the action of a basic species, to take a proton from the activated amide; typically the role is played by the ligand since the most common ligands are amines; when a simple salt is used, the counterion is basic (e.g. acetate) [9]. Few examples exist where there is no apparent base in the reaction media; nevertheless, water can play the role since the proton acidity is increased by the metal and the solvent [10]. The equilibrium is shifted by the addition of the amine to form **B**; this has two purposes, first diminishing the stability of cyclic complex **A** and second increasing the acidity of the amine protons. The equilibrium toward a more stable species **C** is spontaneous and produced after proton exchange. Then the reaction of the amine nitrogen with the highly electrophilic carbonyl results in the formation of a new cyclic intermediate **D**, which readily isomerizes to **E**. The equilibrium **C**-**D**-**E**-**F** will produce a statistical redistribution of the amides, which is the reason why the direct transamidation typically uses an excess of the amine. Once **F** is formed, it can take a proton from the protonated ligand – or in general from the conjugate acid of the base used to produce **A** – to generate **G**. Throughout the reaction of **G** with another molecule of amide and base, the complex **A** is regenerated with the concomitant ejection of ammonia or a volatile amine, which also helps to shift the equilibrium to **F**. Even if less common, the production of a non-nucleophilic amine also helps the shift of the equilibrium, making the ejection the driving force of transamidations.

The transamidation of tertiary amides should follow a different mechanism, since they do not have an acidic proton. Stahl's proposal [11] suggests that simultaneous activation of the amide and the amine is the first step for this transformation (see Scheme 6.6). The Lewis acid playing a dual role led to the formation of intermediate **H**, which has the activated amide and the more nucleophilic amine close to each other. The intramolecular nucleophilic addition of the amine let to the stable intermediate **I**, which is similar in structure to **D** and **E**. **I** is in equilibrium with its isomer **K**, through the acyclic species **J**. The generation of the new complex **L** followed by the interaction of a new pair of molecules (amide and amine) led to the transamidation product and restarts the catalytic cycle.

The transamidation of ureas is slightly atypical; however, the most successful protocols for direct transamidations include this kind of molecules in the study of

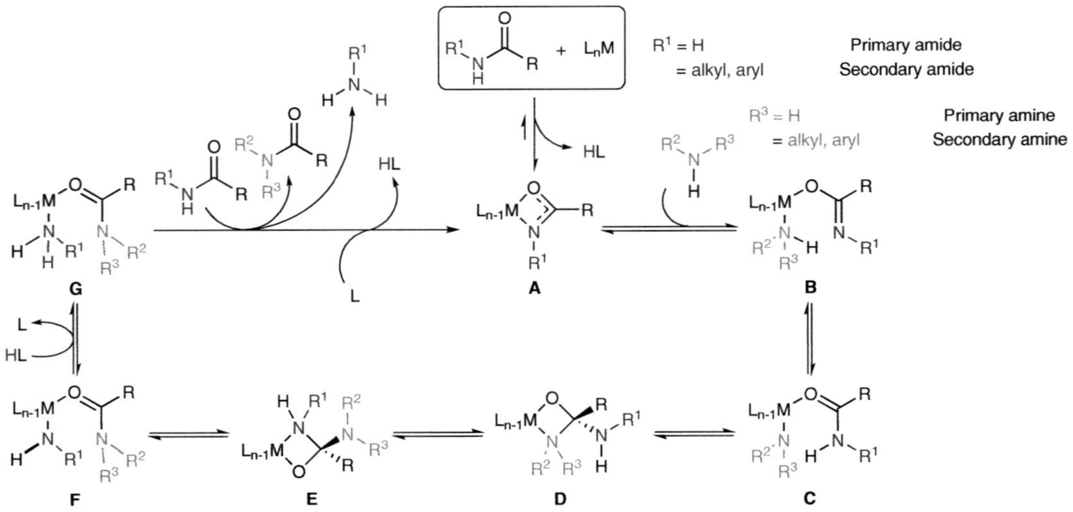

Scheme 6.5 Mechanism of transamidation of primary and secondary amides.

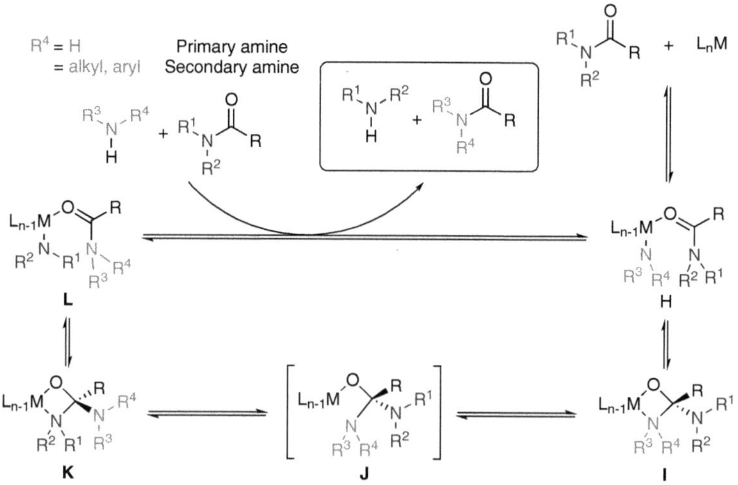

Scheme 6.6 Mechanism of transamidation of tertiary amides.

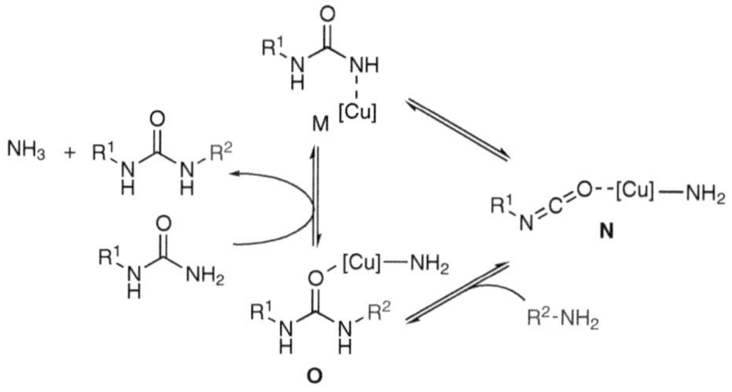

Scheme 6.7 Mechanistic proposal for transamidation of ureas.

the scope and limitations. The so-called urea dearrangement seems to be the most plausible mechanism for transamidation of ureas as depicted in Scheme 6.7. The interaction between the metal and one of the nitrogen atoms in **M** will produce an activated isocyanate **N**; then the nucleophilic addition of the new amine will afford the transamidated urea **O**. Finally, the catalytic cycle restarts ureas exchange [12].

We have presented the mechanistic proposals for the transamidation of amides and ureas. The evidence provided by the authors suggests that activation is the key step of any direct transamidation, so we can assume that activations with heterogeneous catalysts or Brønsted acids should follow a very similar mechanism.

6.3.2 Transition Metal Catalysis

In this section we will summarize the metals that have been used in the transamidation process; we will use a table format since the particular examples have been reviewed in recent literature [13] (Table 6.1).

Table 6.1 Summary of direct transamidation catalyzed by metals.

Metal	Substrates		Observations	
	Amide	Amine	Positive	Negative
$Cu(AcO)_2$ [9]	Aliphatic 1°	Aliphatic 1° and 2°	Works with urea, hydroxy amides are used	Only 1° amides were used
$Fe(NO_3)_3 \cdot 9H_2O$ [10]	Aliphatic 1°, 2°, and 3°. Aromatic 1°	Aliphatic 1°and 2°. Aromatic 1°	Works with urea, phthalimide, excess of amide or amine	Lower yields with aromatic amides and hindered secondary amines
Nb [14]	Aliphatic 1°, 2°, and 3°. Aromatic 1°	Aliphatic 1°and 2°. Aromatic 1°	Works with phthalimide, a slight excess of amine	Only useful with the production of volatile amines
Cp_2ZrCl_2 [15]	Aliphatic 1° and aromatic 1°	Aliphatic 1°and 2°	Works at low temperatures	Low yields with aromatic amines
$Co(AcO)_2$ [16]	Aliphatic 1°, 2°, and 3°. Aromatic 1°	Aliphatic 1°and 2°. Aromatic 1° and 2°	Works with polyfunctional molecules	High temperature and load of catalyst (15 mol%)
CeO_2 [17]	Aliphatic 1° and aromatic 1°	Aliphatic 1°and 2°		Limited substrate scope
Mn(II) [18]	Aliphatic 1°	Aliphatic 1°and 2°. Aromatic 1° and 2°		Limited substrate scope
MnO_2 [19]	Aliphatic 1° and Aromatic 1°	Aliphatic 1°and Aromatic 1°	Works with phthalimide	
MoS_2 [20]	Aliphatic 1°, 2°, and 3°. Aromatic 1°	Aliphatic 1°and 2°		Low yields with acyls other than formyl
$Pd(AcO)_2$ [21]	Aliphatic 1°, 2°, and 3°	Aromatic 1°		Only useful with aromatic amines

In addition to the methods summarized in the table, the use of Fe(III) [22] and Cu(II) [23] has been described under different conditions and with limited substrate scopes. On the other hand, some recent examples include the use of Zn and Ni. We decided to make a special mention of those methods.

The use of direct transamidations in peptide synthesis is pretty rare; only a few successful examples have been described using amino acids; however, Ballet, Maes, and coworkers [24] described an interesting transamidation using zinc as the catalyst. The method is based on the synthesis of a special amide **3** with the *tert*-butyl nicotinate as a directing group; the zinc will coordinate the carbonyl oxygen and the nitrogen of the heterocycle, activating the amide and stabilizing the leaving group (see transition state); the reaction with the amino acid ester yields the corresponding dipeptide **4**. Authors succeeded in the synthesis of peptides up to seven units of amino acids using this method. Interestingly, it can be applied in solid-phase synthesis, and the *tert*-butyl nicotinate is stable to classic deprotection conditions in peptide synthesis (see Scheme 6.8).

A special challenge in direct transamidation is the use of secondary and tertiary amides **5** having a big amine moiety; when those amides react, they produce a nonvolatile amine, and it is difficult to switch the equilibrium; also the use of aromatic amines as nucleophiles has always been difficult. Successful transamidation for these substrates has been described using a mixture of manganese, TMSCl, and Ni/briphos as the catalyst [25]. Remarkably, the reaction is based on the activation of the amine forming the bis-TMS amine, which (according to the authors) is the substrate for the transamidation. The activation of the amide is reached by the nickel complex; however, the role of TMSCl and manganese is still under evaluation, and somehow the mechanism of the transformation is undetermined. A summary of the scope is presented in Scheme 6.9.

6.3.3 Organocatalysis

Organocatalysis has gained attention in the last few decades, and amazing results have been obtained in different fields of organic chemistry. Amide synthesis and, particularly, the transamidation reaction is not an exception; consequently, some reports using organocatalysis should be mentioned.

The first description of a transamidation reaction using an organic catalyst was published in 2013 using hydroxylamine hydrochloride ($NH_2OH{\cdot}HCl$) as the promoter [26]. Several ammonium salts were active in stimulating the transamidation of primary amides; however, the hydroxylamine hydrochloride was the most reactive catalyst. The authors proposed activation of the amide carbonyl by the acidic ammonium moiety and a hydrogen bond between the hydroxy group and one of the hydrogens of the amide (Figure 6.1a). Two feasible pathways are supported by kinetic studies, the formation of hydroxamic acid and the direct attack of the amine. Authors prefer the second one; however, they demonstrate that hydroxamic acids would produce the same amide by reaction with an amine under the same reaction conditions. In summary, the mechanism has some important issues to solve. On the other hand, the activation of the amide carbonyls by an ammonium salt was previously described and mentioned in this chapter; unfortunately, the lack of reactivity for secondary amides cannot be explained by the proposed mechanism. Similar activations were proposed with the use of benzoic acid (Figure 6.1b) as the organocatalyst in a reaction with a highly superior substrate scope [27] and also as described in the case of the use of chitosan (Figure 6.1c) [28].

The activation of amides by organic catalysts was also accomplished using L-Proline [29]. In this case, the mechanism suggests dehydration of the amide to form the intermediate **8**, which is supported by computational calculations [30]. It is noteworthy that authors proved that the amine and carboxylic acid groups are

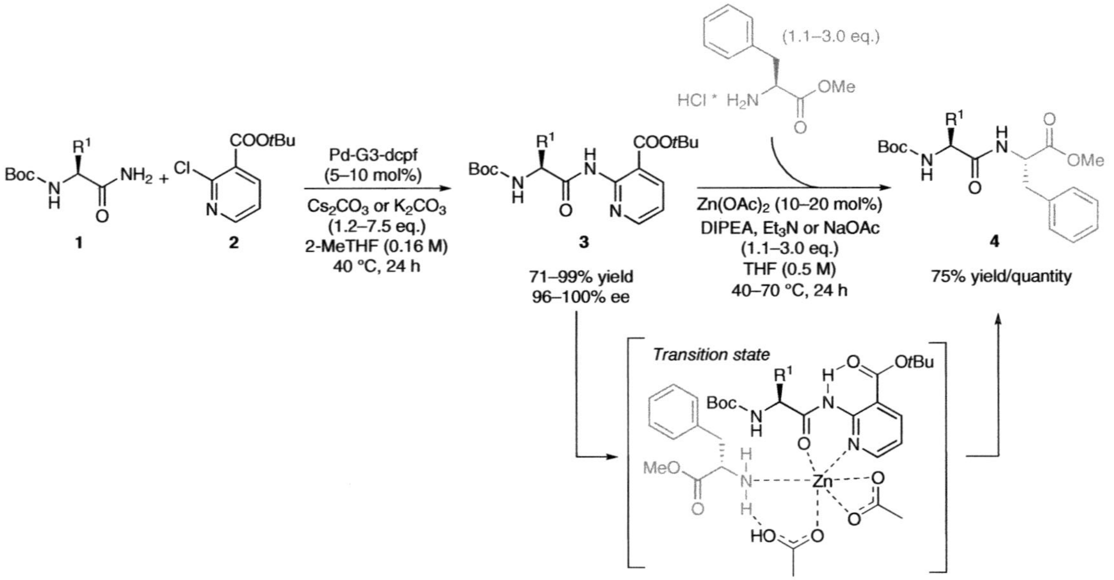

Scheme 6.8 Peptide synthesis by direct transamidation.

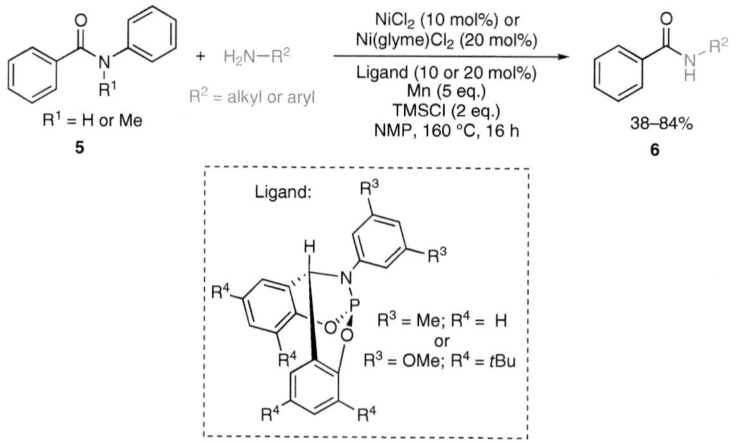

Scheme 6.9 Transamidation of secondary and tertiary amides catalyzed by nickel complexes.

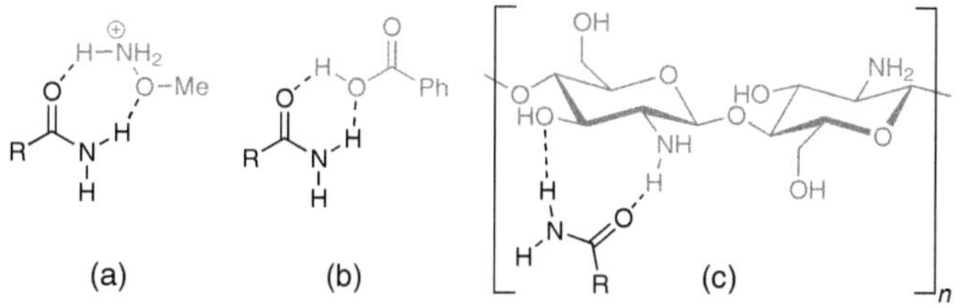

Figure 6.1 Proposed transition states in organocatalytic activation of amides: (a) by ammonium salts, (b) by benzoic acid, and (c) by chitosan.

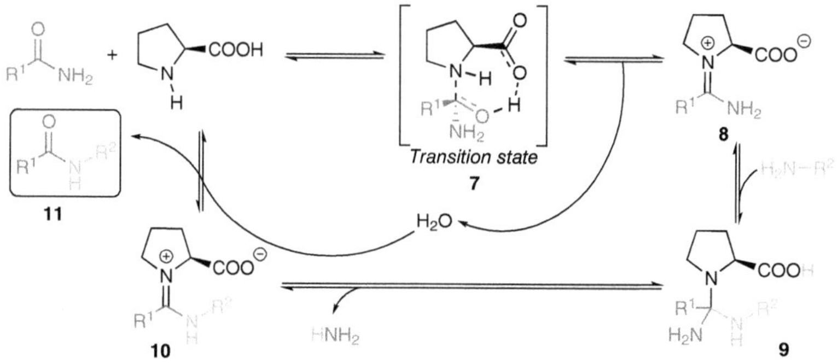

Scheme 6.10 Mechanism of transamidation with L-Proline.

needed for the reaction to proceed, so the formation of **8** is induced by the activation of the carbonyl as shown in Scheme 6.10 (see transition state **7**); the subsequent steps of the mechanism will provide the transamidation product **11**, throughout the intermediates **9** and **10**.

To the best of our knowledge, there are no more reports describing transamidations using organic catalysts; some reports exist using organoboron derivatives;

however, we will treat those compounds in the next section as the proposed mechanism is based on the Lewis acid behavior of boron. In consequence, organocatalytic transamidation is an area of potential growth.

6.3.4 Other Catalytic and Promoted Processes

6.3.4.1 Bases

As we mentioned before, the transamidation reaction needs the activation of the carbonyl group to increase its electrophilicity and make it reactive with weak nucleophiles such as amines; consequently the transamidation promoted by bases is conceptually strange, since bases generally activate the nucleophile instead of the electrophile. Fortunately, some researchers have addressed that question, and some interesting results have been published on transamidations promoted by basic species; notably the use of LiHMDS [31], *t*BuOK [32], and *t*BuONa [33] are the most useful examples.

LiHMDS seems to participate in the reaction with a double role; first as the activator of the amine by deprotonation forming an anionic and more nucleophilic species and second by acting as a Lewis acid in the activation of the carbonyl. The reaction proceeds under amazingly mild conditions; it is noteworthy that the method is focused on the reactivity of tertiary amides **12**, which are the least reactive ones, so this method was an impressive advance in transamidation reactions. The mechanism was studied by computational calculations; the results showed virtually the same results obtained by Stahl as depicted in Scheme 6.6. The main difference is that the metal is coordinated with one or two HMDS moieties, and the authors demonstrated that dimers or trimers have essentially the same reactivity. In Scheme 6.11a some of the most representative examples of this transamidation and the typical reaction conditions are shown. We also illustrate one example using a twisted amide **14** in which the reaction is selective to cleave the C(O)–N bond (Scheme 6.11b) [31].

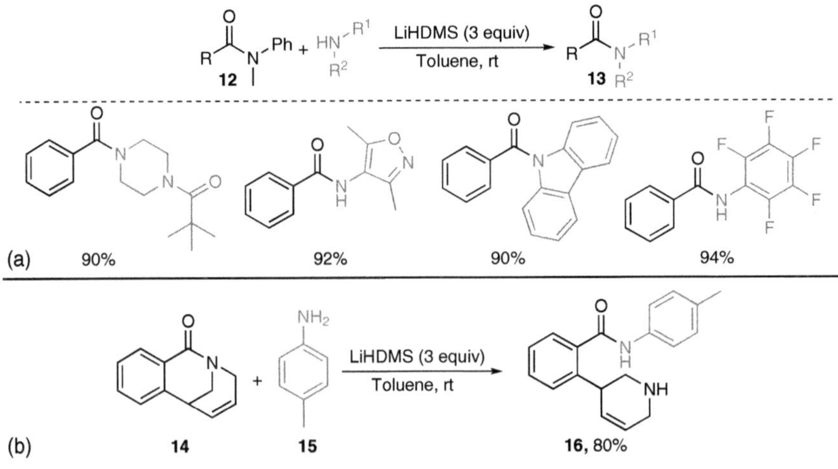

Scheme 6.11 (a) Extract of the substrate scope of transamidation promoted by LiHMDS and (b) Selected example with a twisted amide.

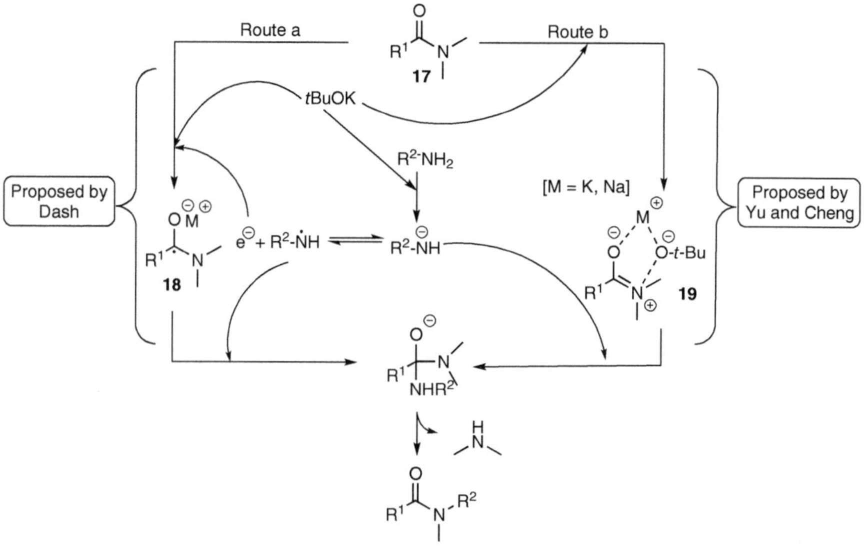

Scheme 6.12 Proposed pathways for transamidation promoted by *tert*-butoxides.

Other strong and bulky bases such as *tert*-butoxides have been used to promote the transamidation of tertiary amides. Notably, the simultaneous reports by Dash [32a] and Yu and Li [32b] showed the transamidation of simple *N,N*-dimethyl amides **17** with a wide range of amines (aromatic and aliphatic); both authors conducted control experiments and showed the possibility to have a radical reaction pathway (route a in Scheme 6.12), and both of them arrive at the same mechanistic proposal; however, Yu and Li also propose a more traditional polar pathway (route b in Scheme 6.12), having the potassium linked to the amide oxygen and the oxygen of the *t*BuOK to the nitrogen (transition state **19**). It is noteworthy that during the studies the control experiments conducted using *t*BuONa showed essentially no effects of radical scavengers, showing that changing the counterion the reaction mechanism may be entirely different. The summary of the mechanisms proposed by Yu and Li as well as that proposed by Dash is showed in Scheme 6.12.

A few months later, Cheng and coworkers reported the use of *t*BuONa [33] with *N,N*-dimethyl amides **17**; the substrate scope is a bit limited since authors only used simple amines; on the other hand, the reaction proceeds under solvent-free conditions, which may be a problem when the two substrates are solids. They adhere to the polar mechanistic proposal made by Yu and Li (route b).

6.3.4.2 Boron Derivatives

The first description of a transamidation using boron compounds was published in 2012 using $B(OH)_3$ [34], which proved active in concentrations as low as 10 mol%. The reaction has a very good substrate scope since it works with primary, secondary, and tertiary amides with a wide range of amines, including aromatic and aliphatic. Interestingly, this method is also useful with phthalimides serving as a mild method to protect primary amines. The authors proposed the activation of the amide based on the Lewis acidity of the boron atom; the boric acid is not strong enough to protonate the amide; in consequence, a transition state similar to those described in Figure 6.1 seems a plausible option. That means an interaction between the carbonyl oxygen with the boron and a hydrogen bond between the nitrogen and the protons of the boric acid, which should activate the amide (intermediate **20**); this should proceed also with borinic acids [35]. However, the hydrogen bond is a secondary interaction, and probably is not essential for the reaction to occur; this is supported by subsequent reports using borates [36] where the formation of the hydrogen bond is impossible. It has to be highlighted that the activation should proceed unexceptionally with primary, secondary, and tertiary amides. Two options for the amine insertion are reasonable; first, the amine can react directly with the activated intermediate to generate **22**, or it can add as a Lewis base on the boron to produce the intermediate **21**, making the addition to the carbonyl easy. Interestingly, the authors suggested that water reacts with intermediate **23** to regenerate the boric acid. That means a proton source is mandatory when the catalyst is boric or borinic acids; in the second case this role is assumed by acetic acid. The reaction using trifluoroethylborate should follow a very similar mechanism in terms of activation, but the subsequent steps and regeneration of the catalyst may proceed differently. See Scheme 6.13 for details of the mechanism with boric and borinic acids.

6.3.4.3 Heterogeneous Catalysis

In addition to transition metals, the heterogeneous catalysts are the most abundant activators for transamidations; this is because the structure of heterogeneous catalysts provides a suitable surface functionalized with Lewis or Brønsted acidic sites; in that way the catalyst may activate the amide as we have seen in the previous sections of this chapter. Unfortunately, the substrate scope of the transamidations promoted by heterogeneous materials is still very limited in comparison with the process using metals or other activators like LiHMDS; however, the use of heterogeneous catalysts

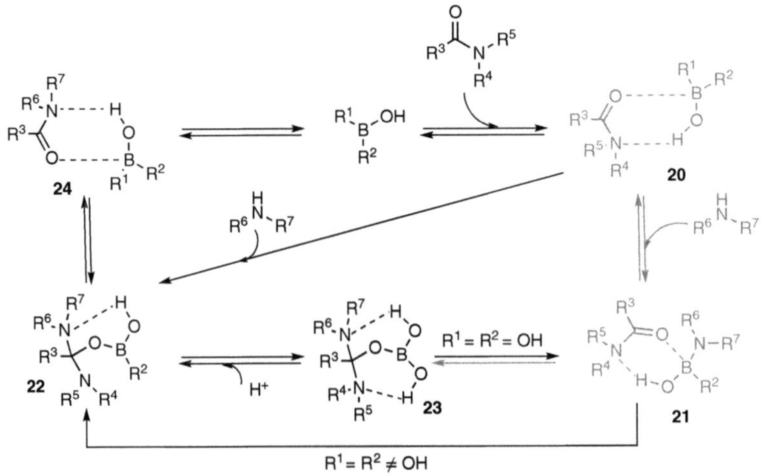

Scheme 6.13 Activation of amides with B–OH derivatives.

Table 6.2 Summary of direct transamidation using heterogeneous catalysts.

Catalyst/ promoter	Substrates		Observations	
	Amide	Amine	Positive	Negative
Mesoporous Nb oxide spheres [37]	Aromatic 1°	Aliphatic 1° and 2°	Good yields	The reaction is performed in a sealed tube at 160 °C
Graphene oxide [38]	Aliphatic 1°and aromatic 1°	Aliphatic 1°and 2°. Aromatic 1°	Works with urea and phthalimide	Long reaction times, around 24 h
Guanidine functionalized nanoparticles [39]	Aliphatic 1° and aromatic 1°	Aliphatic 1°and 2°. Aromatic 1°	Works with phthalimide, urea, and hydrazines, uses stoichiometric amounts of substrates	Only useful with the production of volatile amines
Sulfated tungstate [40]	Aliphatic 1° and aromatic 1°	Aliphatic 1°and 2°. Aromatic 1°	Works with phthalimide and urea uses stoichiometric amounts of substrates	Only useful with the production of volatile amines
H_2SO_4-SiO_2 [41]	*N,N*-Dimethyl amides	Aliphatic 1°and 2°. Aromatic 1°	Brønsted acid activation	
OSU-6 [42]	Aliphatic 1° and aromatic 1°	Aliphatic 1°. Aromatic 1°	Works with hydrazine, Brønsted acid activation	Limited substrate scope

is suitable for industrial applications. In consequence, we will summarize the most important reports on this field in Table 6.2.

In general, the use of heterogeneous catalysis is a powerful alternative when access to calcination and solid characterization techniques is available.

6.3.4.4 Other Promoters

The activation of the amide carbonyl is accomplished by electrophilic reagents. In most cases, these are metals in heterogeneous or homogeneous catalysts; however, some other electrophiles have been used in transamidation processes. The most complete study for transamidation with an electrophilic reagent was published by Singh et al. [43] using hypervalent iodine. The hypervalent iodine reagents are powerful electrophiles [44]; in consequence they can react with the oxygen atom of an amide activating the carbonyl and increasing its electrophilicity. Having this in mind the diacetoxy iodobenzene activates a carbonyl by direct reaction of the oxygen with the iodine atom, producing the highly electrophilic cation **25**, which can react

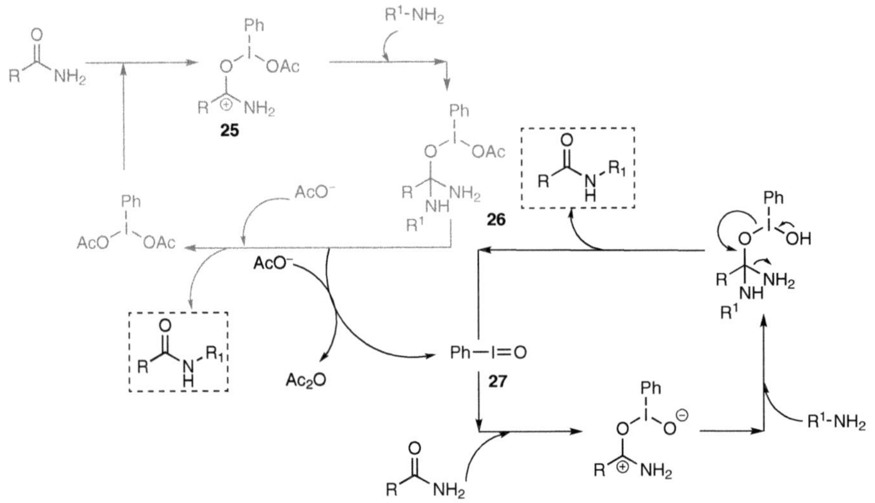

Scheme 6.14 A plausible mechanism for transamidation with hypervalent iodine.

with the amine. According to Singh and coworkers, the tetrahedral intermediate **26** can follow two different pathways; one to regenerate the catalyst and produce the new amide (route a in gray) and another one to produce the iodosobenzene **27**, which may act as a new catalyst (route b in black; see Scheme 6.14).

Transamidation with diacetoxy-iodobenzene has a very good substrate scope and works under relatively mild reaction conditions; however, no solvents are used, and this generates difficulty when both reagents are solids.

The use of other electrophilic reagents is rare for transamidations. Particularly the use of $K_2S_2O_8$ (potassium persulfate) has been described with notable advantages [45]; the reaction uses water as the solvent and may be performed under conventional heating or using microwaves; the substrate scope is very good, and the authors proved that persulfate was superior to other electrophilic oxidants.

Finally, a weak electrophilic reagent such as TMSCl has also been reported [46]. It has to be highlighted that the same reagent was previously used with nickel catalysts, and the authors had proposed the bisTMSamine as the reactive species. In consequence, the reactivity of bisTMSamines and bisTMSamides was evaluated under catalyst-free conditions; unfortunately, the results are insufficient to conclude about the role of TMSCl. We strongly believe that despite the previous descriptions of bisTMSamines as substrates in transamidations, a traditional activation of the carbonyl by the silicon atom is a plausible option.

All the methods discussed to this point try to increase the electrophilicity of the carbonyl, except for methods using strong bases as the promoters; in those cases the deprotonation of the amine serves to increase its nucleophilicity, and consequently the reaction with the amide is easier. In the same context, strong nucleophilic reagents should act as synthetic auxiliaries, which means they will react with the amide substituting the amine moiety, generating a more reactive species. That is the case of imidazolium chloride [47], which has a dual role; its Brønsted acidity

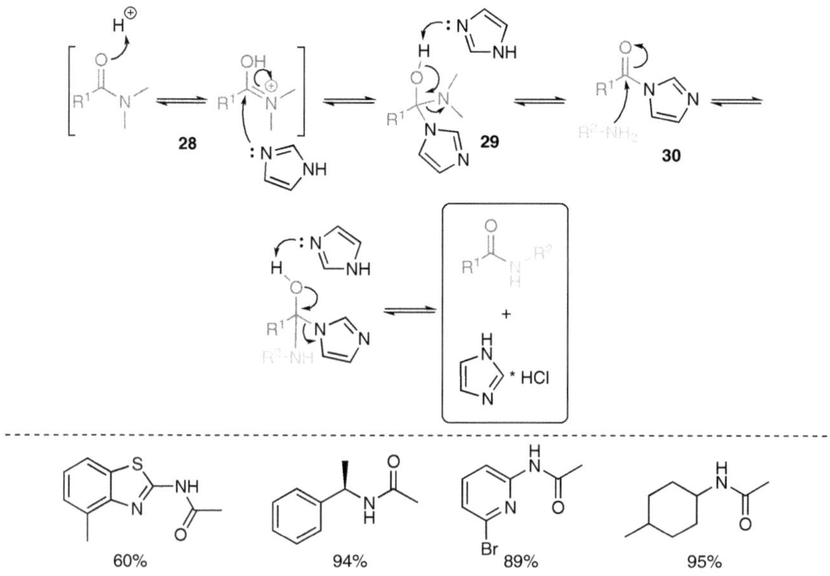

Scheme 6.15 Transamidation with imidazole as auxiliary.

is responsible for the activation of the carbonyl, and the imidazole nucleophilicity induces its reaction with the activated amide, forming the tetrahedral intermediate **29** with the concomitant elimination of the amine, as shown in Scheme 6.15. The imidazole amide **30** is more electrophilic than **28**, so it may react even with fewer nucleophilic aromatic amines.

The last promoter we will discuss for direct transamidations is the *tert*-butylnitrite since it is an electrophilic activator but acts on the nitrogen instead of the oxygen [48]. According to the authors, the formation of the *N*-nitrosoamide **31** increases the reactivity of the carbonyl; it can be seen as a twisted amide. Unfortunately, there is no experimental support for that hypothesis. However, the reaction scope is good, providing a relatively simple method for transamidation of secondary amides, with a wide range of aliphatic amines; the aromatic amines were unreactive under those conditions. In Scheme 6.16, we summarize the proposed mechanism and some representative molecules to illustrate the substrate scope.

6.3.5 Catalyst and Promoter-Free Processes

Thermal transamidations were first reported a long time ago, and they usually provide a distribution of products except when the produced amine was a gas. In that context, thermal transamidation of primary amides and ureas was reinvestigated. One of those reports is based on the hypothesis of the formation of a hydrogen bond between the amine and the amide oxygen, so activating the amide [49]. Given the background, the transamidation with highly reactive amides (formamides, acetamides, among others) was studied under thermal conditions; the reaction was performed on a sealed tube at 150 °C and without solvent. From this angle, the

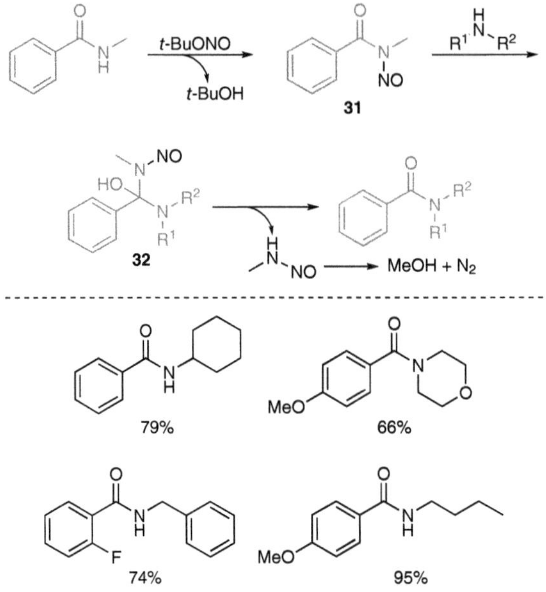

Scheme 6.16 Transamidation through the formation of *N*-nitrosamides.

reaction under microwave heating and using dioxane as the solvent proved useful [50] with similar amides but using aromatic amines, which were unreactive under solvent-free conditions. The transamidation of ureas under similar conditions was also described [51].

6.4 Transamidation by the Previous Functionalization of the Amide Bond

The previous section focused on direct transamidations, which means reactions where the activation is carried out by the interaction of the carbonyl oxygen with an acidic species (Lewis or Brønsted), such interaction increases the carbon electrophilicity making it reactive with some nucleophiles and changing so the natural reactivity of amides, which are the less electrophilic carbonyl compounds. During the last decade a new wave of activation methods, in which the C–N bond in amides is activated changing the amide geometry, has emerged [52]. It is well known that most of the amides are planar, and the consequences of this planarity are studied in general organic chemistry and biochemistry courses; however, deviations to this planarity results in an increase of the electrophilicity of the carbonyl, making it reactive with some nucleophiles and in cross-coupling reactions. The detailed studies on amides structure and geometry are beyond the aim of this chapter; however, the increased reactivity of twisted amides has been reviewed in recent literature [53]. Interestingly, Szostak and coworkers developed a simple scale to measure the electrophilicity of those amides using ^{17}O NMR and ^{15}N MNR [54].

The increased electrophilicity of twisted amides favors their reaction with mild nucleophiles such as amines in transamidation reactions; however, in some cases the use of catalysts is mandatory. In that context, this section is divided into four subsections; in the first one we will present the transamidation of twisted amides under the influence of a metallic species; in the second we will discuss the case of activation with atypical nucleophiles, particularly the use of fluoride as a synthetic auxiliary in transamidations, then we will show some examples where authors succeed a transamidation of activated substrates using other promoters such as oxidants or bases. It is noteworthy that in some of those cases, it is still unclear if the amide is truly twisted. Finally, we will discuss the transamidation of truly twisted amides without the use of an external promoter.

6.4.1 Transamidation of Activated Substrates Using Metallic Catalysts

The most convenient way to obtain a twisted amide is by the insertion of an electron-withdrawing and bulky group, being tosyl and Boc the most frequently used. Typically, the reaction of secondary amides with the appropriate reagents is performed in a separate vessel; the "activated amide" is purified, and then the transamidation is carried out. It is not noteworthy that properly speaking the substrates are not amides, and, more important, the transamidation does not produce an amine; instead, a non-nucleophilic tosylamide or carbamate is obtained, thus avoiding the possibility of an equilibrium.

Taking this into account, changing the geometry of secondary amides by insertion of those groups has demonstrated a wide range of applications, notably in cross-coupling reactions. This is possible by the insertion of a transition metal by cleavage of the C(O)–N bond. The same principle may be used to perform transamidations. The first example of successful transamidation of activated twisted amides with metallic catalysts was described by Lipton and Çalimsiz in 2005 [55], who described the lanthanide-catalyzed transamidation of *N*-Fmoc lactams **33**. This method proved superior to traditional deprotection since epimerization is not observed; unfortunately, only a few lactams were reactive, and, to the best of our knowledge, there are no reports or extension of this methodology to other amides. Scheme 6.17 summarizes the reaction conditions and substrates.

Around a decade later the use of nickel complexes was published by Garg and coworkers [56], and the next year Szostak and coworkers described the use of palladium catalysts [57]. Each team continued working on that subject and published

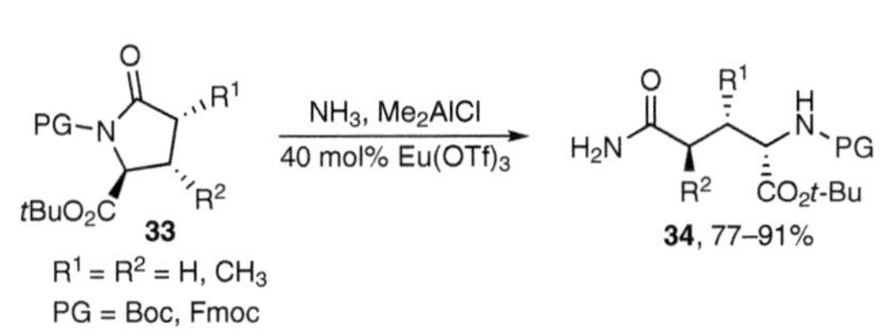

Scheme 6.17 Lanthanide-catalyzed transamidation of twisted amides.

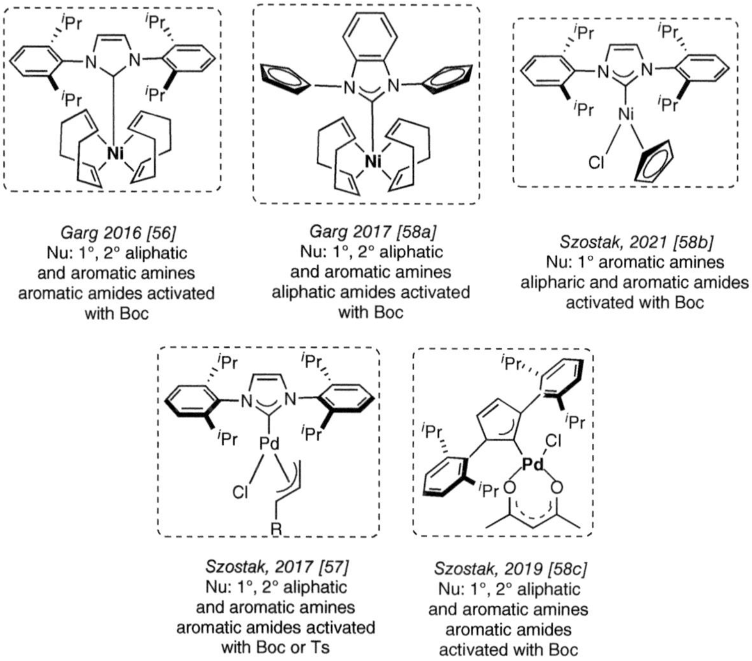

Figure 6.2 Structure of metallic catalysts and scope of transamidation of twisted amides.

the use of other catalysts during the last few years [58]; each of those catalysts has some advantages and disadvantages; also attempts to perform a one-pot process have been described. Interestingly, the transamidation of a wide range of amides (aromatic and aliphatic) was accomplished, and several aromatic and aliphatic amines might be used as nucleophiles. It is easy to assume that those reactions proceed by the insertion of the metal between the carbon and the nitrogen atom, then a nitrogen ligand exchange will produce the non-nucleophilic carbamate or tosylamide. Finally, reconnection of the new amide will liberate the metallic catalyst to restart the cycle. Usually, the Ni catalysts are more active since they react at lower temperatures, but the palladium catalysts are more stable, even commercially available; in summary, there is no superior method or catalysts; all are useful and have very good substrate scopes. Figure 6.2 shows the structure of the catalysts used and summarizes the scope of each catalytic system.

The last example we will present is the use of hydrated cerium chloride as the catalyst [59]. In this case, aromatic *N*-benzyl amides were activated by using trifluoroacetate. It has to be noted that even if this method is similar to the previously described use of lanthanides, it uses milder conditions since trimethyl aluminum is no longer needed, and the reaction is performed under ultrasound irradiation; unfortunately, even if the reaction is cheaper and easy to carry out, the reaction scope is pretty limited; only aromatic amides **35** and amines **36** are suitable substrates, as shown in Scheme 6.18.

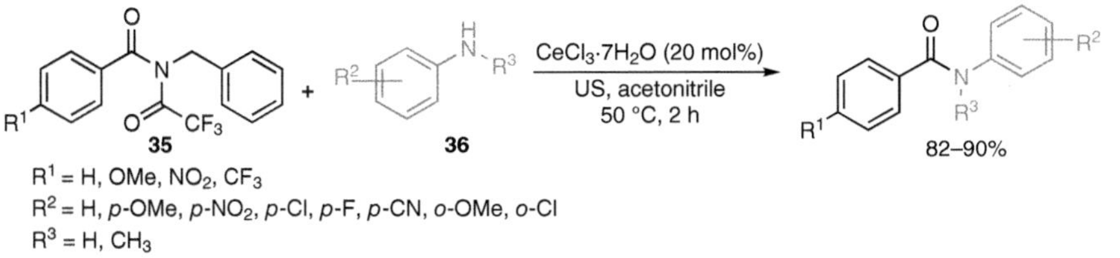

Scheme 6.18 $CeCl_3$ as catalysts for transamidation of twisted amides.

6.4.2 Transamidation of Activated Substrates Using Fluoride as an Auxiliary

The inherent electrophilicity of twisted amides makes them reactive with some mild nucleophiles; however, to the best of our knowledge, this feature has not been properly exploited. The use of atypical nucleophiles like fluorine was described to produce the acyl fluoride *in situ* [60], converting in this way the amide into an acyl source. This transformation could be accomplished with acyl di-*tert*butyl-iminodicarboxylate **37** and with acylisatins **38** using CsF as the nucleophile source. The formation of the acyl fluoride **39** as the intermediate was demonstrated, and it can react with a nucleophilic amine (aliphatic and aromatic). Scheme 6.19 summarizes the transamidations using fluoride as the auxiliary; unfortunately, the reaction scope is still limited, and applications of the same principle with other nucleophiles are still unexplored.

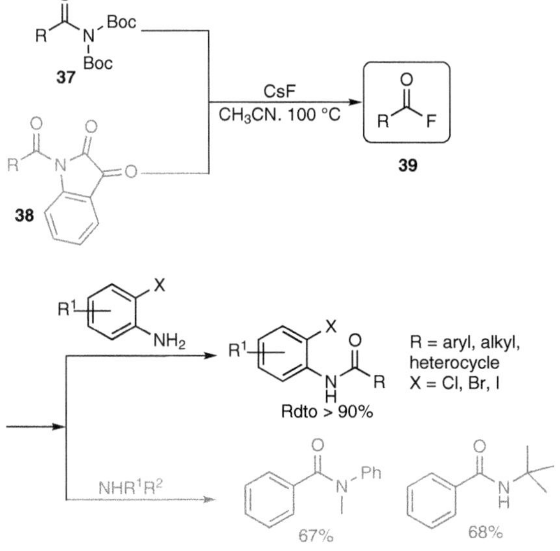

Scheme 6.19 Transamidation of twisted amides via an acyl fluoride.

6.4.3 Transamidation of Activated Substrates Using Other Promoters

Other conditions for transamidation of twisted amides include the use of strong bases and radical initiators. On the other hand, the oxidation of dihydropyridines (DHPs) to pyridinium has also been described for transamidations; the last case is different since there is no evidence of the amide geometry; however, we will discuss that example in this section.

The use of LiHMDS was described for direct transamidations and discussed in a former section of this chapter. As mentioned previously, the use of a strong base should produce the deprotonation of the amine making it more nucleophilic and thus reactive enough to induce an acyl nucleophilic substitution on twisted amides; in that context the less nucleophilic aromatic amines might be reactive [61]. Scheme 6.20 summarizes the most important achievements of this method. It is noteworthy that the reaction is performed using di-*tert*butyl-iminodicarboxylates **37** as the activated twisted amides; also the reaction scope is good, and this achievement is one of the few examples where aromatic amines are useful nucleophiles for transamidations.

The use of *tert*-butylhydroperoxide (TBHP) as a radical initiator in the transamidation of *N*-Boc-substituted amides **40** was described by Srivastava and coworkers [62]. Even if the mechanism has not been elucidated, the authors showed that the introduction of radical scavengers stops the reaction; in consequence, it is presumable that the radical cleavage of the C(O)–N bond is one of the propagation steps in a traditional radical mechanism. The reaction conditions and scope are summarized in Scheme 6.21. The order of amines reactivity is secondary > primary > bulky primary. Traditionally, primary amines are more reactive in transamidation; however, in this case the reaction order is inverse.

Finally, the use of an oxidizing agent demonstrates useful for transamidation of DHP amides [63]. The oxidation of the DHP moiety in **41** produces a pyridinium amide **42**, which certainly contains a good leaving group, making the oxidized

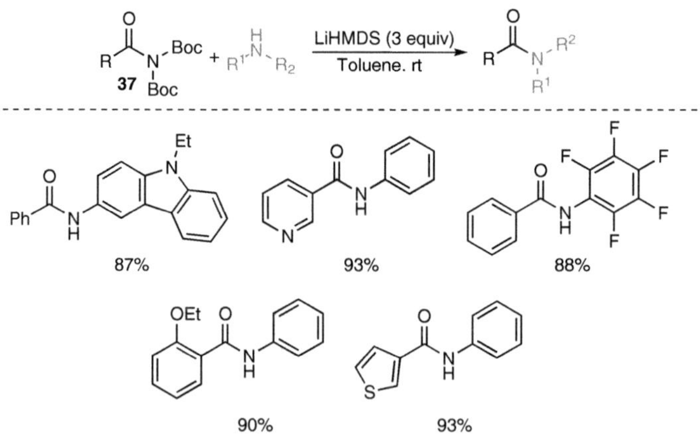

Scheme 6.20 Base promoted transamidation of twisted amides.

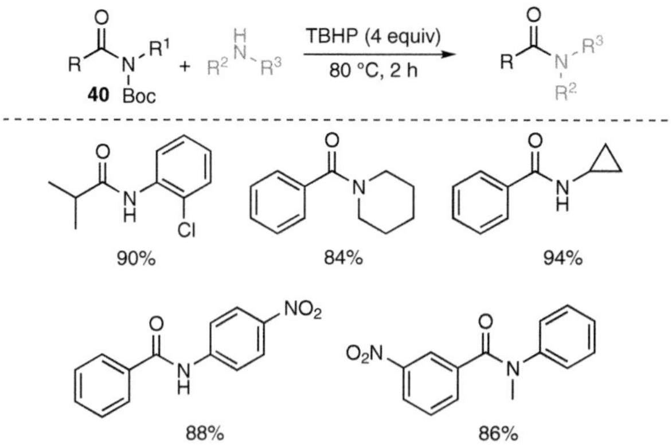

Scheme 6.21 Transamidation of *N*-Boc-substituted amides promoted by TBHP.

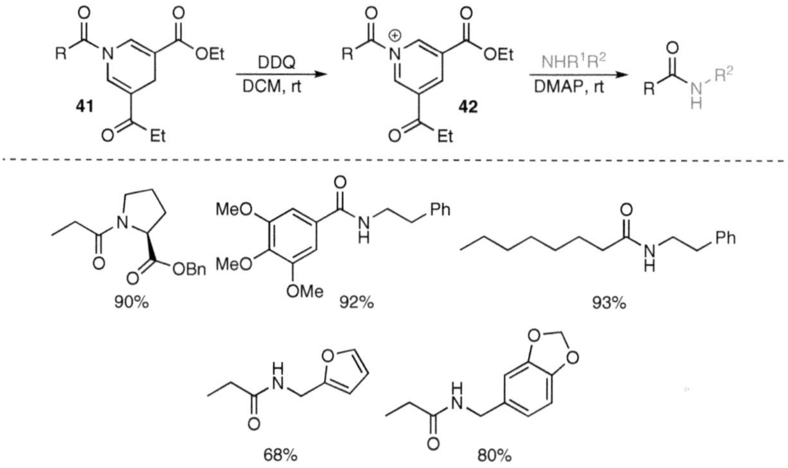

Scheme 6.22 Transamidation of dihydropyridine amides promoted by oxidation.

intermediate highly reactive with weak nucleophiles such as amines, even the less nucleophilic anilines. The reaction is summarized in Scheme 6.22.

6.4.4 Transamidation of Activated Substrates Without Promoters or Catalysts

In this subsection, we summarize the transamidations of twisted amides where no promoters or catalysts are used [64]. The simple heating of a mixture of the amide and the amine with the appropriate solvent and temperature is enough for the reaction to occur. Presumably, the mechanism is a traditional nucleophilic acyl substitution, which means the amine reacts with the electrophilic carbonyl and after a proton transfer a neutral tetrahedral intermediate is formed. The driving force is the elimination of a good leaving group, which is an extremely weak nucleophile. Having this in mind a wide range of highly electrophilic twisted amides have been used;

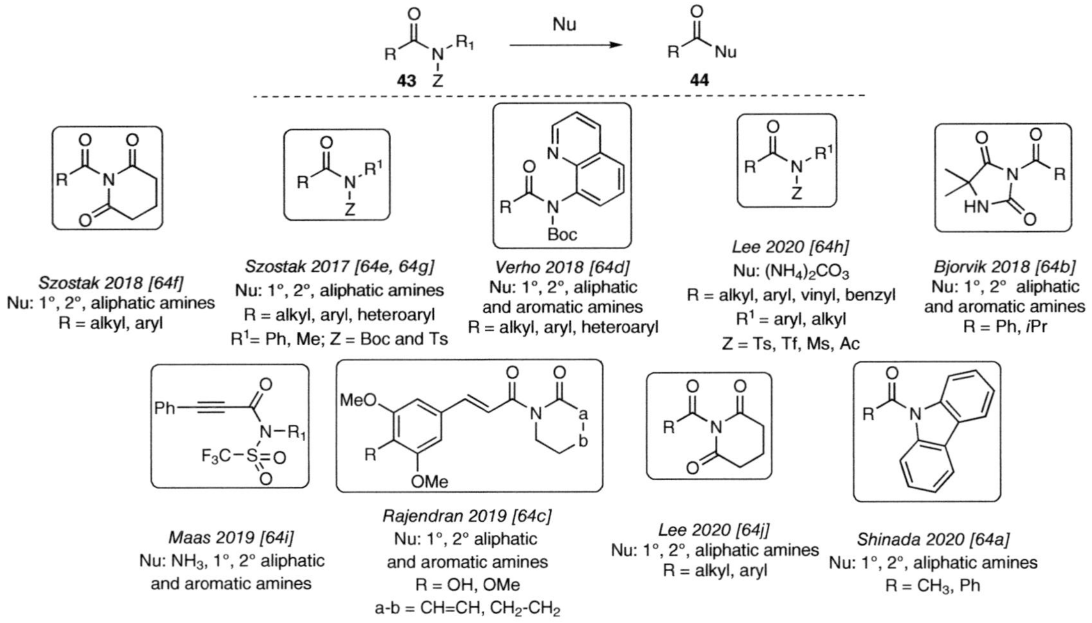

Scheme 6.23 Structure of highly reactive twisted amides.

the structures are shown in Scheme 6.23 where the variety and versatility of twisted amides **43** are exemplified. On the other hand, a wide range of amides **44** was synthesized, including drugs and peptides. Scheme 6.23 shows the twisted amide, the nucleophilic amine used, and the team in which the method was developed.

6.5 Transamidation with Atypical Substrates

In this section we will discuss some transamidations where the substrates are not amides and/or amines; in most of the cases, the traditional reagents are generated *in situ*; however, in some of the examples described herein, the pass throughout the amides or amines is not evident, so a special discussion is needed.

6.5.1 Reductive Transamidation

Reductive transamidation refers to the reaction of an amide with nitrobenzene or a substituted nitroaromatic, under the presence of a reducing agent [65]. It is natural to speculate that the nitrobenzene is reduced to the corresponding amine, and a normal transamidation takes place; however, the authors demonstrated that the diazoaryl compound **46** is the most plausible intermediate instead of the amine. The mechanistic proposal using Ni catalysts and Zn as the reducing agent is depicted in Scheme 6.24. It suggests a reduction of the nitrobenzene **45** with Zn and TMSCl to the corresponding diazo compound **46**, then an insertion of the metallic species will generate a nitrene-type intermediate **47**, which may react with the amide generating the transamidation product and regenerating the catalyst. It is noteworthy that the reaction might be performed using Zn or Mn as the reductants and with or without Ni as the catalyst. Authors demonstrated that the Ni catalyst is needed when an *N*-Boc-substituted amide is used; the method is suitable for two-step transamidation of secondary amides; however, the manganese is reactive enough to promote the reaction with tertiary amides. In that case, we can presume that the Mn plays a dual role, acting as the reducing agent and promoting the transamidation, even by the insertion between the C(O)–N bond or by the generation of a nitrene-type species.

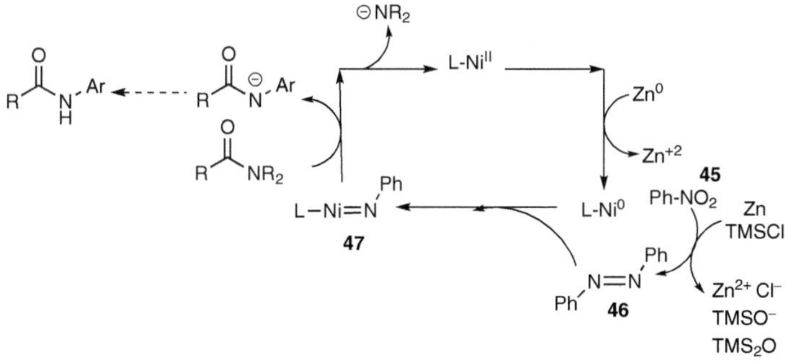

Scheme 6.24 Proposed mechanisms for reductive transamidation.

The use of nitro derivatives over amines has several advantages, notable a lower cost, and easier manipulation. The same principle of this reductive transamidation has served to generate highly nucleophilic nitrogen in other transformations, which are beyond the scope of this chapter, such as amidation of esters [66], among others.

6.5.2 Oxidative Transamidation

Contrary to what was stated in the previous section, oxidative transamidation does not refer to a single kind of reaction. Instead, it refers to any process where the reactive species, amine, or amide equivalent is generated by an oxidation reaction; in that context several transformations are conceivable; however, to the best of our knowledge, only two reports have described transamidations by that kind of sequences.

In 2016 Xu, Liu, and coworkers [67] reported the Cs-catalyzed aminolysis of nitriles. The reaction is performed in water as the solvent, so authors demonstrated that a Cs-catalyzed hydration of the nitrile to the corresponding primary amide (route a in Scheme 6.25) followed by a transamidation is a plausible mechanism, at least more possible over the direct amination of the nitrile followed by oxygenation of the amidine (route b in Scheme 6.25), since the intermediate amide was detected, and the transamidation of the isolated intermediate proceeded with good yield under the same reaction conditions. In that context, the whole transformation is a hydration-transamidation sequence, both reactions catalyzed by Cs. The reaction scope is good since the reaction works with aromatic and aliphatic nitriles and with ammonia, primary and secondary amines. Scheme 6.25 summarizes the probable mechanistic pathways and highlights the most plausible route.

The second report that can be seen as an oxidative transamidation corresponds to the tandem oxidation-transamidation using amines as the acyl source [68]. This reaction is conceptually different from all others showed in this chapter; however, the isolation of the intermediate and computational studies demonstrated that the oxidation of the amine with $Na_2S_2O_8$, catalyzed by $Ru(bpy)_3(PF_6)_2$ using visible light, generates an imine **48**, which is then transformed to the corresponding aldehyde **49**. It suffers the addition of the imidazole moiety to produce an activated amide derivative **50**. This can undergo a transamidation with a wide range of amines; unfortunately, the reaction is limited to the use of benzyl amines as the acyl

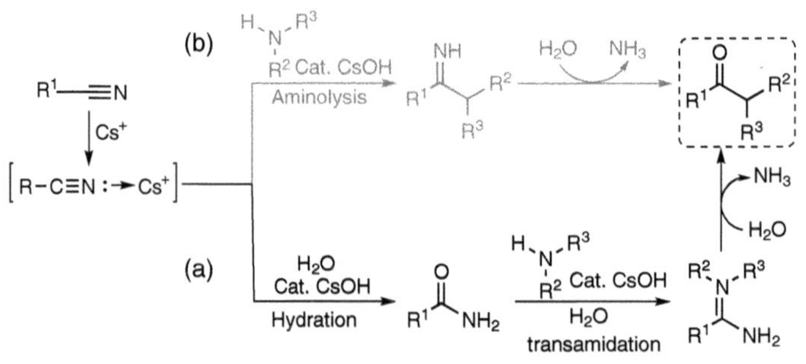

Scheme 6.25 Plausible pathways for amidation of nitriles via a transamidation process.

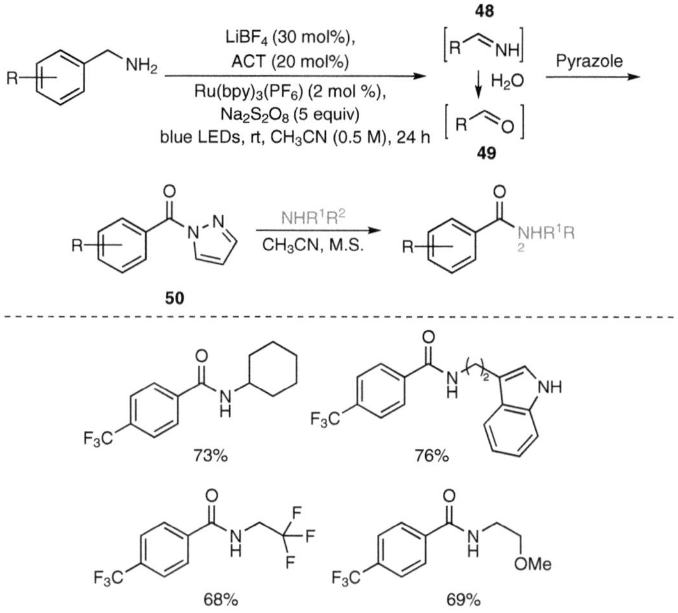

Scheme 6.26 Oxidative transamidation of benzylamines.

source. The general reaction pathway and some representative products obtained by this method are presented in Scheme 6.26.

6.5.3 Using Carbonyl and Thiocarbonyl Heterocycles as Activators

Due to the high significance of the C–C bond formation, the use of easily removable carbonyl heterocycles as auxiliaries has reached amazing developments during the last decades. Probably, the most common derivatives of this type are Evans-type oxazolidinones; however, many other heterocycles have been used as chiral auxiliaries, particularly inducing the diastereoselective addition of electrophiles to enolates. In this context, the cleavage of the auxiliary to an amide might be seen as a transamidation. The well-known method to produce Weinreb amides using $AlMe_3$ and NH_2OH will not be discussed. We will only present two recent examples where amides other than the Weinreb were obtained.

The first example is the ytterbium-catalyzed aminolysis of imides [69]. It is noteworthy that even if imides were discussed as activated substrates, in this case, the system is a C(O)–N–C(O)–X, instead of a C(O)–N–C(O)–C, increasing its stability and thus making the direct production of amides a bit more challenging. Having this in mind, the authors developed a general method to exchange the oxazolidinone moiety in **51** by a wide range of amines as showed in Scheme 6.27.

Less common thiazinanethiones **52** have been proven useful to produce amides under mild reaction conditions even with amino acid derivatives as nucleophiles [70] (see Scheme 6.28).

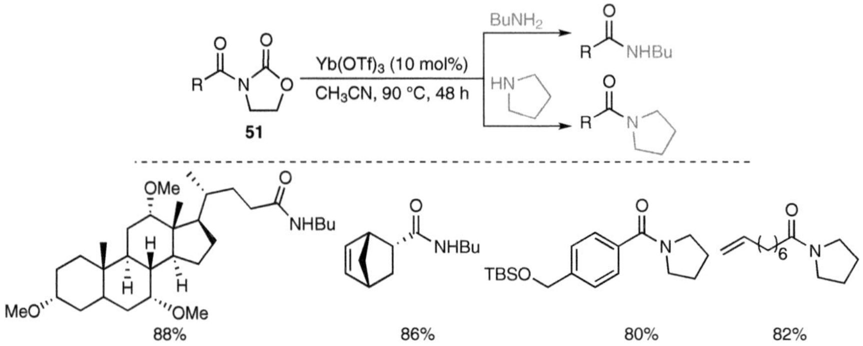

Scheme 6.27 Ytterbium catalyzed transamidation of oxazolidinones.

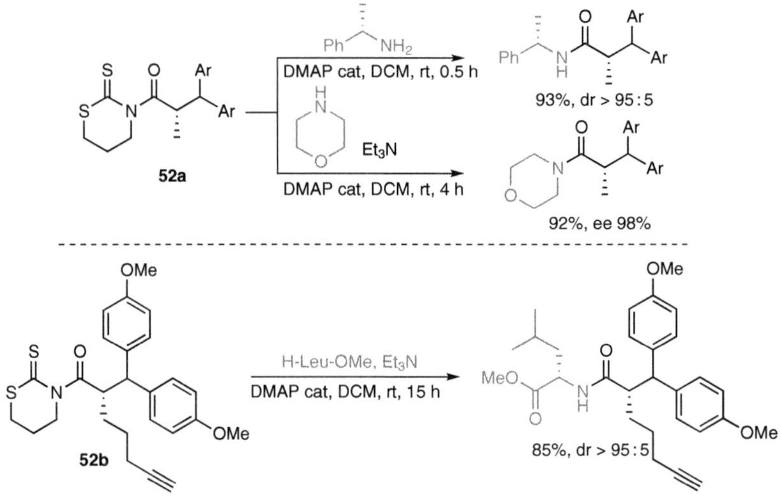

Scheme 6.28 Transamidation of thiazinanethiones.

6.5.4 From Amidines

Amidines might be produced *in situ* from amides; once obtained they may act as acyl donors (route a Scheme 6.29) or be oxidized to produce formyl imides (route b). That means the amidines themselves or the formyl imides might be subjected to a transamidation reaction.

The direct transamidation of amidines **53** may be accomplished using Lewis acid promoters, notably $Sc(OTf)_3$ or $ZrCl_4$, both at room temperature [71]. The reaction may be seen as an activation of the amide by a formyl group, thus making the reaction suitable to obtain complex amides, including highly functionalized amides and dipeptides (see Scheme 6.29).

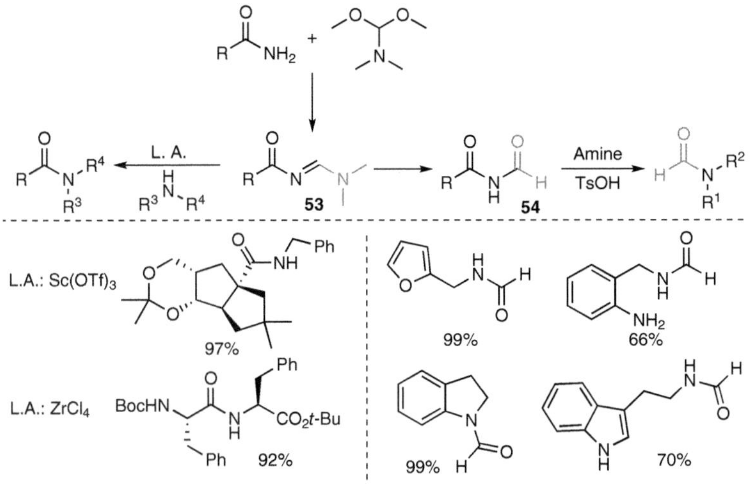

Scheme 6.29 Amidines as intermediates in transamidations.

On the other hand, the oxidation of amidines generated formyl imides **54**, and these have been used as formyl sources [72]. When benzyl formyl imides are used, the reaction is highly chemoselective; unfortunately, there is no data about the selectivity with other asymmetric imides. Scheme 6.29 shows the general reaction pathway and some of the most representative products obtained using these methods.

6.6 Conclusions and Perspectives

Throughout this chapter, we have shown some selected examples of the most representative transamidation methods. We try to cover all different methods of activation of carbonyls in direct transamidation reactions, and, also, we showed how twisted amides may react under soft reaction conditions to produce new amides. In general, the acyl exchange between amines is a useful transformation that has changed the paradigm of amides as extremely stable chemicals and, in consequence, the final products in a reaction sequence. Instead, we (chemists) have now new tools to use amides as acyl sources and profit their stability to carry out acylation reactions with more stable reagents.

In that context, several teams have made amazing developments, and others are currently working to solve some very few issues still present in amides chemistry and more accurately in transamidation chemistry. Examples of those problems are the transamidations of secondary and tertiary amides producing nonvolatile amines or in one step. Even if there are some very effective methods, some further developments are still needed; consequently, the development of more effective and general methods for transamidations will be always welcome, and we predict a brilliant future for this chemistry.

References

1 Hoffmann, C. (1889). *Ber. Dtsch. Chem. Ges.* 22: 2854–2856.
2 (a) Fleischer, A. (1876). *Ber. Dtsch. Chem. Ges.* 9: 992–995. (b) Freund, M. and Goldsmith, B.B. (1888). *Ber. Dtsch. Chem. Ges.* 21: 2456–2469. (c) Hirst, H.R. and Cohen, J.B. (1895). *J. Chem. Soc. Trans.* 67: 829–831. (d) Just, F. (1886). *Ber. Dtsch. Chem. Ges.* 19: 1201–1205. (e) Kelbe, W. (1883). *Ber. Dtsch. Chem. Ges.* 16: 1199–1201.
3 (a) Davis, T.L. and Blanchard, K.C. (1923). *J. Am. Chem. Soc.* 45: 1816–1820. (b) Davis, T.L. and Underwood, H.W. (1922). *J. Am. Chem. Soc.* 44: 2595–2604.
4 Galat, A. and Elion, G. (1943). *J. Am. Chem. Soc.* 65: 1566–1567.
5 Yoshio, O., Noboru, M., and Eiji, I. (1968). *Bull. Chem. Soc. Jpn.* 41: 1485–1485.
6 Kraus, M.A. (1973). *Synthesis* 1973: 361–362.
7 Bon, E., Bigg, D.C.H., and Bertrand, G. (1994). *J. Org. Chem.* 59: 4035–4036.
8 (a) Hoerter, J.M., Otte, K.M., Gellman, S.H., and Stahl, S.S. (2006). *J. Am. Chem. Soc.* 128: 5177–5183. (b) Kissounko, D.A., Hoerter, J.M., Guzei, I.A. et al. (2007). *J. Am. Chem. Soc.* 129: 1776–1783.
9 Zhang, M., Imm, S., Bahn, S. et al. (2012). *Angew. Chem. Int. Ed. Engl.* 51: 3905–3909.
10 Becerra-Figueroa, L., Ojeda-Porras, A., and Gamba-Sánchez, D. (2014). *J. Org. Chem.* 79: 4544–4552.
11 (a) Hoerter, J.M., Otte, K.M., Gellman, S.H. et al. (2008). *J. Am. Chem. Soc.* 130: 647–654. (b) Stephenson, N.A., Zhu, J., Gellman, S.H., and Stahl, S.S. (2009). *J. Am. Chem. Soc.* 131: 10003–10008.
12 Lanigan, R.M. and Sheppard, T.D. (2013). *Eur. J. Org. Chem.* 2013: 7453–7465.
13 (a) Acosta-Guzmán, P., Mateus-Gómez, A., and Gamba-Sánchez, D. (2018). *Molecules* 23: 2382. (b) Ojeda-Porras, A. and Gamba-Sánchez, D. (2016). *J. Org. Chem.* 81: 11548–11555. (c) Kolympadi Marković, M., Marković, D., and Laclef, S. (2020). *Synthesis* 52: 3231–3242. (d) Massolo, E., Pirola, M., and Benaglia, M. (2020). *Eur. J. Org. Chem.* 2020: 4641–4651. (e) Omprakash Rathi, J. and Subray Shankarling, G. (2020). *ChemistrySelect* 5: 6861–6893.
14 Hongting, S., Ruijie, Z., Wenjuan, W. et al. (2017). *Adv. Synth. Catal.* 359: 302–313.
15 Atkinson, B.N., Chhatwal, A.R., Lomax, H.V. et al. (2012). *Chem. Commun.* 48: 11626–11628.
16 Ma, J., Zhang, F., Zhang, J., and Gong, H. (2018). *Eur. J. Org. Chem.* 2018: 4940–4948.
17 Tamura, M., Tonomura, T., Shimizu, K.-i., and Satsuma, A. (2012). *Green Chem.* 14: 717–724.
18 Singh, D.P., Allam, B.K., Singh, K.N., and Singh, V.P. (2014). *RSC Adv.* 4: 1155–1158.
19 Yedage, S.L., D'Silva, D.S., and Bhanage, B.M. (2015). *RSC Adv.* 5: 80441–80449.
20 Zhang, F., Li, L., Ma, J., and Gong, H. (2019). *Sci. Rep.* 9: 2536.
21 Gu, D.-W. and Guo, X.-X. (2015). *Tetrahedron* 71: 9117–9122.

22 Ayub Ali, M., Hakim Siddiki, S.M.A., Kon, K., and Shimizu, K.-i. (2014). *Tetrahedron Lett.* 55: 1316–1319.

23 Sonawane, R.B., Rasal, N.K., Bhange, D.S., and Jagtap, S.V. (2018). *ChemCatChem* 10: 3907–3913.

24 Hollanders, K., Renders, E., Gadais, C. et al. (2020). *ACS Catal.* 10: 4280–4289.

25 (a) Yu, S., Shin, T., Zhang, M. et al. (2018). *Org. Lett.* 20: 7563–7566. (b) Yang, D., Shin, T., Kim, H., and Lee, S. (2020). *Org. Biomol. Chem.* 18: 6053–6057.

26 Allen, C.L., Atkinson, B.N., and Williams, J.M.J. (2012). *Angew. Chem. Int. Ed.* 51: 1383–1386.

27 Wu, J.-W., Wu, Y.-D., Dai, J.-J., and Xu, H.-J. (2014). *Adv. Synth. Catal.* 356: 2429–2436.

28 Nageswara Rao, S., Chandra Mohan, D., and Adimurthy, S. (2014). *Green Chem.* 16: 4122–4126.

29 Rao, S.N., Mohan, D.C., and Adimurthy, S. (2013). *Org. Lett.* 15: 1496–1499.

30 Yang, X., Fan, L., and Xue, Y. (2014). *RSC Adv.* 4: 30108–30117.

31 Li, G., Ji, C.-L., Hong, X., and Szostak, M. (2019). *J. Am. Chem. Soc.* 141: 11161–11172.

32 (a) Ghosh, T., Jana, S., and Dash, J. (2019). *Org. Lett.* 21: 6690–6694. (b) Tan, Z., Li, Z., Ma, Y. et al. (2019). *Eur. J. Org. Chem.* 2019: 4538–4545.

33 Zhang, R., Zhang, J.-C., Zhang, W.-Y. et al. (2020). *Synthesis* 52: 3286–3294.

34 Nguyen, T.B., Sorres, J., Tran, M.Q. et al. (2012). *Org. Lett.* 14: 3202–3205.

35 Dine, T.M.E., Evans, D., Rouden, J., and Blanchet, J. (2016). *Chem. Eur. J.* 22: 5894–5898.

36 Lanigan, R.M., Starkov, P., and Sheppard, T.D. (2013). *J. Org. Chem.* 78: 4512–4523.

37 Ghosh, S.C., Li, C.C., Zeng, H.C. et al. (2014). *Adv. Synth. Catal.* 356: 475–484.

38 (a) Patel, K.P., Gayakwad, E.M., Patil, V.V., and Shankarling, G.S. (2019). *Adv. Synth. Catal.* 361: 2107–2116. (b) Patel, K.P., Kamble, S.S., Boraste, D.R., and Shankarling, G.S. (2020). *Environ. Chem. Lett.* 18: 1731–1735.

39 Kazemi Miraki, M., Arefi, M., Yazdani, E. et al. (2016). *ChemistrySelect* 1: 6328–6333.

40 (a) Wagh, G.D., Pathare, S.P., and Akamanchi, K.G. (2018). *ChemistrySelect* 3: 7049–7053. (b) Pathare, S.P., Jain, A.K.H., and Akamanchi, K.G. (2013). *RSC Adv.* 3: 7697–7703.

41 Rasheed, S., Rao, D.N., Reddy, A.S. et al. (2015). *RSC Adv.* 5: 10567–10574.

42 Nammalwar, B., Muddala, N.P., Watts, F.M., and Bunce, R.A. (2015). *Tetrahedron* 71: 9101–9111.

43 Vanjari, R., Kumar Allam, B., and Nand Singh, K. (2013). *RSC Adv.* 3: 1691–1694.

44 Brown, M., Farid, U., and Wirth, T. (2013). *Synlett* 24: 424–431.

45 Srinivas, M., Hudwekar, A.D., Venkateswarlu, V. et al. (2015). *Tetrahedron Lett.* 56: 4775–4779.

46 Yu, S., Ho Song, K., and Lee, S. (2019). *Asian J. Org. Chem.* 8: 1613–1616.

47 Tian, Q., Gan, Z., Wang, X. et al. (2018). *Molecules* 23: 2234.

48 Sureshbabu, P., Azeez, S., Chaudhary, P., and Kandasamy, J. (2019). *Org. Biomol. Chem.* 17: 845–850.
49 Yin, J., Zhang, J., Cai, C. et al. (2019). *Org. Lett.* 21: 387–392.
50 Vanjari, R., Allam, B.K., and Singh, K.N. (2013). *Tetrahedron Lett.* 54: 2553–2555.
51 Wang, T.C. and Qiao, J.X. (2016). *Tetrahedron Lett.* 57: 1941–1943.
52 (a) Meng, G., Shi, S., Lalancette, R. et al. (2018). *J. Am. Chem. Soc.* 140: 727–734. (b) Liu, C., Shi, S., Liu, Y. et al. (2018). *Org. Lett.* 20: 7771–7774. (c) Rahman, M.M., Liu, C., Bisz, E. et al. (2020). *J. Org. Chem.* 85: 5475–5485.
53 (a) Szostak, M. and Aubé, J. (2013). *Chem. Rev.* 113: 5701–5765. (b) Li, G., Ma, S., and Szostak, M. (2020). *Trends Chem.* 2: 914–928. (c) Li, G. and Szostak, M. (2020). *Chem. Rec.* 20: 649–659. (d) Li, G. and Szostak, M. (2020). *Synthesis* 52: 2579–2599. (e) Szostak, R. and Szostak, M. (2019). *Molecules* 24: 274. (f) Dander, J.E. and Garg, N.K. (2017). *ACS Catal.* 7: 1413–1423.
54 Ielo, L., Pace, V., Holzer, W. et al. (2020). *Chem. Eur. J.* 26: 16246–16250.
55 Çalimsiz, S. and Lipton, M.A. (2005). *J. Org. Chem.* 70: 6218–6221.
56 Baker, E.L., Yamano, M.M., Zhou, Y. et al. (2016). *Nat. Commun.* 7: 11554.
57 Meng, G., Lei, P., and Szostak, M. (2017). *Org. Lett.* 19: 2158–2161.
58 (a) Dander, J.E., Baker, E.L., and Garg, N.K. (2017). *Chem. Sci.* 8: 6433–6438. (b) Buchspies, J., Rahman, M.M., and Szostak, M. (2021). *Molecules* 26: 188. (c) Zhou, T., Li, G., Nolan, S.P., and Szostak, M. (2019). *Org. Lett.* 21: 3304–3309.
59 Mishra, A., Singh, S., and Srivastava, V. (2018). *Asian J. Org. Chem.* 7: 1600–1604.
60 (a) Luo, Z., Wu, H., Li, Y. et al. (2019). *Adv. Synth. Catal.* 361: 4117–4125. (b) Xiong, L., Deng, R., Liu, T. et al. (2019). *Adv. Synth. Catal.* 361: 5383–5391.
61 Li, G. and Szostak, M. (2018). *Nat. Commun.* 9: 4165.
62 Mishra, A., Chauhan, S., Verma, P. et al. (2019). *Asian J. Org. Chem.* 8: 853–857.
63 Funder, E.D., Trads, J.B., and Gothelf, K.V. (2015). *Org. Biomol. Chem.* 13: 185–198.
64 (a) Kang, B., Yasuno, Y., Okamura, H. et al. (2020). *Bull. Chem. Soc. Jpn.* 93: 993–999. (b) Drageset, A. and Bjørsvik, H.-R. (2018). *Eur. J. Org. Chem.* 2018: 4436–4445. (c) Subramani, M. and Rajendran, S.K. (2019). *Eur. J. Org. Chem.* 2019: 3677–3686. (d) Verho, O., Pourghasemi Lati, M., and Oschmann, M. (2018). *J. Org. Chem.* 83: 4464–4476. (e) Rahman, M.M., Li, G., and Szostak, M. (2019). *J. Org. Chem.* 84: 12091–12100. (f) Liu, Y., Achtenhagen, M., Liu, R., and Szostak, M. (2018). *Org. Biomol. Chem.* 16: 1322–1329. (g) Liu, Y., Shi, S., Achtenhagen, M. et al. (2017). *Org. Lett.* 19: 1614–1617. (h) Chen, J., Xia, Y., and Lee, S. (2020). *Org. Lett.* 22: 3504–3508. (i) Fiore, V.A. and Maas, G. (2019). *Tetrahedron* 75: 3586–3595. (j) Idris, M.A. and Lee, S. (2020). *Org. Chem. Front.* 7: 2737–2743.
65 (a) Cheung, C.W., Ma, J.-A., and Hu, X. (2018). *J. Am. Chem. Soc.* 140: 6789–6792. (b) Cheung, C.W., Ploeger, M.L., and Hu, X. (2017). *ACS Catal.* 7: 7092–7096.
66 Cheung, C.W. and Hu, X. (2016). *Nat. Commun.* 7: 12494.
67 Li, Y., Chen, H., Liu, J. et al. (2016). *Green Chem.* 18: 4865–4870.
68 Nandi, J., Vaughan, M.Z., Sandoval, A.L. et al. (2020). *J. Org. Chem.* 85: 9219–9229.

69 Guissart, C., Barros, A., Rosa Barata, L., and Evano, G. (2018). *Org. Lett.* 20: 5098–5102.

70 Kennington, S.C.D., Taylor, A.J., Romea, P. et al. (2019). *Org. Lett.* 21: 305–309.

71 Dineen, T.A., Zajac, M.A., and Myers, A.G. (2006). *J. Am. Chem. Soc.* 128: 16406–16409.

72 Huang, H.-Y., Lin, X.-Y., Yen, S.-Y., and Liang, C.-F. (2020). *Org. Biomol. Chem.* 18: 5726–5733.

7

Amide Bond Esterification and Hydrolysis

Kazushi Mashima, Takahiro Hirai, and Haruki Nagae

Osaka University, Graduate School of Engineering Science, Department of Chemistry, Toyonaka, Osaka 560-8531, Japan

Amide bonds are thermodynamically stable due to conjugation between the carbonyl moiety and the lone pair on the amide nitrogen atom ($n_N \rightarrow \pi^*_{C=O}$ conjugation, Figure 7.1), and its hydrolysis under neutral conditions is very slow: the half-life for the neutral hydrolysis of glycylglycine at 25 °C is estimated to be 350 years [1]. To cleave the C–N bond of amides, it is necessary to break the conjugation of the amide C–N bond. Deformation from the planar conformation of the amide C–N bond is thus a key step for amide hydrolysis and alcoholysis. Enzymes and metalloenzymes use a special amino acid sequence and one or two metal ions at the active site to achieve a deformation suitable for cleaving the amide bond. This chapter summarizes the nonenzymatic stoichiometric and catalytic hydrolysis and esterification of amides [2–6].

7.1 Stoichiometric Reactions

Hydrolysis of a highly twisted amide such as Kirby's amide (**1**) [7–9], in which the nitrogen lone pair is twisted by approximately 90° vs. the π-bond of the carbonyl group, in water at pH 7 is at least 10^{10} times faster than that of ordinary amides (Eq. (7.1)). Another mechanism is the introduction of sterically demanding substituents on the amide nitrogen that triggers a steric repulsion for destabilizing the amide planar configuration. Hydrolysis and alcoholysis of sterically twisted amides **2** bearing a bulky 2,2,6,6-tetramethylpyperizine (TMP) moiety on the amide nitrogen by 2 equiv of water and the corresponding alcohols proceed at room temperature in tetrahydrofuran (THF) (Scheme 7.1) [10]. The bulky TMP group increases the nitrogen basicity to induce a proton switch/amine elimination mechanism via ketene intermediates **3**, which further react with water and alcohols to afford acid and the corresponding esters.

Amide Bond Activation: Concepts and Reactions, First Edition. Edited by Michal Szostak.

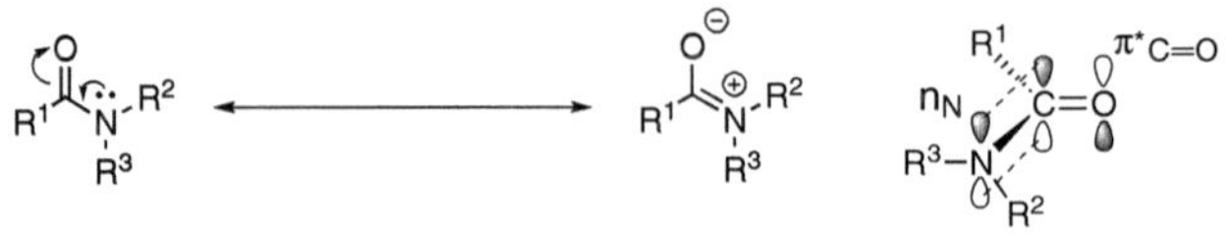

Figure 7.1 Conjugation between the carbonyl moiety and the lone pair on the amide nitrogen atom.

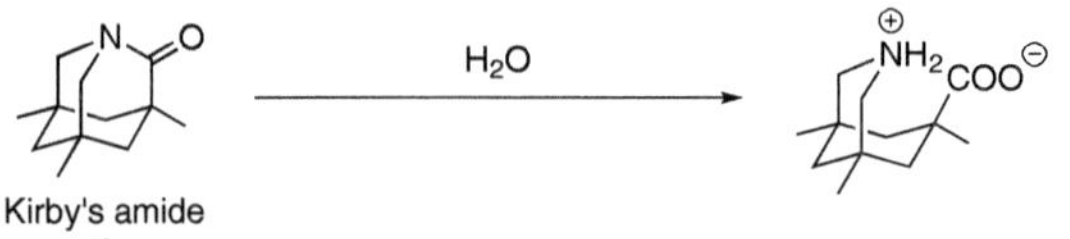

Kirby's amide
1

(7.1)

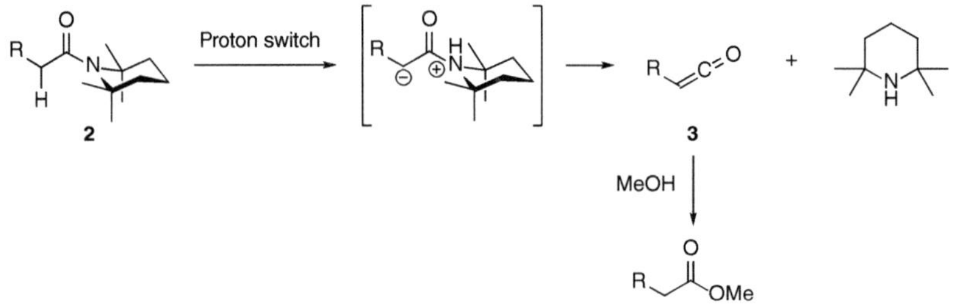

Scheme 7.1 Hydrolysis and alcoholysis of twisted amides **2** bearing a TMP moiety. Source: Adapted from Hutchby et al. [10].

Several activated amides have been converted to the corresponding esters in a stoichiometric manner. *N*-Alkyl-*N*-nitrosoamides undergo an intramolecular rearrangement to give the corresponding esters by elimination of N_2 via an S_Ni mechanism (Scheme 7.2) [11, 12]. This protocol does not work for *N*-aryl amides; however, treatment of *N*-aryl amides with $NaNO_2$ in a mixture of Ac_2O and AcOH (2 : 1 ratio) produced the corresponding esters via *N*-aryl-*N*-nitrosoamides [13]. Primary amides are converted to the corresponding esters upon treatment with *N*,*N*-dimethylformamide dimethyl acetal (DMF-DMA) to give the corresponding *N*-acyl formamidines **6** followed by methanolysis (Scheme 7.3) [14]. Reaction of primary amides with thionyl chloride in the presence of excess amounts of

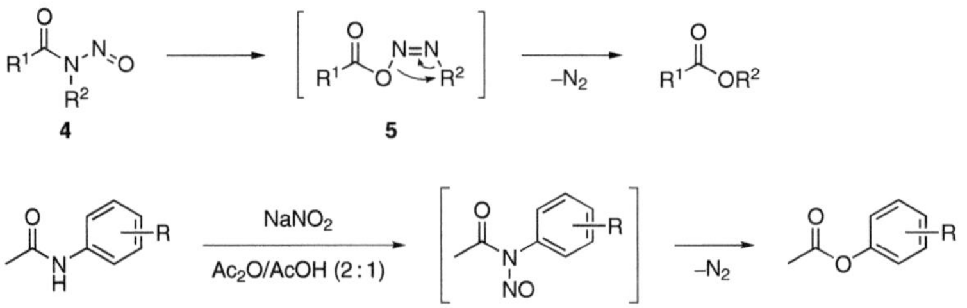

Scheme 7.2 Amide alcoholysis via *N*-nitroamides. Source: Adapted from White et al. [11, 12].

methanol affords the corresponding methyl esters via a sulfuramidite intermediate **7** (Scheme 7.4) [15].

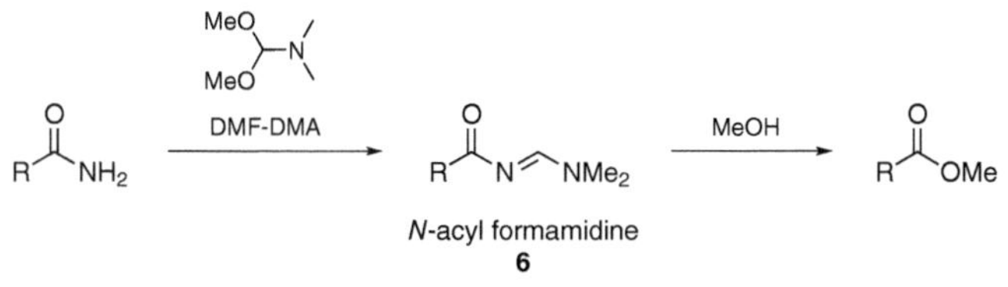

Scheme 7.3 Amide alcoholysis via *N*-acyl formamidine. Source: Adapted from Anelli et al. [14].

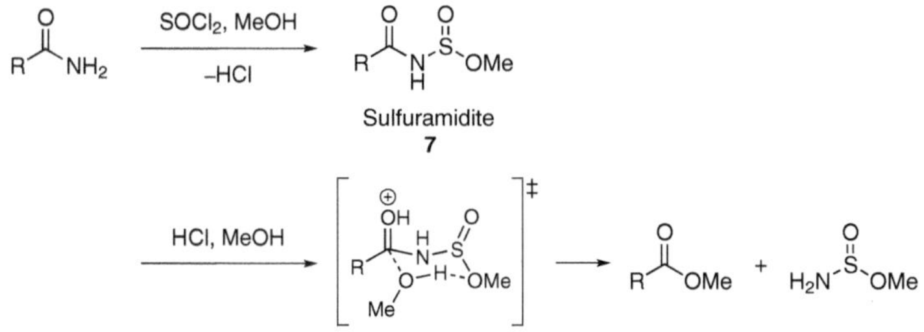

Scheme 7.4 Amide methanolysis via sulfuramidite. Source: Adapted from Li et al. [15].

Highly electrophilic imidoyl species **8** readily react with alcohols to give esters: treatment of *N*-aliphatic secondary and tertiary amides with triflic anhydride (Tf_2O) followed by the addition of alcohols produces the corresponding esters via ortho esters **9** (Scheme 7.5) [16]. Iminochloride intermediates **10** are chemically equivalent to the imidoyl intermediate **8**: treatment of secondary amides with triphenyl phosphite-chlorine adduct ($TPP{\cdot}Cl_2$) followed by treatment with alcohols and then water affords the corresponding esters and amines (Scheme 7.6) [17]. Primary amides are converted to a mixture of nitriles and esters upon treatment with $Te(OEt)_4$ in alcohols (Scheme 7.7) [18]. Activated amides have a protecting group such as Boc and Ts that are treated with 3 equiv of K_3PO_4 to give the corresponding ester (Eq. (7.2)) [19].

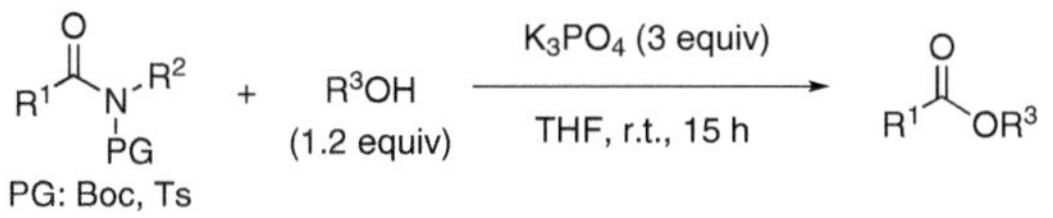

(7.2)

Some metal ions work as Lewis acid catalysts to cleave the amide C–N bond. Mononuclear copper(II) aqua complex coordinated by di(2-pyridyl)amine assists in the hydrolysis of formamide, *N*-methylformamide, and *N*,*N*-dimethylformamide in neutral water at 100 °C [20]. Mononuclear zinc complex $[ZnL(OH_2)]^{2-}$ (L = bis(2-benzimidazolylmethyl)glycinate) exhibits catalytic activity for the hydrolysis of β-lactam [21]. The coordination of metal ions in a chelating manner to amides assists in the a stoichiometric cleavage of the amide C–N bond. Two types of

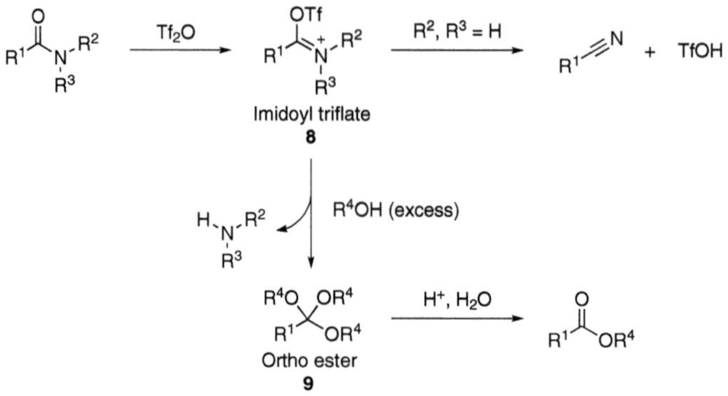

Scheme 7.5 Amide alcoholysis using triflic anhydride (Tf_2O). Source: Adapted from Charette and Chua [16].

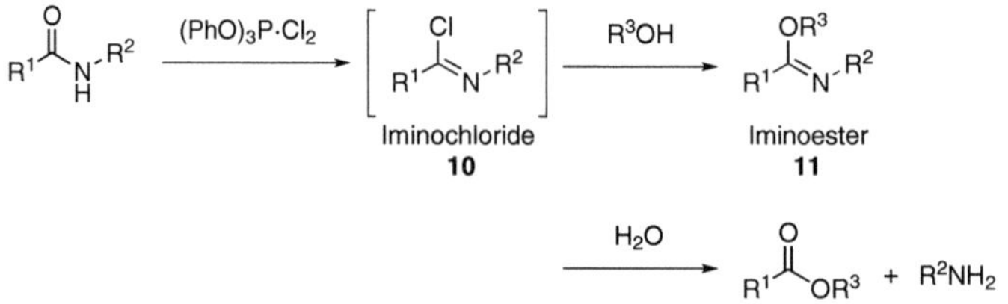

Scheme 7.6 Amide alcoholysis using TPP.Cl_2. Source: Adapted from Spaggiari et al. [17].

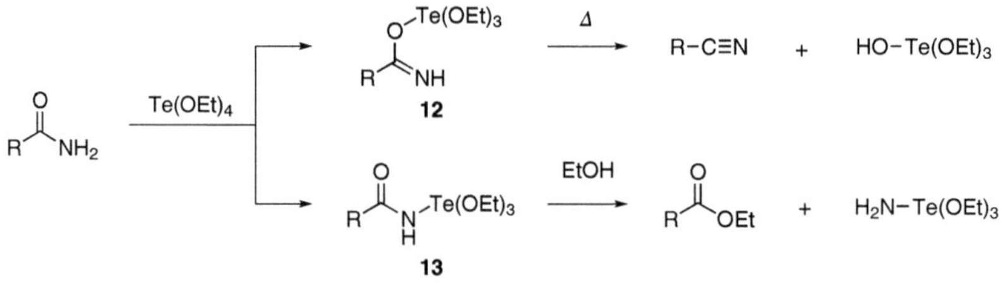

Scheme 7.7 Alcoholysis of primary amides using tellurium tetraethoxide. Source: Adapted from Omote et al. [18].

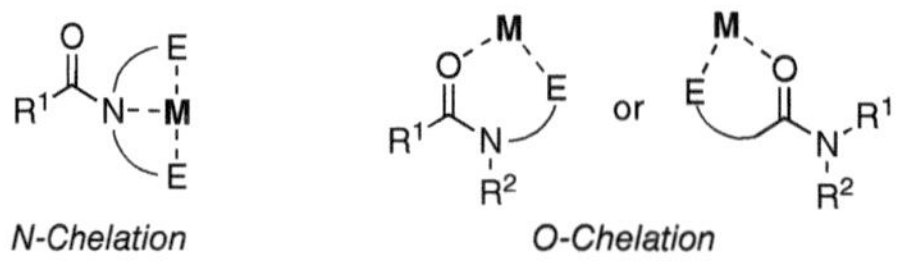

Figure 7.2 Schematic drawings for metal complexation with chelating substituents.

chelations are schematically summarized in Figure 7.2: one is the coordination of the nitrogen atom of the amide directly to metal ions (*N-chelation*) and the other is the coordination of the carbonyl moiety of amides to metal ions without interaction with the amide nitrogen (*O-chelation*).

For *N-chelation*, coordination to metal ions, particularly Cu(II) ions, by the nitrogen atom bearing a chelating auxiliary group accelerates both hydrolysis and alcoholysis. By introducing the coordinating picolinyl moiety on amides, copper ions accelerate the hydrolysis of tertiary (*N*-methyl)amides such as *N*-methyl-picolinanilides and picolinylsarcosine as well as *N*-picolinylimidazoles rather than the corresponding secondary amides [22–25]. The methyl group on the nitrogen atom prevents the formation of the corresponding catalytically inert amido complexes. Accordingly, *N*-acyl bis(picolyl)amines (**14**) treated with $Cu(OTf)_2$ in alcohols and $Cu(OTf)_2/Ba(OH)_2{\cdot}8H_2O$ spontaneously afford methyl esters and carboxylic acids, respectively (Scheme 7.8) [26–29].

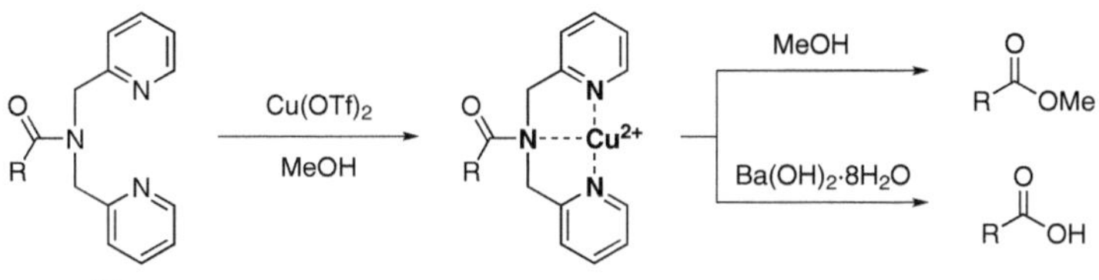

14

Scheme 7.8 Cu-assisted methanolysis and hydrolysis of *N*-acyl bis(picolyl)amides (**14**). Source: Adapted from Houghton and Puttner [26], Bröhmer et al. [27], Mundinger et al. [28], and Jakob et al. [29].

For *O-chelation*, the oxygen atom of the amides coordinates to metal ions to induce smooth hydrolysis and esterification. Anilides bearing 1,10-phenanthrolinyl or 6-carboxypicolinoyl moiety undergo hydrolysis in the presence of some divalent metal ions such as Cu and Zn [30]. Zn(II) and Cd(II) ions assist in the alcoholysis of amides bearing a nitrogen- and sulfur-containing multidentate moiety [31]. Hydrolysis of specially designed cyclic lactams **15** is accelerated by coordination to Cu(II) and Zn(II) (Scheme 7.9) [32–34]. In the presence of lanthanide ions as well as copper ions, amides **16** having 2,2′-bipyridine with two acylamino chains at the 6,6′-positions react with alcohols to give the corresponding esters quantitatively (Eq. (7.3)) [35–37]. When a mixture of **17a** and **17b** derived from Diels–Alder reaction of acrylamide and (2-pyridyl)anthracene is treated with $NiCl_2{\cdot}H_2O$ in ethanol, **17a** is converted to the ethyl esters through a chelating intermediate (Scheme 7.10) [38].

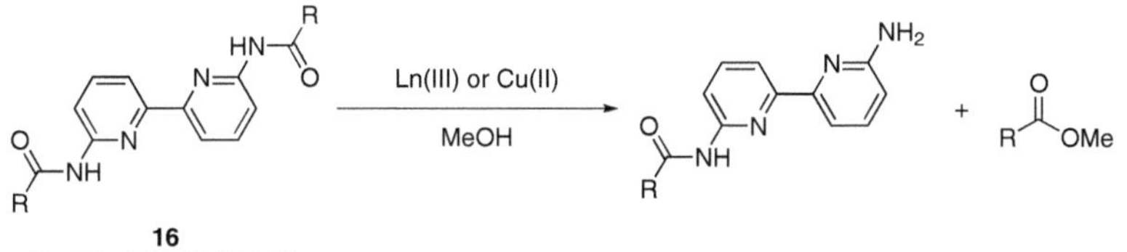

(7.3)

16
R = Ph; $CH(NH_2)CH_2Ph$

Hydrolysis of the amide bond of dipeptides and peptides under neutral conditions in a stoichiometric manner has been investigated as a model reaction of metalloenzymes. Cobalt complexes are used for peptide hydrolysis [39]. Zn(II) cations hydrolyze glycinamide, glycylglycine (Gly–Gly), and triglycine [40]. Copper(II) complexes supported by 1,4,7-triazacyclononane (= [9]aneN_3)

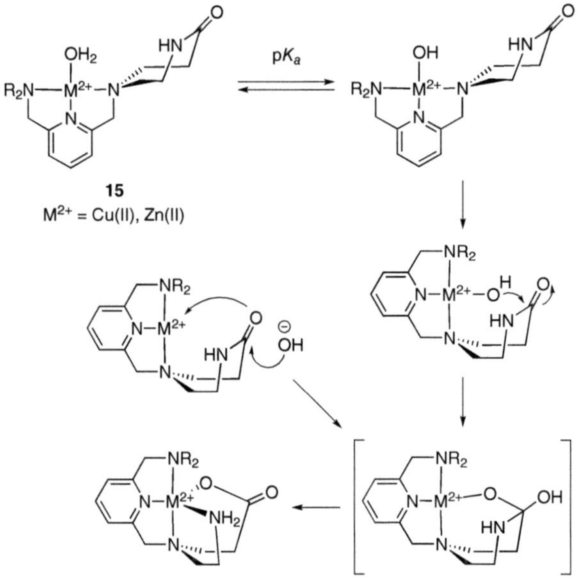

Scheme 7.9 Hydrolysis of cyclic lactams **15** by coordination to Cu(II) and Zn(II). Source: Adapted from Groves et al. [32–34].

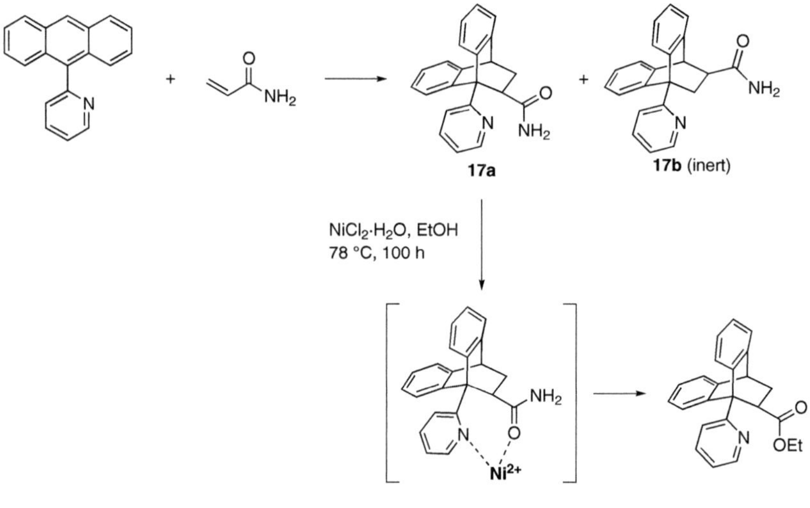

Scheme 7.10 Ni-mediated ethanolysis of an acrylamide Diels–Alder adduct. Source: Adapted from Czarnik [38].

and *cis*,*cis*-1,3,5-triaminocyclohexane (= TACH) ligands hydrolyze Gly–Gly, glycyl-leucine (Gly–Leu), leucylglycine (Leu–Gly), and glycyl-serine (Gly–Ser) [41, 42]. A ternary complex [Cu(TACH)(dipeptide)]$^+$ was crystallographically characterized as an intermediate species [42, 43]. The oxophilicity and Lewis acidity of Zr(IV) have been utilized for the hydrolysis of amide bonds of dipeptides such as Gly–Gly. Zr(IV)-substituted polyoxometalate (POM) efficiently hydrolyzes the amide bonds of dipeptides and oligopeptides [44–51]. Another use of Zr(IV) metal is a metal–organic framework (MOF)-containing $\{Zr_6(\mu\text{-}O)_4(\mu\text{-}OH)_4\}$ cluster units as a stable platform of an effective heterogeneous catalyst for hydrolysis of peptide amide bonds [52]. Hydrolysis of Gly–Gly proceeds at pH 7.4 and 60 °C 10^4 times faster compared with the uncatalyzed reaction. In this reaction, the oxygen atom of the amide moiety and terminal amine nitrogen atom of Gly–Gly bind in a chelating manner to two adjacent Zr(IV) centers of the $\{Zr_6(\mu\text{-}O)_4(\mu\text{-}OH)_4\}$ core [52, 53]. The same MOF-containing $\{Zr_6(\mu\text{-}O)_4(\mu\text{-}OH)_4\}$ cluster units work as a catalyst for the esterification of primary amides with alcohols but are less active for secondary and tertiary amides (*vide infra*) [54]. In the presence of 4,13-diaza-18-crown-6, $ZrCl_4$ hydrolyzes the amide bond of dipeptides such as Gly–Gly [55, 56]. Ce(IV) ions are more active than other lanthanide(III) ions for the hydrolysis of not only amides of α-amino acids such as phenylalaninamide [57] but also oligopeptides such as Gly–Phe and Gly–Gly [58]. The solubility of Ce(IV) ions such as $Ce(NH_4)_2(NO_3)_6$ is increased upon mixing with γ-cyclodextrin, and Ce(IV)-γ-cyclodextrin complexes hydrolyze dipeptides [59].

Selective peptide bond cleavage reactions are achieved based on the nature of the peptide side chains as coordination sites for metals. Hydrolysis of dipeptides having a serine residue at the *C*-terminus, X–Ser, where X is an appropriate amino acid residue, is catalyzed by $ZnCl_2$ at pH 7 [60]. An *N,O*-acyl rearrangement is proposed based on the marked difference between Gly–Ser (83% conversion) and Ser–Gly (4% conversion) (Scheme 7.11). Oxomolybdate, Na_2MoO_4, at pH 7 at 60 °C serves as a catalyst for the hydrolysis of serine-containing dipeptides, such as Gly–Ser, leucylserine (Leu–Ser), and histidylserine (His–Ser) (Eq. (7.4)) [61]. Molybdocene dichloride interacts with cysteine-containing peptides such as Gly–Cys to give an adduct that undergoes amide bond cleavage [62]. Peptides containing a serine residue in the Ser–X–His sequence are readily hydrolyzed in the presence of an Ni(II) complex as much as threonine-containing analogs [63–66].

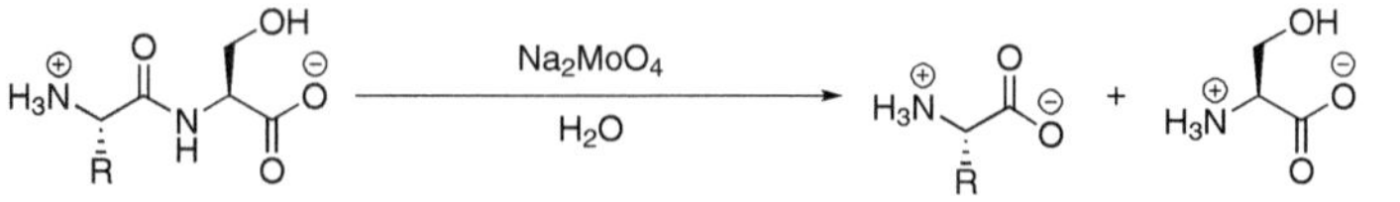

(7.4)

Palladium(II) and platinum(II) complexes favorably coordinate to thioether moieties of methionine as well as nitrogen heterocyclic moieties of histidine and tryptophan of peptides. Hydrolytic cleavage of the amide bond of dipeptides and oligopeptides containing a methionine is assisted by mononuclear palladium(II) and platinum(II) complexes, *cis*-$[M(en)(H_2O)_2]^{2+}$ (M = Pd, Pt; en = ethylenediamine) [67–70]. Dinuclear palladium complexes, such as $[Pd_2(\mu\text{-}SPh)_2(H_2O)_4]^{2+}$ and $[\{Pd(en)(H_2O)\}_2(\mu\text{-}L)]^{4+}$ (L = pyridazine, phthalazine), and mononuclear palladium(II) complexes, such as *cis*-$[Pd(en)Cl_2]$, *cis*-$[Pd(en)(H_2O)_2]^{2+}$, $[Pd(H_2O)_4]^{2+}$, and *cis*-$[Pd(CH_3SCH_2CH_2CH_2SCH_3)(H_2O)_2]^{2+}$, are coordinated by not only the thioether moiety of the methionine residue in methionine-containing dipeptides such as *N*-acetylated methionylglycine (Ac–Met–Gly) but also the

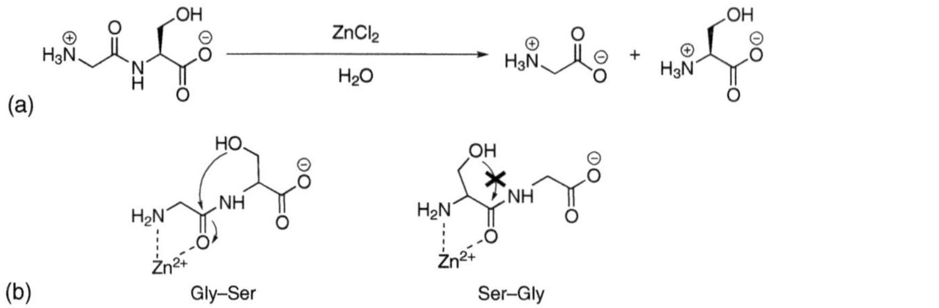

Scheme 7.11 Hydrolysis of dipeptides having a serine residue. (a) Zinc mediated cleavage of the C–N bond on Gly–Ser. (b) A comparison of Gly–Ser and Ser–Gly.

histidine moiety of histidine residue-containing dipeptides such as histidylglycine (Ac–His–Gly) dipeptides, and then the amide bonds of these dipeptides as well as that of some peptides are hydrolyzed [71–82]. Similarly, some dinuclear platinum complexes such as $\{[Pt(en)(H_2O)]_2(\mu\text{-quinoxaline})\}^{4+}$ hydrolyze Ac–Met–Gly and Ac–His–Gly dipeptides [83–85]. Larger cyclic ligands, such as 1,4,7,10,13,16-hexaazacyclooctadecane (= [18]aneN_6), are used for the formation of cationic dinuclear palladium complexes $[Pd_2(\mu\text{-OH})([18]aneN_6)]^{3+}$ and $[Pd_2(H_2O)(OH)([18]aneN_6)]^{3+}$, which efficiently catalyzed the hydrolysis of amide bonds of methionine-containing dipeptides [86]. In addition, palladium(II) complexes such as *cis*-$[Pd(en)(H_2O)_2]^{2+}$ promote the hydrolysis of the second peptide bond upstream from a methionine or a histidine side chain of oligopeptides [87, 88]. Palladium and platinum complexes, *cis*-$[M(en)(H_2O)_2]^{2+}$ (M = Pd, Pt), undergo tryptophan selective hydrolysis of amide bonds of tryptophan-containing peptides such as Ac–Trp–Gly, Ac–Trp–Ala, Ac–Trp–Val, and Ac–Trp–ValOMe [89, 90].

7.2 Catalytic Reactions

Some metal ions act as Lewis acid catalysts for the esterification of amides. Reaction of primary amides in the presence of a catalytic amount (10 mol%) of $TiCl_4$ and 1 equiv of hydrochloric acid in aqueous dioxane affords the corresponding carboxylic acids (Eq. (7.5)) [91]. Notably, hydrolysis of optically pure naproxen amides yields naproxen without any racemization. A catalyst system of $FeCl_3{\cdot}6H_2O$ (20 mol%) in the presence of stoichiometric amounts of HCl also mediates the esterification of not only primary and secondary amides but also tertiary *N*,*N*-dialkylamides **19** (Eq. (7.6)) [92].

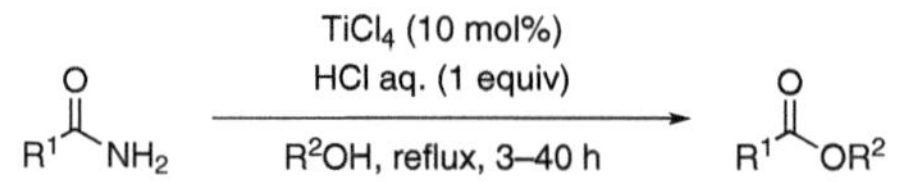

18

(7.5)

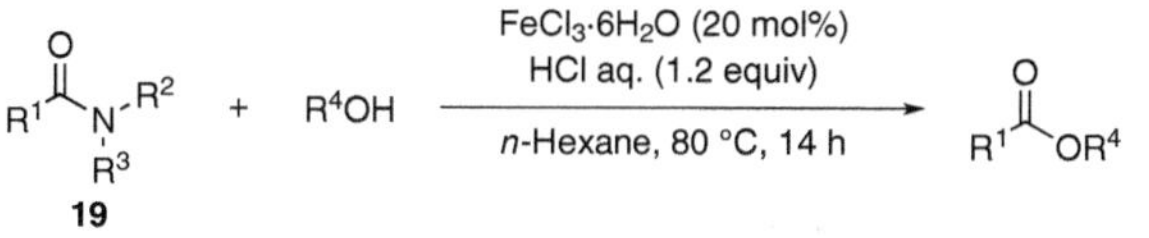

(7.6)

Copper(II) bromide in the presence of 1.5 equiv of 1-(chloromethyl)-4-fluoro-1,4-diazabicyclo[2.2.2]octane-1,4-diium bis(tetrafluoroborate) (Selectfluor) acts as a catalyst for the esterification of tertiary aromatic and aliphatic amides **20** bearing a benzyl moiety on the amide nitrogen atom (Eq. (7.7) and Figure 7.3) [93]. Initially, single electron transfer (SET) from copper(II) to Selectfluor generates a nitrogen-centered radical cation **21** and fluoride anion, and then the radical cation **21** absorbs a hydrogen atom on the benzyl position of **20** (hydrogen atom abstraction process) to afford a benzylic radical species **22**, which is readily oxidized by copper(III) species to form an acyliminium intermediate **23**. The acyliminium intermediate **23** and the fluoride anion give the corresponding acyl fluoride **24**, whose alcoholysis yields the corresponding esters. This catalyst system is advantageous in that the use of amines or thiophene instead of alcohols gives the corresponding amides or 2-acylthiophene

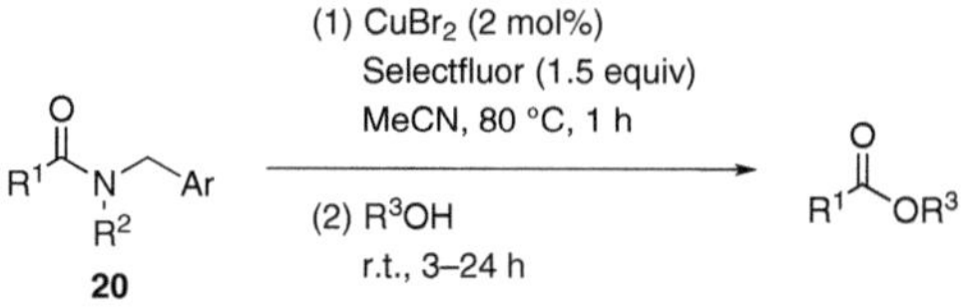

(7.7)

A catalyst system of $Sc(OTf)_3$ and a boronic ester **27** is effective for the alcoholysis of primary aromatic and aliphatic amides **25** as well as deacetylation of secondary acetanilides and their derivatives **26** with alcohols (Eq. (7.8)), in which boronic ester **27** functions to activate the hydroxyl groups of alcohols [94]. Without boronic ester,

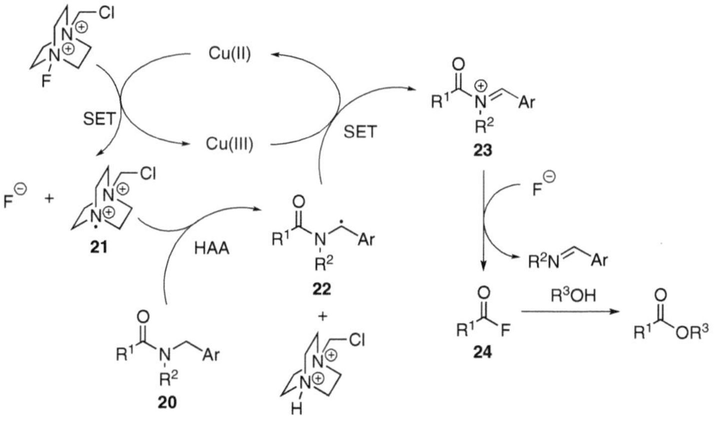

Figure 7.3 Plausible reaction mechanism for $CuBr_2$-catalyzed esterification with Selectfluor.

$Sc(OTf)_3$ itself works as a Lewis acid catalyst for the esterification of aromatic and aliphatic amides **28** with small excess amounts of alcohols (1.2 equiv) (Eq. (7.9)) [95].

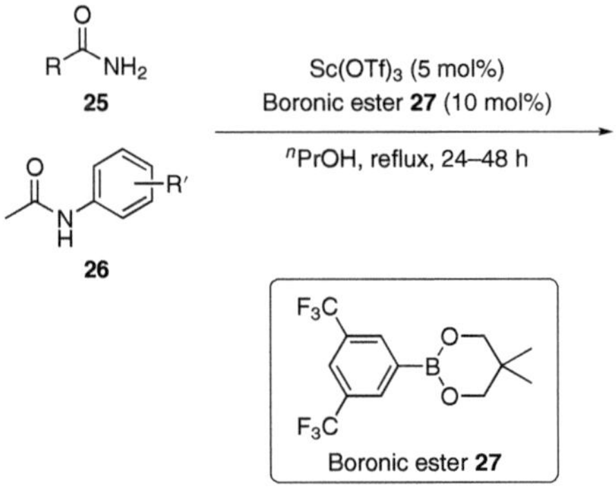

(7.8)

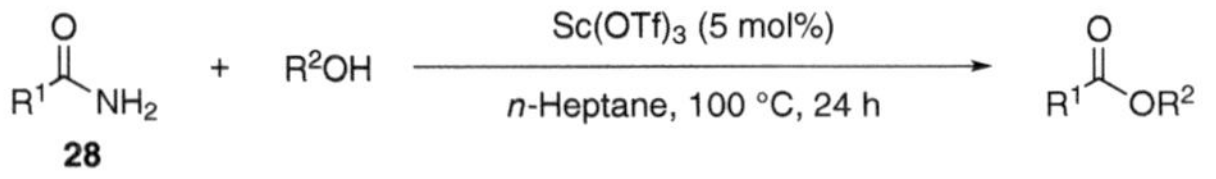

(7.9)

Two metals cooperatively catalyze the esterification of amides; one of two metals activates a carbonyl moiety of amides as a Lewis acid, while an alkoxide on the other metal center attacks the carbonyl moiety. Dinuclear manganese complexes **31** and **32**, derived by treating an alkoxide-bridged manganese tetranuclear cubic complex **30** with 1,10-phenanthroline (phen) or 4,7-bis(dimethylamino)-1,10-phenanthroline (Me_2N-Phen), catalyze the esterification of *N*,*N*-dialkyl aromatic amides **29** in alcohols to give the corresponding esters (Scheme 7.12 and Figure 7.4) [96], while an alkoxide-bridged manganese homodinuclear complex acts as a catalytically active species. In addition, a catalyst system of a mononuclear manganese(II) complex $Mn(acac)_2(Me_2N\text{-Phen})$ (**34**) and KOMe exhibits much higher activity than homodinuclear manganese complex, and catalytic esterification of *N*,*N*-dialkyl aliphatic amides **33**, which is not achieved by the aforementioned manganese homodinuclear system (Eq. (7.10)), occurs via a manganese–potassium heterodinuclear species **35** [97].

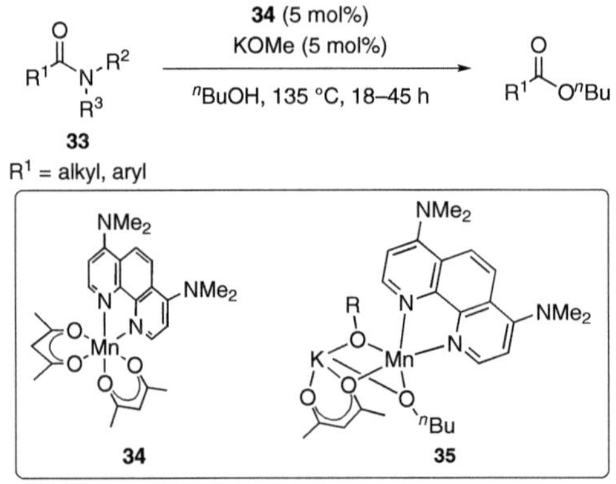

(7.10)

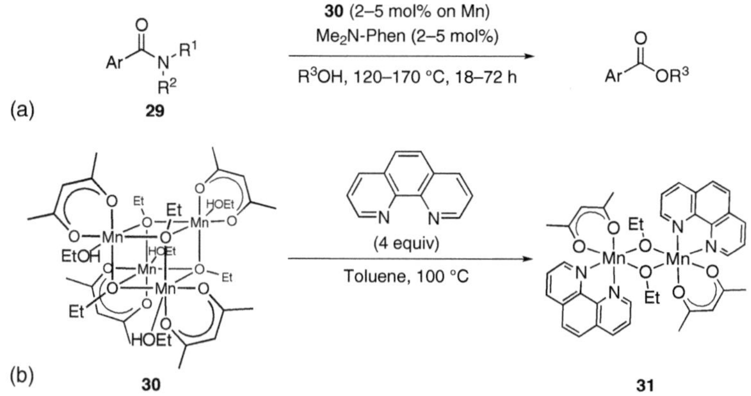

Scheme 7.12 Manganese homodinuclear complex-catalyzed esterification of simple *N,N*-dialkylbenzamides. (a) Alkoxide-bridged Mn_4 cubic complex catalyzed esterification of *N,N*-dialkylbenzamides. (b) Isolation of alkoxide-bridged Mn_2 complex. Source: Adapted from Nagae et al. [96].

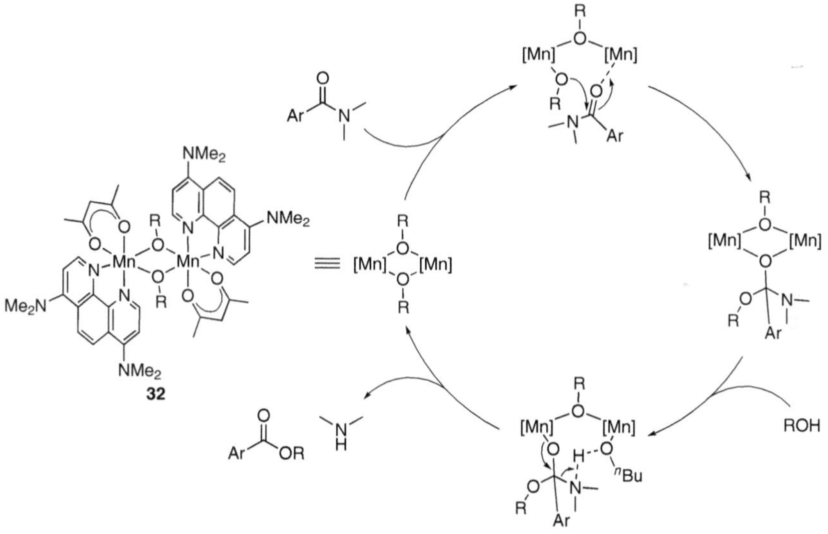

Figure 7.4 Proposed reaction mechanism for manganese dinuclear complex-catalyzed esterification of amides.

Metal oxides were developed as reusable heterogeneous catalysts. At a high temperature of 180 °C, CeO_2 catalyzes the esterification of primary, secondary, and tertiary amides **36** with not only alcohols but also phenol (Eq. (7.11)) [98–100]. The density functional theory calculations revealed that cerium centers on the surface of CeO_2 activate the carbonyl moiety of the amides as a Lewis acid, while oxo moieties on the surface deprotonate the hydroxyl moiety of the alcohols as a Brønsted base. Nb_2O_5 calcined at 500 °C works as a heterogeneous catalyst for the hydrolysis of amides (Eq. (7.12)) [101]. Recently, an MOF such as Zr-MOF-808-P was shown to function as a reusable catalyst at high temperature (150 °C) for the esterification of primary amides **38** along with secondary and tertiary amides (Eq. (7.13)) [54].

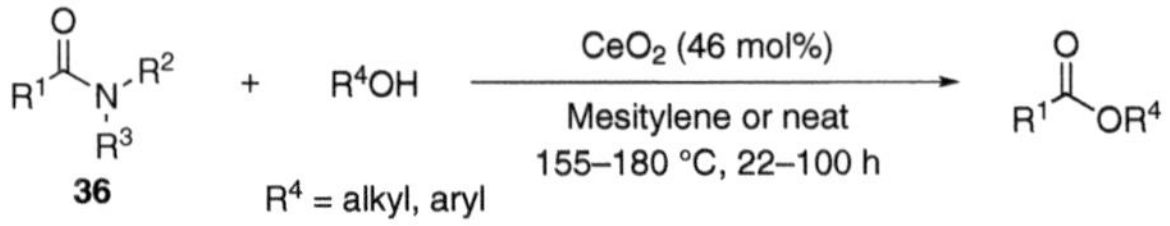

(7.11)

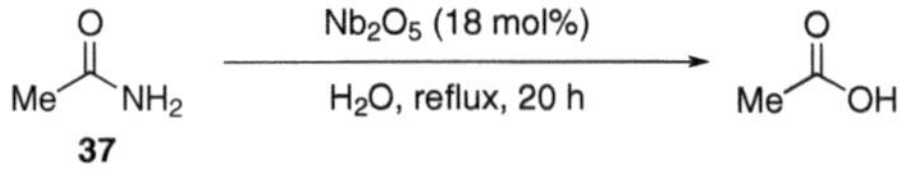

(7.12)

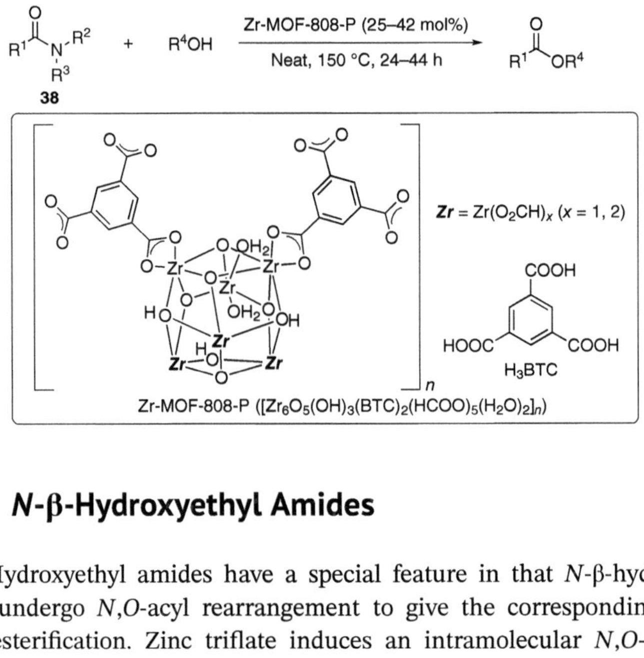

(7.13)

7.3 *N*-β-Hydroxyethyl Amides

N-β-Hydroxyethyl amides have a special feature in that *N*-β-hydroxyethyl moieties undergo *N*,*O*-acyl rearrangement to give the corresponding esters before transesterification. Zinc triflate induces an intramolecular *N*,*O*-acyl rearrangement of *N*-β-hydroxyethyl amides **39** to give the corresponding β-aminoethyl ester intermediate **41**, whose alcoholysis affords the corresponding esters, and a liberated ethanolamine is trapped by diethyl carbonate to give cyclic carbamate **40** (Scheme 7.13a,b) [102]. The formation of cyclic carbamate shifts the equilibrium via *N*,*O*-acyl rearrangement to the β-aminoethyl ester. Another chelation-induced carbonyl activation mechanism was ruled out by the fact that almost no reaction occurred for a methoxyl group-substituted substrate under the same reaction conditions. As an application of this protocol, serine-selective cleavage of dipeptides **43** was achieved using the same zinc catalyst (Scheme 7.13c). Manganese complexes such as $Mn(acac)_2$ in combination with 2,2-bipyridine exhibit higher activity for the esterification of *N*-β-hydroxyethylamides than the zinc system, quantitatively converting an amide **44** into the corresponding ester over six hours (Scheme 7.13d) [103]. It is notable that the manganese system works as a catalyst for the esterification of *N*-methyl-*N*-β-hydroxyethyl amides. The zinc triflate system is applicable to various tertiary *N*-β-hydroxyethyl amides **45** (Scheme 7.13e), leading to one-pot removal of various amide-based directing groups: 8-aminoquinolyl amide **46** is first deprotonated with NaH and then reacted with propylene oxide in the presence of a catalytic amount of $Zn(OTf)_2$ to give the corresponding *N*-β-hydroxyethyl amide, whose subsequent methanolysis in the presence of diethyl carbonate affords the corresponding methyl ester **47** (Scheme 7.14) [104].

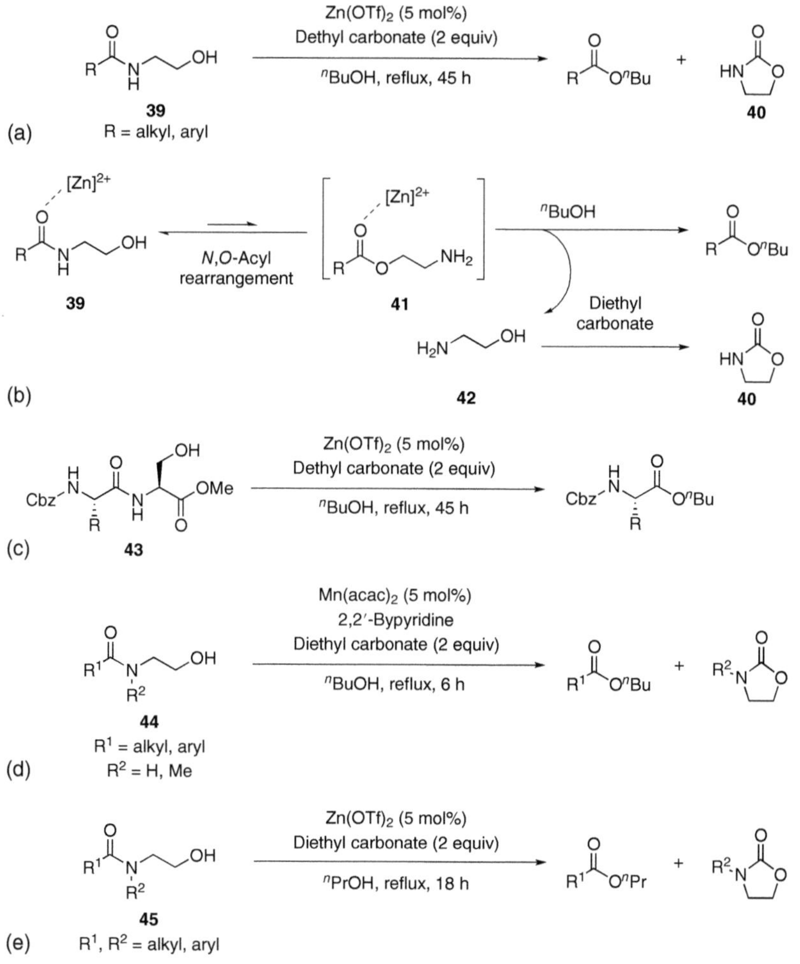

Scheme 7.13 Catalytic alcoholysis of *N*-β-hydroxyethyl amides mediated by zinc and manganese complexes. (a) Zn-catalyzed esterification of *N*-β-hydroxyethyl amides. (b) Plausible reaction mechanism. (c) Zn-catalyzed serine-selective clevage of dipeptides. (d) Manganese catalyzed esterification of *N*-β-hydroxyethylamides. (e) Zinc catalyzed esterification of teritary *N*-β-hydroxyethylamides.

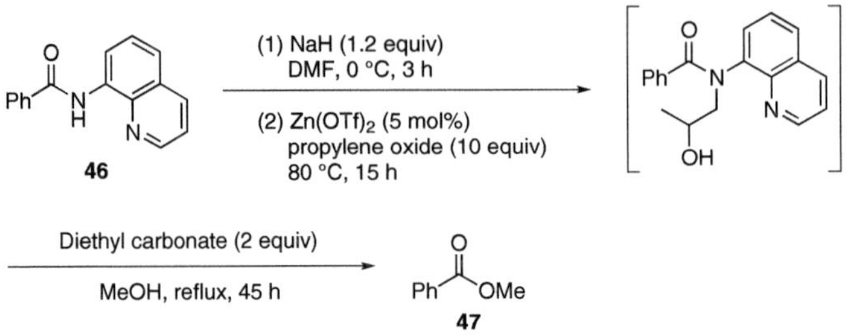

Scheme 7.14 One-pot removal of amide-based directing groups. Source: Nishii et al. [104].

7.4 Chelating Auxiliary at the Nitrogen Atom of Amides

The chelating auxiliary is introduced to the nitrogen atom of amides for activating amide bonds via the coordination of metal complexes. Ni(tmhd)$_2$ (TMHD = 2,2,6,6-tetramethyl-3,5-heptanedionate) is used for the esterification of secondary alkyl amides having an 8-aminoquinoline moiety in methanol (Eq. (7.14)) [105]. Zinc acetate works as a catalyst for alcoholysis of secondary aromatic and aliphatic amides **49** bearing a *tert*-butyl nicotinate moiety on the amide nitrogen atom (Eq. (7.15)) [106], and zinc triflate catalyzes the alcoholysis of tertiary amide **50** bearing a pyridine moiety and an imine moiety on the amide nitrogen atom (Eq. (7.16)) [107]. Yb(OTf)$_3$ is used for the alcoholysis of tertiary *N*-acyloxazolidinones **51** bearing a chiral center on the α-carbon under mild reaction conditions without any racemization of the α-carbon on **51** (Eq. (7.17)) [108].

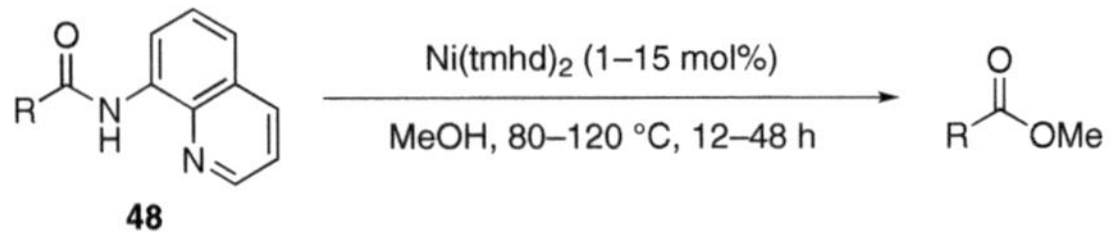

(7.14)

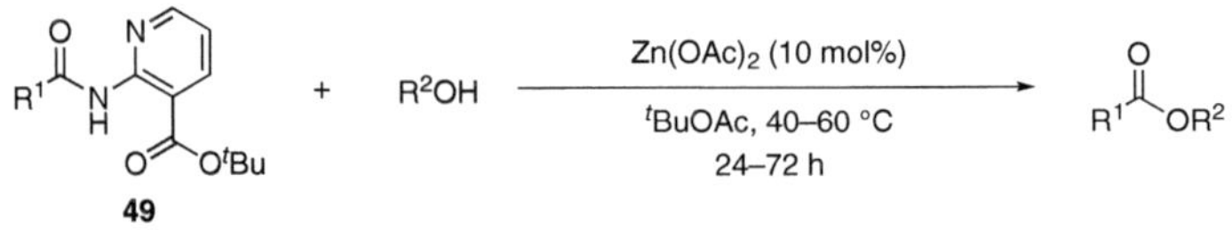

(7.15)

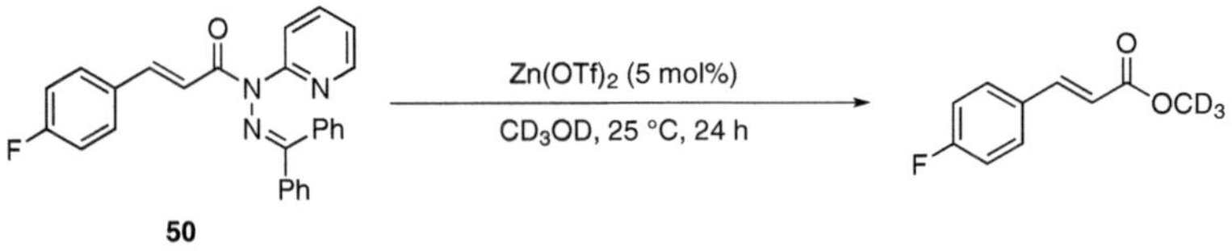

(7.16)

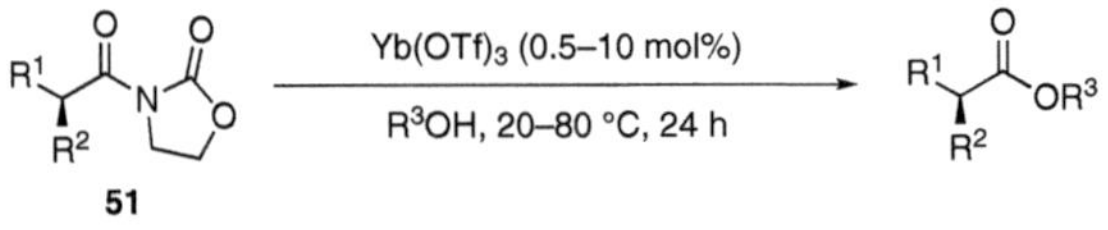

(7.17)

7.5 Activated Amides

Catalytic esterification of activated tertiary amides is achieved using low valent late transition metal complexes through an oxidative addition mechanism. Nickel(0)-SIPr complexes catalyze the esterification of tertiary *N*-phenyl-*N*-methyl aromatic amides **52** (Eq. (7.18)) [109–111]. Nickel acetate in combination with an NHC ligand **54** works as a catalyst for the esterification of bis(Boc) aromatic amides **53** in the presence of stoichiometric amounts of potassium carbonate and triethylamine (Eq. (7.19)) [112]. Terpyridine coordinated to nickel(0) atoms works as a catalyst for the esterification of activated tertiary aliphatic amides such as *N*-benzyl-*N*-Boc tertiary aliphatic amides **55** (Eq. (7.20)) [113]. Bipyridine is another ligand for not only nickel(0) atoms but also low valent cobalt atoms, which act as catalysts for the esterification of *N*-acyloxazolidinons **56** bearing a chiral center on the α-carbon (Eq. (7.21)) [114] and Boc- and Ts-substituted tertiary aromatic and aliphatic amides **57** (Eq. (7.22)), respectively [115].

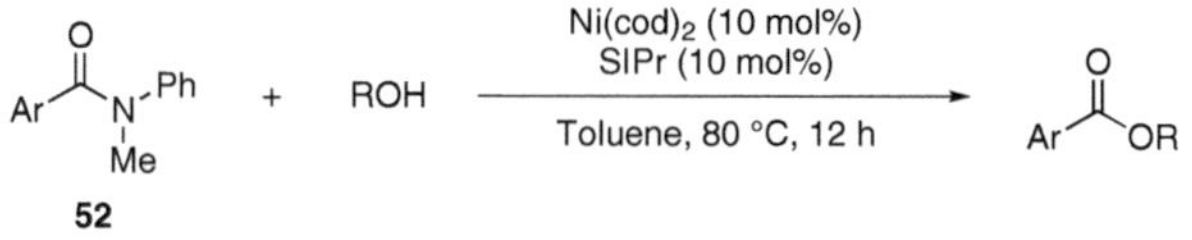

(7.18)

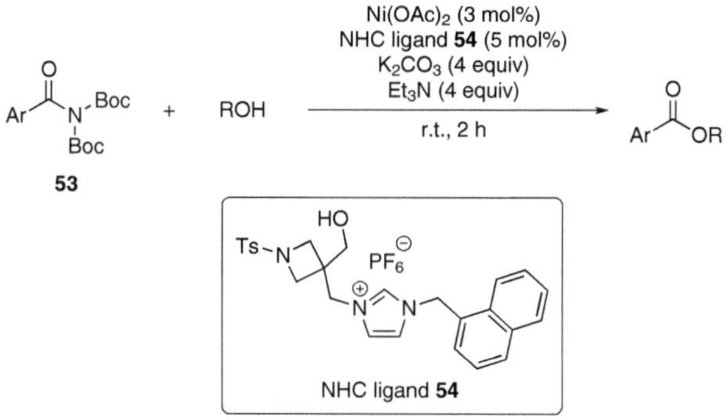

(7.19)

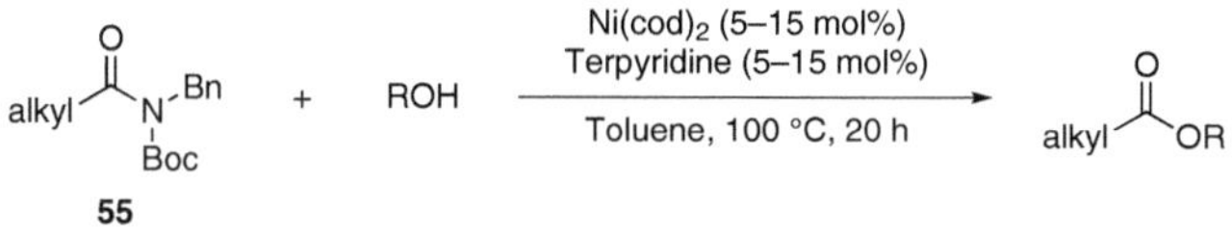

(7.20)

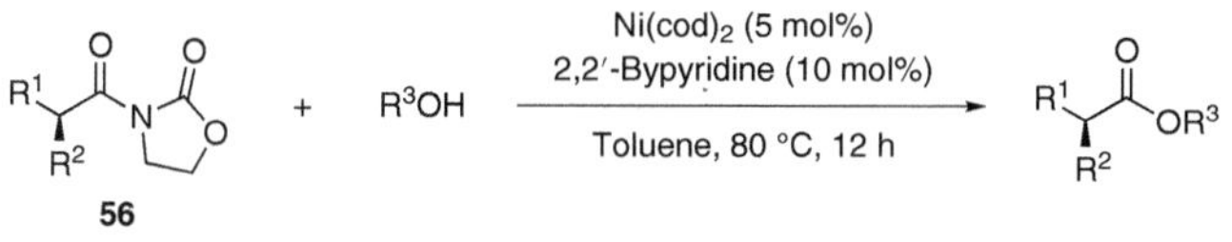

(7.21)

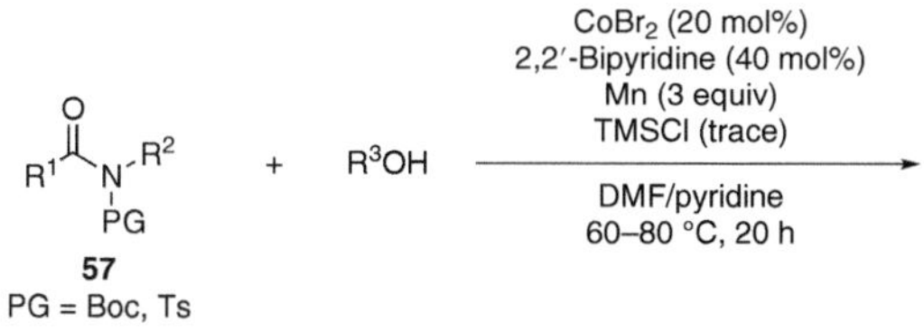

(7.22)

A combination of palladium(II) complexes with fluoride anions acts as a catalyst for the aerobic oxidative coupling of *N*-phenyl-*N*-tosyl amides **58** with arylboronic acids **59** to give aryl esters (Eq. (7.23)) [116]. Cesium salts such as CsF (20 mol%) and Cs_2CO_3 (20 mol%) assist with the esterification of activated tertiary aromatic and aliphatic amides **60** bearing Boc and Ts moieties (Eq. (7.24)) [117, 118]. NaO^{t}Bu (20 mol%) serves as a catalyst for the esterification of Boc-substituted amides **61** with phenol and alcohols at 150 °C to afford the corresponding esters (Eq. (7.25)) [119].

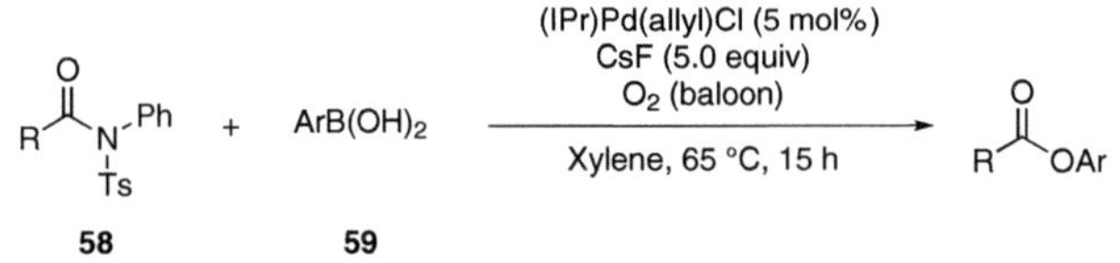

(7.23)

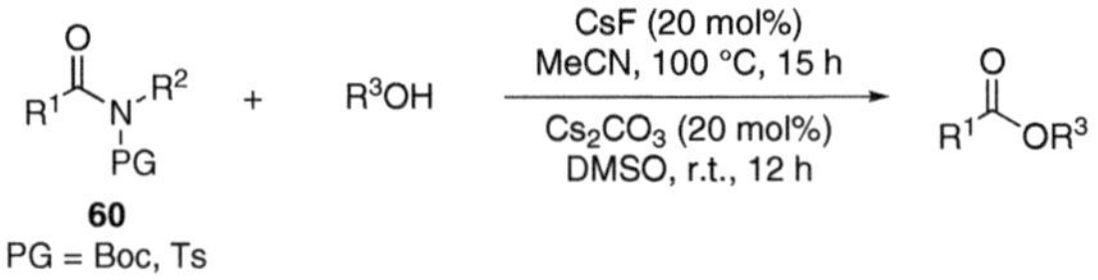

(7.24)

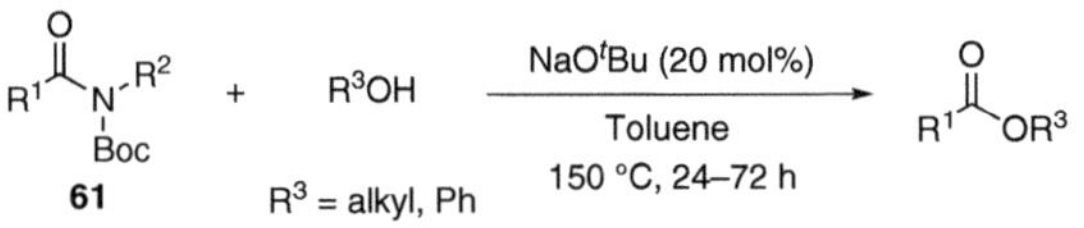

(7.25)

A bifunctional organocatalyst **63** (10 mol%) achieves the catalytic asymmetric esterification of *N*-sulfonyl biaryl lactams **62** to produce axially chiral biaryl amino acid esters (Eq. (7.26)) [120]. Based on control experiments and density functional theory calculations, both N–H moieties on **63** are proposed to cooperatively activate the carbonyl moiety of the amides by the formation of hydrogen bonds, with the quinuclidine moiety on **63** deprotonating the hydroxyl moiety of the alcohol.

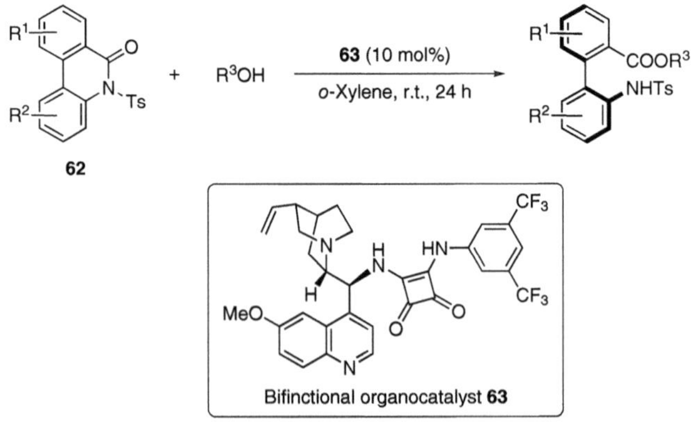

(7.26)

The C–N bond of dipeptide **64** bearing an *N*-troponylglycine moiety is readily cleaved in the presence of the catalytic amount of trifluoroacetic acid to give the corresponding esters and amino acid esters (Eq. (7.27) and Scheme 7.15) [121]. Protonation of the carbonyl group on the troponyl moiety of **64** induces the elimination of the amino acid esters to afford a ketene intermediate **65**, which subsequently reacts with alcohol to give the ester.

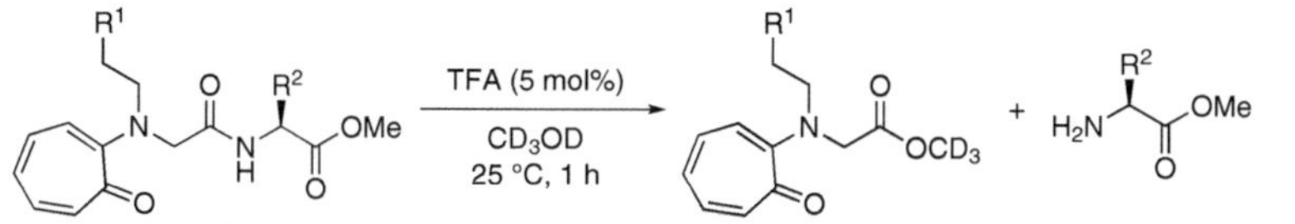

64

(7.27)

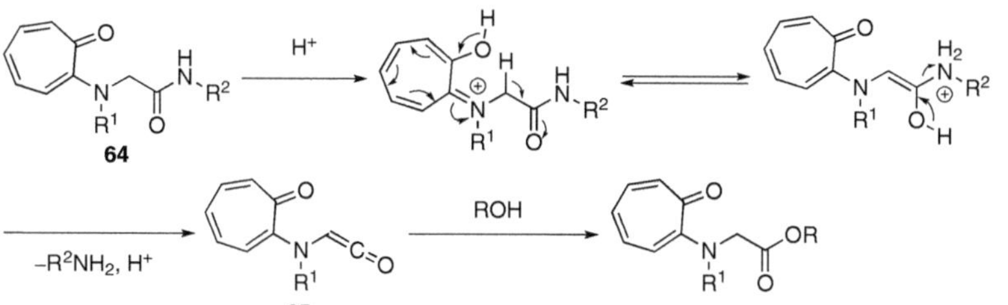

64

65

Scheme 7.15 Plausible reaction mechanism for trifluoroacetic acid-catalyzed esterification of *N*-troponylglycine amides. Source: Balachandra and Sharma [121].

References

1 Radzicka, A. and Wolfenden, R. (1996). *J. Am. Chem. Soc.* 118: 6105–6109.

2 Sigel, A. and Sigel, H. (2001). *Metal Ions in Biological Systems*, vol. 38, CRC Press, Boca Raton, FL.

3 Suh, J. (2003). *Acc. Chem. Res.* 36: 562–570.

4 Suh, J. (1992). *Acc. Chem. Res.* 25: 273–279.

5 Suh, J. and Chei, W.S. (2008). *Curr. Opin. Chem. Biol.* 12: 207–213.

6 Mashima, K., Nishii, Y., and Nagae, H. (2020). *Chem. Rec.* 20: 332–343.

7 Kirby, A.J., Komarov, I.V., Wothers, P.D., and Feeder, N. (1998). *Angew. Chem. Int. Ed.* 37: 785–786.

8 Kirby, A.J., Komarov, I.V., and Feeder, N. (1998). *J. Am. Chem. Soc.* 120: 7101–7102.

9 Kirby, A.J., Komarov, I.V., and Feeder, N. (2001). *J. Chem. Soc. Perkin Trans.* 2: 522–529.

10 Hutchby, M., Houlden, C.E., Haddow, M.F. et al. (2012). *Angew. Chem. Int. Ed.* 51: 548–551.

11 White, E.H., DePinto, J.T., Polito, A.J. et al. (1988). *J. Am. Chem. Soc.* 110: 3708–3709.

12 White, E.H., Field, K.W., Hendrickson, W.H. et al. (1992). *J. Am. Chem. Soc.* 114: 8023–8031.

13 Glatzhofer, D.T., Roy, R.R., and Cossey, K.N. (2002). *Org. Lett.* 4: 2349–2352.

14 Anelli, P.L., Brocchetta, M., Palano, D., and Visigalli, M. (1997). *Tetrahedron Lett.* 38: 2367–2368.

15 Li, L.-C., Ren, J., Liao, T.-G. et al. (2007). *Eur. J. Org. Chem.* 2007: 1026–1030.

16 Charette, A.B. and Chua, P. (1998). *Synlett* 1998, 163–165.

17 Spaggiari, A., Blaszczak, L.C., and Prati, F. (2004). *Org. Lett.* 6: 3885–3888.

18 Omote, K., Aso, Y., Otsubo, T., and Ogura, F. (1994). *Bull. Chem. Soc. Jpn.* 67: 1759–1761.

19 Li, G., Lei, P., and Szostak, M. (2018). *Org. Lett.* 20: 5622–5625.

20 Chin, J., Jubian, V., and Mrejen, K. (1990). *J. Chem. Soc. Chem. Commun.* 1326–1328.

21 Ogawa, K., Nakata, K., and Ichikawa, K. (1998). *Chem. Lett.* **27**: 797–798.

22 Reddy, K.V., Jacobson, A.R., Kung, J.I., and Sayre, L.M. (1991). *Inorg. Chem.* 30: 3520–3525.

23 Sayre, L.M., Reddy, K.V., Jacobson, A.R., and Tang, W. (1992). *Inorg. Chem.* 31: 935–937.

24 Fife, T.H. and Przystas, T.J. (1986). *J. Am. Chem. Soc.* 108: 4631–4636.

25 Tang, W. and Sayre, L.M. (1993). *Inorg. Chem.* 32: 5785–5791.

26 Houghton, R.P. and Puttner, R.R. (1970). *J. Chem. Soc. D* 1270–1271.

27 Bröhmer, M.C., Mundinger, S., Bräse, S., and Bannwarth, W. (2011). *Angew. Chem. Int. Ed.* 50: 6175–6177.

28 Mundinger, S., Jakob, U., and Bannwarth, W. (2014). *Chem. Eur. J.* 20: 1258–1262.

29 Jakob, U., Mundinger, S., and Bannwarth, W. (2014). *Eur. J. Org. Chem.* 2014: 6963–6974.

30 Fife, T.H. and Bembi, R. (1993). *J. Am. Chem. Soc.* 115: 11358–11363.

31 Berreau, L.M., Makowska-Grzyska, M.M., and Arif, A.M. (2000). *Inorg. Chem.* 39: 4390–4391.

32 Groves, J.T. and Dias, R.M. (1979). *J. Am. Chem. Soc.* 101: 1033–1035.

33 Groves, J.T., Rife, R., and Chambers, R. (1984). *J. Am. Chem. Soc.* 106: 630–638.

34 Groves, J.T. and Olson, J.R. (1985). *Inorg. Chem.* 24: 2715–2717.

35 Araki, K., Kajikawa, T., and Kawaguchi, S. (1996). *Chem. Lett.* 295–296.

36 Kawaguchi, S., Kajikawa, T., Kaneko, M. et al. (1999). *Bull. Chem. Soc. Jpn.* 72: 2729–2736.

37 Kawaguchi, S. and Araki, K. (2005). *Inorg. Chim. Acta* 358: 947–956.

38 Czarnik, A.W. (1984). *Tetrahedron Lett.* 25: 4875–4878.

39 Bentley, K.W. and Creaser, E.H. (1973). *Biochem. J* 135: 507–511.

40 Gomez-Reyes, B. and Yatsimirsky, A.K. (2003). *Org. Biomol. Chem.* 1: 866–872.

41 Hegg, E.L. and Burstyn, J.N. (1995). *J. Am. Chem. Soc.* 117: 7015–7016.

42 Fujii, Y., Kiss, T., Gajda, T. et al. (2002). *J. Biol. Inorg. Chem.* 7: 843–851.

43 Shi Tan, X., Fujii, Y., Sato, T. et al. (1999). *Chem. Commun.* 881–882.

44 Ly, H.G.T., Mihaylov, T., Absillis, G. et al. (2015). *Inorg. Chem.* 54: 11477–11492.

45 Absillis, G. and Parac-Vogt, T.N. (2012). *Inorg. Chem.* 51: 9902–9910.

46 Stroobants, K., Absillis, G., Shestakova, P.S. et al. (2014). *J. Cluster Sci.* 25: 855–866.

47 Vanhaecht, S., Absillis, G., and Parac-Vogt, T.N. (2013). *Dalton Trans.* 42: 15437–15446.

48 Ly, H.G.T., Absillis, G., and Parac-Vogt, T.N. (2016). *New J. Chem.* 40: 976–984.

49 Ly, H.G.T., Absillis, G., and Parac-Vogt, T.N. (2015). *Eur. J. Inorg. Chem.* 2015: 2206–2215.

50 Ly, H.G.T., Absillis, G., and Parac-Vogt, T.N. (2013). *Dalton Trans.* 42: 10929–10938.

51 Ly, H.G.T., Absillis, G., Bajpe, S.R. et al. (2013). *Eur. J. Inorg. Chem.* 2013: 4601–4611.

52 Ly, H.G.T., Fu, G.X., Kondinski, A. et al. (2018). *J. Am. Chem. Soc.* 140: 6325–6335.

53 Conic, D., Pierloot, K., Parac-Vogt, T.N., and Harvey, J.N. (2020). *Phys. Chem. Chem. Phys.* 22: 25136–25145.

54 Villoria-del-Alamo, B., Rojas-Buzo, S., Garcia-Garcia, P., and Corma, A. (2021). *Chem. Eur. J.* 27: 4588–4598.

55 Kassai, M., Ravi, R.G., Shealy, S.J., and Grant, K.B. (2004). *Inorg. Chem.* 43: 6130–6132.

56 Kassai, M. and Grant, K.B. (2008). *Inorg. Chem. Commun.* 11: 521–525.

57 Takarada, T., Takahashi, R., Yashiro, M., and Komiyama, M. (1998). *J. Phys. Org. Chem.* 11: 41–46.

58 Takarada, T., Yashiro, M., and Komiyama, M. (2000). *Chem. Eur. J.* 6: 3906–3913.

59 Yashiro, M., Takarada, T., Miyama, S., and Komiyama, M. (1994). *J. Chem. Soc., Chem. Commun.* **6**: 1757–1758.

60 Yashiro, M., Sonobe, Y., Yamamura, A. et al. (2003). *Org. Biomol. Chem.* 1: 629–632.

61 Ho, P.H., Stroobants, K., and Parac-Vogt, T.N. (2011). *Inorg. Chem.* 50: 12025–12033.

62 Erxleben, A. (2005). *Inorg. Chem.* 44: 1082–1094.

63 Krezel, A., Mylonas, M., Kopera, E., and Bal, W. (2006). *Acta Biochim. Pol.* 53: 721–727.

64 Mylonas, M., Krężel, A., Plakatouras, J.C. et al. (2002). *J. Chem. Soc., Dalton Trans.* 4296–4306.

65 Kopera, E., Krężel, A., Protas, A.M. et al. (2010). *Inorg. Chem.* 49: 6636–6645.

66 Protas, A.M., Bonna, A., Kopera, E., and Bal, W. (2011). *J. Inorg. Biochem.* 105: 10–16.

67 Chen, X.H., Zhu, L.G., You, X.Z., and Kostic, N.M. (1998). *J. Biol. Inorg. Chem.* 3: 1–8.

68 Milović, N.M., Dutcă, L.-M., and Kostić, N.M. (2003). *Chem. Eur. J.* 9: 5097–5106.

69 Milović, N.M., Dutcă, L.-M., and Kostić, N.M. (2003). *Inorg. Chem.* 42: 4036–4045.

70 Dutcă, L.-M., Ko, K.-S., Pohl, N.L., and Kostić, N.M. (2005). *Inorg. Chem.* 44: 5141–5146.

71 Milinković, S.U., Parac, T.N., Djuran, M.I., and Kostić, N.M. (1997). *J. Chem. Soc., Dalton Trans.* 2771–2776.
72 Karet, G.B. and Kostic, N.M. (1998). *Inorg. Chem.* 37: 1021–1027.
73 Zivkovic, M.D., Franich, A.A., Asanin, D.P. et al. (2020). *J. Solution Chem.* 49: 1082–1093.
74 Zivkovic, M.D., Rajkovic, S., Glisic, B.D. et al. (2017). *Bioorg. Chem.* 72: 190–198.
75 Zhu, L. and Kostic, N.M. (1993). *J. Am. Chem. Soc.* 115: 4566–4570.
76 Parac, T.N. and Kostić, N.M. (1996). *J. Am. Chem. Soc.* 118: 51–58.
77 Zhu, L., Qin, L., Parac, T.N., and Kostic, N.M. (1994). *J. Am. Chem. Soc.* 116: 5218–5224.
78 Milović, N.M. and Kostić, N.M. (2002). *Inorg. Chem.* 41: 7053–7063.
79 Zhu, L., Bakhtiar, R., and Kostić, N.M. (1998). *J. Biol. Inorg. Chem.* 3: 383–391.
80 Stoffregen, S.A., Griffin, A.K.K., and Kostić, N.M. (2005). *Inorg. Chem.* 44: 8899–8907.
81 Parac, T.N., Ullmann, G.M., and Kostić, N.M. (1999). *J. Am. Chem. Soc.* 121: 3127–3135.
82 Parac, T.N. and Kostić, N.M. (1996). *J. Am. Chem. Soc.* 118: 5946–5951.
83 Rajkovic, S., Warzajtis, B., Zivkovic, M.D. et al. (2018). *Bioinorg. Chem. Appl.* 2018: 3294948.
84 Rajkovic, S., Asanin, D.P., Zivkovic, M.D., and Djuran, M.I. (2013). *Polyhedron* 65: 42–47.
85 Rajkovic, S., Rychlewska, U., Warzajtis, B. et al. (2014). *Polyhedron* 67: 279–285.
86 Yang, G.S., Miao, R., Li, Y.Z. et al. (2005). *Dalton Trans.* 1613–1619.
87 Milović, N.M. and Kostić, N.M. (2002). *J. Am. Chem. Soc.* 124: 4759–4769.
88 Milović, N.M. and Kostić, N.M. (2003). *J. Am. Chem. Soc.* 125: 781–788.
89 Kaminskaia, N.V. and Kostic, N.M. (2001). *Inorg. Chem.* 40: 2368–2377.
90 Kaminskaia, N.V., Johnson, T.W., and Kostić, N.M. (1999). *J. Am. Chem. Soc.* 121: 8663–8664.
91 Fisher, L.E., Caroon, J.M., Stabler, S.R. et al. (1994). *Can. J. Chem.* 72: 142–145.
92 Chen, X., Hu, S., Chen, R. et al. (2018). *RSC Adv.* 8: 4571–4576.
93 Wang, Z., Matsumoto, A., and Maruoka, K. (2020). *Chem. Sci.* 11: 12323–12328.
94 Kita, Y., Nishii, Y., Onoue, A., and Mashima, K. (2013). *Adv. Synth. Catal.* 355: 3391–3395.
95 Atkinson, B.N. and Williams, J.M.J. (2014). *Tetrahedron Lett.* 55: 6935–6938.
96 Nagae, H., Hirai, T., Kato, D. et al. (2019). *Chem. Sci.* 10: 2860–2868.
97 Hirai, T., Kato, D., Mai, B.K. et al. (2020). *Chem. Eur. J.* 26: 10735–10742.
98 Siddiki, S.M.A.H., Touchy, A.S., Tamura, M., and Shimizu, K. (2014). *RSC Adv.* 4: 35803–35807.
99 Toyao, T., Nurnobi Rashed, M., Morita, Y. et al. (2019). *ChemCatChem* 11: 449–456.
100 Rashed, M.N., Siddiki, S., Touchy, A.S. et al. (2019). *Chem. Eur. J.* 25: 10594–10605.
101 Siddiki, S.M.A.H., Rashed, M.N., Touchy, A.S. et al. (2021). *Catal. Sci. Technol.* 11: 1949–1960.

102 Kita, Y., Nishii, Y., Higuchi, T., and Mashima, K. (2012). *Angew. Chem. Int. Ed.* 51: 5723–5726.

103 Nishii, Y., Akiyama, S., Kita, Y., and Mashima, K. (2015). *Synlett* 26: 1831–1834.

104 Nishii, Y., Hirai, T., Fernandez, S. et al. (2017). *Eur. J. Org. Chem.* 2017: 5010–5014.

105 Deguchi, T., Xin, H.-L., Morimoto, H., and Ohshima, T. (2017). *ACS Catal.* 7: 3157–3161.

106 Wybon, C.C.D., Mensch, C., Hollanders, K. et al. (2018). *ACS Catal.* 8: 203–218.

107 Adachi, S., Kumagai, N., and Shibasaki, M. (2017). *Chem. Sci.* 8: 85–90.

108 Stevens, J.M., Parra-Rivera, A.C., Dixon, D.D. et al. (2018). *J. Org. Chem.* 83: 14245–14261.

109 Hie, L., Fine Nathel, N.F., Shah, T.K. et al. (2015). *Nature* 524: 79–83.

110 Weires, N.A., Caspi, D.D., and Garg, N.K. (2017). *ACS Catal.* 7: 4381–4385.

111 Knapp, R.R., Bulger, A.S., and Garg, N.K. (2020). *Org. Lett.* 22: 2833–2837.

112 Li, J.-F., Wang, Y.-F., Wu, Y.-Y. et al. (2020). *Catal. Lett.* 150: 874–880.

113 Hie, L., Baker, E.L., Anthony, S.M. et al. (2016). *Angew. Chem. Int. Ed.* 55: 15129–15132.

114 Huang, P.-Q. and Geng, H. (2018). *Green Chem.* 20: 593–599.

115 Bourne-Branchu, Y., Gosmini, C., and Danoun, G. (2017). *Chem. Eur. J.* 23: 10043–10047.

116 Li, Y., Wu, H., and Zeng, Z. (2019). *Eur. J. Org. Chem.* 2019: 4357–4361.

117 Wu, H., Guo, W., Daniel, S. et al. (2018). *Chem. Eur. J.* 24: 3444–3447.

118 Ye, D., Liu, Z., Chen, H. et al. (2019). *Org. Lett.* 21: 6888–6892.

119 Huang, C., Li, J., Wang, J. et al. (2021). *Sci. China Chem.* 64: 66–71.

120 Wang, G., Shi, Q., Hu, W. et al. (2020). *Nat. Commun.* 11: 946.

121 Balachandra, C. and Sharma, N.K. (2015). *Org. Lett.* 17: 3948–3951.

8

Activation of Amide C–N Bonds by Nickel Catalysis

Liana Hie[1] and Tejas K. Shah[2]

[1] *FMC Stine Research Center, 1090 Elkton Road, Newark, DE 19711, USA*
[2] *Corteva Agriscience, 9330 Zionsville Road, Indianapolis, IN 46204, USA*

8.1 Introduction

Transition-metal-catalyzed cross-couplings have proved to be essential tools in synthetic organic chemistry. Despite the prevalence of palladium catalysis in this field, much interest has recently been devoted to the use of non-precious metal catalysis, specifically to further explore new coupling reactions. In this regard, nickel catalysis has been particularly enabling. In addition to potential cost, toxicity, and environmental benefits relative to precious metal alternatives, nickel catalysis presents opportunities to develop new chemical transformations, notably because of its ability to participate in single-electron processes and to activate strong bonds traditionally considered inert in cross-coupling reactions. In particular, the use of unconventional building blocks (e.g. aniline, phenol, and carboxylic acid derivatives) in cross-couplings offers several advantages because of their availability, bench stability, directing group ability, and orthogonal reactivity to more common aryl and acyl halide electrophiles. In the recent years, the activation of amide C–N bonds using nickel catalysis has emerged as a powerful synthetic tool, enabling the application of amides as acyl electrophiles in cross-couplings. This chapter highlights the efforts in harnessing nickel catalysis for amide activation and application in the C–O bond (esterification and hydrolysis), C–N bond (transamidation), C–C bond (e.g. Suzuki–Miyaura, Negishi, and Heck coupling) formation, and reduction of amides.

8.2 Esterification of Amides

The functional group conversion of amides is essential to numerous biochemical and industrial processes [1]. Particularly, Nature has developed robust enzymes to cleave amides to regulate cellular function. In contrast, the equivalent chemical transformation has been challenging because of the amide's resonance effect and common use of harsh acidic or basic conditions [2]. In recent years, significant attention

Amide Bond Activation: Concepts and Reactions, First Edition. Edited by Michal Szostak.

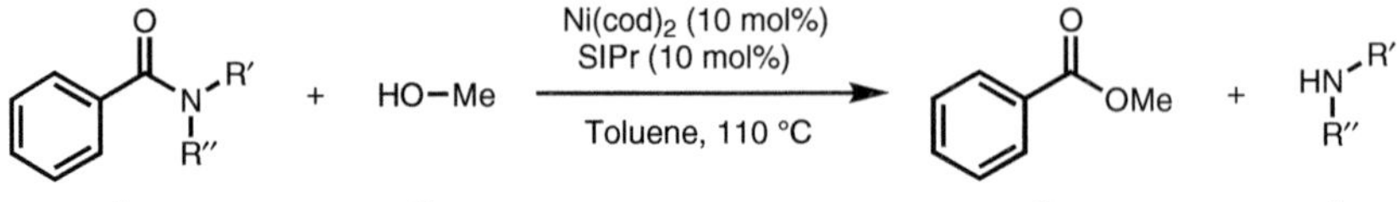

Amide substitution		Calculated ΔG (kcal/mol)	Calculated oxidative addition barrier with Ni/SiPr (kcal/mol)	Yield of ester 10
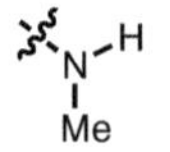	**5**	+3.1	39.0	0%
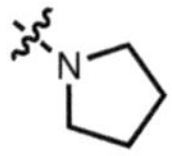	**6**	+2.4	36.8	0%
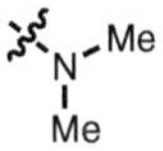	**7**	0.0	36.2	0%
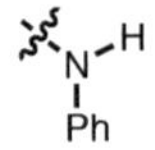	**8**	–4.3	30.6	55%
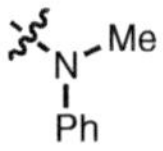	**9**	–6.8	26.0	99%

Figure 8.1 Computational and experimental studies of *N*-substitution effects.

has been focused on developing a catalytic variant using transition metals such as nickel.

In 2015, the Garg and Houk laboratories disclosed the first Ni-catalyzed esterification of amides via acyl C–N bond activation [3]. Initial computational and experimental work started with the examination of amide *N*-substitution in the conversion of benzamide **1** to methyl benzoate (**3**) (Figure 8.1). The change in Gibbs free energy of the overall reaction was calculated for each benzamide and revealed the *N*-substitution significantly influenced the overall reaction. While *N*-alkyl benzamides **5**, **6**, and **7** were calculated to be largely thermodynamically unfavorable or thermoneutral, *N*-phenyl and *N*-phenyl-*N*-methyl benzamides **8** and **9** were favorable with a Δ*G* value of −4.3 and −6.8 kcal/mol, respectively. Additionally, the substitution had a profound effect on the calculated oxidative addition barrier with a nickel catalyst bound to the *N*-heterocyclic carbenes (NHC) ligand, SiPr. *N*-alkyl benzamides **5**, **6**, and **7** had a higher barrier compared to the *N*-phenyl and *N*-phenyl-*N*-methyl substrates **8** and **9**. These computational results were validated by experiments where *N*-phenyl and *N*-phenyl-*N*-methyl benzamides **8** and **9** furnished 55% and 99% yields of ester **3**. Additional mechanistic investigations of the reactivities and properties of non-planar or "twisted" amides have been reported by the Szostak group [4, 5].

With the optimized reaction conditions, the scope of the amide substrate **11** was explored with methanol (**2**) as the nucleophile (Figure 8.2). Benzamide

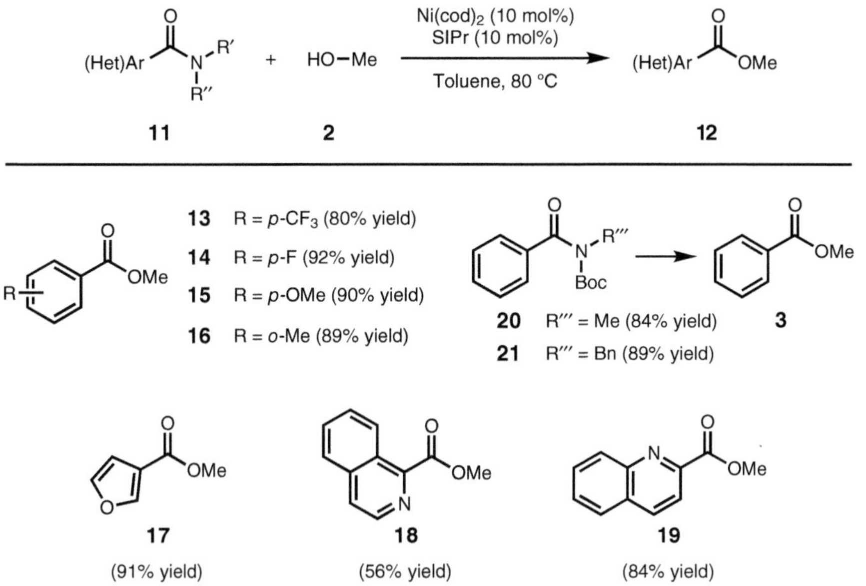

Figure 8.2 Selected examples of amide substrate scope.

substrates containing electron-withdrawing *para*-trifluoromethyl and *para*-fluoro substituents were well tolerated to form methyl esters **13** and **14**, respectively. In addition, electron-donating substrate with a *para*-methoxy substitute afforded methyl ester **15** in high yield. The reaction provided excellent yields for *ortho*- and *meta*-substituted phenyl substrates, **16**, respectively. The methodology was further probed showcasing robustness to heterocycles, such as furan **17**, isoquinoline **18**, and quinoline **19**. Lastly, protected *N*-alkyl benzamides **20** (*N*-Me-*N*-Boc) and **21** (*N*-Bn-*N*-Boc) were able to undergo esterification to provide methyl benzoate (**3**).

Utilizing the *N*-Me-*N*-Ph benzamide **9**, a range of alcohol nucleophiles were subjected to the nickel-catalyzed conditions (Figure 8.3). Simple alcohols such as cyclopropyl carbinol and sterically hindered alcohols such as *t*-butanol coupled smoothly to afford the desired esters **23** and **24** in 90% and 64% yield, respectively. The hindered secondary alcohol, (−)-menthol, also underwent the desired esterification reaction to furnish ester **26** in 88% yield. Furthermore, an unprotected indole with a tethered alcohol and a sugar-containing alcohol were tolerated under the reaction conditions to provide esters **25** and **27** showcasing broad utility.

A notable limitation of the previous methodology was activation of non-aryl amide substrates. The activation of acyl C–N bonds of aliphatic amides posed a significant challenge presumably because of the higher oxidative addition barrier of these substrates. In collaboration with the Catalysis Group at Boehringer-Ingelheim, the Garg laboratory sought to tackle these shortcomings by surveying reaction conditions that included screening ligands [6]. Fortunately, the researchers found a commercially available ligand, terpyridine (**31**) allowed activation of the aliphatic C–N bond and underwent esterification in the presence of an alcohol nucleophile.

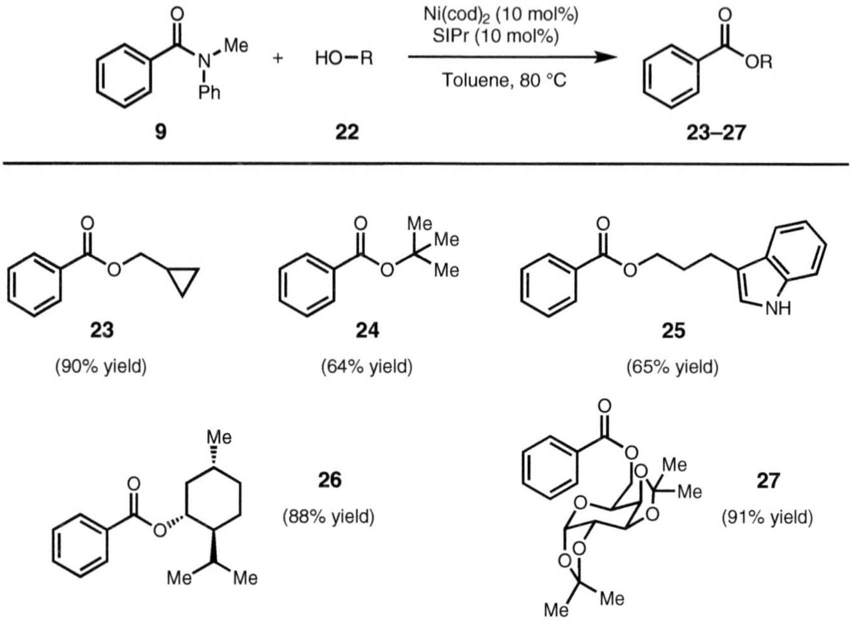

Figure 8.3 Selected examples of alcohol nucleophile scope.

With the appropriate ligand in hand, they sought to explore the generality of the amide substrate using 1-hexanol (**29**) as a nucleophile (Figure 8.4). Non-*alpha* branched and vinyl amides provided esters **32** and **33** in good yield. In addition, *alpha*-branched substrates containing a cyclohexane, dioxane, and cyclopropane were well tolerated as can be demonstrated by the formation of esters **34**, **35**, and **36**. Lastly, heterocyclic substrates containing alpha stereocenters were tested to assess

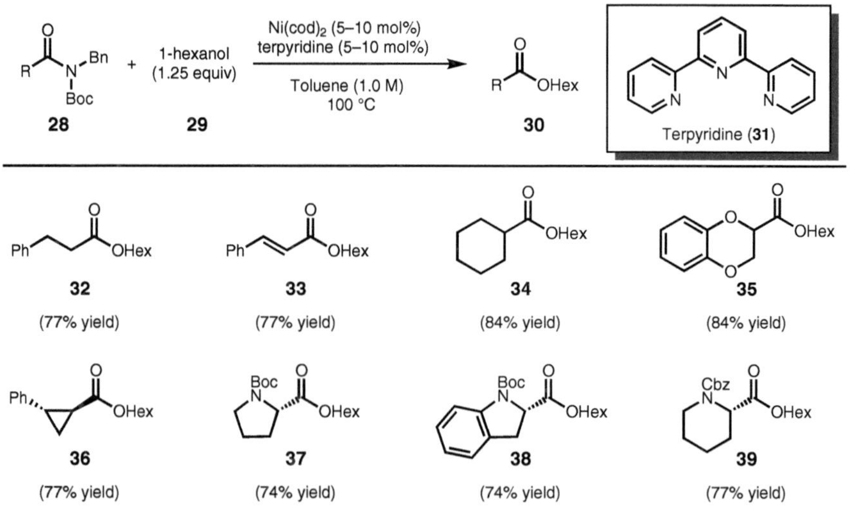

Figure 8.4 Selected examples of amide substrate scope.

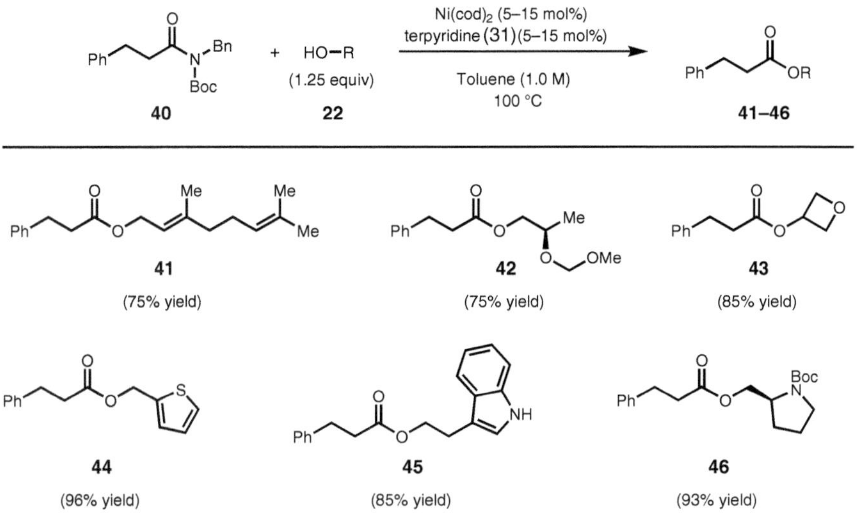

Figure 8.5 Selected examples of alcohol nucleophile scope.

tolerance of epimerizable stereocenters. Esters **37–39** containing a *N*-Boc proline, *N*-Boc indoline, and Cbz-protected piperidine, respectively, were synthesized with no significant loss of enantiopurity, which showcases the mild nature of the reaction conditions.

The breadth of the alcohol nucleophile was also examined and found to be broad (Figure 8.5). Primary alcohols containing olefins and ethers could be employed as seen by esters **41** and **42**. It was also shown that aliphatic and aromatic heterocycles could be tolerated by the formation of esters **43–46** in excellent yields. Of note, ester **45** contains an unprotected indole nitrogen, which speaks to the robustness of this method.

Since these seminal reports, additional advances have been made in decreasing Ni-catalyst loading and expanding the scope of the amide nitrogen functionality. Garg and coworkers, in collaboration with Abbvie, used kinetic modeling to enhance the understanding and practicality of their nickel-catalyzed esterification [7]. This effort led to lower catalyst and ligand loading and reaction temperature. This was exemplified with a 5 g scale coupling of amide **9** and (−)-menthol (**47**) at 45 °C (compared to 80 °C) and less than 1 mol% of $Ni(cod)_2$ and SIPr (compared to 10 mol%) to yield 97% of the ester product **26** (Figure 8.6). In 2017, Morimoto, Ohshima, and coworkers expanded the amide space to include aminoquinoline amides [8], which are well-known directing groups [9]. Utilizing a Ni(II)-catalyst in an alcoholic solvent, both alkyl and aryl aminoquinoline amides underwent esterification. In addition, a variety of functional groups were tolerated (e.g. hydroxy, chloro, cyano, alkyl amides, and *t*-Bu carbamates) to furnish the desired ester products **50–53**. More recently, the Wang group developed the esterification of *N*,*N*-diBoc aryl amide **54** [10], first exploited by Szostak [11]. This reactivity was enabled by a novel NHC ligand and proceeds under room temperature conditions with excellent yields.

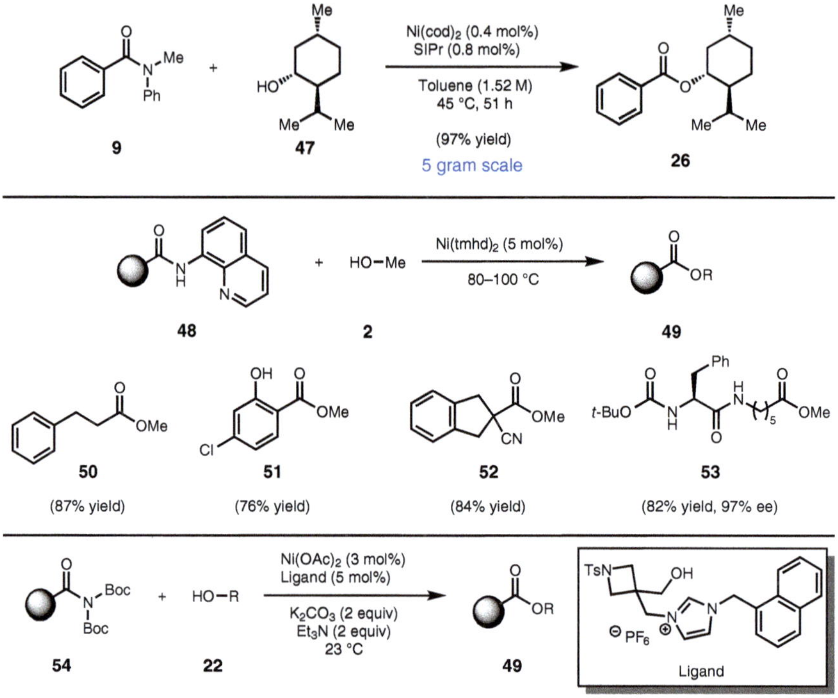

Figure 8.6 Further advances in Ni-catalyzed esterification of amides.

8.3 Hydrolysis of Amides

The hydrolysis of amides to carboxylic acids is a fundamental transformation in organic synthesis [12]. Despite its prominence, it was not until 2020 that a transitional-metal-catalyzed approach was disclosed by Garg and coworkers [13]. Their initial attempts were using the optimized conditions for the transformation of amides to esters. Unfortunately, these conditions using Ni(cod)$_2$, N-heterocyclic carbene ligand SIPr, and water as the nucleophile, led to no formation of the carboxylic acid. An alternative one-pot method (Figure 8.7) was devised using nickel-catalyzed esterification to generate *in situ* protected carboxylic acid **57** followed by deprotection with a mild fluoride source to afford the desired acid **56** in 90% yield.

Several benzamides were examined with this two-step, one-pot approach (Figure 8.8). *N*-Ph-*N*-Me benzamide could undergo the transformation smoothly to provide benzoic acid (**60**) in 73% yield. Benzamides containing electron-withdrawing and electron-donating groups were tolerated to furnish acids **61** and **62** in 84% and 79% yield, respectively. In addition, an amide containing a quinoline heterocycle proceeded to afford acid **63** in 60%, showcasing the tolerance of nitrogen-containing heterocycles.

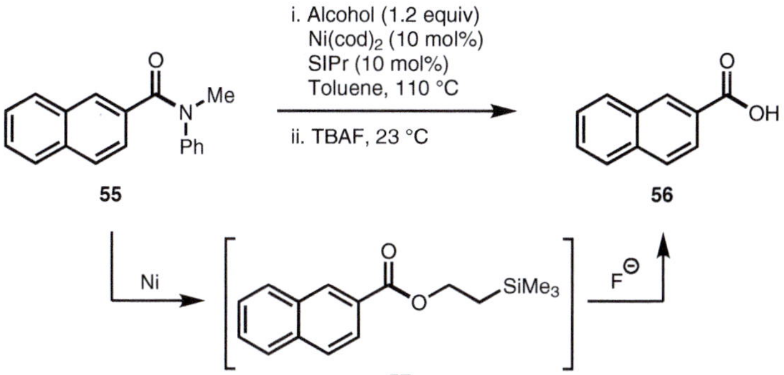

Figure 8.7 Two-step strategy to the synthesis of hydrolysis of amides. TBAF = tetra-*n*-butylammonium fluoride.

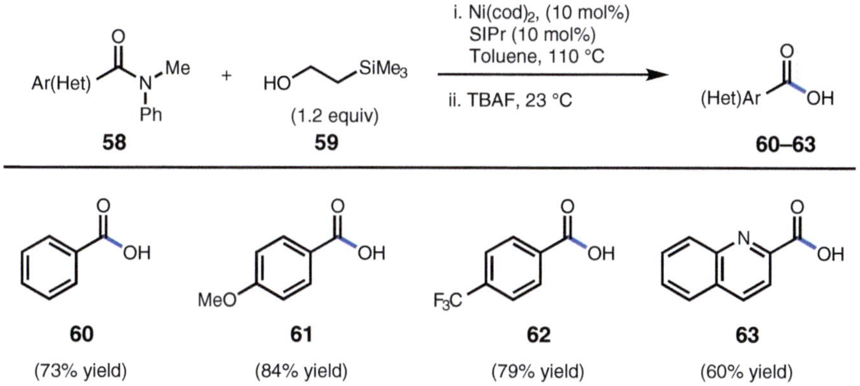

Figure 8.8 Selected examples of hydrolysis of amides. Alcohol = 2-(trimethylsilyl)ethanol.

8.4 Transamidation

Transamidation as an alternative strategy to prepare amides has proven difficult to achieve both synthetically and enzymatically [14]. Besides the high kinetic barrier to break the amide C–N bond, another barrier stems from thermodynamics: the energetics of starting materials and products in transamidation reactions are often comparable, resulting in thermoneutral reactions [1]. The pioneering discovery by Flescher, published in 1876, demonstrated the potential of acyl exchange between an amide and an amine [15]. Transamidation reactions remained hardly explored for another century, with only a few published examples (requiring high temperature and long reaction time) [16]. In 1994, Bertrand and coworkers reported the usage of $AlCl_3$ as a promoter for the transamidation of unfunctionalized aliphatic amides [17]. In 2003, Gellman and coworkers reported the use of a dimeric aluminum

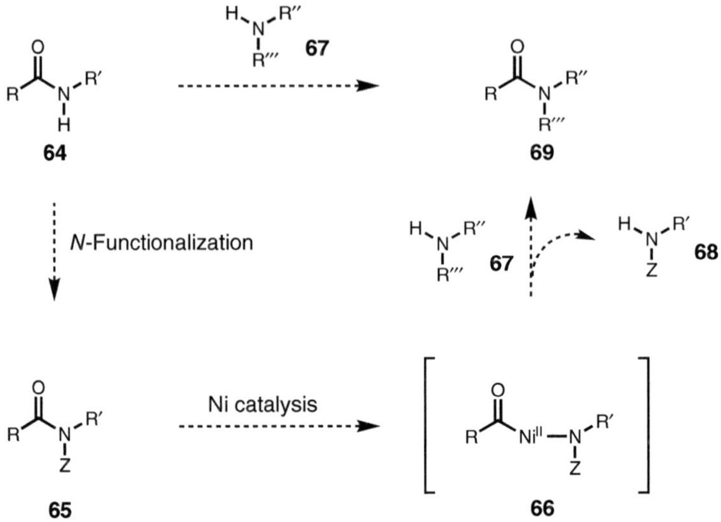

Figure 8.9 Two-step approach to achieve secondary amide transamidation.

complex to promote secondary amide transamidation, which yields equilibrium mixtures [18].

In 2016, Garg and coworkers presented a two-step approach to achieve the net transamidation of secondary amides (Figure 8.9) [19]. Despite the significant progress in utilizing primary amide in transamidation reactions [20], transamidation of secondary amides has remained challenging. The first step relies on activating the amide substrate via *N*-functionalization with an electron-withdrawing group (i.e. Boc group) as a means to weaken the amide C–N bond [21]. The second step is the acyl activation of the *N*-functionalized amide via nickel catalysis. Oxidative addition of the activated species **65** using a nickel catalyst would give acyl metal species **66**. Intercepting this intermediate with amine **67** furnish the desired transamidated product **69**. The process would be driven thermodynamically by the liberation of the *N*-functionalized amine **68**.

Under this approach, a diverse group of amide substrates can undergo transamidation, tolerating electron-donating groups, electron-donating groups, ortho substitutions, and heterocycles as shown by the formation of amides **72–77** (Figure 8.10). Morpholine, aniline, and heterocyclic amine nucleophiles can undergo transamidation under the nickel-catalyzed conditions. The formation of amide **76** also highlighted the tolerance of the method toward steric demand. Moreover, under the reaction conditions, transamidation of a secondary amide substrate using amino-acid-derived nucleophiles was accomplished to forge **77**. In addition, this amino acid containing amide was retained with no appreciable loss of stereochemical fidelity, underlining the mild nature of the methodology.

Garg and coworkers noted that aliphatic amides (e.g. **78**) are not optimal substrates for transamidation under the reaction conditions shown in Figure 8.10.

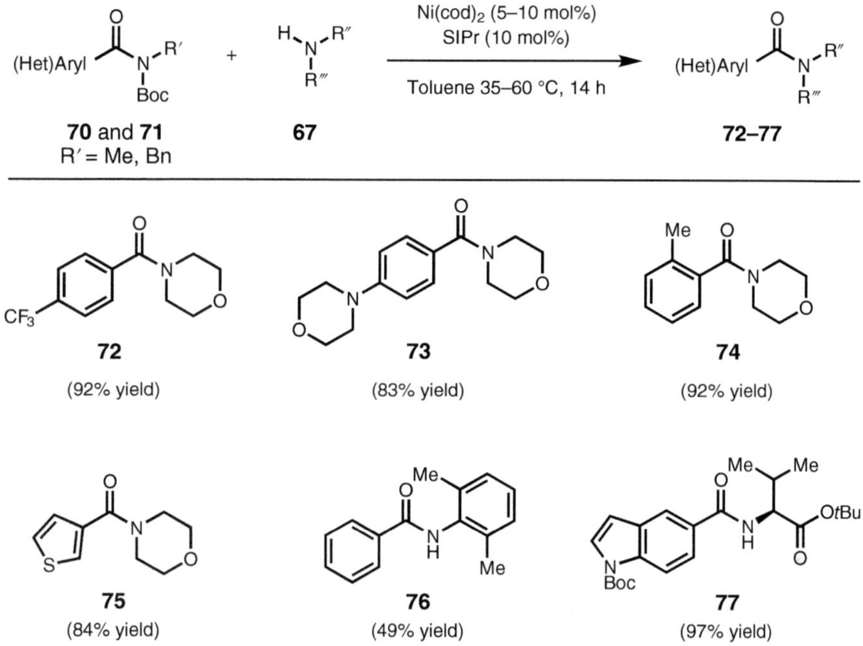

Figure 8.10 Representative examples for transamidation of activated amides.

Subsequent work to tackle this challenge was undertaken starting with a ligand screen (Figure 8.11) [22]. No reaction occurred using terpyridine ligand **31**, which was shown to be an effective ligand in promoting the Ni-catalyzed esterification of aliphatic amide derivatives. Ligand precursor **82**, in combination with NaO*t*Bu for *in situ* free-basing, afforded the desired amide product **80** in quantitative yield. It is hypothesized that this is due to the more electron-rich nature of **82** compared to SIPr ligand **81**, which ultimately renders oxidative addition more facile.

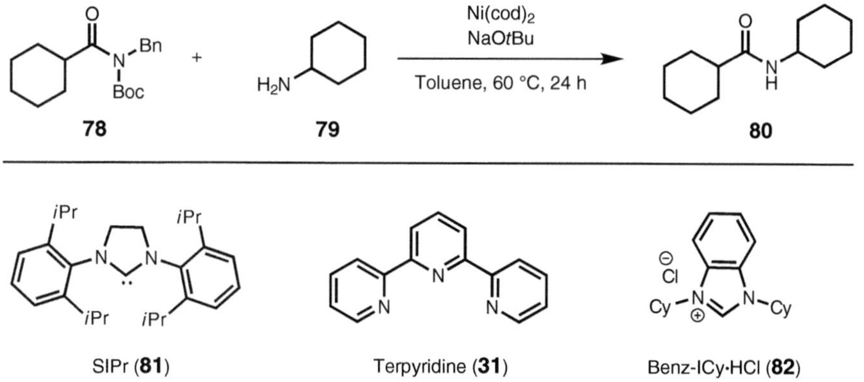

Figure 8.11 Ligand screening.

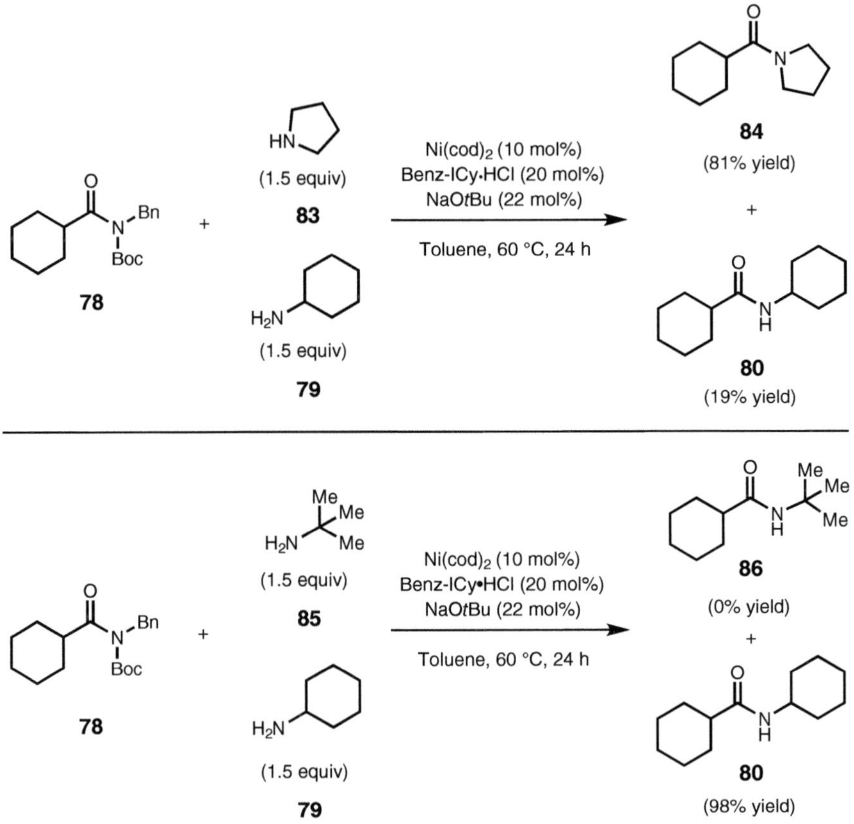

Figure 8.12 Competition experiment.

In a competition experiment, amide **78** was mixed with pyrrolidine (**83**) and cyclohexylamine (**79**) under standard transamidation conditions (Figure 8.12). Pyrrolidine-derived tertiary amide **84** was formed as the major product, rather than secondary amide **80**, consistent with the relative nucleophilicity of the amines being utilized. In another experiment, competition between *t*-butylamine (**85**) and cyclohexylamine (**79**) led to the exclusive formation of cyclohexylamide **80**, presumably driven by steric hindrance.

In 2021, Szostak reported the nickel-catalyzed transamidation of *N*-activated twisted amides, mediated by a half-sandwich Ni(II)–NHC complex [23]. Under the reaction conditions, transamidation of aryl and aliphatic amides with anilines can be achieved. As shown in Figure 8.13, using air- and moisture-stable [CpNi(IPr)Cl] complex, the methodology demonstrated tolerance of anilines containing electron-donating groups (e.g. **89**) and ortho-sterically hindered substituents (e.g. **90** and **91**). However, coupling with anilines containing electron-withdrawing groups is shown to be less effective (e.g. **92**). Aliphatic amide substrate can also be utilized to form amide **93**. Furthermore, cinnamyl amide was shown to be a suitable reaction partner furnishing amide **94**.

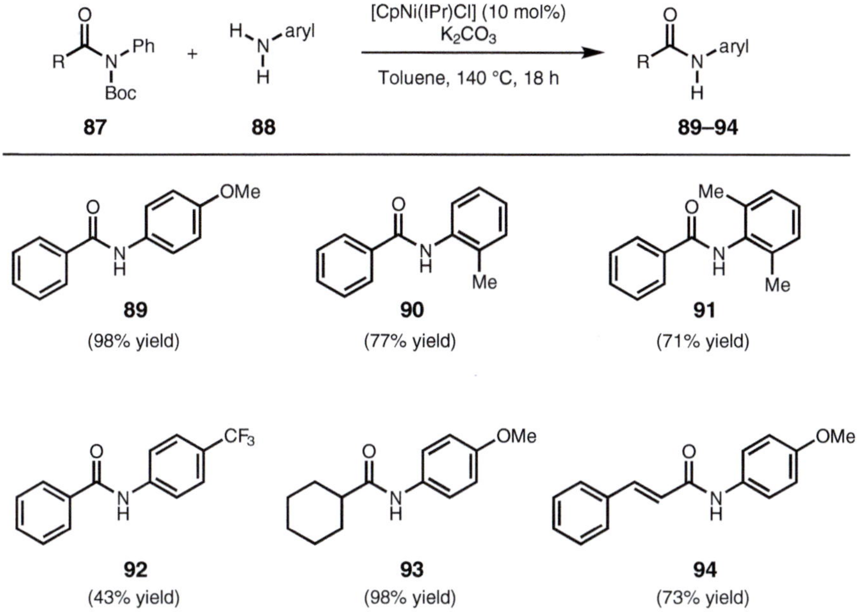

Figure 8.13 Scope of transamidation using a Nickel(II)-NHC complex.

Complimentary to transamidation with amines and anilines, Hu and coworkers reported a two-step reductive transamidation of secondary amides with nitroarenes (Figure 8.14) [24]. Analogous to their report on Ni-catalyzed amidation of esters with nitroarenes [25], the authors identified diazoarene as a probable intermediate, formed by reduction of a nitroarene with zinc in the presence of

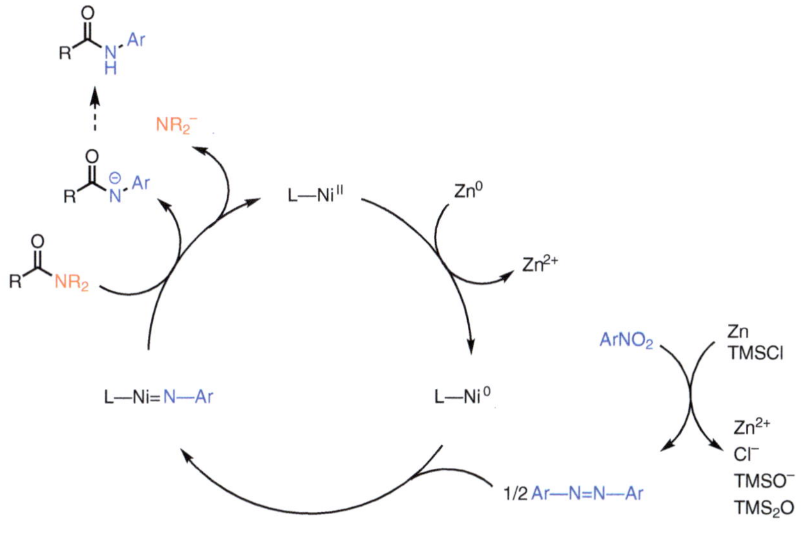

Figure 8.14 Proposed mechanisms for Ni-catalyzed reductive transamidation with nitroarene.

chlorotrimethylsilane (TMSCl). This intermediate could react with a Ni(0) species to give a Ni(II) nitrene-type species. Reaction of the Ni(II) complex with an amide could give a transamidation product.

To probe the nature of the nitrogen-containing intermediates in the reductive transamidation, reactions with Boc-activated alkyl amide **95** and viable intermediates from reduction of nitrobenzene were studied [26] (Figure 8.15). Nitrosobenzene **99**, *N*-phenyl hydroxylamine **100**, and azobenzene **98** were selected as viable intermediates because of their potential to form from nitrobenzene (**97**) under the

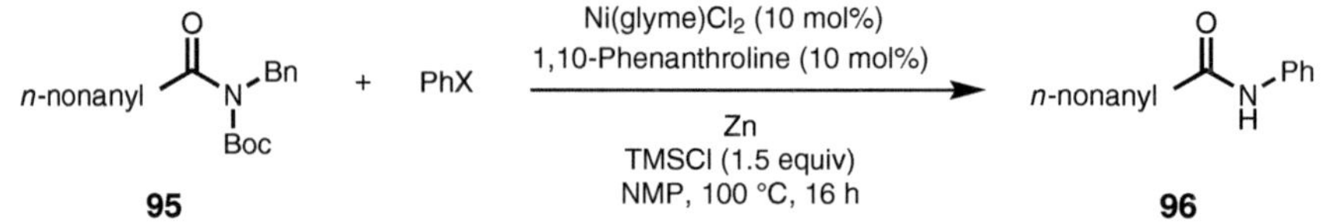

95 **96**

Entry	PhX (1.5 equiv)	Yield of 96	Note
1	$PhNO_2$ (**97**)	80%	5.0 equiv of Zn
2	1/2 Ph–N=N–Ph (**98**)	63%	2.0 equiv of Zn
3	1/2 Ph–N=N–Ph (**98**)	40%	1.0 equiv of Zn
4	Ph–NO (**99**)	0%	2.0 equiv of Zn
5	Ph–NO (**99**)	0%	1.0 equiv of Zn
6	Ph–NHOH (**100**)	0%	2.0 equiv of Zn
7	Ph–NHOH (**100**)	0%	1.0 equiv of Zn

Figure 8.15 Reductive transamidation of alkyl amide derivatives with nitrobenzene.

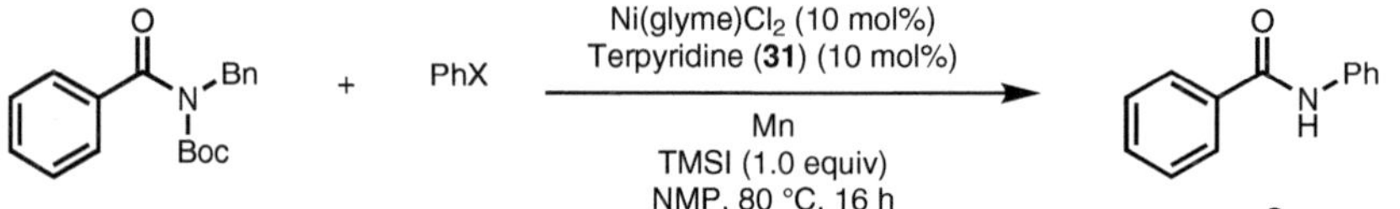

101 **8**

Entry	PhX (1.5 equiv)	Yield of 8	Note
1	$PhNO_2$ (**97**)	88%	5.0 equiv of Mn
2	1/2 Ph–N=N–Ph (**98**)	56%	2.0 equiv of Mn
3	1/2 Ph–N=N–Ph (**98**)	40%	1.0 equiv of Mn
4	Ph–NO (**99**)	35%	2.0 equiv of Mn
5	Ph–NO (**99**)	3%	1.0 equiv of Mn
6	Ph–NHOH (**100**)	0%	2.0 equiv of Mn
7	Ph–NHOH (**100**)	0%	1.0 equiv of Mn

Figure 8.16 Reductive transamidation of aryl amide derivatives with nitrobenzene. TMSI = trimethylsilyl iodide, NMP = *N*-methyl-2-pyrrolidone.

reductive conditions. *N*-Boc-*N*-Bn alkyl amide **95** coupled to nitrobenzene (**97**) to afford the desired amide product in 80% yield (entry 1). Azobenzene in the presence of 2.0 and 1.0 equiv of Zn reacted to give amide **96** in 63% and 40%, respectively (entries 2–3). These results compare to the parent reaction given the inevitable differences in the exact reaction conditions compared to the model reaction. In contrast, reactions with nitrosobenzene (**99**) or *N*-phenyl hydroxylamine (**100**) did not provide the product (entries 4–7).

N-Boc-*N*-Bn aryl amide **101** was also tested with the viable intermediates (Figure 8.16). Using nitrobenzene **97**, the desired amide product **8** formed in 88% yield (entry 1). Azobenzene **98** reacted with **101** to give the amide product **8** in higher yields (entries 2–3) than the reactions with nitrosobenzene (**99**) or *N*-phenyl hydroxylamine (**100**) (entries 4–7). These results suggest that azobenzene is a likely intermediate in the reductive transamidation reactions for both alkyl and aryl amides, which is consistent with the proposed catalytic cycle in Figure 8.14.

8.5 Suzuki–Miyaura Coupling of Amides

The Suzuki–Miyaura coupling reaction has become one of the most prevalent reactions to synthesize a C–C bond in pharmaceuticals, agrochemicals, and materials [27]. Recently, Nickel catalysis has been used to expand the electrophile scope of the Suzuki–Miyaura coupling to phenolic-based substrates (e.g. pivalates, esters, and ethers) [28], thioesters [29], and carboxylic anhydrides [30]. In 2016, Garg and coworkers disclosed the use of aryl amides as electrophiles where the acyl C–N bond would be activated with a nickel catalyst [31]. They utilized *N*-Boc-*N*-Bn (het)aryl amide **102** and (het)aryl boronic ester **103** in the presence of $Ni(cod)_2$, N-heterocycle carbene ligand, SIPr, and 2 equiv of water to afford a biaryl ketone (Figure 8.17). Amides with electron-withdrawing and electron-donating aryl substitution proceeded smoothly to form ketones **104–106**. The scope of the boronic ester coupling partner is equally tolerant of electron-withdrawing and electron-donating as evident by ketones **107** and **108**. In addition, heterocyclic boronic esters were tested including a pyrrole to synthesize ketone **109** in excellent yield.

The robustness of the methodology was further examined by application to a medicinal chemistry-relevant structure. A gram-scale reaction was performed with amide **110** and indolyl boronic ester **111**. The reaction provided ketone **112**, a nanomolar tubulin-binding agent [32] (Figure 8.18). Additional applications were shown, including orthogonal reactivity to Pd-catalysis Suzuki–Miyaura and Ni-catalysis esterification of amides.

In 2017, Huang and Chen furthered the area of Suzuki–Miyura coupling of amides by developing conditions to use *N*-acylpyrroles as electrophiles (Figure 8.19) [33]. These activated amides are valuable substrates because of their ease of synthesis and synthetic utility [34]. It was disclosed that *N*-acylpyrroles (**113**) can undergo a Suzuki–Miyura coupling with (het)aryl boronic neopentylglycol esters (**114**) to form biaryl ketones in the presence of bis-dentate NHC ligand **115** and $Ni(cod)_2$. The scope of the *N*-acylpyrrole was explored and was found to have broad tolerance

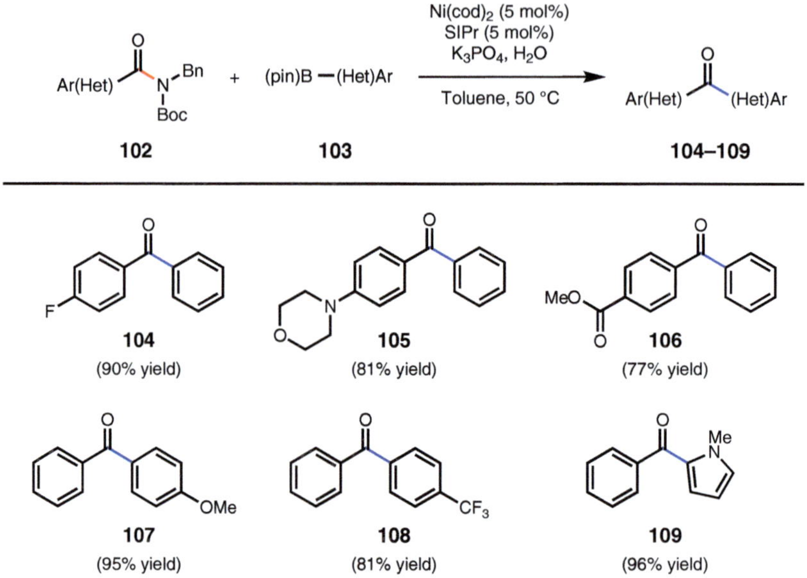

Figure 8.17 Suzuki–Miyaura coupling of aryl amides.

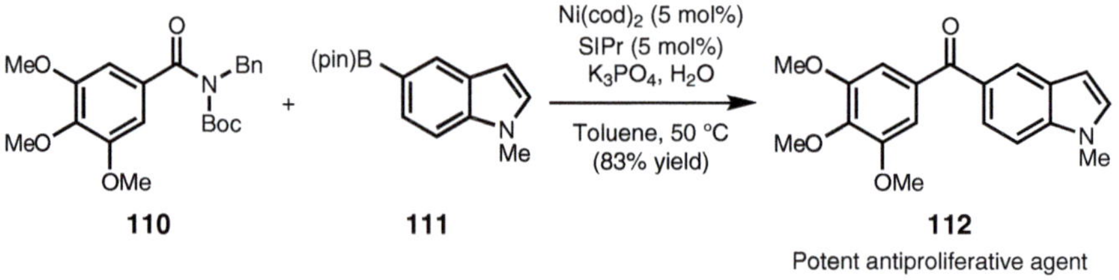

Figure 8.18 Gram-scale synthetic application.

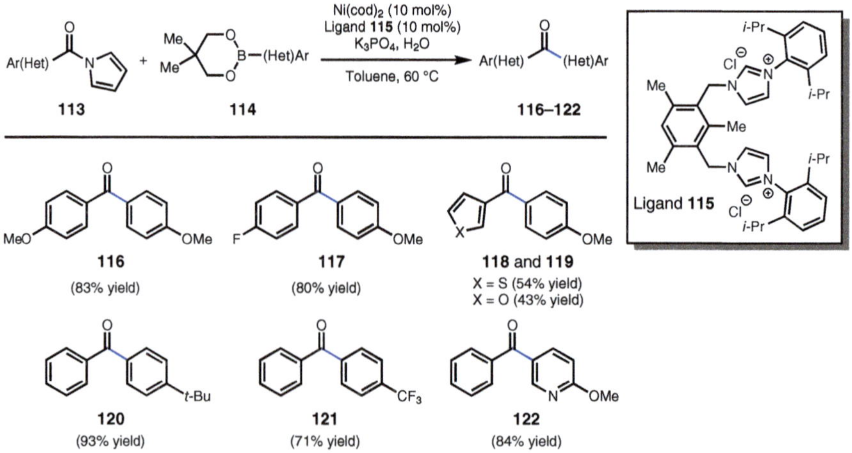

Figure 8.19 Suzuki–Miyaura coupling of *N*-acylpyrroles.

of electron-donating, electron-withdrawing, and oxygen- and sulfur-containing heterocycles to forge ketones **116–119**, respectively. Several boronic ester nucleophiles containing the bulky *t*-butyl group, electron-withdrawing trifluoromethyl, and heterocycles were also surveyed and found to proceed smoothly to give access to ketones **120–122**.

While the initial work focused on aryl amides and the *N*-substitution scope, acyl C–N bond activation of aliphatic amides were still seen as a challenge. In 2018, Garg and coworkers reported the use of an electron-rich NHC ligand precursor, Benz-ICy·HCl (**82**), to activate alkyl amide **123** to undergo a Suzuki–Miyaura coupling in the presence of pincol ester **103** to form ketones **124–129** (Figure 8.20) [35]. Amides containing *N*-Boc piperidine coupled smoothly with heterocyclic boronic acid pincol esters. Fused, five-, and six-membered heterocycles were all tolerated as can be seen by **124**, **125**, and **126**, respectively. Utilizing *N*-Boc-3-proline derived amides furnished ketones **127** in excellent yield. Lastly, amides with oxygen-containing heterocycles and bulky *t*-butyl amides underwent coupling to afford **128** and **129**, respectively, showcasing the broad nature of methodology.

In 2017, Molander and coworkers developed a metallophotoredox methodology to extend acyl cross-coupling of amides to tolerate alkyl boron nucleophiles. In particular, the use of alkyl potassium trifluoroborate salts provides a complementary avenue to alkyl zinc reagents (Figure 8.21) [36]. Utilizing a dual Ir- and Ni-catalytic system, the optimal amide to couple with alkyl potassium trifluoroborate salts **131** to form a ketone was found to be acylsuccinimide **130.** A variety of imides were reacted well under the optimized conditions including aryl, alkyl, and heteroalkyl to furnish ketones **132–137**. Of noteworthy, *alpha*-aminomethyl and *alpha*-alkoxymethyl boron partners provided ketones **135** and **136** in good yields, respectively. Lastly, an

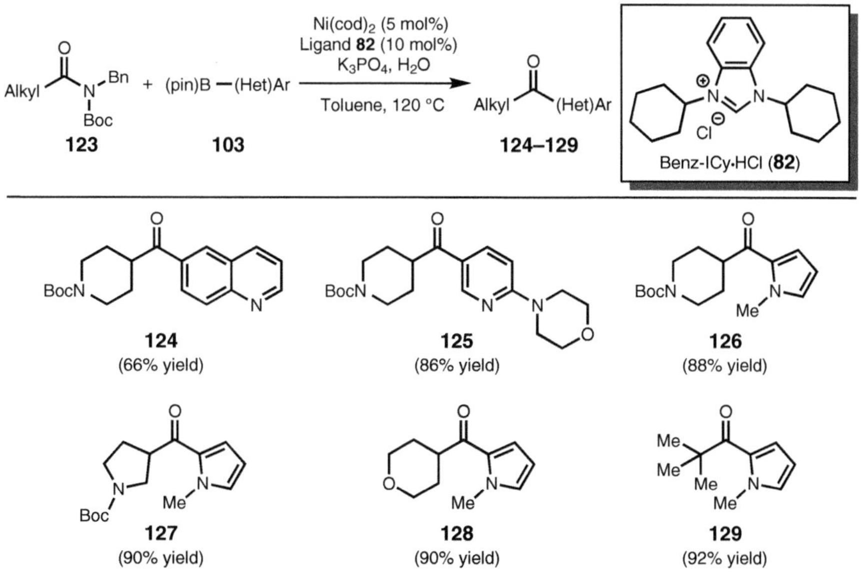

Figure 8.20 Suzuki–Miyaura coupling of alkyl amides.

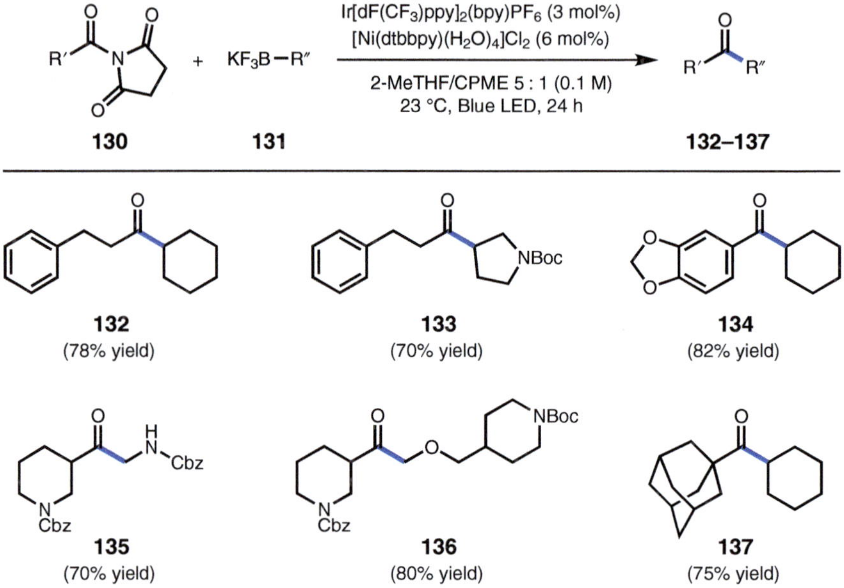

Figure 8.21 Metallophotoredox Suzuki–Miyaura coupling of imides. CPME = cyclopentyl methyl ether, LED = light-emitting diode.

imide containing the bulkier adamantane generated the desired ketone **137** in good yield.

Along with the expansion of the electrophile scope, there have been significant innovations in the utility as well. In 2019, Garg and coworkers in collaboration with Codexis devised a one-pot transformation to convert amides to enantioenriched alcohols using a chemoenzymatic approach [37]. The ability to conduct a Ni-catalyzed Suzuki–Miyaura coupling in aqueous medium allowed for sequential reduction of the ketone products to enantiopure alcohols using a ketoreductase (KRED) enzyme. In 2020, Garg and coworkers tackled the need of a glovebox to handle the air-sensitive $Ni(cod)_2$ that is employed in many of the Suzuki–Miyaura couples of amide bonds [38]. Inspired by Buchwald and Taber's work [39], Garg and coworkers devised a paraffin-encapsulation strategy for $Ni(cod)_2$ catalyst and Benz-ICy·HCl ligand together. These "wax-capsules" could be used on the benchtop and were shown to provide comparable scope and yields to the glovebox chemistry previously published.

8.6 Negishi Coupling of Amides

The Negishi coupling represents one of the most powerful transition-metal-catalyzed coupling reactions [40]. Catalytic acyl couplings [41] with organozinc reagents are well-precedented using acid halides (Pd or Ni) [42], anhydrides (Pd, Ni, or Rh) [43], and thioesters (Pd or Ni) [44]. In 2016, Garg and coworkers reported the first example of nickel-catalyzed alkylation of amide derivatives via Negishi

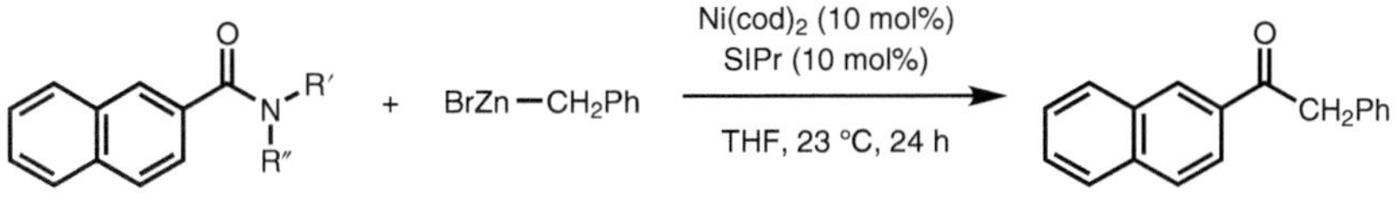

55 and 138–141 **142** **143**

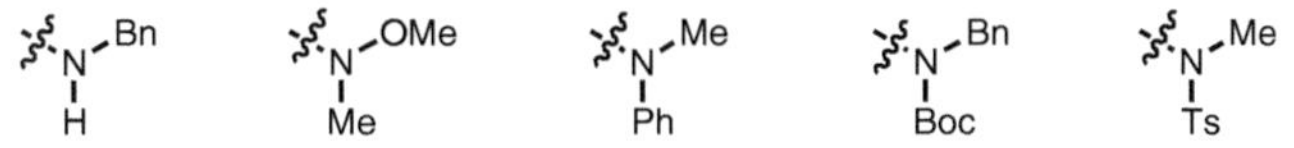

	138	**139**	**55**	**140**	**141**
Yield of ketone (**143**)	(0%)	(0%)	(0%)	(50%)	(81%)
Recovered of amide (**55,138–141**)	(100%)	(51%)	(100%)	(40%)	(17%)

Figure 8.22 Survey of amide *N*-substituents.

coupling (Figure 8.22) [45]. Several of the amide derivatives screened (**55**, **138**, and **139**) failed to undergo the coupling. However, amides with *N*-Alkyl-*N*-Boc and *N*-Alkyl-*N*-Ts (**140** and **141**) derivatives could be utilized. In particular, the organozinc reagent **142** with *N*-Alkyl-*N*-Ts amide provided the highest yield in forming ketone **141**. The alkylative cross-coupling methodology was probed in a synthetic application (Figure 8.23). On a one-gram scale, amide derivative **144** was coupled with cyclohexylzinc iodide (**145**) using the nickel-catalyzed reaction conditions. This transformation provided ketone **146** in 71% yield, which is an intermediate in

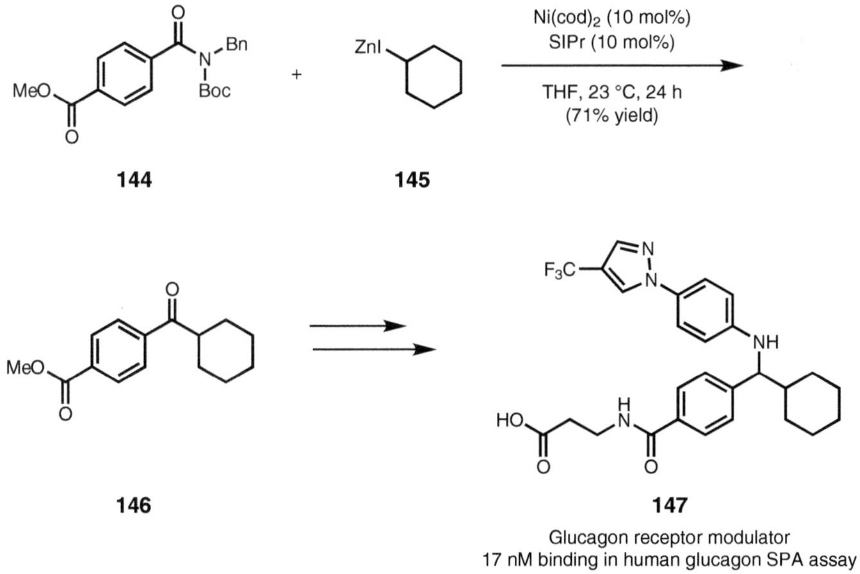

Figure 8.23 Gram-scale coupling to intermediate from Pfizer's route to glucagon receptor modulator.

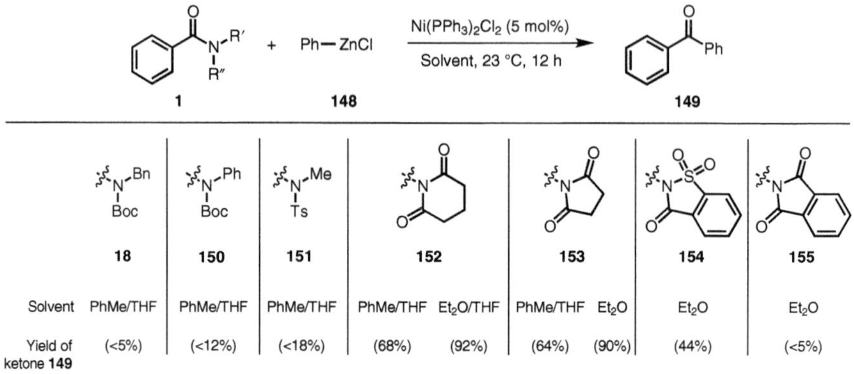

Figure 8.24 Nickel-catalyzed arylation of amide derivatives.

Pfizer's synthesis of the glucagon receptor modulator **147** [46]. The cross-coupling route to **146** provides an alternative to the known Weinreb amide displacement chemistry described in the literature, which proceeds in 34% yield [44].

Progress in nickel-catalyzed arylation of amide derivatives was reported by Shi and Szostak. They discovered the nickel-catalyzed arylation of *N*-acylglutarimides with organozinc reagents (Figure 8.24) [47]. Under the reaction conditions, the arylation of amide derivatives **18**, **150**, and **151** proceeded poorly. However, the arylation of *N*-acylglutarimide **152** and *N*-acylsuccinimide **153** in toluene and tetrahydrofuran (THF) provided ketone **153** in 68% and 64% yields, respectively. *N*-acylglutarimide **153** was chosen for further optimization against a range of the palladium catalysts; however, $[Ni(PPh_3)_2Cl_2]$ was found to be the most effective catalyst. During their condition screening, the authors discovered diethyl ether as a beneficial cosolvent for the arylation reaction. The authors noted the possibility of diethyl ether in stabilizing the putative dihalo-bridged aryl–zinc intermediate [48]. Subsequently, Shi and Szostak were able to improve upon their initial work with *N*-acylsuccinimide **151** by further optimizing conditions to provide high yields of arylated product using diethyl ether as a solvent [49]. To showcase the unique nature of *N*-acylsuccinimides, the authors tested other five-membered *N*-acylimides. While *N*-acylsaccharin **154** furnished ketone **149** in moderate yield, *N*-acylphthalimide **155** provided poor yields.

Furthermore, Shi and Szostak reported the nickel-catalyzed arylation of primary amides with organozinc reagents via double *N*-functionalization of primary amides [50]. The authors proposed that the double *N*-functionalization with an electron-withdrawing group [21a] would decrease the double-bond character of primary amides (barrier to resonance of 15–20 kcal/mol; acetamide, 18.4 kcal/mol) [51], allowing this high-energy precursor to undergo metal insertion into the amidic N–C bond. *N*,*N*-diBoc amides were identified as readily accessible precursors for highly selective nickel insertion into the amide C–N bond.

Intermolecular competition experiments (Figure 8.25) suggested transmetalation to be the rate-limiting step [52], assuming reversible oxidative addition [53].

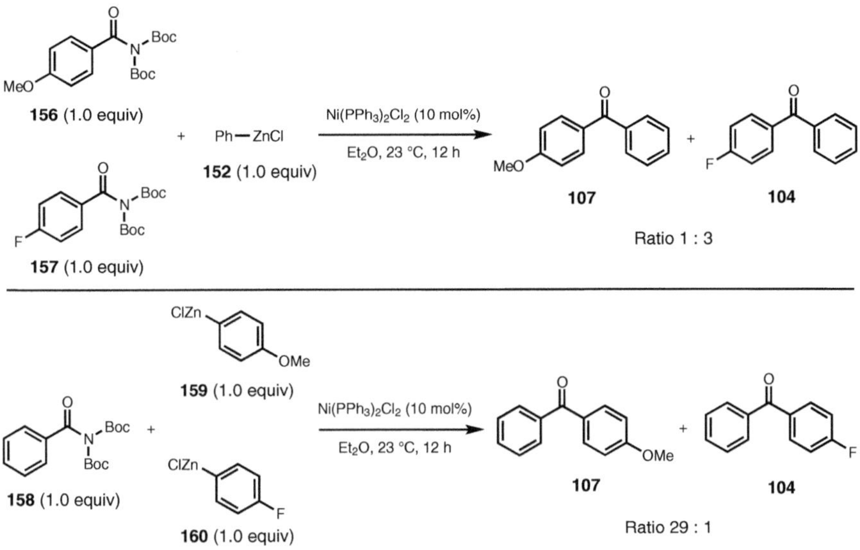

Figure 8.25 Competition experiments for mechanistic studies.

Competition experiments between amide derivatives **156** and **157** revealed the preference for electron-deficient amides consistent with facilitation of metal insertion into the C–N bond [5]. Additionally, competition experiments between organozinc reagents **159** and **160** indicated that electron-rich nucleophiles react preferentially.

8.7 Mizoroki–Heck Coupling of Amides

In 2017, Garg and coworkers utilized their nickel-catalyzed amide activation methodology to construct quaternary carbon centers via Mizoroki–Heck cyclization (Figure 8.26) [54]. In the presence of $Ni(cod)_2$ as the catalyst and Benz-ICy·HCl as the ligand, amide derivatives **161** could be converted into cyclized products **162**, which bear the desired quaternary centers. Following the nickel coordination to the amide derivative, the reaction is proposed to undergo the classical Mizoroki–Heck mechanism, involving oxidative addition (**161** → **163**), olefin coordination and insertion (**163** → **164**), and *beta* hydride elimination (**164** → **162**) (Figure 8.27).

The authors also demonstrated that the diastereoselective Mizoroki–Heck cyclization proceeds for the controlled formation of vicinal, highly substituted sp^3 stereocenters (Figure 8.28). Trisubstituted olefin **165** was treated under the reaction conditions to deliver indanone **166** in 80% yield with a 92 : 8 ratio of diastereomers. The diastereoselectivity of indanone **166** can be rationalized by considering the two competing olefin insertion transition states, TS1 and TS2. In both cases, the olefin insertion event is thought to occur via a standard four-centered transition state, which prompts allylic strain arguments. In TS1, the A(1,3) strain between the two

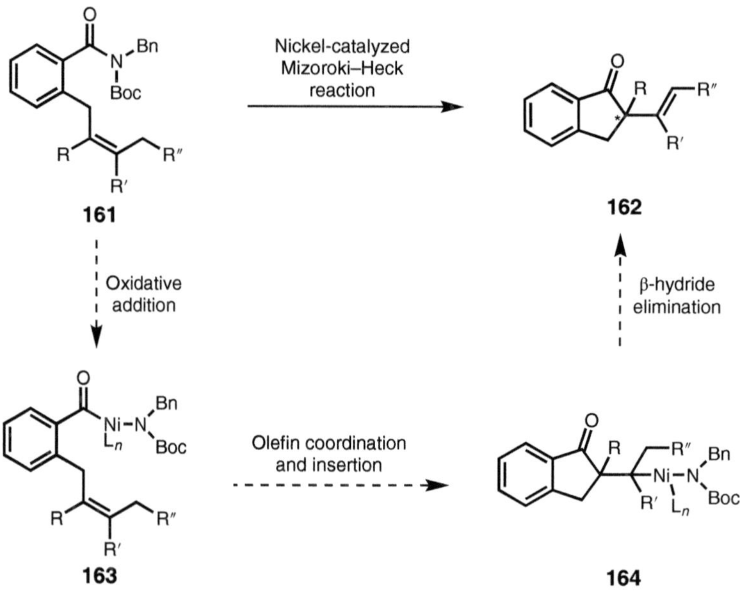

Figure 8.26 Nickel-catalyzed Mizoroki–Heck reaction of amide derivatives to create quaternary centers.

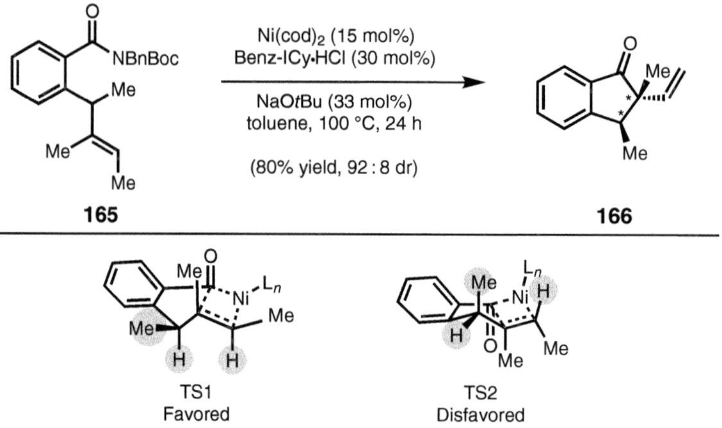

Figure 8.27 Diastereoselective Mizoroki–Heck cyclization for the introduction of vicinal sp^3 stereocenters.

highlighted hydrogens is minimal and the methyl group rests in a pseudo-equatorial disposition. As such, TS1 is favorable and leads to the major diastereomer of **166** shown, with the methyl groups residing in a cis fashion. On the other hand, the minor diastereomer of **166** is thought to arise from TS2, which displays a less favorable A(1,3) interaction between the highlighted hydrogen and the methyl substituent.

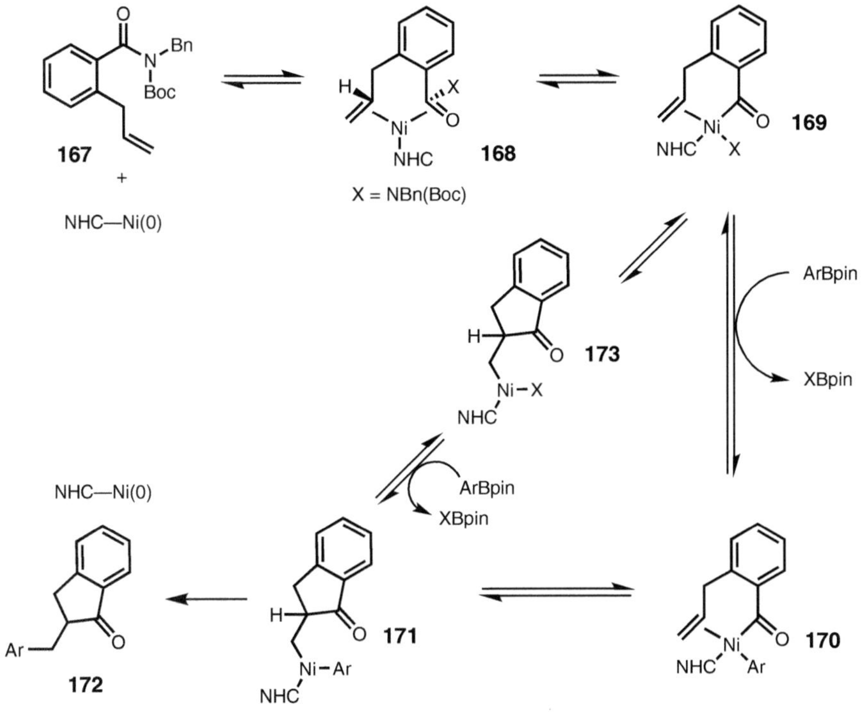

Figure 8.28 Proposed mechanism for nickel-catalyzed alkene carboacylation.

As the Garg laboratory reported the Mizoroki–Heck cyclization reaction of amides to form a single C–C σ bond, the Stanley laboratory independently reported the difunctionalization of an alkene with the formation of two C–C σ bonds [55]. The proposed mechanistic pathways for the carboacylation of amide derivative **167** with arylboronic acid pinacol ester are shown in Figure 8.28. After coordination of the NHC–Ni(0) catalyst to amide derivative **167** to form complex **168**, oxidative addition of the amide C–N bond to the Ni(0) center could form acyl–Ni(II)–amido complex **169**. At this stage, the mechanism of the formal carboacylation may diverge depending on the order of the subsequent transmetalation and migratory insertion events. If transmetalation of ArBpin with **169** occurs first, acyl–Ni(II)–aryl complex **170** would be generated. Subsequent migratory insertion of the tethered alkene into the Ni–C(acyl) bond would form alkyl–Ni(II)–aryl complex **171**. Reductive elimination of **171** provides indanone **172** and NHC–Ni(0). The latter can coordinate with another molecule of **167**, which would close the catalytic cycle. Alternatively, if migratory insertion precedes transmetalation, alkyl–Ni(II)–amido complex **173** would be formed by insertion of the tethered alkene into the Ni–C(acyl) bond of **169**. Subsequent transmetalation of **173** with ArBpin would form **171**, which upon reductive elimination forms indanone **172**.

The authors proposed that transmetalation of **169** with arylboronic acid pinacol ester precedes migratory insertion of the tethered alkene on the basis of two

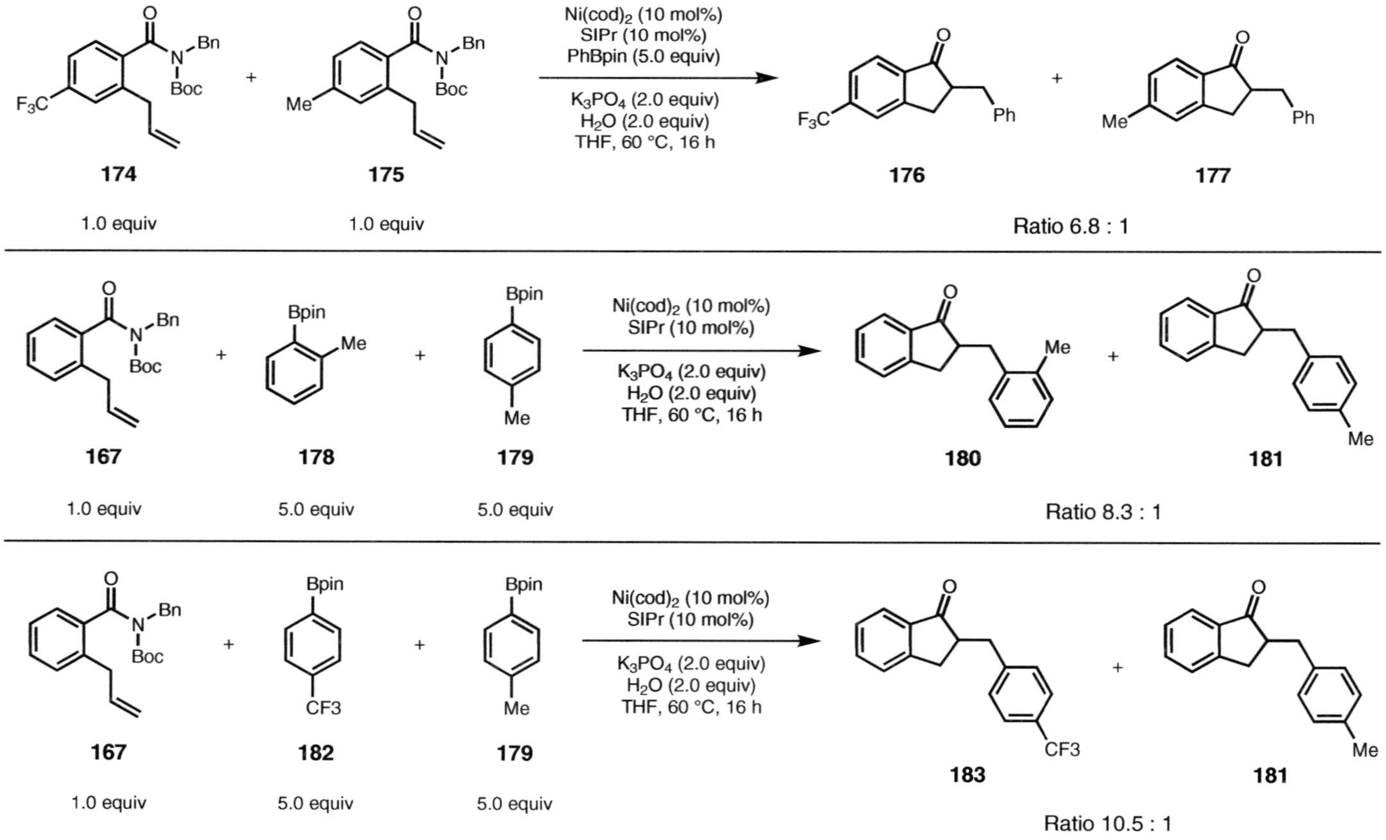

Figure 8.29 Competition experiments.

observations. First, the identity of the arylboronic acid pinacol ester significantly impacts the amount of alkene isomerization observed under the reaction conditions. Second, alkene isomerization is not observed in the absence of arylboronic acid pinacol ester. Taken together, these results are consistent with transmetalation of **169** with arylboronic acid pinacol ester occurring first to form **170** followed by migratory insertion to generate **171**.

The competition experiment between 4-(trifluoromethyl)benzamide **174** and 4-methylbenzamide **175** formed ketones **176** and **177** in a 6.8 : 1 ratio, respectively, which is consistent with the relative reactivity of electron-deficient and electron-rich benzamides in the context of Suzuki–Miyaura and Negishi coupling (Figure 8.29). However, the authors also noted that the finding contrasted with the more facile nature of oxidative addition into electron-rich benzamides **175** versus electron-deficient benzamide **174** because of the increased amidic resonance that would be expected for **174** vs. **175** [56]. In addition, this result suggests that the ratio of the products observed is not determined by the relative rates of oxidative addition. Competition experiments between the 2-tolylboronic acid pinacol ester (**178**) and 4-tolylboronic acid pinacol ester (**179**) formed ketones **180** and **181** in a 8.3 : 1 ratio while competition between 4-(trifluoromethyl)phenylboronic acid pinacol ester (**182**) and 4-tolylboronic acid pinacol ester (**179**) formed ketones **183** and **181** in a 10.5 : 1 ratio. The observation that ketones derived from reactions with sterically hindered and electron-deficient arylboron nucleophiles are favored suggests that transmetalation is fast relative to reductive elimination and that the ratio of products is determined by the relative rates of either reductive elimination or migratory insertion into the Ni–C(acyl) bond. Given that a nearly equimolar ratio of **180** and **181** would be expected from the competition between **178** and **179** if reductive elimination were turnover-limiting, migratory insertion of the alkene into the Ni–C(acyl) bond is the elementary step critical to determining the product ratio.

8.8 Reduction and Reductive Coupling of Amides

Amide reduction is a powerful synthetic tool with notable contributions by Beller (Fe or Ru, 2° or 3° amides; Zn, 3° amides; and Cu, 2° amides) [57], Nagashima (Pt or Fe, 3° amides) [58], and Adolfsson (Mo, 3° amides) [59]. Beller's Fe-catalyzed amide reduction [57a] is the most general of the methodologies that utilize non-precious metal catalysis to reduce tertiary amides. However, there are only scattered reports of nickel-catalyzed amide reduction with no general methodologies having been disclosed. The most known examples of nickel-catalyzed amide reduction require the use of Raney nickel, high temperature, and high pressure of H_2 gas [60, 61] or are selective for double reduction of α-ketoanilides to the corresponding α-hydroxylamines [62].

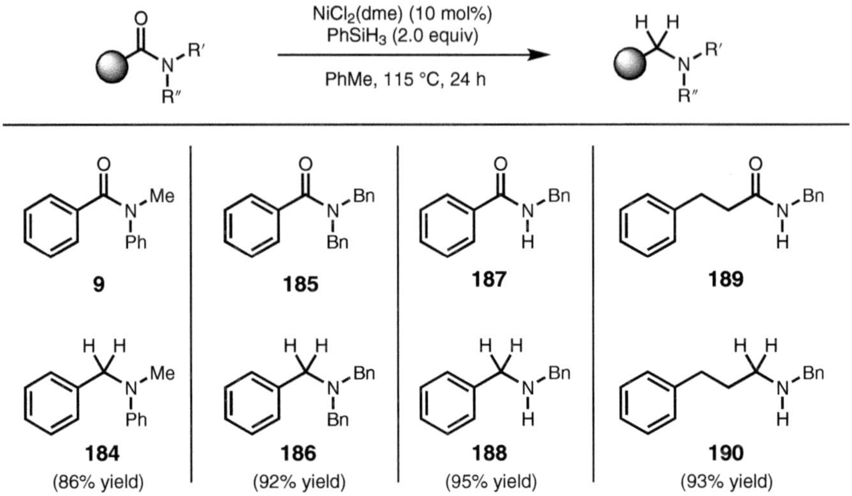

Figure 8.30 Reduction of secondary and tertiary amides.

Garg and coworkers found that tertiary amide **9** underwent efficient reduction to amine **184** in the presence of $PhSiH_3$ and a catalytic amount of the air-stable Ni(II) precatalyst $NiCl_2$(dme) (Figure 8.30) [63]. The authors noted that the use of 2 equiv of $PhSiH_3$ was required for the reduction of most substrates. Altering the silane (e.g. Ph_2SiH_2, Ph_3SiH, and Et_3SiH) proved detrimental to the yield of amine **184**. Numerous nickel(II) precatalysts (e.g. $NiBr_2$, $NiCl_2$, $Ni(acac)_2$, $NiCl_2{\cdot}H_2O$, and $NiCl_2(PPh_3)_2$) were screened in the reduction of tertiary amide **9** with no appreciable decrease in yield observed. A variety of *N*-substitutions were surveyed to understand scope. *N*,*N*-Bn amide **185**, a substrate bearing two aliphatic nitrogen substituents, was reduced to give amine **186** in 92% yield. Secondary aryl amide **187** and aliphatic amide **189** furnished amines **188** and **190**, respectively, in excellent yield.

Using the nickel-catalyzed reduction methodology, a variety of optically enriched amino acid derivatives were reduced using $PhSiD_3$ to give the corresponding medicinally relevant α-deuteroamines (Figure 8.31) [64]. The authors noted that reduction of amino acid derivatives has previously been achieved using lithium aluminum deuteride, which led to global reduction and the formation of α-deutero amino alcohols [65]. Garg and coworkers found that a valine-derived amide bearing a methyl ester could be reduced to give deuteroamine **191**. Similarly, reduction of a tyrosine derivative proceeded in the presence of the free alcohol to furnish **192**. In both cases, the ester remained intact and no epimerization was observed. The authors also examined the reduction of prolinol derivatives, bearing a free alcohol or an acetoxy group. In both cases, the reduction proceeded smoothly to give the corresponding amines **193** and **194**, respectively.

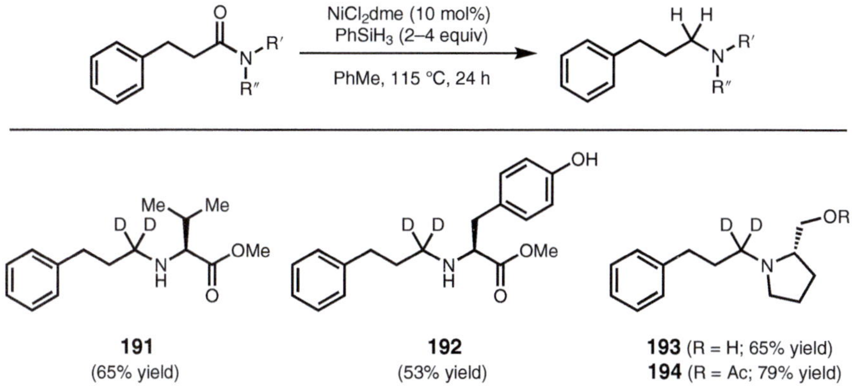

Figure 8.31 Reduction of amino acid derivatives.

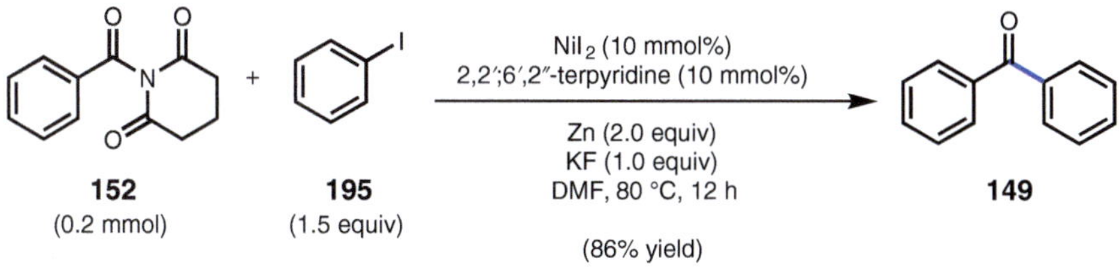

Figure 8.32 Ni-catalyzed reductive cross-coupling of amides with aryl iodide. DMF = dimethylformamide.

Han and coworkers reported a Nickel-catalyzed reductive cross-electrophile coupling between amides and aryl iodides (Figure 8.32) [66]. This work is the first example using an amide as an electrophile to couple with another electrophile.

Han and coworkers proposed a mechanism, which contains C–N bond cleavage, radical oxidative addition, and reductive coupling key steps (Figure 8.33). Starting with the formation of Ni(0) catalyst from reduction of Ni(II), the Ni(0) catalyst then inserts to the C–N bond of amide **152**, resulting in the Ni(II) complex **196**. Iodobenzene **195** can be reduced by the reactive Ni(I) species **198** to give the phenyl radical **200**. Subsequently, phenyl radical **200** can undergo radical oxidative addition to Ni(II) complex **196** to afford the phenyl nickel(III) intermediate **197**. Reductive elimination of Ni(III) intermediate **197** can form the desired cross-coupling ketone product **149** as well as the reactive nickel(I) species **198**. The Ni(I) complex **198** can be oxidized by iodobenzene to give amido Ni(II) iodide intermediate **199**, which could be reduced by Zn power to regenerate the active Ni(0) catalyst for the next catalytic cycle. The generation of aryl radical in the first catalytic cycle could occur by the reaction between Ni(II) intermediate **196** and iodobenzene **195** according to the general mechanism found in atom-transfer radical addition reactions. Alternatively, the catalytic cycle may begin with Ni(0) species and could proceed with the initial oxidative addition of aryl iodide followed by nucleophilic acyl substitution.

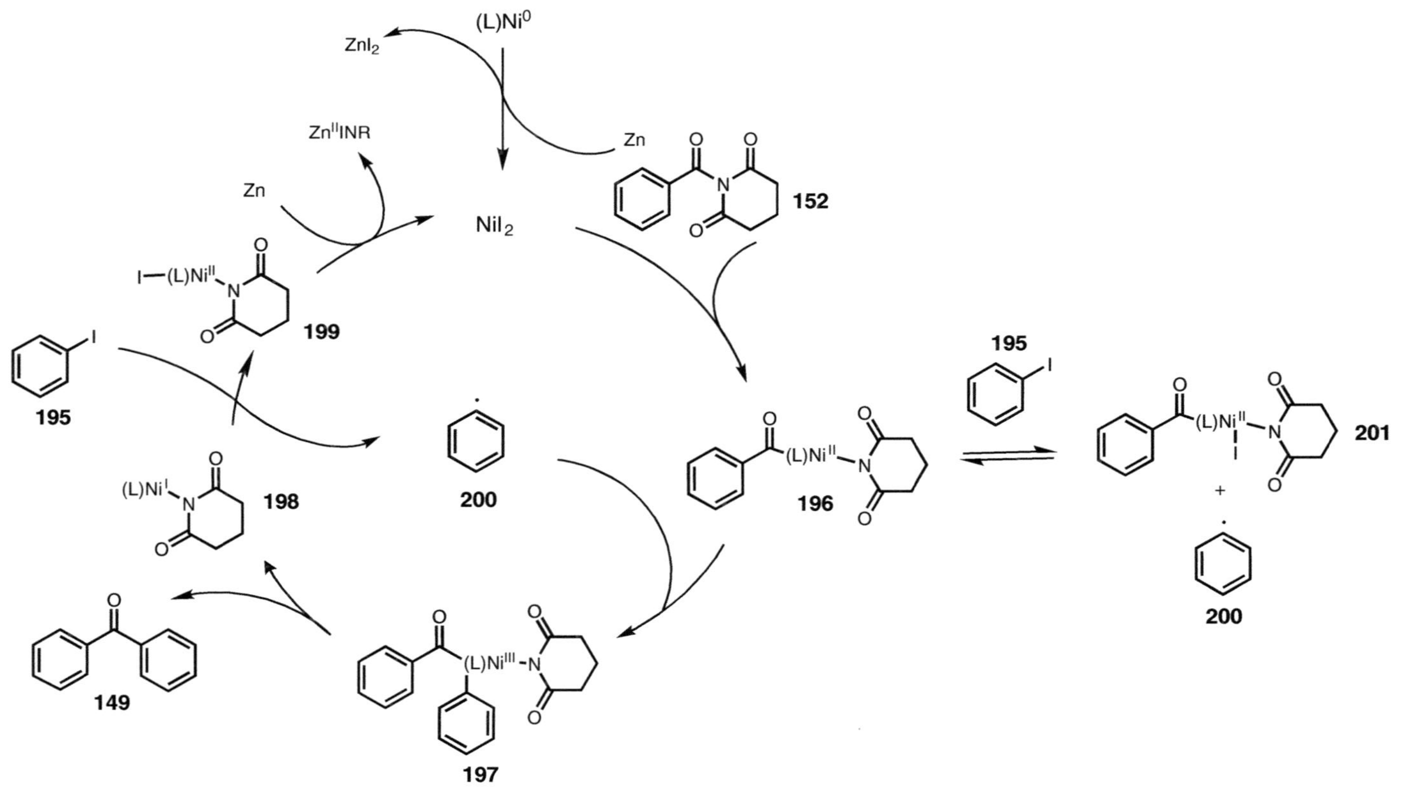

Figure 8.33 Proposed mechanism for nickel-catalyzed reductive cross-coupling of amides with aryl iodide.

References

1 Greenberg, A., Breneman, C.M., and Liebman, J.F. (ed.) (2003). *The Amide Linkage: Structural Significance: Chemistry, Biochemistry and Materials Science*. New York: Wiley-Interscience.

2 Pauling, L., Corey, R.B., and Branson, H.R. (1951). *Proc. Natl. Acad. Sci. U.S.A.* 37: 205–211.

3 Hie, L., Fine Nathel, N.F., Shah, T.K. et al. (2015). *Nature* 524: 79–83.

4 Meng, G., Shi, S., and Szostak, M. (2016). *Synlett* 27: 2530–2540.

5 Liu, C. and Szostak, M. (2017). *Chem. Eur. J.* 23: 7157–7173.

6 Hie, L., Baker, E.L., Anthony, S.M. et al. (2016). *Angew. Chem. Int. Ed.* 55: 15129–15132.

7 Weires, N.A., Caspi, D.D., and Garg, N.K. (2017). *ACS Catal.* 7: 4381–4385.

8 Deguchi, T., Xin, H.-L., Morimoto, H., and Ohshima, T. (2017). *ACS Catal.* 7: 3157–3161.

9 For selected reviews on directing group amides, see:(a)Rouquet, G. and Chatani, N. (2013). *Angew. Chem. Int. Ed.* 52: 11726–11743. (b) Daugulis, O., Roane, J., and Tran, L.D. (2015). *Acc. Chem. Res.* 48: 1053–1064. (c) Liu, J., Chen, G., and Tan, Z. (2016). *Adv. Synth. Catal.* 358: 1174–1194.

10 Li, F.-F., Wang, W.-F., Wu, Y.-Y. et al. (2000). *Catal. Lett.* 150: 874–880.

11 Meng, G., Shi, S., and Szostak, M. (2016). *ACS Catal.* 6: 7335–7339.

12 (a) Steiger, R.E. (1942). *Org. Synth.* 22: 13. (b) Brown, R.S., Bennet, A.J., and Slebocka-Tilk, H. (1992). *Acc. Chem. Res.* 25: 481–488. (c) Smith, M.B. and March, J. (2001). Aliphatic substitution: nucleophilic and organometallic. In: *March's Advanced Organic Chemistry* (ed. M.B. Smith and J. March), 474–477. Wiley. (d) Slebocka-Tilk, H., Neverov, A., and Brown, R.S. (2003). *J. Am. Chem. Soc.* 125: 1851–1858.

13 Knapp, R.R., Ana, S., Bulger, A.S., and Garg, N.K. (2020). *Org. Lett.* 22 (7): 2833–2837.

14 Sergeeva, M.V., Mozhaev, V.V., Rich, J.O., and Khmelnitsky, Y.L. (2000). *Biotechnol. Lett* 22: 1419–1422.

15 Fleischer, A. (1876). *Phys. Chem.* 9: 992–995.

16 (a) Galat, A. and Elion, G. (1943). *J. Am. Chem. Soc.* 65: 1566–1567. (b) Pettit, G., Kalnins, M., Liu, T. et al. (1961). *J. Org. Chem.* 26: 2563–2566. (c) Yoshio, O., Noboru, M., and Eiji, I. (1968). *Bull. Chem. Soc. Jpn.* 41: 1485. (d) Kraus, M.A. (1973). *Synthesis* 361–362.

17 Bon, E., Bigg, D.C.H., and Bertrand, G. (1994). *J. Org. Chem.* 59: 4035–4036.

18 Eldred, S.R., Stone, D.A., Gellman, S.H., and Stahl, S.S. (2003). *J. Am. Chem. Soc.* 125: 3422–4036.

19 Baker, E.L., Yamano, M.M., Zhou, Y. et al. (2016). *Nat. Commun.* 7: 11554.

20 (a) Dineen, T.A., Zajac, M.A., and Myers, A.G. (2006). *J. Am. Chem. Soc.* 128: 16406–16409. (b) Allen, C.L., Atkinson, B.N., and Williams, J.M.J. (2012). *Angew. Chem. Int. Ed.* 51: 1383–1386.

21 (a) Davidsen, S.K., May, P.D., and Summers, J.B. (1991). *J. Org. Chem.* 56: 5482–5485. (b) Mucsi, Z., Chass, G.A., and Csizmadia, I.G. (2008). *J. Phys. Chem. B* 112: 7885–7789.

22 Dander, J.E., Baker, E.L., and Garg, N.K. (2017). *Chem. Sci.* 8: 6433–6438.

23 Buchspies, J., Rahman, M.M., and Szostak, M. (2021). *Molecules* 26: 188–195.

24 Cheung, C.W., Ploeger, M.L., and Hu, X. (2017). *ACS Catal.* 7: 7092–7096.

25 Cheung, C.W., Ploeger, M.L., and Hu, X. (2017). *Nat. Commun.* 8: 14878.

26 Blaser, H.-U. (2006). *Science* 313: 312–313.

27 (a) Diederich, F. and Meijere, A. (2004). *Metal-Catalyzed Cross-Coupling Reactions*. Weinheim: Wiley-VCH. (b) Hassan, J., Se'vignon, M., Gozzi, C. et al. (2002). *Chem. Rev.* 102: 1359–1469. (c) Miyaura, N. (ed.) (2002). *Topics in Current Chemistry*, vol. 219. New York: Springer-Verlag. (d) Corbet, J. and Mignani, G. (2006). *Chem. Rev.* 106: 2651–2710.

28 Rosen, B.M., Quasdorf, K.Q., Wilson, D.A. et al. (2011). *Chem. Rev.* 111: 1346–1416.

29 Blangetti, M., Rosso, H., Prandi, C. et al. (2013). *Molecules* 18: 1188–1213.

30 Chen, Q., Fan, X.-H., Zhang, L.-P., and Yang, L.-M. (2014). *RSC Adv.* 4: 53885–53890.

31 Weires, N.A., Baker, E.L., and Garg, N.K. (2016). *Nat. Chem.* 8: 75–79.

32 Hu, L., Jiang, J.-d., Qu, J. et al. (2007). *Bioorg. Med. Chem. Lett.* 17: 3613–3617.

33 Huang, P.-Q. and Chen, H. (2017). *Chem. Commun.* 53: 12584–12587.

34 Goldish, A.M. and McErlean, C.S.P. (2012). *Eur. J. Org. Chem.* 1877–1888.

35 Boit, T.B., Weires, N.A., Kim, J., and Garg, N.K. (2018). *ACS Catal.* 8: 1003–1008.

36 Amani, J., Alam, R., Badir, S., and Molander, G.A. (2017). *Org. Lett.* 19: 2426–2429.

37 Dander, J.E., Giroud, M., Racine, S. et al. (2019). *Commun. Chem.* 2: 82–90.

38 Mehta, M.M., Boit, T.B., Dander, J.E., and Garg, N.K. (2020). *Org. Lett.* 22: 1–5.

39 (a) Sather, A.C., Lee, H.G., Colombe, J.R. et al. (2015). *Nature* 524: 208–211. (b) Taber, D.F. and Frankowski, K.J. (2003). *J. Org. Chem.* 68: 6047–6048.

40 (a) Haas, D., Hammann, J.M., Greiner, E., and Knochel, P. (2016). *ACS Catal.* 6: 1540–1552. (b) Meijere, A., Bräse, S., and Oestreich, M. (2014). *Metal-Catalyzed Cross-Coupling Reactions and More*. Wiley. (c) Negishi, E., Hu, Q., Huang, Z. et al. (2005). *Aldrichim. Acta* 38: 71–87. (d) Negishi, E. (2002). *Handbook of Organopalladium Chemistry for Organic Synthesis*. Wiley.

41 Scheiper, B., Bonnekessel, M., Krause, H., and Fürstner, A. (2004). *J. Org. Chem.* 69: 3943–3949.

42 (a) Zhang, Y. and Rovis, T. (2004). *J. Am. Chem. Soc.* 126: 15964–15965. (b) Cherney, A.H. and Reisman, S.E. (2014). *Tetrahedron* 70: 3259–3265. (c) Harada, T., Kotani, Y., Katsuhira, T., and Oku, A. (1991). *Tetrahedron Lett.* 32: 1573–1576. (d) Negishi, E.-i., Bagheri, V., Chatterjee, S. et al. (1983). *Tetrahedron Lett.* 24: 5181–5184. (e) Iwai, T., Nakai, T., Mihara, M. et al. (2009). *Synlett* 1091–1094. (f) Grey, R.A. (1984). *J. Org. Chem.* 49: 2288–2289. (g) Sato, T., Naruse, K., Enokiya, M., and Fujisawa, T. (1981). *Chem. Lett.* 10: 1135–1138.

43 (a) Wang, D. and Zhang, Z. (2003). *Org. Lett.* 5: 4645–4648. (b) Cook, M.J. and Rovis, T. (2009). *Synthesis* 335–338. (c) Bercot, E.A. and Rovis, T. (2002). *J. Am.

Chem. Soc. 124: 174–175. (d) Bercot, E.A. and Rovis, T. (2005). *J. Am. Chem. Soc.* 127: 247–254. (e) Johnson, J.B., Cook, M.J., and Rovis, T. (2009). *Tetrahedron* 65: 3202–3210. (f) Rogers, R.L., Moore, J.L., and Rovis, T. (2007). *Angew. Chem. Int. Ed.* 46: 9301–9304. (g) Bercot, E.A. and Rovis, T. (2004). *J. Am. Chem. Soc.* 126: 10248–10249. (h) Johnson, J.B., Yu, R.T., Fink, P. et al. (2006). *Org. Lett.* 8: 4307–4310. (i) Johnson, J.B., Bercot, E.A., Rowley, J.M. et al. (2007). *J. Am. Chem. Soc.* 129: 2718–2725. (j) Cook, M.J. and Rovis, T. (2007). *J. Am. Chem. Soc.* 129: 9302–9303.

44 (a) Tokuyama, H., Yokoshima, S., Yamashita, T., and Fukuyama, T. (1998). *Tetrahedron Lett.* 39: 3189–3192. (b) Mori, Y. and Seki, M. (2004). *Tetrahedron Lett.* 45: 7343–7345. (c) Shimizu, T. and Seki, M. (2002). *Tetrahedron Lett.* 43: 1039–1042. (d) Miyazaki, T., Han-ya, Y., Tokuyama, H., and Fukuyama, T. (2004). *Synlett* 477–480. (e) Shimizu, T. and Seki, M. (2001). *Tetrahedron Lett.* 42: 429–432.

45 Simmons, B.J., Weires, N.A., Dander, J.E., and Garg, N.K. (2016). *ACS Catal.* 6: 3176–3179.

46 Apnes, G.E., Didluk, M.T., Filipski, K.J., et al. (2012). Glucagon receptor modulators. US Patent 20120202834, 09 August 2012.

47 Shi, S. and Szostak, M. (2016). *Chem. Eur. J.* 22: 10420–10424.

48 (a) Liu, Q., Lan, Y., Liu, J. et al. (2009). *J. Am. Chem. Soc.* 131: 10201–10210. (b) Jin, L., Liu, C., Liu, J. et al. (2009). *J. Am. Chem. Soc.* 131: 16656–16657.

49 Shi, S. and Szostak, M. (2017). *Synthesis* 49: 3602–3608.

50 Shi, S. and Szostak, M. (2016). *Org. Lett.* 8: 5872–5875.

51 Glover, S.A. and Rosser, A.A. (2012). *J. Org. Chem.* 77: 5492–5502.

52 Phapale, V.B. and Cardenas, D.J. (2009). *Chem. Soc. Rev.* 38: 1598–1607.

53 Meng, G. and Szostak, M. (2016). *Org. Biomol. Chem.* 14: 5690–5705.

54 Medina, J.M., Moreno, J., Racine, S. et al. (2017). *Angew. Chem. Int. Ed.* 56: 6567–6571.

55 Walker, J.A., Vickerman, K.L. Jr., Humke, J.N., and Stanley, L.M. (2017). *J. Am. Chem. Soc.* 139: 10228–10231.

56 Szostak, R., Meng, G., and Szostak, M. (2017). *J. Org. Chem.* 82: 6373–6378.

57 (a) Zhou, S., Junge, K., Addis, D. et al. (2009). *Angew. Chem. Int. Ed.* 48: 9507–9510. (b) Das, S., Addis, D., Zhou, S. et al. (2010). *J. Am. Chem. Soc.* 132: 1770–1771. (c) Das, S., Join, B., Junge, K., and Beller, M. (2012). *Chem. Commun.* 48: 2683–2685. (d) Cabrero-Antonino, J.R., Alberico, E., Junge, K. et al. (2016). *Chem. Sci.* 7: 3432–3442.

58 (a) Nakatani, N., Hasegawa, J.-Y., Sunada, Y., and Nagashima, H. (2015). *Dalton Trans.* 44: 19344–19356. (b) Sunada, Y., Kawakami, H., Imaoka, T. et al. (2009). *Angew. Chem. Int. Ed.* 48: 9511–9514.

59 Tinnis, F., Volkov, A., Slagbrand, T., and Adolfsson, H. (2016). *Angew. Chem. Int. Ed.* 55: 4562–4566.

60 (a) Smith, M.E. and Adkins, H. (1938). *J. Am. Chem. Soc.* 60: 407–409. (b) Ovakimian, G., Christman, C.C., Kuna, M., and Levene, P.A. (1940). *J. Biol. Chem.* 134: 151–161. (c) Witkop, B. (1949). *J. Am. Chem. Soc.* 71: 2559–2566.

(d) Riebsomer, J.L. (1950). *J. Org. Chem.* 15: 68–73. (e) Robba, M., Rault, S., and Sevricourt, C.D. (1979). *Heterocycles* 12: 1009–1011.

61 For the reduction of imides using Raney nickel and H_2, see:(a)Adkins, H. and Cramer, H.I. (1930). *J. Am. Chem. Soc.* 52: 4349–4358. (b) Paden, J.H. and Adkins, H. (1936). *J. Am. Chem. Soc.* 58: 2487–2499. (c) Arévalo, A., Ovando-Segovia, S., Flores-Alamo, M., and García, J.J. (2013). *Organometallics* 32: 2939–2943.

62 Mamillapalli, N.C. and Sekar, G. (2014). *Chem. Commun.* 50: 7881–7884.

63 Simmons, B.J., Hoffmann, M., Hwang, J. et al. (2017). *Org. Lett.* 19: 1910–1913.

64 For examples of deuterium in pharmaceuticals, see:(a)Foster, A.B. (1984). *Trends Pharmacol. Sci.* 5: 524–527. (b) Kushner, D.J., Baker, A., and Dunstall, T.G. (1999). *Can. J. Physiol. Pharmacol.* 77: 79–88. (c) Nelson, S.D. and Trager, W.F. (2003). *Drug Metab. Dispos.* 31: 1481–1497. (d) Sharma, R., Strelevitz, T.J., Gao, H. et al. (2012). *Drug Metab. Dispos.* 40: 625–634. and references cited therein.

65 Franchi, P., Lucarini, M., Mezzina, E., and Pedulli, G.F. (2004). *J. Am. Chem. Soc.* 126: 4343–4354.

66 Ni, S., Zhang, W., Mei, H. et al. (2017). *Org. Lett.* 19: 2536–2539.

9

Pd–NHC Catalysis in Cross-Coupling of Amides

Faez S. Alotaibi[1], *Michael R. Chhoun*[2], *and Gregory R. Cook*[1]

[1] *North Dakota State University, Department of Chemistry and Biochemistry, 1311 Albrecht Avenue, Fargo, ND 58108-6050, USA*
[2] *American Preclinical Services, 9055 Evergreen Blvd, Minneapolis, MN 55433-9055, USA*

9.1 Introduction

Transition metal-catalyzed cross-coupling reactions are a powerful tool in organic synthesis. Much progress has been made in transition metal-mediated cross-coupling reactions, and it now serves as a powerful methodology for the synthesis of natural products, pharmaceuticals, and polymers. Though many functional groups have been well established in transition metal-mediated cross-coupling reactions, only until recently have amides been identified suitable for cross-coupling reactions. The amide functional group is a privileged motif due to its abundance in nature and its high benchtop stability [1, 2]. Methodology for C–N bond cleavage of amides is scarce due to the high double bond characteristic of the amide functional group. Amide stability is dictated by its structural planarity, which limits the rotation of the N–C(O) bond. The planarity can be explained by the resonance model, which suggests that n_N to $\pi^*_{C=O}$ conjugation is responsible for the planar structure that makes oxidative addition of the C–N bond difficult [3]. A well-developed common methodology in the literature for amide C–N bond cleavage involves the disruption of planarity of the amide via steric or decreasing the resonance ability, utilizing electronics under transition metal catalysis conditions. This involves rotation of the amide N–C(O) bond, which consequently will disrupt the n_N to $\pi^*_{C=O}$ conjugation. This is commonly referred to as a twisted amide. It is important to mention that in order to achieve amide C–N oxidative addition, the amide needs to be destabilized by the exploitation of steric or electronic effects. This is normally achieved by incorporating the nitrogen atom of the amide into a strained cyclic system, leading the amide functional group to adopt a pyramidalized structure [3]. Alternatively, it has been shown that incorporation of electron-withdrawing groups for electronics destabilization under suitable reaction conditions is sufficient for amide cross-coupling. This chapter discusses the Pd–N-heterocyclic carbene (NHC)-catalyzed cross-coupling of activated amides and the advances that have been reported by the literature [4, 5].

Amide Bond Activation: Concepts and Reactions, First Edition. Edited by Michal Szostak.

9.2 Pd(II)–NHC-Catalyzed Cross-Coupling Reactions of Amides

Since their discovery in the 1960s, palladium-catalyzed cross-coupling reactions have had a large impact on organic synthesis and continue to be a key method for the synthesis of natural products, pharmaceuticals, and polymers. Highly efficient Pd catalysts typically require the use of strong donor ligands such as electron-rich phosphines. A major development in cross-coupling came when the use of electron-rich and bulky NHCs were used as ligands. NHCs are strong σ-donor ligands, which can induce facile oxidative addition and reductive elimination [6, 7]. Due to their high reactivity, Pd–NHC complexes have allowed for the cross-coupling of normally inert functional groups at relatively mild conditions. Most recently, this has been displayed in amide cross-coupling where Pd–NHC catalytic systems were shown to easily facilitate cross-coupling reactions of amides via Suzuki–Miyaura and Hartwig–Buchwald coupling with high turnover numbers (TONs) and mild conditions. In this section we discuss the methodologies reported for the coupling of amides via Pd–NHC catalytic systems [8, 9]. Though examples of phosphines have been demonstrated for amide cross-coupling by Zou and Szostak involving *N*-phenyl-*N*-tosyl amides and *N*-glutarimide (Scheme 9.1), it is clear that NHC complexes are significantly more capable of these types of cross-coupling reactions, which are displayed by their lower catalyst loadings, superior yields, broader substrate scope, and milder conditions (Table 9.1) [2, 10–12].

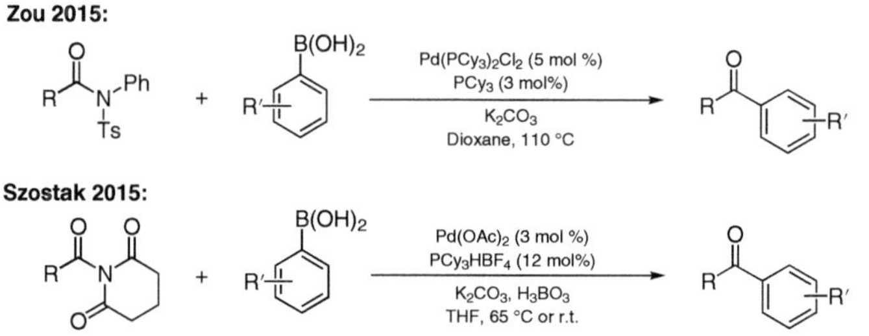

Scheme 9.1 Suzuki–Miyaura cross-coupling of activated amides via Pd/phosphines.

Table 9.1 Comparison of Pd/phosphine and Pd–NHC-catalyzed reactions when reacted with 4-Tol-$(B(OH)_2$ to form benzophenone derivatives.

	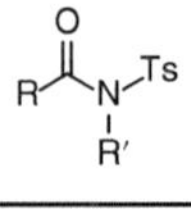	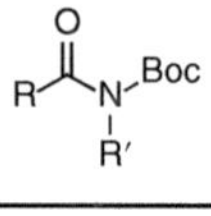	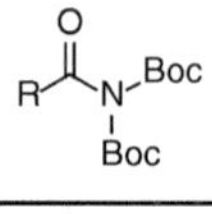	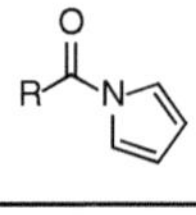	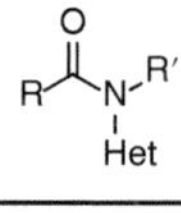
Pd/PCy_3	>95%, 110 °C	>30%, 110 °C	<5%, 110 °C	<5%, 110 °C	<5%, 110 °C
Pd(IPr)(cin)Cl	>95%, 60 °C	>95%, r.t.	>95%, r.t.	>90%, 110 °C	>90%, 110 °C

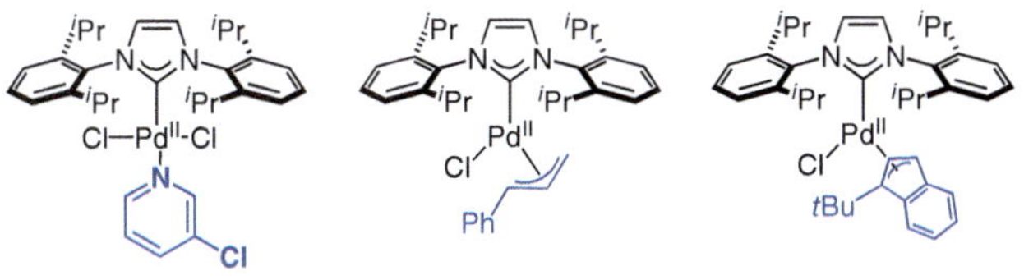

Pd-PEPPSI **Pd(IPr)(cinnamyl)Cl** **Pd(IPr)(indenyl)Cl**

Figure 9.1 Common well-defined precatalyst reported for amide cross-coupling.

Pd/phosphine systems have been shown to be limited in its substrate scope as opposed to Pd–NHC complexes. Pd–NHC complexes have been shown to undergo facile oxidative addition of various types of amides to produce a variety of aryl ketones, whereas Pd/phosphines have been shown to have a limited substrate scope and require harsher conditions (Table 9.1) [1].

Pd–NHC catalyst are often used in the form of well-defined precatalysts. These precatalysts are fully ligated structures in the form of Pd(II) that allow for the use of the metal and ligand in a 1 : 1 ratio. Well-defined precatalysts can be activated in situ to form the active catalyst Pd(0), which facilitates the cross-coupling reaction. The benefits of these types of reactions include having no need for ancillary ligands, low catalyst loadings, easy handling, easy storage, and air stability. Various types of precatalysts have been reported and will be discussed regarding their ability for amide cross-coupling (Figure 9.1) [1]. Herein, we discuss the various transformations reported in the literature that involve the cross-coupling of amides via Pd–NHC complexes.

9.3 Pd(NHC)(Allyl)Cl Precatalyst in Suzuki–Miyaura Cross-Coupling of Amides

Among the various advantages that well-defined precatalysts have (i.e. strong sigma donation, tunability, steric hindrance), Nolan and coworkers have demonstrated the advantages of η^3-allyl and η^3 cinnamyl precatalysts utilizing NHC ligands (Figure 9.2) [1, 6, 13, 14]. Szostak and coworkers have displayed the effectiveness of Pd(NHC)(allyl)Cl precatalysts on amide substrates, revealing them to be highly effective on a variety of amides. The commercial availability and the air stability of these precatalysts make the process of cross-coupling amide substrates simple and efficient. The Pd(NHC)(cinnamyl)Cl scaffold specifically has been incorporated in many cross-coupling reactions such as synthesis of tetra-ortho-substituted biaryls via Suzuki–Miyaura coupling, Buchwald–Hartwig coupling with secondary amines, and recently amide coupling for C–C and C–N bond formations [8, 9].

In 2017, Szostak and coworkers demonstrated that Pd(IPr)(cinnamyl)Cl was an effective catalyst for amide cross-coupling in relatively mild conditions. This versatility of the reaction was displayed by demonstrating its ability to cross-couple *N*-glutarimide, *N*-Ph-*N*-Ts, and *N*-Ph-*N*-Boc amides into the corresponding biaryl ketones [15]. Up to this point only reports of *N*-glutarimide and *N*-Ph-*N*-Ts amides

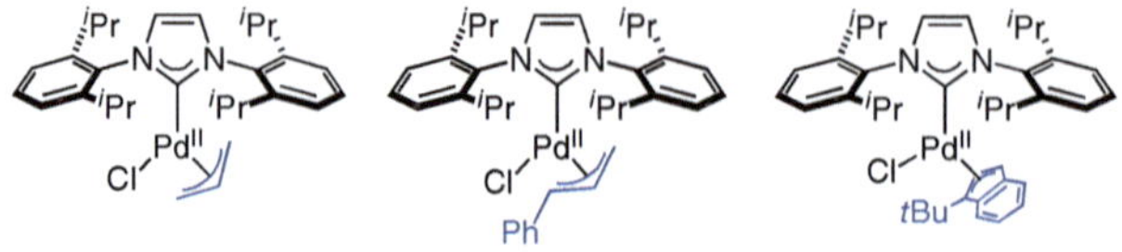

Pd(IPr)(allyl)Cl **Pd(IPr)(cinnamyl)Cl** **Pd(IPr)(indenyl)Cl**

Figure 9.2 Pd(NHC)(allyl)Cl-type precatalysts.

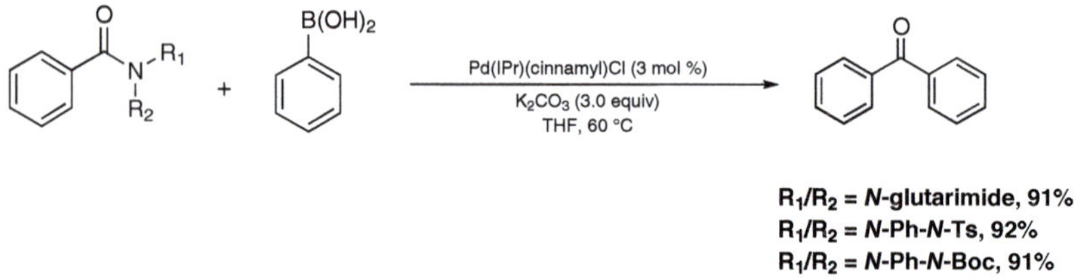

Scheme 9.2 Pd(IPr)(cinnamyl)Cl-mediated Suzuki–Miyaura cross-coupling of activated amides. Source: Adapted from Meng and Szostak [12].

have been reported by Zou and Szostak using a Pd/PCy_3 system. The use of the protecting groups such as *N*-Boc and *N*-Ts is appealing due to its easy installation. Using this methodology, it would potentially be possible to cross-couple a variety of amides. As mentioned previously, the NHC ligand acts as a strong σ-donor that initiates oxidative addition of the C(O)–N bond, which is normally not very reactive and followed up by facile reductive elimination due to the bulky NHC ligand (Scheme 9.2) [12].

The reaction was shown to be effective with not only Suzuki–Miyaura cross-coupling reactions but was also shown to be able to efficiently undergo Buchwald–Hartwig amination using aniline derivatives as the nucleophile. In 2017, Szostak and coworkers also demonstrated that C–N bond formation was also possible under standard conditions using Pd(IPr)(cinnamyl)Cl precatalysts. The methodology allowed for the use of alkyl, aryl, and sterically hindered amides to produce the corresponding amide at good to excellent yields. Later reports showed that the amination reaction worked well using other variations of Pd(NHC)(allyl)Cl precatalysts such as Pd(IPr*)(cinnamyl)Cl, Pd(SIPr)(cinnamyl)Cl, Pd(IMes)(allyl)Cl, Pd(IPr)(allyl)Cl, [Pd(IPr)Cl_2]$_2$, and [Pd(SIPr)Cl_2]$_2$ (Scheme 9.3) [16, 17].

The focus of work up to this point emphasized cross-coupling of nonplanar amides, but that was changed when amides containing planarity were shown to be cross-coupled via Pd(IPr)(cinnamyl)Cl. Twisted amides such as *N*-Ph-*N*-Boc, *N*-Ph-*N*-Ts, and *N*-glutarimide benefit strongly from these twist angle, which decreases its resonance ability and therefore allows for facile oxidative addition. *N*-acylpyrroles and *N*-pyrazole rings do not contain that type of resonance distortion, but Szostak and coworkers were able to cross-couple the amide to synthesize the corresponding biaryl ketone. Utilizing Pd(IPr)(cinnamyl)Cl, the amide substrate was allowed to selectively undergo oxidative addition, which was presumed to

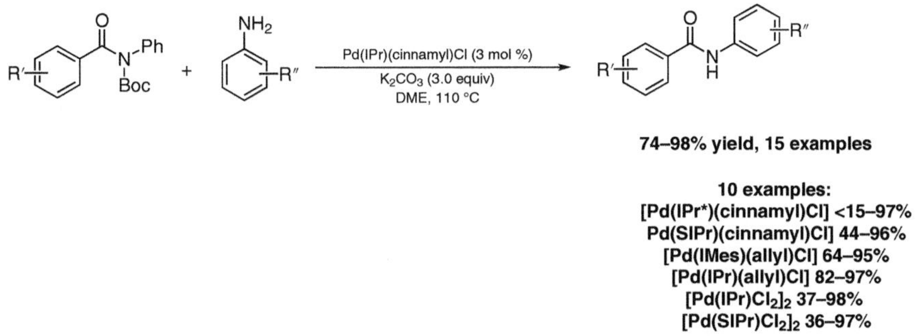

Scheme 9.3 Transamidation of *N*-Ph-*N*-Boc amides catalyzed via Pd–NHC complexes. Source: Adapted from Meng et al. [16] and Zhou et al. [17].

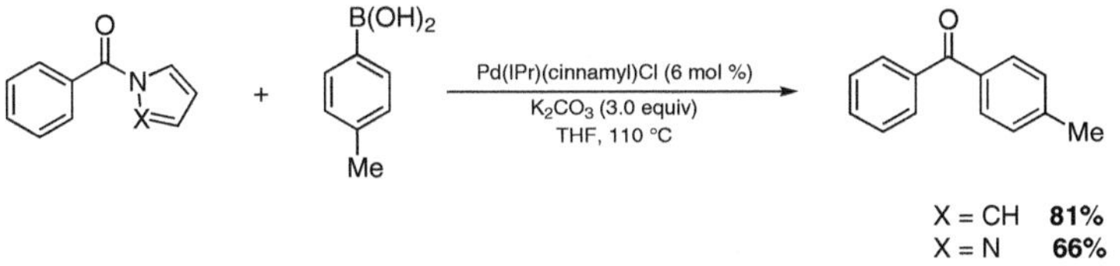

Scheme 9.4 Pd(IPr)(cinnamyl)Cl-mediated Suzuki–Miyaura cross-coupling of pyrrole/pyrazoles. Source: Adapted from Meng et al. [18].

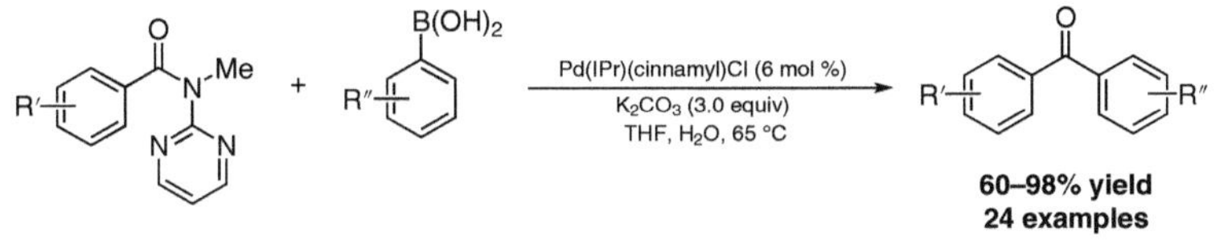

Scheme 9.5 Pd(IPr)(cinnamyl)Cl-mediated Suzuki–Miyaura cross-coupling of *N*-methylaminopyrimidyl amides (MAPA). Source: Adapted from Meng et al. [19] and Li et al. [20].

be allowed by the N_{lp} into the π-electron system of the pyrrole/pyrazole ring that sufficiently destabilized the resonance structure. Optimization of the reaction conditions revealed that the reaction required higher loadings of the precatalysts, which shows that cleavage of non-twisted amides to be difficult, yet possible utilizing Pd(II)–NHC complexes (Scheme 9.4) [18].

Compared with Pd/phosphine systems, Pd(II)–NHC has shown to be the superior catalyst system for the coupling of many different types of twisted or electronically distorted amides. Reports of Suzuki–Miyaura cross-coupling continued to be published with Szostak and coworkers reporting the cross-coupling *N*-methylaminopyrimidyl amides (MAPA) as they showed to be very active for cross-coupling, giving good to excellent yields. Interestingly, it was shown that addition of water was able to assist in increasing reaction yield (Scheme 9.5) [19, 20].

Zeng and coworkers then reported their own reaction conditions using Pd(IPr)(allyl)Cl for the Suzuki–Miyaura cross-coupling of yet another amide derivate. Zeng was able to display that the incorporation of commercially available and inexpensive *N*-acylhydantoins amides is suitable for Suzuki–Miyaura cross-coupling under the conditions outlined in Scheme 9.5. Luo and coworkers carried out the coupling of acylhydantoins using the Pd(IPr)(allyl)cl precatalyst rather than the cinnamyl-derived precatalyst used by others, which demonstrated the tune Pd-NHC precatalysts for individual reactions (Scheme 9.6) [21].

Methods reported so far have led to promising reactions applicable for organic synthesis. However, synthetic modifications to form specific amide substrates can be tedious and limit utility. The use of protecting groups are a good methodology for activating amides easily, but so far only secondary and tertiary amides have been utilized such as *N*-Ph-*N*-Boc, *N*-Ph-*N*-Ts, and *N*-glutarimide amides. Szostak and coworkers reported the use of *N*,*N*-di-Boc-activated amides that can be formed from primary amides. This methodology increases the utility of amide cross-coupling reactions by increasing the substrate scope of amides that can be used in organic synthesis. The methodology utilizes [Pd(IPr)(cinnamyl)Cl]/KF catalyst system for selective metal insertion into the C–N acyl bond. The addition of water (5.0 equiv) was found to be crucial for the reaction compared with low yields or no product observed in the absence of water (Scheme 9.7) [22].

The high catalytic efficiency Pd(IPr)(cinnamyl)Cl was once again displayed by Szostak when he reported the cross-coupling of *N*,*N*-di-Boc-activated amides at room temperature at gram scale (Scheme 9.7). The reaction was able to be scaled up, and the catalyst loading was reduced to 0.5 mol% while still giving excellent yields (Scheme 9.8) [23].

Employing standard conditions utilizing Pd(IPr)(cinnamyl)Cl, Szostak and coworkers were able to cross-couple amides to display C(sp^2)–C(sp^3) bond formation, which has been a challenge in the cross-coupling field. The transformation was achieved utilizing *N*,*N*-di-Boc-activated amides and B-sp^3 alkyl reagents. The reaction was also expanded to various amides such as *N*-Ph-*N*-Boc, *N*-glutarimide, *N*-Ph-*N*-Boc, and *N*-Ph-*N*-Ts resulting in good to excellent yields (Scheme 9.9) [15].

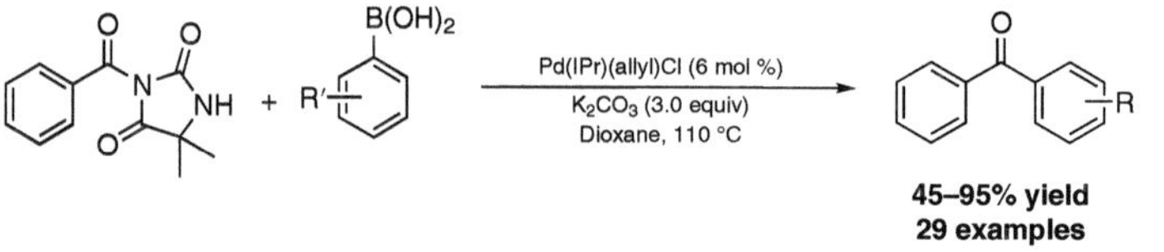

Scheme 9.6 Pd(IPr)(allyl)Cl-mediated Suzuki–Miyaura cross-coupling of *N*-acylhydantoins. Source: Adapted from Luo et al. [21].

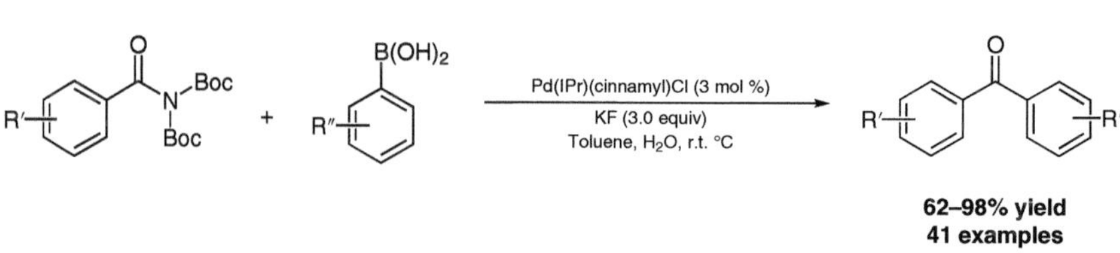

Scheme 9.7 Pd(IPr)(cinnamyl)Cl-mediated Suzuki–Miyaura cross-coupling of *N*,*N*-di-Boc-activated amides. Source: Adapted from Liu et al. [22].

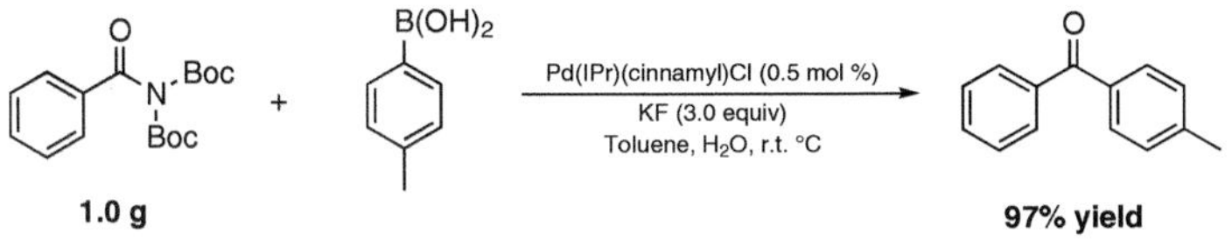

Scheme 9.8 Scale-up study in Suzuki–Miyaura cross-coupling of *N,N*-di-Boc-activated amides. Source: Adapted from Lei et al. [23].

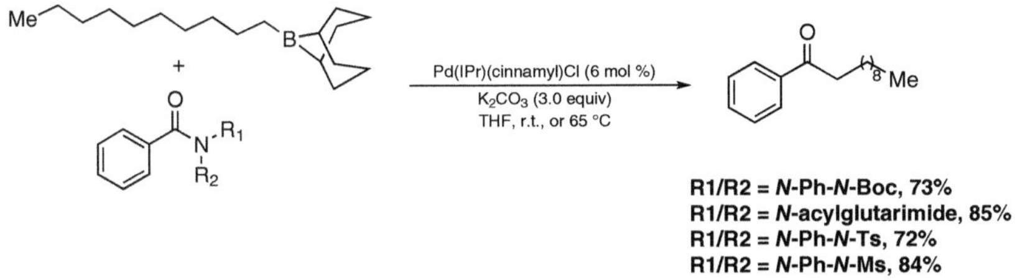

Scheme 9.9 $C(Sp^2)-C(sp^3)$ bond formation from amides via Pd(IPr)(cinnamyl)Cl. Source: Adapted from Meng and Szostak [15].

9.4 Pd(η^3-1-*t*-Bu-Indenyl)(IPr)Cl-Catalyzed Suzuki–Miyaura Cross-Coupling of Amides

The high reactivity of Pd(allyl)(NHC)Cl precatalysts can be attributed to its ability to be used at a 1 : 1 metal to ligand ratio and its ability to quickly reduce to Pd(0) [6]. The activation mechanism of Pd(allyl)(NHC)Cl precatalysts has been shown to be base mediated, which results in the formation of the active Pd(0) catalyst [20]. Hazari and coworkers reported that fast activation of the precatalyst results in higher catalytic efficiency. Comproportionating pathways present in the reaction generate Pd(I) μ-allyl dimers, which reduce the efficiency of the reaction (Scheme 9.9). The mechanistic understanding of the activation mechanism allowed for the design of highly efficient Pd–NHC precatalysts. Reducing the ability of comproportionation of Pd(II) and Pd(0) and invoking rapid activation of Pd(II) to Pd(0) before comproportionating can occur were outlined as important factors for Pd–NHC precatalyst design (Scheme 9.10) [24, 25].

Mechanistic insight on the Pd(IPr)(allyl)Cl-type precatalysts allowed for the Hazari group to effectively design the precatalyst Pd(η^3-1-*t*-Bu-indenyl)(NHC)Cl (Hazari catalyst) [26]. A major feature of the catalyst is its ability to reduce the formation of the less reactive Pd(I) dimers that increases catalytic efficiency. In 2017, Szostak and coworkers demonstrated that the Hazari catalyst was effective at Suzuki–Miyaura cross-coupling-activated *N*-Ph-*N*-Boc amides under standard conditions of amide cross-coupling (Scheme 9.11) [27]. The high reactivity of the catalyst can be observed by the fact that it was done at room temperature while utilizing no additional additives.

In 2019, Szostak and coworkers reported a sustainable methodology for Suzuki–Miyaura cross-coupling of amides and esters utilizing Pd(II)–NHC precatalysts. The incorporation of 2-methyltetrahydrofuran (2-MeTHF) contributes

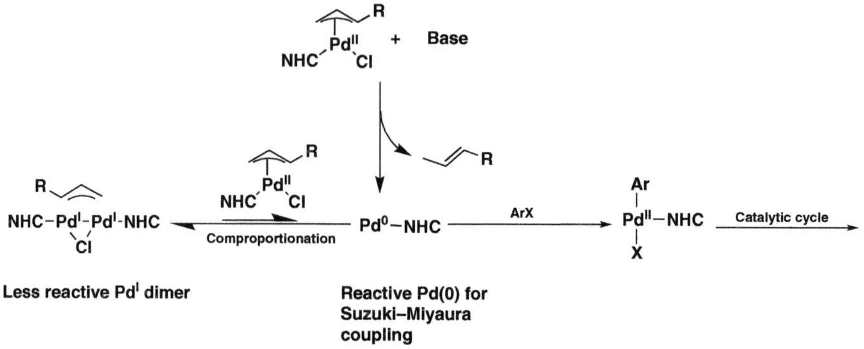

Scheme 9.10 Activation pathway of Pd(NHC)(allyl)Cl precatalysts. Source: Adapted from Hruszkewycz et al. [24, 25].

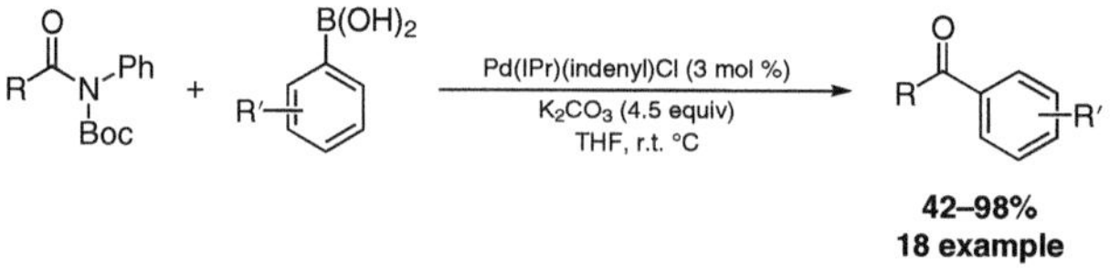

Scheme 9.11 Pd(indenyl)(IPr)Cl-catalyzed (Hazari catalyst) amide cleavage of *N*-Ph-*N*-Boc amides. Source: Adapted from Lei et al. [27].

a sustainable factor to the reaction. The advantage of 2-MeTHF is that it can be generated from renewable resources such as furfural or levilinic acid. The utility of this methodology was demonstrated when Flumorph was synthesized successfully. Flumorph is a bioactive ketone intermediate that highlights this methodology's applications in medicinal chemistry. The Suzuki–Miyaura cross-coupling of *N*-Ph-*N*-Boc secondary amides with Pd–NHC precatalysts were found optimum in conditions that employ Pd(IPr)(cinnamyl)Cl, K_2CO_3, and 2-MeTHF under ambient temperature. Furthermore, 2-MeTHF was found applicable with other commercially available Pd(II)–NHC precatalysts including Pd-PEPPSI-IPr, and Pd(IPr)(indenyl)Cl. The scope of the reaction was found compatible with many boronic acids including neutral, electron-donating, and deactivated electron-withdrawing as well as sterically hindered boronic acids. Various amides were found effective under these conditions including fluorinated amides, heterocyclic amides, electronically deactivated amides, and sterically hindered amides, which are widely employed amides in pharmaceutical entities (Scheme 9.12) [28].

In 2020, Szostak and coworkers reported air-stable NHC–Pd(II) chloro dimers (Pd(NHC)(μ-Cl)Cl]$_2$) for amide cross-coupling. The advantages of this precatalyst is the removal of the allyl "through away ligand" and its easy generation from Pd(OAc)$_2$ and NHC salts. Pd(NHC)(allyl)Cl precatalysts activation mechanism allows for the formation of Pd(I) dimers that hinder the reaction. Removal of the "throw away ligand" allowed for high catalyst reactivity with lower catalyst loadings. Compared with Pd(NHC)(allyl)Cl precatalysts, Pd(NHC)(μ-Cl)Cl]$_2$ precatalysts are

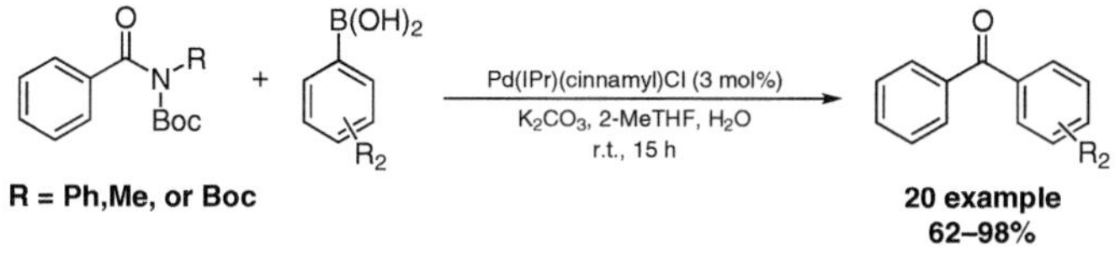

Scheme 9.12 Scope of the Suzuki–Miyaura cross-coupling of amides in sustainable 2-MeTHF. Source: Adapted from Lei et al. [28].

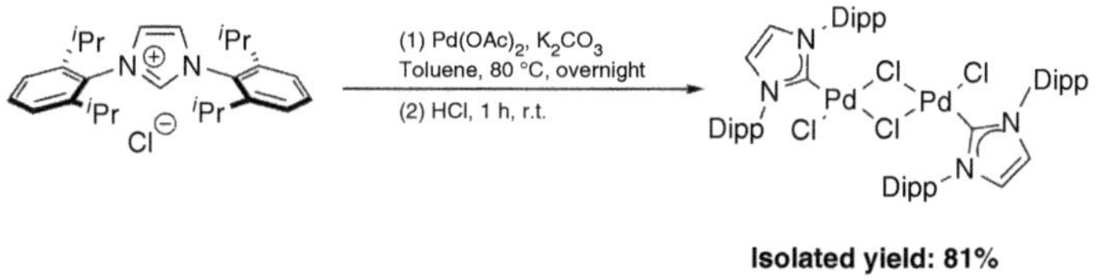

Scheme 9.13 Synthesis of Pd(NHC)(μ-Cl)Cl_2 from IPr and Pd salts.

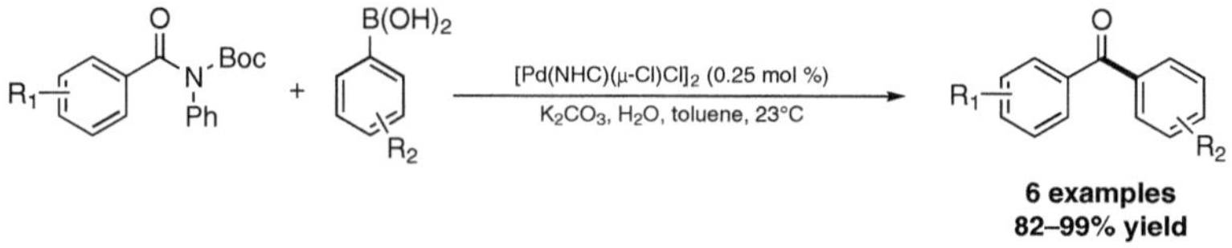

Scheme 9.14 Suzuki–Miyaura cross-coupling of *N*-Ph-*N*-Boc amide via Pd(NHC)(μ-Cl)Cl_2 precatalyst. Source: Adapted from Zhou et al. [29].

readily activated to Pd(0) by dimer dissociation. This class of precatalysts has been shown to be effective in biaryl synthesis and late-stage functionalization to synthesize medicinal compounds. Activation mechanism of this class of precatalysts was provided through computational studies, which demonstrated the less likelihood of comproportionation compared with the Pd(NHC)(allyl)Cl precatalyst. This allows Pd(NHC)(μ-Cl)$Cl]_2$ complexes to be highly efficient in cross-coupling reactions as it is demonstrated to be highly effective at catalyst loadings as low as 0.025 mol% (Scheme 9.13) [29].

Kinetic profiling revealed that Pd(NHC)(μ-Cl)Cl_2 was able to cross-couple *N*-Ph-*N*-Boc amides at a faster rate than Pd(IPr)(cinnamyl)Cl, Pd(IPr)(indenyl)Cl, and Pd-PEPPSI under standard conditions. Substrate screening revealed that the precatalyst successfully cross-coupled a range of amides giving high yields (Scheme 9.14).

9.5 Pd-PEPPSSI Precatalyst in the Suzuki–Miyaura Cross-Coupling of Amides

The use of NHC ligands has resulted in tremendous growth in the development of highly reactive and functional group tolerant catalytic systems. A major development regarding Pd–NHC amide cross-coupling was the use of Pd-PEPPSI precatalyst

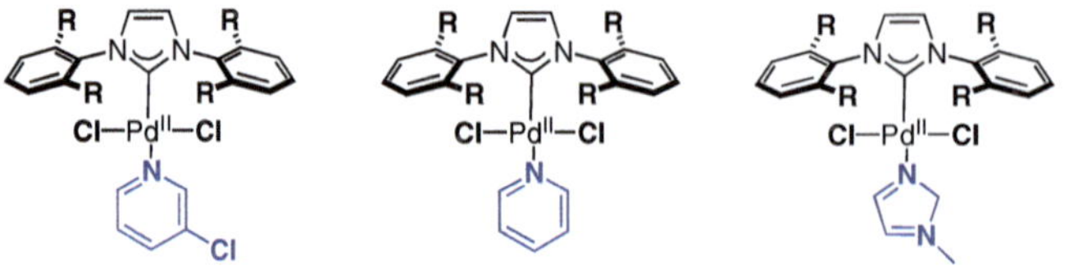

Figure 9.3 Common well-defined Pd-PEPPSI-type precatalysts employed in cross-coupling reactions.

(pyridine-enhanced precatalyst preparation, stabilization, and initiation), developed by the Organ group [30]. Pd-PEPPSI has two major components, the first being the NHC ligand and the second being the pyridine portion which is often referred to as the "throw away ligand" (Figure 9.3). NHC ligands being highly electron rich and bulky give the precatalysts high reactivity in terms of oxidative addition and reductive elimination. The "throw away ligand" allows for precatalyst stabilization, which in this case is the pyridine ring as opposed to an allyl ligand. Pd-PEPPSI has been shown to be active for many types of cross-coupling reactions such as Suzuki–Miyaura, Negishi, Kumada–Tamao–Corriu, and Buchwald–Hartwig amination reactions [31]. The facile synthesis, stability, and efficiency of this precatalyst make it an appealing catalyst to use in a variety of reactions especially when it comes to amide cross-coupling.

Szostak and coworkers reported the use of Pd-PEPPSI precatalysts in the Suzuki–Miyaura cross-coupling of various amides [32]. The precatalyst shows high catalytic activity for the conversion of activated amides to form ketones (Scheme 9.15). The reaction utilized conditions similar to Pd(IPr)(allyl)Cl precatalysts system and resulted in comparable yields.

Szostak and Shi also reported Pd-PEPPSI as a suitable precatalyst for the transamidation of *N*-Ph-*N*-Boc amides. This work was reported alongside Pd-PEPPSI-catalyzed phenyl ester transamidation, which is not discussed in this chapter. However, the work reported transamidation of the activated amide under harsher conditions than what was reported for ketone synthesis (Scheme 9.16) [33].

As a follow-up to Zeng and coworkers report of Suzuki–Miyaura cross-coupling of *N*-acylhydantoins, Szostak and coworkers reported their own variation of *N*-acylhydantoin cross-coupling using Pd-PEPPSI instead of Pd(IPr)(allyl)Cl (Scheme 9.17). They reported that the use of a Pd/phosphine catalyst system was

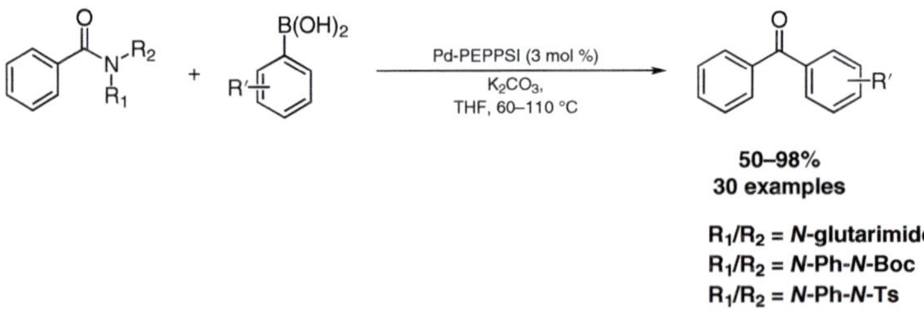

Scheme 9.15 First Pd-PEPPSI-catalyzed amide Suzuki–Miyaura cross-coupling reaction reported.

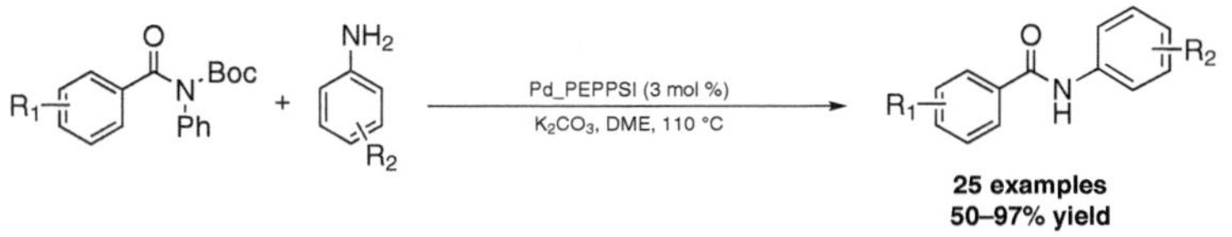

Scheme 9.16 Pd-PEPPSI-mediated transamidation of *N*-Ph-*N*-Boc amides. Source: Adapted from Shi and Szostak [33].

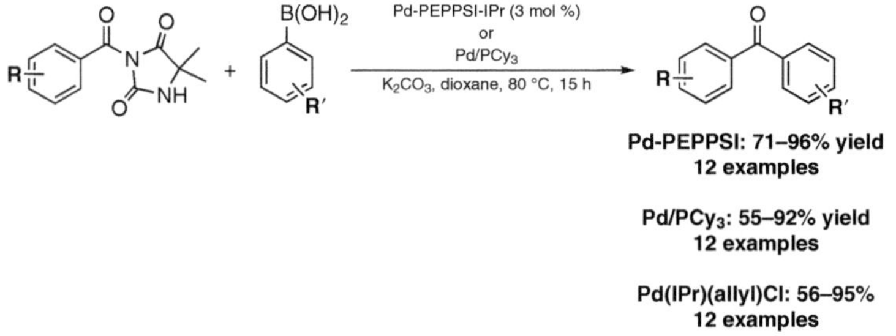

Scheme 9.17 Pd-PEPPSI, Pd/PCy_3, and Pd(IPr)(allyl)Cl-catalyzed Suzuki–Miyaura cross-coupling of *N*-acylhydantoins.

also suitable for successful cross-coupling. They were able to demonstrate that the use of Pd-PEPPSI allowed for milder conditions that included a lower reaction temperature and catalyst loading [34].

Szostak and coworkers also reported the successful Suzuki–Miyaura cross-coupling of *N*-acylcarbazoles and *N*-acylindoles. Similar to his report regarding *N*-acylpyrroles, Szostak was able to exploit the electron-withdrawing effects of the carbazole and indole ring, which have planar geometry to induce selective metal insertion to form the corresponding ketone (Scheme 9.18) [35]. This highlighted Pd-PEPPSI's ability to successfully cross-couple planar amides successfully with good to excellent yields.

Typically, *N*-glutarimide has been shown in the literature to be the more reactive of the twisted amides as it has often been shown to be facilitated using milder conditions with low catalyst loadings. This is most likely due to the fact that it is a very bulky amide, which destabilizes the resonance ability of the amide. This was also demonstrated in Figure 9.4, which shows that the *N*-glutarimide was the most efficiently coupled using Pd(IPr)(cinnamyl)Cl and Pd-PEPPSI. *N*-Ph-*N*-Boc was observed to be comparable, but Pd(IPr)(cinnamyl)Cl was shown to facilitate

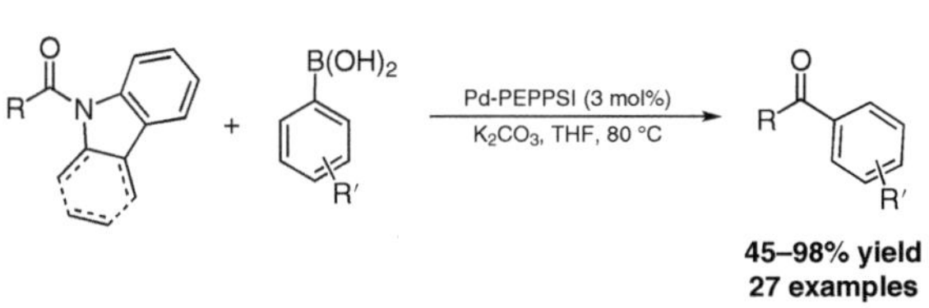

Scheme 9.18 Suzuki–Miyaura cross-coupling of *N*-acylcarbazoles and *N*-acylindoles via Pd-PEPPSI. Source: Adapted from Buchspies et al. [35].

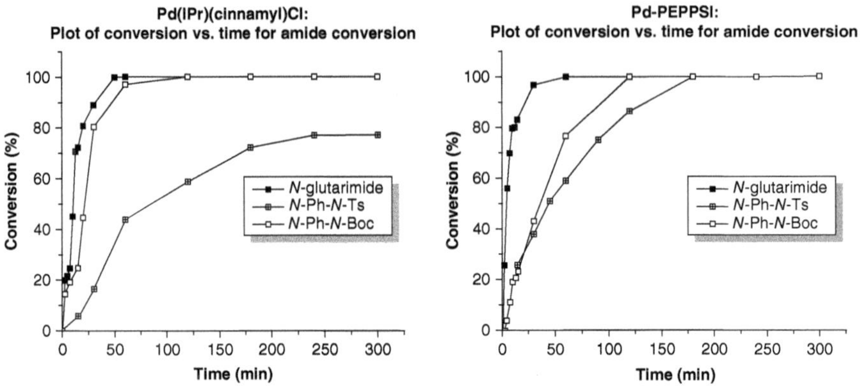

Figure 9.4 Kinetic study displaying the conversion percent of *N*-glutarimide, *N*-Ph-*N*-Ts, and *N*-Ph-*N*-Boc using Pd–NHC precatalyst Pd(IPr)(cinnamyl)Cl (left) and Pd-PEPPSI (right). Source: Reprinted with permission from Lei et al. [32]. Copyright 2017 American Chemical Society.

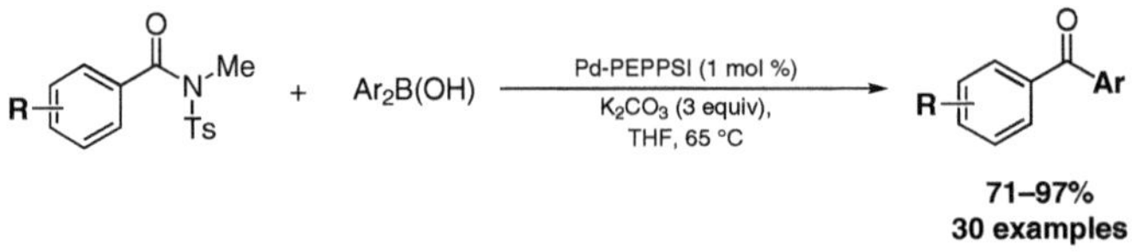

Scheme 9.19 Pd-PEPPSI-catalyzed Suzuki–Miyaura cross-coupling of *N*-alkyl-*N*-Ts amides using diaryl boronic acids. Source: Adapted from Wang et al. [37].

the reaction at a faster rate than Pd-PEPPSI. However, when *N*-Ph-*N*-Ts-type amide were coupled, it was observed that Pd-PEPPSI-type catalyst was superior in the conversion to the corresponding ketone in terms of rate and yield. This data displays the significance of choosing the suitable catalyst for different types of amide substrates [1, 36].

Zou and coworkers reported the Suzuki–Miyaura cross-coupling of *N*-alkyl-*N*-Ts amides while utilizing diarylboronic acids under Pd-PEPPSI conditions (Scheme 9.19) [37]. The work was able to extend the nucleophile substrates to stable diarylboronic acids and uniquely cross-couple a less reactive substrate in *N*-alkyl-*N*-Ts amides. Notably, no product was observed under Pd/phosphine conditions, but the reaction yielded good to excellent yields with a low catalyst loading of 1 mol% of Pd-PEPPSI.

Zou and coworker then expanded on the work by using trialkylboranes as a coupling partner under Pd-PEPPSI conditions to produce alkyl ketones (Scheme 9.20) [38]. The use of trialkyl boranes as the alkyl source allowed for the alkylation of *N*-Me-*N*-Ts amides in good to excellent yields using catalyst loading of 5 mol% Pd-PEPPSI.

Wang, Liu, and coworkers reported a series of Pd–NHC precatalyst using various N-heterocycles as stabilizers to act as the "throw away ligand." The various catalyst was used in a Suzuki–Miyaura cross-coupling reaction for the coupling of

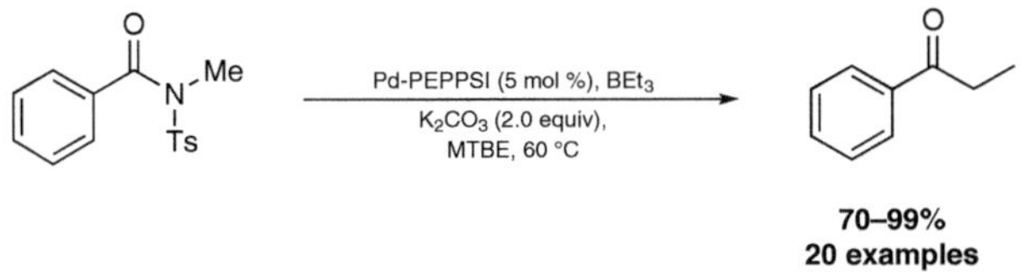

Scheme 9.20 Pd-PEPPSI-mediated alkylation of *N*-Me-*N*-Ts amides via alkyl boranes. Source: Adapted from Shi and Zou [38].

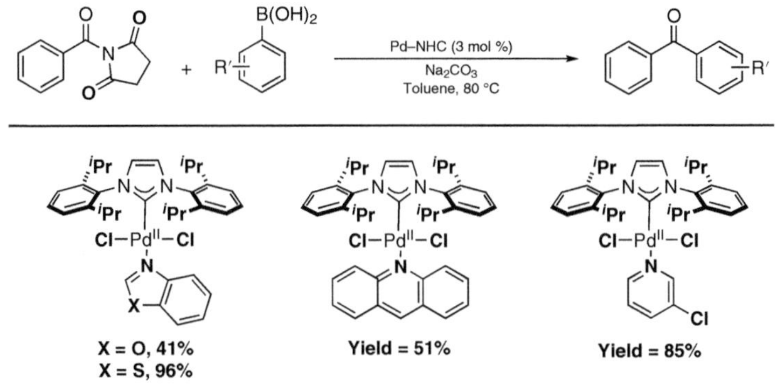

Scheme 9.21 Suzuki–Miyaura cross-coupling reactions of amides meditated by Pd–NHC precatalysts stabilized by N-heterocycles (Pd-PEPPSI derivatives). Source: Adapted from Wang et al. [39].

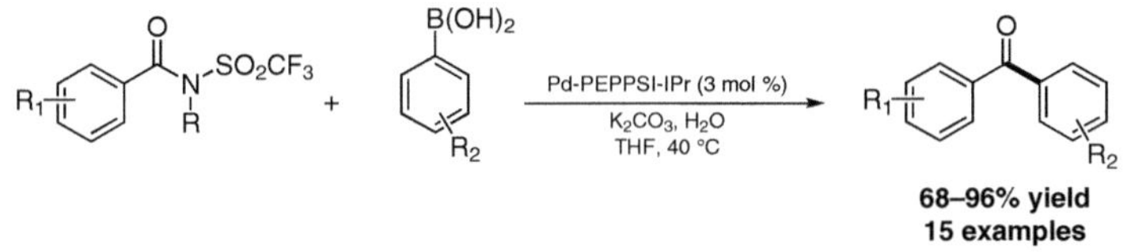

Scheme 9.22 Pd-PEPPSI-mediated Suzuki–Miyaura cross-coupling of triflamides (N-Tf). Source: Adapted from Shi et al. [40].

N-succinimide reactions (Scheme 9.21) [39]. When benzothiazole was used, it was shown to be the most effective out of all the precatalyst including Pd-PEPPSI stabilized via 3-chloro pyridine ligands under the conditions outlined by Zeng.

In 2019, Szostak and coworkers reported Suzuki–Miyaura cross-coupling of inexpensive triflamides (N-Tf) (Scheme 9.22) [40]. They envisioned that the trifluoromethansulfonyl group as an electron-withdrawing group could activate the C–N amide bond for metal insertion. Initial investigation of the amide substrate revealed it was highly reactive for Suzuki–Miyaura cross-coupling reactions via Pd(II)–NHC precatalysts such as Pd-PEPPSI, Pd(IPr)(cinnamyl)Cl, and Pd(IPr)(indenyl)Cl. Though all precatalysts performed well, the reaction was optimized via Pd-PEPPSI for convenience. The reaction was optimized to be done at fairly mild condition. It is interesting to note that Pd/phosphines were able to facilitate the reaction at moderate yields, which reveals the high reactivity of these substrates.

9.6 Various Pd–NHC Precatalysts Suitable for Cross-Coupling of Amides

Pd(IPr)(allyl)Cl and Pd-PEPPSI-type precatalysts have by far been the most reported when it comes to Suzuki–Miyaura cross-coupling and Buchwald–Hartwig amination of activated amides. However, there have been a few other examples that have not been as heavily reported on but have shown promise in the amide cross-coupling field. These precatalysts are also of the Pd–NHC category and show high reactivity toward amide C–N cleavage.

Ghandhi and coworker reported the use of unsymmetrical pyrene-based carbene in the use of their variation of the Pd–NHC precatalysts (Figure 9.5) [41]. Like Pd-PEPPSI, it is stabilized by a pyridine ligand. Using this new type of precatalyst, they were able to cross-couple *N*-acylsaccharin amides with azoles and azole derivatives.

N-acylsaccharin, being a twisted amide, was able to undergo Suzuki–Miyaura cross-coupling utilizing the newly designed precatalyst (Figure 9.5). The work focused on cross-coupling *N*-acylsaccharin and benzoxazole and its derivates via a C–H activation process (Scheme 9.23). *N*-aryl-*N*-Ts were found ineffective and resulted in no product being detected.

Szostak and coworkers recently reported the use of the Pd–NHC precatalysts Pd(NHC)(acac)Cl for amide Suzuki–Miyaura cross-coupling (Scheme 9.24) [42]. The Pd(NHC)(acac)Cl-type precatalyst is a Pd–NHC complex that uses acetoacetate (acac) as the "throw away ligand." The precatalyst could easily be synthesized from Pd(acac)$_2$ and NHC salts. However, Szostak and coworkers developed a protocol that allows for *in situ* formation of the precatalyst, allowing for the reaction to be done in one-pot fashion.

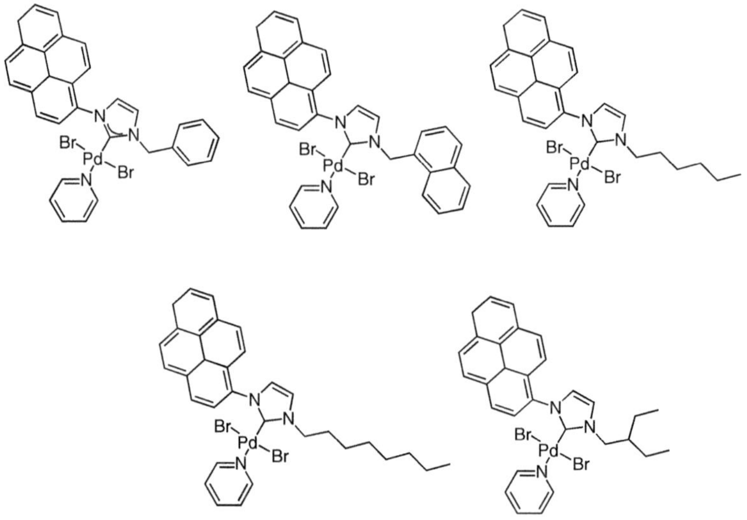

Figure 9.5 Pyrene-based Pd-PEPPSI catalysts employed in the cross-coupling reactions.

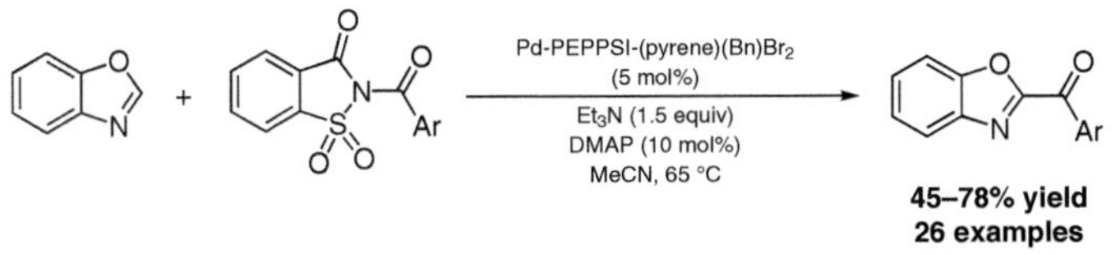

Scheme 9.23 Pd-PEPPSI-pyrene precatalysts employed in the Suzuki–Miyaura cross-coupling of *N*-acylsaccharin with benzoxazole.

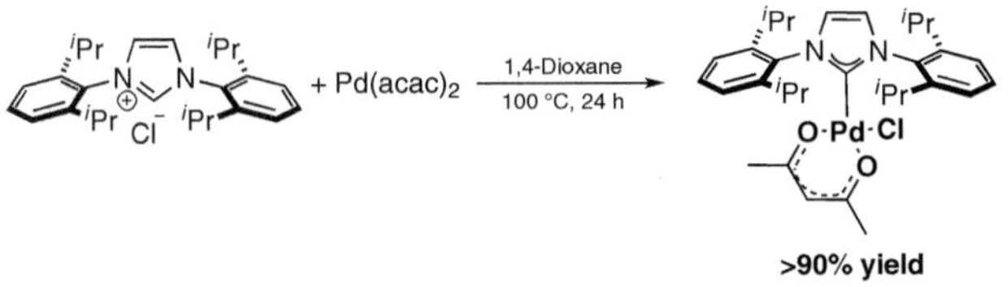

Scheme 9.24 Synthesis of Pd(IPr)(acac)Cl from Pd(acac)$_2$ and IPr salts. Source: Adapted from Zhou et al. [42].

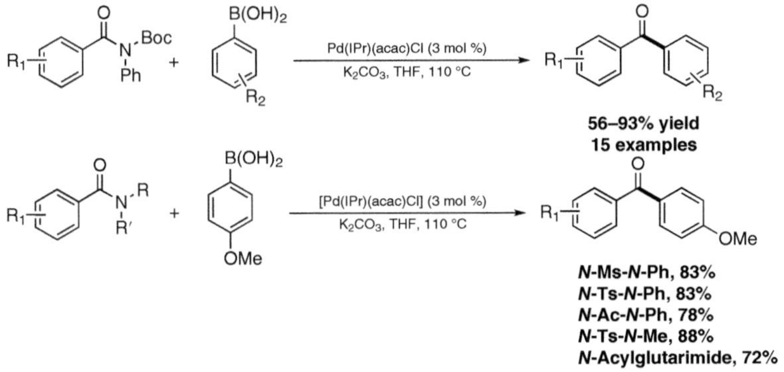

Scheme 9.25 Suzuki–Miyaura cross-coupling of various activated amides via Pd(IPr)(acac)Cl.

The reaction was optimized using *N*-Ph-*N*-Boc amides as the electrophile. Upon optimization of the reaction, it was found that the reaction was tolerant of a variety of electron-rich, electron-deficient, and sterically hindered substrates. They also reported that the reaction was compatible with a range of activated amides (Scheme 9.25). Though reaction temperatures are higher than past examples, it is important to stress the utility of the reaction due to the fact that the precatalyst can be generated in situ [8, 9, 29, 31].

Szostak and coworkers then expanded the utility of the Pd(NHC)(acac)Cl precatalysts by displaying its ability to also facilitate Buchwald–Hartwig amination of activated amides (Scheme 9.26). Pd(NHC)(acac)Cl was shown to have the ability to efficiently undergo transamidation reactions utilizing various anilines alongside activated amide electrophiles. Optimized conditions were identical to Suzuki–Miyaura cross-coupling mentioned earlier for the same Pd–NHC precatalysts. Similarly, the reaction scope was well tolerant of a variety of electron-withdrawing and electron-deficient substrates and activated amide electrophiles.

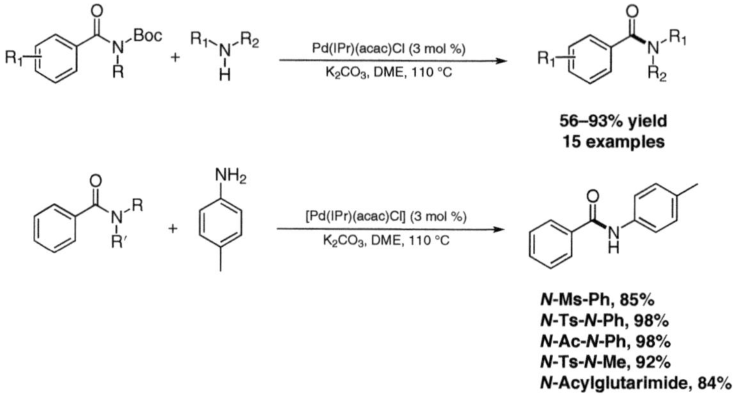

Scheme 9.26 Buchwald–Hartwig amination of various activated amides via Pd(IPr)(acac)Cl.

9.7 Conclusion

Well-defined Pd precatalysts have been proven to be advantageous in challenging cross-coupling reactions of amides electrophiles to yield ketones (Suzuki–Miyaura cross-coupling) and amides (Buchwald–Hartwig transamidation), which is facilitated by the strong σ-bond donation of the NHC to the Pd center, which in turns facilitate the oxidative addition. The scope and generality of the amide bond activation is growing over the last few years and many amide precursors were reported and found utilizable in various amide cross-coupling reactions. Palladium–NHC precatalysts were found exceptionally selective in the activation of C–N bonds of the amide. Well-defined Pd precatalysts became widely applicable due to their air and moisture stability, commercial availability, and ease of usage and continue to be reported on to catalyze difficult cross-coupling reactions under normal circumstances.

References

1 Shi, S., Nolan, S.P., and Szostak, M. (2018). Well-defined palladium(II)-NHC precatalysts for cross-coupling reactions of amides and esters by selective N–C/O–C cleavage. *Acc. Chem. Res.* 51 (10): 2589–2599.

2 Vemula, S.R., Chhoun, M.R., and Cook, G.R. (2019). Well-defined pre-catalysts in amide and ester bond activation. *Molecules* 24 (2): 215.

3 Meng, G., Shi, S., Lalancette, R. et al. (2018). Reversible twisting of primary amides via ground state N–C(O) destabilization: highly twisted rotationally inverted acyclic amides. *J. Am. Chem. Soc.* 140 (2): 727–734.

4 Adachi, S., Kumagai, N., and Shibasaki, M. (2018). Conquering amide planarity: structural distortion and its hidden reactivity. *Tetrahedron Lett.* 59 (13): 1147–1158.

5 Li, G., Ma, S., and Szostak, M. (2020). Amide bond activation: the power of resonance. *Trends Chem.* 2 (10): 914–928.

6 Marion, N. and Nolan, S.P. (2008). Well-defined N-heterocyclic carbenes-palladium(II) precatalysts for cross-coupling reactions. *Acc. Chem. Res.* 41 (11): 1440–1449.

7 Valdés, H., Canseco-González, D., Germán-Acacio, J.M., and Morales-Morales, D. (2018). Xanthine based N-heterocyclic carbene (NHC) complexes. *J. Organomet. Chem.* 867: 51–54.

8 Chartoire, A., Lesieur, M., Falivene, L. et al. (2012). [PdA (IPr*) (cinnamyl)Cl]: an efficient pre-catalyst for the preparation of tetra-ortho-substituted biaryls by Suzuki–Miyaura cross-coupling. *Chem. Eur. J.* 18 (15): 4517–4521.

9 Chartoire, A., Frogneux, X., and Nolan, S.P. (2012). An efficient palladium-NHC (NHC=N-heterocyclic carbene) and aryl amination pre-catalyst: [Pd(IPr*)(cinnamyl)Cl]. *Adv. Synth. Catal.* 354 (10): 1897–1901.

10 Buchspies, J. and Szostak, M. (2019). Recent advances in acyl Suzuki cross-coupling. *Catalysts* 9 (1): 53.

11 Li, X. and Zou, G. (2015). Acylative Suzuki coupling of amides: acyl-nitrogen activation via synergy of independently modifiable activating groups. *Chem. Commun.* 51 (24): 5089–5092.

12 Meng, G. and Szostak, M. (2015). Sterically controlled Pd-catalyzed chemoselective ketone synthesis via N–C cleavage in twisted amides. *Org. Lett.* 17 (17): 4364–4367.

13 Marion, N., Navarro, O., Mei, J. et al. (2006). Modified (NHC)Pd(allyl)Cl (NHC=N-heterocyclic carbene) complexes for room-temperature Suzuki–Miyaura and Buchwald–Hartwig reactions. *J. Am. Chem. Soc.* 128 (12): 4101–4111.

14 Navarro, O., Marion, N., Mei, J., and Nolan, S.P. (2006). Rapid room temperature Buchwald–Hartwig and Suzuki–Miyaura couplings of heteroaromatic compounds employing low catalyst loadings. *Chem. Eur. J.* 12 (19): 5142–5148.

15 Meng, G. and Szostak, M. (2018). Palladium/NHC (NHC=N-heterocyclic carbene)-catalyzed B-alkyl Suzuki cross-coupling of amides by selective N–C bond cleavage. *Org. Lett.* 20 (21): 6789–6793.

16 Meng, G., Lei, P., and Szostak, M. (2017). A general method for two-step transamidation of secondary amides using commercially available, air- and moisture-stable palladium/NHC (N-heterocyclic carbene) complexes. *Org. Lett.* 19 (8): 2158–2161.

17 Li, G., Zhou, T., Poater, A. et al. (2020). Buchwald–Hartwig cross-coupling of amides (transamidation) by selective N–C(O) cleavage mediated by air- and moisture-stable [Pd(NHC)(allyl)Cl] precatalysts: catalyst evaluation and mechanism. *Catal. Sci. Technol.* 10 (3): 710–716.

18 Meng, G., Szostak, R., and Szostak, M. (2017). Suzuki–Miyaura cross-coupling of N-acylpyrroles and pyrazoles: planar, electronically activated amides in catalytic N–C cleavage. *Org. Lett.* 19 (13): 3596–3599.

19 Meng, G., Lalancette, R., Szostak, R., and Szostak, M. (2017). N-methylamino pyrimidyl amides (MAPA): highly reactive, electronically-activated amides in catalytic N–C(O) cleavage. *Org. Lett.* 19 (17): 4654–4659.

20 Li, G., Lei, P., Szostak, M. et al. (2018). Mechanistic study of Suzuki–Miyaura cross-coupling reactions of amides mediated by [Pd(NHC)(allyl)Cl] precatalysts. *ChemCatChem* 10 (14): 3096–3106.

21 Luo, Z., Liu, T., Guo, W. et al. (2018). N-acyl-5,5-dimethylhydantoin, a new mild acyl-transfer reagent in Pd catalysis: highly efficient synthesis of functionalized ketones. *Org. Process Res. Dev.* 22 (9): 1188–1199.

22 Liu, C., Li, G., Shi, S. et al. (2018). Acyl and decarbonylative Suzuki coupling of N-acetyl amides: electronic tuning of twisted, acyclic amides in catalytic carbon-nitrogen bond cleavage. *ACS Catal.* 8 (10): 9131–9139.

23 Lei, P., Meng, G., Ling, Y. et al. (2017). General method for the Suzuki–Miyaura cross-coupling of primary amide-derived electrophiles enabled by [Pd(NHC)(cin)Cl] at room temperature. *Org. Lett.* 19 (24): 6510–6513.

24 Hruszkewycz, D.P., Balcells, D., Guard, L.M. et al. (2014). Insight into the efficiency of cinnamyl-supported precatalysts for the Suzuki–Miyaura reaction: observation of Pd(I) dimers with bridging allyl ligands during catalysis. *J. Am. Chem. Soc.* 136 (20): 7300–7316.

25 Hruszkewycz, D.P., Guard, L.M., Balcells, D. et al. (2015). Effect of 2-substituents on allyl-supported precatalysts for the Suzuki–Miyaura reaction: relating catalytic efficiency to the stability of palladium(I) bridging allyl dimers. *Organometallics* 34 (1): 381–394.

26 Melvin, P.R., Nova, A., Balcells, D. et al. (2015). Design of a versatile and improved precatalyst scaffold for palladium-catalyzed cross-coupling: $(\eta^3\text{-1-}t\text{Bu-indenyl})_2(\mu\text{-Cl})_2\text{Pd}_2$. *ACS Catal.* 5 (6): 3680–3688.

27 Lei, P., Meng, G., Shi, S. et al. (2017). Suzuki–Miyaura cross-coupling of amides and esters at room temperature: correlation with barriers to rotation around C–N and C–O bonds. *Chem. Sci.* 8 (9): 6525–6530.

28 Lei, P., Ling, Y., An, J. et al. (2019). 2-Methyltetrahydrofuran (2-MeTHF): a green solvent for Pd–NHC-catalyzed amide and ester Suzuki–Miyaura cross-coupling by N–C/O–C cleavage. *Adv. Synth. Catal.* 361 (24): 5654–5660.

29 Zhou, T., Ma, S., Nahra, F. et al. (2020). $[\text{Pd(NHC)}(\mu\text{-Cl)Cl}]_2$: versatile and highly reactive complexes for cross-coupling reactions that avoid formation of inactive Pd(I) off-cycle products. *iScience* 23 (8): 101377.

30 Valente, C., Çalimsiz, S., Hoi, K.H. et al. (2012). The development of bulky palladium NHC complexes for the most-challenging cross-coupling reactions. *Angew. Chem. Int. Ed.* 51 (14): 3314–3332.

31 Froese, R.D.J., Lombardi, C., Pompeo, M. et al. (2017). Designing Pd–N-heterocyclic carbene complexes for high reactivity and selectivity for cross-coupling applications. *Acc. Chem. Res.* 50 (9): 2244–2253.

32 Lei, P., Meng, G., Ling, Y. et al. (2017). Pd-PEPPSI: Pd–NHC precatalyst for Suzuki–Miyaura cross-coupling reactions of amides. *J. Org. Chem.* 82 (13): 6638–6646.

33 Shi, S. and Szostak, M. (2017). Pd-PEPPSI: a general Pd–NHC precatalyst for Buchwald–Hartwig cross-coupling of esters and amides (transamidation) under the same reaction conditions. *Chem. Commun.* 53 (76): 10584–10587.

34 Wang, C.A., Liu, C., and Szostak, M. (2020). N-Acyl-5,5-dimethylhydantoins: mild acyl-transfer reagents for the synthesis of ketones using Pd-PEPPSI or Pd/phosphine catalysts. *Org. Process Res. Dev.* 24 (6): 1043–1051.

35 Buchspies, J., Rahman, M.M., Szostak, R., and Szostak, M. (2020). N-Acylcarbazoles and N-acylindoles: electronically activated amides for N–C(O) cross-coupling by Nlpto Ar conjugation switch. *Org. Lett.* 22 (12): 4730–4704.

36 Pace, V., Holzer, W., Meng, G. et al. (2017). Corrigendum: structures of highly twisted amides relevant to amide N—C cross-coupling: evidence for ground-state amide destabilization. *Chem. Eur. J.* 23 (14): 14494–14498.

37 Wang, C., Huang, L., Wang, F., and Zou, G. (2018). Highly efficient synthesis of aryl ketones by PEPPSI-palladium catalyzed acylative Suzuki coupling of amides with diarylborinic acids. *Tetrahedron Lett.* 59 (23): 2299–2301.

38 Shi, W. and Zou, G. (2018). Palladium-catalyzed room temperature acylative cross-coupling of activated amides with trialkylboranes. *Molecules (Basel, Switzerland)* 23 (10): 2412.

39 Wang, T., Guo, J., Wang, H. et al. (2018). N-heterocyclic carbene palladium(II)-catalyzed Suzuki–Miyaura cross coupling of N-acylsuccinimides by C–N cleavage. *J. Organomet. Chem.* 877: 80–84.

40 Shi, S., Lalancette, R., Szostak, R., and Szostak, M. (2019). Triflamides: highly reactive, electronically activated N-sulfonyl amides in catalytic N–C(O) amide cross-coupling. *Org. Lett.* 21 (5): 1253–1257.

41 Karthik, S. and Gandhi, T. (2017). Palladium(II)/N-heterocyclic carbene-catalyzed direct C–H acylation of heteroarenes with N-acylsaccharins. *Org. Lett.* 19 (19).

42 Zhou, T., Li, G., Nolan, S.P., and Szostak, M. (2019). Well-defined, air-stable, and readily available precatalysts for Suzuki and Buchwald–Hartwig cross-coupling (transamidation) of amides and esters by N–C/O–C activation. *Org. Lett.* 21 (9): 3304–3309.

10

Cross-Coupling of Amides Through Decarbonylation

Hong Lu and Hao Wei

Northwest University, College of Chemistry & Materials Science, Key Laboratory of Synthetic and Natural Functional Molecule of the Ministry of Education, Room 425, 1 Xuefu Ave., Chang'an District, Xi'an, China

10.1 Introduction

Modern chemical research has witnessed the development of C(acyl)–N bond activation and functionalization mediated by transition metals [1]. The use of amides as coupling partners in cross-coupling reactions has multiple advantages, such as low cost; bench, air, and moisture stability; the absence of halides; and potential for functional group conversion [2]. Historically, the activation of the amide C(acyl)–N bond by transition metals has been unexplored because of strong $n_N \rightarrow \pi^*_{C=O}$ conjugation. Recent structural studies have revealed that with electronic activation and steric distortion [3, 4], which involve electron deficiency, weak π–π interactions, amide bond twisting, and chelation assistance [1], C(acyl)–N bond activation is, in contrast to conventional strategies, quite straightforward, opening the door for the utilization of amides as electrophiles in cross-coupling reactions in the synthesis of valuable molecules through previously unknown catalytic methods [4, 5].

Notably, a series of carbonyl-retaining acyl cross-coupling reactions involving C(acyl)–N bond cleavage by transition metals have been developed, representing an intriguing class of amide cross-coupling reactions. Unlike in these well-established carbonyl-retaining transformations, in decarbonylative cross-coupling reactions, the entire amide group acts as a leaving group to produce a new chemical bond, allowing synthetically valuable functional group conversion of amides, and these reactions have attracted considerable attention in recent years (Scheme 10.1) [6–8].

Over the past decade, decarbonylative cross-coupling reactions of amides have experienced rapid growth, and significant theoretical and experimental studies have been conducted [6–8]. Mechanistic investigations have revealed that decarbonylative cross-coupling potentially proceeds through C(acyl)–N bond activation (oxidative addition) by a transition metal catalyst (M) to give acyl intermediate **A**. Subsequent activation of a C(acyl)–C bond of intermediate **A** leads to decarbonylation to generate aryl- or alkyl-metal intermediate **B**, which undergoes transmetalation and reductive elimination to deliver the desired cross-coupling

Amide Bond Activation: Concepts and Reactions, First Edition. Edited by Michal Szostak.

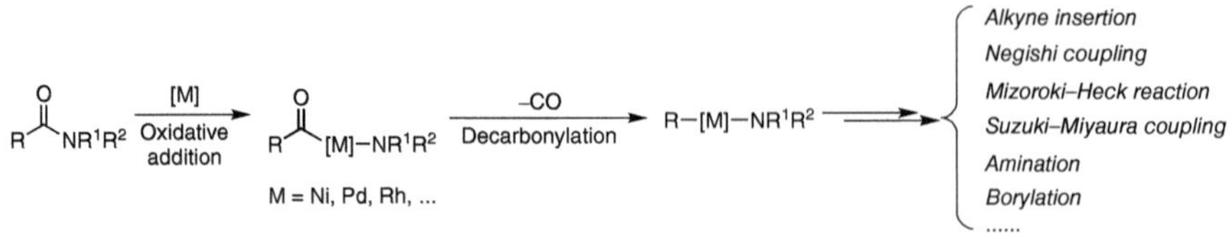

Scheme 10.1 General procedure for transition metal-catalyzed decarbonylation. Source: Adapted from Lu et al. [6], Guo and Rueping [7], and Liu and Szostak [8].

product (Path A). During this process, the order in which CO extrusion and transmetalation occur largely depends on the nature of the metal catalyst and ligand; transmetalation can occur first in some cases. Notably, in addition to cross-coupling reactions, intramolecular recombination is possible for intermediate **B**, when no external nucleophile is involved (Path B), giving the decarbonylative amidation product, R–NR^1R^2 (Scheme 10.2).

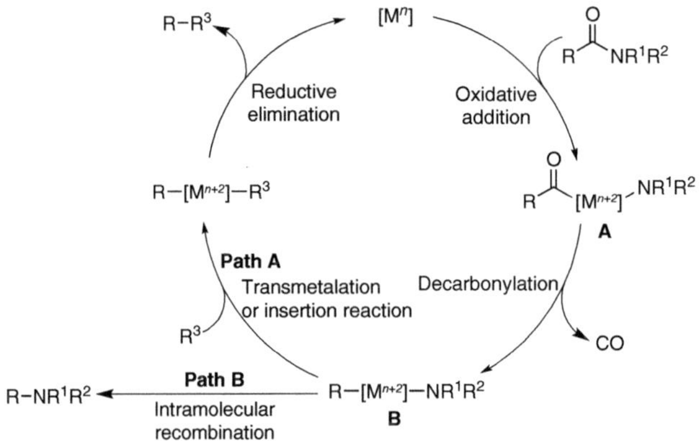

Scheme 10.2 Proposed general reaction mechanism for metal-catalyzed decarbonylative reactions.

While a number of excellent reviews on cross-coupling of amides have been published, this chapter focuses on transition metal-catalyzed decarbonylative cross-coupling using amides as coupling electrophiles, from the discovery of new synthetic strategies to the elucidation of reaction mechanisms, to disclose the great achievements and potential in this field. This chapter is organized according to reaction type based on the two general types of amides, cyclic and acyclic amides, which have been used for decarbonylation processes, wherein the cleavage of amide bonds is mainly driven by electron deficiency or disruption of amide resonance.

10.2 Decarbonylation of Cyclic Amide Derivatives

Since the development of Ni-catalyzed decarbonylative addition using phthalimides in 2008 [9], the decarbonylative cross-coupling of cyclic amides has experienced rapid growth and become an effective instrument for the construction of carbon–carbon and carbon–heteroatom bonds, allowing the conversion of amides to various other functional groups.

10.2.1 Phthalimides

Phthalimides with electron-withdrawing groups on their nitrogen atoms are appealing substrates for decarbonylative processes. Kurahashi and Matsubara first reported phthalimides **1** as "activated amide derivatives" in a decarbonylative ring-opening/alkyne insertion reaction (Scheme 10.3a) [9]. Isoquinolones **3** could be synthesized from alkynes in the presence of a $Ni(cod)_2/PMe_3$ catalyst. Substrate scope investigation revealed that electron-withdrawing arenes on the phthalimide

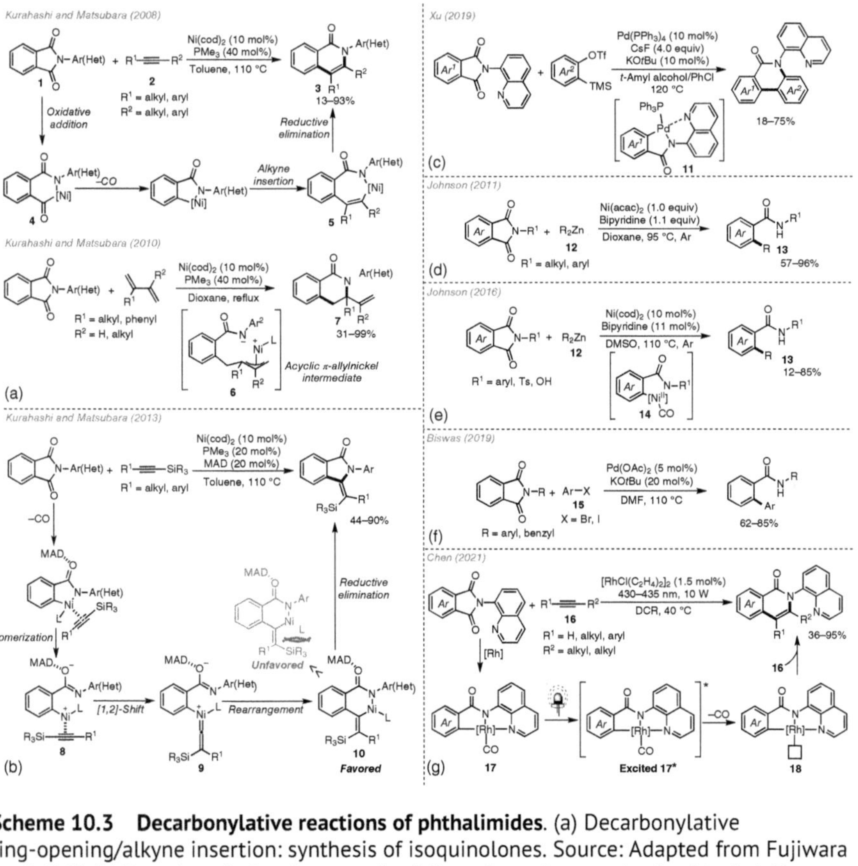

Scheme 10.3 Decarbonylative reactions of phthalimides. (a) Decarbonylative ring-opening/alkyne insertion: synthesis of isoquinolones. Source: Adapted from Fujiwara et al. [10]. (b) Decarbonylative ring-opening/alkyne insertion: synthesis of isoindolinones. Source: Adapted from Shiba et al. [11]. (c) Decarbonylative ring-opening/alkyne insertion: synthesis of phenanthridinones. Source: Adapted from Meng et al. [12]. (d) Stoichiometric decarbonylative coupling of phthalimides and diorganozinc reagents. (e) Catalytic decarbonylative coupling of phthalimides and diorganozinc reagents. (f) Decarbonylative coupling of phthalimides and aryl halides. Source: Adapted from Samanta and Biswas [13]. (g) Visible light-induced decarbonylative coupling. Source: Adapted from Min et al. [14].

nitrogen atom afforded better yields. The use of diarylalkynes **2** yielded multiple arylated isoquinolones.

This decarbonylation was initially believed to proceed through the direct oxidative addition of nickel to the amide C(acyl)–N bond of phthalimide **1** to give six-membered nickelacycle **4**, followed by carbon monoxide extrusion and insertion of alkyne **2** to generate seven-membered nickelacycle **5**, which undergoes reductive elimination to produce the desired product. This was confirmed by density functional theory (DFT) studies conducted by Poater et al. [15].

Kurahashi and Matsubara also reported the decarbonylative cycloaddition of phthalimides with 1,3-dienes catalyzed by nickel (Scheme 10.3a) [10]. They proposed that acyclic π-allylnickel intermediate **6** was formed upon the insertion of a bidentate diene to activate the C–Ni bond, and subsequent nucleophilic addition afforded isoquinolones **7**. Similar to isoquinolone synthesis, an investigation of the substituents on the nitrogen atom of the phthalimides revealed that those with an electron-withdrawing group, especially an *N*-pyrrolyl group, were the most effective. For unsymmetrical phthalimides, the substituents on the aromatic ring had an obvious influence on the regioselectivity. For electron-deficient or easily coordinated groups, such as –F, $-CF_3$, –OMe, or $-NMe_2$, cleavage of the closer C(acyl)–N bond was preferred, owing to either the electrophilic or chelating effect. However, cleavage of the farther C(acyl)–N bond may be favored with electron-donating groups (Me) or substrates with keto-enol tautomerism ($-NH_2$).

Kurahashi and Matsubara also found that trimethylsilyl-substituted alkynes can also participate in the Ni-catalyzed decarbonylative ring-opening/alkyne insertion reaction with phthalimides with a distinct reaction pattern (Scheme 10.3b) [11]. In this process, Lewis acid methylaluminum bis(2,6-di-*tert*-butyl-4-methylphenoxide) (MAD) was added to coordinate to the carbonyl oxygen to facilitate the oxidative addition of Ni to the imide C(acyl)–N bond and the formation of acyclic cationic nickel species **8**. The use of electron-rich phosphine ligand PMe_3, which stabilizes the cationic nature of the nickel center, may also be effective for the formation of intermediate **8**. Subsequent nucleophilic addition to the alkyne accompanied by a [1,2]-shift of the silyl group affords vinylidene complex **9**. The insertion of a carbon–nickel bond into the vinylidene moiety gives six-membered nickelacycle **10**, which undergoes reductive elimination to furnish an isoindolinone. The high stereoselectivity observed in the reaction was attributed to the directed insertion of the vinylidene moiety in intermediate **10**, in which the steric repulsion between the trimethylsilyl group and phosphine ligand on the nickel center was minimal.

Xu and coworkers demonstrated that arynes and phthalimides can also undergo decarbonylative annulation in the presence of $Pd(PPh_3)_4$ as a catalyst (Scheme 10.3c) [12]. They found that quinoline substituents on the N atom of the phthalimides may be crucial for stabilizing the active Pd species **11**.

In addition to decarbonylative additions across π-bonds, Johnson et al. also demonstrated Ni-mediated decarbonylative coupling of phthalimides and diorganozinc reagents **12** [16], producing *ortho*-substituted benzamides **13** in high yields (Scheme 10.3d). They also discovered that the introduction of electron-withdrawing groups, such as Ts, OH, and electron-deficient aromatics, onto the nitrogen atom of the phthalimide led to the catalytic version of this reaction [17]. They suggested that nitrogen-coordinated nickel intermediate **14** may be involved in this decarbonylative process (Scheme 10.3e).

Very recently, Biswas and coworker used aryl halides **15** as coupling partners instead of an organometallic reagent for the Pd-catalyzed decarbonylative cross-coupling of phthalimides (Scheme 10.3f) [13]. Compared to previously reported decarbonylative coupling processes, shorter reaction times and lower temperatures can be used under these ligand-free conditions.

In addition to one based on the thermal dissociation of CO from transition metals, Chen and coworkers developed a visible light-induced Rh-catalyzed decarbonylative coupling of phthalimides with alkynes **16** under ambient conditions (Scheme 10.3g) [14]. The mildness of the reaction conditions was attributed to photosensitive Rh complex **17**; under visible light irradiation, the bond dissociation energy of the Rh–CO bond of excited Rh complex **17*** decreases, owing to electron transfer from the π-backbonding Rh–CO orbital to an antibonding orbital, facilitating the subsequent CO extrusion and giving Rh intermediate **18** a vacant coordination site for coordination and migratory insertion of alkyne. They also found that the quinoline on the nitrogen atom of the phthalimides not only acts as a directing group for C(acyl)–N bond cleavage but also promotes energy transfer.

10.2.2 Saccharins and Other Cyclic Amide Derivatives

Owing to the strong electronegativity and steric effects of the SO_2 group, saccharins, which have emerged as bench-stable, amide-based, electrophilic coupling partners, have also been used as substrates in decarbonylative reactions. In 2015, Bi and coworkers first observed intermolecular decarbonylative C–C homocoupling of saccharins **19** in the presence of stoichiometric amounts of nickel catalyst (Scheme 10.4a) [18]. Notably, using an internal alkyne, a Ni-catalyzed

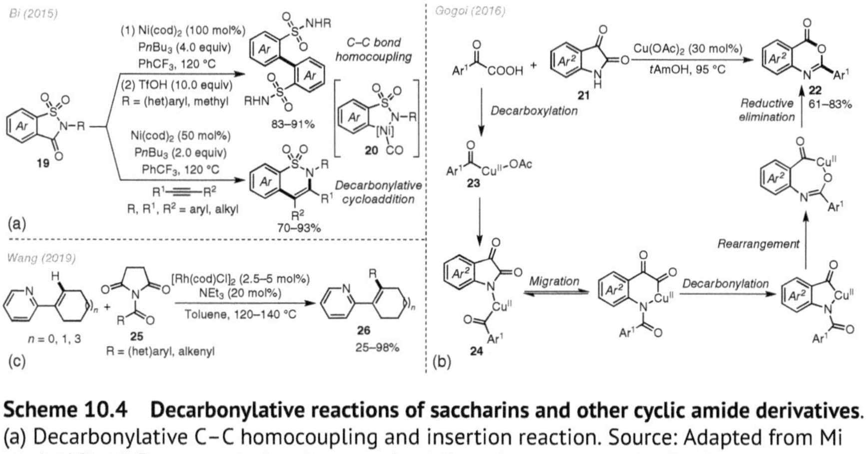

Scheme 10.4 **Decarbonylative reactions of saccharins and other cyclic amide derivatives.** (a) Decarbonylative C–C homocoupling and insertion reaction. Source: Adapted from Mi et al. [18]. (b) Decarbonylative ring-opening/alkyne insertion: synthesis of phenanthridinones. Source: Adapted from Prakash and Gogoi [19]. (c) Decarbonylative alkenyl C–H functionalization of aroyl and acrylamides. Source: Adapted from Huang et al. [20].

decarbonylative cycloaddition could also be achieved. They proposed that decarbonylative intermediate **20** was involved in both cases.

Because of the efforts of others, various other types of substrates can be used in decarbonylative functionalization. In 2016, Gogoi and coworker found that isatin is an applicable electrophilic coupling partner in Cu-catalyzed decarbonylative coupling (Scheme 10.4b) [19]. In this process, isatin **21** was proposed to react with intermediate **23**, which was generated from the decarboxylation of an α-oxocarboxylic acid, to afford intermediate **24**. Subsequent migration of Cu and decarbonylation followed by rearrangement and reductive elimination produced products **22**. The reaction proceeded well for isatin with a free N–H group, but no products were observed with N–H-protected isatin. In 2019, Wang and coworkers found that aroyl and acrylamides **25** can participate in Rh-catalyzed decarbonylative alkenyl C–H bond functionalization to give arylated or alkenylated alkenes **26** (Scheme 10.4c) [20].

10.3 Decarbonylation of Acyclic Amide Derivatives

10.3.1 *N*-Acyl-Glutarimides

Steric activation of amides through the disruption of amide bond resonance has provided a new amide reactivity platform for N–C bond scission. A seminal study by Szostak and coworker reported a Pd-catalyzed decarbonylative Mizoroki–Heck cross-coupling of *N*-acyl-glutarimides **27** (Scheme 10.5a) [21]. The twisted amide can be used to facilitate metal insertion into the C(acyl)–N amide bond, while selective decarbonylation provides valuable aryl/alkenyl-metal intermediate **28** capable of the subsequent migratory insertion, and then β-hydride elimination results in the product. Other amides such as Weinreb amides, *N*-acylpyrrole, and *N*-acyl-succinimide were unreactive.

In 2016, Szostak and coworker first reported the Rh-catalyzed decarbonylative C–H arylation of *N*-acyl-glutarimides using a directing group strategy (Scheme 10.5b) [22]. Various directing groups, such as quinoline, pyridine, pyrimidine, pyrazole, and imine groups, were effective for this reaction. Notably, when excess *N*-acyl-glutarimide (3.0 equiv) was used for these reactions, biaryl products **29** were formed. Recently, Zhou et al. also discovered that *N*-acyl-glutarimides can be used for the C–H functionalization of 1,3-azoles **30** under Pd-catalyzed conditions (Scheme 10.5b) [29]. Notably, the use of a CuBr cocatalyst facilitates the transmetalation by generating Cu species **32** so that insertion occurs instead at the C–N double bond of intermediate **31**.

Applying their established ground-state amide bond destabilization concept, Szostak and coworker disclosed a Pd-catalyzed decarbonylative cyanation of *N*-acyl-glutarimides (Scheme 10.5c) [23]. A wide range of amides bearing highly reactive functional groups, such as esters and aldehydes, can effectively couple with zinc cyanide, affording various aryl, heteroaryl, and alkenyl nitrile derivatives. Rueping and coworkers reported a similar amide-to-nitrile conversion in

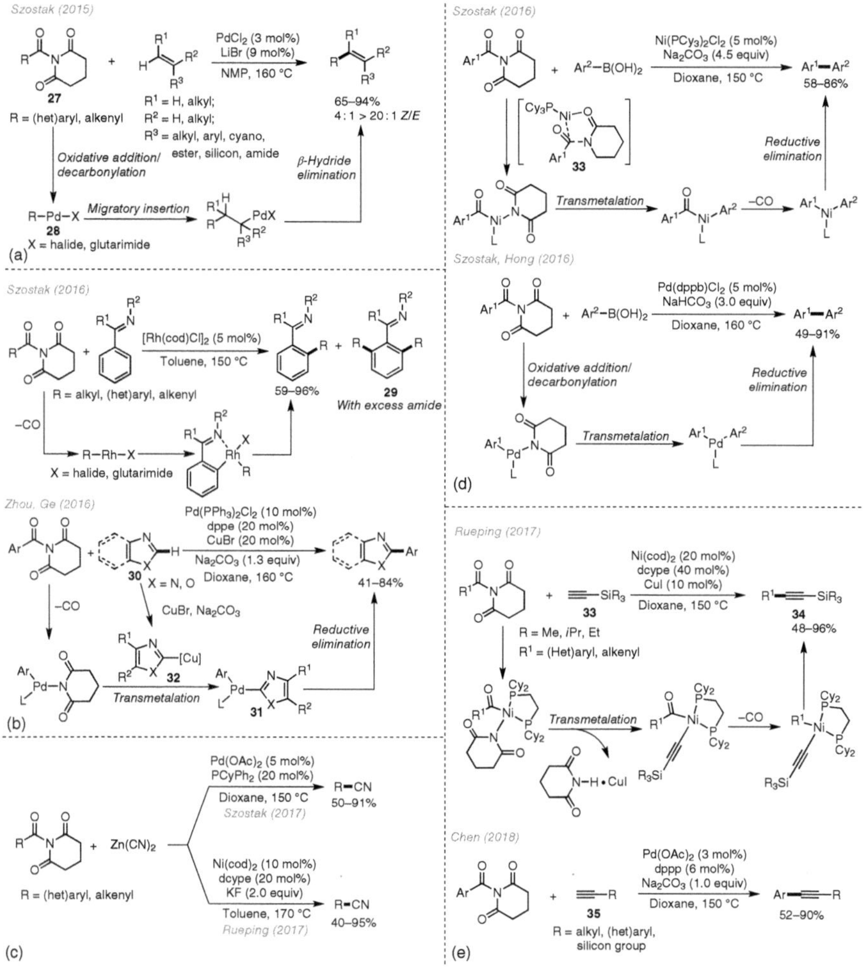

Scheme 10.5 Decarbonylative reactions of *N*-acyl-glutarimides for carbon–carbon bond formation. (a) Decarbonylative Mizoroki–Heck reaction. Source: Adapted from Meng and Szostak [21]. (b) Decarbonylative C–H arylation. Source: Adapted from Meng and Szostak [22]. (c) Decarbonylative Negishi-type reaction. Source: Adapted from Shi et al. [23] and Chatupheeraphat et al. [24]. (d) Decarbonylative Suzuki–Miyaura-type reaction. Source: Adapted from Shi et al. [25] and Zhou et al. [26]. (e) Decarbonylative Sonogashira-type reaction. Source: Adapted from Srimontree et al. [27] and Liu et al. [28].

the presence of $Ni(cod)_2$/dcype (Scheme 10.5c) [24]. They also found that the transformation can be scaled up smoothly by using $Ni(acac)_2$ instead of air-sensitive $Ni(cod)_2$ under microwave (MW) irradiation.

The Suzuki–Miyaura reaction is a useful and versatile method for preparing biaryl compounds. Given the importance of this transformation in the construction of C–C bonds, substantial attention has been focused on extending the scope of the electrophiles toward avoiding the use of aryl halides. In 2016, Szostak et al. demonstrated that sterically distorted *N*-acyl-glutarimides can also be electrophilic coupling partners under Ni-catalyzed conditions (Scheme 10.5d) [25]. A mechanism involving transmetalation prior to decarbonylation was proposed. Computational studies by Hong and coworker revealed that the interaction between nickel and the *N*-glutarimide amide carbonyl provided a crucial directing effect for C(acyl)–N bond cleavage (transition state **33**) [30].

As a result of their continued interest in decarbonylative coupling, Szostak et al. found that Pd(dppb)Cl_2, in the presence of mild base $NaHCO_3$, also enabled the decarbonylative Suzuki–Miyaura cross-coupling of *N*-acyl-glutarimides (Scheme 10.5d) [26]. In contrast to the Ni-catalyzed process, other amides, such as *N*-acetyl (*N*-Ac) amides, *N*-succinimides, *N*-phthalimides, acyclic *N*-Ms, and *N*-Ts sulfonamides, and even challenging *N*-Boc_2 amides, which are prone to deactivation through facile *N*-Boc cleavage, are also applicable under these conditions. A DFT study of the mechanism revealed that decarbonylation occurs prior to transmetalation in the palladium catalytic cycle.

The Sonogashira coupling is one of the most popular and efficient procedures for the synthesis of aryl alkynes. In 2017, Rueping and coworkers first developed a Ni/Cu-cocatalyzed decarbonylative Sonogashira cross-coupling using *N*-acyl-glutarimides as electrophilic coupling partners (Scheme 10.5e) [27]. In the presence of $Ni(cod)_2$ and CuI catalysts and a dcype ligand, a variety of *N*-acyl-glutarimides were successfully converted into alkynes **34** in good to excellent yields. Although this strategy represents an impressive development in this field, the scope of the nucleophilic reagents was, to some extent, restricted to silylacetylenes **33**, owing to the facile polymerization of aryl and alkyl terminal acetylenes. Later, Chen and coworkers disclosed a Pd-catalyzed version of this transformation (Scheme 10.5e) [28]. The authors proposed that decarbonylation occurs prior to transmetalation in the presence of $Pd(OAc)_2$/dppp. In addition to silylacetylenes **33**, the scope of the nucleophilic reagents also included aromatic and aliphatic terminal alkynes **35**.

Rueping and coworkers extended the decarbonylative reaction to borylation using B_2pin_2 as a nucleophile (Scheme 10.6a) [31]. A catalytic system comprising $Ni(cod)_2$ and $PnBu_3$ provided satisfactory results. Recently, Szostak and coworker also reported a Pd-catalyzed direct decarbonylative borylation of amides via cleavage of the amide bond in *N*-acyl-glutarimides (Scheme 10.6a) [32]. It was found that primary and secondary amides did not react under these conditions, and decomposition via N–O cleavage was observed for *N*-methoxy amides.

The first example of decarbonylative conversion of amides to amine derivatives **38** was also reported by the Rueping group [31, 35]; consequently, amides emerged as competent alternatives for aryl halides in classical Buchwald–Hartwig aminations. The reaction proceeded smoothly with aromatic *N*-acyl-glutarimides and benzophenone imine **36** in the presence of a $Ni(cod)_2$/dppf catalytic system (Scheme 10.6b). Notably, corresponding primary amines **38** could also be obtained effectively through acidic hydrolysis of ketimine intermediates **37**.

In 2017, Rueping and coworkers presented a Ni-catalyzed decarbonylative silylation using *N*-acyl-glutarimides as coupling partners (Scheme 10.6c) [31]. The issue of the low nucleophilicity of silicon in the preparation of arylsilanes can be addressed

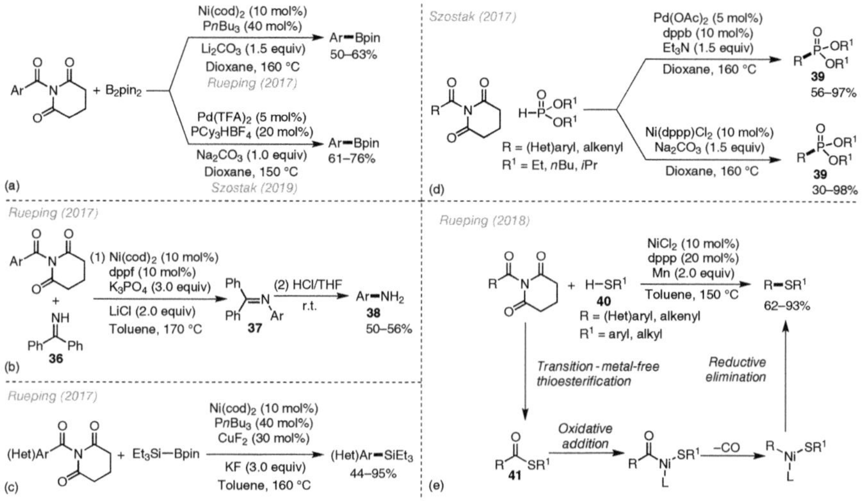

Scheme 10.6 Decarbonylative reactions of *N*-acyl-glutarimides for carbon–heteroatom bond formation. (a) Decarbonylative borylation. Source: Adapted from Lee et al. [31] and Shi and Szostak [32]. (b) Decarbonylative amination. (c) Decarbonylative silylation. Source: Adapted from Lee et al. [31]. (d) Decarbonylative phosphorization. (e) Decarbonylative thioetherification. (d, e) Source: Adapted from Liu and Szostak [33] and Lee et al. [34].

by using Et_3Si–Bpin in the presence of CuF_2 as a selective nucleophilic Si transfer reagent. *N*-Acyl-succinimides and *N*-Boc-carbamates also exhibited promising reactivity in this process.

Decarbonylative carbon–phosphorus bond formation using *N*-acyl-glutarimides as electrophiles has also been reported by Szostak and coworker (Scheme 10.6d) [33]. This new Hirao-type cross-coupling reaction was successfully performed using either a $Pd(OAc)_2$/dppb catalytic system or $Ni(dppp)Cl_2$, affording aryl phosphonates **39** in good to moderate yields. Notably, acyclic amides, such as *N*-Ts and *N*-Ms sulfonamides, are also compatible with the Pd-catalyzed conditions. Mechanistic studies suggested that transmetalation occurs prior to decarbonylation under both Pd- and Ni-catalyzed conditions, which was also supported by recent computational investigations [36]. Other computational studies have disclosed other aspects of the mechanism, including the rate-determining steps of these reactions, the influence of additives on the reactivities, and the chemoselectivity under Pd-catalyzed conditions [37].

The only decarbonylative thioetherification of amides with mercaptans **40** was developed by the Rueping group (Scheme 10.6e) [34]. The utilization of a $NiCl_2$/dppp catalytic system in combination with stoichiometric Mn as a reductant was found to be effective for this transformation. Various amides such as aryl-, heteroaryl-, and alkenyl-substituted *N*-acyl-glutarimides as well as *N*-Ph and *N*-Boc amides were tolerated. It should be noted that the amides were not involved in the decarbonylation process. Transition metal-free thioesterification of the amides followed by Ni-catalyzed decarbonylation of thioesters **41** may have also occurred.

10.3.2 *N*-Acylsaccharin Amides

N-Acylsaccharins **42** can also be used as electrophilic coupling partners in decarbonylative reactions via C(acyl)–N bond cleavage. In 2016, Szostak and coworkers reported a Pd-catalyzed decarbonylative Heck reaction (Scheme 10.7a) [38]. Competition experiments revealed that *N*-acylsaccharins **42** were more reactive than *N*-acyl-glutarimides under these conditions.

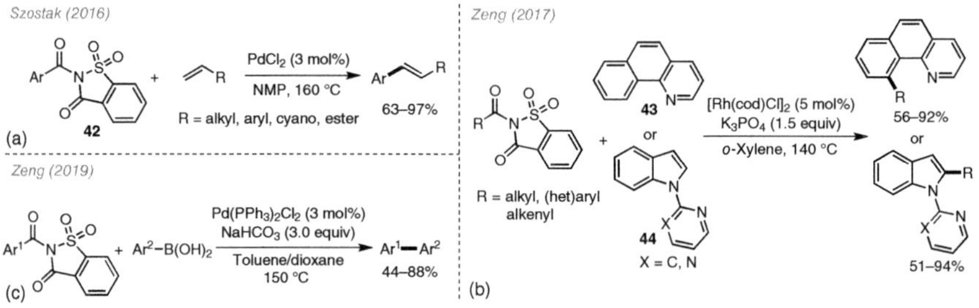

Scheme 10.7 Decarbonylative reactions of *N*-acylsaccharins. (a) Decarbonylative Mizoroki–Heck reaction. Source: Adapted from Liu et al. [38]. (b) Decarbonylative C–H functionalization. Source: Adapted from Wu et al. [39]. (c) Decarbonylative Suzuki–Miyaura coupling. Source: Adapted from Luo et al. [40].

In 2017, Zeng and coworkers developed a Rh-catalyzed C–H arylation using *N*-acylsaccharins as coupling reagents (Scheme 10.7b) [39]. In addition to benzo[*h*]quinoline **43**, this strategy could be used for the direct C2 arylation of indoles **44** by using the *N*-pyridyl/*N*-2-pyrimidyl ring as a directing group. Recently, Zeng and coworkers revealed that *N*-acylsaccharins are applicable substrates in decarbonylative Suzuki–Miyaura coupling (Scheme 10.7c) [40]. Intriguingly, the author proposed that transmetalation occurs prior to decarbonylation, in contrast to the work of Szostak.

10.3.3 Other Acyclic Amides

Decarbonylation involving acyclic amides in which steric and electronic properties are remodeled has also attracted significant interest. In a pioneering study from 2002, Miura et al. used *N*-(2-thenoyl)aniline for decarbonylative coupling with bromobenzene [41]. Wang and coworkers successfully developed a Pd-catalyzed decarbonylative C–H arylation of 1,3-azoles **46** with aromatic amides **45** in the presence of oxidants (Scheme 10.8a) [42]. A wide range of heteroaryls was tolerated under the optimized conditions. This process was proposed to proceed through a Pd(II)/Pd(0) catalytic cycle. Here, elimination of CO_2 (Path B) and decarbonylation (Path A) via activation of the amide C(acyl)–N bond are both possible routes for the generation of Pd-aryl intermediates **48**, and Fourier transform infrared (FT-IR) analysis gave a CO/CO_2 ratio of 29.3 : 1, indicating that decarbonylation (Path A) was the major route. Insertion of intermediates **48** into the C=N double bond of a 1,3-azole followed by β-hydride elimination provided biaryl compounds **47**.

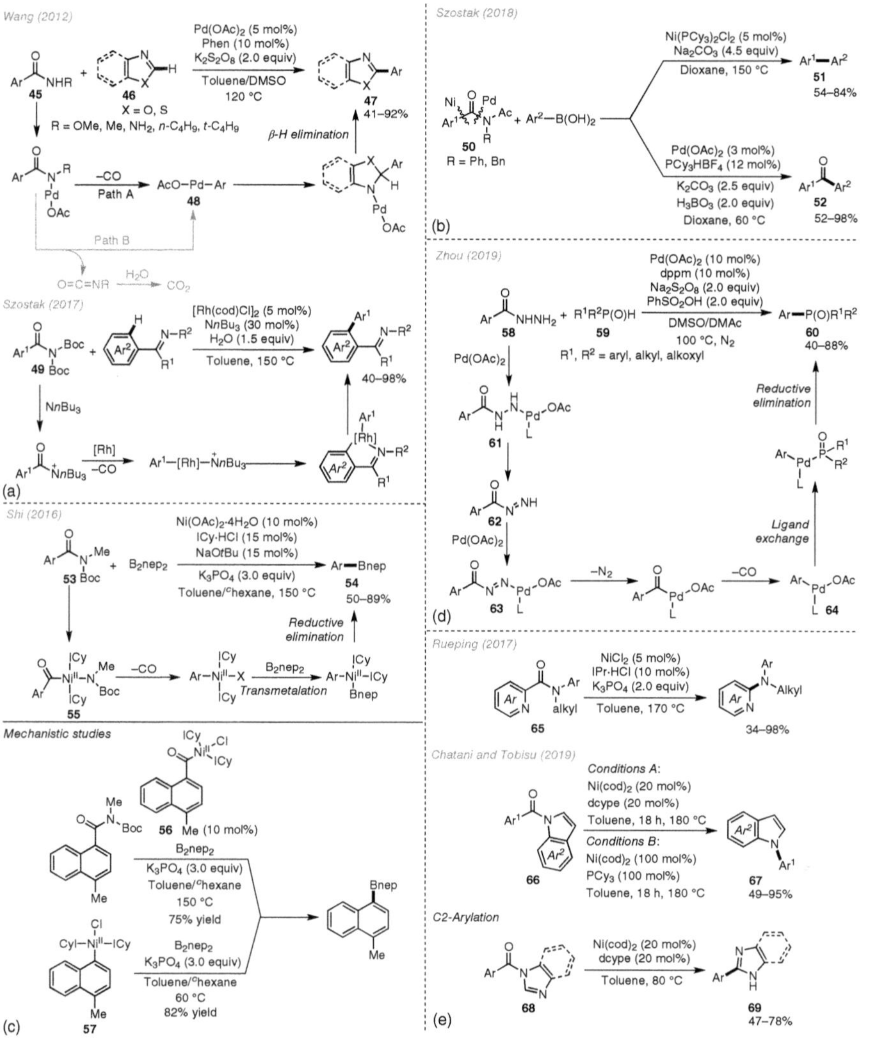

Scheme 10.8 Decarbonylative reactions of acyclic amides. (a) Decarbonylative C–H functionalization. Source: Adapted from Li et al. [42] and Meng and Szostak [43]. (b) Decarbonylative Suzuki–Miyaura coupling. (c) Decarbonylative borylation. Source: Adapted from Hu et al. [44]. (d) Decarbonylative C–P bond formation. Source: Adapted from Dong et al. [45]. (e) Intramolecular decarbonylative reactions. Source: Adapted from Liu et al. [46] and Morioka et al. [47].

Szostak and coworker found that twisted primary amides **49** with *N*-Boc substituents can be used as arylating reagents in the presence of $[Rh(cod)Cl]_2$ and $NnBu_3$ (Scheme 10.8a) [43]. During the reaction, $NnBu_3$, acting as a Lewis base, facilitates metal insertion by converting the inert C(acyl)–N bonds into weak acylammonium bonds.

Another impressive example of catalyst-controlled Suzuki–Miyaura coupling was also discovered by Szostak and coworkers [48]. When $Ni(PCy_3)_2Cl_2$ was used as the catalyst, decarbonylative coupling using amides **50** gave biaryl products **51** (Scheme 10.8b). Substitution of $Ni(PCy_3)_2Cl_2$ with $Pd(OAc)_2$ at lower temperatures led to the formation of non-decarbonylative products **52**. Structural and computational studies revealed that amide twisting, electronic destabilization, and the high resonance energy of the *N*-Ac group are all critical factors for the high reactivity of the C(acyl)–N bond. A DFT computational investigation conducted by Xie and coworkers revealed that distinct rate-determining steps were involved in these two mechanisms [49]. Transmetalation was the rate-determining step for the Ni-catalyzed reactions, while oxidative addition was the rate-determining step with Pd.

In 2016, Shi and coworkers established an intriguing method for the decarbonylative borylation of amides **53** with B_2nep_2 using a Ni/*N*-heterocyclic carbene catalytic system (Scheme 10.8c) [44]. This protocol provided a new method to cleave a C(acyl)–N bond while generating a versatile boron-derived entity. A variety of aryl, alkenyl, and alkyl amides gave corresponding boronate esters **54** in moderate to good yields. The combination of a Boc group and an alkyl or aryl group was necessary for the oxidative addition of Ni to the C(acyl)–N bond to yield acyl-nickel(II) intermediate **55**, and then decarbonylation, transmetalation, and reductive elimination afforded the desired products.

To shed light on the reaction mechanism, relevant mechanistic studies were carried out by the authors (Scheme 10.8c). Catalytic reactions employing intermediate **56** and treatment of a stoichiometric amount of intermediate **57** with B_2nep_2 both gave the borylation product. This indicated that **56** and **57** were both intermediates in this decarbonylative borylation reaction.

Recently, Zhou and coworkers developed a decarbonylative carbon–phosphorus bond formation reaction between aroylhydrazides **58** and P(O)H compounds **59** (Scheme 10.8d) [45]. Various P(O)H compounds, such as secondary phosphine oxides and H-phosphinates, were suitable for the reaction, giving aryl phosphorus compounds **60** in satisfactory yields. Under Pd-catalyzed conditions, aroylhydrazides **58** first underwent oxidative dehydrogenation to produce Pd-azo intermediates **63** via the formation of intermediates **61** and **62**, and subsequent extrusion of molecular nitrogen and decarbonylation afforded Ar-Pd-OAc intermediates **64**, which underwent ligand exchange and reductive elimination to produce aryl phosphorus compounds **60**. The author proposed that the cooperative reaction system featuring the association of the Brønsted acid and the bidentate phosphine ligand can effectively suppress the undesired oxidation and coordination of the P(O)H compounds and activate the C(acyl)–N bonds selectively.

Rueping and coworkers reported a Ni-catalyzed intramolecular decarbonylative recombination of 2-azinecarboxamides **65** using $NiCl_2$ and an *N*-heterocyclic carbene ligand (Scheme 10.8e) [46]. 2-Azinecarboxamides **65** were more reactive than the other amides. An intermolecular crossover reaction conducted by Rueping et al. confirmed that the intramolecular process occurs.

Chatani and Tobisu applied an intramolecular decarbonylative recombination strategy toward preparing *N*-arylated *N*-heteroarenes **67** (Scheme 10.8e) [47]. In this process, the pK_a value of the parent amine of amide **66** was observed to have a significant impact on the reactivity. More basic amines led to stronger C(acyl)–N bonds, which lowered the reactivities of these substrates under catalytic conditions; thus, one equivalent of Ni and PCy_3 was used as an alternative. Interestingly, when *N*-acylated benzimidazoles **68** were used as substrates, unexpected 1,2-migration occurred after decarbonylation, affording 2-arylated benzimidazoles **69**. Preliminary mechanistic studies revealed that distinct intermolecular pathways are likely to be involved.

10.4 Conclusion

While still in the developing stage, transition metal-catalyzed decarbonylative cross-coupling of amides has emerged as a useful synthetic methodology and an attractive alternative to traditional cross-couplings with aryl halides and pseudohalides. The synthetically valuable functional group conversions of amides have led to more than 15 types of reactions that have previously not been achieved with traditional methods, enabling a wide variety of carbon–carbon and carbon–heteroatom bond formation reactions with excellent functional group tolerance, which shows that they have great potential for increasing the number of synthetic methods that can be used in late-stage functionalization and the synthesis of bioactive molecules. Despite these remarkable achievements, some interesting yet challenging problems remain, such as substrate restraints, high catalyst loadings, and harsh reaction conditions; consequently, industrial application has been limited to date. Further advances can be expected, including the development of more efficient catalytic systems and novel reactive yet stable amide precursors, which would lead to milder reaction conditions. More in-depth mechanistic studies, which would be beneficial for refinements and further improvements in the field of catalysis, are also highly desirable.

References

1 Takise, R., Muto, K., and Yamaguchi, J. (2017). *Chem. Soc. Rev.* 46: 5864–5888.
2 Meng, G. and Szostak, M. (2018). *Eur. J. Org. Chem.* 2018: 2352–2365.
3 Szostak, R., Shi, S., Szostak, M. et al. (2016). *J. Org. Chem.* 81: 8091–8094.
4 Lei, P., Meng, G., Szostak, M. et al. (2017). *Chem. Sci.* 8: 6525–6530.
5 Liu, C. and Szostak, M. (2017). *Chem. Eur. J.* 23: 7157–7173.
6 Lu, H., Yu, T.Y., Wei, H. et al. (2021). *Chem. Rev.* 121: 365–411.
7 Guo, L. and Rueping, M. (2018). *Chem. Eur. J.* 24: 7794–7809.
8 Liu, C. and Szostak, M. (2018). *Org. Biomol. Chem.* 16: 7998–8010.
9 Kajita, Y., Matsubara, S., and Kurahashi, T. (2008). *J. Am. Chem. Soc.* 130: 6058–6059.
10 Fujiwara, K., Kurahashi, T., and Matsubara, S. (2010). *Org. Lett.* 12: 4548–4551.
11 Shiba, T., Kurahashi, T., and Matsubara, S. (2013). *J. Am. Chem. Soc.* 135: 13636–13639.

12 Meng, Y.Y., Si, X.J., Xu, F. et al. (2019). *Chem. Commun.* 55: 9507–9510.
13 Samanta, K.P. and Biswas, P. (2019). *J. Org. Chem.* 84: 3968–3976.
14 Min, X.T., Ji, J.W., Chen, Q.A. et al. (2021). *Angew. Chem. Int. Ed.* 60: 1583–1587.
15 Poater, A., Vummaleti, S.V.C., and Cavallo, L. (2013). *Organometallics* 32: 6330–6336.
16 Havlik, S.E., Simmons, J.M., Johnson, J.B. et al. (2011). *J. Org. Chem.* 76: 3588–3593.
17 DeGlopper, K.S., Fodor, S.K., Johnson, J.B. et al. (2016). *Polyhedron* 114: 393–398.
18 Mi, P., Liao, P., Bi, X. et al. (2015). *Chem. Eur. J.* 21: 5332–5336.
19 Prakash, R. and Gogoi, S. (2016). *Adv. Synth. Catal.* 358: 3046–3049.
20 Huang, Y., Pan, W.J., and Wang, Z.X. (2019). *Org. Chem. Front.* 6: 2284–2290.
21 Meng, G. and Szostak, M. (2015). *Angew. Chem. Int. Ed.* 54: 14518–14522.
22 Meng, G. and Szostak, M. (2016). *Org. Lett.* 18: 796–799.
23 Shi, S. and Szostak, M. (2017). *Org. Lett.* 19: 3095–3098.
24 Chatupheeraphat, A., Liao, H.H., and Rueping, M. (2017). *Org. Lett.* 19: 4255–4258.
25 Shi, S., Meng, G., and Szostak, M. (2016). *Angew. Chem. Int. Ed.* 55: 6959–6963.
26 Zhou, T., Hong, X., Szostak, M. et al. (2019). *Chem. Sci.* 10: 9865–9871.
27 Srimontree, W., Chatupheeraphat, A., Rueping, M. et al. (2017). *Org. Lett.* 19: 3091–3094.
28 Liu, L., Zhou, D., and Chen, T. (2018). *Org. Lett.* 20: 2741–2774.
29 Zhou, P.X., Shi, S., and Ge, C. (2019). *Org. Chem. Front.* 6: 1942–1947.
30 Ji, C.L. and Hong, X. (2017). *J. Am. Chem. Soc.* 139: 15522–15529.
31 Lee, S.C., Guo, L., Rueping, M. et al. (2017). *Synlett* 28: 2594–2598.
32 Shi, S. and Szostak, M. (2019). *ACS Omega* 4: 4901–4907.
33 Liu, C. and Szostak, M. (2017). *Angew. Chem. Int. Ed.* 56: 12718–12722.
34 Lee, S.C., Liao, H.H., and Rueping, M. (2018). *Chem. Eur. J.* 24: 3608–3612.
35 Yue, H., Guo, L., Rueping, M. et al. (2017). *Angew. Chem. Int. Ed.* 56: 4282–4285.
36 Xu, Y., Wang, B., Fu, Y. et al. (2019). *J. Org. Chem.* 84: 9474–9479.
37 Tong, W.Y., Wu, Y.B., and Wang, X. (2020). *Chem. Commun.* 56: 113–116.
38 Liu, C., Meng, G., and Szostak, M. (2016). *J. Org. Chem.* 81: 12023–12030.
39 Wu, H., Liu, T., and Zeng, Z. (2017). *Org. Biomol. Chem.* 15: 536–540.
40 Luo, Z., Xiong, L., and Zeng, Z. (2019). *J. Org. Chem.* 84: 10559–10568.
41 Okazawa, T., Satoh, T., Miura, M., and Nomura, M. (2002). *J. Am. Chem. Soc.* 124: 5286–5287.
42 Li, C., Li, P., and Wang, L. (2012). *Chem. Commun.* 48: 4214–4216.
43 Meng, G. and Szostak, M. (2017). *ACS Catal.* 7: 7251–7256.
44 Hu, J., Zhao, Y., and Shi, Z. (2016). *Angew. Chem. Int. Ed.* 55: 8718–8722.
45 Dong, J., Liu, L., Zhou, T. et al. (2019). *Org. Lett.* 21: 3198–3203.
46 Liu, X., Yue, H., and Rueping, M. (2017). *Chem. Eur. J.* 23: 11771–11775.
47 Morioka, T., Chatani, N., Tobisu, M. et al. (2019). *Chem. Sci.* 10: 6666–6671.
48 Liu, C., Li, G., Szostak, M. et al. (2018). *ACS Catal.* 8: 9131–9139.
49 Liu, J., Ling, M., and Xie, H. (2020). *Comput. Theor. Chem.* 1185: 112889.

11

Transition Metal-Catalyzed Radical Reactions of Amides

Taline Kerackian, Didier Bouyssi, Nuno Monteiro, and Abderrahmane Amgoune

Univ Lyon, Université Lyon 1, Institut de Chimie et Biochimie Moléculaires et Supramoléculaires, UMR 5246 du CNRS, 1 rue Victor Grignard, Villeurbanne, 69100, France

11.1 Introduction

The merger of modern radical chemistry and organometallic catalysis has opened new avenues for the design of more resource-efficient and milder cross-coupling methodologies. This new paradigm has notably driven the development of innovative chemical methods for the functionalization of amides or their transformation into other classes of high-value compounds such as ketones. On the one hand, organometallic strategies involving direct activation of the amide substrate by the metal catalyst and subsequent reaction with a catalytically generated radical coupling partner have been explored, which include acylation methods through amide bond activation as well as amide-directed proximal C–H bond functionalization via metalation. On the other hand, approaches combining radical activation of amides with cross-coupling reactions have also been developed, which include remote functionalization via radical translocation, intramolecular carboamination of alkenes, and carbamoylation reactions (Figure 11.1).

This chapter describes representative synthetic transformations of amides with transition metal catalysts involving radical species. The chapter is divided in two main sections organized around the type of amide activation strategy, i.e. via transition metal activation or radical activation and further by the mechanistic aspects of the cross-coupling process. The intent of this chapter is not to provide an exhaustive survey of the literature but rather to highlight important developments in this exciting and rapidly advancing area with special emphasis on carbon–carbon bond-forming strategies. Radical reactions of conjugated amides and enamides will not be discussed here.

Amide Bond Activation: Concepts and Reactions, First Edition. Edited by Michal Szostak.

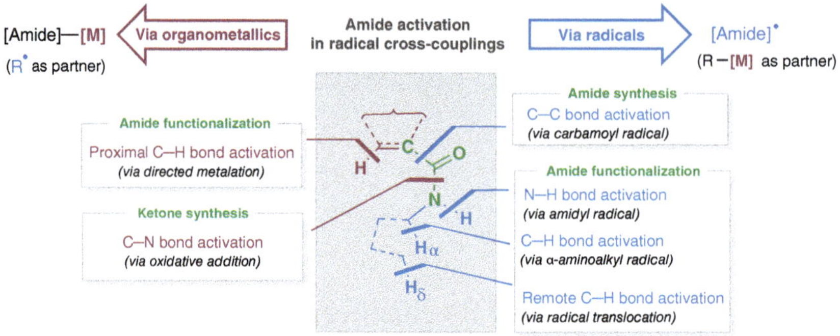

Figure 11.1 Typical activation sites available for metal-catalyzed radical cross-coupling of amides, with associated strategies and synthetic applications.

11.2 Reactions Involving Amides as Precursors to Organometallic Compounds

11.2.1 Radical Reactions of Amides via Metal-Catalyzed C–N Bond Activation

Resonance-destabilized amides have been receiving considerable attention since the mid-2010s as a new type of acyl electrophiles that can successfully participate in many innovative and useful synthetic transformations. For instance, transition metal-catalyzed reactions proceeding via amide C–N bond activation, which are reminiscent of traditional cross-coupling methods (e.g. Suzuki and Negishi reactions), have been highlighted as a very powerful strategy to forge C–C bonds with broad application in the preparation of aryl and alkyl ketones. These reactions mostly employ palladium or nickel catalysts and proceed through two-electron redox processes (polar mechanism) [1]. Alongside these discoveries, radical-based acylative cross-coupling reactions have been developed with nickel catalysts. In addition of being naturally abundant and inexpensive, nickel readily participates in radical pathways via single electron transfer (SET) processes, thus allowing facile access to the Ni(I) and Ni(III) oxidation states [2, 3]. Furthermore, nickel displays increased propensity for oxidative addition as well as increased resistance to β-hydride elimination, thus facilitating catalytic acylation of challenging $C(sp^3)$-hybridized substrates. *N*-acylimides (**1**) featuring a twisted C–N bond, allowing facile insertion of the transition metal, are the privileged motifs in recently developed nickel-catalyzed radical-based acylation methods. A simplified picture of the commonly proposed mechanism is presented in Figure 11.2. The reaction starts with the oxidative addition of the C–N bond to zerovalent nickel catalyst to generate a Ni(II)-acyl species (**A$_1$**). Then, trapping of a transiently generated aryl or alkyl radical species affords a high-valent Ni(III) species (**B$_1$**). An alternative pathway would involve the trapping of the radical by Ni(0) as a first step, followed by oxidative addition of the amide to the organonickel(I) intermediate. Finally, reductive elimination would release the ketone and produce a Ni(I) species (**C$_1$**). In such a process, external metal reductants or visible light photoredox catalysis will

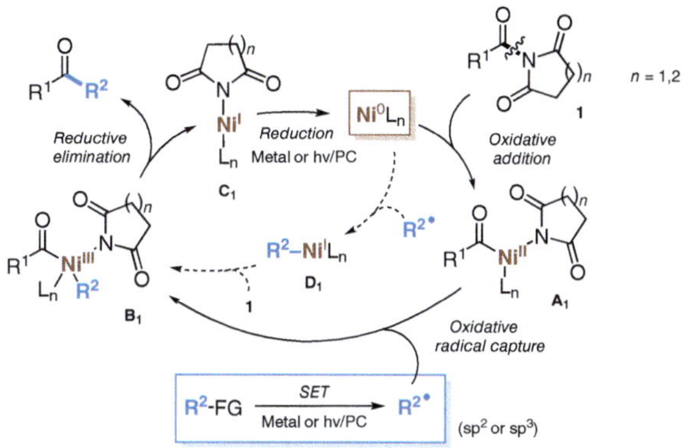

Figure 11.2 Commonly proposed nickel catalytic cycle for radical functionalization of amides via C–N bond activation.

generally be needed as a means of generating radicals from organic substrates via SET, as well as regenerating the Ni(0) active species to restart the process.

In this section, we will discuss the two main radical connective approaches to ketones employing amides as electrophilic acyl-transfer agents, i.e. the nickel-catalyzed reductive cross-coupling and the nickel/photoredox-catalyzed cross-coupling reactions.

11.2.1.1 Reductive Cross-Electrophile Cross-Coupling Reactions

Cross-coupling reactions involving the coupling of two electrophiles, the so-called "cross-electrophile couplings" have rapidly evolved as a straightforward strategy to forge carbon–carbon bonds. In contrast to traditional Pd/Ni-catalyzed cross-coupling reactions that typically involve the coupling of an organo(pseudo)halide electrophile with a nucleophilic partner, this methodology circumvents the need for highly reactive, preformed organometallic reagents. However, excess amounts of external metal powders such as Mn, Zn, or Mg are used as reducing agents for (re)generating the active catalyst [4]. Nickel is the catalyst of choice for reductive coupling of organohalides with acylating reagents, such as acyl chlorides and acid anhydrides [5].

In 2017, the group of Han reported the first example of an acylative cross-electrophile coupling involving an amide derivative [6]. The reductive cross-coupling of aromatic *N*-acyl-glutarimides (**1a**) with aryl iodides was performed using a catalytic system based on NiI_2/terpyridine and in the presence of excess zinc as metal reductant (Figure 11.3). The method was shown to deliver diaryl ketones in excellent yields. As illustrated with ketones **2a–b**, electron-rich aryl iodide electrophiles proved more reactive than electron-deficient ones. Control experiments suggest that the reactions proceed through aryl radical intermediates rather than via *in situ*-generated organozinc intermediates. On the basis of several control experiments, the authors tentatively proposed the intermediacy of a Ni(II) acyl species $\mathbf{A_2}$ ensuing from oxidative addition of *N*-acyl-glutarimides to Ni(0). The Ni(II) complex $\mathbf{A_2}$ is then proposed to capture an aryl radical, generated via reduction of iodoarene by Ni(I) species, to give the phenyl nickel(III) intermediate $\mathbf{B_2}$, which would then undergo reductive elimination to release the ketone and amido-nickel(I) specie $\mathbf{C_2}$. The latter would reduce the iodoarene to generate the aryl radical and

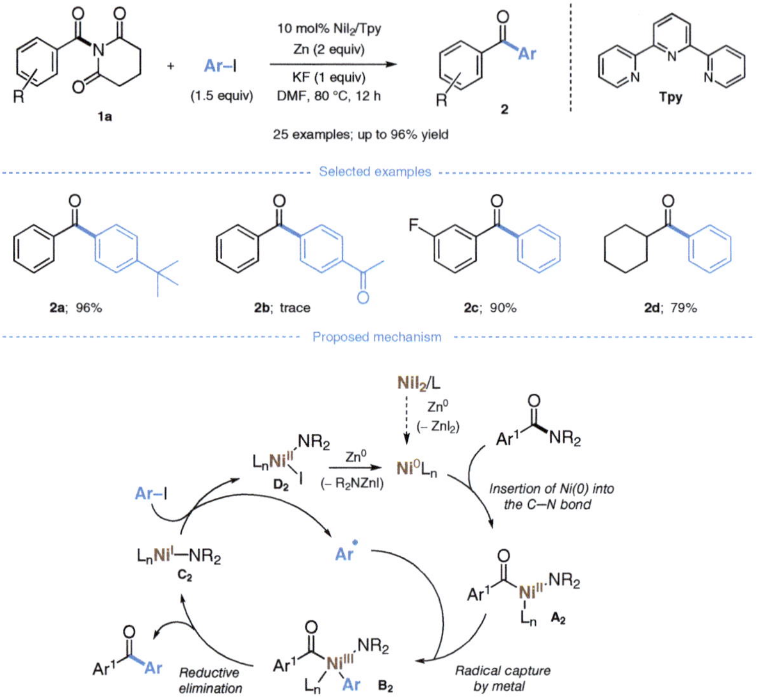

Figure 11.3 Reductive cross-electrophile coupling of *N*-acyl-glutarimides with aryl iodides.

the amido-Ni(II) complex $\mathbf{D_2}$. Zn metal would ensure regeneration of the active zerovalent nickel catalyst (Figure 11.3). An alternative pathway involving initial oxidative addition of iodoarene to nickel followed by nucleophilic acyl substitution could not be excluded.

In 2020, Matsuo and coworker described a nickel-catalyzed deaminative acylation of primary or secondary aliphatic amines with aromatic *N*-acyl-succinimides (**1b**) enabling the preparation of alkyl ketones via C(acyl)–C(sp^3) bond formation [7]. The amines are pre-activated in the form of redox-active 2,4,6-triphenylpyridinium salts (**3**, Katritzky salts) by simple condensation with the corresponding pyrylium salt. While Mn(0) was shown to be essential for the reaction, it was also noticed that additional amount of $MgCl_2$ had a beneficial effect on the reaction. The authors proposed a mechanism similar to that proposed by Han. Here, the alkyl radical species issued from precursor **3** may be either generated via reduction by Mn(0) (initiation step) or catalytically by SET from a Ni(I) species (Figure 11.4).

A nickel-catalyzed cross-coupling of *N*-acyl-imidazoles (**5**) with alkyl or aryl halides was also developed in the same year by the group of Li [8]. The method is applicable for a wide range of primary, secondary, or sterically hindered tertiary acyl-imidazoles, and a large scope of alkyl and aryl bromides were investigated. Notably, nonsymmetrical dialkyl ketones (**6**) are produced in excellent yields under mild conditions. Again large amounts of Zn powder are needed as the reductant. Interestingly, it was observed that symmetrical ketones (**8**) derived from

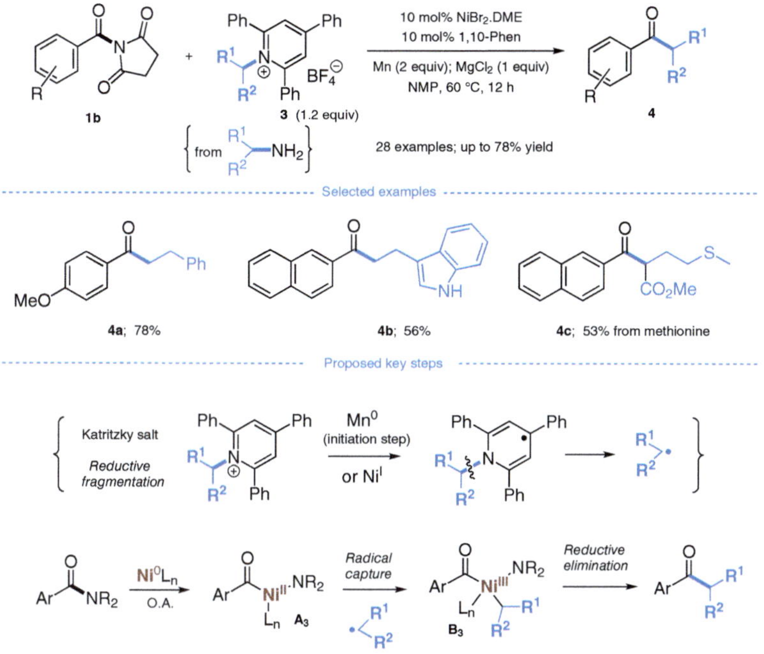

Figure 11.4 Reductive deaminative cross-electrophile coupling of *N*-acyl-succinimides with Katritzky salts.

the homocoupling of the organobromide partners were common by-products of the cross-coupling reactions, suggesting a decarbonylative pathway of acyl-imidazoles. The authors proposed two distinctive mechanisms depending on the steric parameters of the acyl-imidazoles (Figure 11.5). Sterically hindered acyl-imidazoles were proposed to follow pathway A. As shown by stoichiometric experiments, Ni(0) complexes preferably react with the organobromide instead of the sterically hindered acyl-imidazole to form Ni(II) species $\mathbf{A_4}$. The latter would be reduced by Zn(0) to form Ni(I) complex $\mathbf{B_4}$, which would then undergo direct oxidative addition to generate acyl-Ni(III) complex $\mathbf{C_4}$. Reductive elimination would then deliver the corresponding ketones (**6**/**7**). Less hindered primary and secondary acyl-imidazoles were proposed to follow the alternative pathway B. The C–N bond would be activated by single electron reduction by a Ni(I) complex (that may be derived from reduction of Ni(II) precatalyst) to give a radical anion that would undergo fragmentation to acyl radical **9**. This species would afford an alkyl radical through extrusion of carbon monoxide that can be trapped by Ni(0). Reaction of the Ni(0) carbonyl species with the organobromide, followed by CO insertion, and radical trapping would give acyl-Ni(III) complex $\mathbf{H_4}$. Reductive elimination would afford ketones **6**/**7**. The symmetrical ketone by-products **8** were proposed to be issued from reaction of the acyl-Ni(II) complex $\mathbf{G_4}$ with Ni(II)-carbonyl complex $\mathbf{F_4}$. Since these reactions have been shown to benefit from $ZnBr_2$ or $MgCl_2$ salts as additives, it was suggested that these Lewis acids might interact with the carbonyl group, thus strengthening the oxidative power of acyl-imidazoles. It is noteworthy

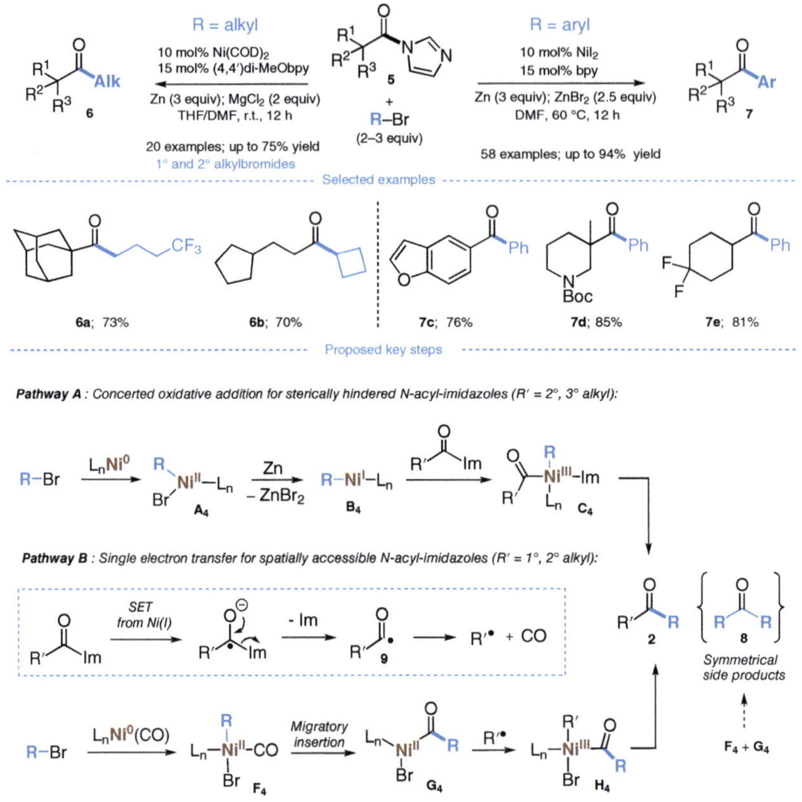

Figure 11.5 Reductive cross-electrophile coupling of *N*-acyl-imidazoles with organobromides.

that tertiary acyl-imidazoles might also be able to be reduced by Ni(I) complex to generate tertiary radicals, but the authors suggest that the steric hinderance of such radicals prevents from further reaction with Ni complexes (Figure 11.5).

11.2.1.2 Photoredox Cross-Coupling Reactions

In parallel to the nickel-catalyzed reductive cross-coupling methodologies, nickel/photoredox dual catalysis has rapidly advanced as a more sustainable strategy for the alkylation of amide derivatives. In this approach, photoredox cocatalysts promote photo-induced SET processes enabling the generation of a variety of alkyl radical species under mild conditions. The photocatalyst also participates in the reduction of nickel species, thereby obviating the mandatory need for excess amounts of metal reductants [9, 10]. Iridium-based complexes and nickel bipyridine complexes are the catalysts of choice in these methodologies. In 2017, the group of Molander developed a first alkylation process of *N*-acyl-succinimides using alkyltrifluoroborate nucleophiles (**10**) as radical precursors via single-electron transmetalation with the aid of visible light-mediated photoredox catalysis [11]. The method employs readily available, bench-stable cooperative catalysts: Ir[dF(CF_3)ppy]$_2$(bpy)PF_6 (**12**) [bpy = 2,2′-bipyridine] as the photocatalyst (under blue light irradiation) and [Ni(d*t*bbpy)-(H_2O)$_4$]Cl_2 (**13**) [d*t*bbpy = 4,4′-di-*tert*-butyl-2,2′-bipyridine] as the source of active nickel catalyst. Primary and secondary alkyl fragments could be incorporated. Interestingly, amino groups and nitrogen heterocycles were tolerated, which could be problematic with more electrophilic acylating agents (e.g. acyl chlorides). Notably, ketones bearing α-amino (or α-alkoxy) primary alkyl groups (e.g. **11a** and **11c**) have been made readily accessible by this method. The mechanistic proposal underlines the cooperative action of the Ir photoredox and Ni catalyst. SET oxidation of the alkyltrifluoroborate by the excited state of Ir photoredox catalyst yields a C-centered radical that is captured by the oxidative addition acyl-Ni(II) complex $\mathbf{A_5}$ to form Ni(III) intermediate $\mathbf{B_5}$. The cross-coupled product and Ni(I) complex $\mathbf{C_5}$ are then generated via irreversible reductive elimination. Finally, single-electron reduction of Ni(I) by the iridium complex simultaneously regenerates the ground state photocatalyst and the Ni(0) catalyst. As an alternative mechanistic pathway, the oxidative addition and radical trapping steps can be inverted, and the reaction proceed via alkyl-Ni(I) intermediate $\mathbf{D_5}$ (Figure 11.6).

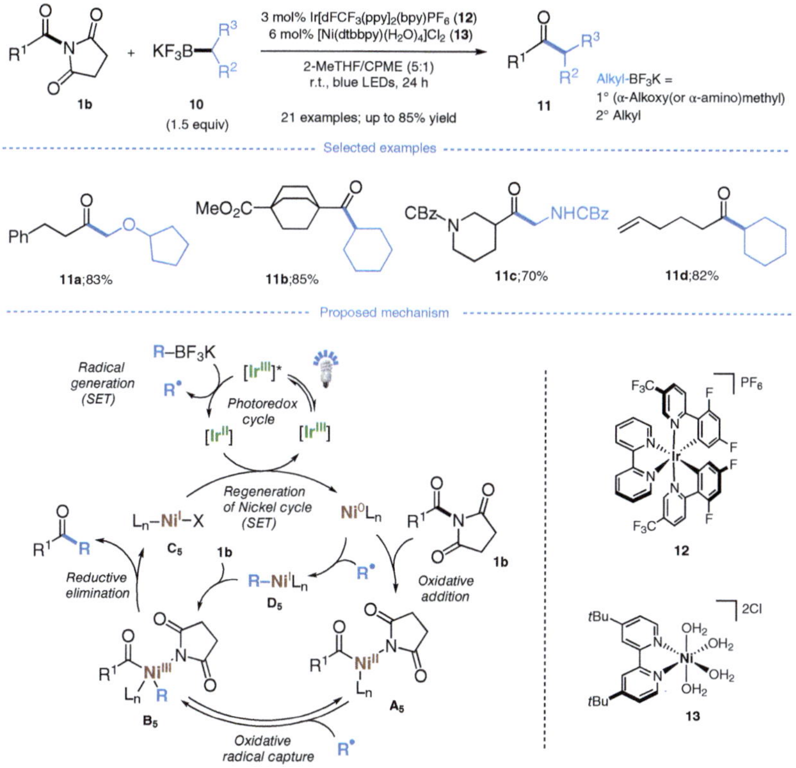

Figure 11.6 Nickel/photoredox-catalyzed cross-coupling of *N*-acyl-succinimides with potassium trifluoroborates.

Amgoune and coworkers subsequently developed a silyl radical-mediated cross-electrophile coupling process enabling the intermolecular alkylation of amides with nonactivated alkyl halides [12]. The key point of the mechanistic design relied on the ability of silyl radical intermediates (**15**), which can be photocatalytically generated, to activate an alkyl bromide via halogen–atom abstraction [13]. The reactions were conducted with a dual catalytic system based on [Ni(d*t*bbpy)(H_2O)$_4$]Cl_2 complex (**13**) and Ir[(dF(CF_3)ppy)$_2$(d*t*bbpy)]PF_6 complex (**16**) as the photocatalyst, in the presence of stoichiometric amounts of (Me_3Si)$_3$SiH. The scope of alkyl bromides that underwent the transformation encompassed a range of primary and secondary nonactivated derivatives. The method failed to provide ketones bearing primary α-alkoxy groups but allowed access to sensitive β-alkoxy-substituted ketones (e.g. **14e**). A mechanistic pathway similar to that proposed by Molander and coworkers was proposed to account for the formation of the alkyl ketones. Concerning the generation of the nucleophilic alkyl radical, it was suggested that (Me_3Si)$_3$Si radical intermediate – generated directly via oxidative scission of the (Me_3Si)$_3$Si–H bond by the strongly oxidizing photoexcited *Ir(III) catalyst in the presence of the base – would activate the alkyl bromide via bromine–atom abstraction with concomitant production of bromotris(trimethylsilyl)silane (Figure 11.7). Of note, *N*-silylated-succinimide could be identified as a by-product of the reaction, most likely deriving from nucleophilic substitution of the silyl bromide by succinimide anion released from the nickel catalytic cycle.

Later on, the same authors [14] investigated another, economically more attractive, strategy to access alkyl ketones that relies on the selective C(sp^3)–H bond activation of simple alkanes through a radical hydrogen atom transfer (HAT) process (Figure 11.8). Chlorine radical had previously been demonstrated as a potent

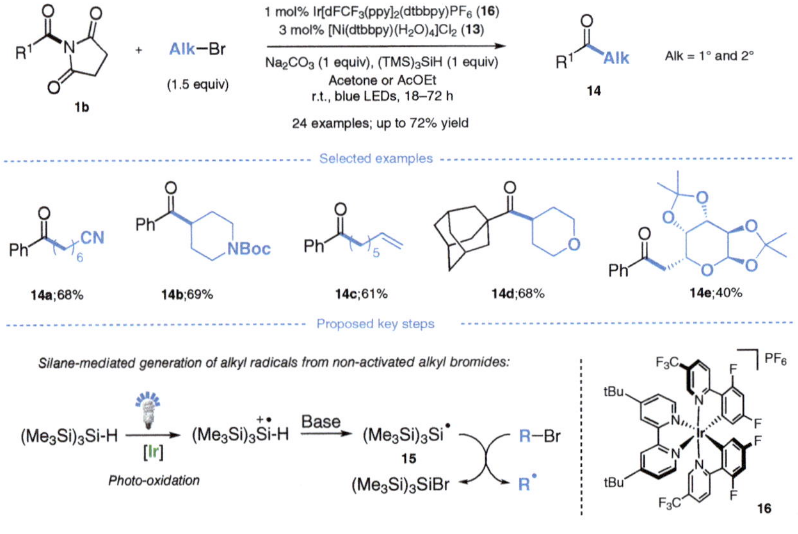

Figure 11.7 Nickel/photoredox-catalyzed cross-coupling of *N*-acyl-succinimides with alkyl bromides.

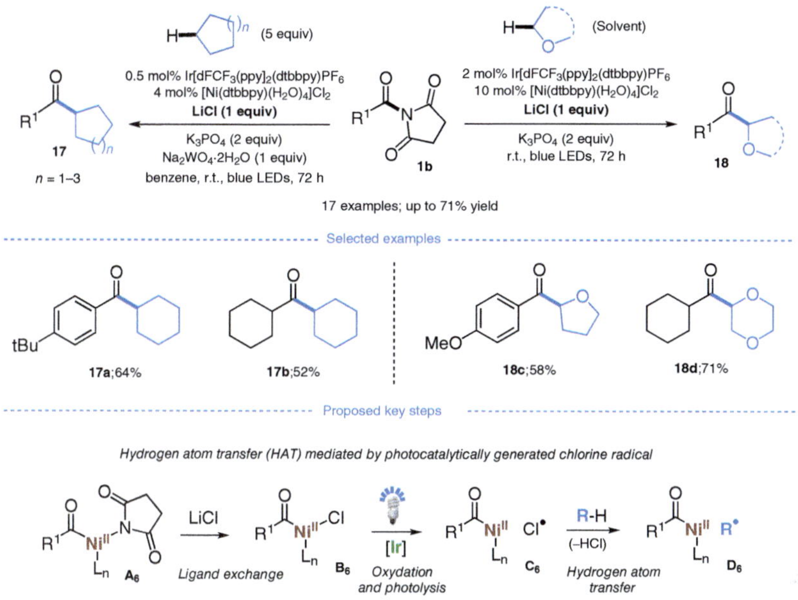

Figure 11.8 Nickel/photoredox-catalyzed cross-coupling of *N*-acyl-imides with simple alkanes or ethers.

H-abstracting agent in Ni/photoredox-catalyzed alkylations of organochloride electrophiles with aliphatic compounds. In these reactions, the chlorine radical species is presumably generated photocatalytically via Ni–Cl bond homolysis and mediates the generation of a substrate-derived alkyl radical [15]. Here, it was shown that in the presence of an exogenous source of chloride atom like lithium chloride, this strategy could also be applicable to the cross-coupling of amide electrophiles. It was proposed that oxidative addition of the amide to nickel (**A$_6$**) and subsequent succinimide to chloride ligand exchange occurs readily to generate an intermediate acyl nickel chloride (**B$_6$**) that could possibly serve as source of chlorine radicals upon photo-oxidation (**C$_6$**) and thereby lead to the generation of the alkyl radical species via HAT (**D$_6$**). The reactivity of different acyl succinimides and potential HAT substrates, including a selection of simple cyclic alkanes and ethers, was reported using the same dual catalyst system (**13**/**16**) as previously mentioned. Lithium chloride was found to be the most effective source among other salts tested (e.g. NaCl, KCl, nBu$_4$NCl, Et$_3$BnNCl, NH$_4$Cl). The opportunity of using low boiling point dialkyl ethers both as alkylating reagents and as solvents was considered as a safer alternative to benzene that proved to be the only efficient solvent for this type of transformation. Based on stoichiometric and catalytic evaluation of a series of isolated nickel intermediates with different oxidation states, the authors suggested that the C–N bond activation step might proceed at Ni(I) during catalysis.

Noteworthy, in the same time, a very similar protocol for the alkylation of *N*-acyl-succinimides was reported [16]. In this case, a polar mechanism involving C–H bond activation of the alkane prior to the oxidative addition step was proposed.

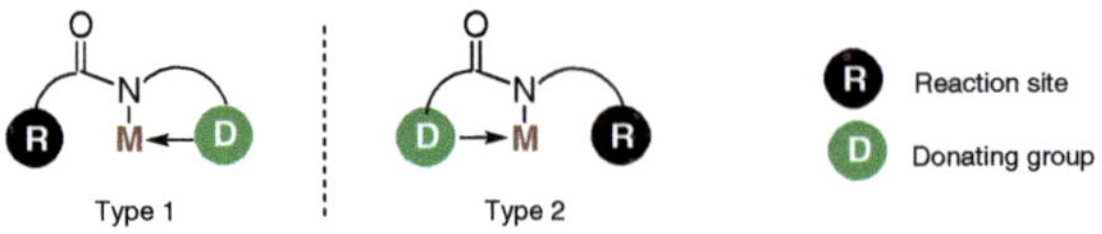

Figure 11.9 Bidentate chelation modes for assisted functionalization of amides.

11.2.2 Chelation-Assisted Radical Reactions of Amides

Amides are commonly used as versatile chelating auxiliaries in a variety of transition metal-catalyzed processes, most notably in the activation of C–H bonds [17, 18]. In these reactions, secondary amides featuring another nitrogen-donating group are typically employed, enabling dual chelation of the metal center (Figure 11.9). Two different modes of chelation assistance can be envisioned for amide functionalization depending on the location of the N-donating group, either directly attached to the amide nitrogen atom (type 1) or part of the acyl moiety backbone (type 2). Type 1 strategies are the ones explored most frequently probably due to the relative ease of installation and removal of the auxiliaries. The amide chelation assistance strategy has been successfully applied to design radical transformations. In this section, a selection of significant achievements in this field, including amide ortho C–H bond functionalization and remote functionalization of pendant alkenes, is presented.

11.2.2.1 Amide-Directed C–H Bond Functionalization

Amide-directed C(sp^2)–H and C(sp^3)–H bond functionalization is a broad field of research that has been thoroughly reviewed in recent years [19, 20]. However, compared to conventional cross-coupling methods, radical processes remain far less developed. The exploration of radical pathways has been focused on the use of 8-aminoquinoline (AQ) amides (**19**), a very popular class of directing groups (first introduced by Daugulis [21]), which are easily removable, allowing further downstream synthetic transformations. A commonly proposed radical capture mechanism is presented in Figure 11.10. The metallic center is first chelated by the aminoquinoline moiety in a *N,N*-bidentate fashion, followed by intramolecular C–H bond cleavage to form relatively stable five-membered metallacycle intermediate $\mathbf{A_7}$. Oxidative capture of a transiently generated radical species generates higher oxidation state complex $\mathbf{B_7}$, which undergoes reductive elimination and decoordination that releases the product (**20**). The reaction design requires further metal oxidation step to enable catalytic turnover.

Among the repertoire of cross-coupling methods that involve the cleavage of C–H bonds, the cross-dehydrogenative couplings of two different C–H bonds have emerged as an environmentally attractive strategy. In 2014, the group of Chatani, reported a nickel(II)-catalyzed C(sp^2)–H benzylation of aromatic amides **20** with toluene derivatives (**21**), leading to *ortho*-benzylated benzamides **22** [22] (Figure 11.11a). Notably, the 5-position in the quinoline ring was blocked by a chlorine atom to avoid undesired functionalization at this site. The most important feature of the method was the use of a fluoroalkyl iodide, *i*-C_3F_7I, to activate the benzylic C(sp^3)–H bond, which supposedly occurs through formation of the corresponding perfluoroalkyl radical via base-promoted SET, followed by hydrogen abstraction from toluene to generate a benzyl radical. The perfluoroalkyl iodide

Figure 11.10 Typical reaction pathway for catalytic amide-directed metalation/radical functionalization of C–H bonds.

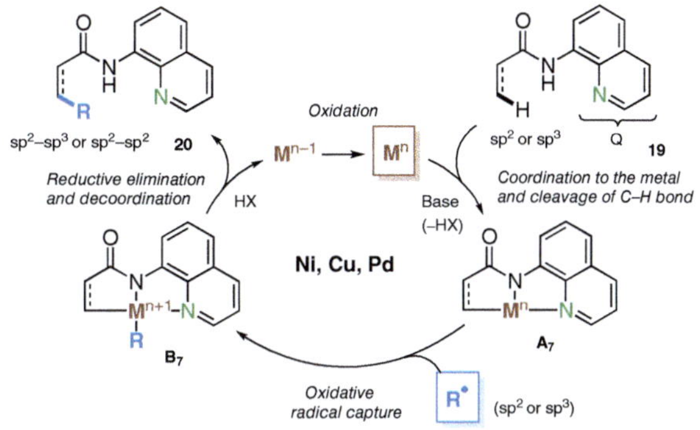

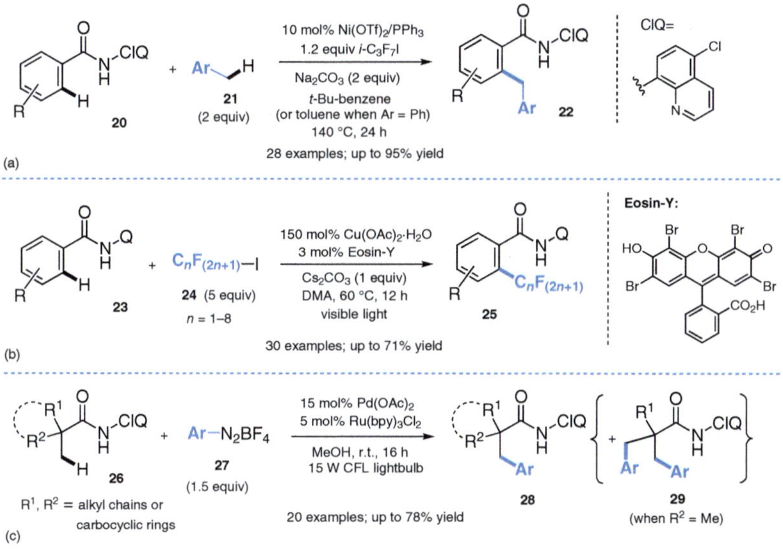

Figure 11.11 Illustrative examples of radical-type amide-directed C–H bond functionalizations. (a) Nickel-catalyzed C(sp^2)-H benzylation. (b) Photocatalytic Cu-mediated C(sp^2)-H perfluoroalkylation. (c) Photoredox/Pd-catalyzed C(sp^3)-H arylation.

would also be involved as an oxidant in the regeneration of the active Ni(II) catalyst through SET, obviating the need for peroxides as strong oxidants. Subsequently, the same group developed another nickel-catalyzed process allowing direct *ortho*-methylation of benzamides using dicumyl peroxide as a source of methyl radical [23] (not shown).

Perfluoroalkyl radicals can also be generated from perfluorohalides (**24**) when exposed to visible light in the presence of a photocatalyst. Interestingly, it was demonstrated that directed *ortho*-perfluoroalkylation of benzamides **23** was possible under such conditions using eosin-Y as a cost-effective photoredox catalyst. However, more than stoichiometric amounts of a copper salt are needed to promote the reaction [24] (Figure 11.11b).

The amide-directed radical functionalization of C(sp^3)–H bonds has also been investigated. In 2017, the group of Polyzos described the dual photoredox/palladium(II)-catalyzed β-arylation of ClQ-derived quaternary alkyl amides **26** (system that lacks β-hydrogen atoms) (Figure 11.11c) [25]. In this redox-neutral process, the aryl radical is produced by SET from diazonium tetrafluoroborates (**27**) to the excited state of the photocatalyst Ru(bpy)$_3$Cl$_2$. The reaction proved highly selective for β-methyl C(sp^3)–H bonds, with no coupling being observed for β-methylene C(sp^3)–H bonds. However, substantial amounts of diarylated compounds **29** are formed as side products where two methyl groups are made available.

11.2.2.2 Amide-Directed Functionalization of Unactivated Alkenes

Chelating assistance strategies using the 8-aminoquinoline group as *N,N*-bidentate auxiliary have also proved very effective for challenging transition metal-catalyzed mono- or difunctionalization of nonconjugated alkenes, although this area is still less explored [26]. The chelation of the 8-aminoquinoline group facilitates coordination of the alkene π-bond to the metal (**A$_8$**), which increases the chemical reactivity of the electronically unbiased olefin and helps controlling the regio- and stereoselectivity of the functionalization process. Olefin activation can facilitate alkyl radical (Wacker-type) additions to internal alkenes, which are known to have poor radical trapping properties compared with terminal olefins. β,γ-Unsaturated amides are the privileged substrates in such catalytic-directed alkene functionalizations, owing to the much higher stability of five-membered metallacycles (**B$_8$** and **C$_8$**, $n = 1$) compared with six-membered or larger rings. Various catalytic radical pathways such as 1,2-dialkylation (**30**), Heck-type alkylation (**31**), carboamination (**32**), and hydroalkylation (**33**) have been designed to give access to a diverse range of alkylated products (Figure 11.12). Radical additions to conjugated amides involving Lewis acid activation are not covered here.

In 2018, the group of Engle described a nickel-catalyzed 1,2-dialkylation of β,γ- or γ,δ-unsaturated amides **34** via three-component conjunctive cross-coupling of

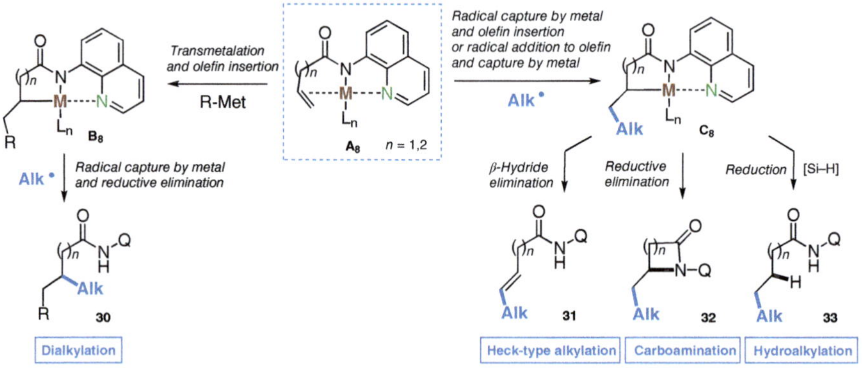

Figure 11.12 Explored strategies toward catalytic amide-directed radical functionalization of nonconjugated alkenes.

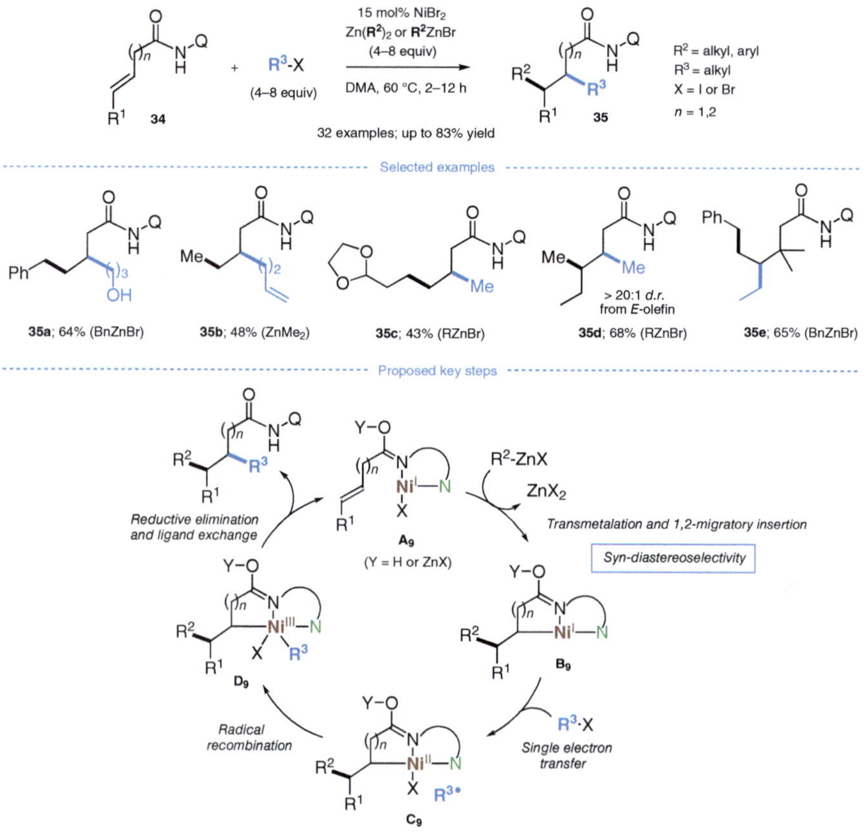

Figure 11.13 Nickel-catalyzed amide-directed 1,2-dialkylation of nonconjugated alkenes.

an organozinc nucleophile and an organohalide electrophile [27]. Importantly, this method delivers two different alkyl fragments across the nonconjugated alkene allowing the forging of two vicinal C(sp^3) stereocenters (Figure 11.13). The reaction is proposed to follow a Ni(I)/Ni(III) redox catalytic cycle. The strong chelation of the 8-aminoquinoline group was proposed to control the regiochemical outcome through five-(or six-) membered nickelacycle stabilization and to prevent competitive β-hydride elimination from alkylmetal intermediates. The first step involves transmetalation of a *N,N*-ligated Ni(I) catalyst $\mathbf{A_9}$ with the alkylzinc nucleophile, followed by *syn*-migratory insertion. SET from $\mathbf{B_9}$ to the alkyl halide followed by radical recombination generates a Ni(III) intermediate $\mathbf{D_9}$. Finally, reductive elimination and ligand exchange liberate the product and restart the catalytic cycle. Importantly, internal alkenes provided the *syn*-1,2-difunctionalized products with high diastereomeric ratios (>20 : 1; e.g. **35d**), which is consistent with the olefin insertion proposal.

In the same year, Fu and Bi reported the Heck-type radical alkylation of β,γ-unsaturated amides **36** under copper(I) catalysis (Figure 11.14). Functionalized alkyl halide electrophiles were used as a source of stabilized carbon radical

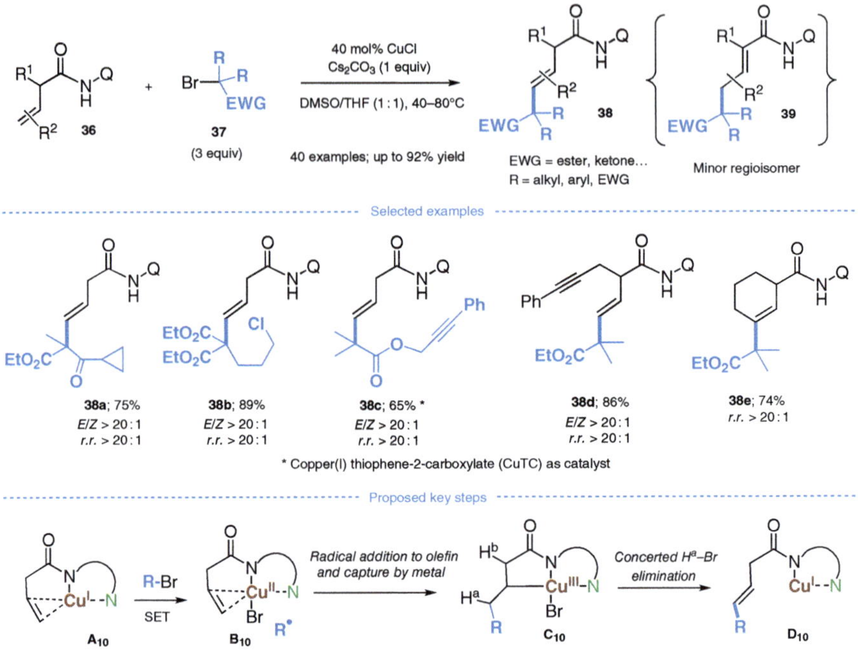

Figure 11.14 Copper-catalyzed amide-directed Heck-type alkylation of alkenes with organobromides.

intermediates that ensure efficient capture by the weakly reactive olefin partner [28]. The reaction tolerates 1°, 2°, and 3° alkyl halides **37** and affords the coupled olefins **38** with excellent regio- and *E*/*Z*-stereoselectivity. Tri- or tetra-substituted olefins showed no reactivities. The regioisomeric conjugated amides (**39**) have been identified as minor side products of the reaction that do not originate from alkene isomerization. The use of dimethyl sulfoxide as a cosolvent was shown to have a strong impact on yields and regioisomeric ratios. The authors proposed a mechanistic pathway involving electron transfer from directing group-coordinated copper(I) $\mathbf{A_{10}}$ to the alkyl halide to give a Cu(II) intermediate and an alkyl radical ($\mathbf{B_{10}}$). Then, addition of the alkyl radical to the olefin moiety and recombination of the resulting carbon radical with the nearby Cu(II) species were proposed to generate a five-membered cupracycle $\mathbf{C_{10}}$, which then undergoes β-hydride elimination. At this stage, the origin of regioselectivity can be understood in terms of discriminating between hydrogen atoms H_a and H_b. Interestingly, DFT calculations indicated a dimethyl sulfoxide-assisted concerted H_a–Br elimination event of a conformationally strained Cu(III) cyclic transition state. The lack of reactivity observed with γ,δ-unsaturated substrates may be explained by a less favorable six-membered metallacycle intermediate. Later on, Fu applied the same strategy to the cyanoalkylation of unactivated alkenes with cyclic ketone oximes [29] (not shown; see Figure 11.16 for another illustration of the use of oximes as cyanoalkyl radical precursors).

A different trend in the copper-catalyzed amide-directed radical addition to alkenes has been documented, showing that these processes could also be terminated by reductive elimination instead of β-hydride elimination pathway. This provided the opportunity to access β-lactam derivatives through efficient intramolecular radical carbo-amination of terminal alkenes. For instance, Wang, Chen, and He developed an enantioselective version of such reaction using 4-alkyl Hantzsch esters (alkyl-DHPs, **41**) as precursors of alkyl radicals via oxidative fragmentation, BINAPO-type diphosphine oxides (L1/L2) as chiral ligands, and AdO_2CO_2tBu (**43**) as an oxidant (Figure 11.15). The nucleophilic alkyl radical is generated by reaction of *tert*-butoxy radical with the alkyl-DHP. The reaction occurs under mild conditions and tolerates well internal olefins, which offers an efficient asymmetric access to β-substituted β-lactams bearing two contiguous stereocenters (**42**). Notably, (*Z*)-olefins performed more efficiently than (*E*)-olefins: diminished yield and enantioselectivity were observed with these substrates (compare **42a** vs. **42a′**), and *N*-alkylation occurred as a side reaction. Mechanistic studies indicated that a nucleophilic addition of the alkyl radical to Cu(II)-coordinated alkene ($\mathbf{A_{11}}$) was the enantio-determining step. The observed *anti*-diastereoselectivity dismisses the possible radical capture by the metal and subsequent *syn*-olefin insertion as an alternative reaction pathway [30]. It is of note that an access to β-lactams proceeding

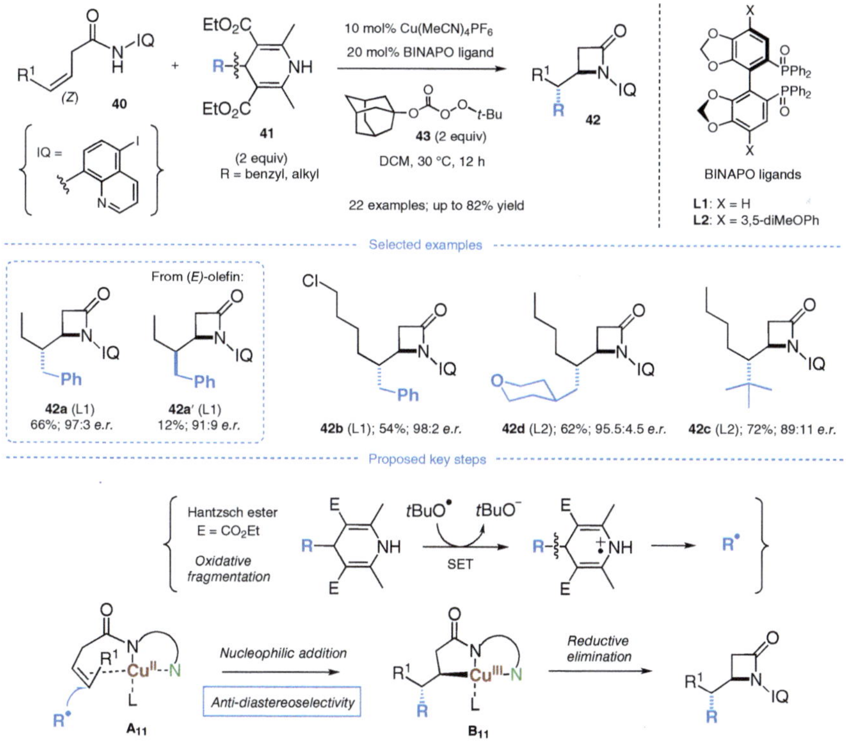

Figure 11.15 Copper-catalyzed amide-directed asymmetric carboamination of alkenes with alkyl Hantzsch esters.

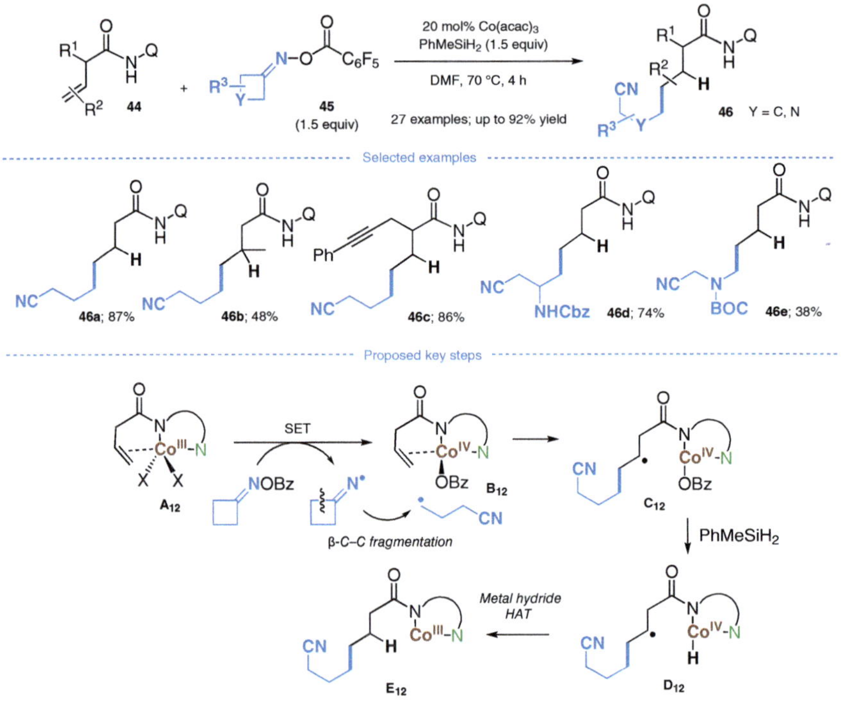

Figure 11.16 Cobalt-catalyzed amide-directed *anti*-Markovnikov hydroalkylation of alkenes.

via benzylation reactions of (*E*)-olefins involving C(benzyl)–H radical abstraction had previously been investigated [31].

The radical hydroalkylation of olefinic substrates is another interesting strategy to achieve C(sp^3)–C(sp^3) bond formation. In 2020, Niu and Song reported a cobalt(III)-catalyzed hydroalkylation of unactivated alkenes using cyclobutanone *O*-perfluorobenzoyl oximes (**45**) as effective precursors of non-stabilized cyanoalkyl radicals via β-carbon–carbon radical fragmentation. $PhMeSiH_2$ was used as the reductant [32] (Figure 11.16). The general pathway was proposed to proceed through SET reaction between the amide-chelated Co(III) intermediate $\mathbf{A_{12}}$ and the cyclic oxime to give the Co(IV)–OBz species ($\mathbf{B_{12}}$), followed by addition of the cyanoalkyl radical to the alkene moiety. The resulting intermediate $\mathbf{C_{12}}$ is converted to the Co–H species $\mathbf{D_{12}}$ in the presence of $PhMeSiH_2$, which is followed by intramolecular metal hydride HAT. Isotope-labeling experiments using $PhMeSiD_2$ indicated that $PhMeSiH_2$ is likely to be the source of hydrogen atom for the formation of cobalt hydride intermediates. Metal hydride HAT reactions with alkenes typically exhibit Markovnikov selectivity [33]. Here, the use of an amide directing group strategy allowed exclusive *anti*-Markovnikov selectivity in the radical hydroalkylation of nonconjugated alkenes.

11.3 Reactions Involving Amides as Precursors to Nitrogen- or Carbon-Centered Radicals

11.3.1 Reactions of Amides via Amidyl Radicals

Amidyl radicals (**49**) represent a very useful class of reactive intermediates that have found numerous applications, notably in the development of inter- and intramolecular functionalization methods for substituted amides. These *N*-centered radicals can be generated through SET processes directly from secondary amides via homolysis of highly stable N–H bonds (bond-dissociation free energy [BDFE] of N–H bonds: 107–110 kcal/mol) or from prefunctionalized N–X bonds [**48**, X = halide, OR, NR_2] featuring a traceless redox handle that often requires multistep synthesis and leads to substrates with poor stability. Remarkable recent progress toward the generation and use of amidyl radicals has also taken advantage of photoredox catalysis as a means of inducing SET processes under mild conditions (*vide infra*). Amidyl radicals may subsequently be integrated in cross-coupling reactions to provide a set of powerful methods for the creation of carbon–nitrogen and carbon–carbon bonds [34] (Figure 11.17). In this section, recent applications of amidyl radicals in intramolecular 1,2-difunctionalization reactions of alkenes, i.e. the carboamination–cyclization processes (**50**), and radical-directed remote alkylation reactions of nonactivated aliphatic C–H bonds (**51**) are described. The intermolecular N–H/C–H oxidative cross-coupling reactions that lead to the direct N–H functionalization of the amides (**52**) will not be discussed herein. These processes featuring transition metal chelation-assisted C–N radical–radical couplings of *N*-alkoxyamides are somewhat beyond the scope of this chapter, but readers

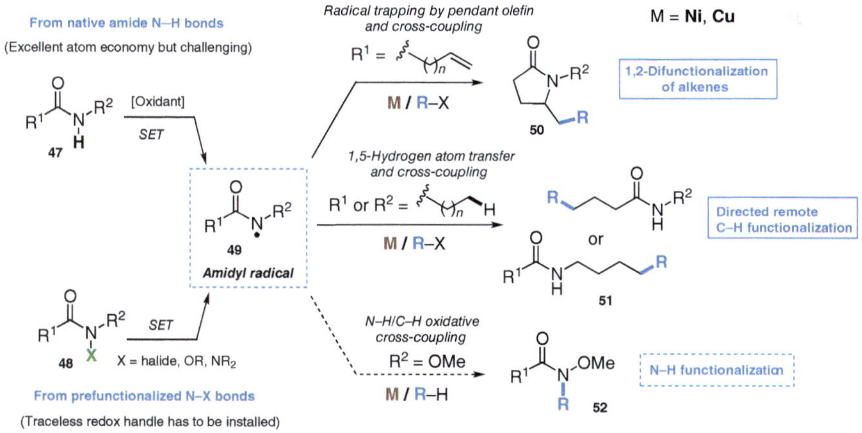

Figure 11.17 General pathways explored for amidyl radical-based catalytic functionalization of amides.

can refer to specific leading articles by Lan and Lei on this otherwise important emerging topic [35–37].

11.3.1.1 Vicinal Difunctionalization of Pendant Olefins

Radical difunctionalization of unsaturated carbon–carbon bonds via visible light photoredox catalysis has developed into a robust tool for the construction of two different vicinal chemical bonds [38]. Intramolecular carboamination of alkenes is one important goal in this field as it grants access to valuable nitrogen-containing heterocycles. Remarkably, it has been shown that iridium-based photocatalysts are capable of generating amidyl radicals directly from N–H bonds in the presence of phosphate base under mild conditions through concerted oxidative proton-coupled electron transfer (PCET). The phosphate base engages in hydrogen bonding with the amide and lowers the redox potential of the amide [39], thereby obviating the need for stoichiometric amounts of strong oxidants and high temperatures.

Within this context, the group of Molander recently developed new protocols merging visible light Ir-photoredox PCET and nickel catalysis to achieve cross-coupling of alkenyl amides **53** with organohalides, including (hetero)aryl bromides [40] and acyl chlorides [41], enabling selective access to functionalized pyrrolidinones **54** and **55**, respectively (Figure 11.18). The reactions are conducted in the presence of superstoichiometric amounts of the phosphate base $Bu_4N[OP(O)(OBu)_2]$. The use of *t*BuOH as cosolvent was found to have a beneficial effect on the coupling of amides with (hetero)aryl bromides, presumably due to acidity enhancement of the N–H bond. Remarkably, the reactions showed excellent

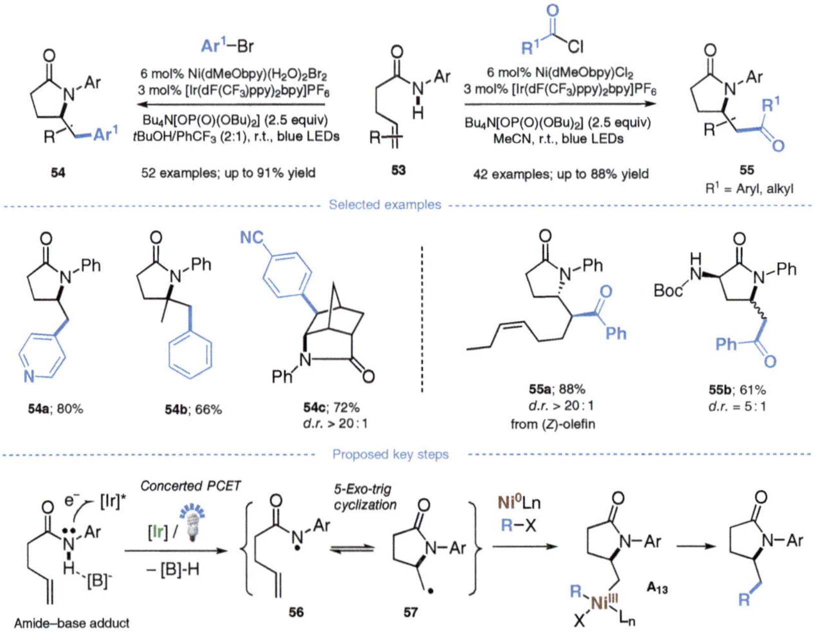

Figure 11.18 Photoredox PCET/Ni dual-catalyzed carboamidation of nonactivated olefins.

regioselectivities, affording exclusively 5-*exo*-trig cyclization products. In addition, high *trans*-diastereoselectivities were generally observed in reactions involving internal alkenes (e.g. **54c** and **55a**). Mechanistically, the process is proposed to proceed through concerted PCET process, followed by fast intramolecular amidyl radical addition across the pendant alkene to give a carbon-centered radical (**57**) that can subsequently be captured by a nickel species (diastereo-determining step). At this point, radical trapping by Ni(0) before oxidative addition of the organohalide or by Ni(II) after oxidative addition can be envisioned. Ultimately, the high-valent Ni(III) complex $\mathbf{A}_{13}$ would be formed, thus delivering the desired product.

11.3.1.2 Distant C–H Bond Functionalization

A classical approach to the functionalization of nonactivated C(sp^3)–H bonds involves the use of heteroatom-centered radicals, by harnessing their ability to undergo translocation processes through 1,5-hydrogen atom transfer (1,5-HAT) [20, 42]. Such reactivity is facilitated with amidyl radicals, owing to their inherent electrophilic character [43, 44]. In this area, remarkable examples of regioselective amidyl radical-directed carbon–carbon bond formation at distal C(sp^3) sites have been enabled by intercepting the transposed radical with catalytically generated organometallic species to afford δ-functionalized amides. For instance, in 2019, Rovis and coworkers reported the regioselective alkylation of aliphatic trifluoroacetamides (**58**) through Ni/photoredox-catalyzed cross-coupling with alkyl halides in the presence of a tribasic phosphate base [45] (Figure 11.19). The trifluoroacetamide group proved to be the only productive directing group, and primary alkyl bromides

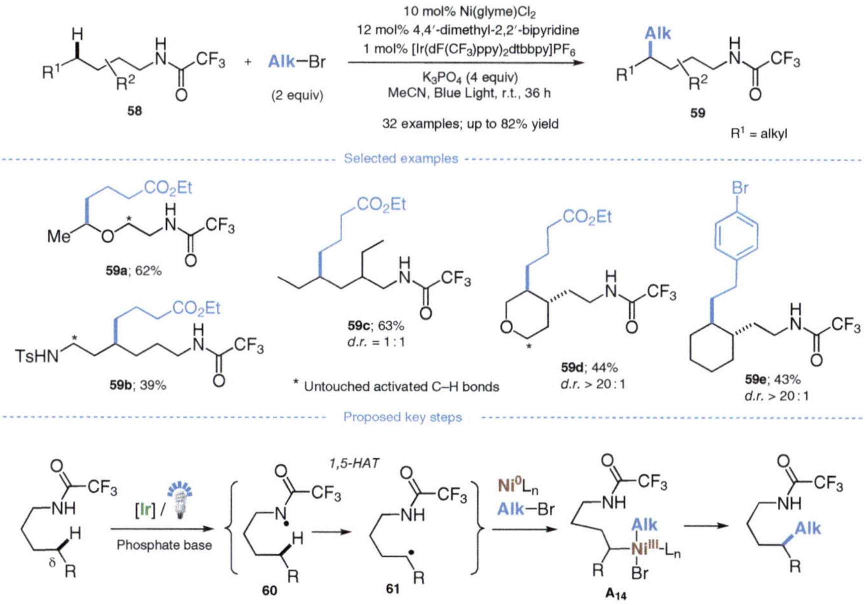

Figure 11.19 Amide-directed Ni/photoredox-catalyzed δ-alkylation of C(sp^3)–H bonds with alkyl bromides.

were found to be the most suitable coupling partners. High selectivity was observed toward monoalkylation at the δ-position of the amide, even in the presence of more activated functionalizable sites. Furthermore, alkylation was shown to proceed preferably at secondary δ $C(sp^3)$–H bonds, thus tolerating the presence of other primary carbon centers at δ-positions (e.g. **59c**). Remarkably, the low reactivity of sterically hindered tertiary reactive sites impedes overfunctionalization and thus favors mono-selective alkylation. High levels of *trans*-diastereoselectivity were also generally observed in the alkylation of cyclic systems (e.g. **59d–e**). From a mechanistic point of view, the trifluoroacetamide directing group was considered sufficiently acidic to be deprotonated by the phosphate base and oxidized by the excited state of the photocatalyst to generate the amidyl radical (**60**). 1,5-Abstraction of a hydrogen atom by this radical intermediate from a δ-methylene would afford the corresponding carbon-centered translocated radical (**61**) that can then be captured by nickel.

In the same year, Tambar and coworker reported the allylation of unactivated $C(sp^3)$–H bonds via a similar trifluoroacetamide-directed 1,5-HAT/photoredox strategy [46] (Figure 11.20). Here, allylation reactions proceeded best at δ-methine (vs. δ-methylene) centers, and allyl chlorides (**64**) proved to be more suitable coupling partners compared with the corresponding bromides and iodides that generate undesired *N*–H allylation coproducts. Importantly, a notable feature of this process is that the nickel catalyst improves the coupling process but is not essential in product formation. Besides, the bis(2-oxazoline) ligand (L3) proved more effective than the classical bipyridines. Deuterium-labeled experiments suggested that the presence of the nickel complex does not significantly impact the reaction pathway and dismissed the intermediacy of a π-allyl-Ni complex. It was proposed that, instead, the Ni(0) complex improves the yields by stabilizing the allylic chloride substrates (**A$_{15}$**) to photochemical double bond isomerization and to decomposition through allylic radical dimerization.

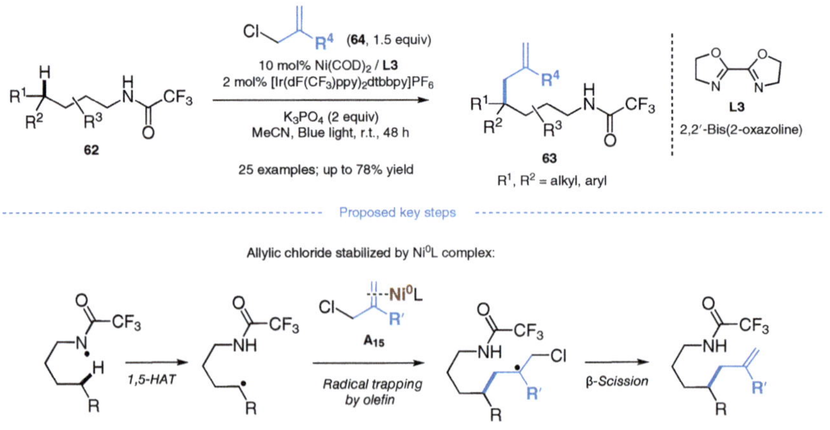

Figure 11.20 Amide-directed photoredox-catalyzed δ-allylation of $C(sp^3)$–H bonds with allyl chlorides.

Remote δ-functionalization of amides has been also achieved with copper catalysts enabling both the generation of the amidyl radical from a prefunctionalized amide substrate and the subsequent radical coupling process. A first example of such strategies has been reported by the group of Zhu in 2018, describing the cross-coupling of *N*-fluorocarboxamides (**65** or **66**) with (hetero)arylboronic acids under mild conditions [47] (Figure 11.21). Reactions were performed at room temperature in trifluorotoluene in the presence of copper(II) trifluoroacetylacetonate/2,2′-bipyridine as catalyst system and sodium *tert*-butoxide as base. Interestingly, two sets of different substrates proved competent coupling partners; the *N*-alkyl-*N*-fluorobenzamides **65** allowed arylation of the *N*-substituent of the amide, while *N*-(*tert*-butyl)-*N*-fluoro-2-methylbenzamides **66** directed the functionalization process on the acyl fragment to yield 1,1-diarylmethanes (**68**). The mechanism of the reaction was proposed to proceed via the generation of amidyl radical by Cu(I)-mediated reductive cleavage of the N–F bond, followed by 1,5-HAT, and radical capture of the subsequent carbon radical by copper intermediate ArCu(II)X ensuing from transmetalation with arylboronic acid. Reductive elimination from the resulting aryl-Cu(III) species $\mathbf{A_{16}}$ would deliver the arylated product and recycle the Cu(I) active catalyst. Subsequently to this work, the ability

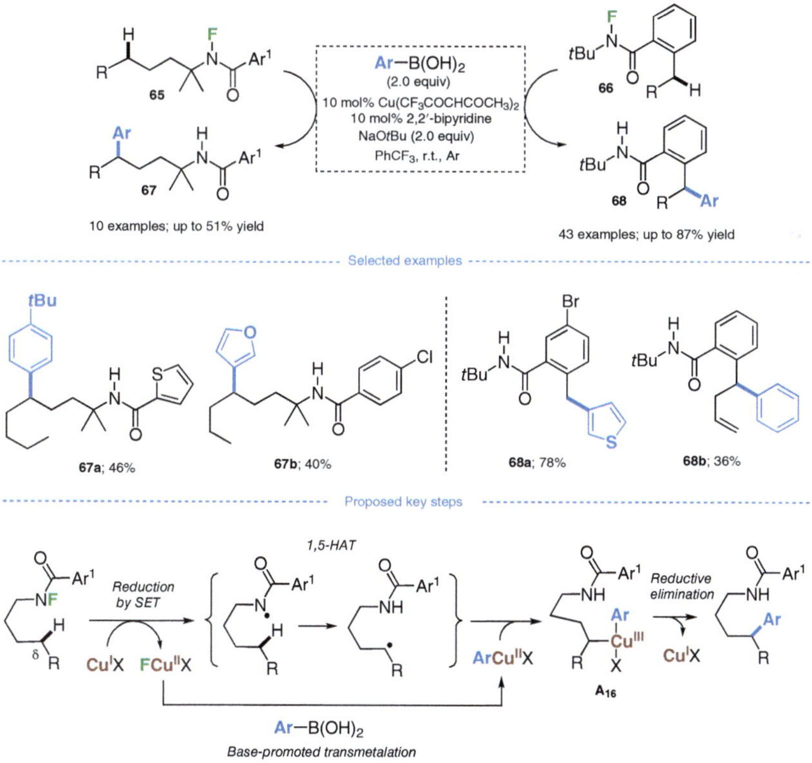

Figure 11.21 Amide-directed copper-catalyzed C(sp^3)–H δ-arylation of *N*-fluorocarboxamides with boronic acids.

of other *N*-functionalized amidyl radical precursors, including benzohydrazides (via N–N bond cleavage) [48] and *O*-acyl hydroxamides (via N–O bond cleavage) [49], to undergo copper-catalyzed remote C–H functionalizations has also been documented.

11.3.2 Reactions of Amides via α-Aminoalkyl Radicals

The α-amino C–H bonds of amides can undergo H-atom abstraction to generate α-aminoalkyl radicals (**69**), which normally requires the combination of a peroxide and metal catalyst at elevated temperatures but can also be mediated by a photocatalyst [50]. Subsequent radical capture processes that have been explored to functionalize the amide within the context of transition metal catalysis include intermolecular additions to olefins and cross-coupling reactions with organohalides (Figure 11.22).

11.3.2.1 C–H Bond Functionalization via Radical Addition to Alkenes

The functionalization of amides via decarboxylative cross-coupling reactions with α,β-unsaturated carboxylic acids (**71**) has been investigated, yielding *N*-allyl amides (**72**) and *N*-(3-oxoalkyl) amides (**73**) (Figure 11.23). Simple acyclic and cyclic amides that can be used as solvents (e.g. *N,N*-dimethylacetamide, *N*-methyl-2-pyrrolidone) have been employed as the source for α-aminoalkyl radicals, and reactions were promoted by the presence of peroxides at elevated temperatures.

A first method, developed by Mao and coworkers in 2014, allows direct alkenylation of the amides under copper(II) oxide catalysis and in the presence of di-*tert*-butyl peroxide as radical promoter [51] (Figure 11.23, left). The process is proposed to generate a copper(II) carboxylate intermediate ($\mathbf{A}_{17}$), which undergoes addition of the acetamido methyl radical across the double bond to form radical species $\mathbf{B}_{17}$. Decarboxylative elimination would then produce the desired product (**72**) and recycle the copper catalyst. A second cross-coupling procedure, reported concomitantly by the group of Wang, is conceptually very similar but affords *N*-(3-oxoalkyl) amides (**73**) as the coupling products [52] (Figure 11.23, right). The method employs Ni(II) acetate as the most efficient catalyst among other active metal acetates, and *tert*-butyl hydroperoxide was the most effective radical promoter. Again, the reaction is believed to proceed via the intermediacy of a radical metal alkylcarboxylate ($\mathbf{C}_{17}$) that would here ultimately combine with a hydroxyl radical (generated from homolysis of the hydroperoxide) to afford the corresponding hydroxyl-alkylated product **74**. A final oxidation step would afford the desired carbonyl compound. It is of note that Mao and coworkers also described in their aforementioned paper [51] an alternative access to *N*-(3-oxoalkyl) amides

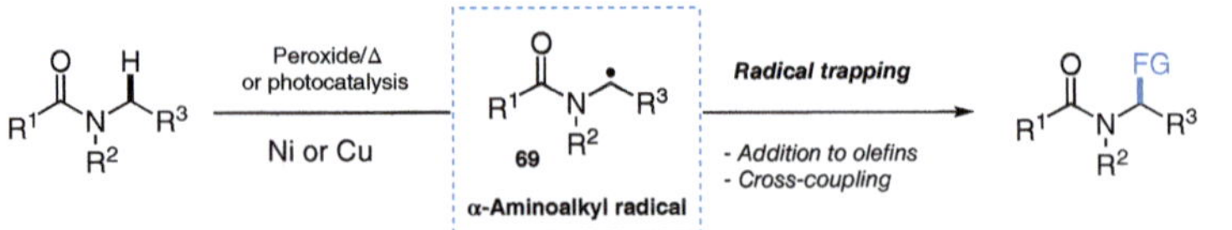

Figure 11.22 Strategies developed for α-C(sp^3)–H radical functionalization of amides.

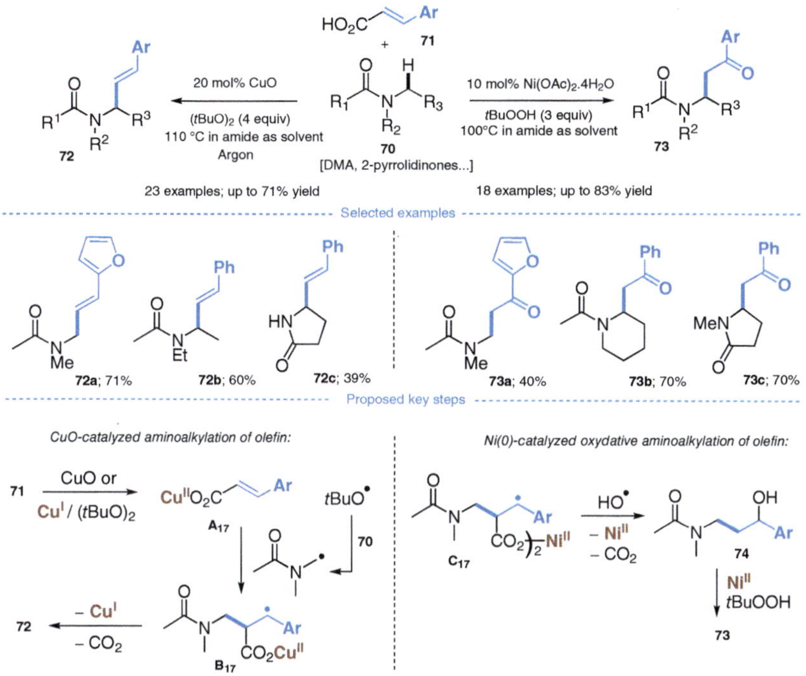

Figure 11.23 α-C(sp^3)–H functionalization of amides with α,β-unsaturated carboxylic acids.

using simple vinylarenes as radical acceptors and copper(II) fluoride as a catalyst. In addition, a synthesis of homologous *N*-(4-oxoalkyl) amides through a Ni-mediated process that involves allylic alcohols as radical acceptors and is terminated by a neophyl-type rearrangement was documented in 2015 [53] (not shown).

11.3.2.2 C–H Bond Functionalization via Cross-coupling

Exploration of radical α-C–H functionalization of amides using organohalides as coupling partners with the aid of a metal catalyst began nearly at the same time, with notable examples of nickel-mediated intramolecular arylations reported in 2013 [54]. However, significant advances to intermolecular reactions only appeared in the early 2020s, benefiting from the advent of visible light metallaphotoredox catalysis. For instance, König and coworkers described the (hetero)arylation of tertiary amides catalyzed by a combination of a heterogeneous organic semiconductor, mesoporous graphitic carbon nitride (mpg-CN, **76**), as a cost-effective recyclable photocatalyst and a nickel catalyst [55]. Simple amides were essentially used as both C–H precursor and solvent (Figure 11.24). Salient features of this method include a high robustness, a very good regio- and chemoselectivity, a compatibility with a large set of diversely substituted (hetero)aryl bromides and chlorides, and facile separation and recycling (up to five cycles) of the heterogeneous photocatalyst.

On the basis of mechanistic investigations and kinetic studies, the authors proposed a mechanism in favor of an energy transfer process (EnT), rather than SET, from the organic semiconductor to the Ni(II) oxidative addition complex $\mathbf{A_{18}}$ that

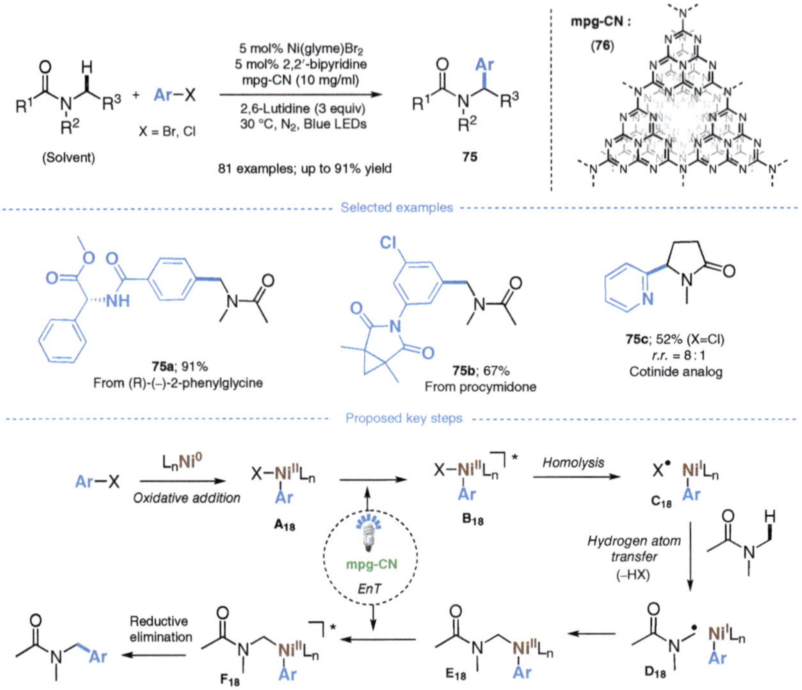

Figure 11.24 α-C(sp^3)–H functionalization of tertiary amides with aryl halides.

would generate electronically excited Ni(II) species $\mathbf{B_{18}}$. Ni(II)–X bond homolysis would then produce a halogen radical that can rapidly abstract a hydrogen atom from the amide to form the α-aminoalkyl radical species $\mathbf{D_{19}}$, which would rebound with Ni(I) to give the recombined Ni(II) complex $\mathbf{E_{18}}$. Reductive elimination from the electronically excited species $\mathbf{F_{18}}$, promoted by energy transfer with the organic semiconductor, would furnish the desired product and regenerate the active Ni(0) catalyst.

Protocols allowing chemo- and regioselective α-(hetero)arylations and α-alkylations of secondary amides have also been reported by Montgomery and Martin using a dual Ni/photoredox catalysis with iridium-based photocatalysts [56] (Figure 11.25). A notable feature of the method is the involvement of secondary amides having long *N*-alkyl chains susceptible to undergoing concurrent δ-substitution via amidyl radical formation. Remarkably, the reactions proved selective for activation of C(sp^3)–H bonds adjacent to the nitrogen atom of the amides, and only monosubstitution was observed. Although no certainty was gained regarding the mode of generation of key alkyl radical intermediate **79**, possible pathways include direct intermolecular hydrogen atom abstraction by a bromine radical (possibly issued from photochemical homolysis of Ni–Br bond of $\mathbf{B_{19}}$) or amidyl radical formation (**80**) (via base-assisted PCET) and subsequent 1,2-HAT. Noteworthy, amides bearing smaller *N*-alkyl chains were also amenable to α-arylations (e.g. **78d**), ruling out a mechanism involving a 1,5-HAT process

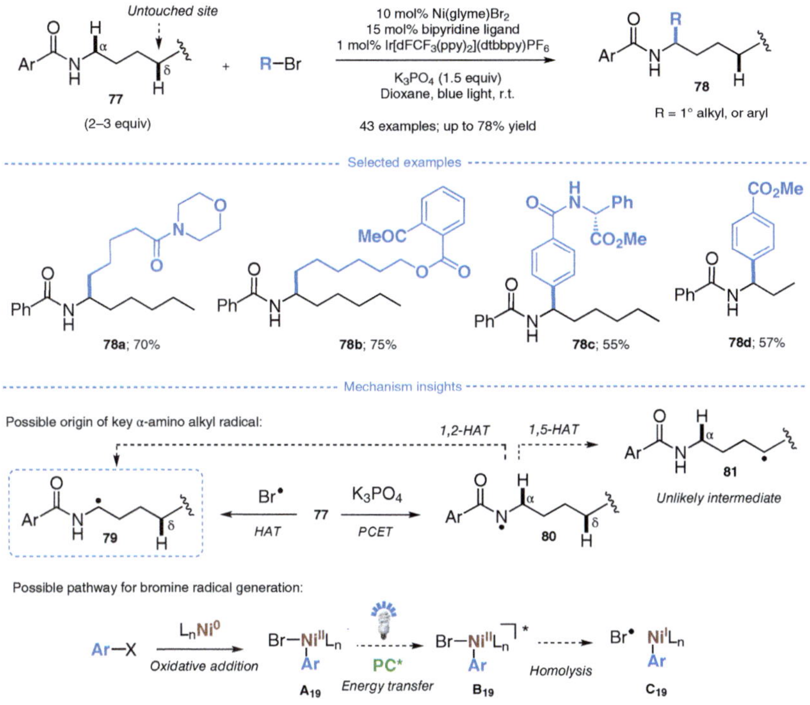

Figure 11.25 α-C(sp^3)–H functionalization of secondary amides with organohalides.

(via **81**), followed by migration of nickel along the alkyl chain. It is interesting to note also that an enantioselective version of the α-arylation reactions was developed using a chiral 2,2′-bis(2-oxazoline) ligand.

11.3.3 Reactions of Amides via Carbamoyl Radicals

Carbamoyl radicals (**85**) represent another useful class of reactive intermediates that have found interesting applications, most naturally in the synthesis of amides through direct carbamoylation of unsaturated systems, including (hetero)arenes and alkenes. In this area, visible light-mediated photoredox catalysis has provided convenient synthetic routes to access these valuable carbon-centered radicals from readily available precursors [57, 58] and combine them with cross-coupling reactions. For instance, Shang and Fu reported in 2015 a Pd/photoredox decarboxylative cross-coupling of oxalate monoamide salts (**82**) with (hetero)aryl halides [59] (Figure 11.26). In the process, a carboxyl radical (**84**) would be first generated through oxidation of the carboxylate anion by the excited iridium complex (*Ir^{III}), and rapidly extrude carbon dioxide to deliver the carbamoyl radical. The latter would be trapped by the Pd(II) oxidative addition complex $\mathbf{A_{20}}$ to form Pd(III) complex $\mathbf{B_{20}}$. From this point, DFT calculations suggested that the relay between the two catalytic cycles was likely to take place on this Pd(III) intermediate and that reductive elimination would thus occur from reduced Pd(II) species $\mathbf{C_{20}}$. Of note,

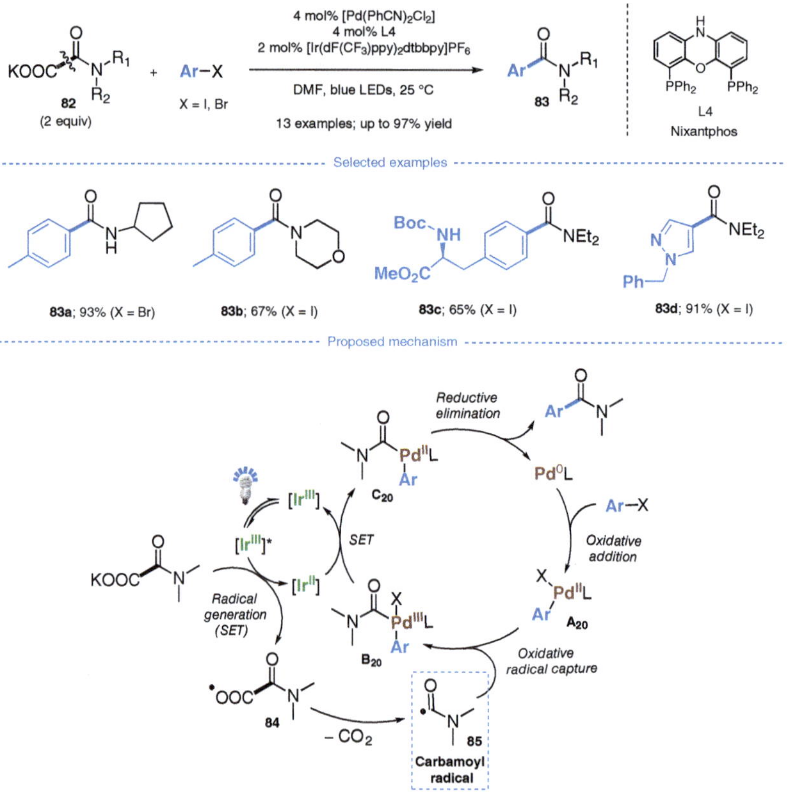

Figure 11.26 Carbamoylation of arenes with oxamic acids.

oxalate monoamides had been previously employed to achieve the Pd-catalyzed carbamoylation of potassium aryltrifluoroborates in the presence of potassium persulfate as an oxidant [60].

An alternative method for the direct carbamoylation of (hetero)aryl bromides was reported in 2020 by the group of Melchiorre [61]. The method is also based on the combination of transition metal and photoredox catalysis, here nickel and the organic photocatalyst 2,4,5,6-tetra(9*H*-carbazol-9-yl)isophthalonitrile (4CzIPN, **88**), but uses 4-carbamoyl-1,4-dihydropyridines **86** as precursors of carbamoyl radicals. These reagents are readily available through direct condensation between the corresponding 4-carboxyl-1,4-dihydropyridine and the selected amines and allowed the preparation of secondary and tertiary amides as well as the installation of amide scaffolds within biologically relevant heterocycles. As a possible mechanism, it was proposed that the 4-carbamoyl-1,4-dihydropyridine can easily generate carbamoyl radicals upon SET oxidation by the photoexcited organic photocatalyst. The nucleophilic carbamoyl radical would be intercepted by Ni(II) oxidation complex $\mathbf{A_{21}}$ to form the Ni(III) intermediate $\mathbf{B_{21}}$, from which the desired amide would be delivered. Recycling of both catalysts would proceed through SET reduction of Ni(I) species $\mathbf{C_{21}}$ by the reduced form of the photoredox catalyst (Figure 11.27).

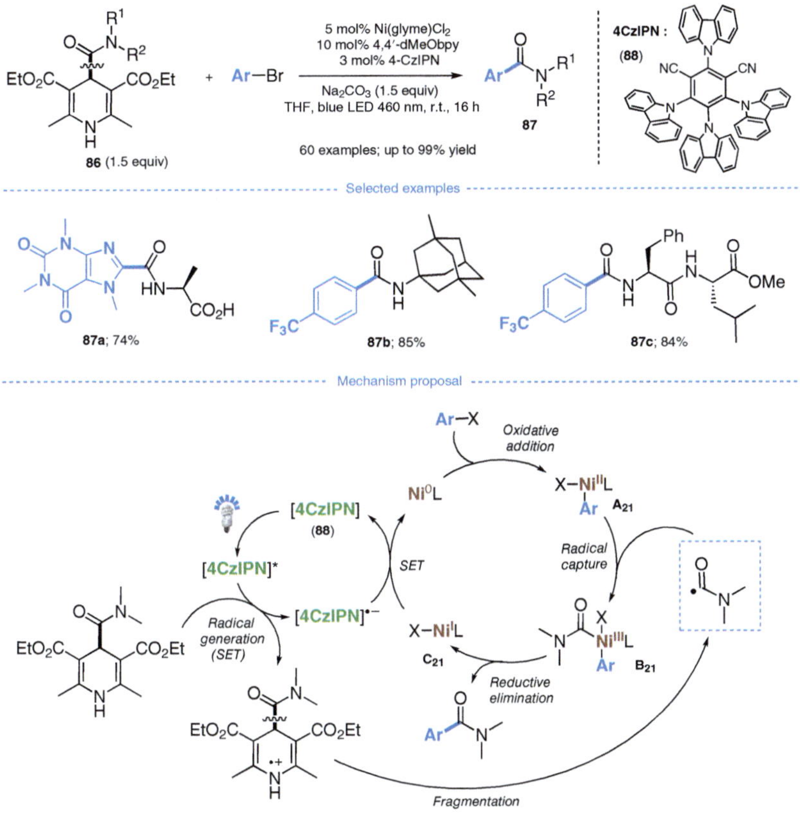

Figure 11.27 Carbamoylation of arenes with 4-carbamoyl-1,4-dihydropyridines.

11.4 Conclusion

The transformation of amides via radical cross-coupling processes has attracted renewed attention in recent years, benefitting greatly from the emergence of new photochemical strategies for the mild generation of synthetically valuable radical intermediates, and their possible combination with organometallic catalysis. Amides can also be easily diversified by taking advantage of their inherent abilities to act as effective directing groups in regioselective radical bond-forming processes, *i.e.* distant functionalization via radical translocation or proximal functionalization via transition metal chelation. A very remarkable feature of these modular substrates – classically considered inert to substitution reactions – is their recently demonstrated ability to serve as stable, mild acylating reagents upon transition metal-induced activation of the amide linkage enabled by ground state destabilization. In addition, amides can serve as convenient precursors to a variety of nitrogen- and carbon-centered radicals, e.g. amidyl, α-aminoalkyl, or carbamoyl radicals, which can engage in several radical capture mechanisms to deliver useful functionalized structures.

Overall, various strategies featuring selective cleavage of C–N, C–C, C–H, or N–H bonds have been implemented, particularly thanks to the use of dual metal/photoredox catalytic systems. The field is still progressing very rapidly, and the design of new catalytic systems and methodologies enabling the activation and functionalization of simple amides will certainly provide powerful tools for the synthesis and late-stage modification of complex molecular scaffolds of pharmaceutical relevance.

References

1 Li, G., Ma, S., and Szostak, M. (2020). Amide bond activation: the power of resonance. *Trends Chem.* 2 (10): 914–928.

2 Tasker, S.Z., Standley, E.A., and Jamison, T.F. (2014). Recent advances in homogeneous nickel catalysis. *Nature* 509: 299–309. (review).

3 Diccianni, J.B. and Diao, T. (2019). Mechanisms of nickel-catalyzed cross-coupling reactions. *Trends Chem.* 1: 830–844. (review).

4 Everson, D.A. and Weix, D.J. (2014). Cross-electrophile coupling: principles and selectivity. *J. Org. Chem.* 79: 4793–4798. (review).

5 Moragas, T., Correa, A., and Martin, R. (2014). Metal-catalyzed reductive coupling reactions of organic halides with carbonyl-type compounds. *Chem. Eur. J.* 20: 8242–8258. (review).

6 Ni, S., Zhang, W., Mei, H. et al. (2017). Ni-catalyzed reductive cross-coupling of amides with aryl iodide electrophiles via C–N bond activation. *Org. Lett.* 19 (10): 2536–2539.

7 Yu, C.-G. and Matsuo, Y. (2020). Nickel-catalyzed deaminative acylation of activated aliphatic amines with aromatic amides via C–N bond activation. *Org. Lett.* 22 (3): 950–955.

8 Zhuo, J., Zhang, Y., Li, Z. et al. (2020). Nickel-catalyzed direct acylation of aryl and alkyl bromides with acylimidazoles. *ACS Catal.* 10 (6): 3895–3903.

9 Twilton, J., Le, C., Zhang, P. et al. (2017). The merger of transition metal and photocatalysis. *Nat. Rev. Chem.* 1: 0052. (review).

10 Milligan, J.A., Phelan, J.P., Badir, S.O. et al. (2019). Alkyl carbon-carbon bond formation by nickel/photoredox cross-coupling. *Angew. Chem. Int. Ed.* 58: 6152–6163. (review).

11 Amani, J., Alam, R., Badir, S. et al. (2017). Synergistic visible-light photoredox/nickel-catalyzed synthesis of aliphatic ketones via N–C cleavage of imides. *Org. Lett.* 19 (9): 2426–2429.

12 Kerackian, T., Reina, A., Bouyssi, D. et al. (2020). Silyl radical mediated cross-electrophile coupling of N-acyl-imides with alkyl bromides under photoredox/nickel dual catalysis. *Org. Lett.* 22 (6): 2240–2245.

13 Li, J.-S. and Wu, J. (2018). Recent developments in the photo-mediated generation of silyl radicals and their application in organic synthesis. *ChemPhotoChem* 2: 839–846. (review).

14 Kerackian, T., Reina, A., Krachko, T. et al. (2021). C(sp^3)–H bond acylation with N-acyl-imides under photoredox/nickel dual catalysis. *Synlett* 32: 1531–1536.

15 Kariofillis, S.K. and Doyle, A.G. (2021). Synthetic and mechanistic implications of chlorine photoelimination in nickel/photoredox C(sp^3)–H cross-Coupling. *Acc. Chem. Res.* 54: 988–1000.

16 Lee, G.S., Won, J., Choi, S. et al. (2020). Synergistic activation of amides and hydrocarbons for direct C(sp^3)–H acylation enabled by metallaphotoredox catalysis. *Angew. Chem. Int. Ed.* 59 (39): 16933–16942.

17 Rej, S., Ano, Y., and Chatani, N. (2020). Bidentate directing groups: an efficient tool in C–H bond functionalization chemistry for the expedient construction of C–C Bonds. *Chem. Rev.* 120 (3): 1788–1887. (Review).

18 Sambiagio, C., Schönbauer, D., Blieck, R. et al. (2018). A comprehensive overview of directing groups applied in metal-catalysed C–H functionalisation chemistry. *Chem. Soc. Rev.* 47: 6603–6743. (Review).

19 Zheng, Q., Liu, C.F., Chen, J. et al. (2020). C–H functionalization of aromatic amides. *Adv. Synth. Catal.* 362 (7): 1406–1446. (Review).

20 Tan, P.W. and Seayad, J. (2019). Advances in amide and thioamide assisted C(sp^3)-H functionalization. *Tetrahedron Lett.* 60 (50): 151338. (review).

21 Daugulis, O., Do, H.-Q., and Shabashov, D. (2009). Palladium- and copper-catalyzed arylation of carbon-hydrogen bonds. *Acc. Chem. Res.* 42: 1074–1086.

22 Aihara, Y., Tobisu, M., Fukumoto, Y. et al. (2014). Ni(II)-catalyzed oxidative coupling between C(sp^2)-H in benzamides and C(sp^3)-H in toluene derivatives. *J. Am. Chem. Soc.* 136: 15509–15512.

23 Kubo, T. and Chatani, N. (2016). Dicumyl peroxide as a methylating reagent in the Ni-catalyzed methylation of ortho C–H bonds in aromatic amides. *Org. Lett.* 18: 1698–1701.

24 Chen, X., Tan, Z., Gui, Q. et al. (2016). Photocatalytic/Cu-promoted C–H activations: visible-light-induced ortho-selective perfluoroalkylation of benzamides. *Chem. Eur. J.* 22: 6218–6222.

25 Czyz, M.L., Lupton, D.W., and Polyzos, A. (2017). Auxiliary-directed C(sp^3)–H arylation by synergistic photoredox and palladium catalysis. *Chem. Eur. J.* 23: 14450–14453.

26 Lin, C. and Shen, L. (2019). Recent progress in transition metal-catalyzed regioselective functionalization of unactivated alkenes/alkynes assisted by bidentate directing groups. *ChemCatChem* 11: 961–968. (Review).

27 Derosa, J., van der Puyl, V.D., Tran, V.T. et al. (2018). Directed nickel-catalyzed 1,2-dialkylation of alkenyl carbonyl compounds. *Chem. Sci.* 9: 5278–5283.

28 Tang, C., Zhang, R., Zhu, B. et al. (2018). Directed copper-catalyzed intermolecular Heck-type reaction of unactivated olefins and alkyl halides. *J. Am. Chem. Soc.* 140: 16929–16935.

29 Deng, Y., Zhao, C., Zhou, Y. et al. (2020). Directing-group-based strategy enabling intermolecular Heck-type reaction of cycloketone oxime esters and unactivated alkenes. *Org. Lett.* 22: 3524–3530.

30 Bai, Z., Zhang, H., Wang, H. et al. (2021). Enantioselective alkylamination of unactivated alkenes under copper catalysis. *J. Am. Chem. Soc.* 143: 1195–1202.

31 Shi, P., Wang, J., Gan, Z. et al. (2019). A practical copper-catalyzed approach to β-lactams via radical carboamination of alkenyl carbonyl compounds. *Chem. Commun.* 55: 10523–10526.

32 Yang, D., Huang, H., Li, M.-H. et al. (2020). Directed cobalt-catalyzed *anti*-Markovnikov hydroalkylation of unactivated alkenes enabled by "Co–H" catalysis. *Org. Lett.* 22: 4333–4338.

33 Crossley, S.W.M., Obradors, C., Martinez, R.M. et al. (2016). Mn-, Fe-, and Co-catalyzed radical hydro-functionalizations of olefins. *Chem. Rev.* 116: 8912–9000.

34 Kärkäs, M.D. (2017). Photochemical generation of nitrogen-centered amidyl, hydrazonyl, and imidyl radicals: methodology developments and catalytic applications. *ACS Catal.* 7 (8): 4999–5022. (review).

35 Zhou, L., Yi, H., Qi, X. et al. (2015). Tuning the reactivity of radical through a triplet diradical Cu(II) intermediate in radical oxidative cross-coupling. *Sci. Rep.* 5: 15934.

36 Zhou, L., Tang, S., Qi, X. et al. (2014). Transition-metal-assisted radical/radical cross-coupling: a new strategy to the oxidative $C(sp^3)$–H/N–H cross-coupling. *Org. Lett.* 16: 3404–3407.

37 Song, L., Qi, X., Bai, R. et al. (2019). Theoretical study of Ni-catalyzed C–N radical–radical cross coupling. *J. Org. Chem.* 84: 3321–3327.

38 Badir, S.O. and Molander, G.A. (2020). Developments in photoredox/nickel dual-catalyzed 1,2-difunctionalizations. *Chem* 6: 1327–1339. (review).

39 Choi, G.J. and Knowles, R.R. (2015). Catalytic alkene carboaminations enabled by oxidative proton-coupled electron transfer. *J. Am. Chem. Soc.* 137: 9226–9229.

40 Zheng, S., Gutiérrez-Bonet, A., and Molander, G.A. (2019). Merging photoredox PCET with Ni-catalyzed cross-coupling: cascade amidoarylation of unactivated olefins. *Chem* 5: 339–352.

41 Zheng, S., Zhang, S.-Q., Saeednia, B. et al. (2020). Diastereoselective olefin amidoacylation via photoredox PCET/nickel-dual catalysis: reaction scope and mechanistic insights. *Chem. Sci.* 11: 4131–4137.

42 Kumar, G., Pradhan, S., and Chatterjee, I. (2020). *N*-Centered radical directed remote C-H bond functionalization via hydrogen atom transfer. *Chem. Asian J.* 15: 651–672. (review).

43 Choi, G.J., Zhu, Q., Miller, D.C. et al. (2016). Catalytic alkylation of remote C–H bonds enabled by proton-coupled electron transfer. *Nature* 539: 268–271.

44 Chu, J.C.K. and Rovis, T. (2016). Amide-directed photoredox-catalysed C–C bond formation at unactivated Sp^3 C–H bonds. *Nature* 539: 272–275.

45 Thullen, S.M., Treacy, S.M., and Rovis, T. (2019). Regioselective alkylative cross-coupling of remote unactivated $C(sp^3)$–H bonds. *J. Am. Chem. Soc.* 141: 14062–14067.

46 Xu, B. and Tambar, U.K. (2019). Remote allylation of unactivated $C(sp^3)$–H bonds triggered by photogenerated amidyl radicals. *ACS Catal.* 9: 4627–4631.

47 Li, Z., Wang, Q., and Zhu, J. (2018). Copper-catalyzed arylation of remote C(sp^3)-H bonds in carboxamides and sulfonamides. *Angew. Chem. Int. Ed.* 57: 13288–13292.

48 Bao, X., Wang, Q., and Zhu, J. (2019). Copper-catalyzed remote C(sp^3)–H azidation and oxidative trifluoromethylation of benzohydrazides. *Nat. Commun.* 10: 769.

49 Chen, H., Jin, W., and Yu, S. (2020). Enantioselective remote C(sp^3)–H cyanation via dual photoredox and copper catalysis. *Org. Lett.* 22 (15): 5910–5914.

50 Nakajima, K., Miyake, Y., and Nishibayashi, Y. (2016). Synthetic utilization of α-aminoalkyl radicals and related species in visible light photoredox catalysis. *Acc. Chem. Res.* 49: 1946–1956.

51 Yan, H., Lu, L., Rong, G. et al. (2014). Functionalization of amides via copper-catalyzed oxyalkylation of vinylarenes and decarboxylative alkenylation of sp^3 C-H. *J. Org. Chem.* 79: 7103–7111.

52 Zhang, J.-X., Wang, Y.-J., Wang, N.-X. et al. (2014). Nickel-catalyzed sp^3 C–H bond activation from decarboxylative cross-coupling of α,β-unsaturated carboxylic acids with amides. *Synlett* 25: 1621–1625.

53 Song, R.-J., Tu, Y.-Q., Zhu, D.-Y. et al. (2015). A nickel-mediated oxidative α-C(sp^3)–H functionalization of amides with allylic alcohols terminated by radical 1,2-aryl migration. *Chem. Commun.* 51: 749–752.

54 Li, Z.-L., Sun, K.-K., and Cai, C. (2018). Nickel-catalyzed cross-dehydrogenative coupling of α-C(sp^3)–H bonds in *N*-methylamides with C(sp^3)–H bonds in cyclic alkanes. *Org. Lett.* 20: 6420–6424.

55 Das, S., Murugesan, K., Villegas Rodríguez, G.J. et al. (2021). Photocatalytic (het)arylation of C(sp^3)–H bonds with carbon nitride. *ACS Catal.* 11: 1593–1603.

56 Rand, A.W., Yin, H., Xu, L. et al. (2020). Dual catalytic platform for enabling sp^3 α C–H arylation and alkylation of benzamides. *ACS Catal.* 10: 4671–4676.

57 Raviola, C., Protti, S., Ravelli, D. et al. (2019). Photogenerated acyl/alkoxycarbonyl/carbamoyl radicals for sustainable synthesis. *Green Chem.* 21: 748–764. (review).

58 Banerjee, A., Lei, Z., and Ngai, M.-Y. (2019). Acyl radical chemistry via visible-light photoredox catalysis. *Synthesis* 51: 303–333. (review).

59 Cheng, W.-M., Shang, R., Yu, H.-Z. et al. (2015). Room-temperature decarboxylative couplings of α-oxocarboxylates with aryl halides by merging photoredox with palladium catalysis. *Chem. Eur. J.* 21: 13191–13195.

60 Li, M., Wang, C., Fang, P. et al. (2011). Pd(II)-catalyzed decarboxylative cross-coupling of oxamic acids with potassium phenyltrifluoroborates under mild conditions. *Chem. Commun.* 47: 6587–6589.

61 Alandini, N., Buzzetti, L., Favi, G. et al. (2020). Amide synthesis by nickel/photoredox-catalyzed direct carbamoylation of (hetero)aryl bromides. *Angew. Chem. Int. Ed.* 59: 5248–5253.

12

Weinreb Amide as a Multifaceted Directing Group in C–H Activation

Jayabrata Das and Debabrata Maiti

Department of Chemistry, Indian Institute of Technology Bombay, Mumbai, Powai-400076, India

12.1 Introduction

N-methoxy-*N*-methyl amides, popularly known as Weinreb amides, are one of the most well-known and versatile synthetic precursor in organic synthesis. The first synthesis of Weinreb amide was reported in 1981 by Steven M. Weinreb and Steven Nahm [1]. Weinreb amide has served as an excellent acylating agent for organolithium or organomagnesium reagents and as an effective equivalent for an aldehyde group [2]. Direct addition of organolithium or organomagnesium reagents to acyl chloride results in the formation of tertiary alcohols due to overaddition. This occurs even if the equivalents of nucleophile are closely controlled. Weinreb amides solve this problem of overaddition by forming a stable five-membered tetrahedral intermediate upon addition of the first equivalent of the organometallic species or reducing agent (Scheme 12.1). This stability of tetrahedral cyclic intermediate precludes the collapse to a ketone or aldehyde under the reaction conditions and thus prevents the formation and subsequent possibility of addition to the ketone or aldehyde. Further, an aqueous workup ensures the collapse of tetrahedral intermediate and quenches any excess organometallic species or reducing agent present in the reaction mixture. In recent years, there has been a spike in usage of Weinreb amides, and this is primarily due to four reasons: (i) ease of preparation by *in situ* activation of the carboxyl group or through its stable ester derivative, even in a highly functionalized molecule; (ii) its success as an effective acylating agent for organometallics, facilitating convenient access to highly functionalized ketones, particularly in the total synthesis of complex natural products; (iii) easy scale of preparation of Weinreb amides in their routine usage in organic synthesis; and, finally, (iv) stability of Weinreb amides. Their facile reduction to aldehydes with hydride reducing agents, coupled with all the previously mentioned attractive features, makes them serve as an excellent precursor to sensitive aldehyde in the masked form. Assembling valuable organic compounds in green and atom economic approaches has long been the core goal of synthetic chemistry. On that aspect, direct C–H functionalization has been developed into a powerful synthetic tool that could meet this goal by excluding

Amide Bond Activation: Concepts and Reactions, First Edition. Edited by Michal Szostak.

the requirement for undesired prefunctionalization [3, 4]. The use of an exogenous directing groups to carry out C–H functionalization reactions has become a great success [5]. However, the laborious installation/removal of the external directing group decreases the overall step as well as atom economy of such reactions. Therefore, it is more desirable that the directing group either should be part of the target molecule or should be easily modifiable/removable after the C–H functionalization step is over. Weinreb amides fit perfectly in this category. No wonder, in recent years Weinreb amides have received a great attention from researchers over the globe to direct C–H functionalization of not only C(sp^2)–H but also more inert C(sp^3)–H bonds [6]. This short chapter focuses on recent advancement on Weinreb amide-assisted C(sp^2)–H and C(sp^3)–H functionalizations. The relevant reports are sorted by the type of C–H bond functionalized and transition metals utilized.

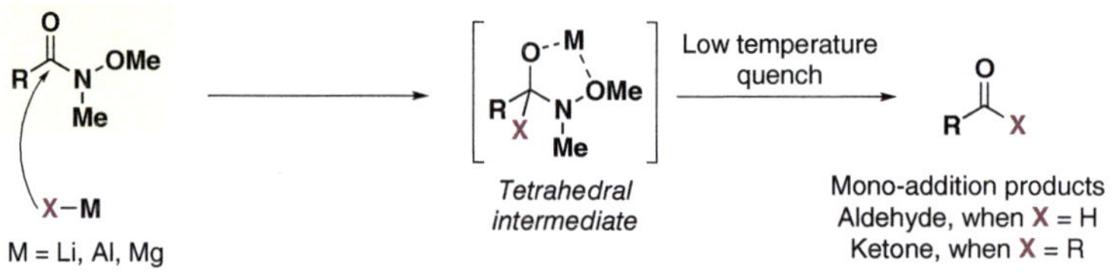

Scheme 12.1 Application of Weinreb amide: utilization as an effective acylating reagent.

12.2 Weinreb Amide-Directed C(sp^2)–H Activation

12.2.1 Ru-Catalyzed Reactions

The lower price of Ru compared with second and third row transition metals makes it an appealing choice for carrying out C–H functionalization. In recent years, Ru has been utilized extensively for C–H functionalization in accordance with amide directing groups [7, 8]. In 2013, Ackermann group reported a Ru-catalyzed Weinreb amide-directed *ortho*-oxygenation of arenes (Scheme 12.2) [9]. $[RuCl_2(p\text{-}cymene)]_2$

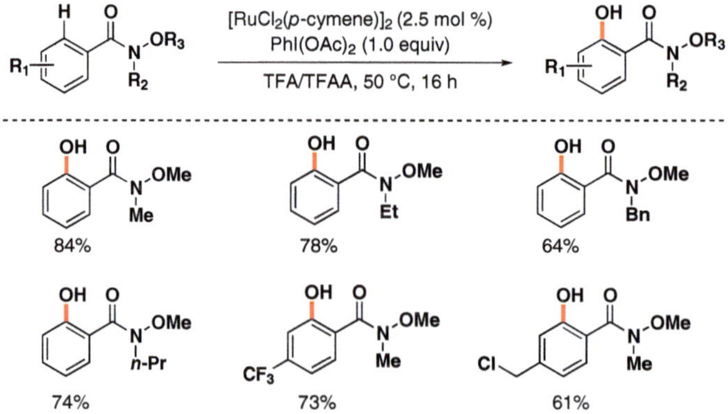

Scheme 12.2 Ru-catalyzed *ortho*-hydroxylation of Weinreb amides. Source: Adapted from Yang et al. [9].

was utilized as the catalyst precursor and $PhI(OAc)_2$ as the external oxidant in the reaction. Notably, Ag(I) or Cu(II) oxidants were ineffective in this transformation. The reaction was carried out at lower temperature 50 °C. Changing the substituents on the N of Weinreb amide did not affect the efficacy of the reaction. Further, a number of substituted arene moieties including *meta*-substituted arenes provided *ortho*-oxygenated protocol in good to excellent yields. An intermolecular competition experiment showed that electron-rich arenes were more susceptible to the reaction compared with electron-deficient arenes.

A reversibility experiment with $[D_5]$-labeled Weinreb amide substrate suggests an irreversible nature of the reaction. A higher value in primary kinetic isotope experiment confirms that C–H activation is the rate-determining step of the reaction [10]. To showcase a practical application of the reaction, *ortho*-oxygenated Weinreb amide was converted to *ortho*-hydroxy benzaldehyde (Scheme 12.3).

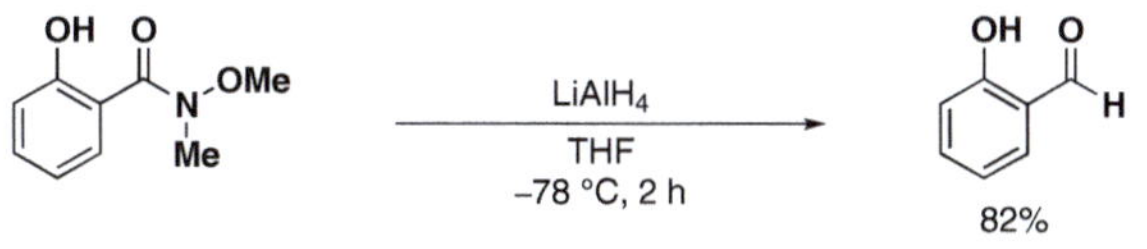

Scheme 12.3 Weinreb amide to aldehyde in one step.

In 2015, Kapur and coworker disclosed a Ru-catalyzed olefination of various Weinreb amides including cyclic amides (Scheme 12.4) [11]. The methodology resembles Fujiwara–Moritani reaction [12]. A range of styrenes and activated olefins were compatible with the protocol and provided *ortho*-olefinated products in good to excellent

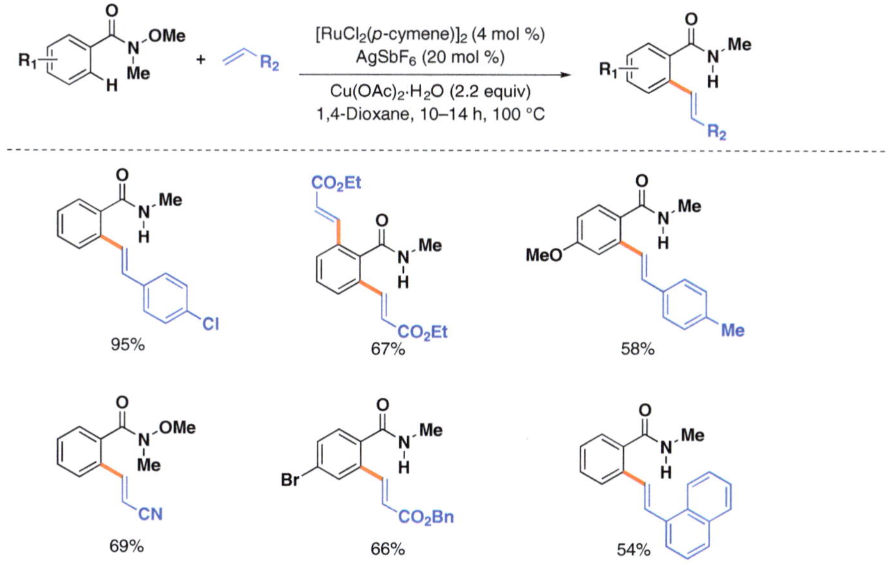

Scheme 12.4 Ru-catalyzed *ortho*-olefination of Weinreb amides. Source: Adapted from Das et al. [11].

yields. Various substituted arenes including halogens in the arene moiety were also tolerated in the reaction. Several five-membered Weinreb amides were also subjected to the reaction.

Olefination occurred efficiently in these cyclic amides with ring opening of the isoxazolidine ring (Scheme 12.5). Interestingly, in case of most entries, cleavage of Weinreb amide N–O bond occurred. However, the authors were not sure about whether N–O bond of Weinreb amide acts as internal oxidant, and thus cleavage of N–O bond occurs.

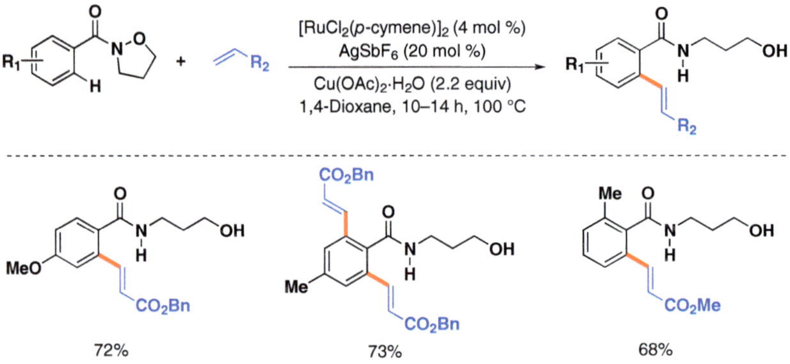

Scheme 12.5 Ru-catalyzed *ortho*-olefination of Weinreb amides with ring opening of cyclic Weinreb amides.

The authors proposed a tentative mechanism for the reaction where a five-membered ruthanacycle formation is followed by coordination of olefin. Next, migratory insertion of the olefin and then beta hydride elimination resulted in the product formation. Ru(0) then reoxidized by Cu^{II} (Scheme 12.6). Nonetheless, the reason for N–O bond cleavage is not clear in the mechanism. Although a number of recent reports outlined the role of Weinreb amide *N*-alkoxy bond cleavage as internal oxidant, a control experiment by the authors showed that even without the olefin partner, Weinreb amide N–O bond cleavage took place, which is suggestive of the fact that N–O bond cleavage is not acting as internal oxidant in this protocol. $[Ru^{II}\text{-}H]^+$ formed after beta hydride elimination in the reaction could also play a role in N–O bond cleavage. All in all, the reason for cleavage of N–O bond in the reaction is not clarified by the authors.

Later, to prevent the cleavage of N–O bond, the same group showed that N–O bond cleavage could be prevented by judicious substrate design (Scheme 12.7). When they utilized six- and seven-membered Weinreb amide, olefination occurred without the loss of Weinreb amide functionality [13]. The authors put forward the instability of higher size ruthanacycle that would form if Ru inserts into N–O bond of six- or seven-membered Weinreb amide as the reason for resistance of N–O bond breakage in higher membered Weinreb amides. Electronic effects on both arene and olefin coupling partner were well tolerated. Further synthetic utility of the formed product was showcased by transforming the Weinreb amide to aldehyde and ketone.

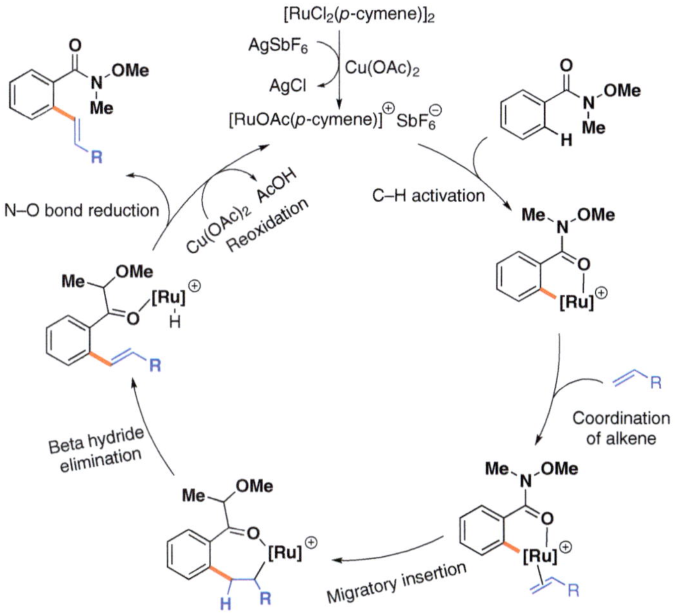

Scheme 12.6 Catalytic cycle for Ru-catalyzed *ortho*-olefination.

A single example of Weinreb amide was shown to work by Jeganmohan group as part of their study on *ortho*-benzoxylation of various tertiary amides (Scheme 12.8) [14]. $[RuCl_2(p\text{-}cymene)]_2$ was utilized as the catalyst in the reaction along with $AgSbF_6$, $(NH_4)S_2O_8$ as terminal oxidant, and 1,2-dichloroethane as solvent. The reaction closely resembles Ackermann's Ru-catalyzed hydroxylation reaction discussed previously except a higher temperature and usage of $(NH_4)S_2O_8$ as an external oxidant.

In 2020, Ackermann and coworkers disclosed an electrocatalyzed oxygenation of Weinreb amides (Scheme 12.9) [15]. In recent years, electrosynthesis has received a great attention of chemists due to storability and sustainable properties [16]. A platinum plate at the anode was found to be most effective for the transformation. The optimum current was found to be 2.67 mA/cm^2. Iodobenzene was utilized in the protocol as a cocatalyst that enabled C–H oxygenation. Notably, the electrocatalyzed C–H oxygenation protocol is enabled by orchestrating the catalytic generation of hypervalent iodine(III) reagent with electricity as the only terminal oxidant. Molecular hydrogen is formed as the only by-product in this reaction. Furthermore, control experiments established the crucial role electricity in the reaction. A range of aromatic amides containing *meta* and *para* substituents were transformed into *ortho*-hydroxylated products. Halogens like chloro, bromo, and iodo groups in the arene were all compatible with the protocol. Reversibility experiment and primary kinetic isotope data suggests that ruthanacycle formation is fast and reversible, and C–H activation is not the rate-determining step of the reaction.

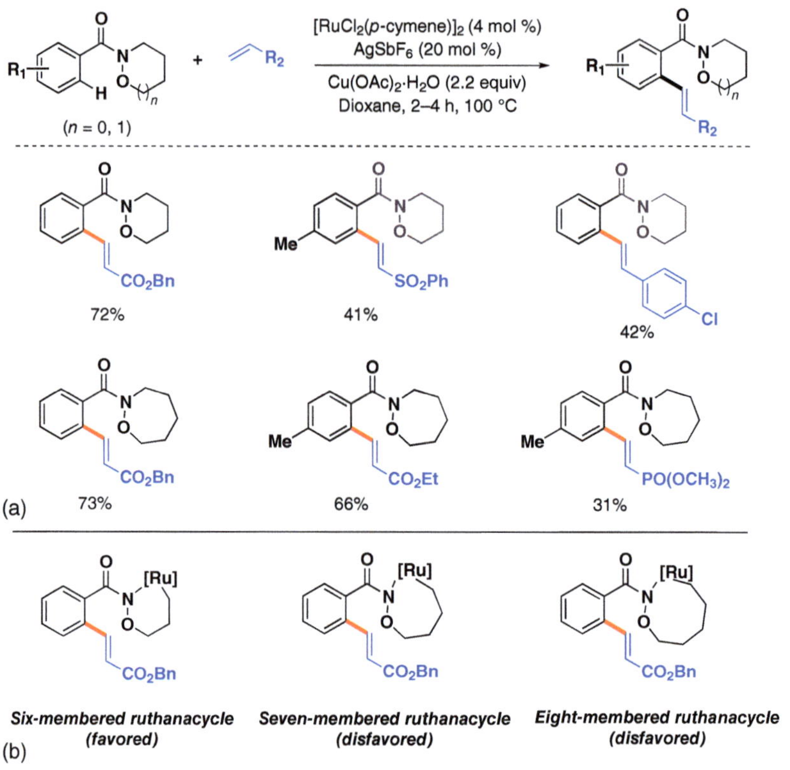

Scheme 12.7 Ru-catalyzed *ortho*-olefination of cyclic Weinreb amides without N–O bond cleavage.

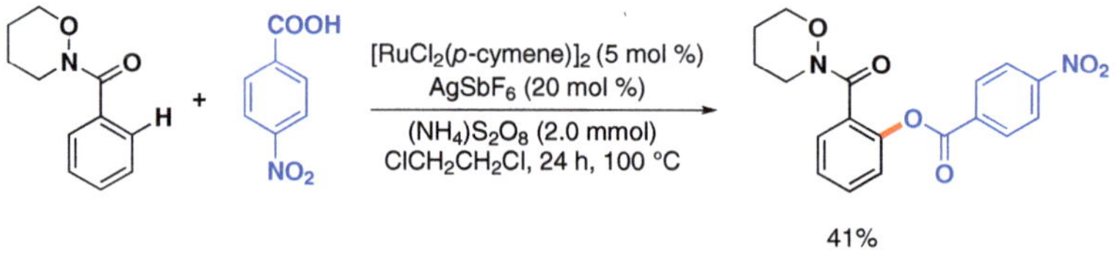

Scheme 12.8 Ru-catalyzed Weinreb amide-directed *ortho*-benzoxylation of arenes. Source: Adapted from More et al. [14].

12.2.2 Co-Catalyzed Reactions

In recent years, Co has drawn significant attention in the realm of C–H functionalization due to its less expense and abundance compared with second and third row transition metals. A number of C–H functionalization reactions have been developed using various Co catalysts [17, 18]. In 2013, Yoshino and Matsunaga reported the first Co-catalyzed *ortho*-functionalization of arenes [19, 20]. Later, the same group in a subsequent study demonstrated a single example of allylation reaction on Weinreb amide (Scheme 12.10) [22]. Bench stable Cp*Co(CO)I_2

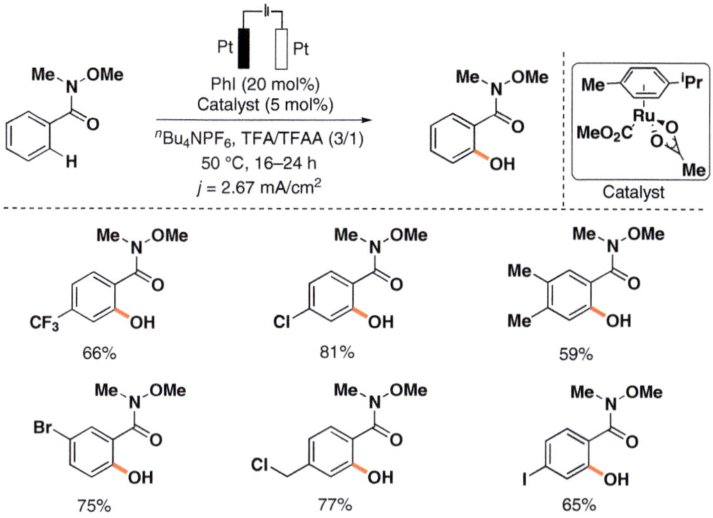

Scheme 12.9 Ru-catalyzed *ortho*-oxygenation of Weinreb amides utilizing electrocatalysis. Source: Adapted from Massignan et al. [15].

was utilized as the precatalyst in the reaction. Along with $AgNTf_2$ and AgOAc as external oxidants, hexafluoroisopropanol (HFIP) was used as the solvent in the protocol. An acetate-assisted concerted metalation-deprotonation (CMD) mechanism thought to be operative in the reaction [21]. The moderate yield in case of Weinreb amide is due to weaker coordination of Weinreb amide group compared with strong coordination of benzamide or purine N_{sp2} as mentioned in the book.

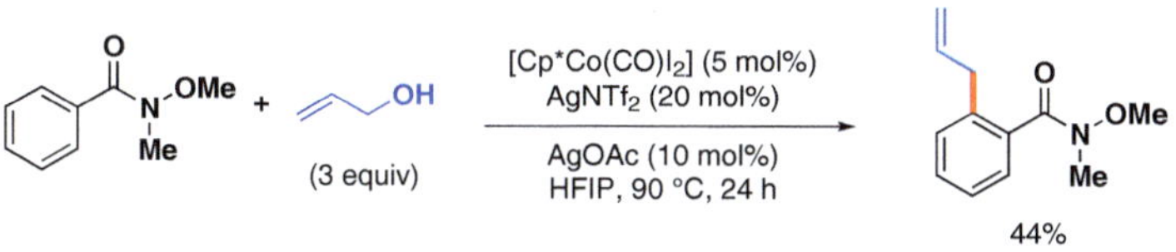

Scheme 12.10 An initial example of Co-catalyzed Weinreb amide-directed *ortho*-allylation. Source: Adapted from Planas et al. [21].

In 2018, Matsunaga and coworkers further explored the Weinreb amides and carried out a number of Co-catalyzed functionalizations at *ortho*-position of Weinreb amides (Scheme 12.11) [23]. The functionalizations included allylation with allyl carbonate, alkenylation with acrylates, iodination with *N*-iodo succinimide, and amidation with dioxazolones. The reaction condition in all these cases involved $Cp^*Co(CO)I_2$ as catalyst, AgOAc and a cationic silver salt ($AgNTf_2$ or $AgSbF_6$), and dichloroethane (DCE) or HFIP as solvent. Various aromatic, heteroaromatic, and α,β-unsaturated Weinreb amides were employed as substrates in this protocol. Competition experiment and kinetic isotope effect experiments revealed that C–H activation step is irreversible and rate limiting step in case of C–H allylation reaction. This supported the authors earlier work on Weinreb amide where they proposed an acetate-assisted CMD mechanism [22].

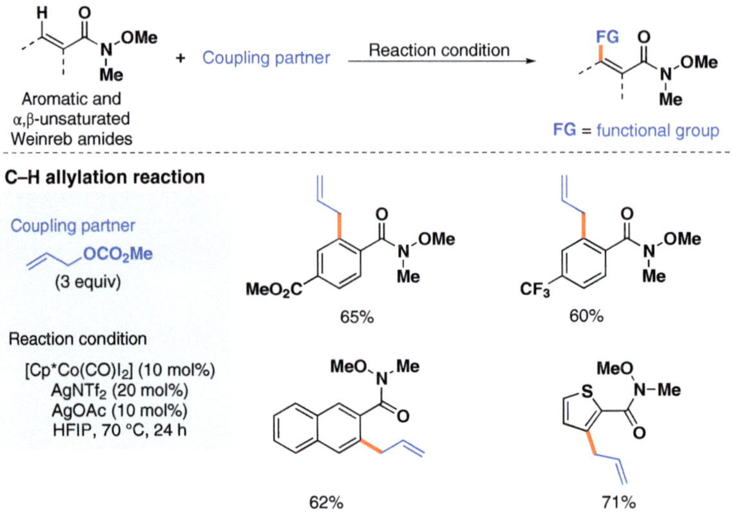

Scheme 12.11 Co(III)-catalyzed Weinreb amide-directed C–H transformations of various arenes and heteroarenes. Source: Adapted from Kawai et al. [23].

A tentative mechanism for allylation has been proposed by the authors (Scheme 12.12). A coordinatively unsaturated species Cp*CoIII(OAc)$^+$ was formed from Cp*Co(CO)I_2, $AgNTf_2$, and AgOAc. Next, coordination of the substrate occurs with Cp*CoIII(OAc)$^+$. A C–H activated intermediate then forms, which is the rate limiting step of the mechanism. Further, coordination and migratory insertion of allyl carbonate followed by β- elimination of carbonate group affords the allylated product.

12.2.3 Pd-Catalyzed Reactions

Pd is the most established metal in C–H functionalization chemistry among all the transition metals. Many Pd catalysts are utilized regularly in cross-coupling and C–H activation reactions [24–26]. The well-defined properties of Pd catalysts allow systematic tuning of their properties, which in turn makes them suitable catalysts in C–H functionalization reactions. Due to their unique properties, these Pd catalysts work via various mechanisms such as Pd(0)/Pd(II), Pd(II)/Pd(IV), Pd(II)/Pd(0), etc. [25, 27, 28]. Despite the prevalence of Pd catalyst in C–H functionalization reactions, there are handful reports of Weinreb amide-catalyzed C–H functionalization reactions catalyzed by Pd.

The first report on Pd-catalyzed *ortho*-functionalization of Weinreb amides came from group of Wang in 2015 (Scheme 12.13) [29]. The reaction took place in presence of $Pd(OAc)_2$ as catalyst, AgOTf as oxidant, and DCE as solvent. Notably, addition of

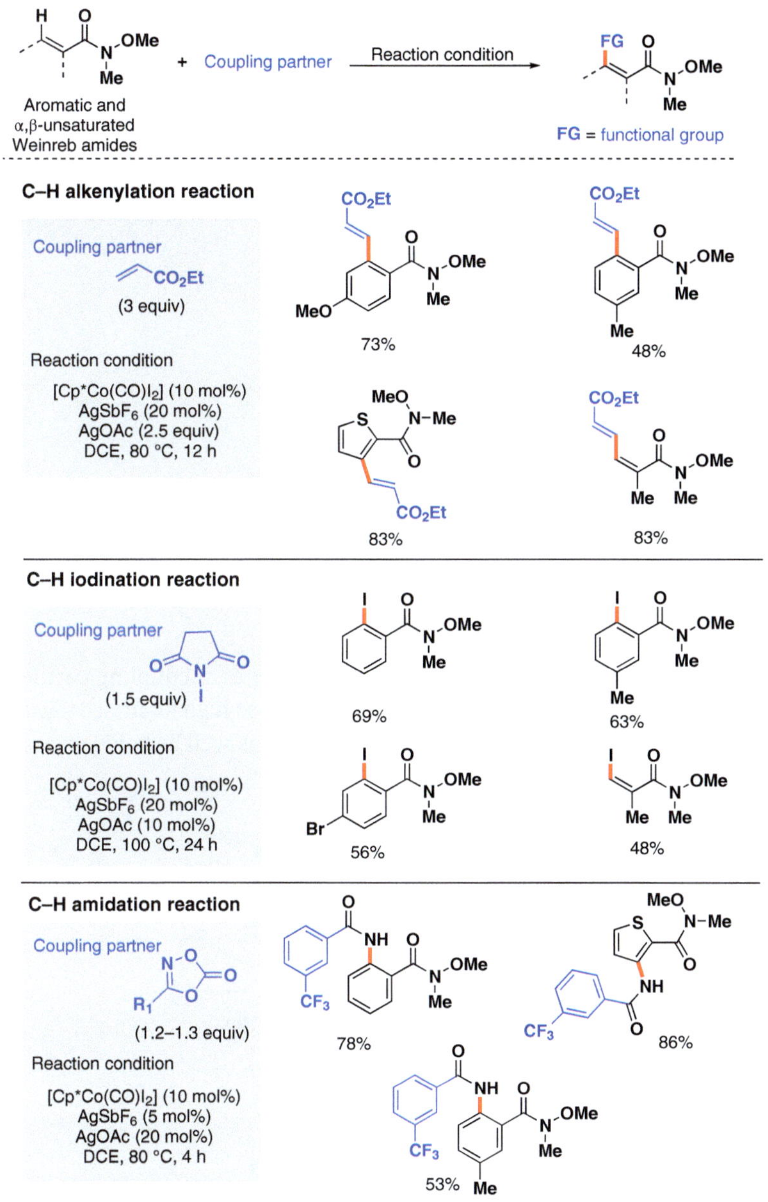

Scheme 12.11 (*Continued*)

TfOH was found to be fruitful in the reaction. Formation of *in situ* cationic $Pd(OTf)_2$ is critical in the reaction as it facilitated the C–H activation by easily coordinating to the Weinreb amide for its enhanced Lewis acidity. To further favor the formation of $Pd(OTf)_2$, TfOH was added in the reaction, and it helped to increase the yield further. Both aryl Weinreb amides and benzyl Weinreb amides were found to be

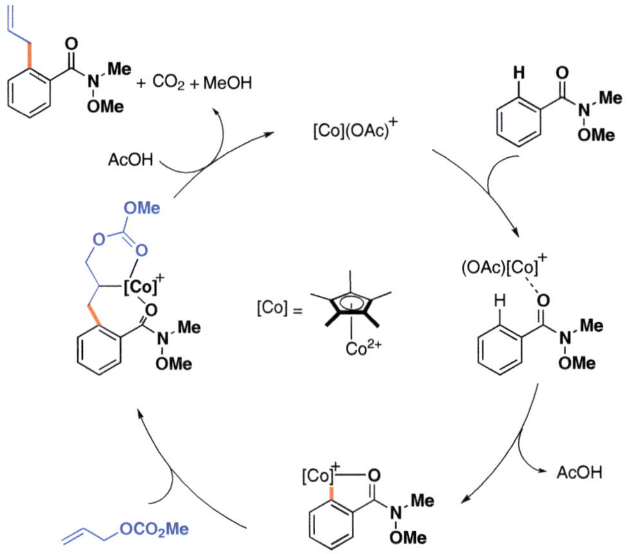

Scheme 12.12 Mechanism of Co(III)-catalyzed Weinreb amide-directed C–H allylation.

successful substrates in the reaction in presence of iodoarenes as coupling partners (Schemes 12.13 and 12.14). Both Weinreb amide types showed high tolerance toward electron-donating and electron-withdrawing substituents, as well as halogens with moderate to good yields.

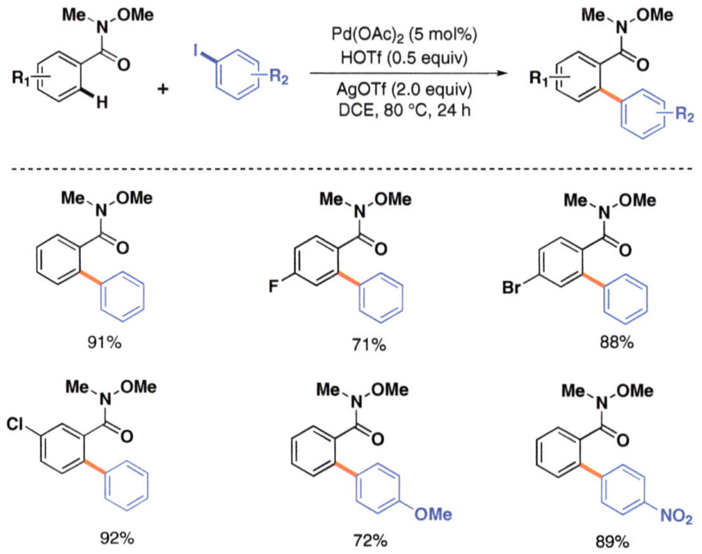

Scheme 12.13 Pd(II)-catalyzed C–H arylation of aryl Weinreb amides. Source: Adapted from Bonney et al. [28].

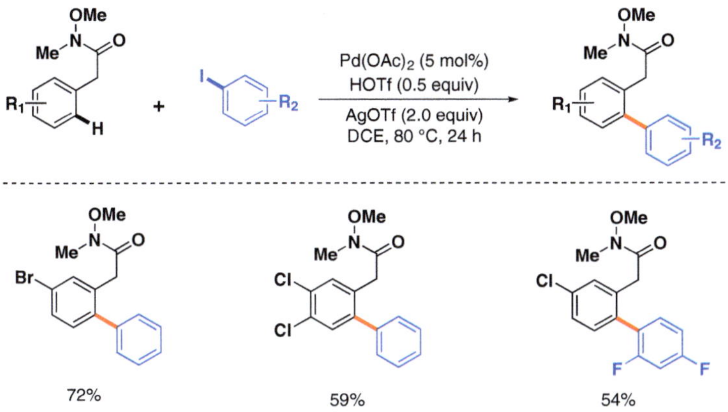

Scheme 12.14 Pd(II)-catalyzed C–H arylation of benzyl Weinreb amides.

The authors proposed a Pd(II)/Pd(IV) catalytic cycle as the mechanism for the reaction (Scheme 12.15). $Pd(OTf)_2$ first formed from or by anion exchange of $Pd(OAc)_2$ and HOTf afforded the ArPd(II) species. Then carbonyl oxygen of Weinreb amide coordinates with electrophilic palladium to form an *ortho*-activated palladacycle. Further, oxidative addition of ArI to Pd(II) forms a Pd(IV) intermediate, which then undergoes reductive elimination to release the product. The Pd(II) species C then regenerates the active $Pd(OTf)_2$ catalyst by anion exchange with AgOTf [30]. However, the authors did not disclude the possibility of a Pd(0)/Pd(II) mechanism.

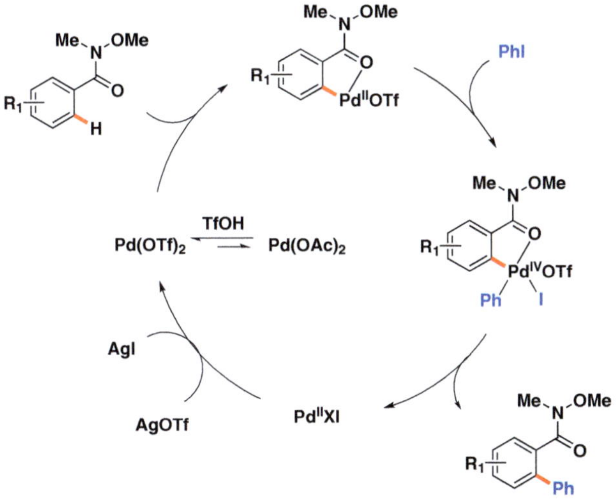

Scheme 12.15 Mechanism for *ortho*-C–H arylation of Weinreb amides.

Another example of Pd-catalyzed arylation was disclosed by Bhanage and coworker in 2016 [31]. In their work on *ortho*-arylation of *N*-methoxybenzamide, the authors reported a single example of Weinreb amide *ortho*-arylation utilizing aniline as arylating reagent (Scheme 12.16). Under the reaction condition, aniline undergoes *in situ* oxidation by ᵗBuONO to generate an electrophilic coupling partner. Nonetheless, yield of the reaction was poor and indicates that the reaction is sluggish in case of Weinreb amides compared with other amide substrates.

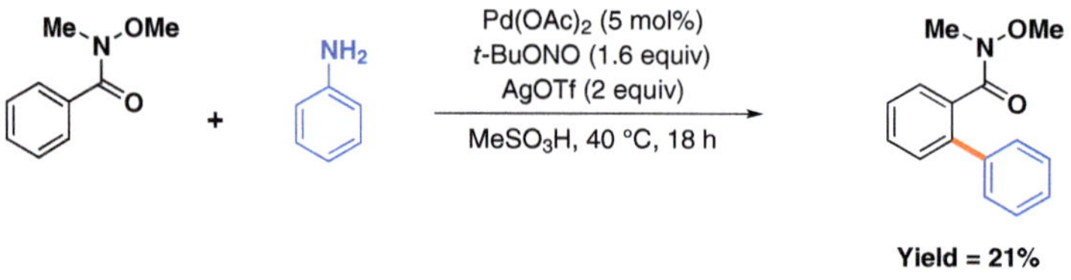

Scheme 12.16 Usage of aniline as an arylating reagent for Pd-catalyzed C–H arylation.

In 2015, Yu and coworkers developed an efficient protocol for *ortho*-olefination and acetoxylation of phenylacetyl Weinreb amides, overcoming the limitations of entropic barrier possessed by extra methylene unit in substrate (Scheme 12.17) [32]. The longer distance between target C–H bond and directing group (amide) did not hamper the outcome of the reaction. The interplay between distal weak coordination (by carbonyl of amide) and ligand acceleration is thought to be crucial for the success of the reaction. Ac-Gly-OH was found to be the most effective ligand for *ortho*-olefination of phenylacetyl Weinreb amides [33], whereas *ortho*- and *meta*-substituted benzyl Weinreb amides generated mono-exclusive olefinated products in excellent yields; reduced selectivity was observed (both mono and di product formed) in case of *para*-substituted Weinreb amides.

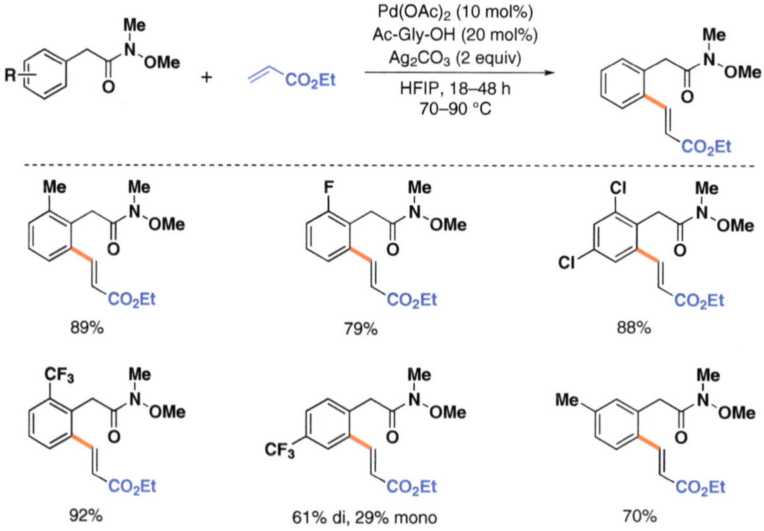

Scheme 12.17 Pd(II)-catalyzed *ortho*-olefination of benzyl Weinreb amides. Source: Adapted from Li et al. [32].

Utilizing $PhI(OAc)_2$ as an oxidant in presence of Ac_2O, acetoxylation of phenylacetyl Weinreb amides was developed (Scheme 12.18). Various electronic substitutions in the phenyl ring were well tolerated under the reaction condition. Moreover, it was possible to perform the reaction in presence of inexpensive ^{t}butyl peroxyacetate as an oxidant instead of $PhI(OAc)_2$ with slightly longer reaction times.

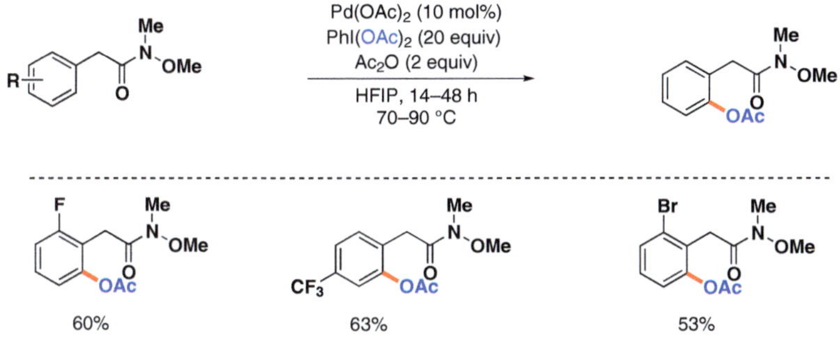

Scheme 12.18 Pd(II)-catalyzed *ortho*-C-H acetoxylation of benzylic Weinreb amides.

The insertion of halogen group into organic substrates is of great importance due to versatility of C–halogen bonds [34]. The newly formed C–halogen bond then could be utilized for a number of transformations. In 2017, group of Kapur disclosed a Pd-catalyzed *ortho*-halogenation method for Weinreb amides. Iodination, bromination, and chlorination have been carried out using *N*-halo succinimide as the coupling partner (Scheme 12.19) [35]. A combination of $Pd(OAc)_2$ and $Cu(OTf)_2$ was found to be crucial for the success of the reaction along with the coupling partner. A variety of substituents on the aryl amide were compatible with the

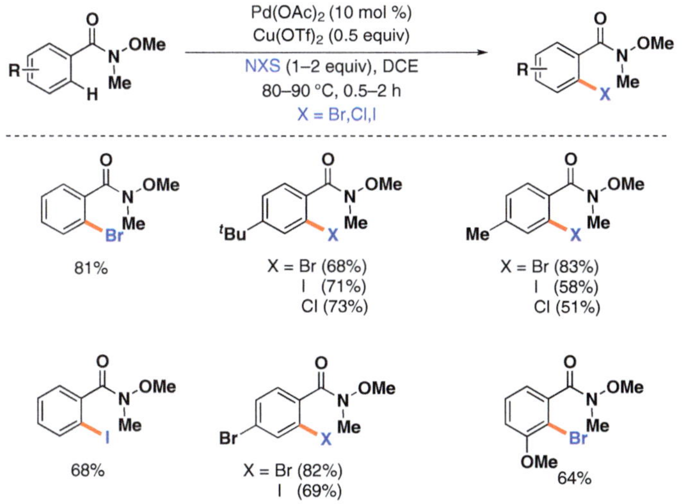

Scheme 12.19 *Ortho*-C–H halogenation of aryl Weinreb amides. Source: Adapted from Das et al. [35].

reaction. Nonetheless, the nitro group switched off the reaction from either *ortho* or *para* position in aryl ring.

Some preliminary mechanistic experiments revealed that C–H activation is irreversible, but not the rate limiting step of the reaction. The catalytic cycle of the reaction starts with carbonyl-directed cyclometalation to form a five-membered palladacycle A. Oxidative addition with *N*-halo succinimide generates a Pd^{IV} intermediate B (a Pd^{III} bimetallic intermediate is also possible [36]), which after reductive elimination forms the *ortho*-halogenated product (Scheme 12.20).

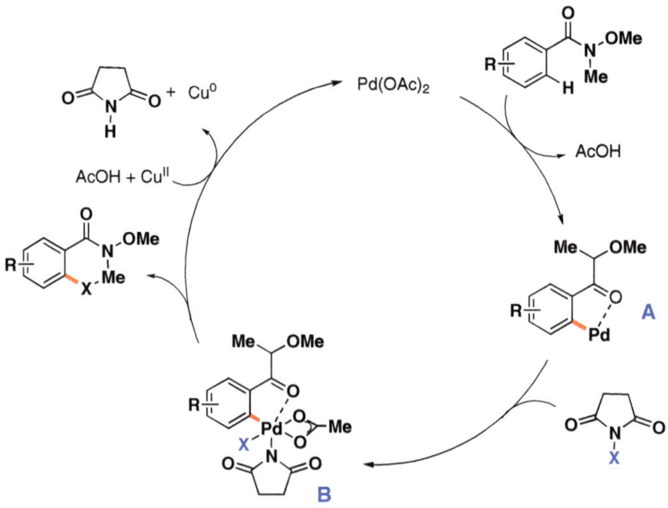

Scheme 12.20 Mechanistic cycle for *ortho*-C–H halogenation.

12.2.4 Rh-Catalyzed Reactions

In recent years, there has been upsurge in the usage of Rh(III) catalysts for C–H functionalization. Substantial advancement has been made in both scope and mechanistic understanding in Rh-catalyzed C–H functionalization reactions [37–39]. Rh(III)-catalyzed C–H activation features mild reaction conditions, good functional group tolerance, high reaction efficiency, good regioselectivity, etc. In 2013, in the quest of knowing the key questions regarding regioselectivity, Rovis group carried out intramolecular alkene hydroarylation reaction directed by Weinreb amide in excellent yield (Scheme 12.21) [40].

Wang et al. disclosed an Rh(III)-catalyzed *ortho*-olefination of aryl Weinreb amides (Scheme 12.22) [41]. Notably, $Cu(OAc)_2$ was utilized as an catalytic oxidant that regenerated under air in the catalytic cycle. *N*-Butyl acrylate was employed as the olefin partner with a number of substituted aromatic Weinreb amides. Both electron-donating and withdrawing substituents at *ortho*, *meta*, and *para* position of phenyl ring were well compatible under the reaction condition and formed olefinated product in good to excellent yield. Further, competition experiments revealed that electron-rich arenes reacted faster than electron withdrawing arenes.

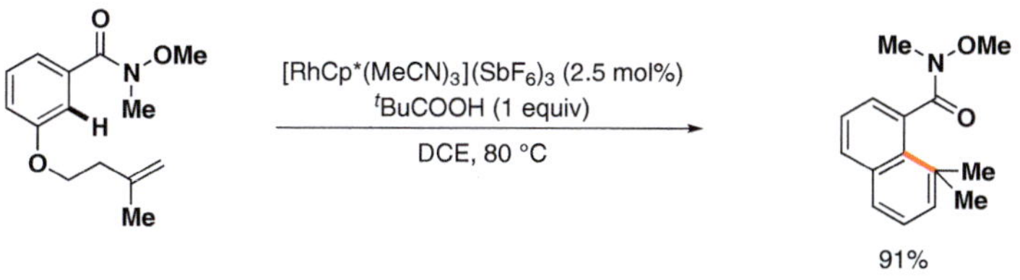

Scheme 12.21 Rh(III)-catalyzed Weinreb amide-directed alkene hydroarylation. Source: Adapted from Davis et al. [40].

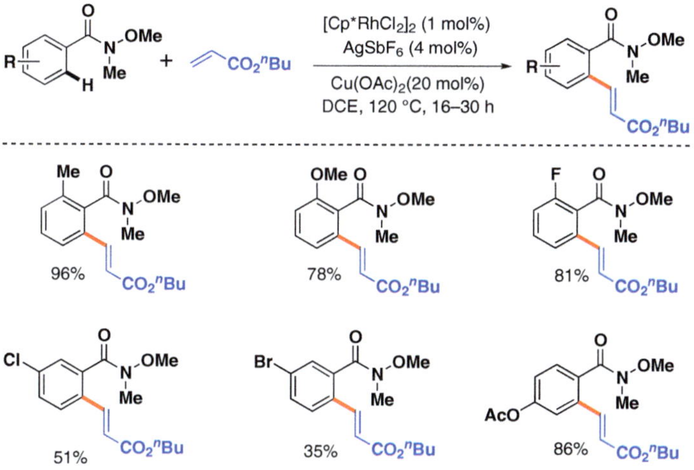

Scheme 12.22 Rh-catalyzed C–H alkenylation directed by Weinreb amides. Source: Adapted from Wang et al. [41].

An intermolecular kinetic isotope effect value of 4.6 suggested that C–H activation is involved in the rate-determining step of the reaction. The authors also proposed a reaction mechanism of the olefination reaction (Scheme 12.23). Anion exchange of $[RhCp^*Cl_2]_2$ with $AgSbF_6$ forms a more reactive Rh catalyst. Next, the Weinreb amide generates a C–H five-membered rhodacycle A after coordination with reactive Rh catalyst. Coordination of alkene is followed by insertion into the rhodacycle that forms C. Finally, elimination from C generates the olefinated product and Rh(I) species, which is reoxidized by Cu(II) to Rh (III). Cu(I) is then reoxidized by air to Cu(II).

In 2018, Qin and coworkers disclosed another Rh-catalyzed olefination protocol that was similar to Wang's condition previously stated (Scheme 12.24) [42]. Ethane sulfonyl fluoride (ESF) was utilized as the coupling partner and 1,4-dioxane as the solvent in the reaction unlike Wang's reaction condition. A number of substituted aryl Weinreb amides were shown to be compatible with the protocol with good to excellent yields in most cases. When the authors increased the $AgSbF_6$ loading to 1 equiv, cyclic lactone was formed instead of acyclic *ortho*-olefinated product. Residual

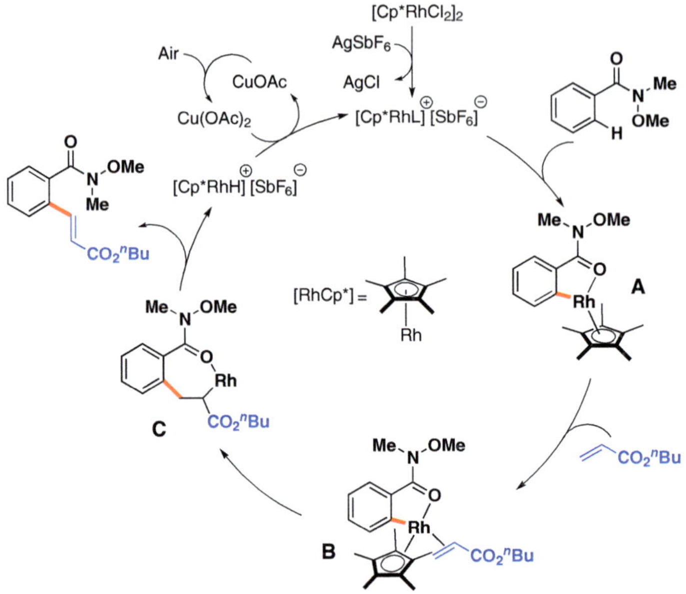

Scheme 12.23 Proposed mechanism for Rh-catalyzed C–H alkenylation.

water formed from hygroscopic $AgSbF_6$ was proposed to promote the hydrolysis of Weinreb amide and thereby formed lactone, although the authors were not sure of the mechanism.

Recent years have witnessed an upsurge of photocatalyzed C–H functionalization reactions. Such reactions provide a greener route approach to some extent by eliminating the need of using external oxidants [43–45]. A photocatalyst whose oxidative power is obtained via photoexcitation replaces the exogenous oxidant in photocatalyzed reactions. Rueping group in 2014 applied this approach for alkenylation of Weinreb and other amides, driven by a visible light-regenerated Ru-based photocatalyst, $[Ru(bpy)_3][PF_6]_2$ (Scheme 12.25) [46]. The extra oxidant $Cu(OAc)_2$ in Wang's condition (mentioned previously) was replaced here with the Ru photocatalyst. This protocol is not just efficient for Weinreb amides but also for other amides. Various olefin partners including acrylates, sulfone, and silyl ethers were well tolerated under this reaction condition. In absence of light or the photocatalyst, trace amount of product formed, which is evident of the role of light and photocatalyst in the reaction. The mechanism of the reaction resembles that of Wang's catalytic cycle except the fact that Rh(III) catalyst regeneration in this case was attributed to photoredox catalyst.

12.2.5 Ir-Catalyzed Reactions

Ir-catalyzed C–H functionalizations got more popularity in the past decade. Boronates are one of the most versatile groups in organic synthesis as they can be transformed to a number of other functionalities through different mechanisms. Among other methods of introducing boronates into organic compounds,

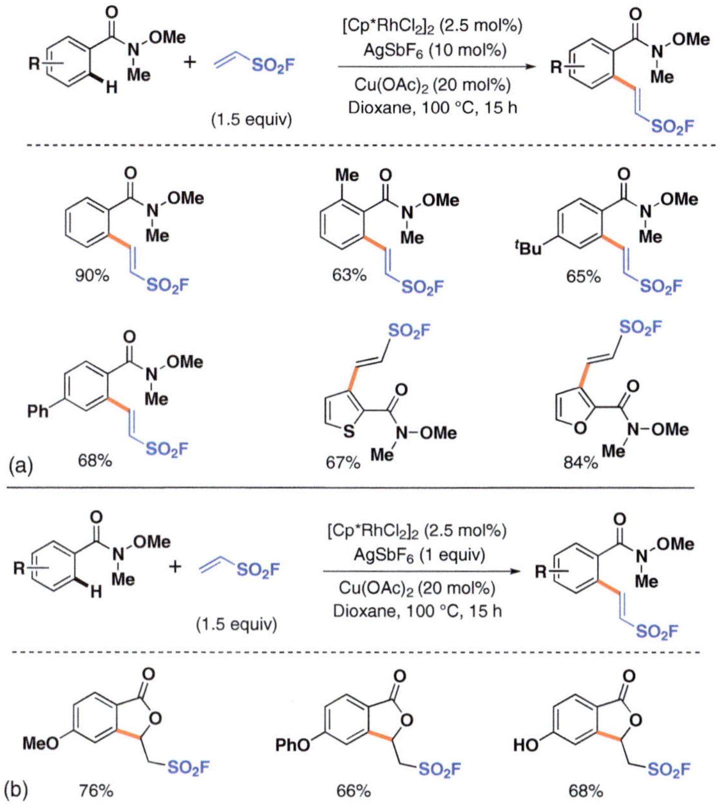

Scheme 12.24 Rh-catalyzed *ortho*-olefination utilizing ethane sulfonyl fluoride as olefinating partner. Source: Adapted from Wang et al. [42].

Ir-catalyzed borylation holds a special place as it provides one of the mildest condition and gives control over regioselectivity through sterics, directing groups, or inherent electronic property of substrates [47–52].

In 2012, Wang and coworkers disclosed an Ir-catalyzed hydrogenation procedure of α,β-unsaturated Weinreb amides in an enantioselective manner (Scheme 12.26) [53]. An Ir complex bearing a spiro P^N ligand spinPhos was utilized for enantioselective hydrogenation of Weinreb amides. 50 atm pressure was essential for better conversion of the hydrogenated product. Reducing pressure resulted in lower yield as well as enantioselectivity. Changing substituents at β,β′- position did not affect the yield and enantioselectivity of the reaction. Almost full conversions were observed for most substrates along with excellent enantioselectivities.

Krska, Maleczka, and Smith in 2014 reported an Ir-catalyzed *ortho*-borylation of Weinreb amide utilizing di-tertiary bipyridine (dtbpy) as ligand (Scheme 12.27) [54]. Although a single example was disclosed, the successful reaction on Weinreb amides demonstrated the competence of Weinreb amide as directing group in borylation reaction. Also, the other advantages of the reaction such as high functional group tolerance could be paired with Weinreb amide substrates. [Ir(μ-OMe)(COD)]$_2$ was

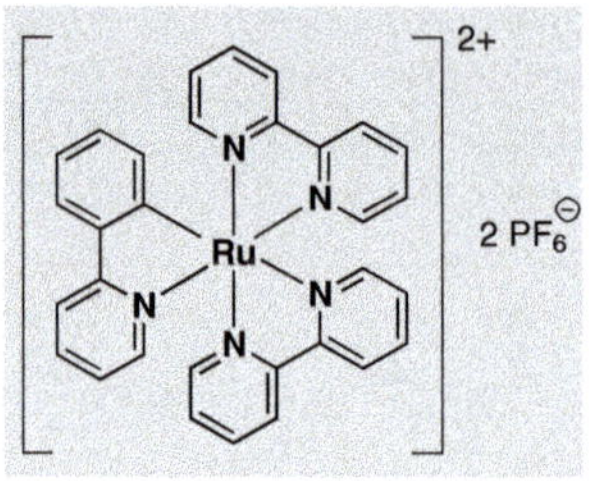

$[Ru(bpy)_3](PF_6)_2$

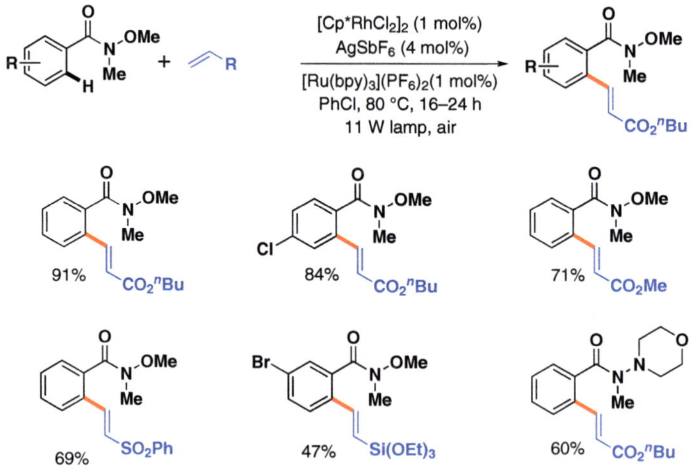

Scheme 12.25 Rh-catalyzed olefination of aryl Weinreb amides enabled by Ru photocatalysis. Source: Adapted from Fabry et al. [46].

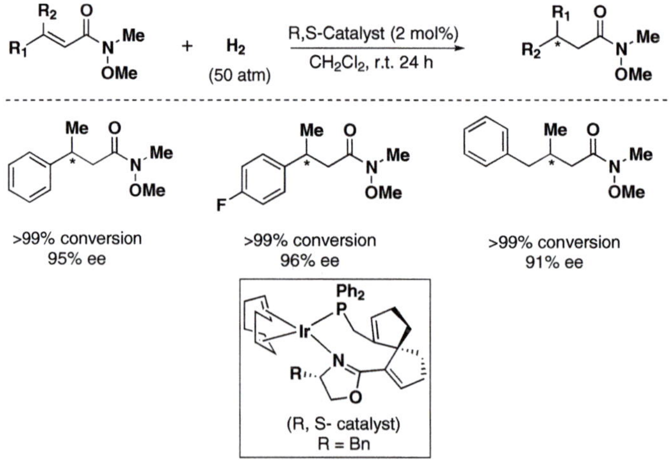

Scheme 12.26 Enantioselective hydrogenation of α,β-unsaturated Weinreb amides using a chiral Ir catalyst. Source: Adapted from Shang et al. [53].

utilized as the catalyst in the reaction along with dtbpy as ligand and $B_2(Pin)_2$ as the boron source. The single entry of Weinreb amides formed *ortho*-borylated product in 84% yield.

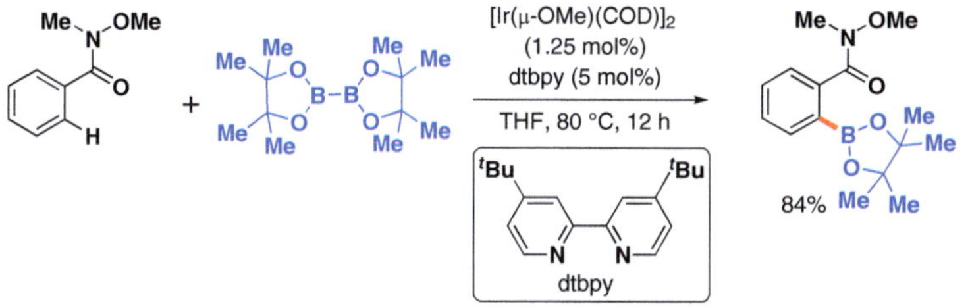

Scheme 12.27 Ir-catalyzed Weinreb amide-directed *ortho*-borylation of arene. Source: Adapted from Ghaffari et al. [54].

In 2015, Yamamoto and Shirai outlined an Ir-catalyzed protocol for asymmetric hydroarylation of bicycloalkenes (Scheme 12.28) [55]. A single entry of Weinreb was included in the paper with 64% yield and 99% ee. Although a single was described, this reaction was representative for Weinreb amides and could be extended to other aryl Weinreb amides as well. A cationic Ir catalyst was utilized for this reaction. The enantioselectivity was induced by a chiral bis(phosphoramidite) ligand (R,R)-S-Me-(R,R′)-Me-BIPAM.

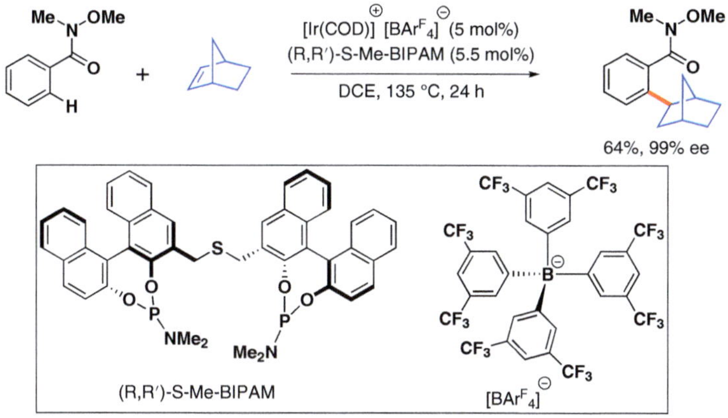

Scheme 12.28 Ir-catalyzed enantioselective C–H hydroarylation of an olefin, directed by a Weinreb amide. Source: Adapted from Shirai et al. [55].

Kinetic studies suggested the C–H cleavage step as the rate-determining step of the reaction. The mechanism of the reaction involves formation of a metallacycle (Ir catalyzed) via C–H cleavage followed by coordination and migratory insertion of bicycloalkene into Ir–C bond. Next, reductive elimination generates the product along with regeneration of the Ir catalyst (Scheme 12.29).

Formation of C–N bonds is of special importance in organic synthesis due to their ubiquitous presence in nature [56]. Although transition metal-catalyzed

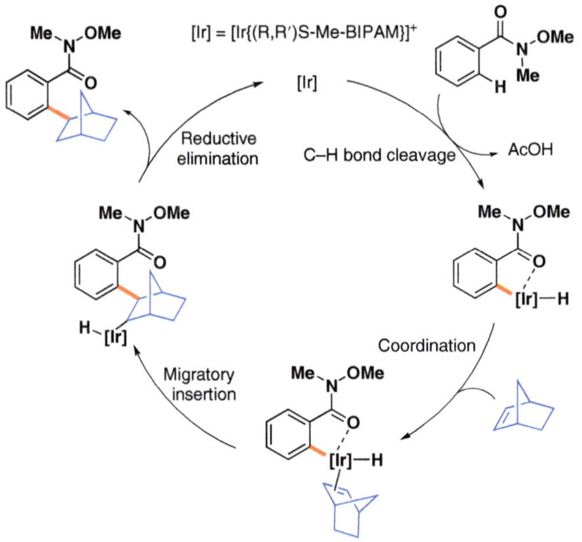

Scheme 12.29 Mechanism of Ir-catalyzed enantioselective hydroarylation of bicyclic olefin.

C–H amination reactions have been explored in recent years, amination in case of Weinreb amides remained unexplored until Lu group disclosed an Ir-catalyzed *ortho*-amination of aryl Weinreb amides using 2,2,2-trichloroethoxycarbonyl (Troc) azide as an aminating agent (Scheme 12.30) [57]. Organic azides are excellent aminating reagents as they are easy to prepare and generate N_2 as the only by product. The authors utilized $TrocN_3$ as the aminating reagent because of the electron-withdrawing property of Troc group that increases reactivity of the azide. A combination of catalytic AgOTf and 1 equiv of KOAc was employed to increase the yield of products. A number of electron-donating and electron-withdrawing functional groups on the Weinreb amide were found to be well compatible with the reaction and provided aminated products in good to excellent yields.

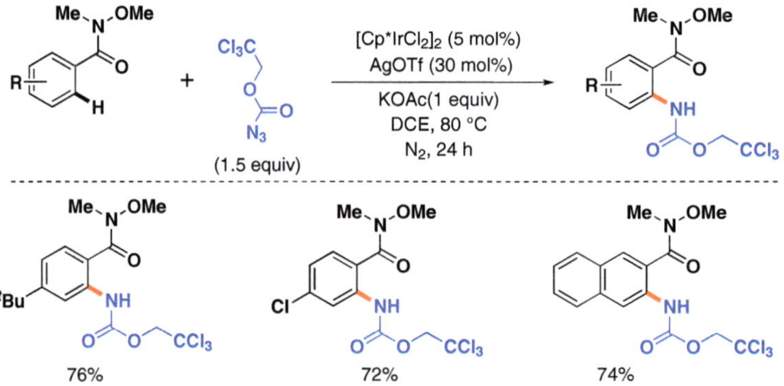

Scheme 12.30 Ir-catalyzed Weinreb amide-directed *ortho*-C–H amination of arenes. Source: Adapted from Dong et al. [57].

Further, the NHTroc group in the products was hydrolyzed to $-NH_2$ group by treating with K_2CO_3 in dimethyl sulfoxide (Scheme 12.31a). Thus a simple hydrolysis provided various substituted anilines. The carbonyl group in Weinreb amide remained unaffected by the hydrolysis condition due to less electrophilicity compared with carbonyl group of Troc moiety. The authors utilized the aminating products for further synthetic manipulation by reacting it with primary amines. Using N,N-diisopropylethylamine (DIPEA) as a base, primary amine, which acts as nucleophile, reacts with the NHTroc group to form quinazolin-2,4-dione (Scheme 12.31b). Various primary amines like 5-methoxyindole-3-ethylamine, furan-2-methylamine, cyclohexylamine, isobutylamine, ammonia, etc. have been successfully employed in this cascade reaction for the formation of quinazolin-2,4-diones. Notably, quinazolin-2,4-dione is also a key fragment in many natural alkaloids. Naturally occurring alkaloids such as pelanserine and goshuyuamide II have been prepared by utilizing this methodology.

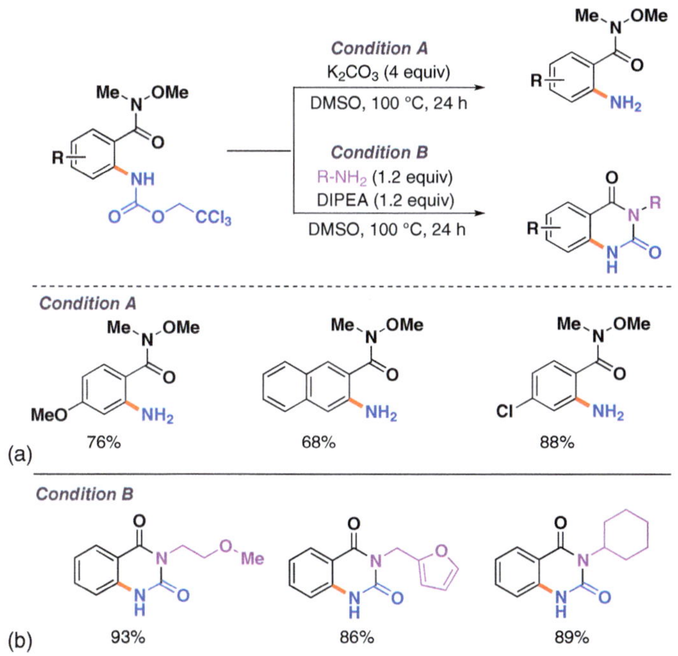

Scheme 12.31 (a) A selective deprotection procedure for the synthesis of *ortho*-NH_2 aryl Weinreb amides. (b) Quinolin-2,4-diones synthesis via a cascade aminocyclization approach.

12.3 Weinreb Amide-Directed C(sp³)–H Activation

Although C(sp^2)–H functionalization assisted by Weinreb amide have been explored, C(sp^3)–H functionalization remains a significant challenge. The inert and fluxional nature of C(sp^3)–H bonds makes their activation extremely challenging [4, 58]. On top of that, Weinreb amide in this context presents unique challenges. First, inductive effect of –OMe weakens the binding ability of carbonyl group (of Weinreb amide group) with transition metal, and, secondly, bi coordination with both carbonyl group and OMe group could slow down the catalytic activity of the catalyst. Instead of these challenges, a few reports have been disclosed in recent years. In 2018, Yu and coworkers disclosed the first report of Weinreb amide-directed C(sp^3)–H activation [59]. Using a pyridine sulfonic acid as ligand, β-C(sp^3)–H arylation of Weinreb amides have been carried out. The non-coordinating anion sulfonate on pyridine promotes dissociation of acetate ion via electrostatic repulsion from anionic complex C, and thus an active intermediate forms, which then activates the β-C(sp^3)–H bond (Scheme 12.32). In absence of sulfonate ligand on pyridine, dissociation of acetate from complex A is disfavored due to high energy associated with cationic complex B. Thus pyridine sulfonate ligand stabilizes the active Pd intermediate D as a zwitter ionic species.

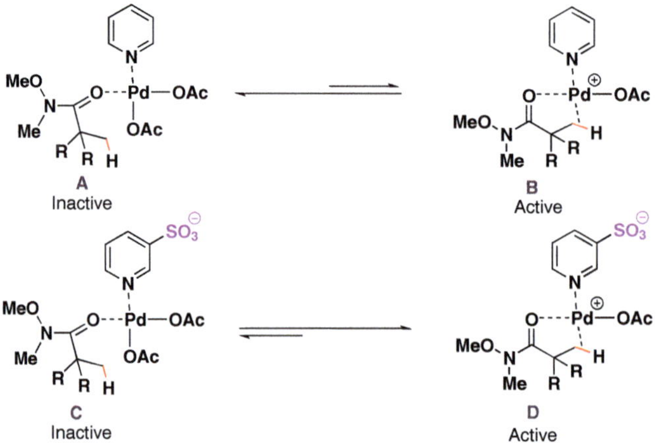

Scheme 12.32 Working hypothesis of pyridine sulfonic acid ligand.

A number of different Weinreb amides derived from aliphatic acids and cyclopropane carboxylic acids were compatible with the protocol. Nonetheless, α-H-containing Weinreb amides did not work presumably due to lack of Thorpe–Ingold effect. Similarly, various substituted aryl iodides were also tolerated in the reaction and provided arylated products in good to excellent yields (Scheme 12.33). Further, computational studies suggested the role of pyridine sulfonic acid in stabilizing the substrate bound Pd species (Scheme 12.32), which is likely due to the enhanced cationic character of the Pd center.

The same group later reported a single example of Weinreb amide-directed olefination in their study of native amide-directed C(sp^3)–H olefination (Scheme 12.34) [60]. Utilizing an electron-deficient pyridine sulfonic acid ligand (the authors established pyridine sulfonic acid ligand in their previous arylation work) olefination was carried out in excellent yield. Notably, olefination was also accomplished on a substrate that do not contain α-quaternary center.

In 2020, Yu group devised a [3+2] cycloaddition reaction utilizing the same ligand between a Weinreb amide containing substrate and maleimide via a 1,3-twofold

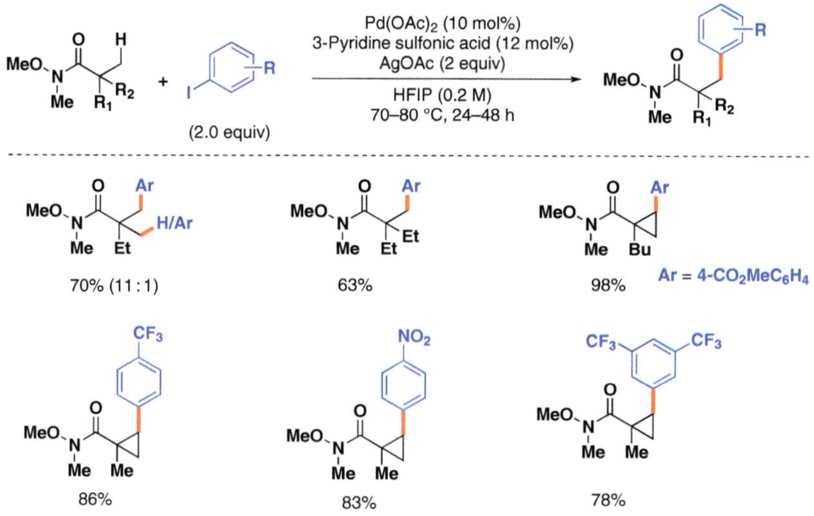

Scheme 12.33 Weinreb amide-directed Pd-catalyzed β-C(sp^3)–H arylation.

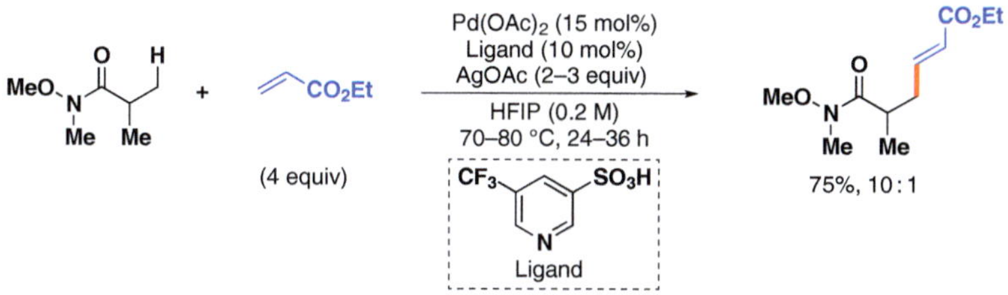

Scheme 12.34 Weinreb amide-directed β-C(sp^3)–H olefination of aliphatic acids. Source: Adapted from Park et al. [60].

C(sp^3)–H activation [61]. A single entry of Weinreb amide substrated was showcased, which formed the cycloaddition product in excellent yield (Scheme 12.35). The role of weakly coordinating directing group was proposed to be essential in carrying out this double activation protocol as otherwise strong directing group would have given Heck-type or alkylation product.

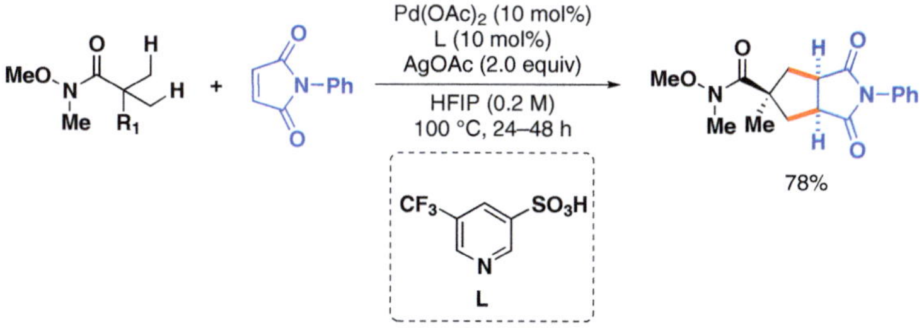

Scheme 12.35 Weinreb amide-assisted [3+2] cycloaddition via a twofold β-C(sp^3)–H activation.

12.4 Conclusions and Outlook

Recent developments in C–H functionalization methodology have greatly expanded the synthetic toolbox of chemists. Weinreb amide has come up as one of the privileged motifs to carry out C–H bond transformations in aromatic as well as aliphatic substrates. Strong coordinating directing groups have mainly dominated the C–H activation field in the past decade. Nonetheless, weaker coordinating directing groups are getting more attention recently due to their unique properties. A weaker coordination increases the diversity of reactions by increasing the selectivity and reactivity in a number of reactions. Although Weinreb amides have been well explored by synthetic chemists to get synthons such as aldehyde or ketone, they are still unexplored in the domain of C–H activation. Even when Weinreb amides are showcased in C–H activation reactions, they are shown as special examples in studies outlining a general scope. There are not many examples where Weinreb amides are presented as a substrate class. Also, compared with other amides, Weinreb amides often provide different product under the same reaction condition due to reactivity of their N–O bond. The examples provided in the chapter is evident of the fact that Weinreb amide could serve as an excellent directing group with a number of transition metal-catalyzed reactions. Considering this and the other synthetic utility of Weinreb amides, more exploration is necessary to unleash the potential of Weinreb amides as a directing group in C–H activation.

Notably, most Weinreb amide-directed reactions involve use of toxic halogenated solvents. Moving toward greener solvents would be a challenge in the future. Mostly, second or third row transition metals have worked well with Weinreb amide-directed reactions. Usage of cheap and abundant metals like Cu, Ni, Al, Fe, and Mn would be another area to investigate. Weinreb amide-assisted reactions are also compatible with newer strategies of carry out C-H activation reactions such as photocatalysis and electrocatalysis. More research is expected on these in the coming days.

Finally, C(sp^3)–H activation examples directed by Weinreb amide are extremely rare. A handful of examples in the literature already established that Weinreb amides could also activate inert and more challenging sp^3 C–H bonds. Further honing is required to discover the potential of Weinreb amides in sp^3 C–H bond activation reactions. Although Weinreb amides remain a less-explored auxiliary in C–H activations, remarkable advances made in recent years will stimulate further developments and eventually make it a valuable motif to accomplish unique C–H bond transformations.

References

1 Nahm, S. and Weinreb, S.M. (1981). *Tetrahedron Lett.* 22: 3815.
2 Balasubramanium, S. and Aidhen, I.S. (2008). *Synthesis* 23: 3707.
3 Gensch, T., Hopkinson, M., Glorius, F. et al. (2016). *Chem. Soc. Rev.* 45: 2900.

4 Das, J., Guin, S., and Maiti, D. (2020). *Chem. Sci.* 11: 10887.
5 Sambiagio, C., Schönbauer, D., Blieck, R. et al. (2018). *Chem. Soc. Rev.* 47: 6603.
6 Kalepu, J. and Pilarski, L.T. (2019). *Molecules* 24: 830.
7 Arockiam, P.B., Bruneau, C., and Dixneuf, P.H. (2012). *Chem. Rev.* 112: 5879.
8 Ruiz, S., Villuendas, P., and Urriolabeitia, E.P. (2016). *Tetrahedron Lett.* 57: 3413.
9 Yang, F. and Ackermann, L. (2013). *Org. Lett.* 15: 718.
10 Simmons, E.M. and Hartwig, J.F. (2012). *Angew. Chem. Int. Ed.* 51: 3006.
11 Das, R. and Kapur, M. (2015). *Chem. Asian J.* 10: 1505.
12 Ferreira, E.M., Zhang, H., and Stoltz, B.M. (2009). Oxidative Heck-Type Reactions (Fujiwara–Moritani Reactions). In: *The Mizoroki–Heck Reaction* (ed. M. Oestreich). Chichester: Wiley. 9: 345–382.
13 Das, R. and Kapur, M. (2016). *Chem. Eur. J.* 22: 16986.
14 More, N.Y., Padala, K., and Jeganmohan, M. (2017). *J. Org. Chem.* 82: 12691.
15 Massignan, L., Tan, X., Meyer, T.H. et al. (2019). *Angew. Chem. Int. Ed.* 59: 3184.
16 Yan, M., Kawamata, Y., and Baran, P.S. (2017). *Chem. Rev.* 117: 13230.
17 Moselage, M., Li, J., and Ackermann, L. (2016). *ACS Catal.* 6: 498.
18 Yoshikai, N. (2014). *Bull. Chem. Soc. Jpn.* 87: 843.
19 Yoshino, T., Ikemoto, H., Matsunaga, S. et al. (2013). *Chem. Eur. J.* 19: 9142.
20 Yoshino, T., Ikemoto, H., Matsunaga, S. et al. (2013). *Angew. Chem. Int. Ed.* 52: 2207.
21 Planas, O., Chirila, P.G., Whiteoak, C.J. et al. (2018). *Adv. Organomet. Chem.* 69: 209.
22 Bunno, Y., Murakami, N., Suzuki, Y. et al. (2016). *Org. Lett.* 18: 2216.
23 Kawai, K., Bunno, Y., and Yoshino, T. et al. (2018). *Chem. Eur. J.* 24: 10231.
24 Johansson Seechurn, C.C.C., Kitching, M.O., Colacot, T.J. et al. (2012). *Angew. Chem. Int. Ed.* 51: 5062.
25 Sehnal, P., Taylor, R.J.K., and Fairlamb, I.J.S. (2010). *Chem. Rev.* 110: 824.
26 Wang, D., Weinstein, A.B., White, P.B. et al. (2018). *Chem. Rev.* 118: 2636.
27 Topczewski, J.J. and Sanford, M.S. (2015). *Chem. Sci.* 6: 70.
28 Bonney, K.J. and Schoenebeck, F. (2014). *Chem. Soc. Rev.* 43: 6609.
29 Wang, Y., Zhou, K., Lan, Q. et al. (2015). *Org. Biomol. Chem.* 13: 353.
30 Mudarra, Á.L., Martínez de Salinas, S., and Pérez-Temprano, M.H. (2018). *Org. Biomol. Chem.* 17: 1655.
31 Yedage, S.L. and Bhanage, B.M. (2016). *J. Org. Chem.* 81: 4103.
32 Li, G., Wan, L., Zhang, G. et al. (2015). *J. Am. Chem. Soc.* 137: 4391.
33 Engle, K.M., Wang, D.-H., and Yu, J.-Q. (2010). *J. Am. Chem. Soc.* 132: 14137.
34 Das, R. and Kapur, M. (2018). *Asian J. Org. Chem.* 7: 1524.
35 Das, R. and Kapur, M. (2017). *J. Org. Chem.* 82: 1114.
36 Powers, D.C., Lee, E., Ariafard, A. et al. (2012). *J. Am. Chem. Soc.* 134: 12002.
37 Chatani, N. and Rej, S. (2018). *Angew. Chem. Int. Ed.* 58: 8304.
38 Colby, D.A., Bergman, R.G., and Ellman, J.A. (2010). *Chem. Rev.* 110: 624.
39 Vásquez-Céspedes, S., Wang, X., and Glorius, F. (2018). *ACS Catal.* 8: 242.
40 Davis, T.A., Hyster, T.K., and Rovis, T. (2013). *Angew. Chem. Int. Ed.* 52: 14181.
41 Wang, Y., Li, C., Li, Y. et al. (2013). *Adv. Synth. Catal.* 355: 1724.
42 Wang, S.-M., Li, C., Leng, J. et al. (2018). *Org. Chem. Front.* 5: 1411.

43 Wang, C.-S., Dixneuf, P.H., and Soulé, J.-F. (2018). *Chem. Rev.* 118: 7532.
44 Romero, N.A. and Nicewicz, D.A. (2016). *Chem. Rev.* 116: 11075.
45 Prier, C.K., Rankic, D.A., and MacMillan, D.W.C. (2013). *Chem. Rev.* 113: 5322.
46 Fabry, D.C., Zoller, J., Raja, S. et al. (2014). *Angew. Chem. Int. Ed.* 53: 10228.
47 Hartwig, J.F. (2011). *Acc. Chem. Res.* 45: 864.
48 Smith, M.R., Bisht, R., Haldar, C. et al. (2018). *ACS Catal.* 8: 6216.
49 Li, H.-L., Kanai, M., and Kuninobu, Y. (2017). *Org. Lett.* 19: 5944.
50 Crawford, K.M., Ramseyer, T.R., Daley, C.J.A. et al. (2015). *Angew. Chem. Int. Ed.* 53: 7589.
51 Hoque, M.E., Bisht, R., Haldar, C. et al. (2017). *J. Am. Chem. Soc.* 139: 7745.
52 Larsen, M.A. and Hartwig, J.F. (2014). *J. Am. Chem. Soc.* 136: 4287.
53 Shang, J., Han, Z., Li, Y. et al. (2012). *Chem. Commun.* 48: 5172.
54 Ghaffari, B., Preshlock, S.M., Plattner, D.L. et al. (2014). *J. Am. Chem. Soc.* 136: 14345.
55 Shirai, T. and Yamamoto, Y. (2015). *Angew. Chem. Int. Ed.* 54: 9894.
56 Park, Y., Kim, Y., and Chang, S. (2017). *Chem. Rev.* 117: 9247.
57 Dong, X., Ma, P., Zhang, T. et al. (2020). *J. Org. Chem.* 85: 13096.
58 He, C., Whitehurst W. G., Gaunt, M.J. (2019). *Chem.* 5, 1031.
59 Park, H., Chekshin, N., Shen, P.-X. et al. (2018). *ACS Catal.* 8: 9292.
60 Park, H., Li, Y., and Yu, J.-Q. (2019). *Angew. Chem. Int. Ed.* 58: 1.
61 Park, H. and Yu, J.-Q. (2020). *J. Am. Chem. Soc.* 142: 16552.

13

Computational Studies of Amide C–N Bond Activation

Xin Hong, Pei-Pei Xie, Zhi-Xin Qin, and Shuo-Qing Zhang

Center of Chemistry for Frontier Technologies, Department of Chemistry, State Key Laboratory of Clean Energy Utilization, Zhejiang University, 38 Zheda Road, Hangzhou, 310027, China

13.1 Introduction

As a ubiquitous functional moiety in organic chemistry, amide group plays an important role in a wide array of areas [1]. The distinctive chemical properties of amide give this special C–N linkage the extra stability, the planar geometry, and the rich possibilities for intermolecular interactions and organic transformations, which leads to the remarkable biologic, material, and synthetic functions of amide molecules [2]. The decades of fruitful discoveries of functional amide molecules resulted in its wide appearance in drugs, polymers, and synthetic building blocks, making it one of the backbone molecules for human society [3]. Figure 13.1 presents a few highlighted cases. Of the new therapeutic small molecule drugs approved by FDA in 2020 [4], 17 of them have amide linkages, including the transformative Rukobia for HIV disease [5]. Polyamide, or nylon, is an outstanding polymer molecule with desirable strength and elasticity. The manufacturing of polyamide fiber is a massive industry process and still under extensive research efforts and competitions. As a stable functional group, one of the typical applications of amide in synthetic chemistry is to serve as a protecting group [6]. Phthalimide is a classic protecting group for amine moiety and still widely applied in modern total synthesis [7].

The amide C–N bonds are generally very strong. The n(N)-π*(CO) orbital interaction stabilizes the C–N linkage and results in the planar geometry of amide moiety. This is the reason for the typical resonance forms of amide (Figure 13.2a), in which the zwitterionic form has a double C=N bond. This resonance stability ranges from 15 to 20 kcal/mol in typical amides [8], creating the additional barrier requirement for amide C–N bond cleavage as compared with the other typical C–N single bonds. Figure 13.2b lists the C–N bond dissociation energies (BDEs) of a few representative amides and the corresponding amines from the *i*Bond database [9]. The C–N BDEs of amides are generally higher than those of amines by 20 kcal/mol (Figure 13.2b). The C–N BDE of benzamide is 96.4 kcal/mol [10], which is close to a typical C–H bond. The intrinsic stability of amide C–N bond results in its chemical

Amide Bond Activation: Concepts and Reactions, First Edition. Edited by Michal Szostak.

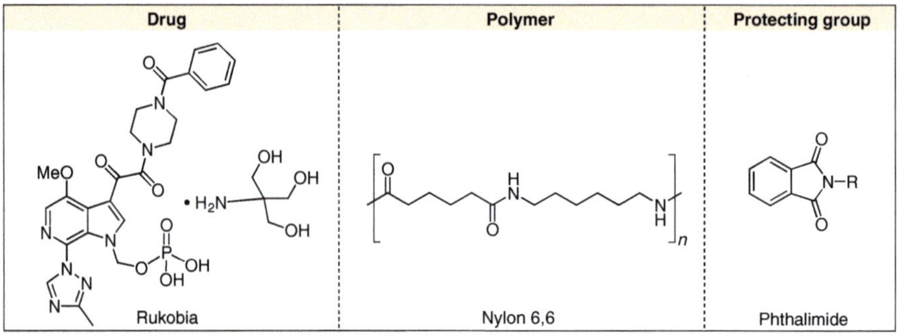

Figure 13.1 Representative applications of amide molecules.

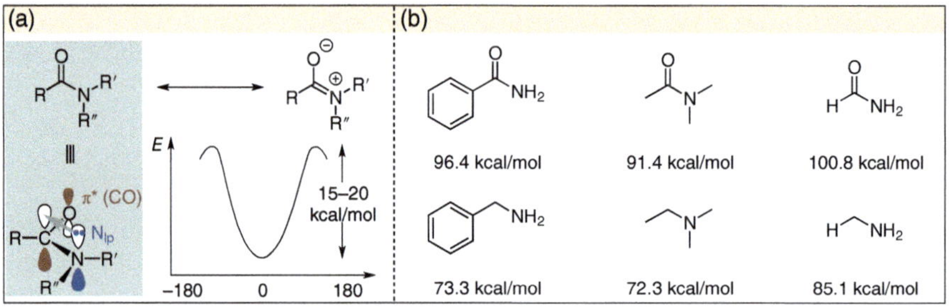

Figure 13.2 Resonance and typical bond dissociation energies of amide C–N bond. (a) Amide bond resonance. (b) C–N BDEs of representative amides and amines.

inertness. Amide is the most inert member in the family of carboxylic acid derivatives. Although similar acyl-LG bond cleavage and derivatization mechanisms are in principle operative for all members of carboxylic acid derivatives, the transformations with amides under the same conditions generally have limited or no productivity [11]. This leads to the fact that the synthetic transformations involving amide C–N bond derivatization are significantly less productive as compared with acyl halide, anhydride, acid or ester. Therefore, the synthetic value of amide molecules has been limited due to the lack of efficient approaches to break amide C–N bond.

In order to break and transform amides, organic chemists have dedicated a tremendous amount of efforts toward this synthetic challenge. Particularly in the recent few decades, the field has witnessed the development of a series of exciting new strategies to realize the catalytic or stoichiometric transformations of amides, which remarkably broadened the scope of amide chemistry [12]. In addition to the experimental advances of these innovative reactions, which are summarized and discussed in the other chapters of this book, the physical organic questions on the amide C–N bond activation mechanism and controlling factors of amide reactivity also emerged. Computations have played a critical role in this regard [13]. Through quantum chemical computations, the molecular level mechanistic understandings are revealed for the target amide transformations. These computational explorations not only explained the experimental phenomena from an energetic and molecular perspective but also provided the mechanistic basis to understand the structure–activity relationship in the C–N bond activation of amide. The synergy between experiment and computation leads to the strong momentum that drives the design and engineering of the next generation of amide transformations [14], pushing the frontier of amide chemistry. In this chapter, we will focus on the computational studies of amid C–N bond activation. The general activation modes and the origins of the reactivity and selectivity trend in related amide transformations are discussed. The coverage of the selected computational studies is by no means comprehensive, but this selection provides a reasonable overview of the amide C–N bond activation mechanism and how the collaborations between experiment and computation supported the understanding and development of amide transformations.

13.2 General Mechanisms of Amide C–N Bond Cleavage and Derivatization

Despite the abundant experimental explorations and computational studies on the amide C–N bond activation and functionalization reactions, the highlighted computational reports in this chapter and the discovered reaction mechanisms can be generalized into two typical mechanistic models, nucleophilic substitution and transition metal catalysis (Figure 13.3). These two typical mechanistic models are quite distinctive from each other, which reflect the nature of the wide applied synthetic strategies for amide C–N bond cleavage. Computational studies and mechanistic knowledge of these two models provide a fundamental basis to comprehend and design synthetic transformations with amides.

The mechanistic model of nucleophilic substitution of amides is elaborated in Figure 13.3a. Under stoichiometric or sub-stoichiometric Lewis acid additives, the carbonyl group of amide moiety is activated, leading to the nucleophilic addition of amide carbonyl group by external nucleophile. This step breaks the amide C–N resonance and generates the tetrahedral intermediate. Subsequently, the elimination of the amino leaving group from the tetrahedral intermediate produces the substituted acyl derivative. Due to the intrinsic stability of amide functional group, this type of

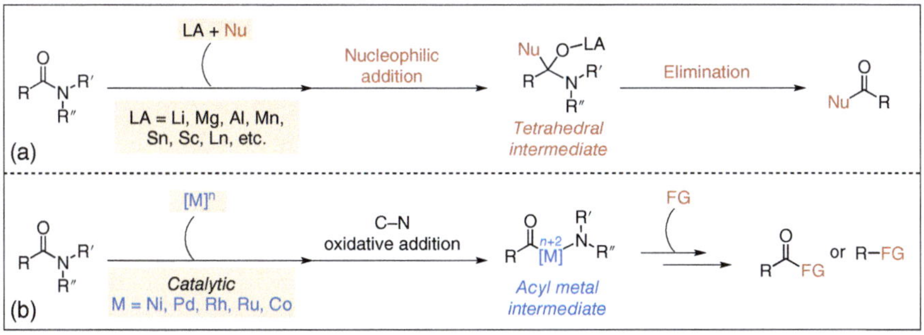

Figure 13.3 Typical reaction mechanisms of C–N bond activation of amides. (a) Nucleophilic substitution. (b) Transition metal catalysis.

nucleophilic substitution generally requires the usage of fairly strong Lewis acid (Li, Mg, Al, Mn, Sn, Sc, Ln, and others) [15] and certain engineering of the substituents of amide (e.g. Weinreb amide [16]), in order to boost the overall reactivity of C–N cleavage.

In addition to the classic nucleophilic substitution mechanism for functionalizing the carboxylic acid derivatives, the transition metal catalysis recently emerges as a powerful strategy for breaking amide C–N bond [17]. This type of transformation generally proceeds via a typical cross-coupling reaction mechanism, in which amide acts as an electrophile (Figure 13.3b). From the amide substrate, the low valent transition metal (Ni [18], Pd [19], Rh [20], Ru [21], or Co [22]) can break the strong amide C–N bond via oxidative addition, leading to the acyl metal intermediate. This acyl metal intermediate can undergo either the decarbonylative pathway or the carbonyl retention pathway to produce the observed functionalized products. Both pathways are operative, and the competition relies on the substrate and transition metal catalyst. The decarbonylative functionalization pathway involves the decarbonylation of the acyl metal intermediate, which generates the LM(R)(amino) species. Subsequent transmetalation and reductive elimination lead to the formation of the decarbonylative coupling product and regenerate the low valent active catalyst. Alternatively, the carbonyl retention pathway excludes the decarbonylation step. The acyl metal intermediate directly undergoes the sequential transmetalation and reductive elimination to give the acyl product.

Although this chapter focuses on these two general reaction mechanisms of amide C–N bond activation, we want to emphasize that these two mechanisms are *not* the only working pathways for amide C–N bond cleavages. These two mechanisms are widely seen in recent computational studies of amide C–N bond activations mainly because there are quite a number of experimental reports using these two synthetic strategies, and these two mechanistic models are suitable for readily available computational tools. Other approaches including single electron reduction [23] were also proposed in related experimental studies on amide C–N bond activation and functionalization, but computational studies and mechanistic elucidations are still limited for these reaction mechanisms.

13.3 Computational Studies on the Mechanism and Selectivity of Lewis Acid-Mediated Nucleophilic Substitution of Amides

Lewis acid is a typical catalyst or activating reagent for the nucleophilic substitution of carboxylic acid derivatives. This strategy also works for amide derivatization. Synthetic chemists have developed a series of powerful transformations that can achieve productive and selective amide C–N bond activation and functionalization reactions, even with excellent control of enantioselectivity. Computations have served a prominent role in establishing the reaction mechanism and the origins of the structure–activity relationship for these transformations, which is the focus of this section.

13.3.1 Computational Studies on the Mechanism and Selectivity of LiHMDS-Mediated Transamidation

Using LiHMDS reagent, Szostak, Hong, and coworkers reported a general and highly chemoselective approach for the transamidation of unactivated tertiary amides under mild conditions (Figure 13.4a) [24]. This transition metal-free transformation can efficiently break and functionalize the strong C–N bond of a wide scope of amides, offering a useful synthetic route for the amide synthesis. In this work, density functional theory (DFT) calculations were performed to elucidate the reaction mechanism and reactivity trend. A clear and concise mechanistic picture is present, which provides a general guideline to understand and predict amide reactivity in this transformation.

Under the M06-2X/6-311+G(d,p)-SMD(toluene)//B3LYP-D3(BJ)/6-31G(d) level of theory, DFT calculations provided a complete free energy diagram for the LiHMDS-mediated transamidation between *N*-methyl-*N*-phenylbenzamide and aniline (Figure 13.4b,c). LiHMDS is known to have the dimer and trimer forms, which were both considered in the computations. From the LiHMDS dimer **int1**, the deprotonation of aniline is exergonic by 0.3 kcal/mol, which leads to the intermediate **int2**. This intermediate complexes with the amide substrate through the Li^+-carbonyl interaction, generating the on-cycle resting intermediate **int3**. From **int3**, the intramolecular nucleophilic addition of the deprotonated aniline to the Li^+-activated amide occurs via **TS4**. This elementary step breaks the carbonyl-amino resonance in amide and produces the key tetrahedral intermediate **int5** of nucleophilic substitution mechanism. The DFT-computed barrier for the nucleophilic addition step is 19.2 kcal/mol from **int3** to **TS4**, which is considered to be the limiting step for dimer pathway. The tetrahedral intermediate **int5** undergoes the isomerization via **TS6** to have the eliminating amino group coordinated to lithium in intermediate **int7**. In **int7**, the original C–N bond of amide is significantly weakened due to the lack of resonance stability and the presence of three carbon–heteroatom bonds. Thus, the subsequent elimination via **TS8** only requires barrier of 4.9 kcal/mol to break the original C–N bond of amide and completes the transamidation. They found that the further deprotonation of the transamidation product to intermediate **int9** produces the critical thermodynamic driving force for the favorable product formation. This formation of amidate anion is important for the thermodynamic equilibrium of the transamidation reaction since the tertiary amide substrate does not contain the acidic N–H, which also explains the requirement of overstoichiometric requirement of LiHMDS reagent.

The LiHMDS trimer-mediated pathway is similar to the LiHMDS dimer-mediated pathway (Figure 13.4c). The same reaction mechanism applies, and the change from LiHMDS dimer to trimer results in minor changes on the computed free energy profile. Both pathways have the limiting nucleophilic addition step (**TS4** and **TS13**) and the facile elimination step (**TS8** and **TS17**). Comparing the two DFT-computed free energy profiles, the LiHMDS dimer-mediated pathway is 2.1 kcal/mol more favorable than the LiHMDS trimer-mediated pathway (**TS4** vs. **TS13**). This free energy difference is not that significant, which suggests that both pathways could be operative if the amide substrate varies. In the authors'

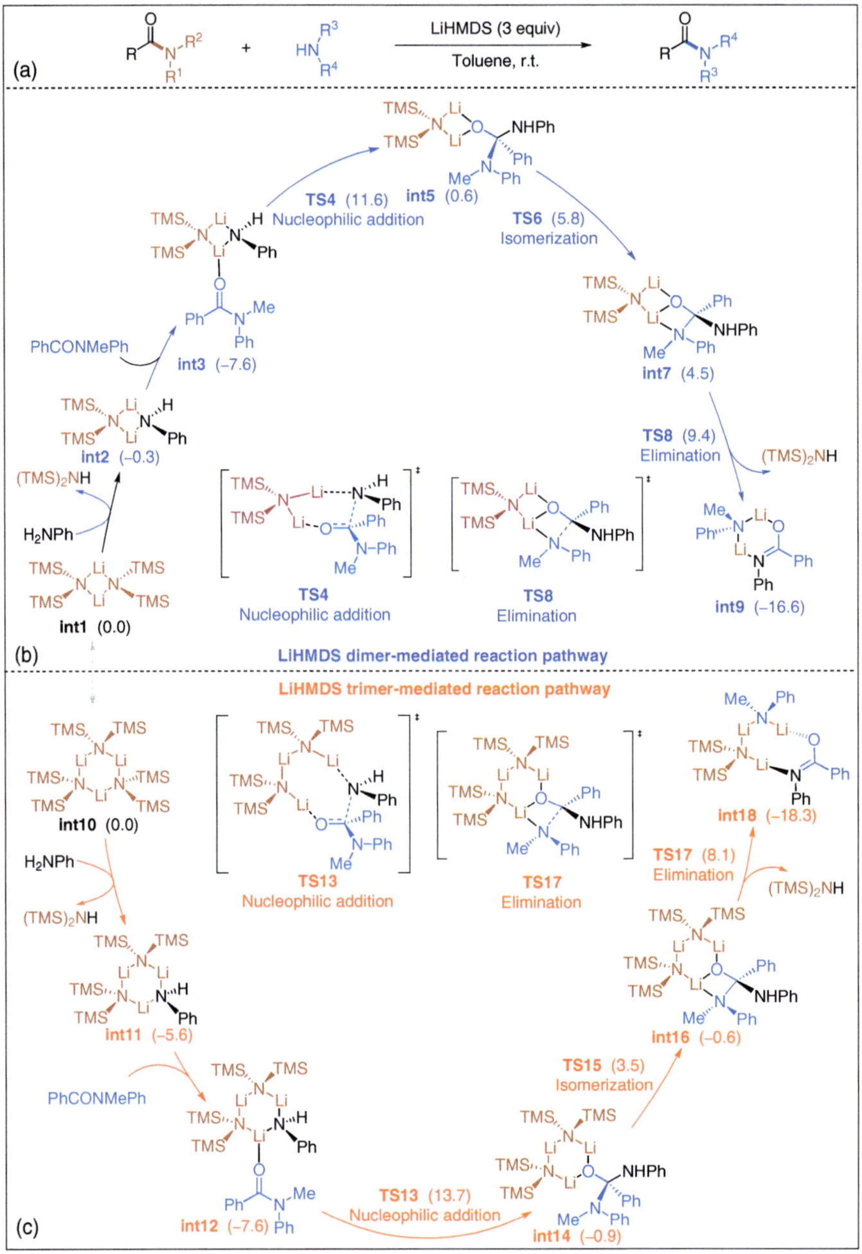

Figure 13.4 a) LiHMDS-mediated transamidation of unactivated tertiary amides. b) DFT-computed free energy profile of LiHMDS dimer-mediated transamidation. c) DFT-computed free energy profile of LiHMDS trimer-mediated transamidation. Source: From Li, G. et al. [24], © 2019 American Chemical Society. Reprinted with permission of American Chemical Society.

computations of the reactivity trend (*vide infra*), they found that the preference between dimer and trimer pathways can vary. Therefore, both pathways should be considered for reaction designs in this transformation. Due to the possible deprotonation of tertiary amide with α-C–Hs under the basic conditions, the authors also examined the possibility of ketene formation (Scheme 13.1). From the deprotonated tertiary amides, the ketene formations are endergonic by about 17 kcal/mol, whose possibility was excluded.

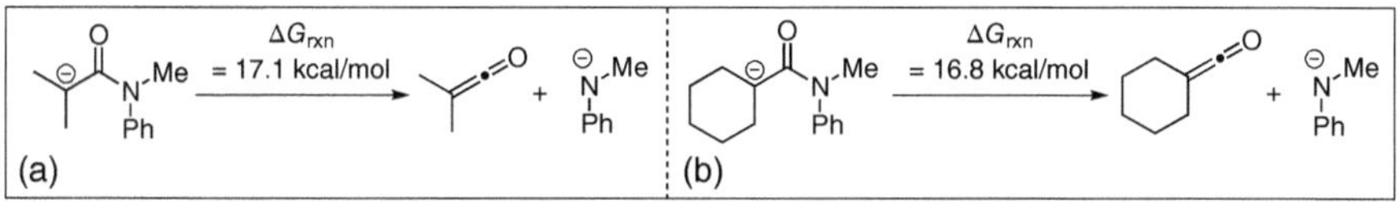

Scheme 13.1 DFT-computed free energy changes of the ketene formation from the deprotonated amides.

The elucidation of reaction mechanism provides the basis to rationalize the reactivity trend. The authors explored the barriers of the rate-limiting nucleophilic addition step for two representative twisted amides **19** and **26** (Figure 13.5). Both amides have lower barrier for the nucleophilic addition step as compared to *N*-methyl-*N*-phenylbenzamide, especially the fully twisted amide **26.** These computations are consistent with the experimentally observed high reactivities of these twisted amides. This change of reactivity was rationalized based on the geometric structures of the twisted amides. **26** is more twisted than **19**, which is reflected in the geometric indexes (τ and χ_N) of the corresponding X-ray crystal structures (Figure 13.5c). The geometric twisting also leads to the elongation of the amide C–N bond and the mitigation of the amide resonance stability energy (E_R). These comparisons highlight the role of geometric twisting in the determination of amide C–N bond activation barrier. The authors further extended the reactivity comparison of two amides to a quantified reactivity scale of a series of typical amides in this transformation (Scheme 13.2). This computational scale agrees well with the experimental observations, which provides a useful guideline for related reaction designs.

Under the same conditions, the authors reported that the C–N bond cleavage of ester is also feasible, leading to the amidation of a series of unactivated esters (Figure 13.6a) [24]. Through DFT calculations, they found that the same reaction mechanism applies to the amidation of ester (Figure 13.6b). Both LiHMDS dimer and trimer pathways were considered for the mechanistic computations of the model reaction between methyl benzoate and aniline. From the LiHMDS dimer **int1**, the deprotonation of aniline leads to the intermediate **int2**, as in the transamidation mechanism. Subsequent complexation of ester generates the intermediate **int33**. This intermediate undergoes the rate-limiting nucleophilic addition from the deprotonated aniline to the Lewis acid-activated ester, which produces the tetrahedral intermediate **int35**. From **int35**, the elimination of methoxy group is facile through **TS36**, leading to the Li^+-amide complex **int37**. Similar to the transamidation reaction, the formation of the N–H deprotonated amidation product

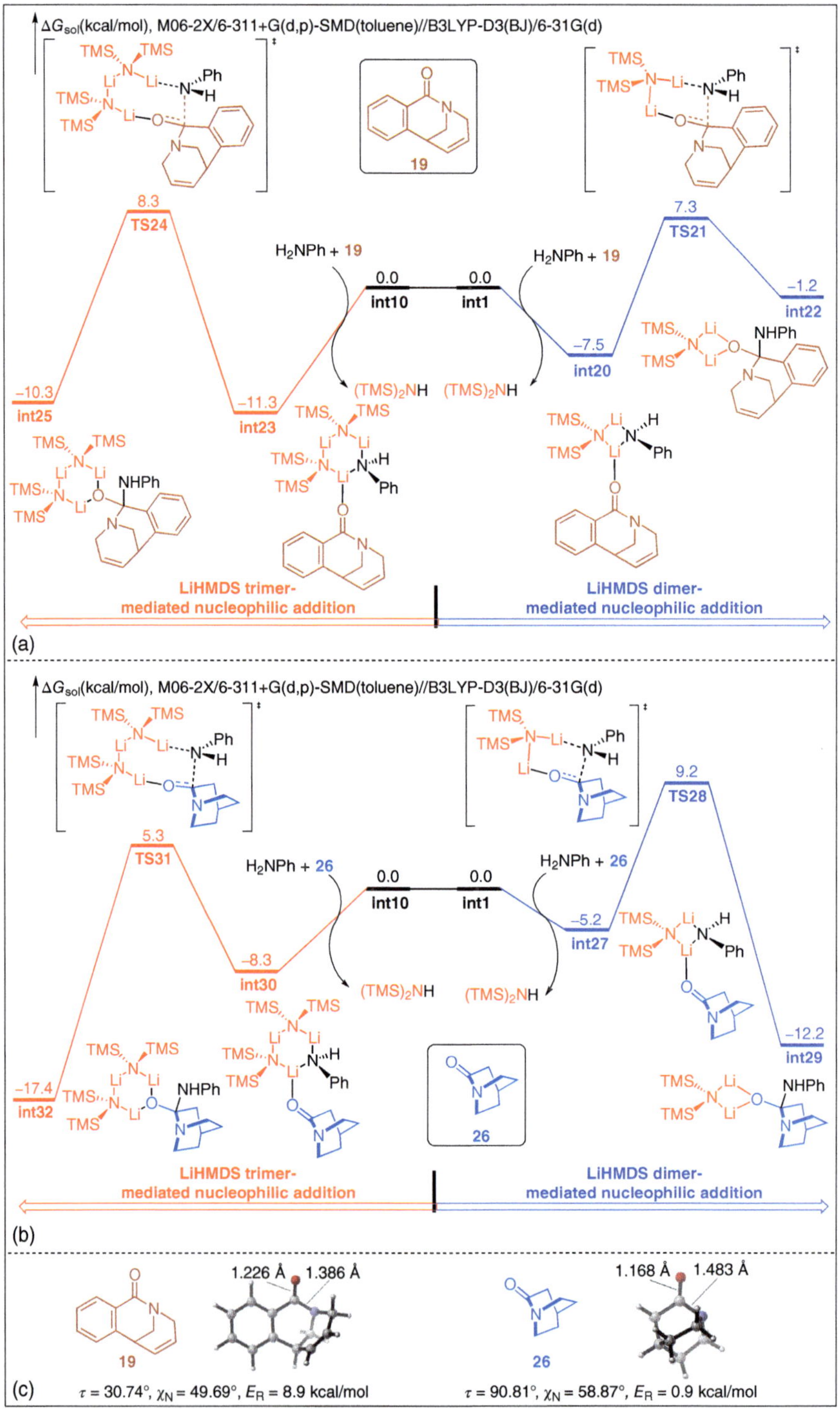

Figure 13.5 DFT-computed free energy changes of LiHMDS-mediated nucleophilic addition of selected twisted amides. Source: From Li, G. et al. [24], © 2019 American Chemical Society. Reprinted with permission of American Chemical Society.

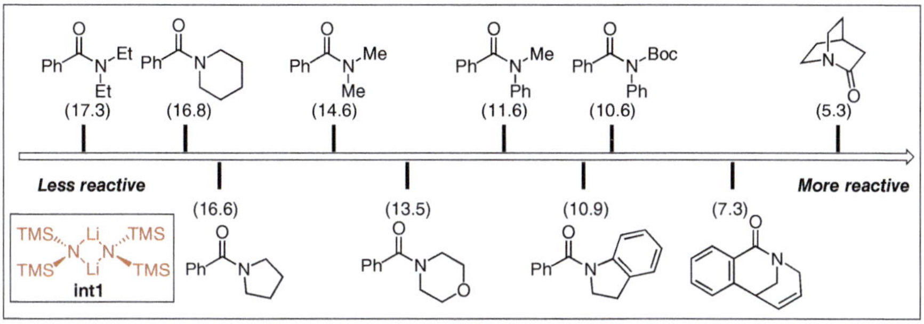

Scheme 13.2 Computational reactivity scale of amides in LiHMDS-mediated transamidation. Free energy barriers are in kcal/mol. The amide reactivities are ranked from the same LiHMDS dimer intermediate **int1**. Source: From Li, G. et al. [24], © 2019 American Chemical Society. Reprinted with permission of American Chemical Society.

int39 is needed for the favorable thermodynamic equilibrium. The alternative LiHMDS trimer-mediated amidation pathway is similar to the dimer pathway. The overall barrier of the rate-limiting nucleophilic addition in trimer pathway is very close to that of the dimer pathway (**TS34** vs. **TS41**), which indicates that both pathways are operative. These computations revealed that a general mechanistic model applies to the LiHMDS-mediated nucleophilic substitution of carboxylic acid derivatives.

13.3.2 Computational Studies on the Mechanism and Reactivity of Zn-Catalyzed Esterification of Amides

To realize the challenging alcoholysis of primary amides, Maes and coworkers developed a powerful Zn-catalyzed transamidation reaction (Figure 13.7) [25]. This two-step approach first derivatizes the primary amide to install the designed *tert*-butyl nicotinate directing group on the amide nitrogen under palladium catalysis, and subsequent Zn-esterification of the *N*-substituted amide occurs smoothly under weak base and mild conditions. As a biomimic of metallo-exopepetidase activity, this amide activation utilizes the judicious design of the directing group to allow the favorable bis-chelation to Zn(II) catalyst, and the hydrogen-bonding between acetate and alcohol facilitates the desired nucleophilic addition to activate amide. The authors thoroughly explored the effect of directing group and found that the conformational preference is essential for the success of directing group design. With the *tert*-butyl nicotinate directing group that favors the bis-ligating conformation, a wide scope of amides and alcohols are tolerated in this transformation.

To understand the reaction mechanism and origins of directing group effect, the authors performed DFT calculations using B3LYP functional and 6-31G(d,p) basis set [25]. The computed free energy profiles of the Zn(II)-catalyzed esterification of benzamide derivatives are shown in Figure 13.8. For the favorable *tert*-butyl nicotinate directing group, zinc acetate, methanol, and amide substrate first complex to form the complex **int49**. Subsequent addition of the hydrogen bonded methanol to the Zn(II)-activated amide occurs via **TS50**, generating the tetrahedral intermediate

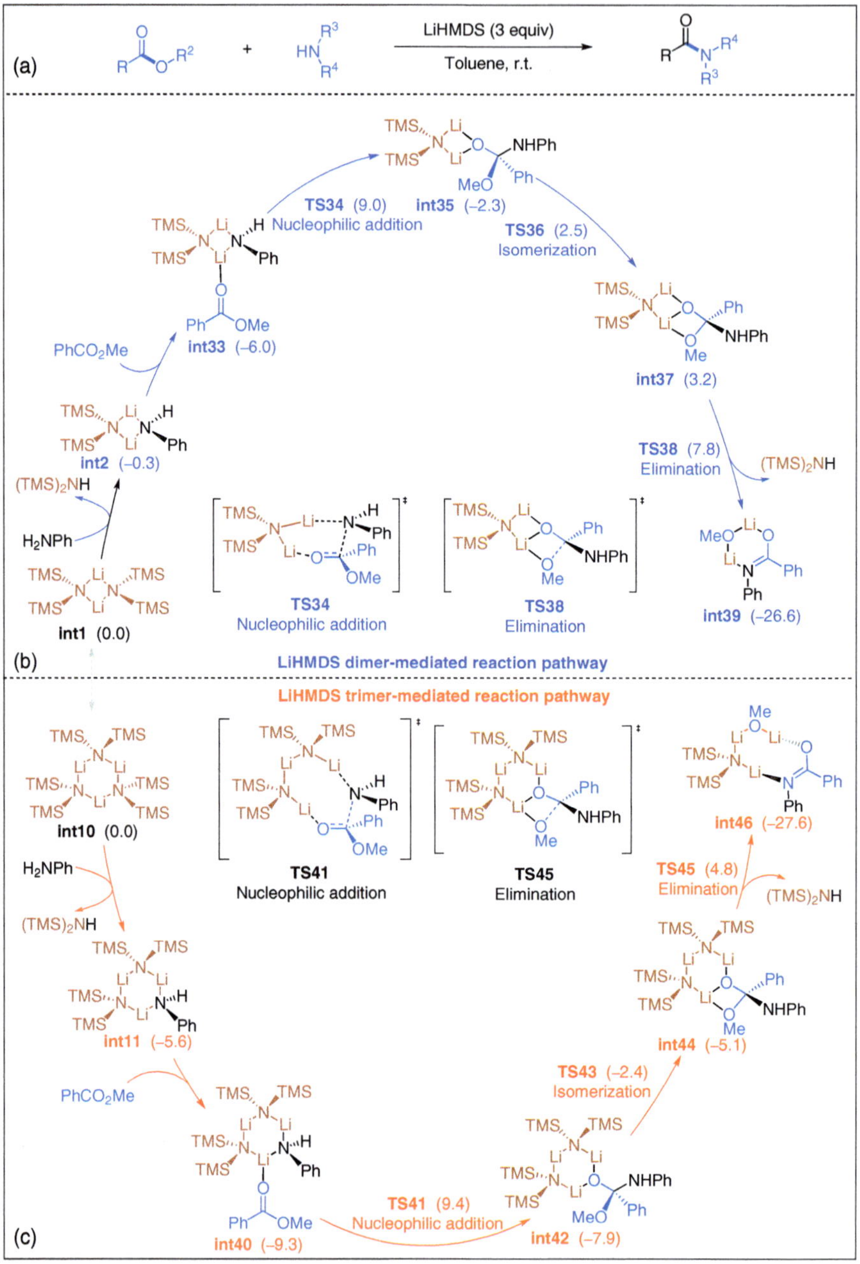

Figure 13.6 DFT-computed free energy profiles of LiHMDS-mediated amidation of esters. Source: From Li, G. et al. [24], © 2019 American Chemical Society. Reprinted with permission of American Chemical Society.

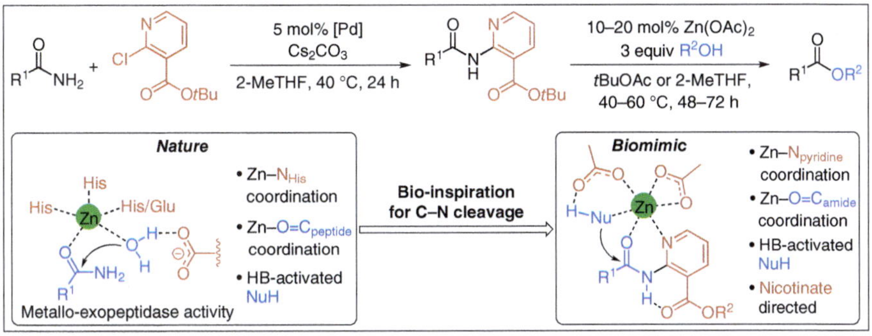

Figure 13.7 Bio-inspired Zn-catalyzed esterification of amides.

int51. This addition requires a barrier of 19.7 kcal/mol and is endergonic by 19.4 kcal/mol. From the tetrahedral intermediate **int51**, the elimination step through **TS52** produces the esterification product and regenerates the zinc acetate catalyst. Due to the endergonic nucleophilic addition, the rate-determining step of the overall esterification is the elimination step via **TS52**, which requires barrier of 23.8 kcal/mol.

The comparisons between the computed free energy profiles of the *tert*-butyl nicotinate and simple pyridine directing groups elucidated the directing group effect on amide reactivity (Figure 13.8a vs. b). The same reaction mechanism applies to the *N*-pyridine-substituted benzamide, but the overall barrier of the rate-determining elimination is 26.2 kcal/mol from intermediate **int56** to **TS60**. The increasing barrier for the pyridine directing group is attributed to the stability of the resting state **int56**, in which the substrate coordinates to zinc via the pyridine nitrogen only. Although both the pyridine nitrogen and carbonyl group of the *N*-pyridine-substituted benzamide can provide coordination, intermediate **int56** (inactive form) is 5.3 kcal/mol more stable than intermediate **int57** (active form). This is reasoned by the intrinsic conformational preference of the *N*-pyridine-substituted benzamide. In contrast, the productive *tert*-butyl nicotinate directing group has the favorable conformation suitable for bis-ligation in intermediate **int49** (Figure 13.8a), which avoids the energy penalty for conformational change and eventually leads to the observed higher reactivity.

13.3.3 Computational Studies on the Mechanism of Ammonium Salt-Mediated Hydrazinolysis of Amides

Ohshima, Morimoto, and coworkers reported a combined experimental and computational mechanistic study on the ammonium salt-mediated hydrazinolysis of unactivated amides [26]. Under the acceleration of simple ammonium iodide, the hydrazinolysis of unactivated secondary and tertiary amides proceeds smoothly with a wide scope of tolerated functional groups (Figure 13.9a). In this mechanistic exploration, the authors performed experimental kinetic studies and found that the initial rate of the hydrazinolysis reaction has first-order dependence on the

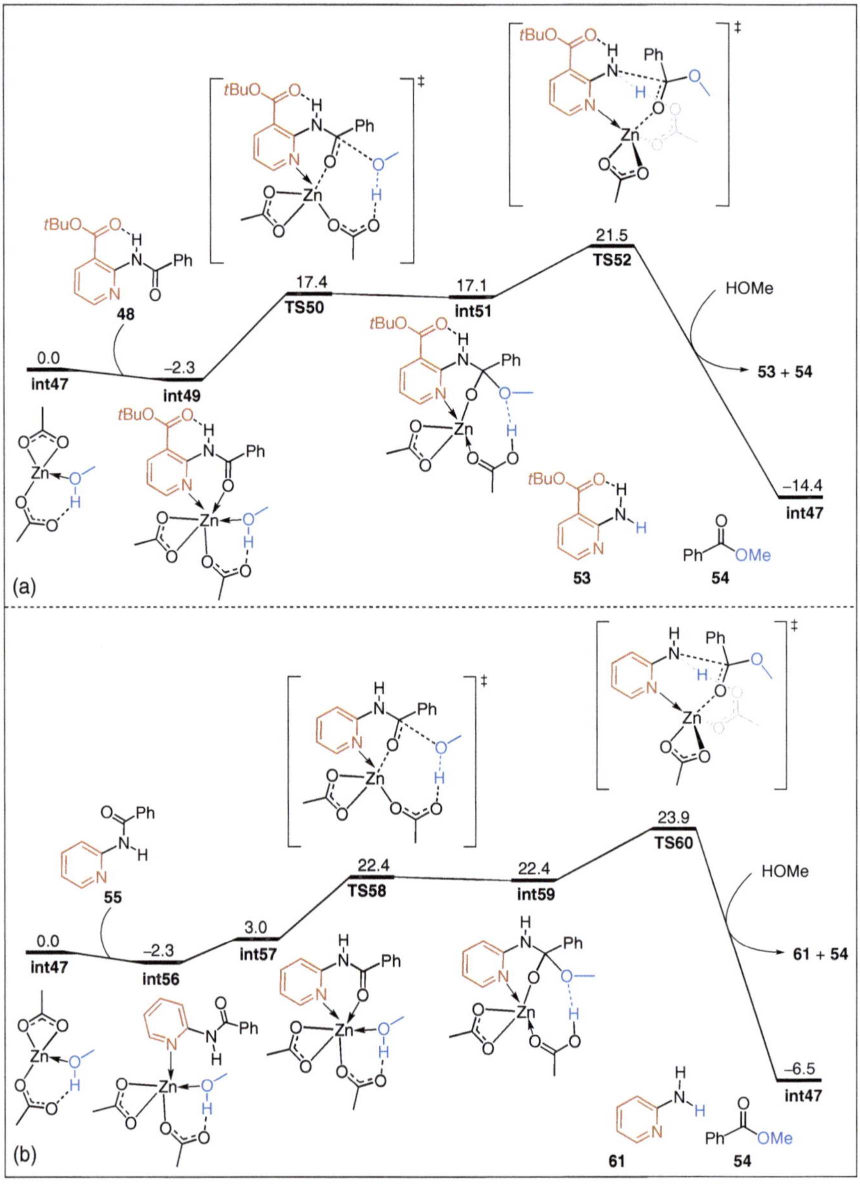

Figure 13.8 DFT-computed free energy changes of Zn-catalyzed esterification of benzamides with *N*-directing group. a) Tert-butyl nicotinate directing group. b) Pyridine directing group.

concentration of amide, hydrazinium bromide, and hydrazine monohydrate, which suggests that the three molecules are involved in the rate-determining step. In addition, they were able to quantify the activation parameters of this transformation ($\Delta H^{\ddagger}$ = 17.5 kcal/mol, $\Delta S^{\ddagger}$ = −33.3 cal/mol/K, and $\Delta G^{\ddagger}$ = 27.4 kcal/mol). The large entropy penalty also agrees with the involvement of three molecules in the rate-determining step.

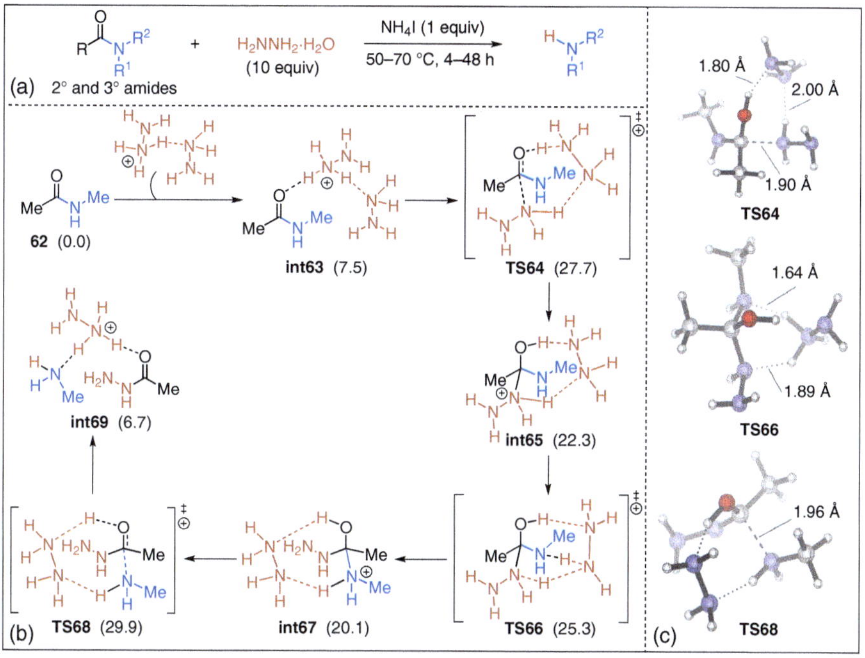

Figure 13.9 Experimental and computational results of ammonium salt-mediated hydrazinolysis of unactivated amides. Source: From Noshita, M. et al. [26], © 2019 American Chemical Society. Reprinted with permission of American Chemical Society.

DFT calculations under the M06-2X/6-311g(d,p)-SMD/B3LYP/6-31g(d) level of theory were performed to understand the reaction mechanism and the origins of ammonium salt acceleration [26]. The DFT-computed free energy profile in the presence of hydrazinium salt is shown in Figure 13.9b. Comparing with the separated starting materials, the hydrogen bonding complex is 7.5 kcal/mol higher in free energy. Subsequent addition of hydrazine occurs via **TS64** to generate the key tetrahedral intermediate **int65**. In **TS64**, a hydrogen bonding network was identified, in which the protonated hydrazine activates the carbonyl group of amide. From the tetrahedral intermediate **int65**, the proton transfer through **TS66** to the eliminating amino group is necessary, leading to the pre-elimination intermediate **int67**. **int67** undergoes the elimination step via **TS68** to generate the hydrazinolysis product **int69**. Based on the computed free energy profile, the elimination step is the rate-limiting step, which requires an overall barrier of 29.9 kcal/mol (from **62** to **TS68**). This barrier is mainly due to the 20.1 kcal/mol endergonicity of the generation of tetrahedral intermediate **int67**. The DFT-computed free energy barrier is in nice agreement with the experimentally measured activation parameters.

Without the assistance of the ammonium salt, similar mechanism is operative but requires a significantly higher overall barrier (Figure 13.10). Without the extra proton in the hydrogen bonding network, the formation of the hydrogen bonding complex **int70** between amide and two hydrazine molecules is now endergonic by 21.1 kcal/mol. The lack of Lewis acid promotion also increases the intrinsic barrier for hydrazine addition, leading to the 44.4 kcal/mol overall barrier via

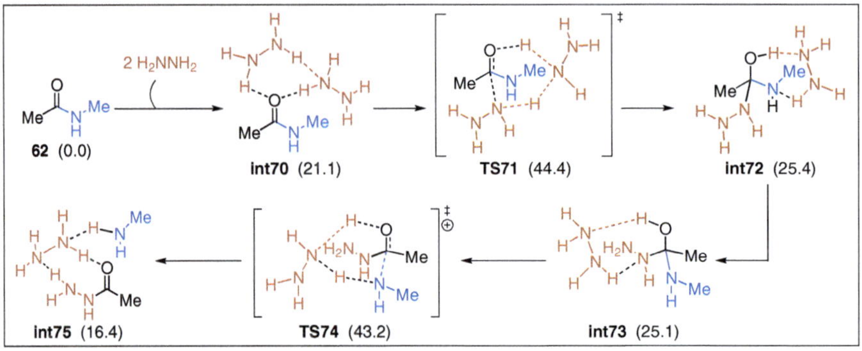

Figure 13.10 DFT-computed free energy changes of hydrazinolysis of amides without hydrazinium salt assistance. Source: From Noshita, M. et al. [26], © 2019 American Chemical Society. Reprinted with permission of American Chemical Society.

TS71. It should be noted that the elimination step also becomes less efficient without the presence of proton in the hydrogen bonding network. The generation of pre-elimination intermediate **int73** is 25.1 kcal/mol (Figure 13.10), which is similar to the generation of pre-elimination intermediate **int67** under ammonium salt (20.1 kcal/mol, Figure 13.9b). The intrinsic barrier for the elimination step, however, increases to 18.1 kcal/mol from **int73** to **TS74**. The same step only requires a barrier of 9.8 kcal/mol under the hydrazinium salt (**int67** to **TS68**, Figure 13.9b). These computations elucidated the origins of the ammonium salt acceleration and provides useful mechanistic insights on the acid effect in the C–N bond cleavage of amide.

13.3.4 Computational Studies on the Mechanism of Organocatalytic Asymmetric Alcoholysis of *N*-Sulfonyl Amide

In addition to the previously mentioned transition metal-free C–N bond activation and functionalization of achiral amides, an interesting organocatalytic asymmetric alcoholysis of pre-chiral *N*-sulfonyl amide was reported by Huang, Fu, Wei, and coworkers [27]. This transformation provides a novel approach to access axially chiral biaryl amino acids via amide C–N bond activation (Figure 13.11a). Under the bifunctional organocatalysts **Cat-1** with both cinchona alkaloid and squaramide moieties, the desired asymmetric alcoholysis of the pre-chiral amides occurs smoothly with benzyl alcohol, generating the axially chiral product with excellent yield and enantioselectivity. This transformation tolerates a wide scope of amides and alcohols, leading to good to excellent enantioselectivities in an array of substrates. This direct organocatalytic asymmetric version of the amide C–N bond activation provides a useful synthetic strategy for asymmetric amide derivatization.

To understand the reaction mechanism and origins of enantioselectivity of this interesting organocatalyzed asymmetric amide C–N bond activation, the authors conducted DFT calculations at the M06-2X-GD3/6-311++G(2d,2p)/IEF-PCM(*o*-xylene)//M06-2X/6-31G(d,p)/IEF-PCM(*o*-xylene) level of theory. The DFT-computed free energy profile is shown in Figure 13.11b. From the

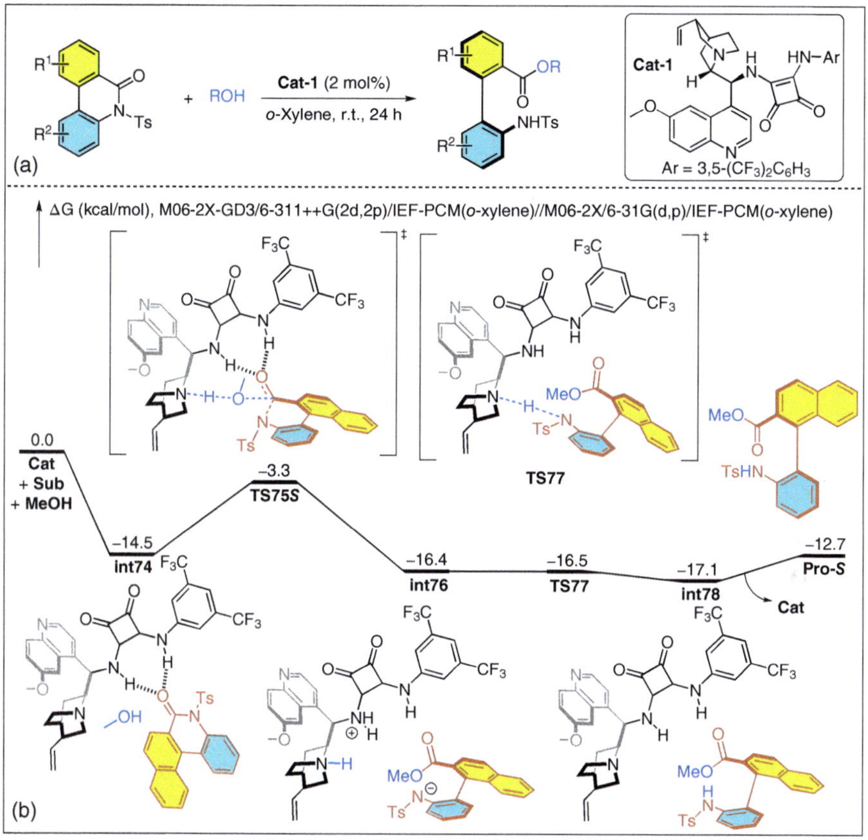

Figure 13.11 Organocatalytic asymmetric alcoholysis of *N*-sulfonyl amides to access axial chirality. Source: Wang et al. [27]/Springer Nature/CC BY 4.0.

catalyst-amide species and alcohol, the favorable complexation leads to the intermediate **int74**. This intermediate undergoes the organocatalytic C–N bond activation via **TS75*S*** to the intermediate **int76**. **TS75*S*** is distinctive from the previously mentioned reaction mechanisms of transition metal-free amide C–N bond cleavages. This step is proposed to be a concerted process, which involves the deprotonation of methanol by the tertiary amine, the nucleophilic addition of deprotonated methanol to carbonyl group of amide, and the elimination of the amino leaving group. Therefore, this reaction mechanism does *not* involve the typical tetrahedral intermediate as in the classic mechanism of nucleophilic substitution of carboxylic acid derivatives, according to the authors' computations. This concerted substitution of amide is interesting since concerted S_N2-fashion substitution generally occurs in sp^3-hybridized carbon. This step was found to be the rate- and enantioselectivity-determining step of the overall transformation. After this step, the axially chiral intermediate **int76** undergoes the subsequent facile proton transfer from the protonated organocatalyst to the amidate, eventually producing the observed axially chiral products.

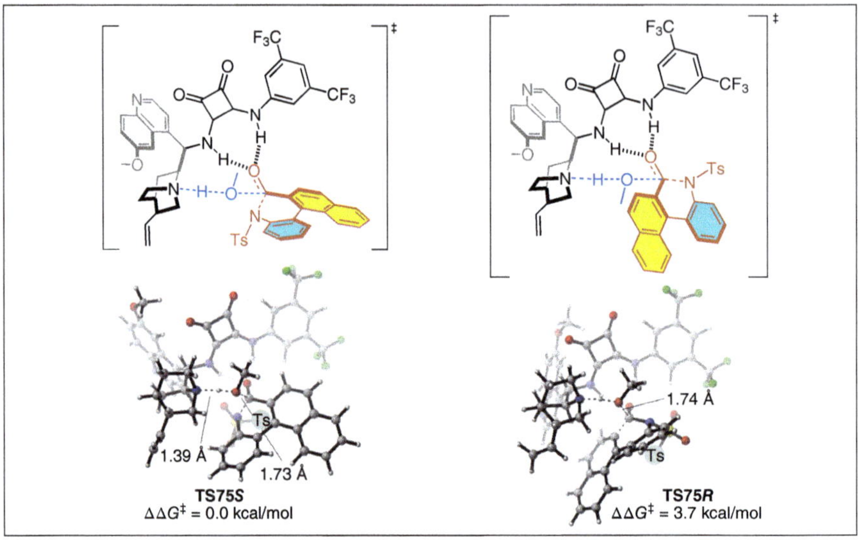

Figure 13.12 Optimized structures and relative free energies of the enantioselectivity-determining substitution transition states in organocatalytic asymmetric alcoholysis of *N*-sulfonyl amides. Source: Wang et al. [27]/Springer Nature/CC BY 4.0.

Based on the DFT-computed reaction mechanism, the authors further explored the origins of enantioselectivity of the concerted substitution step (Figure 13.12). The squaramide of the bifunctional catalyst activates the carbonyl group of the amide, and the tertiary amine moiety of the bifunctional catalyst deprotonates the methanol to promote its nucleophilicity. This bifunctional linkage creates a chiral pocket and allows the efficient chiral induction in the concerted substitution step. **TS75*S*** is 3.7 kcal/mol more stable than the competing **TS75*R***, which agrees with the experimental enantioselectivity. This chiral induction was attributed to the non-covalent interaction (NCI). NCI analysis by the authors identified that the number and strength of the hydrogen bonding interactions are more favorable in the major transition state **TS75*S***, which results in the observed enantioselectivity.

13.4 Computational Studies on the Mechanism and Selectivity of Transition Metal-Catalyzed Cross-coupling of Amides

In addition to the classic Lewis acid-catalyzed nucleophilic substitution, transition metal catalysis emerges as a useful strategy to activate the inert amides [17]. The C–N bond activation of amide is realized by the C–N oxidative addition of low valent transition metal, which generates the acylmetal species. This species is similar to the arylmetal intermediate in the classic cross-coupling mechanism, which can undergo subsequent transmetalation and reductive elimination to achieve a productive cross-coupling catalytic cycle. Therefore, most of the current transition metal-catalyzed amide C–N bond activation and functionalization reactions are cross-coupling. This section focuses on the computational studies on the reaction mechanism and origins of reactivity and selectivity of transition metal-catalyzed cross-couplings of amides.

13.4.1 Computational Study on the Mechanism and Reactivity of Ni-Catalyzed Esterification of Amides

Garg, Houk, and coworkers reported a pioneering study on the Ni-catalyzed C–N bond activation and functionalization of amides (Figure 13.13a) [28]. In this transformation, the Ni/SIPr catalyst is able to efficiently cleave the amide C–N bond and realize the esterification of amides. Both the reaction kinetics and thermodynamics are intriguing considering the cleavage of strong amide C–N bond and the overall transformation from stable amide to more reactive ester. To understand the reaction mechanism and origins of reactivity, DFT calculations were performed at the M06/SDD-6-311+G(d,p)-SMD(toluene)//B3LYP/6-31G(d)-LANL2DZ level of theory.

The DFT-computed free energy profile and the optimized structures of key transition states are shown in Figure 13.13b and c. From the catalyst resting state $Ni(SIPr)_2$ **int79**, the oxidative addition via **TS80** breaks the amide C–N bond and generates the LNi(acyl)(amino) intermediate **int81**. This step requires a barrier of 26.0 kcal/mol and is endergonic by 15.6 kcal/mol. Subsequent proton transfer from methanol to

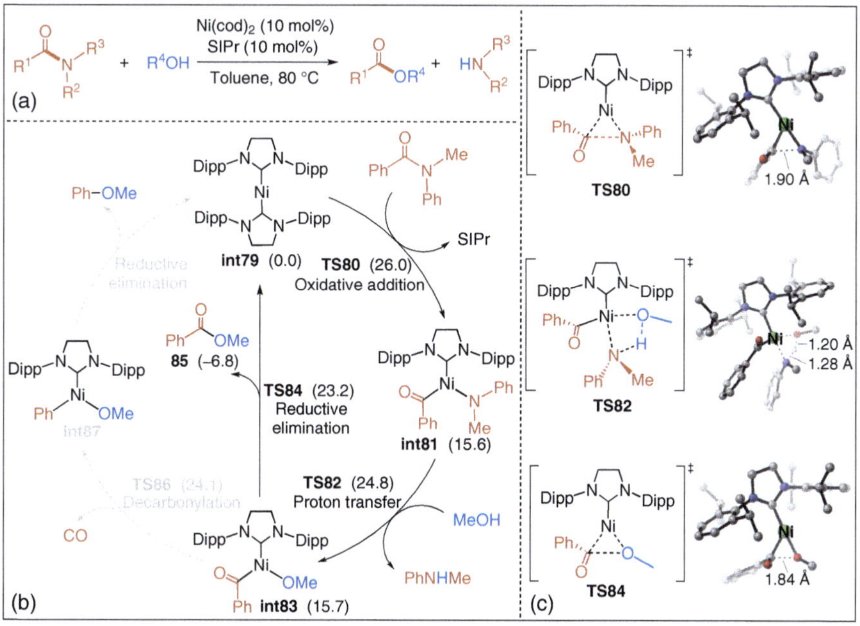

Figure 13.13 Experimental and computational results of Ni/SIPr-catalyzed esterification of amides.

the coordinated amino group occurs via **TS82**, generating the pre-reductive elimination intermediate **int83**. This intermediate undergoes the C–O reductive elimination through **TS84** to produce the esterification product **85** and regenerates the active nickel(0) catalyst. Based on the computed free energy changes of the catalytic cycle, the oxidative addition step is the rate-limiting step. However, it should be noted that all three transition states have close free energy barriers (**TS80**, **TS82**, and **TS84**), which suggests that the related reaction designs should involve the rationalization of all three steps. The authors also examined the alternative decarbonylative functionalization pathway. The LNi(acyl)(methoxy) intermediate could undergo the decarbonylation via **TS86** to generate the LNi(phenyl)(methoxy) intermediate, which may eventually lead to the decarbonylative cross-coupling ether product. This step is found less favorable as compared to the C–N reductive elimination (**TS84** vs. **TS86**), but the close competition reveals that the decarbonylative pathway is a competitive reaction in Ni-catalyzed amide transformations, which was indeed observed in related catalytic systems (*vide infra*).

In addition to the computational study on the reaction mechanism, the authors further explored the *N*-substituent effect on the amide reactivity. The *N*-substituents turned out to have a phenomenal effect on the amide reactivity, and excellent consistency between the computation and experiment was identified (Table 13.1). Only with the *N*-phenyl-*N*-methyl benzamide (entry 8), the overall esterification has reasonable exergonicity (−6.8 kcal/mol) and surmountable barrier for oxidative addition (26.0 kcal/mol), which meets both the thermodynamic and kinetic requirements for reaction efficiency. The other tested amides have

Table 13.1 Experimental and computational results of Ni/SIPr-catalyzed esterification of *N*-substituted amides.

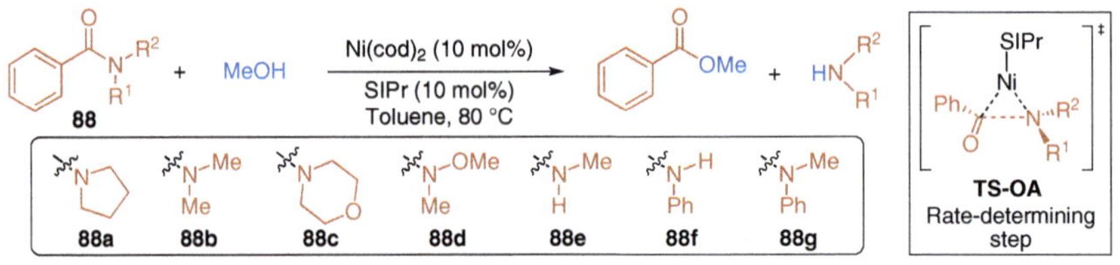

Entry	N(R¹)–R²	ΔG_{rxn} (kcal/mol)	$\Delta G^{\ddagger}_{(TS\text{-}OA)}$ (kcal/mol)	Temperature (°C)	Equivalents of MeOH	Yield of ester
1	**88a**	+2.4	36.8	110	2.0	0%
2	**88b**	0.0	36.2	110	2.0	0%
3	**88c**	−1.1	34.0	110	2.0	23%
4	**88d**	−6.1	31.9	110	2.0	22%
5	**88e**	+3.1	39.0	110	2.0	0%
6	**88f**	−4.3	30.6	110	2.0	55%
7	**88g**	−6.8	26.0	110	2.0	>99%
8				80	1.2	>99%

Source: Adapted from Hie et al. [28].

either significantly higher barrier for amide bond cleavage and/or unfavorable thermodynamic equilibriums, which results in much lower observed yield for ester product. These computations emphasized that the synergistic design of amide *N*-substitution and transition metal catalysis is critical to have a productive amide transformation.

The collaboration between Garg and Houk further revealed that the Ni-catalyzed esterification of amides can be reversed by adding stoichiometric amount of $Al(OtBu)_3$ Lewis acid, leading to the Ni/SIPr-catalyzed amidation of ester (Figure 13.14a) [29]. Through DFT calculations at M06/SDD-6-311+G(d,p)-SMD(toluene)//B3LYP/6-31G(d)-LANL2DZ level of theory, the authors elucidated the reaction mechanism and the kinetic and thermodynamic effects of Lewis acid. The computed free energy profile of the catalytic cycle is shown in Figure 13.14b, and the optimized structures of the key transition states are shown in Figure 13.14c. From the $Ni(SIPr)_2$ catalyst resting state **int89**, the oxidative addition of Lewis acid-bonded ester substrate proceeds via **TS91**, generating the acylnickel(II) intermediate **int92**. This step requires barrier of 26.8 kcal/mol, and the generated acylnickel(II) intermediate is 20.0 kcal/mol higher in free energy as compared with **int89**. From **int92**, subsequent proton transfer between the aniline and coordinated methoxy group leads to the (NHC)Ni(acyl)(amino) intermediate **int94**. This intermediate undergoes the C–N reductive elimination via **TS95** to produce the amidation product, instead of the decarbonylation via **TS97**. The C–N reductive elimination transition state **TS95** is essentially the C–N oxidative addition transition

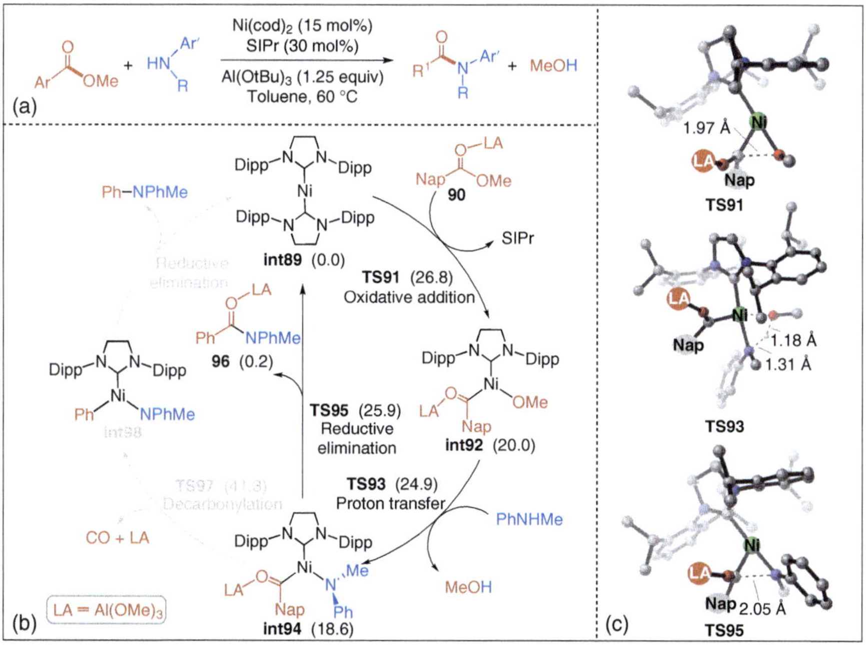

Figure 13.14 Experimental and computational results of Ni/SIPr-catalyzed amidation of esters. Source: Hie et al. [29].

state for amide activation. The rate-determining step is the oxidative addition step based on the DFT-computed free energy profile.

DFT computations provided a clear picture of the Lewis acid effect. Without the Lewis acid, the amidation reaction is endergonic by 4.9 kcal/mol (Scheme 13.3a). This suggests that the amide activation and esterification is thermodynamically more favorable, as found in the authors' previous collaboration [28]. Through the addition of external Lewis acid, the complexation with Lewis acidic $Al(OtBu)_3$ is more favorable for amide than ester, due to the Lewis basicity change of the carbonyl group. This additional complexation drives the thermodynamic equilibrium toward the formation of the amidation product, leading to the favorable amidation reaction. The authors also noted that the Lewis acid is able to lower the activation barrier for C–O bond cleavage (Scheme 13.3b). The liberation of strong coordinating NHC ligand is unfavorable, which results in the unsurmountable barrier of C–O oxidative addition via **TS100**. The Lewis acid complexation promotes the substrate's ability to accept the d-back donation from nickel, which significantly lowers the C–O bond cleavage barrier and kinetically promotes the desired amidation. These calculations provided a useful mechanistic basis to understand and design the Lewis acid additive in the context of Ni-catalyzed transformation of amides.

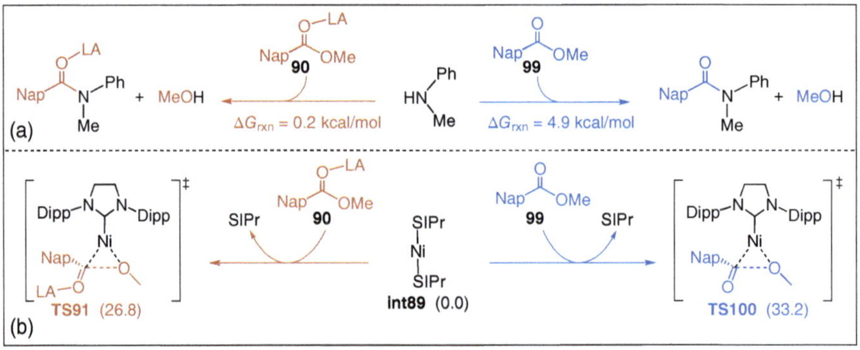

Scheme 13.3 Computational studies on the thermodynamic and kinetic effects of Lewis acid on the Ni-catalyzed amidation of ester. Source: Hie et al. [29].

Newman and coworkers later disclosed an Ni/IPr-catalyzed amidation of ester under Lewis acid-free conditions (Figure 13.15a) [31]. Hong and coworkers explored the reaction mechanism and the substituent effects on the thermodynamic equilibrium using DFT calculations at the M06/6-311+G(d,p)-SDD-SMD(toluene)//B3LYP-D3(BJ)-SMD(toluene)/6-31G(d)-LANL2DZ level of theory [30]. They found that the amidation reaction involves the sequential oxidative addition, proton transfer, reductive elimination, and product liberation, which is similar to the above cases (Figure 13.15b). The rate-limiting step is the oxidative addition step that breaks the C–O bond (**TS102**), but the reductive elimination step for C–N bond formation has similar free energy barrier. To guild the future reaction designs,

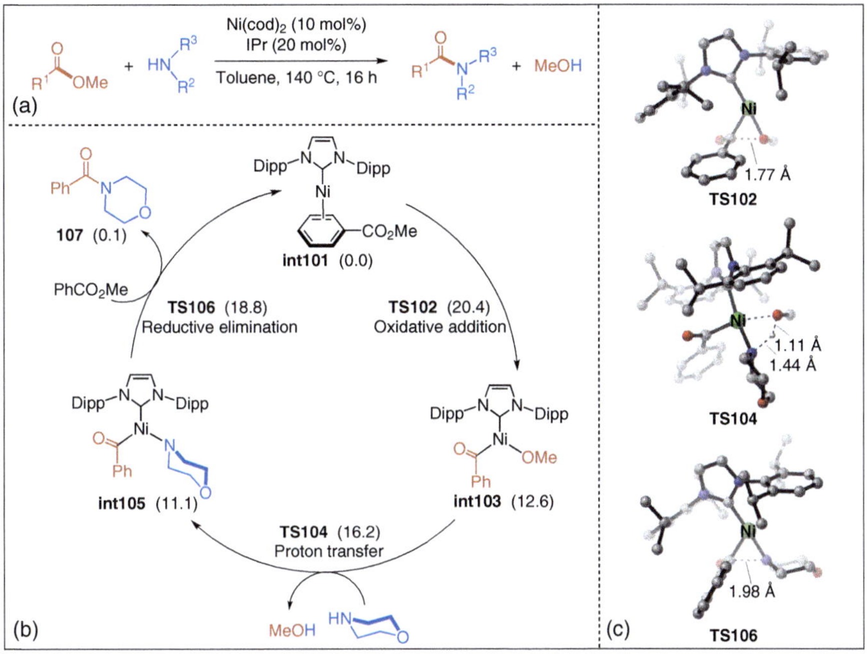

Figure 13.15 Computational study on the Ni/IPr-catalyzed amidation of esters without the Lewis acid assistance. Source: Ji et al. [30]/MDPI/CC BY 4.0.

Table 13.2 DFT-computed thermodynamic reaction free energies of amidation of esters.

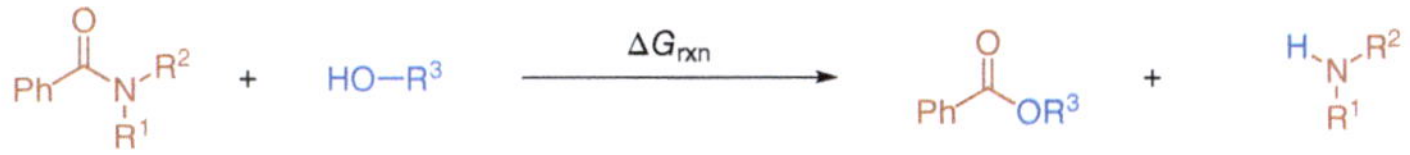

		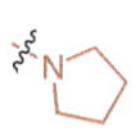 **Aliphatic ring amide**	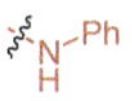 **Primary amide**	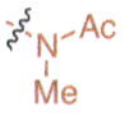 **Electronically activated amide**	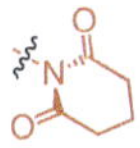 **Twisted amide**
Primary alcohol		2.6	−4.5	−15.9	−19.3
Secondary alcohol	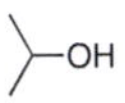	6.6	−0.5	−11.9	−15.4
Tertiary alcohol	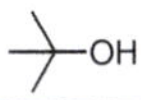	6.7	−0.4	−15.8	−15.3

Source: Ji et al. [30]/MDPI/CC BY 4.0.

the authors provided a thorough thermodynamic landscape of the amide-ester transformation for a wide array of amides and esters. A few selected key cases are summarized in Table 13.2. They discovered that the *O*-substitution of ester has limited effects on the thermodynamic equilibrium, while the *N*-substitution of

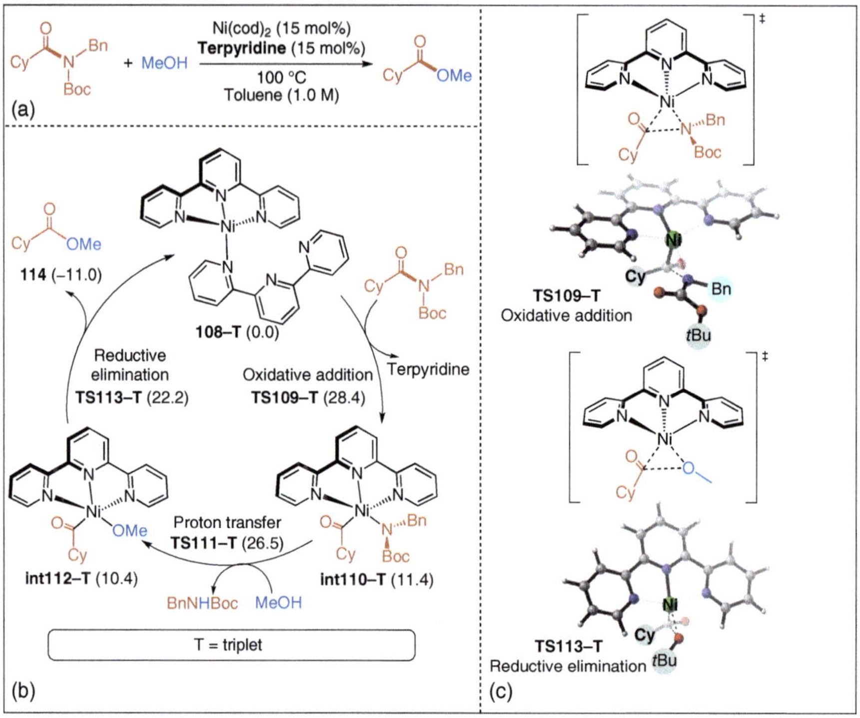

Figure 13.16 Experimental and computational studies of Ni-catalyzed esterification to aliphatic amides. Source: From Chu et al. [32], © 2018 American Chemical Society. Reprinted with permission of American Chemical Society.

amide is critical for the thermodynamic stability of amides. Electron-activating substituents or geometric twisting can provide strong driving force for the amide transformation.

Following the success of Ni/NHC-catalyzed esterification of aromatic amides, Garg and coworkers discovered a new catalytic system for the same reaction with aliphatic amides (Figure 13.16a) [18]. Terpyridine ligand was identified as the key ligand that can promote this transformation for aliphatic amides [18]. Dang and coworker performed DFT studies on the esterification mechanism and ligand effect at the M06/6-311++G(2d,p)-MDF10-SMD(toluene)//ωB97XD/6-31G(d)-LANL2DZ level of theory [32]. The DFT-computed free energy profile is shown in Figure 13.16b, and the optimized structures of the key transition states are shown in Figure 13.16c. From the Ni(terpyridine)$_2$ catalyst resting state **108-T**, the oxidative addition of aliphatic amide breaks its C–N bond via the monoligated transition state **TS109-T**. This step requires a barrier of 28.4 kcal/mol, generating the (terpyridine)Ni(acyl)(amino) species **int110-T**. Subsequent proton transfer from methanol to the coordinating amino group produces the pre-elimination intermediate **int112-T**. Eventually the C–O reductive elimination via **TS113-T** leads to the aliphatic ester product. Based on the DFT-computed free energy profile, the reaction mechanism and rate-determining step are similar to the previously mentioned studies.

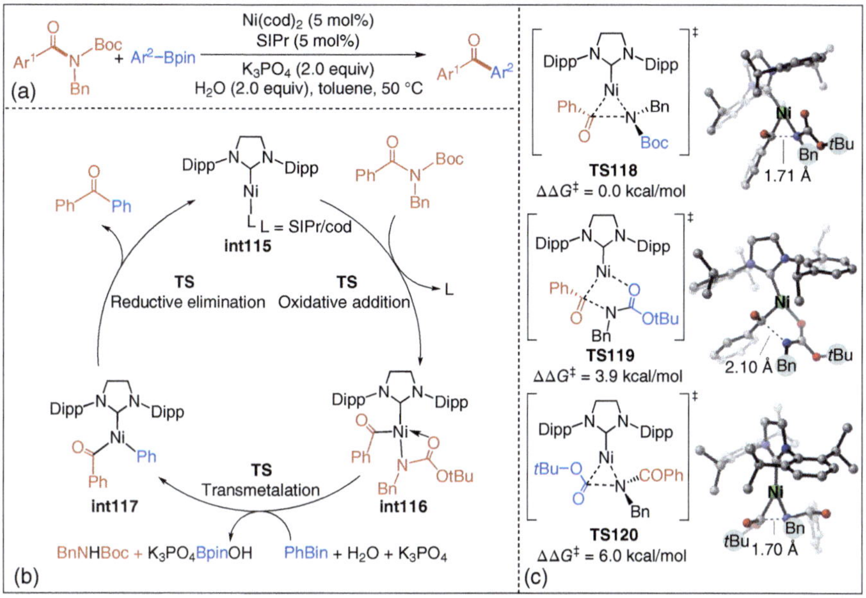

Figure 13.17 Experimental and computational studies of Ni-catalyzed Suzuki–Miyaura cross-coupling of amides. Source: From Liu et al. [33], © 2016 American Chemical Society. Reprinted with permission of American Chemical Society.

13.4.2 Computational Study on the Mechanism of Ni-Catalyzed Suzuki–Miyaura Coupling of Amides

Based on the powerful nickel-catalyzed C–N bond activation of amide, the Garg group extended the synthetic applications of their amide activation and developed the Ni-catalyzed Suzuki–Miyaura cross-coupling reaction of amides (Figure 13.17a) [34]. This transformation realized the C–C bond formation between aromatic amides and aromatic boronic esters under the Ni/SIPr catalyst and basic conditions, leading to the formation of ketone product. The authors discovered that the *N*-phenyl-*N*-methyl benzamide has limited productivity in this transformation, and the installation of electronically activating Boc group leads to the desired efficiency (Figure 13.17a). A wide substrate scope was tolerated in this useful transformation, and several applications of the developed Suzuki–Miyaura coupling of amides were present to highlight its synthetic potential. This first nickel-catalyzed synthetic method of C–C bond formation from amides further revealed the hidden value of amide as a synthon, highlighting the broad future of Ni-catalyzed amide transformation.

Based on the experimental results from Garg's report, Fu, Yu, and coworker [35], as well as Zhao, Zhu, and coworkers [33], studied the reaction mechanism and chemoselectivity of C–N bond activations using DFT calculations. Consistent mechanistic pictures were elucidated by these two independent computational studies, and the simplified reaction mechanism is shown in Figure 13.17b. From the nickel(0) catalyst, the oxidative addition of *N*-Boc benzamide breaks the amide

C–N bond and generates the LNi(acyl)(amino) intermediate **int116**. Because the Boc group can provide bidentate coordinate, the acylnickel(II) intermediate **int116** has tetra-coordination, which stabilizes this post-intermediate and provides additional thermodynamic driving force for amide C–N bond cleavage. Subsequent base-assisted transmetalation generates the LNi(acyl)(aryl) intermediate **int117**. The K_3PO_4 base was found to play an important role in promoting the transmetalation process by both computational studies. From the LNi(acyl)(aryl) intermediate **int117**, the acyl–aryl reductive elimination leads to the observed ketone product. These computational studies further validated the putative Ni-catalyzed amide C–N bond activation via oxidative addition. The classic cross-coupling reaction mechanism is applicable for amide substrate, in which the amide substrate acts as an alternative electrophile to replace the classic aryl halides.

In these computational studies, the authors also explored the chemoselectivities of amide C–N bond cleavage in the *N*-Boc protected benzamide [35]. Under the M06/SDD-6-311+G(d,p)-SMD(toluene)//B3LYP/6-31G(d)-LANL2DZ level of theory, the free energy differences of the competing oxidative addition transition states are shown in Figure 13.17c. The Ni(SIPr) catalyst is able to break the acyl-nitrogen bond via three-membered ring transition state **TS118** and five-membered ring transition state **TS119**. **TS118** is 3.9 kcal/mol more favorable than **TS119**, which indicates that the Boc group does not provide the carbonyl coordination during the amide C–N bond cleavage. The competing oxidative addition transition state **TS120** is 6.0 kcal/mol less favorable than **TS118**, which suggests that the *N*-Boc bond cleavage is unfeasible under the Ni/SIPr catalyst and is consistent with the experimental chemoselectivity. Distortion/interaction analysis indicated that both distortion energy and interaction energy contribute to the preference of **TS118**.

13.4.3 Computational Study on the Mechanism and Selectivity of Ni-Catalyzed C–N Bond Activation of Twisted Amides

Distinctive to the previously mentioned transition metal-catalyzed C–N bond activation of amides, Szostak and coworkers developed a novel family of *N*-glutarimide amides and related derivatives, namely, the twisted amides (Scheme 13.4a) [36]. This type of twisted amides generally has intrinsic geometric twisting around the acyl-nitrogen axis and exhibits remarkable reactivities toward amide C–N bond activation and functionalization. The series of powerful transformations with twisted amides, developed by the Szostak group, provide a general synthetic platform using twisted amides as a central synthon. The twisted amides have distinctive chemical behavior as compared with the electronically activated amides. Taking Suzuki–Miyaura coupling as an example, the Ni/PCy_3-catalyzed Suzuki–Miyaura coupling with the *N*-glutarimide amides leads to the decarbonylative biaryl coupling product (Scheme 13.4b) [37], while the similar transformation with electronically activated *N*-Boc amides results in the nondecarbonylative ketone product (Scheme 13.4c) [34]. The intriguing experimental phenomena with the twisted amides raised the physical organic questions on the underlying reaction mechanism and the controlling factors of the reactivity of twisted amides.

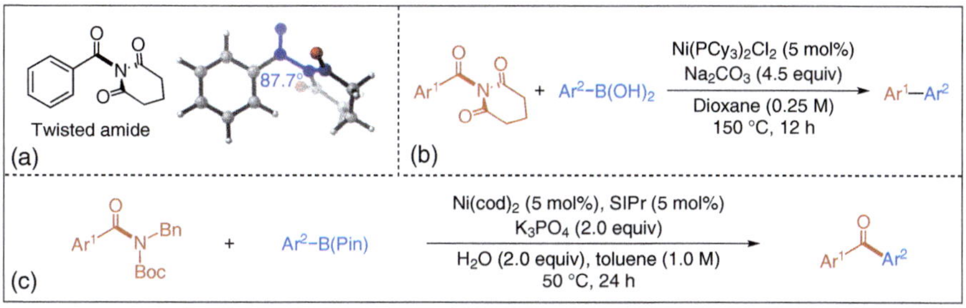

Scheme 13.4 Twisted amides and its application in Ni-catalyzed Suzuki–Miyaura coupling. (a) Source: Adapted from Liu and Szostak [36]. (b) Source: Adapted from Shi et al. [37]. (c) Source: Adapted from Weires et al. [34].

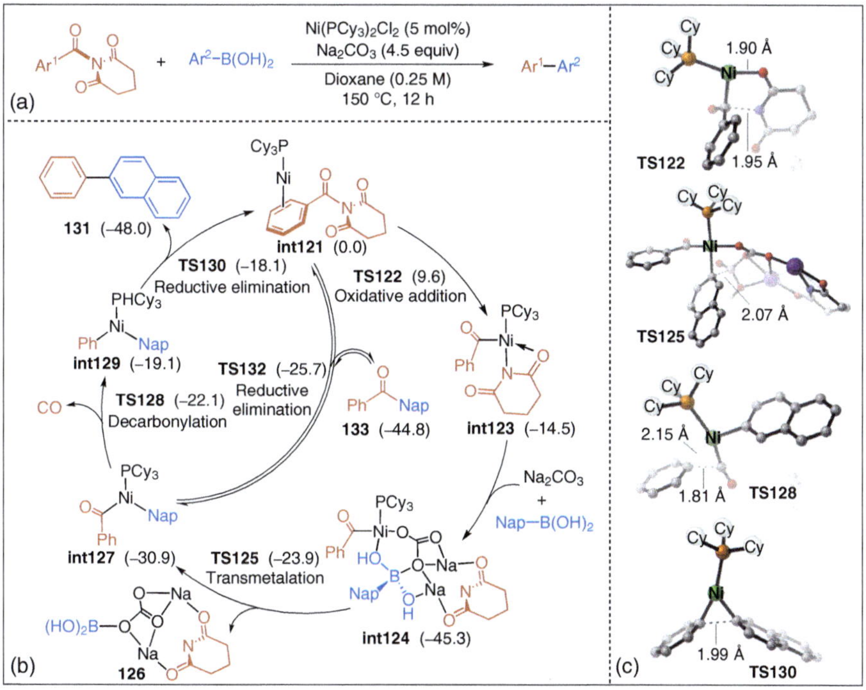

Figure 13.18 Computational study on the Ni-catalyzed Suzuki–Miyaura coupling of *N*-glutarimide amides. Source: From Ji, C.L. et al. [38], © 2017 American Chemical Society. Reprinted with permission of American Chemical Society.

To understand the elements that are responsible for the reactivity and chemoselectivity of *N*-glutarimide amides, Hong and coworkers studied the Ni/Pcy$_3$-catalyzed Suzuki–Miyaura coupling of *N*-glutarimide amides using DFT calculations [38]. Figure 13.18b shows the DFT-computed free energy profile at the M06/6-311+G(d,p)-SDD-CPCM(1,4-dioxane)//B3LYP/6-31G(d)-LANL2DZ level of theory, and the optimized structures of the key transition states are shown in Figure 13.18c. From the substrate-coordinated nickel(0) complex **int121**, the oxidative addition of twisted amide occurs via **TS122**, generating the tetra-coordinated nickel(II) intermediate **int123**. This amide C–N bond cleavage step is quite facile and thermodynamically favorable; the kinetic barrier of oxidative addition is only 9.6 kcal/mol, and the generated nickel(II) intermediate **int123** is 14.5 kcal/mol more stable than the pre-oxidative addition nickel(0) intermediate **int121**. From **int123**, subsequent complexation with base and boronic acid leads to the intermediate **int124**, which further undergoes the base-assisted transmetalation via **TS125** to produce the LNi(acyl)(aryl) intermediate **int127**. **Int127** can either undergo the acyl-aryl reductive elimination through **TS132** to produce the ketone product or the decarbonylation (via **TS128**) and aryl–aryl reductive elimination (via **TS130**) to produce the decarbonylated biaryl product. The computations indicated that the formation of ketone product is kinetically more favorable than that of biaryl product (**TS132** vs. **TS130**). It is the thermodynamic stability of the biaryl product **131** that drives the kinetic ketone product **133** to the thermodynamic biaryl product **131**, leading to the observed decarbonylative Suzuki–Miyaura coupling. These computations are also consistent with Chatani's concurrent experimental report of Ni-mediated decarbonylation of biaryl ketone [39].

The authors also explored the origins of the high reactivities of twisted amides toward Ni-catalyzed amide C–N bond cleavage [38]. The C–N bond cleavage barriers of *N*-glutarimide amide **134** and *N*-Boc amide **136** are compared in Figure 13.19. Two factors were revealed that contribute to the high reactivity of twisted amides. First, the twisted amide has a suitable geometry for the bidentate coordination in the pre-oxidative addition intermediate, which means that the twisted amide is pre-distorted for the nickel complexation. This pre-distortion results in the low energy requirement for the isomerization from the most stable form of nickel(0)-amide complex to the pre-oxidative addition intermediate (**int121** to **int135**, 1.9 kcal/mol). The Boc-protected amide requires 3.8 kcal/mol free energy for the same process (**int137** to **int138**), which increases the overall oxidative addition barrier of *N*-Boc amide **136**. In addition, the geometric twisting weakens the intrinsic strength of amide C–N bond, which lowers the intrinsic oxidative addition barrier. **134** only has a 7.7 kcal/mol intrinsic barrier for the nickel catalyst to break the amide C–N bond (**int135** to **TS122**), while the same process requires a barrier of 13.1 kcal/mol for the Boc-protected amide **136** (**int138** to **TS139**). These computations further elucidated the origins of the high reactivities of twisted amides, providing useful guidelines for the rationalization and design of new amide substrates.

Based on Szostak's experimental report of Ni-catalyzed decarbonylative phosphorylation of twisted amides [40], Fu and Yu studied the reaction mechanism of this transformation with DFT calculations [41]. Under the M06/6-311G (d,p)-SDD-IEF-PCM(1,4-dioxane)//B3LYP/6-311G(d,p)-SDD-IEF-PCM(1,4-dioxane) level of theory, the DFT-computed free energy profile is shown in Figure 13.20. This transformation has a bidentate phosphine ligand, 1,3-bis-(diphenylphosphino) propane (dppp), which leads to noticeable change in the free energy profile. From the *N*-glutarimide amide-coordinated nickel(0) species **int141**, the amide C–N bond cleavage via **TS142** requires a barrier of 26.0 kcal/mol. This high barrier is mainly due to the three-membered ring oxidative addition transition state, in which the glutarimide does not provide the additional coordination for nickel to stabilize the

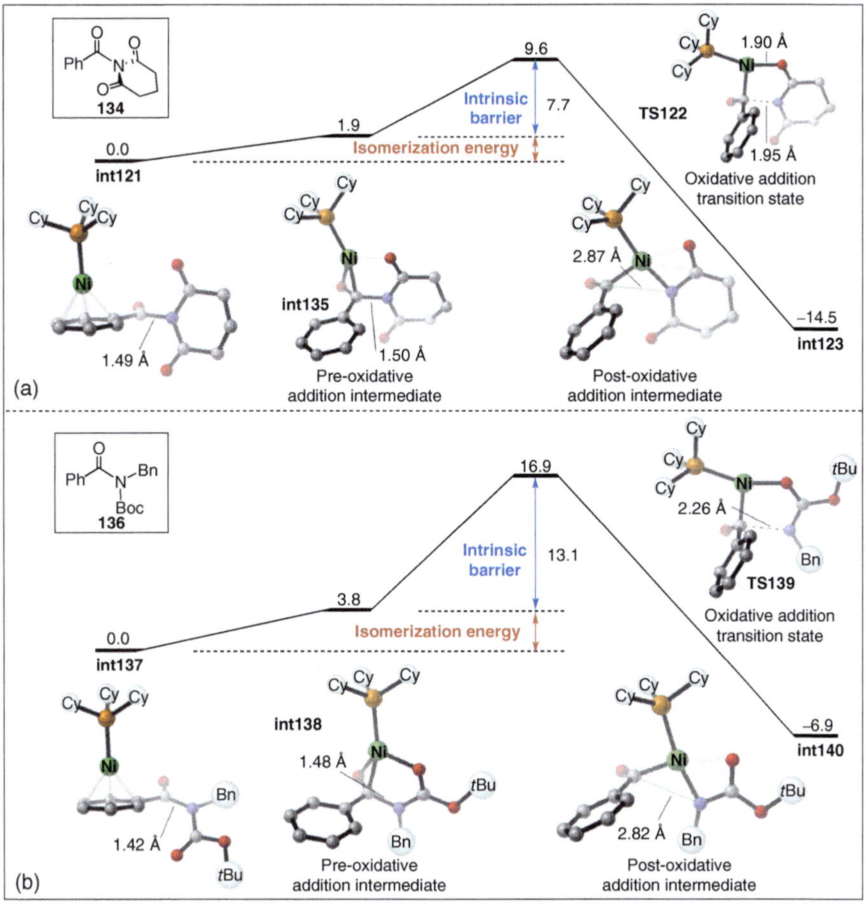

Figure 13.19 DFT-computed free energy barriers of Ni-catalyzed C–N bond activation of representative amides. Source: From Ji, C.L. et al. [38], © 2017 American Chemical Society. Reprinted with permission of American Chemical Society.

C–N bond cleavage transition state. Subsequent deprotonation of diethyl phosphite and dissociation of glutarimidate leads to intermediate **int146**. **Int146** undergoes the decarbonylation via **TS137** and the C–P reductive elimination via **TS149** to produce the observed phosphorylation product. Based on the DFT-computed free energy profile, the amide C–N bond cleavage was identified as the rate-limiting step.

13.4.4 Computational Study on the Structure–Activity Relationship of Ni-Catalyzed C–N Bond Activation of Amides

The rich chemistry of nickel(0)-catalyzed C–N bond activation of amides and the critical effect of *N*-substitution on the amide reactivity raised the question on the structure–activity relationship of the amide C–N bond activation barrier. To provide a general guideline on the feasibility of amide C–N bond cleavage under nickel catalysis, the Hong group computationally studied the activation barriers

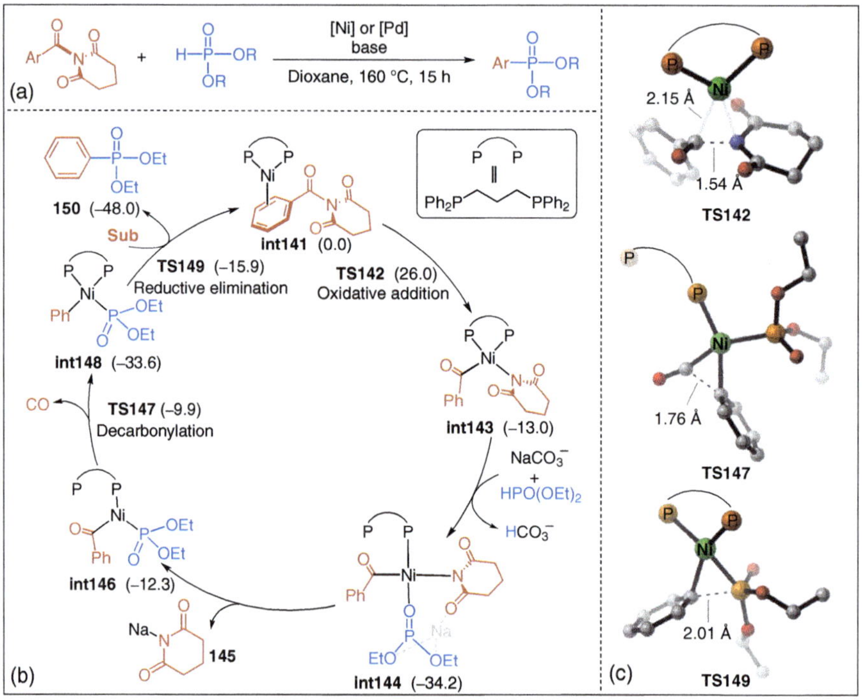

Figure 13.20 Experimental and computational studies on the Ni-catalyzed decarbonylative phosphorylation of twisted amides. Source: From Xu, Y. et al. [41], © 2019 American Chemical Society. Reprinted with permission of American Chemical Society.

of Ni/PCy$_3$-catalyzed C–N bond cleavages of a series of typical amides in the related transformations and analyzed the controlling factors of the amide activation barriers using distortion/interaction analysis [42].

Predictive structure–activity relationships of amide C–N bond cleavage was identified by their DFT calculations. The DFT-computed activation barriers of Ni/PCy$_3$-catalyzed amide C–N bond cleavage, at the M06/SDD-6-311G(d,p)-CPCM (1,4-dioxane)//B3LYP/6-31G(d)-LANL2DZ level of theory, are summarized in Figure 13.21. These amides, with coordinating *N*-substituent or not, all undergo the oxidative addition to break the amide C–N bond via the three-membered ring transition state. In this type of amide activation transition state, the distortion of amide substrate is found to be the controlling factor that differentiates the reactivity of the 12 amides. Due to the nature of heterolytic C–N bond cleavage during the amide activation, the authors found a linear correlation between the amide activation barriers and the heterolytic C–N bond dissociation energies (HBDEs) of the amides (Figure 13.21c). The widely used homolytic BDEs of amides actually have no noticeable correlation with their reactivities in C–N bond activation. These results revealed that a general structure–activity relationship is applicable, and the C–N HBDE can serve as a predictive physical organic descriptor for amide reactivity in the nickel(0)-catalyzed oxidative addition of amides.

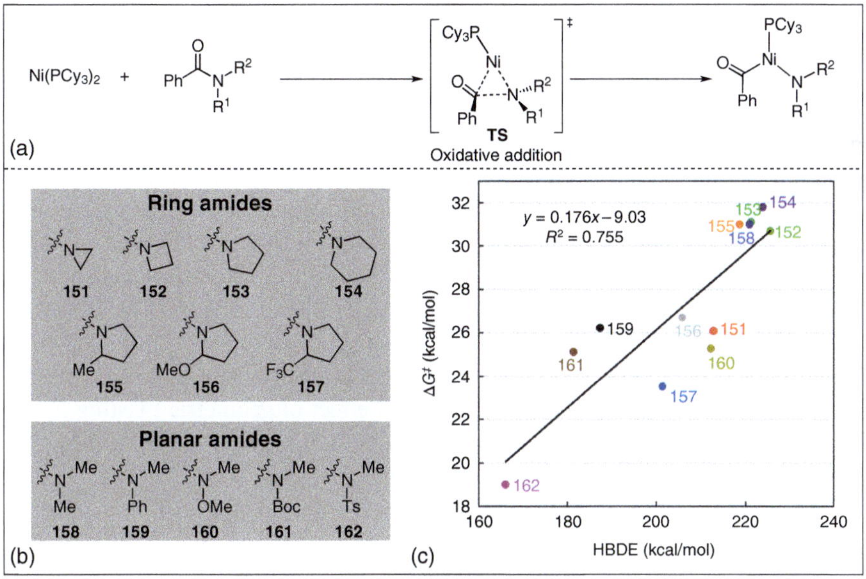

Figure 13.21 HBDE-related structure–activity relationship of Ni-catalyzed amide C–N bond activation of amides.

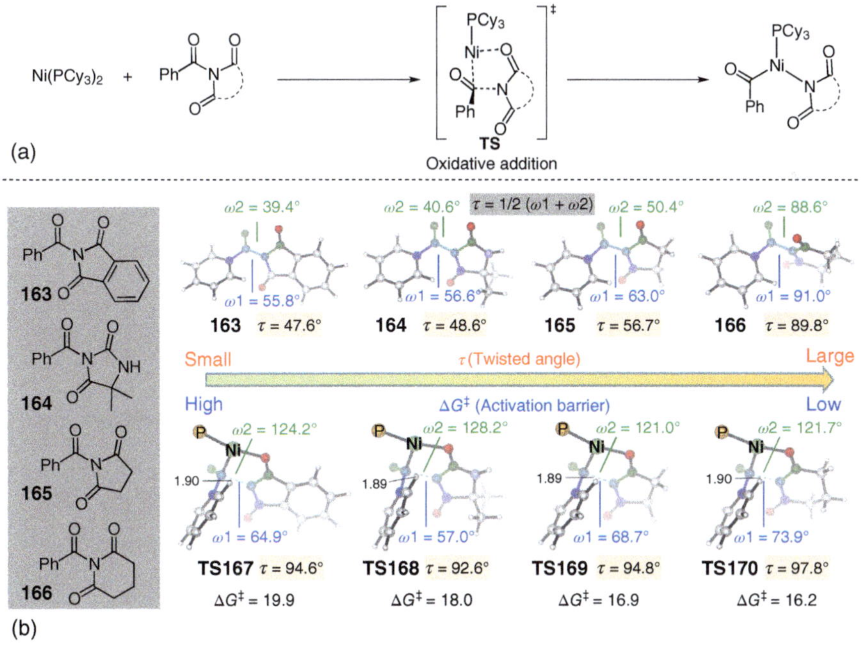

Figure 13.22 Geometric twisting-controlled Ni-catalyzed C–N bond activation of twisted amides.

Unlike the HBDE-controlled reactivity of the 12 amides in Figure 13.21, the authors noted that the twisted amides follow a different structure–activity relationship for the amide activation barrier. Under the same Ni/PCy_3 catalyst, the C–N bond activation barriers of the computed twisted amides are controlled by the distortion of amide substrate, but the substrate distortion of twisted amides is not due to the heterolytic C–N bond stretching. Therefore, the authors found that the C–N HBDEs are not related to the activation barriers for the twisted amides. Instead, the twisted angle (τ) is closely related to the reactivity of twisted amides in Ni-catalyzed C–N bond activation (Figure 13.22). All the twisted amides undergo the C–N bond activation via the chelation-assisted transition state, in which the acyl and amino planes are almost perpendicular to each other (τ = 90°). The higher the τ is, the lower the amide activation barrier would be. These mechanistic insights further corroborated in proposed promotional effect of geometric twisting for the reaction of twisted amides and offer helpful guidance for the rational design of amide reagents in related nickel catalysis.

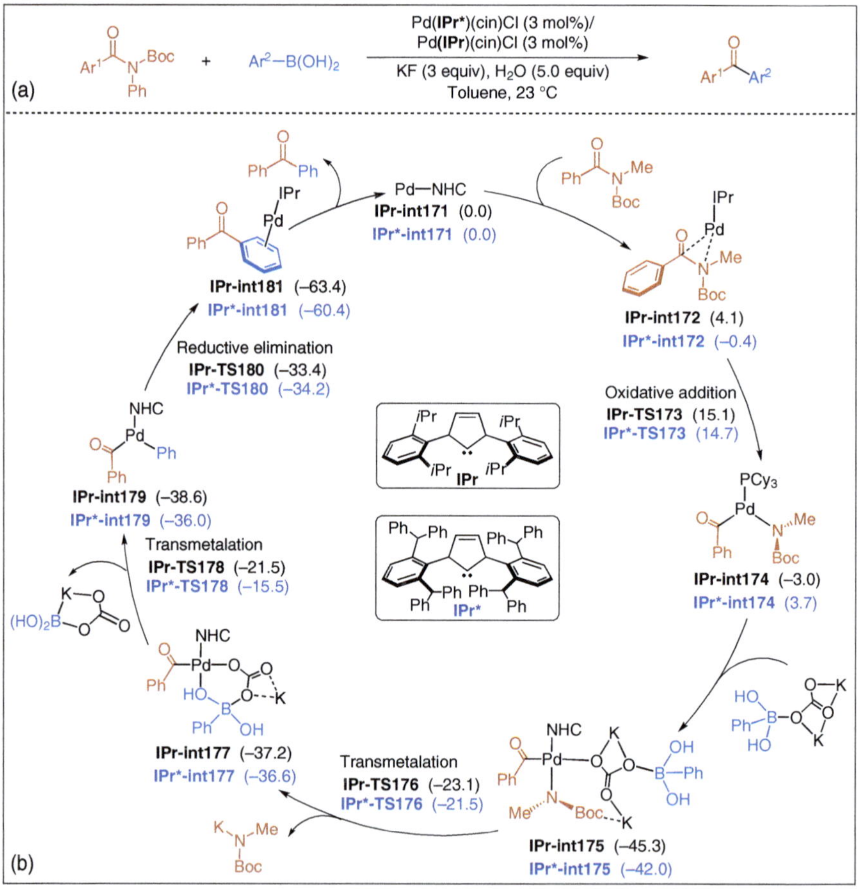

Figure 13.23 Experimental and computational studies of Pd-catalyzed Suzuki–Miyaura cross-coupling of amides.

13.4.5 Computational Study on the Mechanism of Pd-Catalyzed Suzuki–Miyaura Coupling of Amides

Palladium complexes have also been found catalytically active in the amide C–N bond activation and functionalization reactions. A number of cross-coupling reactions with amides under the palladium catalysis were developed, which provides useful alternatives to the previously mentioned nickel catalysis. In 2018, Nolan, Poater, Szostak, and coworkers conducted a combined experimental and computational study on the palladium-catalyzed Suzuki–Miyaura coupling of *N*-Boc-*N*-phenyl aryl amides (Figure 13.23a) [43]. The reaction mechanism was elucidated by DFT calculations at the M06/SDD-Def2-TZVP-PCM(dimethyl ether)//BP86-D3(BJ)/Def2-SVP-SDD level of theory. A Pd(0)-Pd(II) catalytic cycle was discovered, and the DFT-computed free energy profiles of two NHC ligands are shown in Figure 13.23b. From the amide substrate and palladium catalyst, the oxidative addition via **TS173** breaks the amide C–N bond and generates the

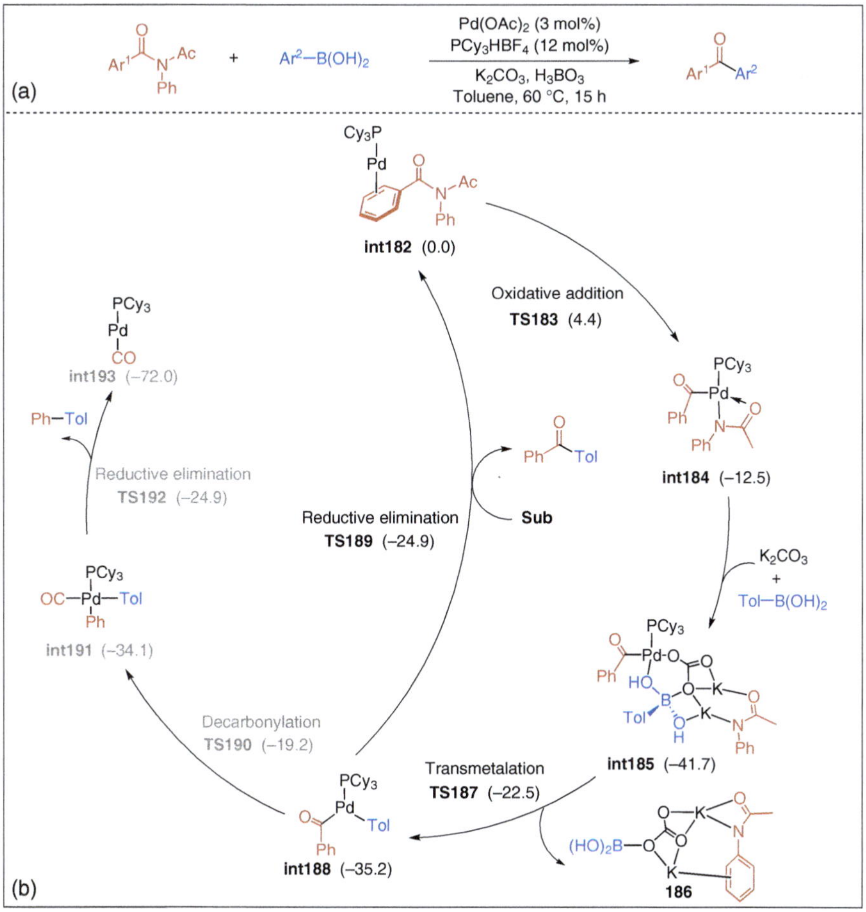

Figure 13.24 Computational study of the Pd-catalyzed Suzuki–Miyaura cross-coupling of electronically activated amides.

palladium(II) species **int174**. This species favorably complexes with base and aryl boronic acid, leading to the intermediate **int175**. Subsequent dissociation of amino group and transmetalation leads to the LPd(acyl)(aryl) intermediate **int179**, which further undergoes the acyl-aryl reductive elimination to produce the observed ketone product. These calculations provided an important mechanistic insight that the metal(0)–metal(II) mechanism applies in both nickel and palladium catalysis.

Dang and coworkers also computationally explored the reaction mechanism of Suzuki–Miyaura coupling of electronically activated amides, with a focus on the origins of chemoselectivity change between palladium and nickel catalysis (Figure 13.24) [44]. Their computations at the M06-L/SDD-6-311++G(d,p)-SMD//B3LYP/SDD-6-31G(d) level of theory are consistent with the previously presented computations on similar transformation, which further corroborated that the Pd(0)–Pd(II) cross-coupling mechanism is operative for electronically activated amides (Figure 13.24b). In addition, Dang et al. also found that the PCy_3Pd(acyl)(aryl) **int188** species prefers to undergo the acyl-aryl reductive elimination to produce the ketone product (**TS189** vs. **TS190**), which is consistent with the observed chemoselectivity. This chemoselectivity is rationalized by the d(palladium) to π(CO)* back donation and natural population analysis of the involved palladium species.

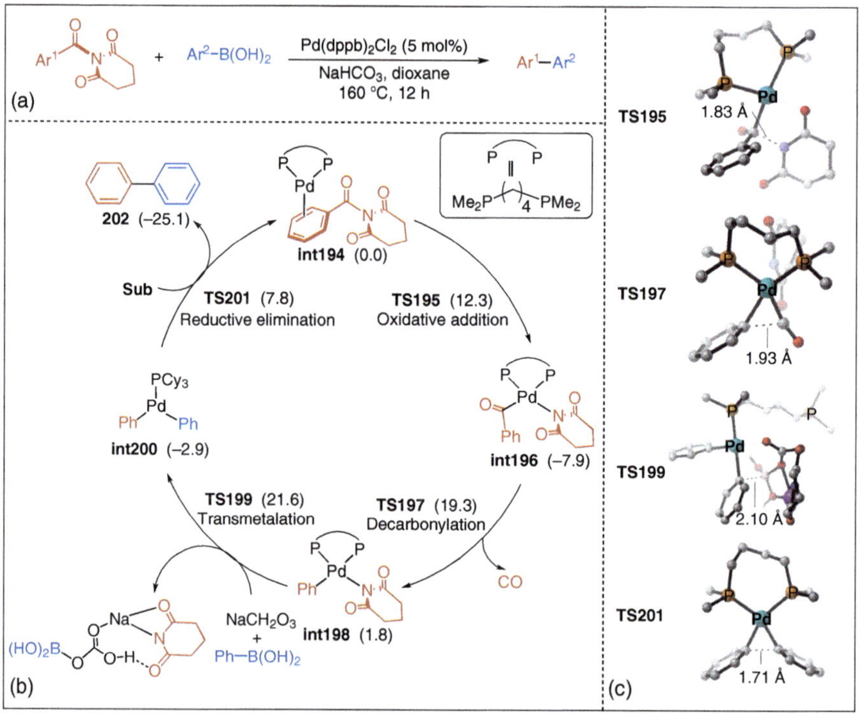

Figure 13.25 Experimental and computational studies on the Pd-catalyzed decarbonylative Suzuki–Miyaura cross-coupling of twisted amides.

The collaboration between Szostak and Hong reported a combined experimental and computational study on the Pd/dppb-catalyzed decarbonylative Suzuki–Miyaura coupling of twisted amides [45]. The DFT-computed free energy profile is shown in Figure 13.25b. From the twisted amide-coordinated complex **int194**, the oxidative addition of twisted amides breaks the C–N bond via **TS195**, generating the LPd^{II}(acyl)(amino) intermediate **int196**. This step only requires a barrier of 12.3 kcal/mol and is exergonic by 7.9 kcal/mol. Subsequently, the decarbonylation via **TS197** occurs to produce the decarbonylated intermediate **int198**. This intermediate **int198** further undergoes the transmetalation through **TS199** to generate the biaryl palladium(II) species **int200**. The reductive elimination via **TS201** produces the decarbonylative cross-coupling product and regenerates the palladium(0) active catalyst. Based on the DFT-computed free energy profile of the catalytic cycle, the transmetalation is the rate-limiting step with a 21.6 kcal/mol overall barrier. These computations elucidated the catalytic behavior of bisligated palladium catalyst in C–N bond activation of amides.

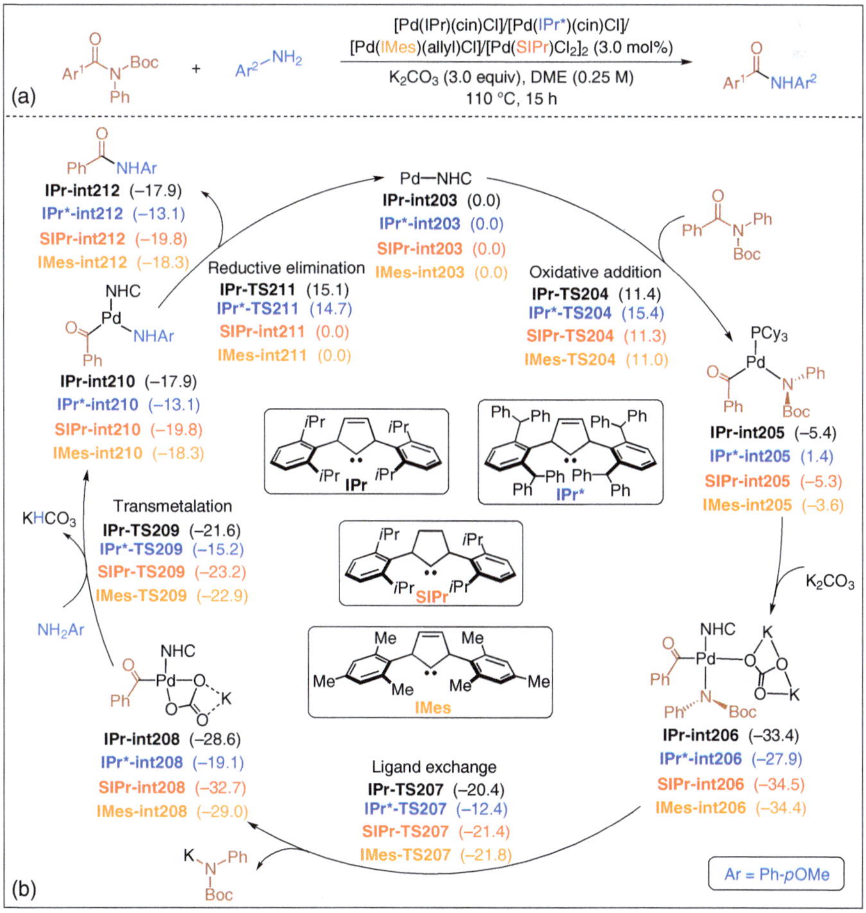

Figure 13.26 Computational study of Pd/NHC-catalyzed transamidation of amides.

13.4.6 Computational Study on the Mechanism of Pd-Catalyzed Transamidation of Amides

The collaboration between Szostak, Nolan, and Poater also revealed that palladium catalysis can achieve transamidation of electronically activated amides (Figure 13.26a) [46]. The *N*-Boc-*N*-phenyl aryl amides undergo smooth C–N coupling with aniline derivatives to produce the transamidation product. In addition to the synthetic advances, the authors also explored the mechanism of the generation of the active palladium(0) catalyst and the Pd-catalyzed transamidation reaction using DFT calculations. The typical cross-coupling mechanism was supported by the authors' computations at the M06/Def2-TZVP-SDD-PCM(DME)//BP86-D3/Def2-SVP-SDD level of theory (Figure 13.26b). The overall transformation involves sequential C–N oxidative addition, base-assisted proton transfer, and C–N reductive elimination. The calculations also revealed how the steric effects of *N*-heterocyclic carbene ligands affect the free energy profile and eventually the observed reactivity.

13.5 Outlook

In this chapter, we briefly reviewed and discussed the computational studies on amide C–N bond activation reactions. The general mechanistic model, the free energy profile, and the structure–activity relationship would not be available if without the highlighted computations. It is without any doubt that the synergy between computation and experiment significantly accelerated the discovery of amide C–N bond activation reactions, providing a strong and continuing momentum that pushes the frontiers of amide chemistry. The joint force between computation and experiment will pave the way for the emerging and dynamic field of amide activation and transformation, bringing the synthetic utilization of amides to the next level.

Looking at the future of amide activation, there are still a number of critical issues that will certainly benefit from computations. First, the mechanistic understanding of external force-assisted bond activations is significantly underexplored as compared to the thermocatalysis. Photo- and electrocatalysis have been a strong support to open new reaction channels for a wide array of traditionally inert molecules. Their utilizations in amide transformation, however, are still elusive. Rational reaction designs and mechanistic understandings in this regard are highly desirable. In addition, the predictive ability of the available structure–activity relationships in amide chemistry is still limited. Most of the current computational studies focus on the understanding, instead of the rational design and even prediction of new and better catalyst for amide transformation. The trending catalytic systems in amide transformation are mainly discovered by experimental serendipity, not prediction. Data-driven approaches can serve as useful strategies to transfer the mechanistic knowledge to synthetic predictions. The combination between computational mechanistic knowledge and transformative machine learning approaches will facilitate the development of next generation of amide transformations. In summary, over the years in computational exploration of the chemical behavior of amides, the computations have never stopped surprising me. We hope and encourage more colleagues can join this intriguing field and experience the joy and fun in amide chemistry.

References

1 (a) Greenberg, A., Breneman, C.M., and Liebman, J.F. (2003). *The Amide Linkage: Structural Significance in Chemistry, Biochemistry, and Materials Science*. Wiley-Interscience. (b) Pattabiraman, V.R. and Bode, J.W. (2011). *Nature* 480: 471–479.

2 Zabicky, J. (1970). *The Chemistry of Amides*. Wiley.

3 (a) Hughes, A.B. (2011). *Amino Acids, Peptides and Proteins in Organic Chemistry*. Wiley-VCH. (b) Marchildon, K. (2011). *Macromol. React. Eng.* 5: 22–54. (c) Roughley, S.D., Jordan, A.M., and Med, J. (2011). *Chem* 54: 3451–3479.

4 Mullard, A. (2021). *Nat. Rev. Drug Discovery* 20: 85–90.

5 Chen, K., Risatti, C., Bultman, M. et al. (2014). *J. Org. Chem.* 79: 8757–8767.

6 Theodora, W. (1999). *Protective Groups in Organic Synthesis*. Wiley.

7 Green, T.W. and Wuts, P.G.M. (1999). *Protective Groups in Organic Synthesis*, vol. 518-525, 736–739. New York: Wiley-Interscience.

8 Pauling, L. (1940). *The Nature of the Chemical Bond*. Oxford University Press.

9 The C–N BDEs of amides listed in Figure 2 can be found from iBonD 2.0. Internet Bond-Energy Databank (p*K*a and BDE)-*i*BonD Home Page. http://ibond.chem.tsinghua.edu.cn or http://ibond.nankai.edu.cn.

10 Linstrom, P., Mallard, W. (1998). *The NIST Chemistry WebBook: A Tool for Chemical Data Access on the Internet*.

11 (a) Szostak, M. and Li, G. (2020). *Synthesis* 52: 2579–2599. b Kemnitz, C.R. and Loewen, M.J. (2007). *J. Am. Chem. Soc.* 129: 2521–2528.

12 (a) Adachi, S., Kumagai, N., and Shibasaki, M. (2018). *Tetrahedron Lett.* 59: 1147–1158. (b) Campeau, L.-C. and Hazari, N. (2019). *Organometallics* 38: 3–35. (c) Dander, J.E. and Garg, N.K. (2017). *ACS Catal.* 7: 1413–1423. (d) Li, G., Ma, S., and Szostak, M. (2020). *Trends Chem.* 2: 914–928. (e) García-Cárceles, J., Bahou, K.A., and Bower, J.F. (2020). *ACS Catal.* 10: 12738–12759. f Boit, T.B., Bulger, A.S., Dander, J.E., and Garg, N.K. (2020). *ACS Catal.* 10: 12109–12126. g Shi, S., Nolan, S.P., and Szostak, M. (2018). *Acc. Chem. Res.* 51: 2589–2599.

13 Wang, H., Zhang, S.-Q., and Hong, X. (2019). *Chem. Commun.* 55: 11330–11341.

14 (a) Zhang, X., Chung, L.W., and Wu, Y.D. (2016). *Acc. Chem. Res.* 49: 1302–1310. (b) Sperger, T., Sanhueza, I.A., Kalvet, I., and Schoenebeck, F. (2015). *Chem. Rev.* 115: 9532–9586. (c) Cheng, G.J., Zhang, X., Chung, L.W. et al. (2015). *J. Am. Chem. Soc.* 137: 1706–1725.

15 (a) Cheung, C.W., Ma, J.A., and Hu, X. (2018). *J. Am. Chem. Soc.* 140: 6789–6792. (b) Nagae, H., Hirai, T., Kato, D. et al. (2019). *Chem. Sci.* 10: 2860–2868. (c) Fairley, M., Bole, L.J., Mulks, F.F. et al. (2020). *Chem. Sci.* 11: 6500–6509.

(d) Taillier, C., Bellosta, V., Meyer, C., and Cossy, J. (2004). *Org. Lett.* 6: 2145–2147. (e) Li, G. and Szostak, M. (2018). *Nat. Commun.* 9: 4165. (f) Senatore, R., Ielo, L., Monticelli, S. et al. (2019). *Synthesis* 51: 2792–2808. (g) Hoerter, J.M., Otte, K.M., Gellman, S.H. et al. (2008). *J. Am. Chem. Soc.* 130: 647–654. (h) Bon, E., Bigg, D.C.H., and Bertrand, G. (1994). *J. Org. Chem.* 59: 4035–4036. (i) Sheng, H., Zeng, R., Wang, W. et al. (2017). *Adv. Synth. Catal.* 359: 302–313. (j) Stephenson, N.A., Zhu, J., Gellman, S.H., and Stahl, S.S. (2009). *J. Am. Chem. Soc.* 131: 10003–10008. (k) Shirokane, K., Kurosaki, Y., Sato, T., and Chida, N. (2010). *Angew. Chem. Int. Ed.* 49: 6369–6372. (l) Kurosaki, Y., Shirokane, K., Oishi, T. et al. (2012). *Org. Lett.* 14: 2098–2101. (m) Oda, Y., Sato, T., and Chida, N. (2012). *Org. Lett.* 14: 950–953. (n) Yanagita, Y., Nakamura, H., Shirokane, K. et al. (2013). *Chem. Eur. J.* 19: 678–684. (o) Shirokane, K., Wada, T., Yoritate, M. et al. (2014). *Angew. Chem. Int. Ed.* 53: 512–516. (p) Yoritate, M., Meguro, T., Matsuo, N. et al. (2014). *Chem. Eur. J.* 20: 8210–8216. (q) Nakajima, M., Oda, Y., Wada, T. et al. (2014). *Chem. Eur. J.* 20: 17565–17571. (r) Heller, S.T., Newton, J.N., Fu, T., and Sarpong, R. (2015). *Angew. Chem. Int. Ed.* 54: 9839–9843. (s) Jäkel, M., Qu, J., Schnitzer, T., and Helmchen, G. (2013). *Chem. Eur. J.* 19: 16746–16755. (t) Giannerini, M., Vila, C., Hornillos, V., and Feringa, B.L. (2016). *Chem. Commun.* 52: 1206–1209. (u) Nugent, J. and Schwartz, B.D. (2016). *Org. Lett.* 18: 3834–3837. (v) Vincent, G., Guillot, R., and Kouklovsky, C. (2011). *Angew. Chem. Int. Ed.* 50: 1350–1353.

16 Balasubramaniam, S. and Aidhen, I.S. (2008). *Synthesis* 23: 3707–3738.

17 (a) Meng, G., Shi, S., and Szostak, M. (2016). *Synlett* 27: 2530–2540. (b) Takise, R., Muto, K., and Yamaguchi, J. (2017). *Chem. Soc. Rev.* 46: 5864–5888. (c) Liu, C. and Szostak, M. (2018). *Org. Biomol. Chem.* 16: 7998–8010. (d) Kaiser, D., Bauer, A., Lemmerer, M., and Maulide, N. (2018). *Chem. Soc. Rev.* 47: 7899–7925. (e) Bourne-Branchu, Y., Gosmini, C., and Danoun, G. (2019). *Chem. Eur. J.* 25: 2663–2674. (f) Chaudhari, M.B. and Gnanaprakasam, B. (2019). *Chem. Asian J.* 14: 76–93. (g) Buchspies, J. and Szostak, M. (2019). *Catalysts* 9: 53–75. (h) Gao, Y., Ji, C.-L., and Hong, X. (2017). *Sci. China Chem.* 60: 1413–1424.

18 For representative studies:(a)Boit, T.B., Mehta, M.M., Kim, J. et al. (2021). *Angew. Chem. Int. Ed.* 60: 2472–2477. (b) Hie, L., Baker, E.L., Anthony, S.M. et al. (2016). *Angew. Chem. Int. Ed.* 55: 15129–15132. (c) Baker, E.L., Yamano, M.M., Zhou, Y. et al. (2016). *Nat. Commun.* 7: 11554. (d) Shi, S. and Szostak, M. (2016). *Chem. Eur. J.* 22: 10420–10424. (e) Buchspies, J., Rahman, M.M., and Szostak, M. (2020). *Catalysts* 10: 372–382. (f) Shi, S. and Szostak, M. (2016). *Org. Lett.* 18: 5872–5875. (g) Medina, J.M., Moreno, J., Racine, S. et al. (2017). *Angew. Chem. Int. Ed.* 56: 6567–6571. (h) Dander, J.E., Giroud, M., Racine, S. et al. (2019). *Commun. Chem.* 2: 82–90. (i) Hu, J., Wang, M., Pu, X., and Shi, Z. (2017). *Nat. Commun.* 8: 14993. (j) Boit, T.B., Weires, N.A., Kim, J., and Garg, N.K. (2018). *ACS Catal.* 8: 1003–1008.

19 a Meng, G. and Szostak, M. (2015). *Org. Lett.* 27: 4364–4367. b Meng, G. and Szostak, M. (2015). *Angew. Chem. Int. Ed.* 54: 14518–14522. c Meng, G., Shi, S., and Szostak, M. (2016). *ACS Catal.* 6: 7335–7339. d Li, X. and Zou, G. (2015). *Chem. Commun.* 51: 5089–5092. e Liu, C., Meng, G., and Szostak, M. (2016).

J. Org. Chem. 81: 12023–12030. f Meng, G. and Szostak, M. (2016). *Org. Biomol. Chem.* 14: 5690–5707. g Lei, P., Meng, G., Ling, Y. et al. (2017). *Org. Lett.* 19: 6510–6513. h Shi, S. and Szostak, M. (2017). *Org. Lett.* 19: 3095–3098. i Meng, G., Lei, P., and Szostak, M. (2017). *Org. Lett.* 19: 2158–2161. j Liu, C., Liu, Y., Liu, R. et al. (2017). *Org. Lett.* 19: 1434–1437. k Lei, P., Meng, G., and Szostak, M. (2017). *ACS Catal.* 7: 1960–1965. l Shi, S. and Szostak, M. (2017). *Chem. Commun.* 53: 10584–10587. m Lei, P., Meng, G., Shi, S. et al. (2017). *Chem. Sci.* 8: 6525–6530. n Liu, P., Rong, G., Ling, Y. et al. (2017). *J. Org. Chem.* 82: 6638–6646. o Liu, C., Li, G., Shi, S. et al. (2018). *ACS Catal.* 8: 9131–9139. p Meng, G. and Szostak, M. *Org. Lett.* 20: 6789–6793. q Zhou, P.X., Shi, S., Wang, J. et al. (2019). *Org. Chem. Front.* 6: 1942–1947. r Zhou, T., Li, G., Nolan, S.P., and Szostak, M. (2019). *Org. Lett.* 21: 3304–3309. s Ma, S., Zhou, T., Li, G., and Szostak, M. (2020). *Adv. Synth. Catal.* 362: 1887–1892.

20 (a) Meng, G. and Szostak, M. (2016). *Org. Lett.* 18: 796–799. (b) Meng, G. and Szostak, M. (2017). *ACS Catal.* 7: 7251–7256.

21 Li, W., Zhang, S., Feng, X. et al. (2021). *Org. Lett.* 23: 2521–2526.

22 Bourne-Branchu, Y., Gosmini, C., and Danoun, G. (2017). *Chem. Eur. J.* 23: 10043–10047.

23 (a) Jiao, J. and Wang, X. (2021). *Angew. Chem. Int. Ed.* 60: 17088–17093. (b) Zhuo, J., Zhang, Y., Li, Z., and Li, C. (2020). *ACS Catal.* 10: 3895–3903. (c) Seebach, D. (1979). *Angew. Chem. Int. Ed. Engl.* 18: 239–258. (d) Seebach, D. (1979). *Angew. Chem.* 91: 259–278. (e) Bugaut, X. and Glorius, F. (2012). *Chem. Soc. Rev.* 41: 3511–3522. (f) Streuff, J. (2013). *Synthesis* 45: 281–307. (g) Just-Baringo, X. and Procter, D.J. (2015). *Acc. Chem. Res.* 48: 1263–1275. (h) Szostak, M., Spain, M., Eberhart, A.J., and Procter, D.J. (2014). *J. Am. Chem. Soc.* 136: 2268–2271. (i) Szostak, M., Spain, M., and Procter, D.J. (2014). *J. Am. Chem. Soc.* 136: 8459–8466. (j) Szostak, M., Spain, M., Eberhart, A.J., and Procter, D.J. (2014). *J. Org. Chem.* 79: 11988–12003. (k) Parmar, D., Duffy, L.A., Sadasivam, D.V. et al. (2009). *J. Am. Chem. Soc.* 131: 15467–15473. (l) Huq, S.R., Shi, S., Diao, R., and Szostak, M. (2017). *J. Org. Chem.* 82: 6528–6540.

24 Li, G., Ji, C.L., Hong, X., and Szostak, M. (2019). *J. Am. Chem. Soc.* 141: 11161–11172.

25 Wybon, C.C.D., Mensch, C., Hollanders, C. et al. (2018). *ACS Catal.* 8: 203–218.

26 Noshita, M., Shimizu, Y., Morimoto, H. et al. (2019). *Org. Process Res. Dev.* 23: 588–594.

27 Wang, G., Shi, Q., Hu, W. et al. (2020). *Nat. Commun.* 11: 946–955.

28 Hie, L., Fine Nathel, N.F., Shah, T.K. et al. (2015). *Nature* 524: 79–83.

29 Hie, L., Fine Nathel, N.F., Hong, X. et al. (2016). *Angew. Chem. Int. Ed.* 55: 2810–2814.

30 Ji, C.L., Xie, P.P., and Hong, X. (2018). *Molecules* 23: 2681–2690.

31 Ben Halima, T., Masson-Makdissi, J., and Newman, S.G. (2018). *Angew. Chem. Int. Ed.* 57: 12925–12929.

32 Chu, C.Q. and Dang, L. (2018). *J. Org. Chem.* 83: 5009–5018.

33 Liu, L.L., Chen, P., Sun, Y. et al. (2016). *J. Org. Chem.* 81: 11686–11696.

34 Weires, N.A., Baker, E.L., and Garg, N.K. (2016). *Nat. Chem.* 8: 75–79.

35 Xu, Z.Y., Yu, H.Z., and Fu, Y. (2017). *Chem. Asian J.* 12: 1765–1772.

36 (a) Liu, C. and Szostak, M. (2017). *Chem. Eur. J.* 23: 7157–7173. (b) Szostak, M. and Aube, J. (2013). *Chem. Rev.* 113: 5701–5765. (c) Szostak, R. and Szostak, M. (2019). *Molecules* 24: 274–290.

37 Shi, S., Meng, G., and Szostak, M. (2016). *Angew. Chem. Int. Ed.* 55: 6959–6963.

38 Ji, C.L. and Hong, X. (2017). *J. Am. Chem. Soc.* 139: 15522–15529.

39 Morioka, T., Nishizawa, A., Furukawa, T. et al. (2017). *J. Am. Chem. Soc.* 139: 1416–1419.

40 Liu, C. and Szostak, M. (2017). *Angew. Chem. Int. Ed.* 56: 12718–12722.

41 Xu, Y., Wang, B., Jiang, J. et al. (2019). *J. Org. Chem.* 84: 9474–9479.

42 Xie, P.P., Qin, Z.X., Zhang, S.Q., and Hong, X. (2021). *ChemCatChem* 13: 3536–3542.

43 Li, G., Lei, P., Szostak, M. et al. (2018). *ChemCatChem* 10: 3096–3106.

44 Li, R., Xu, H., Zhao, N. et al. (2020). *J. Org. Chem.* 85: 833–840.

45 Zhou, T., Ji, C.L., Hong, X., and Szostak, M. (2019). *Chem. Sci.* 10: 9865–9871.

46 Li, G., Zhou, T., Poater, A. et al. (2020). *Catal. Sci. Technol.* 10: 710–716.

14

Esters as Viable Acyl Cross-Coupling Electrophiles

Omid Daneshfar and Stephen G. Newman

University of Ottawa, Centre for Catalysis Research and Innovation, Department of Chemistry and Biomolecular Sciences, 10 Marie Curie, Ottawa, ON K1N 6N5, Canada

14.1 Introduction

Transition metal-catalyzed cross-coupling reactions are among the most frequently run reactions as they provide an efficient and reliable way to construct C–C and C–heteroatom bonds [1]. Traditionally, these reactions are often run with aryl halides as the electrophilic reaction partner, generating halide salt by-products. Ongoing research efforts have led to the expansion of the types of pseudohalides that can act as surrogates, avoiding the halide salt waste and enabling a wider variety of synthetic pathways. For example, sulfamates, pivalates, carbamates, and aryl ethers have all been engaged in cross-coupling reactions [2].

In tandem to the expansion of the type of (pseudo)halide leaving groups that could be cleaved, great progress has been made to diversify the carbon scaffolds that can participate. For example, carboxylic acid derivatives can be activated to make a range of product classes. Pioneering work in this area was realized by Yamamoto in 1976, when the stoichiometric reaction between Ni(0) and a phenyl ester was disclosed [3]. Ni(0) was proposed to oxidatively insert into the acyl(C)–O, as evidenced by the decarbonylated ester product and Ni(0)–CO complex that were obtained. Around the same time, acyl chlorides were first used as cross-coupling electrophiles to give ketones [4]. By the end of the twentieth century, anhydrides [5] and thioesters [6] were similarly demonstrated to be viable reaction partners. A key mechanistic step in coupling of these acyl electrophiles is the oxidative addition into the C(acyl)–X bond, which is facilitated by selecting a (pseudo)halide that acts as a good leaving group. More robust carboxylic acid derivatives such as esters and amides can be difficult to engage in cross-coupling because of the relative stability of the C(acyl)–O and C(acyl)–N bonds, which have substantial double bond character. The cross-coupling of amides was ultimately achieved by two key strategies: choosing powerful catalysts capable of cleaving the robust amide bond or identifying amides that are more readily activated. This topic is covered in depth in the chapter 'cross-coupling of amides'. The cross-coupling of esters shares key similarities along with many

Amide Bond Activation: Concepts and Reactions, First Edition. Edited by Michal Szostak.

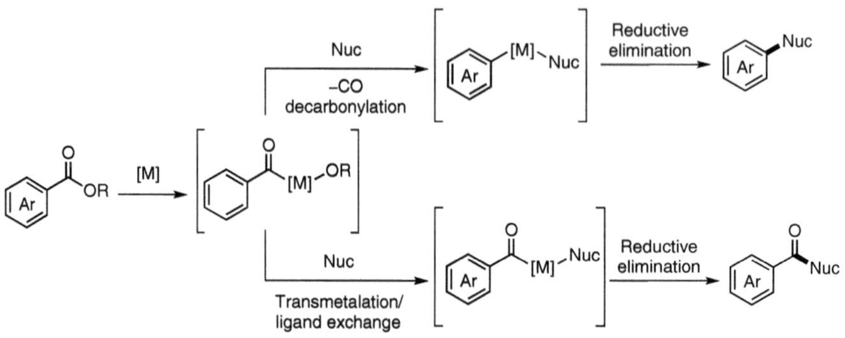

Figure 14.1 General mechanism for the cross-coupling of esters.

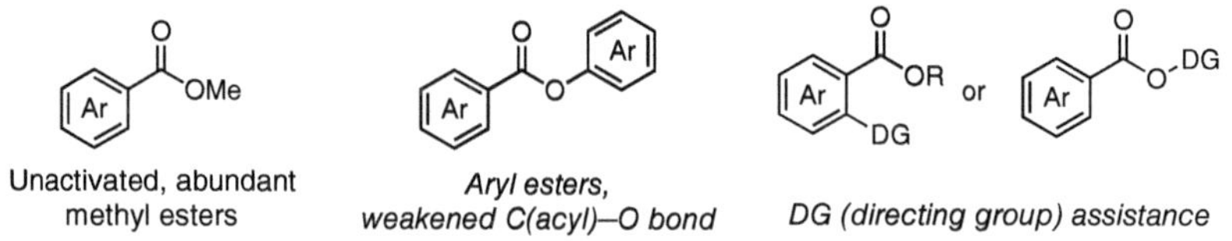

Figure 14.2 Classes of esters that can be activated.

distinct differences. The key step involves oxidative addition of a transition metal catalyst into the C(acyl)–O bond (Figure 14.1). The complex resulting from oxidative addition can react with nucleophiles in a decarbonylative fashion, or in a carbonyl retentive manner. Two different product classes can arise from the use of a single electrophile. Controlling this process provides access to a wide range of products.

Section 14.2 of this chapter aims to first provide a very brief overview of the pioneering work in cross-coupling of activated carboxylic acid derivatives – acyl chlorides, anhydrides, and thioesters. Afterward, three main classes of esters that can be engaged in cross-coupling will be discussed: aryl esters (Sections 14.3–14.4), methyl esters (Sections 14.5–14.6), and esters with directing group assistance (Sections 14.4.1 and 14.6.1) (Figure 14.2). Detailed discussion in each section will discuss dates and timelines to provide contextualization.

14.2 Early Work in the Cross-coupling of Carboxylic Acid Derivatives

This section aims to take a brief look at some of the pioneering examples concerning the cross-coupling of activated carboxylic acid derivatives such as acyl chlorides, anhydrides, and thioesters. This is not an exhaustive list of all work concerning the aforementioned electrophiles, nor does it include all varieties of electrophiles studied before, during, and after this field began to make use of esters. Rather, this section seeks to contextualize the early research and follow its natural progression to esters. For more details on carboxylic acid derivative (including amide and other C–O electrophile) cross-couplings, we direct readers to leading reviews [2, 7].

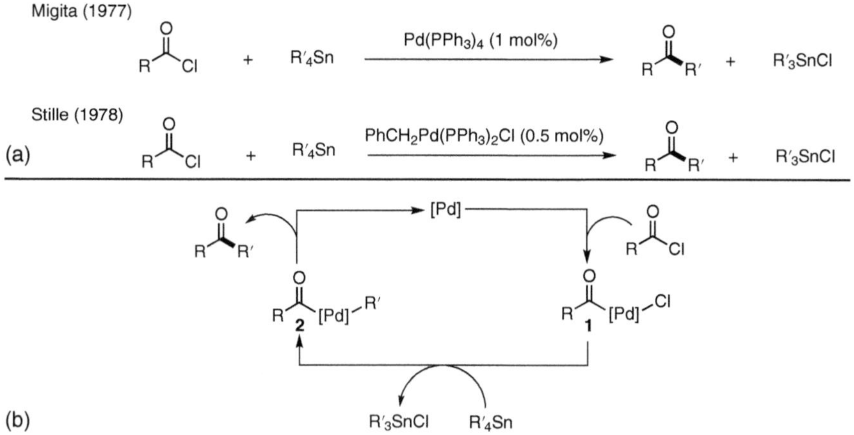

Figure 14.3 (a) First examples of acyl electrophiles used for cross-coupling catalysis. (b) Proposed mechanism of the Pd-catalyzed cross-coupling of acyl chlorides with organotin reagents.

Acyl chlorides are highly electrophilic species that readily react with carbon-based nucleophiles (e.g. Grignard or organolithium reagents) in the absence of transition metal catalysts. Ketones are difficult to access this way due to the electrophilicity of the resulting product, which renders them susceptible to a second attack from the organometallic reagent, giving alcohol products. In the late 1970s, Migita and Stille independently reported the first cross-coupling reactions that made use of a carboxylic acid derivative as the electrophile [4]. In the presence of Pd catalysts, it was demonstrated that acyl chlorides could undergo cross-coupling reactions with alkyl or aryl organotin reagents, forming a variety of different ketone products (Figure 14.3a). The importance of this work was substantial as it demonstrated the power of cross-coupling catalysis by allowing the use of mild nucleophiles, avoiding overaddition of aggressive organometallic nucleophiles, and allowing for the efficient synthesis of ketones. Both reports proposed a similar mechanism, which began with Pd oxidatively inserting into the C(acyl)–Cl bond (Figure 14.3b). This intermediate **1** could then undergo transmetalation with the organotin species, yielding intermediate **2**. Reductive elimination releases the product and the palladium catalyst, allowing for propagation of the catalytic cycle.

Notably, the Weinreb ketone synthesis was developed at approximately the same time as this cross-coupling approach to make ketones, which features carefully chosen amide electrophiles that overcome the overaddition problem [8]. While this method became and has remained powerful in synthetic organic chemistry, the need for harsh organometallic reagents and multistep preparation of the Weinreb amide electrophile left room for improvement for future coupling methods.

In 1998, the Fukuyama group developed coupling of thioesters, providing an advantage over the use of acid chlorides due to their robustness (Figure 14.4a) [6a]. In the so-called Fukuyama coupling, thioesters and organozinc reagents are coupled in the presence of a Pd catalyst to furnish ketone products under mild

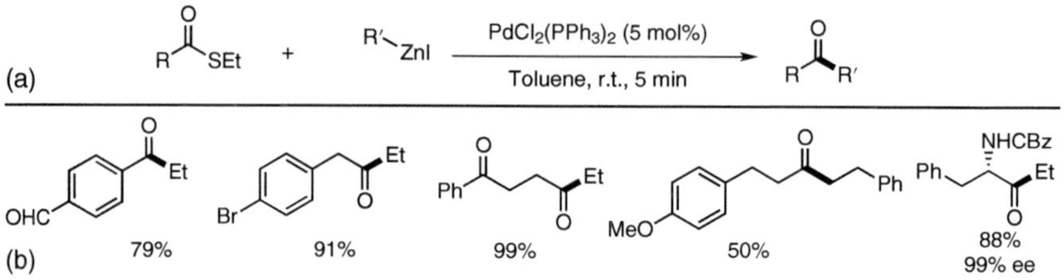

Figure 14.4 (a) Pd-catalyzed carbonyl retentive Negishi-type coupling of thioesters. (b) Representative scope examples.

reaction conditions. Encouragingly, a variety of sensitive functional groups such as ketones, aldehydes, aryl chlorides, and bromides were tolerated, demonstrating high chemoselectivity for ketone formation. Within this study, oxidative addition of the Pd catalyst with aryl halides and deleterious nucleophilic addition of the organozinc reagents to other electrophilic functionality were both avoided (Figure 14.4b).

In many ways, boronic acids are ideal cross-coupling nucleophiles due to their stability, commercial availability, and mildness. In 2000, Liebeskind and Srogl developed a ketone-forming reaction via the coupling of thioesters with boronic acids under strictly nonbasic conditions (Figure 14.5) [6b]. This reaction was facilitated by a Pd catalyst. Critically, Cu(I) thiophene-2-carboxylate (CuTC) was used as a stoichiometric additive. Although the mechanism was not fully understood, it was proposed that the transmetalation from boron to palladium was mediated by CuTC, which was bound to the acyl palladium thiolate complex. This was unique as traditional Suzuki–Miyaura couplings often require base, while the use of base in this chemistry was detrimental due to binding to boron, blocking the necessary coordination of the Cu(I) carboxylate to the trivalent boron for transmetalation. In 2003, Liebeskind and Srogl reported a similar base-free, Cu- and Pd-mediated ketone synthesis via the coupling of thioesters with organotin reagents [6c]. The following year, this was extended to include alkyl nucleophiles through the use of B-alkyl-9-BBN reagents [6d]. A Pd catalyst was used again, along with a Cu(I) mediator; however, this reaction required the use of base.

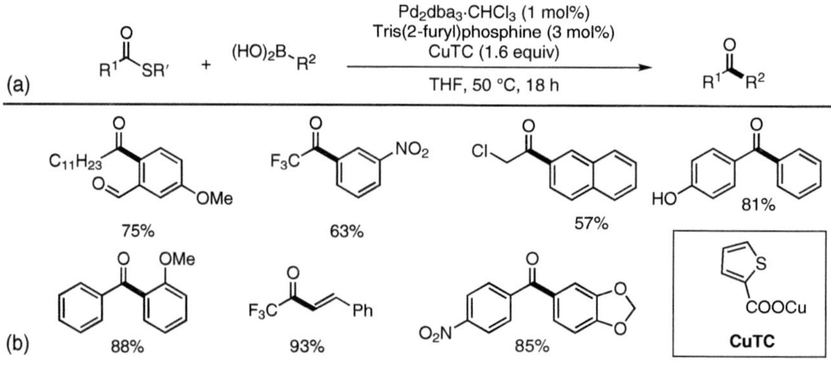

Figure 14.5 (a) Pd-catalyzed carbonyl retentive Suzuki–Miyaura-type coupling of thioesters. (b) Representative scope examples.

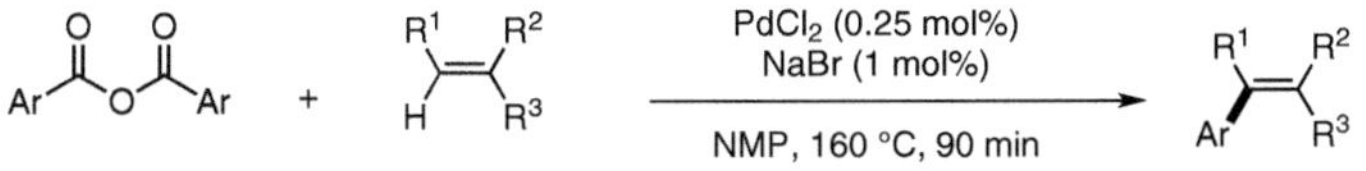

Figure 14.6 Pd-catalyzed decarbonylative Mizoroki–Heck-type cross-coupling of aromatic carboxylic anhydrides.

Around the same time as thioesters, cross-coupling reactions of acid anhydrides were developed. The first such report was in 1998 by de Vries and coworkers (Figure 14.6) [5a]. They demonstrated that in the presence of a Pd catalyst, aromatic carboxylic anhydrides could decarbonylate and be used to arylate olefins. In traditional Mizoroki–Heck reactions with aryl halides, a stoichiometric amount of base is required to neutralize the acid generated during the reaction. Furthermore, a stoichiometric amount of halide salt is generated as waste. Both of these drawbacks were circumvented in this report with anhydrides acting as organohalide surrogates. However, a high reaction temperature was required, ranging from 140 to 190 °C. As a result, mixtures of olefins isomers were formed. While this issue limited the synthetic usefulness, this seminal work demonstrated that both the power of acyl electrophiles to overcome issues associated with existing chemistry and that acyl electrophiles are not limited to ketone-forming coupling reactions.

Gooßen and coworker envisioned that ketone products could be accessed from anhydrides via a mechanistic pathway that mirrors the cross-coupling of acid chlorides (Figure 14.3b). In 2001, Gooßen reported the first Pd-catalyzed coupling of aromatic carboxylic anhydrides with boronic acids in a carbonyl retentive manner (Figure 14.7a) [5b]. In contrast to the decarbonylative reactions developed by de Vries and coworkers described previously, the carbonyl group was retained to provide the ketone product. Free carboxylic acids were also converted to acid anhydrides *in situ* with the use of stoichiometric pivalic anhydride (Figure 14.7b) to form various ketone products (Figure 14.7c). A similar anhydride cross-coupling reaction was also reported by Yamamoto and coworkers in 2002 [5c].

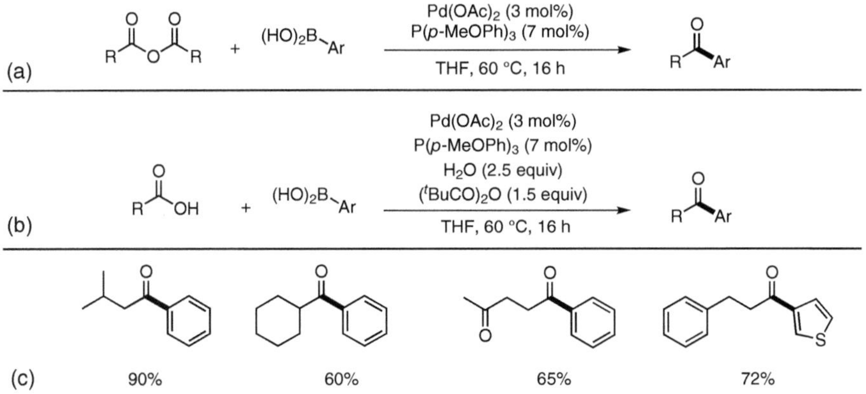

Figure 14.7 (a) Pd-catalyzed carbonyl retentive Suzuki–Miyaura coupling of acid anhydrides. (b) *In situ* formation and coupling of acid anhydrides. (c) Representative scope of *in situ* anhydride coupling.

Although the coupling reactions with acyl electrophiles discussed thus far are of both fundamental and practical importance, they are not without their drawbacks. Acid chlorides and anhydrides are reactive species and thus are seldom useful in late-stage functionalization. Relative to acyl chlorides and anhydrides, thioesters are much more robust; however, they too need to be prepared prior to coupling, and they often require the use of stoichiometric copper salts to be activated. Continued progress in this field calls for the ability to functionalize more abundant and, more importantly, more robust electrophiles. Esters have the potential to satisfy both of these demands. The acyl cross-coupling reactions described thus far paved the way for future achievements with esters that will be described throughout the remainder of this chapter. In particular, the power of esters to form both decarbonylative and carbonyl retentive products will be highlighted.

14.3 Decarbonylative Coupling of Aryl Esters

Cross-coupling reactions of esters that selectively form decarbonylated products provide an alternative route to more traditional reactions with aryl halides. Beyond providing a distinct synthetic disconnection, these reactions avoid the formation of halide salt waste products and display useful chemoselectivity. Furthermore, esters are a common, relatively robust motif that makes them useful in complex molecule synthesis. The majority of these decarbonylative couplings feature esters that are modestly activated by a phenyl ring on oxygen. These so-called aryl esters feature a weaker C(acyl)–O bond than alkyl esters, rendering them more susceptible to cleavage. Leading examples of these transformations are given in the following section.

14.3.1 Mizoroki–Heck-Type Coupling

Since its discovery, the Mizoroki–Heck reaction has been demonstrated as a reliable way to arylate olefins efficiently and often regioselectively [9]. Decarbonylative couplings using acid anhydrides as aryl halide surrogates were previously described (Figure 14.6), which obviate the need for a stoichiometric base. The first example using aryl esters was reported by Gooßen and coworker, who coupled *p*-nitrophenyl esters with olefins using a palladium catalyst (Figure 14.8a) [10]. As was the case in

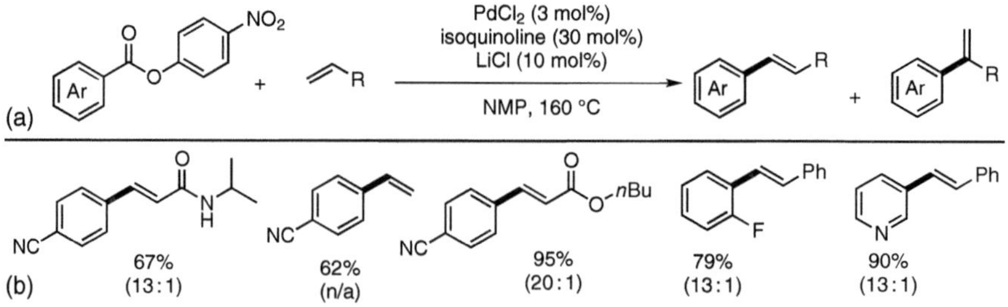

Figure 14.8 (a) Mizoroki–Heck-type reaction of *p*-nitrophenyl esters. (b) Representative scope examples where ratio in brackets represent linear:branched selectivity.

the work of de Vries, $PdCl_2$ was identified as the optimal precatalyst, and the addition of alkali metal halides, most notably LiCl, helped to increase the effectiveness of the catalyst. To diminish catalyst deactivation, nitrogen ligands, particularly isoquinoline, were identified as useful stabilizing additives. A strongly coordinating phosphine ligand like PPh_3 led to diminished yields of product. The *p*-nitro group on the arene was critical; in contrast to the high yields obtained, the use of a simple phenyl ester gave trace product formation. Linear products were formed preferentially using a range of terminal olefin coupling partners (Figure 14.8b). A "waste-free" procedure was also demonstrated, with formation of the aryl ester by condensation and recovery of the *p*-nitrophenol after coupling enabling a closed recycle loop.

14.3.2 C–H Biaryl Coupling

The cross-coupling reactions described thus far feature traditional nucleophilic coupling partners, including olefins, boronic acids, and organozincs. The use of unfunctionalized arenes provides an appealing alternative for C–C bond construction but adds the additional challenge of necessitating a C–H activation step. Toward this goal, Itami and Yamaguchi reported the first decarbonylative C–H arylation of phenyl esters in 2012 (Figure 14.9a) [11]. This involved a nickel catalyst system in a net redox-neutral transformation. The authors were particularly interested

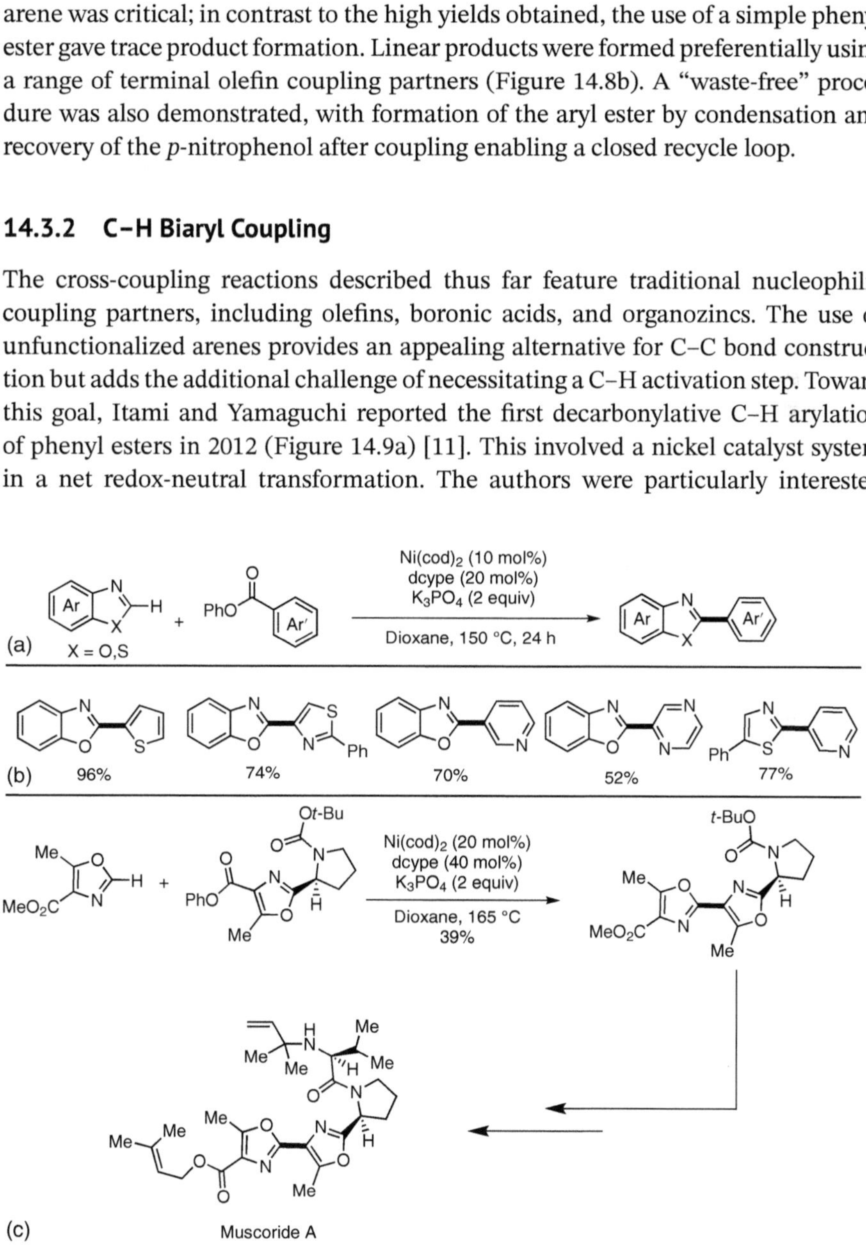

Figure 14.9 (a) Ni-catalyzed decarbonylative C–H coupling of azoles with phenyl esters. (b) Representative scope examples. (c) Formal synthesis of muscoride A.

in investigating the coupling of heteroaromatic esters as they would lead to the formation of heteroarylated azoles (Figure 14.9b), which are important scaffolds in pharmaceutically relevant compounds and natural products. This method was applied toward the formal synthesis of muscoride A using the decarbonylative C–H coupling for access to a key intermediate (Figure 14.9c). Under similar conditions, a decarbonylative alkenylation of azoles with α,β-unsaturated esters has also been demonstrated [12]. While the use of Ni/dcype was limited to the coupling of benzoxazoles and benzothiazoles, Itami and Yamaguchi later expanded the reaction to include benzothiazoles and benzimidazoles by switching to a Pd/dcypt or dcype catalyst system [13].

A Ni(0)/Ni(II) catalytic cycle was proposed for this C–H functionalization, which starts with Ni(0) oxidatively adding into the phenyl ester C–O bond to form complex **3** (Figure 14.10). Decarbonylation occurs to generate intermediate **4**, followed by the key C–H arylation to give intermediate **5**. This Ni(II) intermediate then reductively eliminates the product and generates Ni(0) species **6**, which can finally be turned over via thermal extrusion of CO. Ni(dcype)$(CO)_2$ was prepared and demonstrated to be an effective catalyst, suggesting that CO extrusion is a viable step. Lastly, more user-friendly conditions for this arylation were demonstrated with $NiCl_2$ and Zn powder replacing air-sensitive $Ni(COD)_2$, albeit in lower yields.

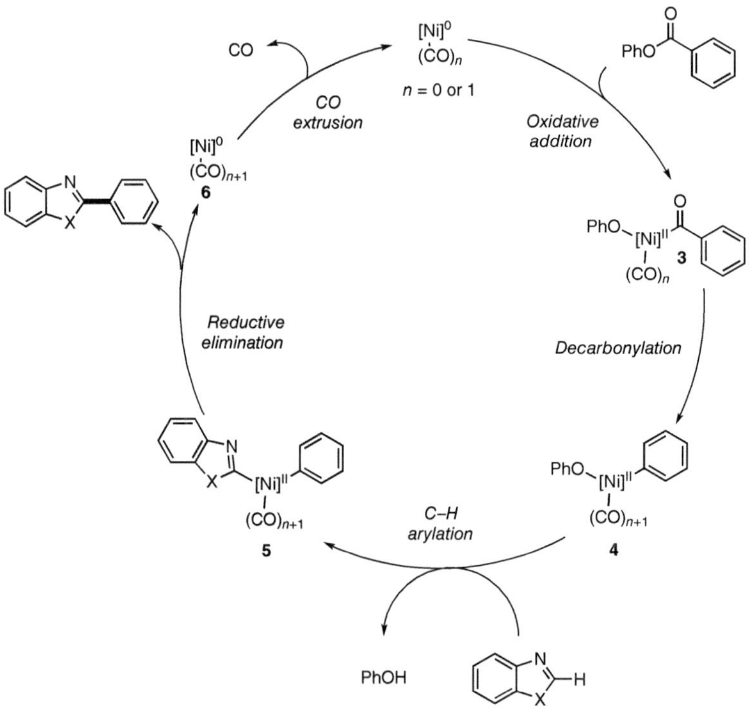

Figure 14.10 Proposed mechanism for decarbonylative C–H coupling of azoles with phenyl esters.

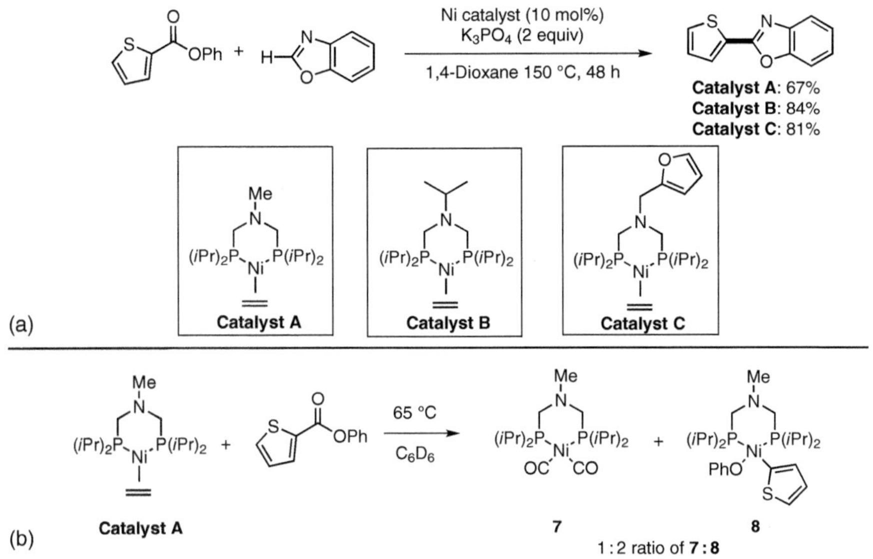

Figure 14.11 (a) New Ni complexes for application in C–H arylation reactions. (b) Isolation of Ni complexes following oxidative addition and decarbonylation.

In 2013, Gade and coworkers investigated new Ni complexes for application in the decarbonylative C–H arylation of phenyl esters with azoles (Figure 14.11a) [14]. At the time of this report, there had been no detailed mechanistic studies for this transformation. The authors were interested in the synthesis of potential intermediates in the proposed catalysis. With this goal in mind, nickel Ni(0) catalyst **A** was treated with phenyl-2-thiophenecarboxylate (Figure 14.11b). NMR was used to monitor consumption of the Ni complex while tracking the gradual formation of products **7** and **8**, which were produced in a 1 : 2 ratio. Product **8** was proposed to arise from the oxidative addition of the Ni complex into the C(acyl)–O bond of the ester followed by rapid carbonyl deinsertion. The CO that is liberated in the process is proposed to react with Ni complex **A**, which leads to the formation of product **7**.

In 2018, Yamaguchi and coworkers extended their decarbonylative C–H arylation of phenyl esters beyond reaction with 1,3-azoles [15]. Attempts at intermolecular C–H arylation of simple arenes had thus far proven unsuccessful, so intramolecular reactions were explored toward accessing dibenzofurans (Figure 14.12). While nickel-catalyzed conditions could not be identified, palladium was very successful for the desired transformation, in particular with $Pd(OAc)_2$/dcype. Control experiments determined that the phenyl ester was necessary for reactivity (Figure 14.12c); methyl ester activation, decarboxylative coupling, and C–H/C–H coupling were all ineffective. Although a precise mechanism is not known, the authors proposed a mechanism analogous to their Ni-catalyzed C–H arylations, with Pd oxidatively inserting into the C(acyl)–O bond followed by subsequent decarbonylation, intramolecular C–H activation, and reductive elimination to release the final product (Figure 14.12d).

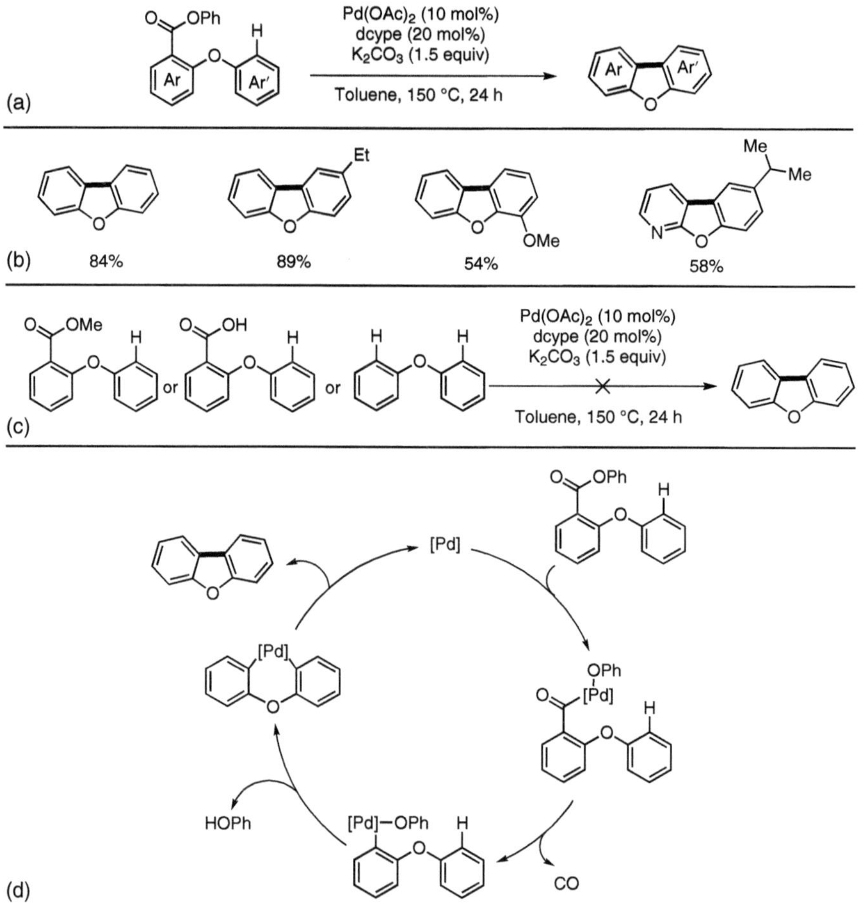

Figure 14.12 (a) Pd-catalyzed decarbonylative intramolecular C–H arylation of phenyl esters. (b) Representative scope examples. (c) Control experiments. (d) Proposed mechanism.

14.3.3 Suzuki–Miyaura Coupling

In modern organic synthesis, the Suzuki–Miyaura coupling has streamlined the synthesis of many important compounds, establishing itself as one of the most frequently used reactions [16]. Some of the earliest reports of the Suzuki–Miyaura coupling of esters led to ketone products, particularly with the use of a directing group [17], and are described in Section 14.4.1 (see also Section 14.6.1). Decarbonylative Suzuki–Miyaura coupling of phenyl esters was not reported until 2015, when Itami, Musaev, and coworkers disclosed a Ni(0)-catalyzed procedure for this transformation (Figure 14.13a) [18]. In their work, extensive ligand screening showed that monodentate ligands were particularly effective with $Ni(OAc)_2/P^nBu_3$ serving as the optimized catalyst. A broad substrate scope was demonstrated for both the ester and the boronic acid coupling partners (Figure 14.13b). $C(sp^3)$–$C(sp^2)$ bond formation was also possible when arylacetic acid phenyl esters

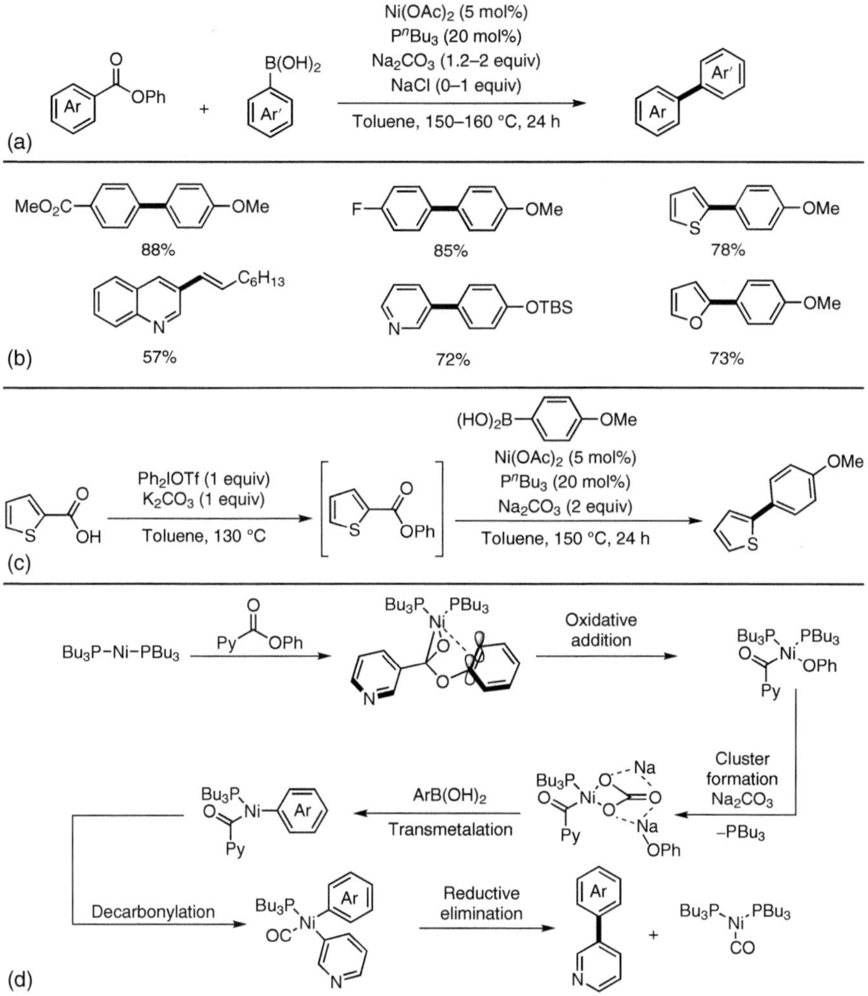

Figure 14.13 (a) Ni/P^{n}Bu$_3$-catalyzed decarbonylative Suzuki–Miyaura cross-coupling of phenyl esters. (b) Representative scope examples. (c) One-pot protocol using carboxylic acids. (d) Important intermediates identify by DFT studies.

were treated with boronic acids, although this necessitated the use of catalytic 4-Dimethylaminopyridine (DMAP). The use of moderately activated phenyl esters was critical, with unactivated alkyl esters being unsuccessful. Interestingly, the reaction provided product (albeit in lower yields) in the absence of an exogenous base, which is usually necessary for Suzuki–Miyaura reactions that implement aryl halides as the electrophile. A one-pot protocol was also developed, which involved making the phenyl ester *in situ* via the corresponding carboxylic acid and diphenyliodonium triflates (Figure 14.13c). A by-product of the esterification was iodobenzene, which was subsequently removed under reduced pressure.

DFT calculations were conducted in order to gain a stronger appreciation for the key mechanistic features associated with the reaction (Figure 14.13d).

π-Complexation between the aryl ester moiety and the empty d-orbitals of Ni was proposed as the first step. This interaction was predicted to be important for a subsequent oxidative addition into the C(acyl)–O bond, as alkyl esters were unreactive under the reaction conditions. Na_2CO_3 reacts with the oxidative addition intermediate to form a cluster complex, which lowers the energy barrier for transmetalation. Following transmetalation is a decarbonylation step, which was proposed to be rate determining. A slightly higher energy pathway in the absence of base was possible, where the transmetalation was the rate-determining step, explaining why the reaction occurs without base in lower yields.

The same year, the Love group reported a similar transformation making use of a $Ni(cod)_2$ and PCy_3 catalyst system (Figure 14.14a) [19]. By running the reaction under a constant flow of N_2, the authors could mitigate the catalyst poisoning effect of CO in the flask. Phenyl esters could be coupled with arylboronic acids to form biaryl products (Figure 14.14b). These reaction conditions were mostly limited to heterophenyl esters. Methyl nicotinate was not tolerated under the reaction conditions, demonstrating that this chemistry could not be extended to simple methyl esters. While 3- and 4-pyrididine carboxylates could be coupled effectively under the reaction conditions, aryl 2-pyridinecarboxylates and methyl esters were not tolerated (Figure 14.14c).

Under Love's standard conditions, non-heteroaryl esters could also be coupled, albeit in low yields (Figure 14.15). A mixture of biaryl and ketone products was obtained with non-heteroaryl esters. The authors proposed that the ketone products stemmed from transmetalation with the boronic acid prior to decarbonylation.

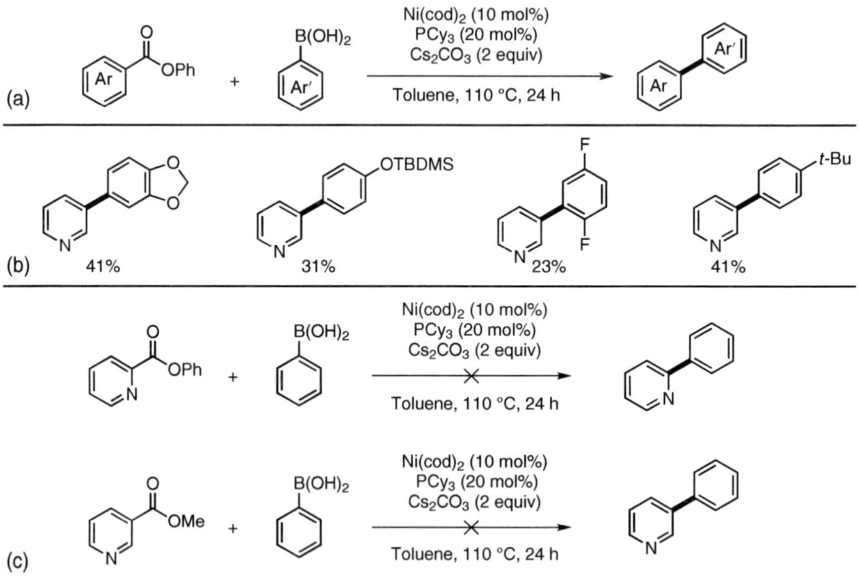

Figure 14.14 (a) Ni/PCy_3-catalyzed decarbonylative Suzuki–Miyaura coupling of heterophenyl esters. (b) Representative scope examples. (c) Scope limitations.

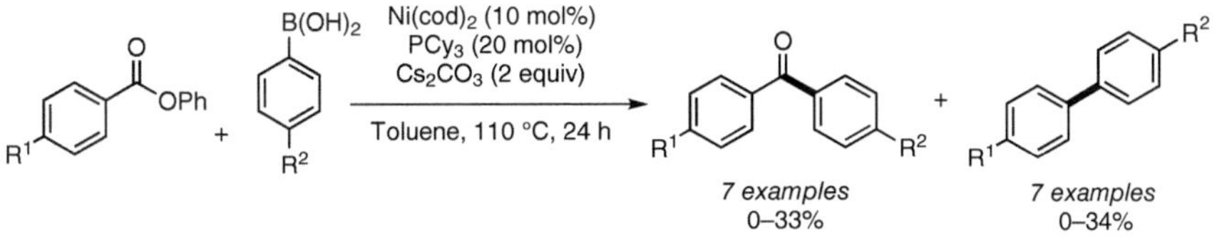

Figure 14.15 Coupling of non-heterophenyl esters under Ni/PCy_3-catalyzed Suzuki–Miyaura conditions.

In 2016, the Yamaguchi group developed a method to functionalize aryl 2-pyridinecarboxylates using Suzuki–Miyaura coupling with Pd catalysis [20]. Application of this method was later demonstrated through a decarbonylative aryl 2-pyridinecarboxylate coupling [20], decarbonylative C–H coupling [11], and a decarboxylative [4+2] cycloaddition to access triarylpyridines toward the formal synthesis of antibiotic compounds [21].

Although the Suzuki–Miyaura coupling is a powerful transformation, it is often used to make C–C bonds between two C(sp^2) fragments [1a]. C(sp^2)–C(sp^3) bond formation is less common, largely due to issues associated with synthesis and use of alkyl boron nucleophiles. For example, they are prone to protodeboration, are more reluctant to transmetalate relative to their aryl nucleophile counterparts, and may undergo β-hydride elimination after transmetalation with a transition metal catalyst [22]. In 2018, the Rueping group expanded the decarbonylative Suzuki–Miyaura-type couplings to include C(sp^3)-hybridized organoboron nucleophiles (Figure 14.16a; also see Section 14.3.5.3) [23]. Interestingly, it was observed that the outcome of the reaction was ligand controlled. When dcype was used, an alkylated arene was obtained selectively. When P^nBu_3 or PCy_3 was used, ketones were formed (see Section 14.4.1). DFT studies suggested that the Ni/dcype system favors oxidative addition into the C(aryl)–C bond, which ultimately decarbonylates and thus leads to the formation of alkylated arenes. A variety of aryl and heteroaryl esters were demonstrated to participate smoothly in the reaction (Figure 14.16b). Substitutions on the organoborane could also be tolerated, including silyl groups

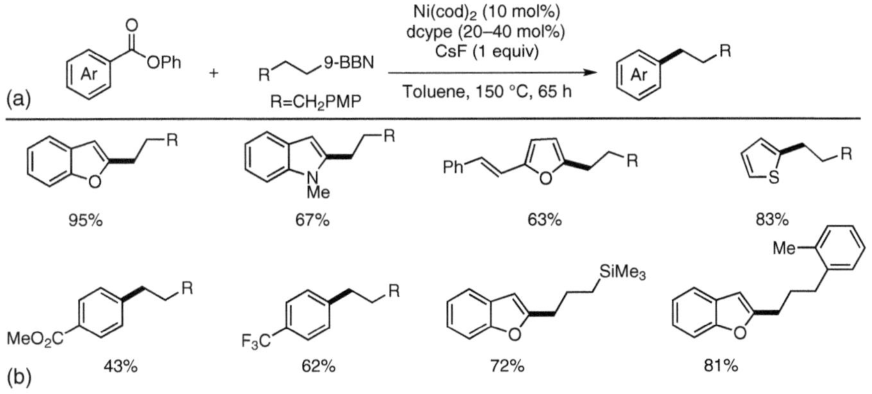

Figure 14.16 (a) Ligand controlled decarbonylative alkylation of phenyl esters. (b) Select scope examples of the decarbonylative alkylation.

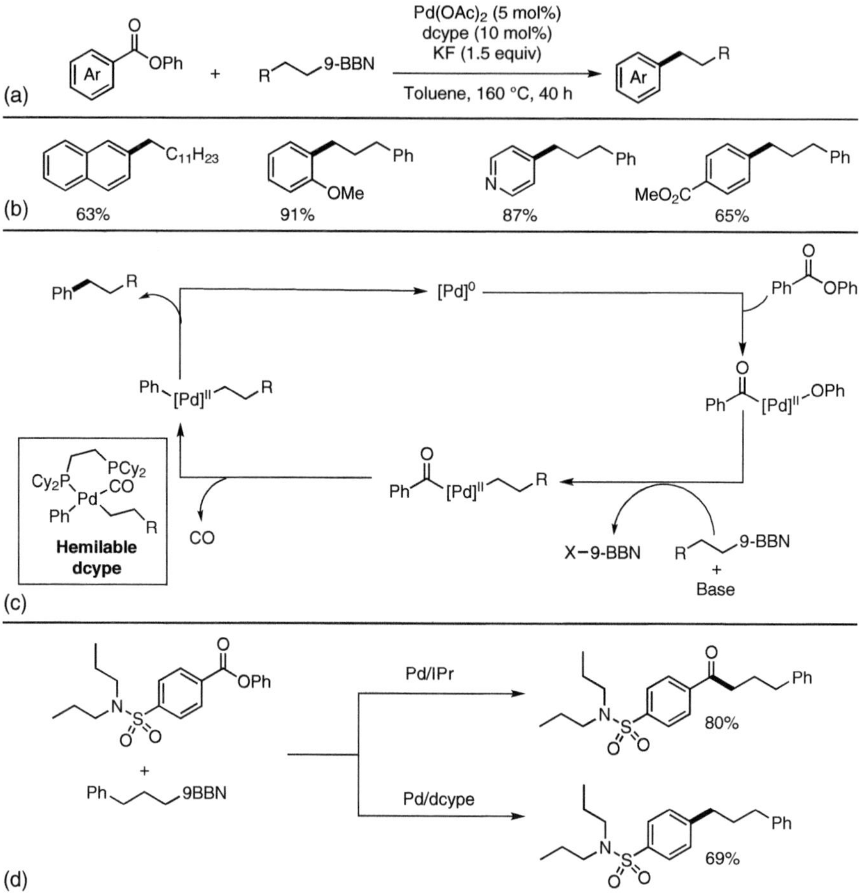

Figure 14.17 (a) Pd-catalyzed decarbonylative alkyl Suzuki–Miyaura coupling of phenyl esters. (b) Representative scope examples. (c) Proposed mechanism. (d) Bioactive molecule derivatization.

and esters. While the B-alkyl-9-BBN reagents were prepared *in situ*, commercially available BEt_3 was also shown to be a suitable alkylating agent. Although not extensively investigated, it was also shown that amides were also capable of undergoing a similar decarbonylative alkylation reaction.

In the same year, the Newman laboratory disclosed a Pd-catalyzed coupling of phenyl esters with B-alkyl-9-BBN reagents (Figure 14.17a) [24]. Several aryl and heteroaryl esters were well tolerated (Figure 14.17b). Under these conditions, the formation of the products also depended on the choice of ligand. A bulky *N*-heterocyclic carbene (NHC) catalyst favored the formation of alkylated ketones (see Section 14.4.1). The use of the bidentate phosphine dcype instead led to the formation of alkylated arenes. The authors proposed that dcype acts as a hemilable ligand, capable of opening coordination sites on the metal that allow decarbonylation to occur, a key step that is prevented by bulky monodentate NHCs (Figure 14.17c). The ability to functionalize a single starting material into different classes of products was also demonstrated by transforming the hyperuricemia treatment Probenecid into either an alkyl arene or alkyl ketone by changing the choice of ligand (Figure 14.17d).

14.3.4 Silylation and Borylation

Organosilicon reagents are important as they can be used toward the synthesis of valuable natural products and can be incorporated into pharmaceutical compounds [25]. Routes to organosilanes often use aggressive organometallic nucleophiles [26] or necessitate the cross-coupling between organohalides and disilanes [27]. Organoboron compounds are also highly useful, particularly as the nucleophilic reactant in the Suzuki–Miyaura coupling. Like organosilanes, they can be prepared by reaction of an organometallic nucleophile with an electrophilic boron source [28] or by cross-coupling of aryl halides and diborons [29]. Constructing C–Si and C–B bonds by decarbonylative coupling reactions of esters provides similar advantages to the C–C forming reactions discussed previously, including a distinct synthetic route, avoidance of metal halide salts, and potential chemoselectivity advantages.

In 2016, the Rueping group developed a decarbonylative silylation of aromatic phenyl esters with silylboranes in the presence of nickel/copper cocatalysis (Figure 14.18a) [30]. The authors demonstrated a wide substrate scope for tolerated functionalities on the ester (Figure 14.18b). The reaction was not limited to Et_3Si-Bpin, as a variety of silylborane coupling partners could be used. The first step in this transformation was proposed to be an oxidative addition of Ni(0) into the C(acyl)–O bond of the ester, yielding an acyl nickel(II) intermediate **9**. Intermediate **9** undergoes transmetalation with the copper silane species that is generated *in situ*, forming complex **10**. Complex **10** then undergoes decarbonylation to form nickel complex **11.** Reductive elimination releases the silylated product and regenerates Ni(0) to further propagate the catalytic cycle (Figure 14.18c).

In the same year, the Shi group reported a similar decarbonylative silylation reaction of aromatic phenyl esters (Figure 14.19a) [31]. This reaction also made use of nickel/copper cocatalysis using dcype as the ligand. The silylation reaction exhibited tolerance for diverse functional groups and heteroaromatic compounds. In the same publication, a decarbonylative borylation of aromatic phenyl esters using B_2nep_2 was also reported (Figure 14.19b). Key to the success of the borylation reaction was the use of the NHC ligand ICy. To demonstrate the synthetic utility of these methods, a phenyl ester analog of a complex molecule was shown to be capable of being borylated or silylated selectively (Figure 14.19c).

Although significant advances had been made in the field of ester cross-coupling, at the time of these publications, the isolation of an acyl nickel(II) species remained an elusive goal. The Shi group were the first to synthesize and isolate an acyl nickel(II) species derived from a phenyl ester (Figure 14.19d). Phenyl 1-naphthoate was treated with a stoichiometric amount of $Ni(cod)_2$, NHC ligand, and NaO^tBu, yielding the corresponding nickel(II) oxidative addition adduct **12**. By heating the complex at 60 °C, decarbonylation occurred to yield nickel (II) complex **13**. When complex **13** was heated at 60 °C under 1 atm of CO, it reverted back to the acyl

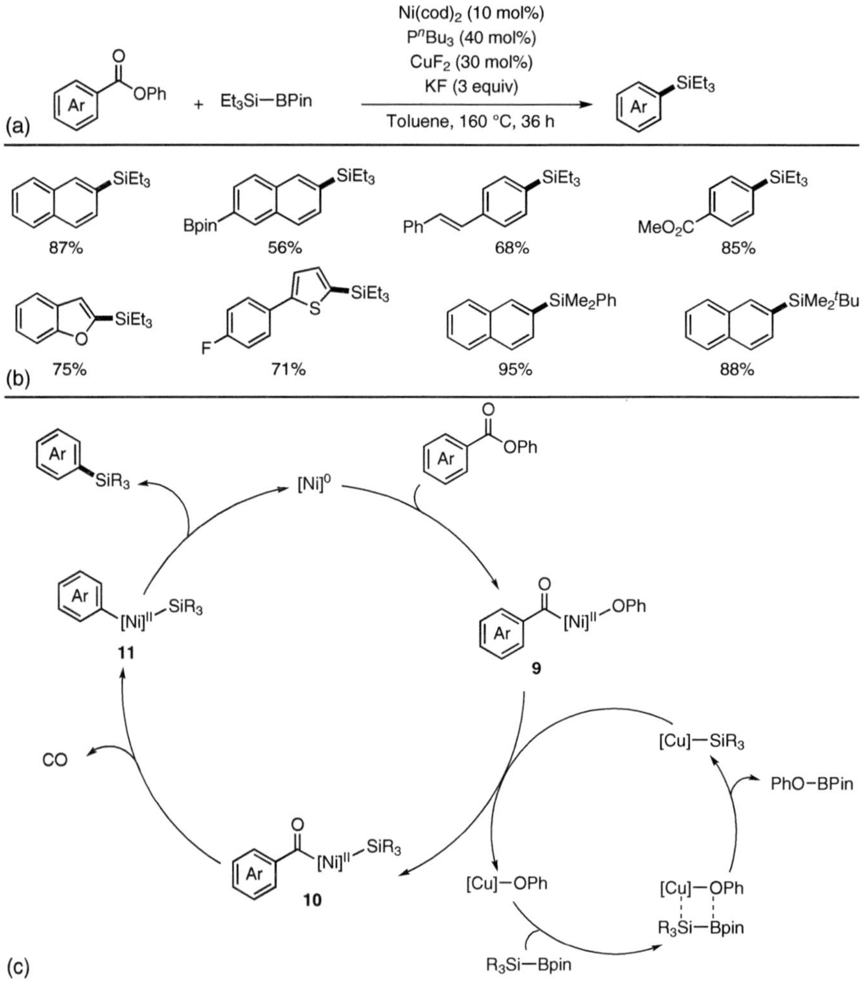

Figure 14.18 (a) Ni-catalyzed decarbonylative silylation of phenyl esters. (b) Representative scope examples. (c) Proposed mechanism.

nickel (II) complex **12**. These results suggest that the decarbonylation process is fast and reversible. It was also found that both the aryl complex **13** and acyl complex **12** were capable of catalyzing the standard reaction in near identical yields, indicating that both species can enter catalytic cycle.

Rueping also reported a decarbonylative borylation reaction of aromatic phenyl esters with B_2pin_2 by making use of a $Ni(cod)_2/P^nBu_3$ catalytic system (Figure 14.20a) [32]. This reaction demonstrated a wide substrate scope and was additionally shown to be capable of functionalizing some α-alkenyl and α-alkyl esters (Figure 14.20b). Gram-scale synthesis was also demonstrated. The synthetic utility of such a transformation was demonstrated when performing sequential couplings, using a carboxylate as a directing group prior to installing and activating a phenyl ester for borylation (Figure 14.20c).

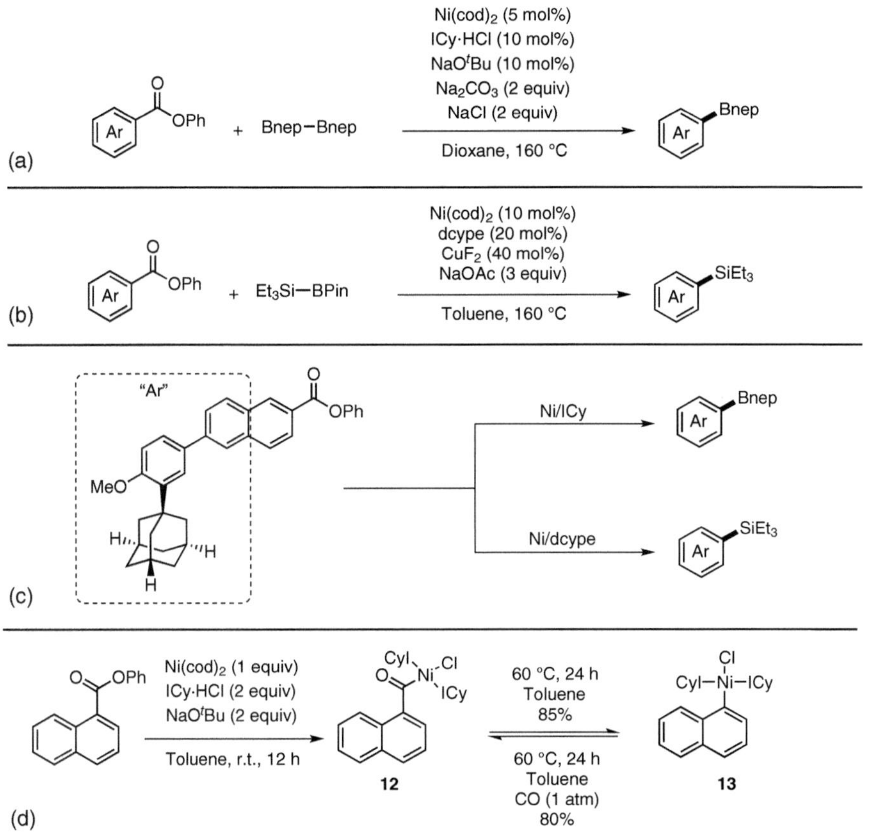

Figure 14.19 (a) Standard conditions A: Ni-catalyzed decarbonylative borylation of phenyl esters. (b) Standard conditions B: Ni-catalyzed decarbonylative silylation of phenyl esters. (c) Complex molecule derivatization. (d) Isolation of nickel intermediates.

14.3.5 Other C–C/C–H Bond Forming Reactions

This section describes miscellaneous impactful phenyl ester cross-couplings that forge C–C or C–H bonds using nucleophilic coupling partners not captured in the prior sections.

14.3.5.1 Sonogashira-Type Couplings

Arylalkynes are important scaffolds that are commonly used as versatile intermediates toward the synthesis of valuable organic compounds [33]. The Sonogashira coupling is an efficient transformation that accesses arylalkynes, traditionally from aryl halides and terminal alkynes [34]. Leading up to 2017, substantial progress was made in research concerning ester cross-coupling. However, decarbonylative C–C bond forming reactions of esters were still limited to C(sp^2)–C(sp^2) bond formation at that time. In 2017, the first example of decarbonylative C(sp^2)–C(sp) bond formation was reported by Itami and Yamaguchi by coupling aromatic phenyl esters with alkynes in the presence of palladium/copper cocatalysts to furnish a variety

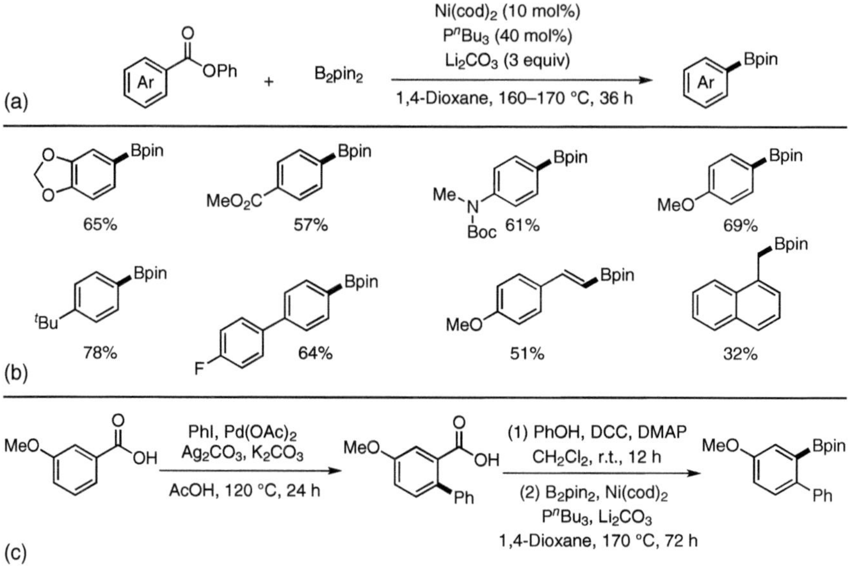

Figure 14.20 (a) Ni/P^{n}Bu$_3$-catalyzed decarbonylative borylation of phenyl esters. (b) Representative scope examples. (c) Sequential cross-couplings.

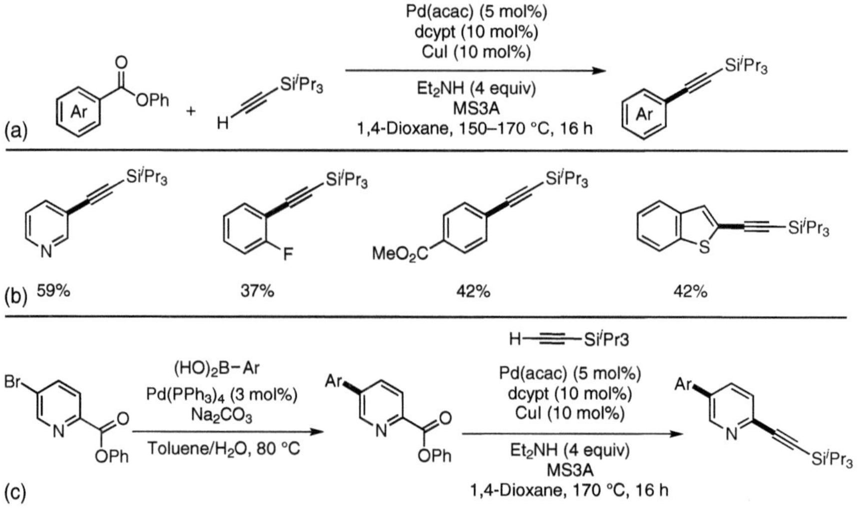

Figure 14.21 (a) Pd-catalyzed decarbonylative Sonogashira-type cross-coupling. (b) Representative scope examples. (c) Sequential cross-coupling sequence.

of alkynylated arene products (Figure 14.21a and b) [35]. This protocol was limited to the use of silylacetylenes, as aryl and alkylacetylenes were not applicable for this reaction. To demonstrate the synthetic utility of such a transformation, sequential cross-couplings were conducted. Notably, traditional Suzuki–Miyaura couplings with aryl halides were shown to leave phenyl esters untouched, which allowed further derivatization (Figure 14.21c).

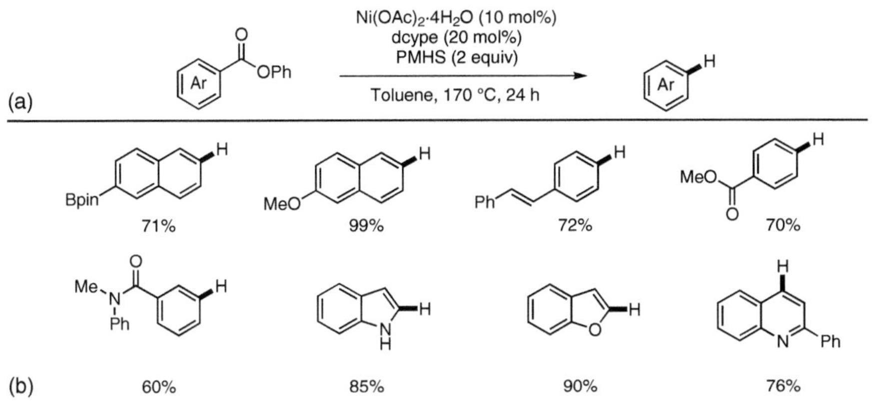

Figure 14.22 (a) Ni-catalyzed decarbonylative reduction of phenyl esters using PMHS. (b) Representative scope examples.

14.3.5.2 Reduction

The reduction of carboxylic acid derivatives such as esters is a staple transformation in organic synthesis, often achieved by the use of metal hydrides [36] or via hydrosilylation [37]. In contrast, the replacement of an ester with a hydrogen atom ($RCO_2R' \rightarrow R\text{-}H$) is an interesting alternative reduction product. Rueping and coworkers were the first to develop a reductive defunctionalization of phenyl esters (Figure 14.22a) [38]. This was the first of such reports to achieve this without the use of directing groups (see Section 14.6.1). The catalyst was an inexpensive combination of $Ni(OAc)_2{\cdot}4H_2O$ and dcype and made use of polymethylhydrosiloxane (PMHS) as the reducing agent. Under the reaction conditions, the authors were capable of reducing a variety of aryl and heterophenyl esters (Figure 14.22b). The reaction was also readily extended to aryl and heteroaryl amides. As for other reduction protocols, triazine esters have also been shown to reduce down to aldehydes in the presence of a Pd-catalyst and a silane [39].

14.3.5.3 Negishi-Type Coupling

Up until 2017, several different nucleophiles had been demonstrated to undergo coupling with esters. Some of these include 1,3-azoles, alkynes, boronic acids and silylboranes. Yet to be reported was a decarbonylative aryl–alkyl cross-coupling of esters. Rueping and coworkers were the first to develop a decarbonylative alkylation of phenyl esters using organozinc reagents (Figure 14.23a) [40]. As is often seen in ester activation, the choice of ligand was particularly important, with the bidentate phosphine ligand dcype being ideal. Under the optimized reaction conditions, a variety of aryl and heterophenyl esters could be coupled efficiently with alkyl organozinc reagents (Figure 14.23b). Interestingly, the reaction was not limited to the use of phenyl esters. Pivaloyl esters could be converted to the same alkylated products via cleavage of the C(aryl)–O bond (Figure 14.23c).

Chen and coworkers also described the Ni-catalyzed cross-coupling of 2-pyridyl esters with alkyl zinc reagents, followed by Ni-H reduction of the corresponding ketones to yield benzylic alcohols in a single step [41].

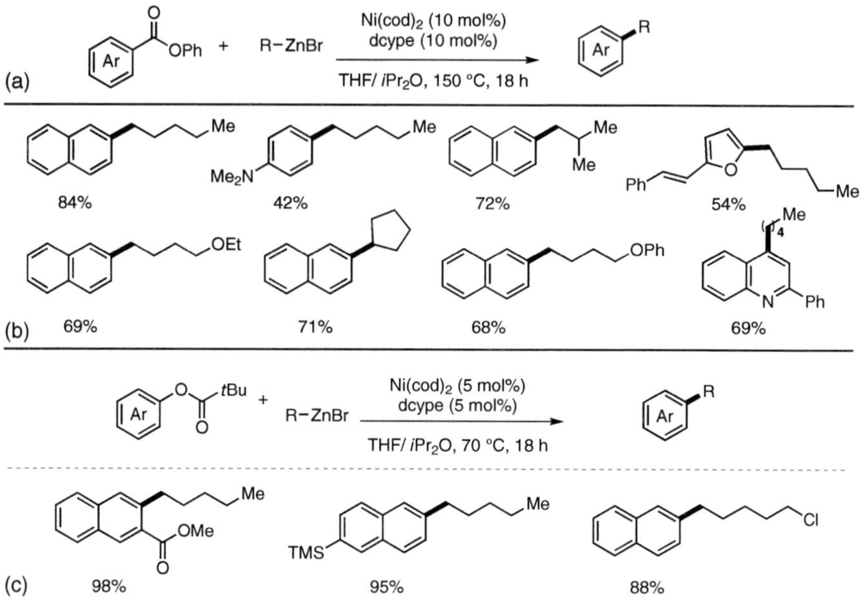

Figure 14.23 (a) Ni-catalyzed decarbonylative Negishi-type coupling. (b) Representative scope examples. (c) Pivalate ester cleavage and scope examples.

14.3.5.4 Cyanation

Traditionally, the introduction of cyano groups into aromatic compounds is achieved by either the Sandmeyer [42] or Rosenmund–von Braun reactions [43]. Unfortunately, both of these methods suffer from relatively limited substrate scopes. The Rueping group were the first to report the nickel-catalyzed decarbonylative cyanation of phenyl esters with zinc cyanide, which seeks to address the substrate scope issue associated with the existing protocols (Figure 14.24a) [44]. Ligand screening with $Ni(cod)_2$ determined that monodentate phosphines and NHC ligands were not useful for this transformation. Instead, success was realized with a $Ni(cod)_2$/dcype catalyst system. A variety of aryl and heterophenyl esters were demonstrated to undergo coupling with zinc cyanide smoothly (Figure 14.24b). The authors also demonstrated that their protocol was capable of being extended to amide bond activation. Decarbonylative cyanation could also be done in an intramolecular fashion, with acyl cyanides directly furnishing the desired aryl nitrile products (Figure 14.24c). The Yamaguchi group have also reported the decarbonylative synthesis of aryl nitriles by coupling phenyl esters with nonmetal cyanating reagents [45].

14.3.5.5 Methylation

The majority of decarbonylative coupling reactions discussed this far feature formation of $C(sp^2)$–$C(sp^2)$ bonds. Toward developing new ester alkylation methods, Yamaguchi and coworkers described a Ni-catalyzed decarbonylative methylation of aromatic esters using methylaluminum reagents (Figure 14.25a) [46]. Dimethylaluminum chloride was particularly effective. Interestingly, this reagent

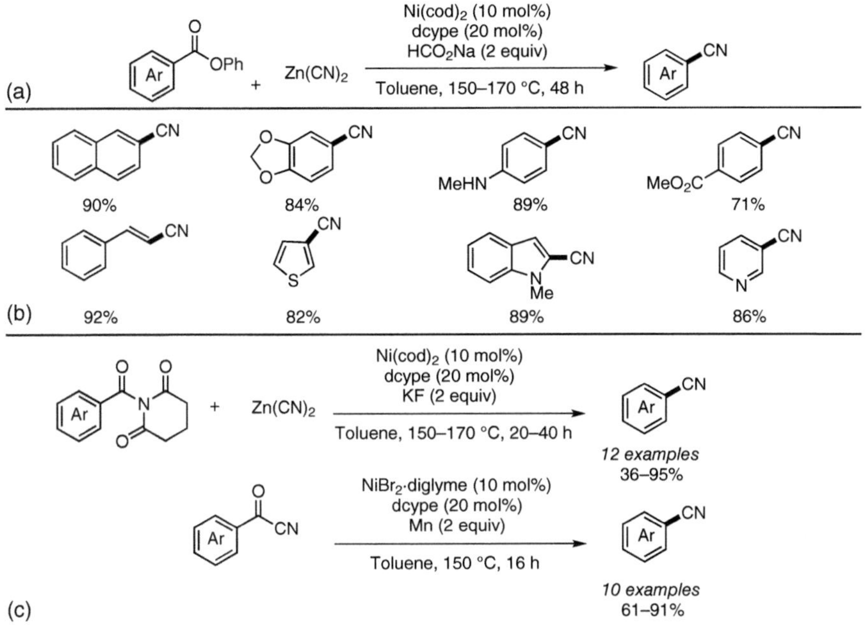

Figure 14.24 (a) Ni-catalyzed decarbonylative aryl nitrile synthesis. (b) Representative scope examples. (c) Cyanation of amides and acyl cyanides.

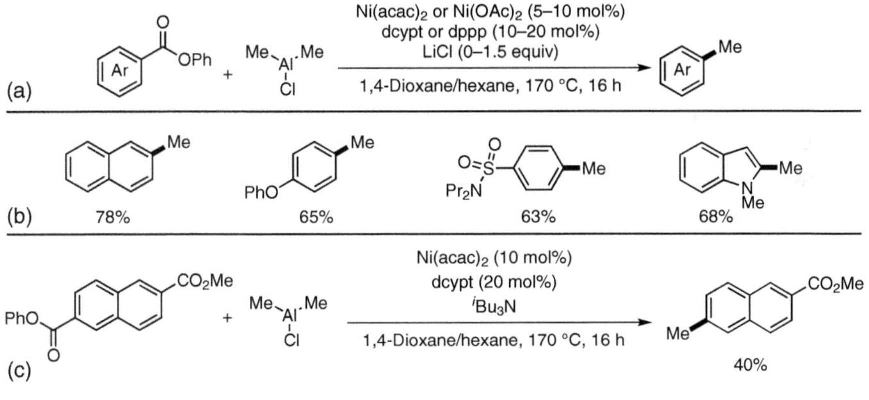

Figure 14.25 (a) Ni-catalyzed decarbonylative methylation of phenyl esters. (b) Representative scope examples. (c) Chemoselective alkyl coupling.

was suggested to play a dual role, acting both as a source of alkyl nucleophile and simultaneously acting as a Lewis acid, allowing for the functionalization of seemingly inert methyl esters (see Section 14.6.2 for further details). Ultimately, Ni(acac)$_2$/dcypt or dppp was determined to be optimal catalyst systems, which allowed for the functionalization of several aryl and heterophenyl esters (Figure 14.25b). The authors also demonstrated that a phenyl ester could be functionalized selectively in the presence of a methyl ester, which is achieved when a tertiary amine is used in the reaction to suppress the Lewis acidity of the methylating agent (Figure 14.25c). In the absence of the tertiary amine, both the phenyl ester and methyl ester were functionalized.

14.3.6 Other C-Heteroatom Bond Forming Reactions

This section will describe miscellaneous impactful phenyl ester cross-coupling that makes C-heteroatom bonds using nucleophilic coupling partners not captured in the prior sections.

14.3.6.1 Etherification

Diaryl ethers are scaffolds commonly seen in pharmaceuticals and natural products. Due to their importance, several methods have been developed for the synthesis of diaryl ethers, including the use of Pd [47] or Cu [48] catalysts to bring together aryl halides and phenols. Although the established protocols are powerful, extending the work to enable ester derivatization would be synthetically useful particularly when late-stage molecule functionalization is concerned.

Itami and Yamaguchi demonstrated that aryl esters such as **14** can be used to form diaryl ethers (Figure 14.26a and b) [49]. This C(acyl)–O bond activation was achieved by either Ni or Pd catalysis without need for an external nucleophile. Mechanistically, an oxidative addition, decarbonylation, and reductive elimination sequence was proposed (Figure 14.26c). The ligands dcypt and dcppt were found to be optimal, regardless of which metal they were paired with. While both Ni and Pd were effective for most substrates, some were observed to only work with one of the two. The reaction conditions were mostly limited to 2-azinecarboyxlates as the ester substrates. The selectivity of this catalyst system was demonstrated when a molecule bearing two phenyl esters was subjected to sequential cross-couplings (Figure 14.26d). First a decarbonylative etherification was achieved on at the pyridine 2-position followed by a decarbonylative Suzuki–Miyaura coupling at the 4-position.

14.3.6.2 Amination

Aryl amines are often found in pharmaceuticals, natural products, and other important organic compounds such as dyes [50]. An efficient way of accessing C(sp^2)–N bonds is the Buchwald–Hartwig amination between aryl halides and amines. Given the value of these products, new protocols that can access C(sp^2)–N scaffolds are still sought, particularly without the use of precious metals such as Pd. In 2017, the Rueping group developed a new aryl amine synthesis via the decarbonylative cross-coupling of phenyl esters with imines (Figure 14.27a) [51]. A Ni/dcype catalyst system, as well as the use of K_3PO_4, was crucial to the success of the reaction. After decarbonylative amination to form the arylated imine, the intermediate was treated with acid to hydrolyze the imine and release the aniline product.

In 2020, the Sanford group developed a protocol for aryl amine synthesis via a decarbonylative cross-coupling between phenyl esters and silylated amines (Figure 14.28a) [52]. By masking the amine with a main group element, these nucleophiles would not react directly with the phenyl ester to form the amide,

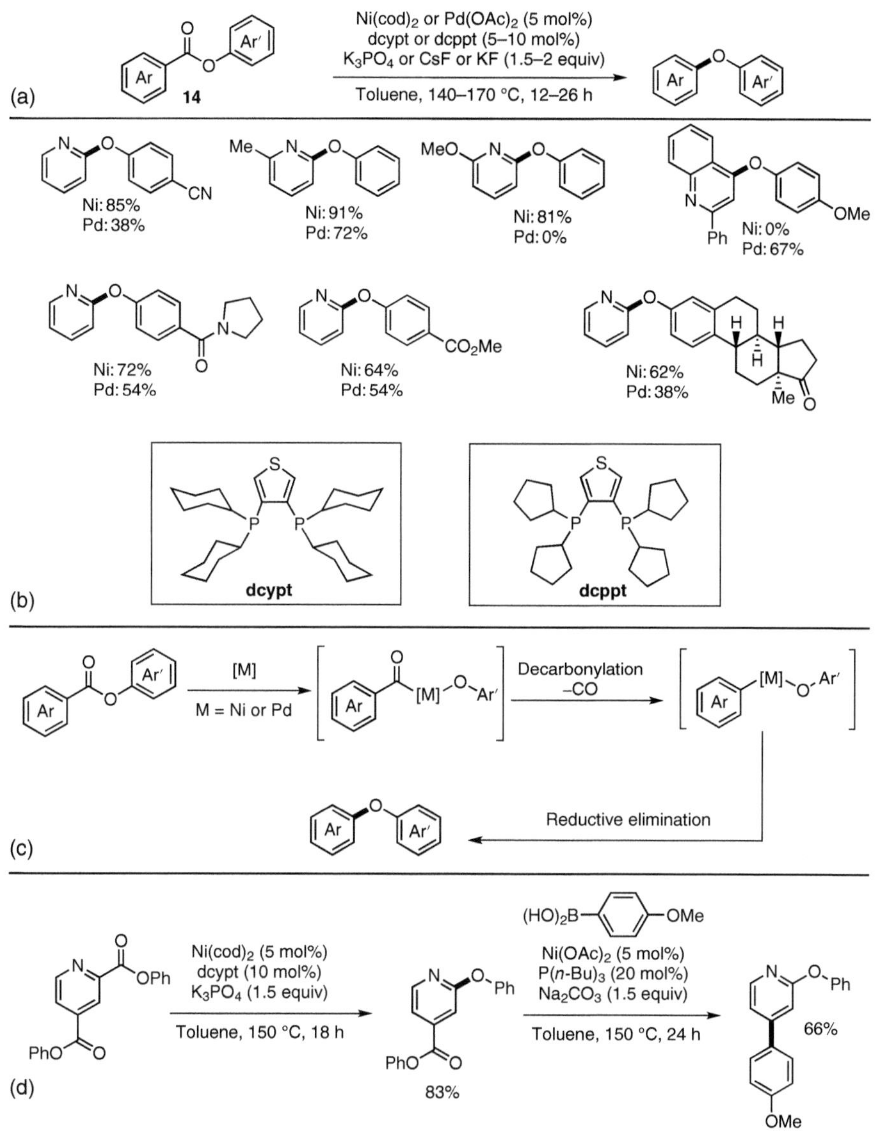

Figure 14.26 (a) Decarbonylative etherification of heterophenyl esters. (b) Representative scope examples. (c) Mechanistic hypothesis. (d) Chemoselectivity/sequential cross-coupling study.

and would instead selectively interact with the catalyst by transmetalation with Ni(II), after a decarbonylation event. In the presence of free amines, low yields of amide products would be obtained as transmetalation with free amines was faster than decarbonylation. By using *N*-methyl-*N*-(trimethylsilyl)trifluoroacetamide (MSTFA), the amines could be silylated *in situ* (Figure 14.28b). A variety of primary and secondary aryl and alkyl amines were demonstrated to be effective coupling partners. It was also demonstrated that $Ni(CO)_2(PPh_3)_2$, a commercial and air stable

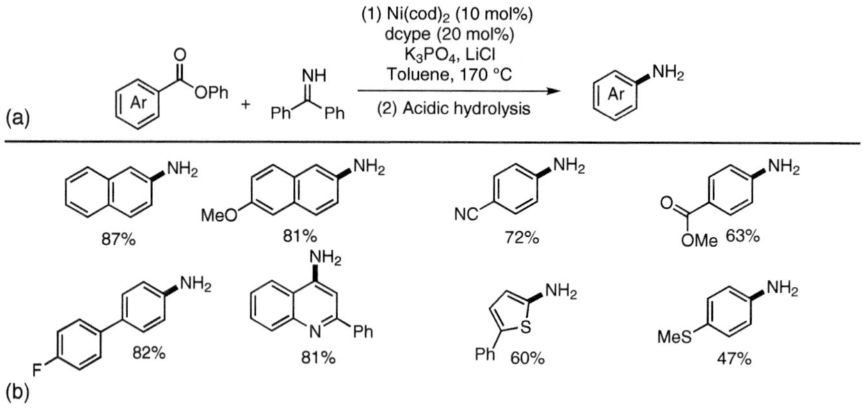

Figure 14.27 (a) Ni-catalyzed decarbonylative aryl amine synthesis. (b) Representative scope examples.

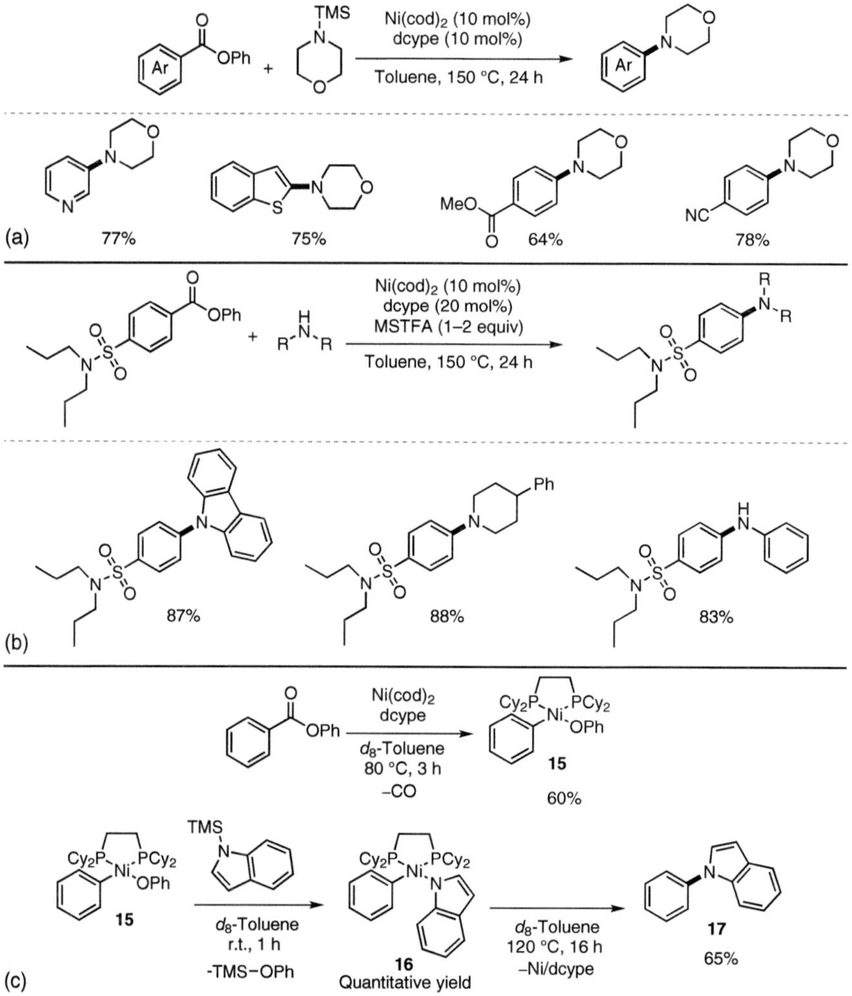

Figure 14.28 (a) Representative silyl amine scope. (b) Representative scope of *in situ* amine silylation. (c) Isolation of nickel intermediates.

source of Ni(0), could be utilized in place of $Ni(cod)_2$. Mechanistic investigation of the reaction led to the isolation of Ni(II) complex **15** formed by oxidative addition and subsequent decarbonylation. No acyl Ni(II) complexes were observed during the studies, likely indicating that decarbonylation occurs rapidly under the reaction conditions. The treatment of **15** with Trimethylsilyl (TMS) indole resulted in transmetalation to form Ni(II) complex **16**, which was stable and isolable at room temperature. Once complex **16** was heated at 120 °C, it allowed the formation of the corresponding aryl amine product **17** (Figure 14.28c).

14.3.6.3 Thioetherification

Thioethers are important scaffolds [53] that can be synthesized via the Chan–Evans–Lam, Ullmann, and related transition metal couplings [54]. In 2018, Rueping and coworkers investigated a decarbonylative thioetherification reaction, with the added goal of developing a glovebox-free protocol (Figure 14.29a) [55]. Investigation of air and moisture stable metal sources and ligands led to the identification of $NiCl_2$/dppp as the optimal catalyst system. A range of aromatic and heteroaromatic esters could be coupled with thiols to provide the corresponding sulfanes (Figure 14.29b). Nonaromatic thiols were also demonstrated to be viable coupling partners. Prompted by their success with esters, the authors also investigated other classes of electrophiles for decarbonylative thioetherification. With

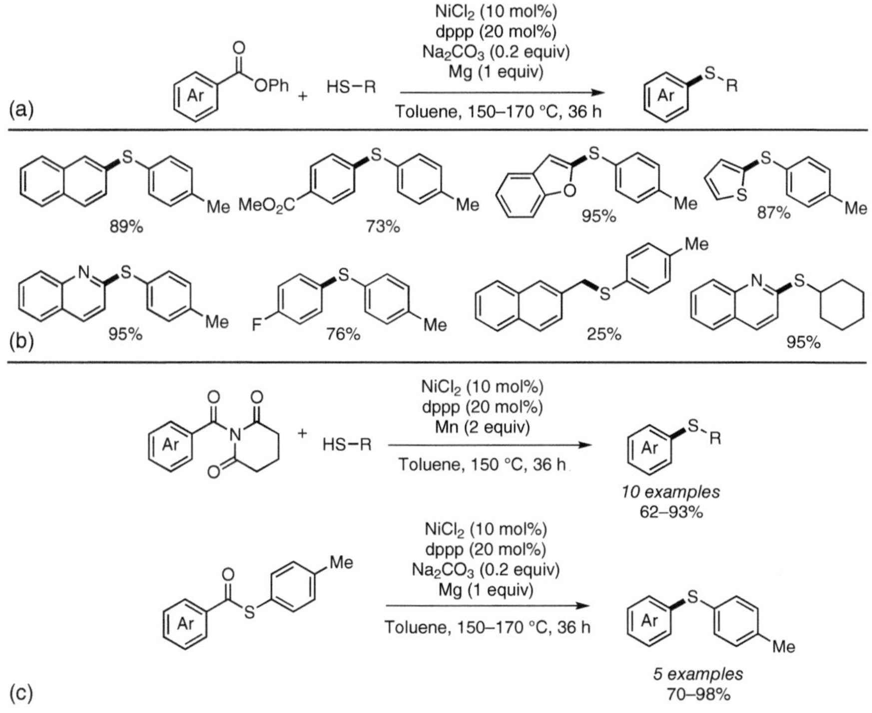

Figure 14.29 (a) Ni-catalyzed decarbonylative thioetherification. (b) Representative scope examples. (c) Amide and thioester thioetherification.

their own respective optimized conditions, aryl amides and thioesters were also shown to be viable cross-coupling electrophiles (Figure 14.29c).

14.3.6.4 Carbon–Phosphorus Bond Formation

The use of organophosphorus nucleophiles in cross-coupling is less common than more traditional coupling partners but can provide a useful route to access valuable derivatives. The Yamaguchi group were the first to develop a catalytic decarbonylative C–P bond forming reaction using phenyl esters as the electrophile (Figure 14.30a) [56]. Initial studies of the phosphorus containing nucleophile used diphenyl phosphine oxide. Robust ligands were screened to avoid the possibility that diphenyl phosphine oxide would poison the catalyst. Common ligands for ester activation, particularly PCy_3 and bulky NHCs, were found to be completely ineffective. Ultimately, a $Ni(OAc)_2$/dcypt catalytic system was most successful. Under the reaction conditions, a variety of aryl and heteroaryl esters were tolerated in modest to good yields (Figure 14.30b). Different phosphates could be used as a nucleophilic coupling partner by using NaF as an additive in toluene. To demonstrate the synthetic utility of this transformation, a sequential cross-coupling was conducted where molecule **18** was first functionalized via Suzuki–Miyaura conditions, followed by C–P bond formation (Figure 14.30c).

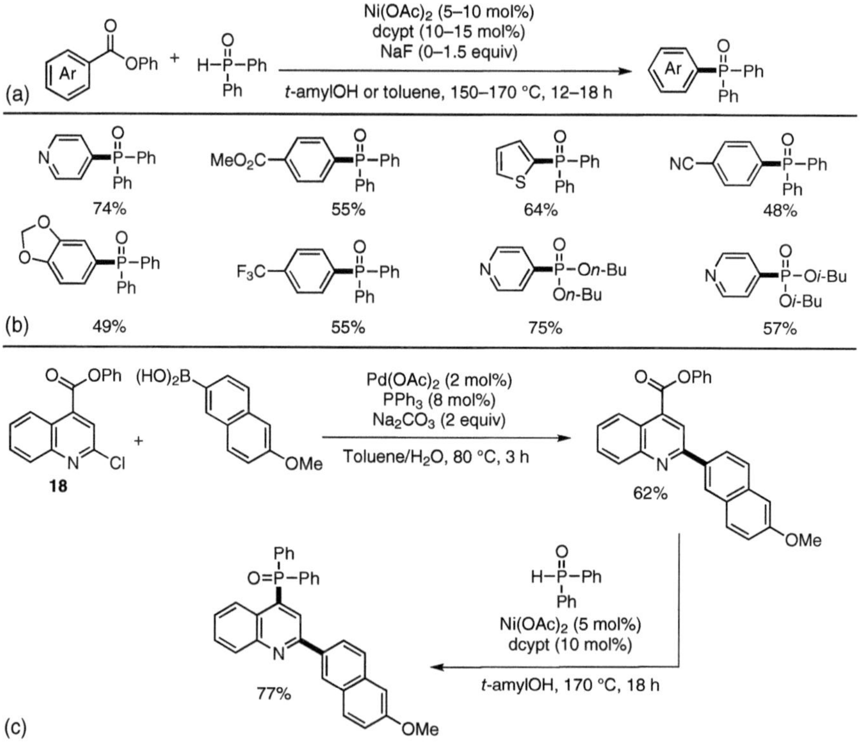

Figure 14.30 (a) Ni-catalyzed decarbonylative C–P bond formation. (b) Representative scope examples. (c) Sequential cross-coupling.

14.4 Carbonyl Retentive Coupling of Phenyl Esters

Section 14.3 described numerous instances where esters act as aryl halide surrogates, losing the carbonyl group in the process. While mechanistically more analogous to traditional cross-couplings, reactions where the carbonyl group is retained are somewhat less common. These so-called carbonyl-retentive couplings allow for the formation of various ketone products when using carbon-based nucleophiles. Ketones are highly desirable scaffolds to access via cross-coupling due to challenges associated with synthesizing them from traditional nucleophile–electrophile chemistry. Organometallic reagents can attack carboxylic acid derivatives; however, due to the harsh nature of these nucleophiles, the ketone products are still susceptible to further attack. To address this issue, the Weinreb ketone synthesis was developed, which prefunctionalizes the desired carbonyl fragment into a stable species referred to as a Weinreb amide [8]. By exploiting a transition metal catalyst, prefunctionalization of the molecule can be avoided, and the catalyst can directly activate the acyl electrophile. In doing so, milder organometallic nucleophiles can be used to access the desired ketones, which will not react with the products. If the nucleophile is an amine, amide products can be obtained. The two common strategies for preparing amides from esters rely on either activation of the electrophile with Lewis acids [57] or activation of the amine nucleophile with harsh organometallic base [58]. Both methods suffer from significant scope limitations. Novel methods of synthesizing amides will also be valuable, given the prevalence of nitrogen in so many bioactive molecules [59]. This section describes routes that maintain the carbonyl group to provide access to ketones and amides.

14.4.1 Suzuki–Miyaura Coupling

The use of phenyl ester electrophiles in acyl Suzuki cross-coupling is advantageous given that they are relatively stable species amenable to orthogonal cross-coupling strategies. One of the earliest breakthroughs was realized by the Chatani group in 2004, who exploited the reactivity of 2-pyridyl esters (Figure 14.31a) (see also Section 14.6.1) [17]. Extensive screening of potential chelating groups was conducted, with 2-pyridyl esters determined to be the most efficient. A Pd/PPh_3 catalyst system allowed for the efficient coupling of aryl and alkyl 2-pyridyl esters under relatively mild reaction conditions (Figure 14.31b). Although not extensively studied, it was also demonstrated that under slightly modified reaction conditions, benzylic B-alkyl-9-BBN reagents could also be coupled, giving rise to alkylated ketone products.

Toward the goal of carbonyl retentive ester coupling, in 2017 the Newman laboratory developed the first catalytic method of aryl phenyl ester activation to provide ketone products selectively and in high yields (Figure 14.32a) [60]. Crucial to the success of this transformation was the use of a Pd catalyst and a bulky NHC ligand. A variety of aryl and heteroaryl substituents were tolerated (Figure 14.32b). Chemoselectivity studies demonstrated the synthetic utility of such a method, as selective functionalization of the phenyl esters could occur in the presence of other

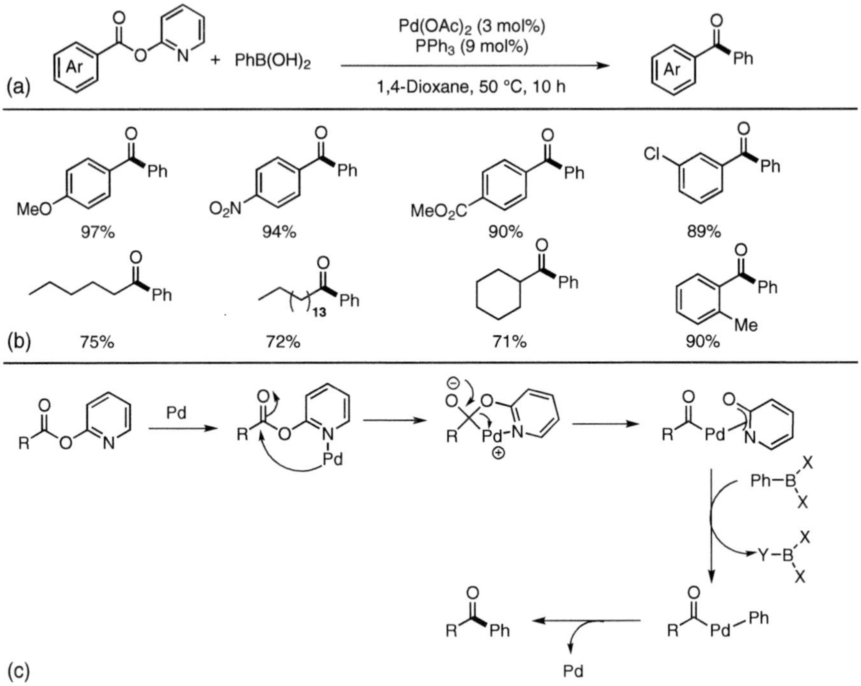

Figure 14.31 (a) Pd-catalyzed carbonyl retentive Suzuki–Miyaura coupling of 2-pyridyl esters. (b) Representative scope examples. (c) Proposed chelation assistance mechanism.

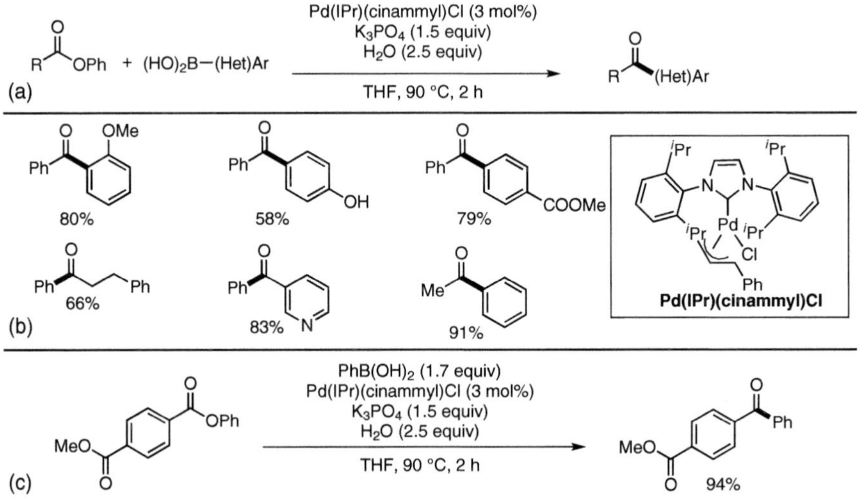

Figure 14.32 (a) Pd-catalyzed carbonyl retentive Suzuki–Miyaura coupling of phenyl esters. (b) Representative scope examples. (c) Chemoselective C–O cleavage.

bonds susceptible to cleavage (Figure 14.32c). Prior to this study, most reported cross-couplings of phenyl esters followed a decarbonylative pathway. Mechanistic investigation provided an explanation for this stark contrast in reactivity. The bulky NHC ligand hinders decarbonylation by blocking open coordination sites on the acyl-Pd(II) intermediate, leaving the carbonyl-retentive pathway as the only accessible mechanism.

An ongoing challenge in catalysis is to enable reactions to occur under mild conditions using low catalyst loading. Toward achieving this goal, the groups of Hazari and Szostak independently demonstrated the use of bench stable Pd precatalysts capable of activating phenyl esters under mild conditions (Figure 14.33). These precatalysts feature 1-^{t}Bu-indenyl groups and activate to form monoligated L_1-Pd(0) species without forming inactive Pd(I) dimers. Using 1 mol% of these precatalysts, Suzuki–Miyaura reactions of phenyl esters and *N*-Boc-protected amides can occur at room temperature. It was also demonstrated that the precatalysts were useful for Buchwald–Hartwig-type couplings of phenyl esters, which could occur near room temperature (see Section 14.4.2) [61–64]. Szostak and coworkers also demonstrated that Pd-PEPPSI-IPr [65] and [Pd(NHC)(acac)Cl] [66] catalyst systems were capable of coupling phenyl esters under similarly mild conditions. Catalyst design was not the sole method of achieving milder conditions. The green solvent 2-methyltetrahydrofuran was also demonstrated to be substantially effective for amide and ester Suzuki–Miyaura coupling [67].

Thus far, all work concerning carbonyl retentive coupling of phenyl esters had made use of NHC ligands. The Szostak group demonstrated in 2018 that phosphine ligands could also be applied to this chemistry successfully (Figure 14.34). Critical to the success of Pd/phosphine catalyst systems was the use of electronically activated pentafluorophenyl esters [68].

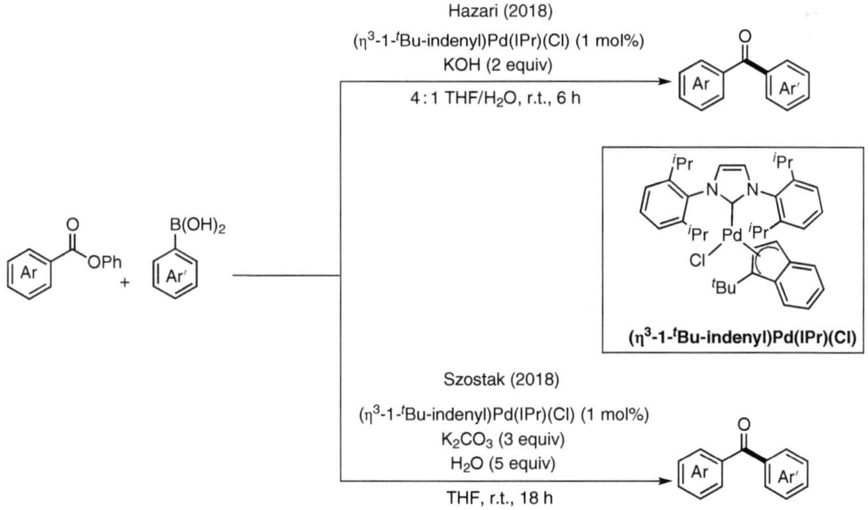

Figure 14.33 Pd catalysts for mild phenyl ester activation.

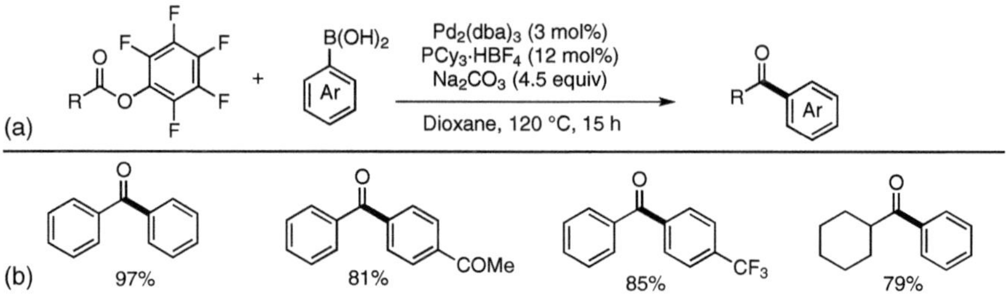

Figure 14.34 (a) Pd-catalyzed carbonyl retentive Suzuki–Miyaura coupling of pentafluorophenyl esters.

Accessing C(sp^2)–C(sp^3) bonds in Suzuki–Miyaura cross-coupling is relatively challenging because of the sensitivity of alkylboranes to protodeboration as well as the potential for undesired β-hydride elimination to occur from key intermediates. The Rueping group expanded the scope of nucleophiles capable of performing Suzuki–Miyaura-type couplings with phenyl esters by developing a method to selectively form alkyl ketones using B-alkyl-9-BBN reagents by utilizing Ni catalysts. The selectivity for ketone product formation was reliant on the choice of ligand and reaction conditions (Figure 14.35a). The use of elevated temperatures and the bidentate ligand dcype leads to the formation of decarbonylated products (see Figure 14.16, Section 14.3.3). DFT studies suggested that the use of monodentate PCy_3 or P^nBu_3 ligands would lead to the cleavage of the C(acyl)–O bond, allowing formation of the ketone products. Under a Ni/PCy_3 or P^nBu_3 catalyst system, a variety of aryl and heteroaryl products could be synthesized efficiently (Figure 14.35b) [23].

Concurrently, the Newman laboratory reported a similar alkylative transformation, which was a Pd-catalyzed process (Figure 14.36a) [24]. Again, crucial to the success of this reaction was the careful selection of reaction parameters. Elevated temperatures and a Pd/dcype catalyst system led to the formation of decarbonylated products (Figure 14.17, Section 14.3.3). In contrast, the use of a bulky Pd-IPr catalyst led to the formation of carbonyl-retentive alkylated ketones. Under the reaction conditions, aryl and heteroaryl esters could be coupled smoothly (Figure 14.36b).

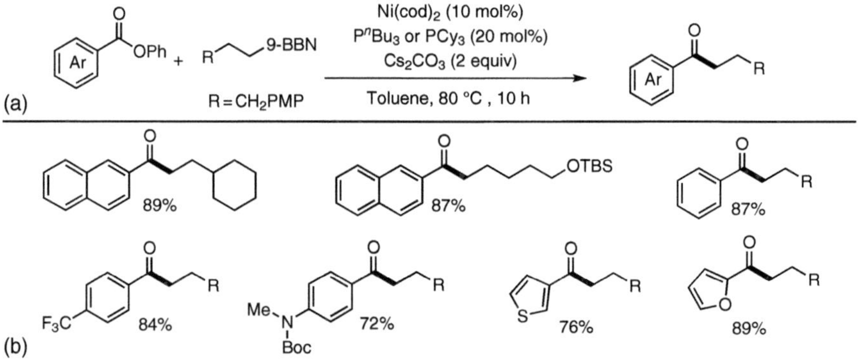

Figure 14.35 (a) Ligand controlled carbonyl retentive Suzuki–Miyaura coupling of phenyl esters. (b) Representative scope examples.

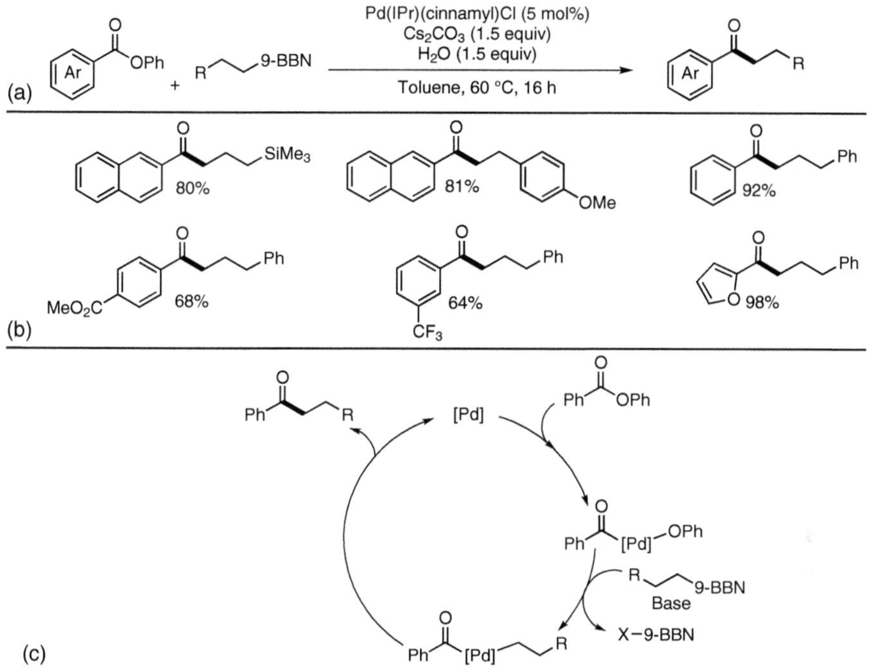

Figure 14.36 (a) Pd-catalyzed carbonyl retentive alkyl Suzuki–Miyaura coupling of phenyl esters. (b) Representative scope examples. (c) Proposed mechanism.

Using a bulky ligand was proposed to block the decarbonylation of the acyl-Pd intermediate, favoring the transmetalation and thus the formation of ketones (Figure 14.36c).

14.4.2 Amidation

Cross-coupling catalysis has opened the door toward accessing new heteroatom containing molecules [59]. The Buchwald–Hartwig reaction is particularly important, bringing together aryl halides and amines, allowing access to new C–N bonds. The strength of such a reaction is demonstrated by its ability to functionalize two molecules that would otherwise be inert to each other. While both acylative cross-coupling reactions and Buchwald–Hartwig amination reactions are independently important, the overlap between these two families is not obvious. For instance, acid chlorides require a Pd catalyst to react with organotin nucleophiles, but not with amines. Phenyl esters, however, are only modestly electrophilic, and do not react with aniline derivatives in the absence of strong base. With this in mind, the Newman laboratory developed a Pd-catalyzed amide bond formation from phenyl esters and anilines (Figure 14.37a) [69]. A variety of aryl, heteroaryl, and alkyl amide products were obtained (Figure 14.37b). This method did not require deprotonation of the aniline derivative with a strong base, so the use of potassium carbonate was sufficient for the coupling. As a result, enantiopure esters did not undergo significant racemization under the reaction conditions (Figure 14.37c).

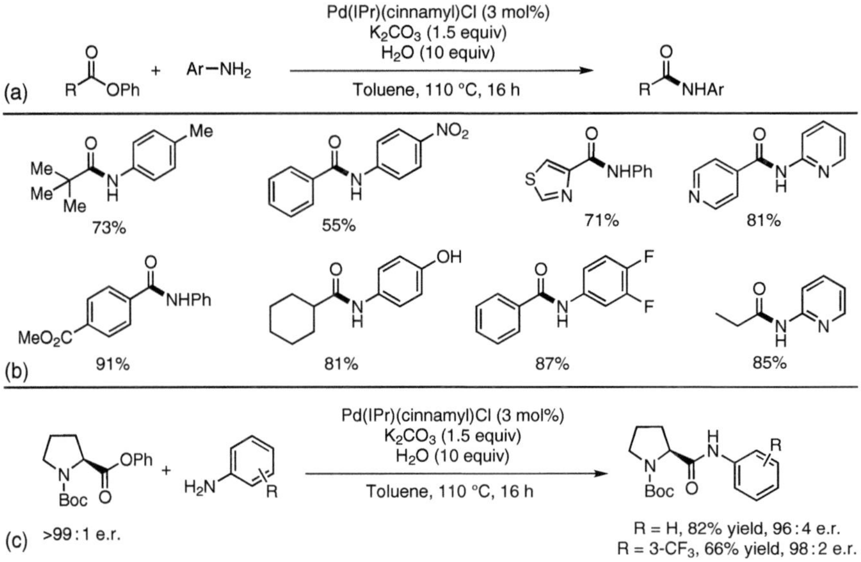

Figure 14.37 (a) Pd-catalyzed amide bond formation from phenyl esters. (b) Representative scope examples. (c) Enantiopure ester coupling.

14.4.3 Cross-Electrophile Coupling

Ni-catalyzed cross-electrophile coupling between acylating agents and aryl or alkyl halides has become a very useful method for preparing ketones while avoiding the use of stoichiometric organometallic reagents [70]. Thus far, the cross-coupling reactions of phenyl esters discussed are net redox neutral. In 2020, Wang and coworkers hypothesized that the lack of phenyl esters in cross-electrophile coupling chemistry is due to the difficulty associated with the activation of the C(acyl)–O bond. They reported the first cross-electrophile coupling of phenyl esters by exploiting the presence of an ortho directing group (Figure 14.38a) [71]. Esters were coupled with primary and secondary alkyl bromides with a Ni/1,10-phenanthroline catalyst system to give a wide variety of ketone products (Figure 14.38b). Tertiary alkyl halides, benzyl halides, and aryl halides were all incompatible with the reaction. The authors demonstrated that the hydroxyl directing group could be substituted for sulfonamides, and a similar outcome with regard to scope would be achieved, albeit with lower yields. The presence of the directing group was absolutely necessary for reactivity, as when the hydroxyl group is methylated or moved to the *para* position, the phenyl ester remains unreactive. Control reactions also ruled out the possibility of *in situ* organozinc formation, thus excluding a Negishi cross-coupling pathway. Even using stoichiometric Ni(cod)$_2$, the authors did not find any product formation, signifying that a catalytic cycle initiated by Ni(0) without an intermediate reduction is not possible. Thus the authors proposed a mechanism featuring oxidative addition of Ni(I) into the C(acyl)–O bond, facilitated by the directing

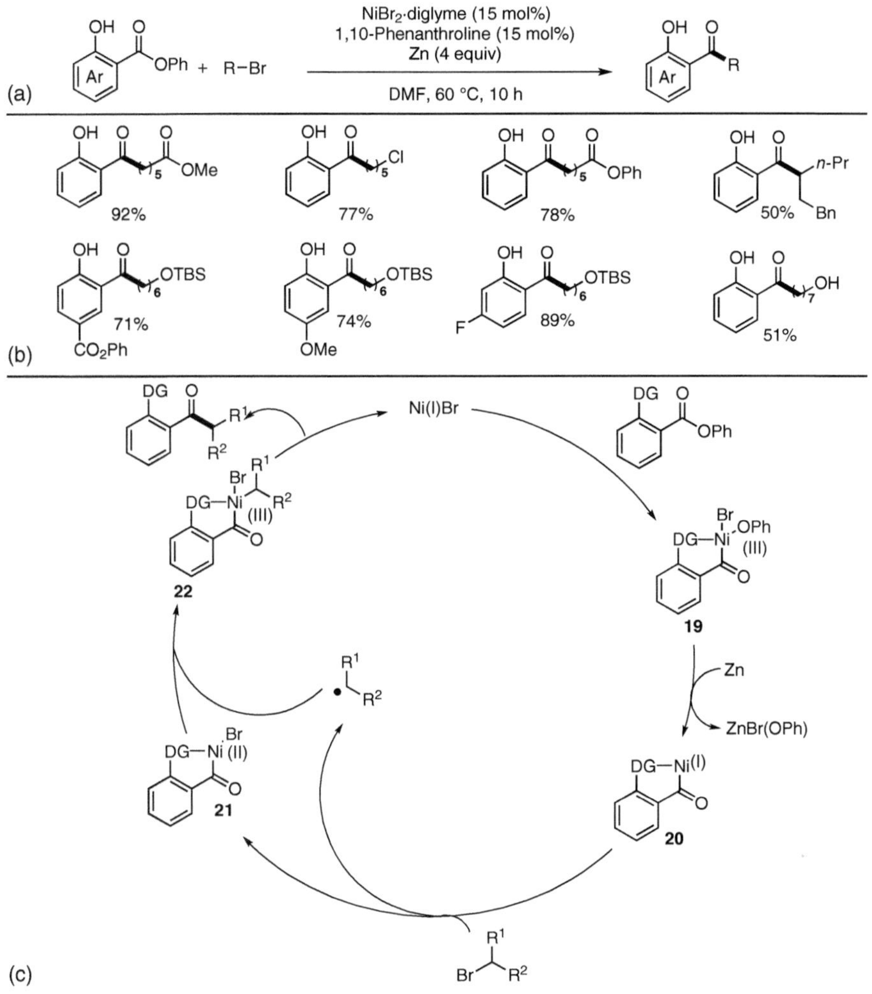

Figure 14.38 (a) Ni-catalyzed cross-electrophile coupling of phenyl esters. (b) Representative scope examples. (c) Proposed mechanism.

group, to give complex **19**. This intermediate is reduced by Zn to afford a Ni(I) species **20**, which can interact with the alkyl bromide to generate a free radical and complex **21**. Recombination with the alkyl radical generates Ni complex **22** and releases the product and regenerates the Ni(I) catalyst after reductive elimination (Figure 14.38c).

14.4.4 Ester Transfer and Ester Dance

Substitution reactions on aromatic substrates are incredibly important reactions in organic chemistry. To extend this chemistry to phenyl esters, the Yamaguchi group have developed ester transfer [72] and ester dance reactions [73]. The ester

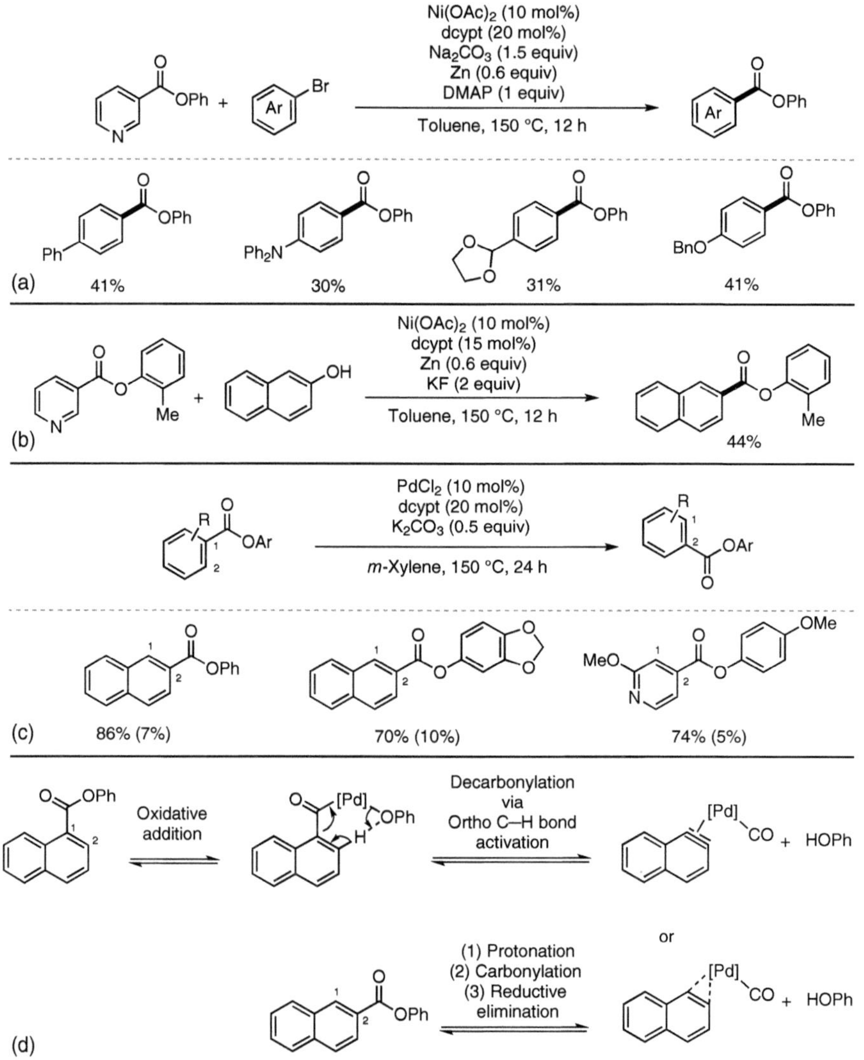

Figure 14.39 (a) Ester transfer reaction with representative scope examples. (b) Ester transfer with phenol-based electrophile. (c) Ester dance reaction from position 1 to position 2 with representative scope examples. Yields in brackets represent starting material recovery. (d) Proposed mechanism of ester dance reaction.

transfer reaction used Ni/dcypt catalysis to transfer aryl groups between aryl halides and aromatic esters (Figure 14.39a). Aryl chlorides and bromides were suitable for this chemistry, whereas aryl iodides were not. As an alternative to aryl halides, phenol-based aryl electrophiles could also form ester transferred product (Figure 14.39b).

The ester dance reaction was catalyzed by Pd/dcypt (Figure 14.39c), and the nature of the base played an important role as the reactions were almost shut down when bases other than K_2CO_3 were used. The nature of the base was proposed to be important for an ortho C–H bond activation step (Figure 14.39d). A variety of aryl and heteroaryl substrates could undergo this ester dance reaction, generating good yields of the product along with relatively small quantities of recovered starting material due to the modest thermodynamic driving force of this equilibration.

14.4.5 Deoxygenative Organophosphorus Coupling

Cross-coupling reactions with esters most commonly form either carbonyl retentive or decarbonylative products (see Figure 14.1). Yamaguchi and coworkers hypothesized that the combination of carbonyl-retentive cross-coupling and carbonyl reduction could lead to unusual deoxygenative coupling products (Figure 14.40a). With this challenge in mind, the authors reported a C–P bond forming reaction using a Pd/dcype catalyst and HCOONa as a stoichiometric reducing agent to give benzyl phosphine oxides (Figure 14.40b) [74]. A wide variety of aryl and heteroaryl compounds were tolerated under the reactions (Figure 14.40c). Control experiments demonstrated that the carbonyl-retentive product was an intermediate in the reaction, suggesting that the carbonyl is reduced to the corresponding methylene via a Pd/HCOONa catalytic system. This reaction could also be applied to carboxylic acids, thanks to *in situ* formation of an anhydride that can react similarly to a phenyl ester (Figure 14.40d).

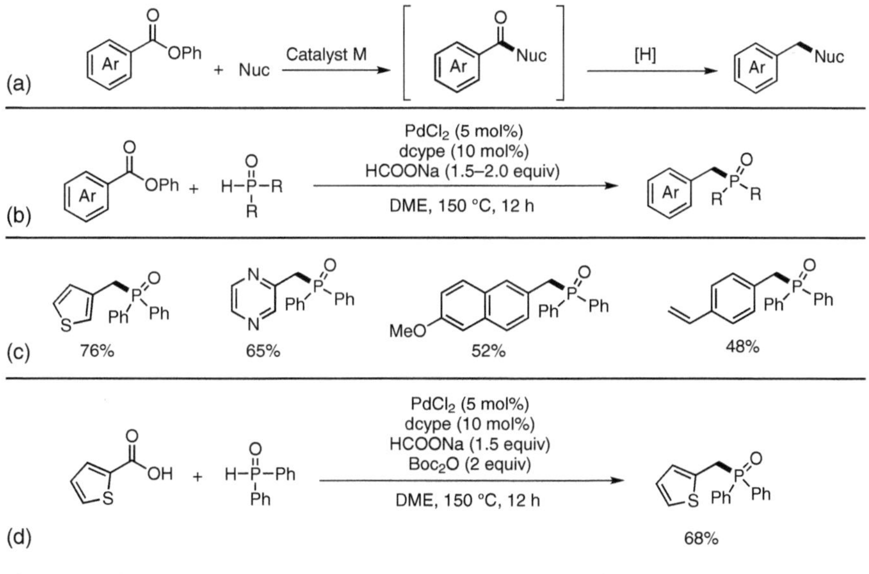

Figure 14.40 (a) Hypothesis for carbonyl retentive coupling followed by reduction. (b) Pd-catalyzed deoxygenative coupling with organophosphorus compounds. (c) Representative scope examples. (d) Deoxygenative coupling of carboxylic acids.

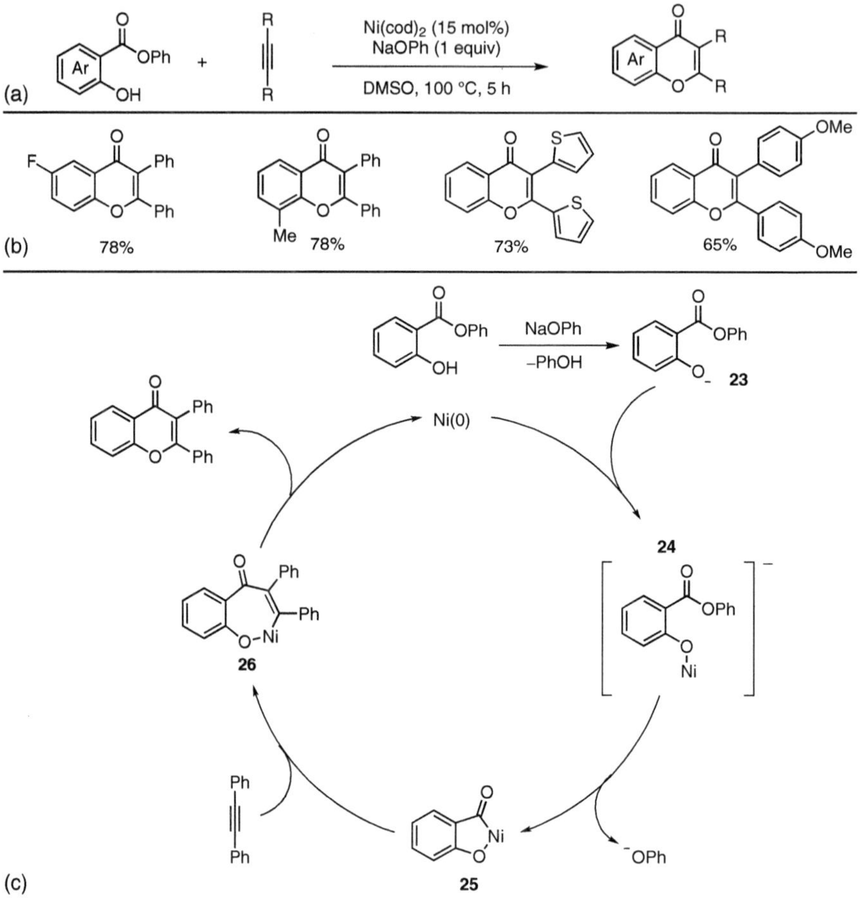

Figure 14.41 (a) Ni-catalyzed annulation reaction of salicylate phenyl esters with alkynes. (b) Representative scope examples. (c) Proposed mechanism.

14.4.6 Alkyne Insertion

In 2008, Matsubara and Kurahashi reported a Ni-catalyzed alkyne insertion/addition reaction of phthalic anhydrides and internal alkynes [75]. This reaction was decarbonylative in nature and formed isocoumarin products. In 2021, the Chatani group reported analogous reactivity between salicylate phenyl esters and alkynes using a nickel catalyst (Figure 14.41a) [76]. Unlike the alkyne insertion of phthalic anhydrides, alkyne insertion of salicylate phenyl esters was found to be carbonyl retentive. C–O bond activation was achieved with $Ni(cod)_2$ in the absence of ligand. The presence of base was necessary for reactivity. The ester was proposed to react with NaOPh to generate the anion **23** (Figure 14.41c). This intermediate can react with the Ni catalyst to give the anionic nickel complex **24**. Oxidative addition into the C(acyl)–O bond gives complex **25**, followed by insertion into the alkyne, forming intermediate **26**. Reductive elimination forms the product and regenerates the catalyst.

14.5 Carbonyl Retentive Coupling of Alkyl Esters

In the past two decades, significant advances have been made in the field of ester cross-coupling through the use of phenyl esters and related compounds as the electrophile. However, these substrates are seldom commercially available and thus need to be synthesized, often from the corresponding carboxylic acid or other acyl electrophiles such as acyl chlorides. In contrast, direct derivatization of unactivated esters via cross-coupling can potentially obviate the need for multistep processes for preparing, e.g. ketones, amides, and aldehydes. However, alkyl esters are much less activated compared with their phenyl ester counterparts, and they lack any significant coordinating ability. Furthermore, unactivated esters are generally left untouched in traditional coupling reactions of aryl halides. As a result, oxidative addition of metal catalysts into the strong C(acyl)–O bond of alkyl esters was not a commonly considered pathway. In this section, transformations of simple alkyl esters will be described.

14.5.1 Amidation

Despite amides being among the most frequently synthesized molecules, converting simple esters to amides can be problematic. Existing methods rely on activation of the electrophile with Lewis acids [57] or activation of the amine nucleophile with harsh organometallic bases [58]. Due to scope limitations of both these methods, esters are often converted to amides by first hydrolizing to the corresponding carboxylic acid, followed by subsequent activation with species such as thionyl chloride or peptide coupling reagents. Direct activation of alkyl esters for amide bond formation via cross-coupling has the potential to shortcut this pathway and provide an acid- and base-free alternative to other direct methods.

In 2016, Garg and Houk were the first to demonstrate that methyl esters can be viable cross-coupling electrophiles in Ni-catalyzed amidation reactions (Figure 14.42a), marking a significant milestone in the field [77]. Using a Ni/SIPr catalyst system, amide products were obtained, although the scope was largely limited to 1-napthyl esters (Figure 14.42b). Vital to this reaction was the addition of $Al(O^tBu)_3$ as a stoichiometric Lewis acid additive. DFT studies suggested that $Al(O^tBu)_3$ was capable of coordinating to the carbonyl group of the ester, lowering the kinetic barrier of the rate-determining oxidative addition step. 1-Naphthyl esters were more successful compared with their 2-naphtyl and methyl benzoate counterparts. The exceptional reactivity of 1-naphthyl esters was also attributed to $Al(O^tBu)_3$. The complex resulting from Lewis acid coordination causes steric repulsion from the bulky 1-naphthyl group, leading to a distortion that makes ester cleavage more thermodynamically favorable. Theoretically, a decarbonylation could occur at steps proceeding oxidative addition. However, it was shown that both the ligand exchange and the reductive elimination are much more energetically favorable than decarbonylation, explaining why amide bond formation takes place. To demonstrate the powerful synthetic utility of this method, orthogonal

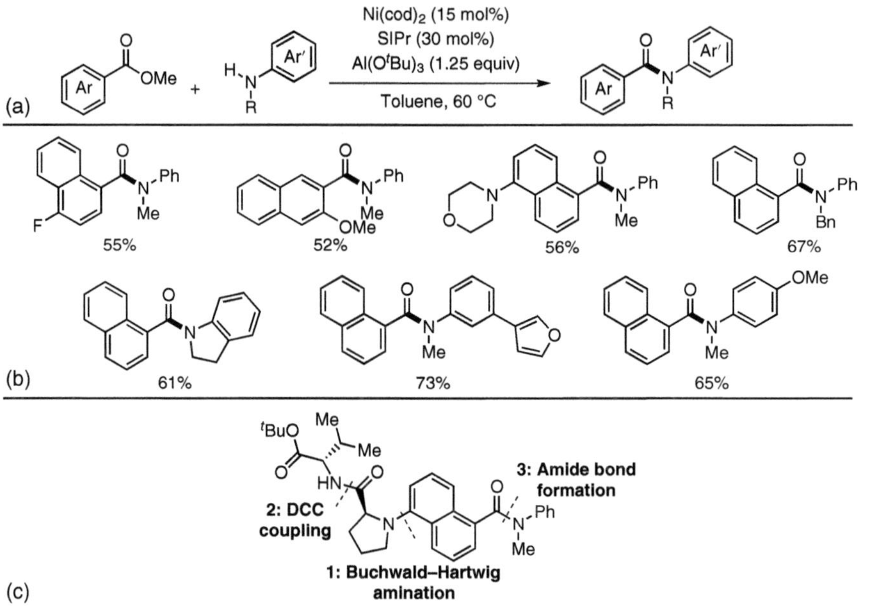

Figure 14.42 (a) Ni-catalyzed activation of methyl naphthoates. (b) Representative scope examples. (c) Sequential cross-couplings.

cross-couplings via Dicyclohexylcarbodiimide (DCC), Buchwald–Hartwig, and methyl ester cross-couplings were performed (Figure 14.42c).

In 2018, the Newman laboratory expanded the scope of ester and amine functionalities that could be tolerated in Ni-catalyzed amide bond formation [78]. Using only a Ni/IPr catalyst, methyl esters were activated toward amide bond formation in the absence of stoichiometric activating agents, with methanol being the sole stoichiometric by-product (Figure 14.43a). The use of elevated temperatures that

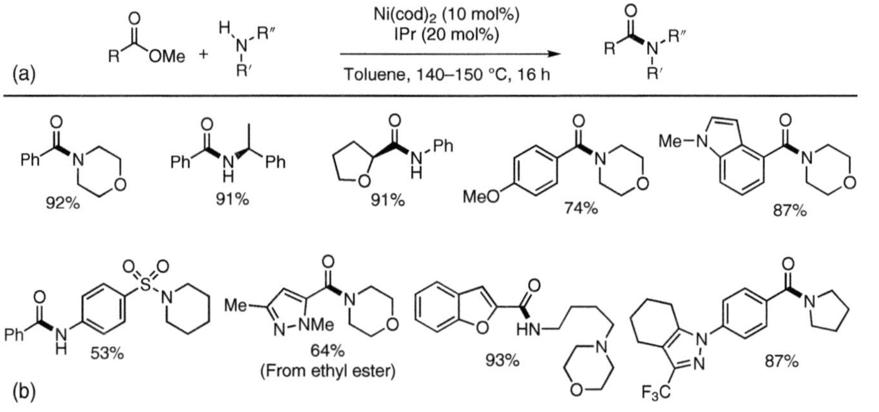

Figure 14.43 (a) Additive-free Ni-catalyzed methyl ester activation toward amide synthesis. (b) Representative scope examples.

drive methanol out of solution was found to be critical, both to prevent catalyst deactivation and to drive the equilibrium forward. A wide variety of aryl, heteroaryl, and alkyl esters were demonstrated to couple smoothly with aniline derivatives as well as aliphatic amines (Figure 14.43b). This method was also shown to be capable of efficiently accessing several bioactive molecules. In all cases, amides were formed selectively, likely due to the role of the bulky NHC ligand, which may prevent decarbonylation. Notably, the Hu group published a mechanistically distinct Ni-catalyzed approach to amide bond formation using nitrobenzene derivatives that are reduced *in situ* with stoichiometric Zn [79]. A similar reaction has also been realized using Cr catalysis [80].

Mechanistic studies by Hong and coworkers suggest that all key steps involved in the Ni-catalyzed amide bond formation from esters and amines – oxidative addition, proton transfer, and reductive elimination – have similar transition state energies [81]. These findings suggest that any of these three elementary steps could be turnover limiting, which implies that different ester and amine pairs may benefit from different conditions. With this in mind, the Newman laboratory reinvestigated several coupling reactions that proved difficult using the conditions described in Figure 14.43, including sterically hindered substrates, certain heterocycles (e.g. furans and coumarins), and dialkylamines. An extensive substrate-dependent ligand optimization study was conducted (Figure 14.44a) in order to couple these challenging substrates [82]. The ligand screening demonstrated that most phosphine ligands were entirely ineffective except for select bidentate phosphine ligands such as dcype. A bidentate nitrogen ligand, Me_4-1,10-Phen, was also shown to be effective for the transformation. Several classes of privileged NHC ligands were also identified (Figure 14.44c). Interestingly, selectivity studies demonstrated that esters bearing α-aryl groups react preferentially to those bearing α-alkyl groups, with the exception of α-benzylic esters, which are the most reactive of the three. In 2021, Szostak and coworkers demonstrated that [CpNi(IPr)Cl] complexes were also capable of coupling amines with phenyl and methyl esters in a carbonyl-retentive fashion to provide amides [83].

14.5.2 Mizoroki–Heck-Type Domino Reactions

With the exception of the initial report of methyl ester activation by Garg and coworkers that required specific substrates, cross-couplings of methyl esters generally require highly elevated temperatures (see also Section 14.6). DFT studies performed by Hong and coworkers [81] suggest that Ni(0) catalysts are capable of oxidatively adding into the C(acyl)–O bond of a methyl ester with reasonable activation energy, making use of milder temperatures feasible. However, the formation of the acyl-Ni complex is calculated to be thermodynamically uphill and reversible, complicating intermolecular couplings as the concentration of the reactive Ni(II) species will be extremely low at any given time.

In 2017, the groups of Garg [84] and Stanley [85] independently reported the trapping of *N*-Boc amide-derived acyl-Ni intermediates with tethered allyl fragments. Using these reports as inspiration, the Newman laboratory hypothesized that

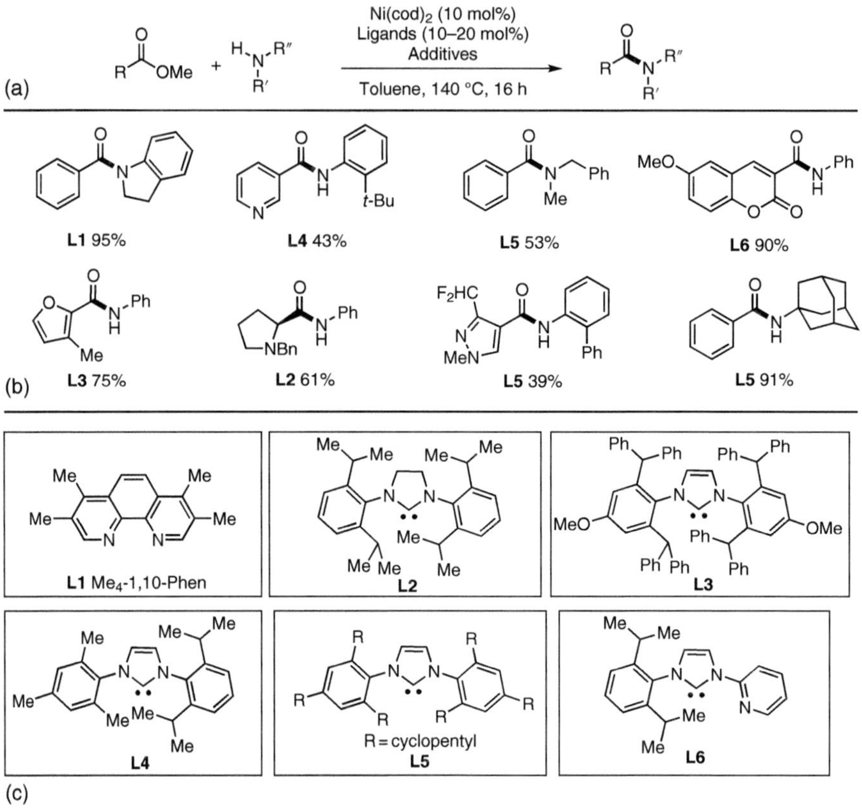

Figure 14.44 (a) Extensive ligand optimization to overcome issues in Ni-catalyzed amide bond formation from methyl esters. (b) Representative examples of expanded scope. (c) Examples of privileged ligands.

trapping the oxidative addition complex of a methyl ester with a tethered allyl coupling partner would be thermodynamically favorable and could lead to synthetically valuable products. In comparison to their *N*-Boc amide counterparts, methyl esters are often commercially available, and their methyl 2-allylbenzoate derivatives necessary for intramolecular coupling can be easily accessed in a single step via allylation from the organohalide. With this in mind, a Ni-catalyzed cyclization with tethered allyl groups was developed to prepare formal carboacylation or hydroacylation products (Figure 14.45a) [86]. A Ni/SIPr system was found to activate the methyl ester, which was proposed to generate acyl-Ni species **27**. This complex inserts the alkene of the tethered allyl group, generating intermediate **28**. Subsequent transmetalation with an arylboronic acid forms a domino Mizoroki–Heck/Suzuki–Miyaura coupling product. Alternatively, addition of an alcohol in the reaction leads to a domino Mizoroki–Heck/reduction product (Figure 14.45b). This cyclization was successful at 80 °C, which is significantly lower than most other reports on similar transformations, suggesting that the intramolecular trapping of the acyl-Ni(II) species may overcome challenges with intermolecular trapping.

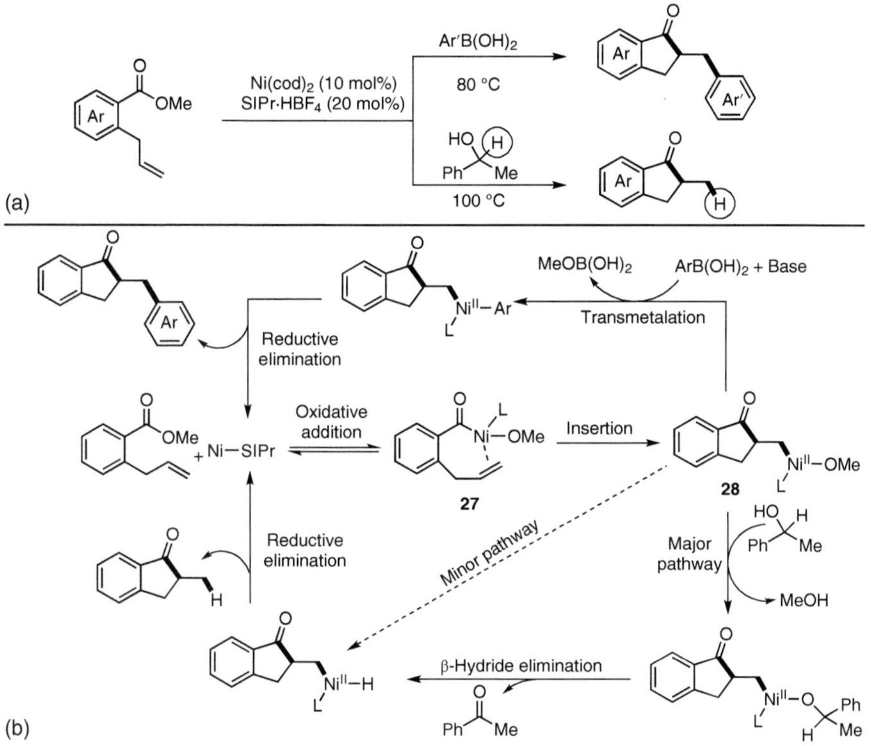

Figure 14.45 (a) Ni-catalyzed domino Mizoroki–Heck-type reactions using methyl esters. (b) Proposed mechanism.

14.5.3 Suzuki–Miyaura Coupling

The coupling of methyl esters with carbon-based nucleophiles can in principle occur in a carbonyl-retentive fashion to provide ketone products, circumventing the need to use a multistep approach such as the Weinreb ketone synthesis [8]. However, such intermolecular couplings have proven more challenging than the previously described amidation and cyclization reactions. In 2021, the Newman laboratory identified a Ni-catalyzed synthesis of ketones from methyl esters and alkyl-9-BBN reagents (Figure 14.46a) [87]. Crucial to the success of the reaction was the use of a 2,4,6-tricyclopentylaniline-derived N-heterocyclic carbene, IPr^{Cyp}, which in conjunction with Ni, provided access to a wide variety of ketone products (Figure 14.46b). The use of alkylboranes was necessary, as arylboronic acids and other boron nucleophiles were unsuccessful. This reaction also showed impressive orthogonality to other Ni-catalyzed cross-coupling reactions (Figure 14.46c). DFT studies were conducted to gain understanding into the challenges of this reaction. With the particularly bulky NHC catalyst, oxidative addition was calculated to be relatively facile, making transmetalation the rate-determining step. The aryl substituents found on the alkyl-9-BBN nucleophiles were found to interact with the para-cyclopentyl ring on the NHC ligand in this step, explaining the nonintuitive importance of these two remote groups.

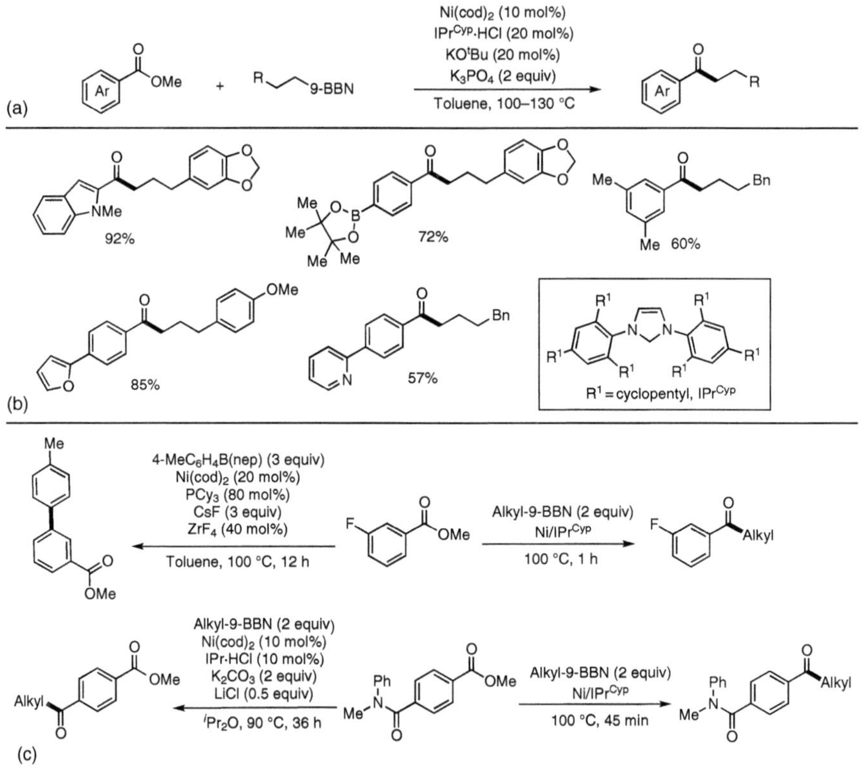

Figure 14.46 (a) Ni-catalyzed synthesis of ketones from methyl esters and alkyl-9-BBN reagents. (b) Representative scope examples. (c) Chemoselectivity studies.

14.6 Decarbonylative Couplings of Alkyl Esters

Most reports on the cross-coupling of simple alkyl esters thus far have given carbonyl-retentive products. This contrasts sharply with the coupling of phenyl esters, which have been primarily used for decarbonylative coupling. In this final section, the few examples of decarbonylative alkyl ester coupling are described.

14.6.1 Directing Group Assistance

The first example of a catalytic reaction involving ester C(acyl)–O bond cleavage followed by decarbonylation was reported in 2001 by the Murai group (Figure 14.47a) [88]. For this reaction, the authors opted to use Ru catalysis. At the time, Yamamoto and coworkers had reported the first Pd-catalyzed Suzuki–Miyaura coupling of electronically activated esters to form ketone products [89]. The decision to use Ru was driven based on earlier studies from their laboratory [90]. Pyridylmethyl esters were used, with the pyridine acting as a directing group to render the Ru more nucleophilic. In the presence of ammonium formate, it was demonstrated that the esters could be reduced. Related work in 2004 from the Chatani laboratory

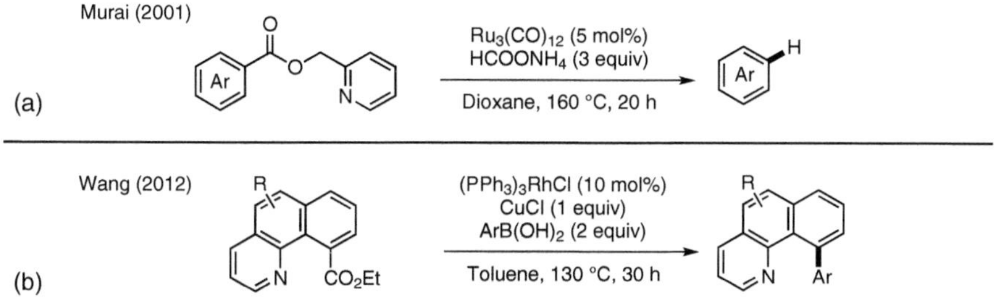

Figure 14.47 (a) Alkyl ester cross-coupling achieved via chelation assistance. (b) Rh-catalyzed ethyl ester activation mediated by chelation assistance.

demonstrated the ability of these substrates to give both carbonyl-retentive and decarbonylative products [91].

In 2012, Wang and coworkers reported a related chelation-assisted decarbonylative Suzuki–Miyaura cross-coupling of ethyl esters (Figure 14.47b) [92]. Chatani has also reported the reduction of alkyl esters in the absence of an external reductant by using a directing group [93]. These early reports on ester activation with carefully selected directing groups were seminal toward future developments in the activation of simpler, more accessible alkyl esters.

14.6.2 Methylation

The decarbonylative methylation of aryl esters via Ni catalysis was outlined in Section 14.3.5.5 [46]. The methylating agent used was dimethylaluminum chloride, which may also act as a Lewis acid to render the ester more reactive. In the same report, a selection of simple alkyl esters was also shown to be amenable using slightly modified conditions (Figure 14.48). While a mechanism was not proposed, formation of an acyl-Ni intermediate may be occurring.

14.6.3 Organostannane Formation

Arylstannanes are very valuable compounds due to the role they play in the Stille coupling, which allows access to complex molecules [94, 95]. In 2018, the Rueping group developed a decarbonylative stannylation reaction of methyl esters, which could also be extended to ethyl, cyclohexyl, benzyl, and phenyl esters

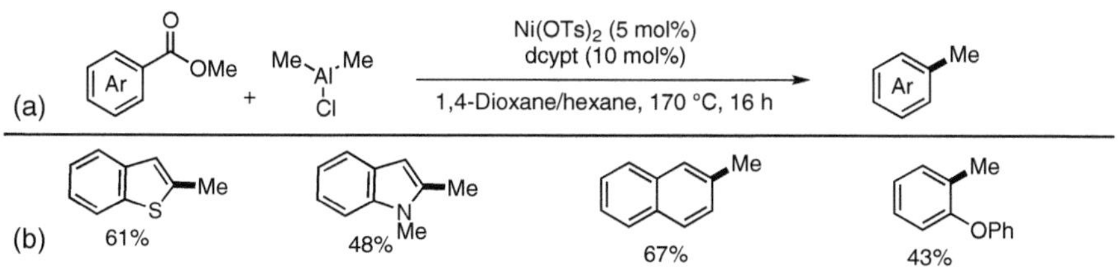

Figure 14.48 (a) Ni-catalyzed decarbonylative methylation of methyl esters. (b) Representative scope examples.

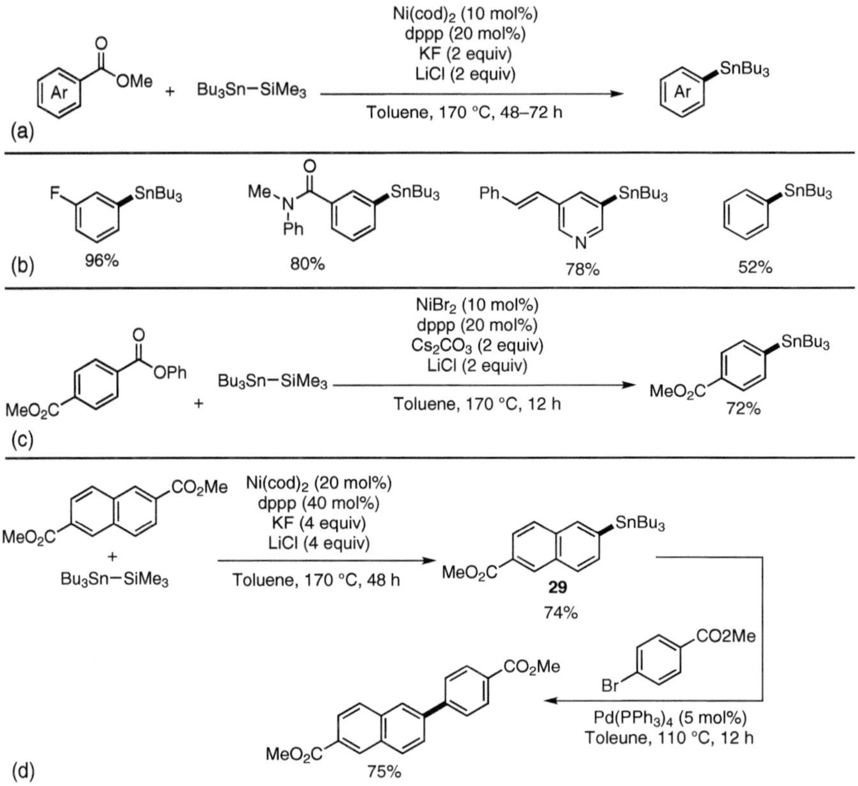

Figure 14.49 (a) Ni-catalyzed decarbonylative ester to stannane conversion. (b) Representative scope examples. (c) Selective stannylation. (d) Monostannylation and sequential cross-coupling.

(Figure 14.49a) [96]. In the presence of a Ni/dppp catalyst system, methyl esters were coupled with trimethyl(tributylstannyl)silane to give a range of arylstannanes (Figure 14.49b). If both a methyl and phenyl ester are present, selective reaction occurs at the more reactive phenyl ester (Figure 14.49c). Monostannylation of substrates bearing two methyl ester groups was also demonstrated, with the resulting product **29** being capable of further derivatization by Stille coupling (Figure 14.49d). No mechanism was proposed, but one possible pathway may involve formation of a Ni-acyl intermediate followed by transmetalation, decarbonylation, and reductive elimination.

14.7 Conclusion and Outlook

The use of esters for cross-coupling reactions has numerous benefits. They are robust functional groups capable of being carried through multistep synthesis and used in late-stage functionalization. They are ubiquitous and, in the case of simple alkyl esters, are often commercially available. Their reactions avoid the generation of halide salt waste formed in the coupling of aryl halides. Most importantly, these coupling reactions can serve as appealing alternatives to traditional routes based on the availability of the starting material or the efficiency by which they can be converted into the desired products. Generally speaking, the cross-coupling of esters proceeding via metal-acyl intermediates can provide two classes of products – those that retain the carbonyl and those that lose the carbonyl. In some cases, control of these two pathways is possible, further highlighting the synthetic utility of esters as cross-coupling electrophiles. Initial reports in ester activation relied on directing groups to facilitate the difficult oxidative addition into the C(acyl)–O bond. Shortly after, ester chemistry was then directed toward the coupling of phenyl esters, as the aromatic ring weakened the C(acyl)–O bond, rendering them modestly activated toward oxidative addition. Most recently, progress has been made in the activation of abundant methyl esters. While not discussed in this chapter, progress in this field shares numerous parallels with the activation of amides, which is generally achieved by identifying substrates with a relatively weak C(acyl)–N bond and/or identifying powerful catalysts capable of directly cleaving stronger C(acyl)–N bonds. These two different strategies have several complementary features. By designing substrates with particularly weak amide or ester bonds, cleavage can occur with mild conditions, relatively low catalyst loadings, and with a range of catalyst choices. However, these substrates are often less abundant and less robust. In contrast, more abundant esters and amides feature stronger bonds that necessitate the use of more aggressive and carefully tuned conditions, which may result in more limited substrate scopes. In the future, we expect further mechanistic understanding, and catalyst design enables the continued expansion of acylative and decarbonylative cross-coupling reactions to provide complementarity to traditional organohalide coupling reactions.

References

1 (a) Miyaura, N. and Suzuki, A. (1995). *Chem. Rev.* 95: 2457. (b) Johansson Seechurn, C.C., Kitching, M.O., Colacot, T.J., and Snieckus, V. (2012). *Angew. Chem. Int. Ed.* 51: 5062. (c) Surry, D.S. and Buchwald, S.L. (2008). *Angew. Chem. Int. Ed.* 47: 6338. (d) Forero-Cortés, P.A. and Haydl, A.M. (2019). *Org. Process Res. Dev.* 23: 1478.

2 (a) Qiu, Z. and Li, C.J. (2020). *Chem. Rev.* 120: 10454. (b) Campeau, L.C. and Hazari, N. (2019). *Organometallics* 38: 3.

3 (a) Ishizu, J., Yamamoto, T., and Yamamoto, A. (1976). *Chem. Lett.* 10: 1091. (b) Yamamoto, T., Ishizu, J., Kohara, T. et al. (1980). *J. Am. Chem. Soc.* 102: 3758.

4 (a) Kosugi, M., Shimizu, Y., and Migita, T. (1977). *Chem. Lett.* 1423. (b) Milstein, D. and Stille, J.K. (1978). *J. Am. Chem. Soc.* 100: 3636.

5 (a) Stephan, M.S., Teunissen, A.J., Verzijl, G.K., and de Vries, J.G. (1998). *Angew. Chem. Int. Ed.* 37: 662. (b) Gooßen, L.J. and Ghosh, K. (2001). *Angew. Chem. Int. Ed.* 40: 3458. (c) Kakino, R., Yasumi, S., Shimizu, I., and Yamamoto, A. (2002). *Bull. Chem. Soc. Jpn.* 75: 137.

6 (a) Tokuyama, H., Yokoshima, S., Yamashita, T., and Fukuyama, T. (1998). *Tetrahedron Lett.* 39: 3189. (b) Liebeskind, L.S. and Srogl, J. (2000). *J. Am. Chem. Soc.* 122: 11260. (c) Wittenberg, R., Srogl, J., Egi, M., and Liebeskind, L.S. (2003). *Org. Lett.* 5: 3033. (d) Yu, Y. and Liebeskind, L.S. (2004). *J. Org. Chem.* 69: 3554.

7 (a) Hirschbeck, V., Gehrtz, P., and Fleischer, I. (2018). *Chem. Eur. J.* 24: 7092. (b) Buchspies, J. and Szostak, M. (2019). *Catalysts* 9: 53. (c) Blangetti, M., Rosso, H., Prandi, C. et al. (2013). *Molecules* 18: 1188. (d) Dzik, W.I., Lange, P.P., and Gooßen, L.J. (2012). *Chem. Sci.* 3: 2671. (e) Dermenci, A. and Dong, G.B. (2013). *Sci. China Chem.* 56: 685. (f) Guo, L. and Rueping, M. (2018). *Chem. Eur. J.* 24: 7794. (g) Takise, R., Muto, K., and Yamaguchi, J. (2017). *Chem. Soc. Rev.* 46: 5864. (h) Liu, C. and Szostak, M. (2018). *Org. Biomol. Chem.* 16: 7998.

8 Nahm, S. and Weinreb, S.M. (1981). *Tetrahedron Lett.* 22: 3815.

9 Heck, R.F. (1982). *Org. React.* 27: 345.

10 Gooßen, L.J. and Paetzold, J. (2002). *Angew. Chem. Int. Ed.* 41: 1237.

11 Amaike, K., Muto, K., Yamaguchi, J., and Itami, K. (2012). *J. Am. Chem. Soc.* 134: 13573.

12 Meng, L., Kamada, Y., Muto, K. et al. (2013). *Angew. Chem.* 125: 10232.

13 Matsushita, K., Takise, R., Hisada, T. et al. (2018). *Chem. Asian J.* 13: 2393.

14 Kruckenberg, A., Wadepohl, H., and Gade, L.H. (2013). *Organometallics* 32: 5153.

15 Okita, T., Komatsuda, M., Saito, A.N. et al. (2018). *Asian J. Org. Chem.* 7: 1358.

16 Miyaura, N. (ed.) (2002). *Cross-Coupling Reactions: A Practical Guide*, Topics in Current Chemistry, vol. 219. Berlin: Springer.

17 Tatamidani, H., Kakiuchi, F., and Chatani, N. (2004). *Org. Lett.* 6: 3597.

18 Muto, K., Yamaguchi, J., Musaev, D.G., and Itami, K. (2015). *Nat. Commun.* 6: 7508.

19 Laberge, N.A. and Love, J.A. (2015). *Eur. J. Org. Chem.* 5546.

20 Muto, K., Hatakeyama, T., Itami, K., and Yamaguchi, J. (2016). *Org. Lett.* 18: 5106.

21 Amaike, K., Itami, K., and Yamaguchi, J. (2016). *Chem. Eur. J.* 22: 4384.

22 Lennox, A.J.J. and Lloyd-Jones, G.C. (2014). *Chem. Soc. Rev.* 43: 412.

23 Chatupheeraphat, A., Liao, H.-H., Srimontree, W. et al. (2018). *J. Am. Chem. Soc.* 140: 3724.

24 Masson-Makdissi, J., Vandavasi, J.K., and Newman, S.G. (2018). *Org. Lett.* 20: 4094.

25 (a) Denmark, S.E., Liu, J., and J.H.-C. (2010). *Angew. Chem. Int. Ed.* 49: 2978. (b) Ramesh, R. and Reddy, D.S. (2018). *J. Med. Chem.* 61 (9): 3779.

26 Manoso, A.S., Ahn, C., Soheili, A. et al. (2004). *J. Org. Chem.* 69: 8305.

27 Denmark, S.E. and Kallemeyn, J.M. (2003). *Org. Lett.* 5: 3483.

28 Pintaric, C., Olivero, S., Gimbert, Y. et al. (2010). *J. Am. Chem. Soc.* 132: 11825.

29 Molander, G.A., Trice, S.L.J., and Dreher, S.D. (2010). *J. Am. Chem. Soc.* 132: 17701.

30 Guo, L., Chatupheeraphat, A., and Rueping, M. (2016). *Angew. Chem.* 128: 11989.

31 Pu, X., Hu, J., Zhao, Y., and Shi, Z. (2016). *ACS Catal.* 6: 6692.

32 Guo, L. and Rueping, M. (2016). *Chem. Eur. J.* 22: 16787.

33 Chinchilla, R. and Nájera, C. (2014). *Chem. Rev.* 114: 1783.

34 Chinchilla, R. and Nájera, C. (2011). *Chem. Soc. Rev.* 40: 5084.

35 Okita, T., Kumazawa, K., Takise, R. et al. (2017). *Chem. Lett.* 46: 218.

36 Seyden-Penne, J. (1997). *Reductions by the Alumino- and Borohydrides in Organic Synthesis*, 2e. New York: Wiley-VCH.

37 Marciniec, B. (2009). *Hydrosilylation: A Comprehensive Review on Recent Advances*. Heidelberg: Springer.

38 Yue, H., Guo, L., Lee, S.-C. et al. (2017). *Angew. Chem.* 129: 4030.

39 He, Z., Wang, Z., Ru, J. et al. (2020). *Adv. Synth. Catal.* 362: 5794.

40 Liu, X., Jia, J., and Rueping, M. (2017). *ACS Catal.* 7: 4491.

41 Wu, X., Li, X., Huang, W. et al. (2019). *Org. Lett.* 21: 2453.

42 Beletskaya, I.P., Sigeev, A.S., Peregudov, A.S., and Petrovskii, P.V. (2004). *J. Organomet. Chem.* 689: 3810.

43 Rosenmund, K.W. and Struck, E. (1919). *Ber. Dtsch. Chem. Ges.* 2: 1749.

44 Chatuphreeraphat, A., Liao, H.-H., Lee, S.-C., and Rueping, M. (2017). *Org. Lett.* 19: 4255.

45 Iizumi, K., Kurosawa, M.B., Isshiki, R. et al. (2020). *Synlett.* 31: https://doi.org/10.1055/s-0040-1705943.

46 Okita, T., Muto, K., and Yamaguchi, J. (2018). *Org. Lett.* 20: 3132.

47 (a) Aranyos, A., Old, D.W., Kiyomori, A. et al. (1999). *J. Am. Chem. Soc.* 121: 4369. (b) Burgos, C.H., Barder, T.E., Huang, X., and Buchwald, S.L. (2006). *Angew. Chem. Int. Ed.* 45: 4321.

48 (a) Maiti, D. and Buchwald, S.L. (2010). *J. Org. Chem.* 75: 1791. (b) Altman, R.A., Shafir, A., Choi, A. et al. (2008). *J. Org. Chem.* 73: 284.

49 Takise, R., Isshiki, R., Muto, K. et al. (2017). *J. Am. Chem. Soc.* 139: 3340–3343.

50 (a) Liang, M. and Chen, J. (2013). *Chem. Soc. Rev.* 42: 3453. (b) Ricci, A. (ed.) (2008). *Amino Group Chemistry: From Synthesis to the Life Sciences*. Weinheim: Wiley-VCH.

51 Yue, H., Guo, L., Liao, H.-H. et al. (2017). *Angew. Chem. Int. Ed.* 56: 4282.

52 Malapit, C.A., Borrell, M., Milbauer, M.W. et al. (2020). *J. Am. Chem. Soc.* 142: 5918.

53 Ilardi, E.A., Vitaku, E., and Njardarson, J.T. (2014). *J. Med. Chem.* 57: 2832.

54 Bichler, P. and Love, J.A. (2010). *Top. Organomet. Chem.* 31: 39.

55 Lee, S.-C., Liao, H.H., Chatupheeraphat, A., and Rueping, M. (2018). *Chem. Eur. J.* 24: 3608.

56 Isshiki, R., Muto, K., and Yamaguchi, J. (2018). *Org. Lett.* 20: 1150.

57 (a) Ishihara, K., Kuroki, Y., Hanaki, N. et al. (1996). *J. Am. Chem. Soc.* 118: 1569. (b) Han, C., Lee, J.P., Lobkovsky, E., and Porco, J.A. (2005). *J. Am. Chem. Soc.* 127: 10039. (c) Tsuji, H. and Yamamoto, H. (2016). *J. Am. Chem. Soc.* 138: 14218. (d) Morimoto, H., Fujiwara, R., Shimizu, Y. et al. (2014). *Org. Lett.* 16: 2018.

58 (a) Basha, A., Lipton, M., and Weinreb, S.M. (1977). *Tetrahedron Lett.* 18: 4171. (b) Wang, W.B. and Roskamp, E.J. (1992). *J. Org. Chem.* 57: 6101. (c) Ohshima, T., Hayashi, Y., Agura, K. et al. (2012). *Chem. Commun.* 48: 5434.

59 (a) Ruiz-Castillo, P. and Buchwald, S.L. (2016). *Chem. Rev.* 116: 12564. (b) Bariwal, J. and Van der Eycken, E. (2013). *Chem. Soc. Rev.* 42: 9283.

60 Ben Halima, T., Zhang, W., Yalaoui, I. et al. (2017). *J. Am. Chem. Soc.* 139: 1311.

61 Dardir, A.H., Melvin, P.R., Davis, R.M. et al. (2018). *J. Org. Chem.* 83: 469.
62 Melvin, P.R., Nova, A., Balcells, D. et al. (2015). *ACS Catal.* 5: 3680.
63 Lei, P., Meng, G., Shi, S. et al. (2017). *Chem. Sci.* 8: 6525.
64 Shi, S. and Szostak, M. (2017). *Chem. Commun.* 53: 10584.
65 (a) Shi, S., Lei, P., and Szostak, M. (2017). *Organometallics* 36: 3784. (b) Li, G., Shi, S., and Szostak, M. (2018). *Adv. Synth. Catal.* 360: 1538.
66 Zhou, T., Li, G., Nolan, S.P., and Szostak, M. (2019). *Org. Lett.* 21: 3304.
67 Lei, P., Ling, Y., An, J. et al. (2019). *Adv. Synth. Catal.* 361: 5654.
68 Buchspies, J., Pyle, D.J., He, H., and Szostak, M. (2018). *Molecules* 23: 3134.
69 Ben Halima, T., Vandavasi, J.K., Shkoor, M., and Newman, S.G. (2017). *ACS Catal.* 7: 2176.
70 (a) Everson, D.A. and Weix, D.J. (2014). *J. Org. Chem.* 79: 4793. (b) Moragas, T., Correa, A., and Martin, R. (2014). *Chem. Eur. J.* 20: 8242. (c) Gu, J., Wang, X., Xue, W., and Gong, H. (2015). *Org. Chem. Front.* 2: 1411. (d) Weix, D.J. (2015). *Acc. Chem. Res.* 48: 1767. (e) Wang, X., Dai, Y., and Gong, H. (2017). *Ni- and Fe-Based Cross-Coupling Reactions* (ed. A. Correa), 61–89. Cham: Springer International Publishing. (f) Richmond, E. and Moran, J. (2018). *Synthesis* 50: 499. (g) Poremba, K.E., Dibrell, S.E., and Reisman, S.E. (2020). *ACS Catal.* 10: 8237.
71 Yang, F., Ding, D., and Wang, C. (2020). *Org. Lett.* 22: 9203.
72 Isshiki, R., Inayama, N., Muto, K., and Yamaguchi, J. (2020). *ACS Catal.* 10: 3490.
73 Matsushita, K., Takise, R., Muto, K., and Yamaguchi, J. (2020). *Sci. Adv.* 6: eaba7614.
74 Kurosawa, M.B., Isshiki, R., Muto, K., and Yamaguchi, J. (2020). *J. Am. Chem. Soc.* 142: 7386.
75 Kajita, Y., Kurahashi, T., and Matsubara, S. (2008). *J. Am. Chem. Soc.* 130: 17226.
76 Iyori, Y. and Chatani, N. (2021). *Chem. Lett.* 50: 510. https://doi.org/10.1246/cl.200885.
77 Hie, L., Nathel, N.F.F., Hong, X. et al. (2016). *Angew. Chem. Int. Ed.* 55: 2810.
78 Ben Halima, T., Masson-Makdissi, J., and Newman, S.G. (2018). *Angew. Chem. Int. Ed.* 57: 12925.
79 Cheung, C.W., Ploeger, M.L., and Hu, X. (2017). *Nat. Commun.* 8: 14878.
80 Ling, L., Chen, C., Luo, M., and Zeng, X. (2019). *Org. Lett.* 21: 1912.
81 Ji, C.L., Xie, P.P., and Hong, X. (2018). *Molecules* 23: 2681.
82 Zheng, Y.-L. and Newman, S.G. (2019). *ACS Catal.* 9: 4426.
83 Buchspies, J., Rahman, M.M., and Szostak, M. (2021). *Molecules* 26: 188.
84 Medina, J.M., Moreno, J., Racine, S. et al. (2017). *Angew. Chem. Int. Ed.* 56: 6567.
85 (a) Walker, J.A. Jr., Vickerman, K.L., Humke, J.N., and Stanley, L.M. (2017). *J. Am. Chem. Soc.* 139: 10228. (b) Kadam, A., Metz, T., Qian, Y., and Stanley, L.M. (2019). *ACS Catal.* 9: 5651.
86 Zheng, Y.-L. and Newman, S.G. (2019). *Angew. Chem. Int. Ed.* 58: 18159.
87 Zheng, Y.-L., Xie, P.-P., Daneshfar, O. et al. (2021). *Angew. Chem. Int. Ed.* https://doi.org/10.1002/anie.202103327.
88 Chatani, N., Tatamidani, H., Ie, Y. et al. (2001). *J. Am. Chem. Soc.* 123: 4849.

89 (a) Nagayama, K., Shimizu, I., and Yamamoto, A. (1999). *Bull. Chem. Soc. Jpn.* 72: 799. (b) Kakino, R., Shimizu, I., and Yamamoto, A. (2001). *Bull. Chem. Soc. Jpn.* 74: 371.

90 (a) Chatani, N., Fukuyama, T., Tatamidani, H. et al. (2000). *J. Org. Chem.* 65: 4039. (b) Chatani, N., Le, Y., Kakiuchi, F., and Murai, S. (1999). *J. Am. Chem. Soc.* 121: 8645. (c) Ishii, Y., Chatani, N., Yorimitsu, S., and Murai, S. (1998). *Chem. Lett.* 157.

91 Tatamidani, H., Yokota, K., Kakiuchi, F., and Chatani, N. (2004). *J. Org. Chem.* 69: 5615.

92 Wang, J., Liu, B., Zhao, H., and Wang, J. (2012). *Organometallics* 31: 8598.

93 Iyori, Y., Takahashi, K., Yamazaki, K. et al. (2019). *Chem. Commun.* 55: 13610–13613.

94 (a) Ragan, J., Raggon, J., Hill, P. et al. (2003). *Org. Process Res. Dev.* 7: 676. (b) Alonso, E., Fuwa, H., Vale, C. et al. (2012). *J. Am. Chem. Soc.* 134: 7467. (c) Valot, G., Regens, C.S., O'Malley, D.P. et al. (2013). *Angew. Chem. Int. Ed.* 52: 9534. (d) Li, J., Yang, P., Yao, M. et al. (2014). *J. Am. Chem. Soc.* 136: 16477. (e) Mailhol, D., Willwacher, J., Kausch-Busies, N. et al. (2014). *J. Am. Chem. Soc.* 136: 15719. (f) Logan, M.M., Toma, T., Thomas-Tran, R., and Du Bois, J. (2016). *Science* 354: 865.

95 (a) Soderquist, J.A. and Hassner, A. (1980). *J. Am. Chem. Soc.* 102: 1577. (b) Gosmini, C. and Perichon, J. (2005). *Org. Biomol. Chem.* 3: 216.

96 Yue, H., Zhu, C., and Rueping, M. (2018). *Org. Lett.* 20: 385.

15

Cross-Coupling of Aromatic Esters by Decarbonylation

Kei Muto[1] *and Junichiro Yamaguchi*[2]

[1] *Waseda University, Waseda Institute for Advanced Study, 513 Wasedatsurumakicho, Shinjuku, Tokyo, 162-0041, Japan*
[2] *Waseda University, Department of Applied Chemistry, 513 Wasedatsurumakicho, Shinjuku, Tokyo, 162-0041, Japan*

15.1 Introduction

Metal-catalyzed cross-coupling is recognized as a reliable method for bond formation and can be applied to the synthesis of natural products, pharmaceuticals, and organic materials in versatile fashion [1]. This method typically uses organic halides as electrophiles with nucleophilic counterparts. With recent advancements in this arena, alternative electrophiles such as phenols [2], anilines [3], and thiophenols [4] can now be used. The development of these advanced methods often brings about benefits not only in avoiding the generation of corrosive halogen-containing chemical wastes but also in providing chemists new synthetic strategies and routes.

Recently, the use of arenecarboxylic acids derivatives such as esters ($ArCO_2R$) and amides ($ArCONR_2$) in cross-coupling has gained considerable attention because they are abundant and useful chemical feedstocks [5, 6]. So far, two types of couplings using arenecarboxylates are known: decarbonylative coupling (aryl–X bond formation) and non-decarbonylative couplings (aroyl–X bond formation). In this book chapter, methods of decarbonylative coupling of aromatic esters through metal catalysis are presented. This topic is subdivided into C–C bond formations, C–X bond formations, and other types of transformations. Strictly speaking, acid anhydrides are not esters; however, this book chapter includes the reactions of acid anhydrides because of its high relevance to this topic. Related decarbonylative couplings of aromatic amides are described elsewhere in this book [7].

15.2 Overview of Decarbonylative Coupling

It has long been known that aroyl chlorides [8], thioesters [9], and aroyl cyanides [10] can participate in metal-catalyzed decarbonylative coupling. Compared to these

Amide Bond Activation: Concepts and Reactions, First Edition. Edited by Michal Szostak.

examples, the development of similar reactions of more ubiquitous esters has been hampered mainly due to the inertness of the C–O bond toward metal catalysis. The decarbonylative coupling of esters proceeds through a C(aroyl)–O bond oxidative addition to a metal catalyst to generate intermediate **A** (Scheme 15.1). This intermediate **A** then excludes CO and reacts with a nucleophilic component to give intermediate **B**, followed by a reductive elimination to release coupling products.

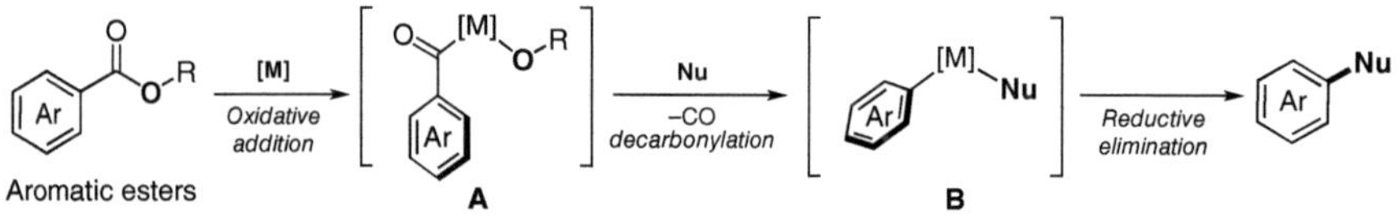

Scheme 15.1 General scheme of decarbonylative coupling of aromatic esters.

In a fundamentally important work, in 1976, Yamamoto discovered that a stoichiometric amount of nickel(0) complex can oxidatively add into the C(acyl)–O bond of esters (Scheme 15.2) [11]. This report unveiled that Ni^0/PPh_3 **C** can react with phenylpropionate to give complex **D**. Decarbonylation of **D** next occurred (complex **E**), after which a β-hydride elimination gave complex **F** with the release of ethylene. Finally, a reductive elimination of phenol and capturing carbon monoxide furnished nickel carbonyl complex **G**. This example provided a significant insight where a nickel complex inserts into a seemingly inert C(acyl)–O bond. However, the implementation of this process into a catalytic coupling reaction was not achieved until close to the twenty-first century.

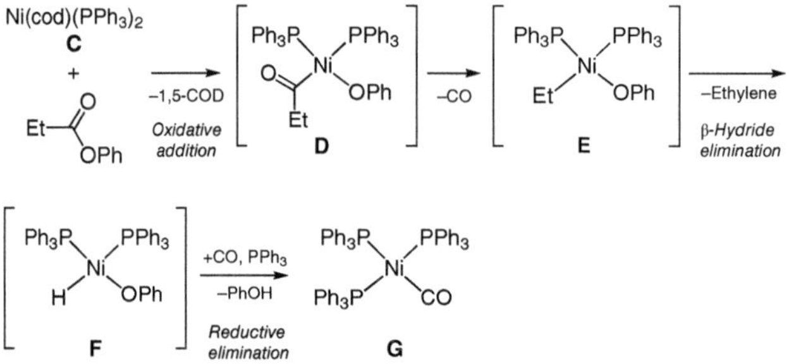

Scheme 15.2 Yamamoto's decarbonylation of phenylpropionate with stoichiometric nickel. Source: Adapted from Ishizu et al. [11].

A chronological progress of decarbonylative transformations of aromatic esters is shown in Figure 15.1. In 1998, de Vries and Stephan developed a Pd-catalyzed decarbonylative Mizoroki–Heck reaction of acid anhydrides with alkenes [12]. This is the first catalytic example of decarbonylative coupling of aromatic "esters." In 2002, Gooßen showed that a similar reaction can be achieved by directly using carboxylic acids through an *in situ* anhydride formation [13]. The first catalytic decarbonylative reaction of aromatic esters was achieved by Murai [14]. After these findings,

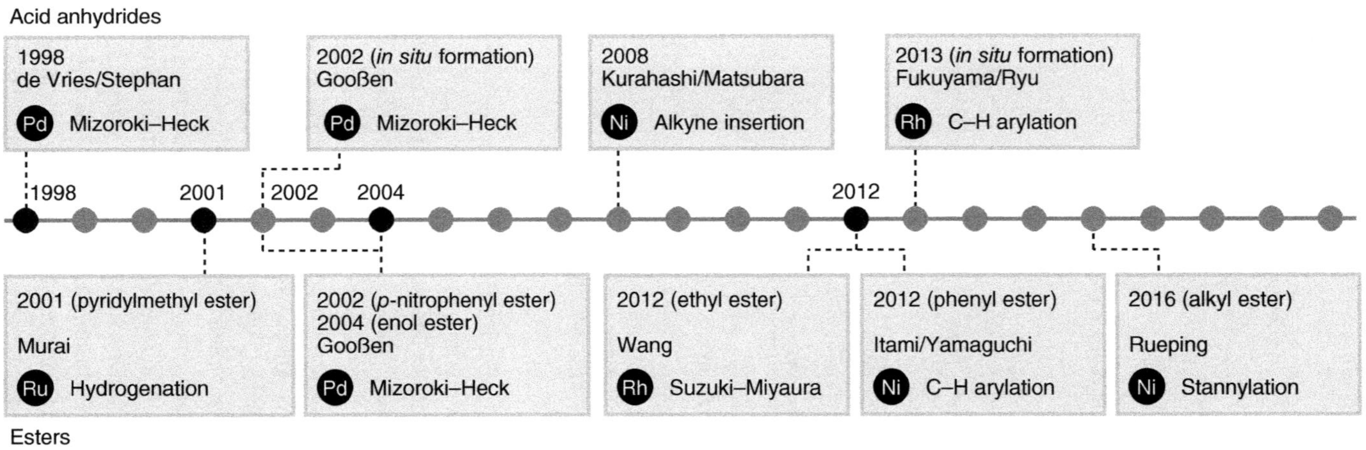

Figure 15.1 A timeline of catalytic decarbonylative couplings of aromatic acid anhydrides and esters (representative examples are given).

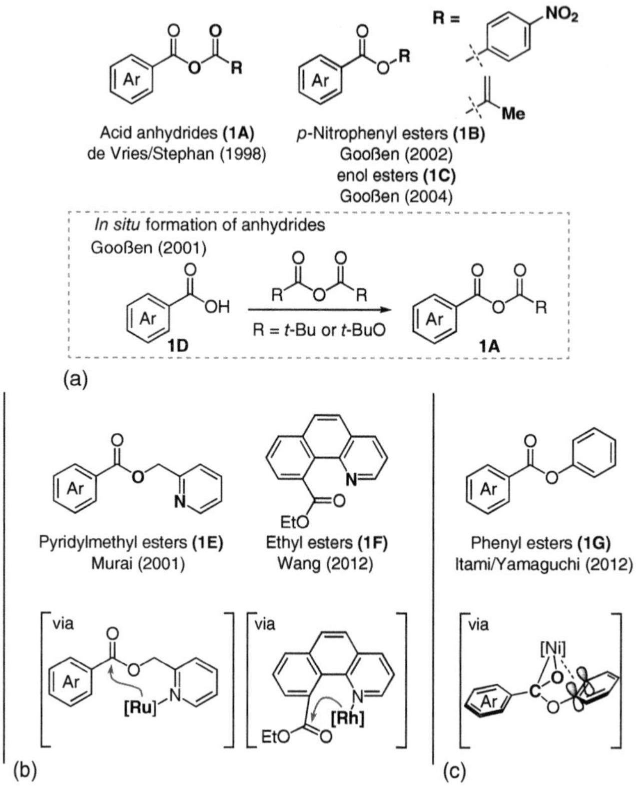

Figure 15.2 A series of esters that can be used in decarbonylative coupling. (a) Driven by strong leaving group. (b) Driven by chelation assistance. (c) Driven by weak chelation.

the decarbonylative coupling of aromatic esters has been extended by using various nucleophiles to form C–C bonds and C–heteroatom bonds.

Generally, the following issues should be taken into account in decarbonylative coupling: (i) the choice of ester (including acid anhydrides), (ii) the driving force to activate C–O bonds by transition metals, and (iii) ligands to be employed for the metal catalysts. Acid anhydrides **1A** were employed in early decarbonylative couplings because they have good leaving ability, forcing the metal catalysts to insert into the aroyl C–O bond (Figure 15.2a) [12]. Similarly, *p*-nitrophenyl- (**1B**) and enol esters (**1C**) can be classified in the same category [15, 16]. Another strategy was shown by Murai's achievement where pyridylmethyl esters **1E** were used (Figure 15.2b) [14]. Driven by chelation assistance of the pyridine moiety, a metal catalyst can undergo the oxidative addition of aroyl C–O bonds. Of note, this strategy enabled the utilization of ethyl esters **1F** as shown in Wang's Rh-catalyzed decarbonylative coupling in 2012 [17]. Phenyl esters **1G**, which have now become rather standard substrates for decarbonylative coupling, can work through a weak chelation of the phenyl ring toward metal catalysts (Figure 15.2c) [18]. Phenyl esters **1G** have been used first in nickel catalysis; however, it was later revealed that this can be

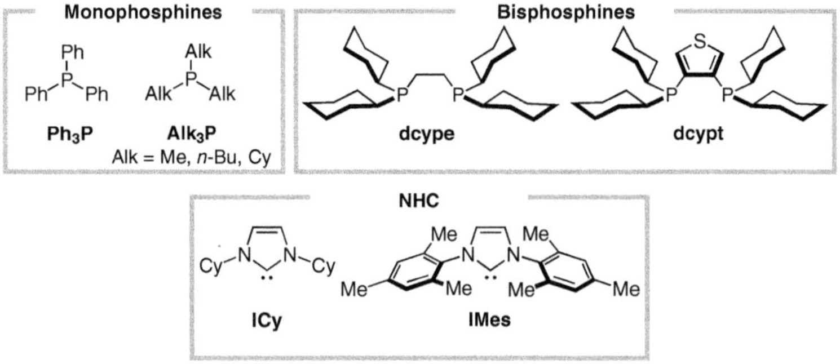

Figure 15.3 Effective ligands in decarbonylative couplings.

applied to palladium catalysis and now extended to a broad range of decarbonylative couplings.

The choice of ligands for catalysts in decarbonylative coupling is also important (Figure 15.3). In earlier developments, simple triphenylphosphine (or isoquinoline) could be used as a ligand because acid anhydrides as well as nitrophenyl ester were reactive enough toward a C–O bond oxidative addition. Later, more electron-rich monodentate phosphines were found to be more suitable for the catalysis [19]. One of the turning points in this area was the discovery of dcype (1,2-bis(dicyclohexylphosphino)ethane), an electron-rich diphosphine, by Yamaguchi and Itami in 2012 [18]. Including structurally similar dcypt (3,4-bis(dicyclohexylphosphino)thiophene), these types of ligands have been recognized to function broadly and efficiently in decarbonylative couplings [20]. Besides phosphines, *N*-heterocyclic carbene ligands such as ICy have been used [21].

Although the overview of this chemistry was briefly documented in this section, the following sections will introduce more details of decarbonylative reactions by categorizing the nucleophiles used in the reactions.

15.3 Decarbonylative Mizoroki–Heck Reaction

The Mizoroki–Heck reaction of haloarenes and alkenes is known as a powerful method to construct aryl–alkene frameworks [22]. A decarbonylative version of this reaction was first realized by de Vries and Stephan (Scheme 15.3) [12]. They used aromatic carboxylic acid anhydrides **1A** as an aryl source with alkenes **2** under

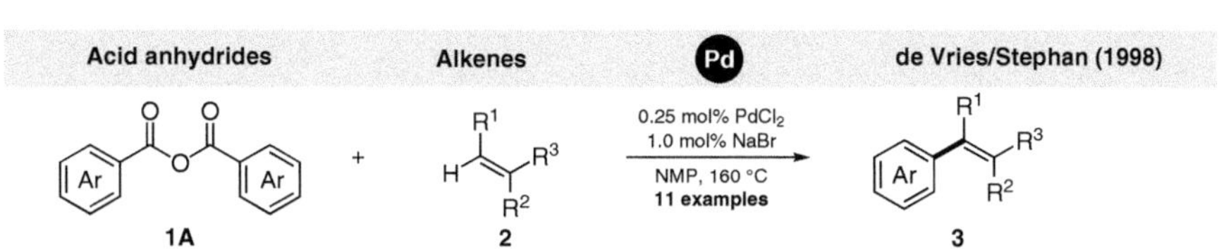

Scheme 15.3 Pd-catalyzed decarbonylative Mizoroki–Heck reaction of benzoic acid anhydrides. Source: Adapted from Stephan et al. [12].

the influence of $PdCl_2$/NaBr catalysis at 160 °C to afford aryl alkenes **3** in good to moderate yields. Of note, this reaction does not require a base (which is commonly required for Mizoroki–Heck reactions), successfully avoiding the generation of salt wastes during the reaction.

An alternative methodology through an *in situ* acid anhydride formation was developed by the Gooßen group in 2002, avoiding the preparation of acid anhydrides **1A** from carboxylic acids before the coupling (Scheme 15.4) [13]. The group established a $PdCl_2$/4-picoline/LiCl catalysis using a stoichiometric amount of Boc_2O as an activator, achieving a decarbonylative Mizoroki–Heck reaction of arenecarboxylic acids **1D** with alkenes. The Chen group reported a similar protocol by using Piv_2O instead of Boc_2O [23].

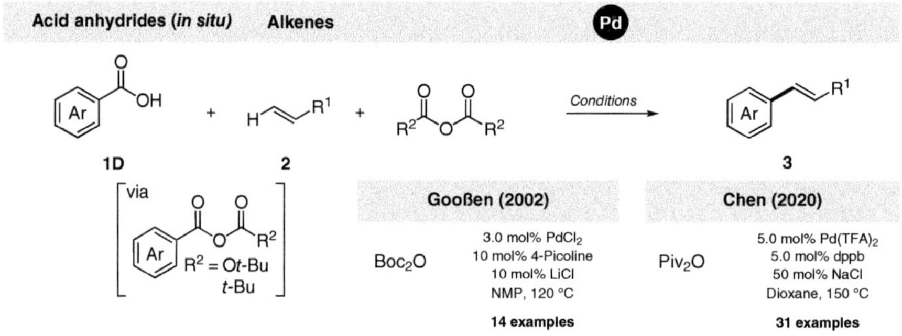

Scheme 15.4 Pd-catalyzed decarbonylative Mizoroki–Heck reaction of benzoic acids **1D** through an *in situ* formation of anhydrides. Source: Adapted from Gooßen et al. [13].

Later, Gooßen reported the first Mizoroki–Heck reaction of aromatic esters **1E** (Scheme 15.5) [15]. Unlike the high reactivity of anhydrides, the C–O bond of esters is generally not reactive toward an oxidative addition to metal centers. To overcome this, the group utilized 4-nitrophenyl esters as an activated ester

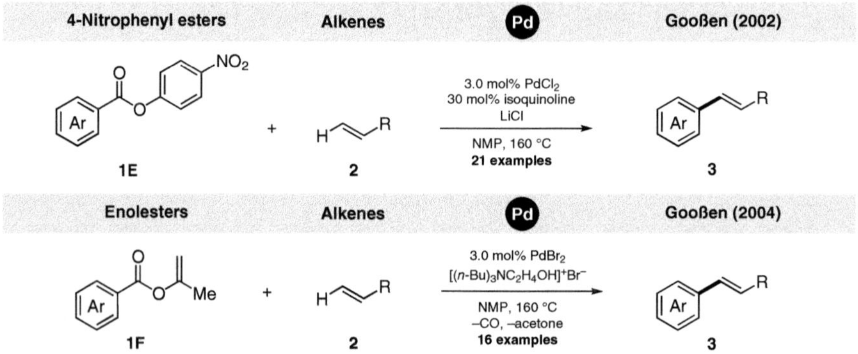

Scheme 15.5 Pd-catalyzed decarbonylative Mizoroki–Heck reaction of aromatic esters. Source: Adapted from Gooßen and Paetzold [15].

for a decarbonylative Mizoroki–Heck reaction through a $PdCl_2$/isoquinoline/LiCl catalysis. The importance of the 4-nitrophenyl moiety was highlighted by the fact that the simple phenyl esters resulted in no reaction. Similarly, enol esters were also reported to be usable as activated esters, with the added advantage of generating only volatile coproducts (acetone and CO) [16].

15.4 Decarbonylative Alkyne Insertions

Metal-catalyzed alkyne insertions have been often utilized to synthesize not only alkenes but also ring structures [24]. In this regard, Matsubara and Kurahashi succeeded in the development of the decarbonylative alkyne insertion of phthalic anhydrides **1H** (Scheme 15.6) [25]. $Ni(cod)_2/PMe_3$ catalyst in the presence of $ZnCl_2$ efficiently promoted this reaction under the relatively mild conditions to generate isocoumarins **5**. The reaction generally uses symmetric alkynes, but unsymmetric silyl acetylenes can also react with high regioselectivity. The reaction proceeds through the following mechanism. The oxidative addition of C(aroyl)–O bonds of **1H** to an Ni(0) species, followed by a decarbonylation, gives a five-membered nickelacycle. To this nickel species, alkyne **4** undergoes an insertion reaction. Finally, reductive elimination with the support of $ZnCl_2$ releases isocoumarin product **5** with the regeneration of the active nickel species.

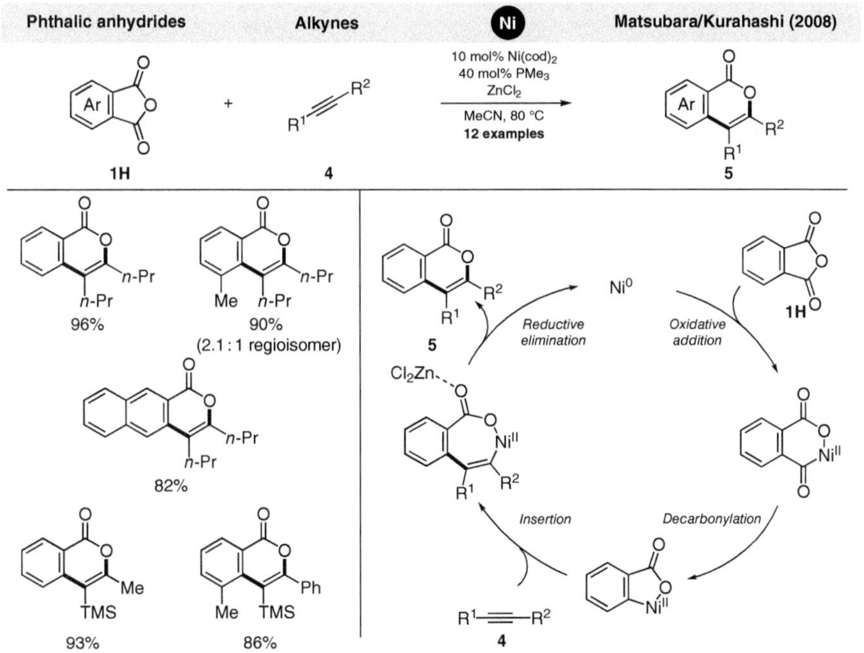

Scheme 15.6 Ni-catalyzed decarbonylative alkyne insertion of phthalic anhydrides **1H**. Source: Adapted from Kajita et al. [25].

15.5 Suzuki–Miyaura Coupling

Suzuki–Miyaura coupling is undoubtedly one of the most powerful C–C bond formations, taking place between aryl halides and boronic acids with a palladium catalyst [1]. To render Suzuki–Miyaura couplings cleaner as a synthetic method, studies to replace aryl halides with other aryl electrophiles have been conducted all over the world. Suzuki–Miyaura-type decarbonylative couplings first appeared in 2004 (Scheme 15.7) [26]. Gooßen found that a catalytic amount of $[Rh(ethylene)Cl]_2$ in the presence of KF realizes the decarbonylative coupling between acid anhydrides **1A** with arylboroxins **7**, giving biaryls **6**. In 2020, Zhu found that an Ni/PCy_3 catalyst in the presence of Cs_2CO_3 efficiently works for the decarbonylative coupling of **1A** with arylboronic acids **8** [27]. Similar to the decarbonylative Mizoroki–Heck reaction, an *in situ* anhydride formation protocol was also amenable to Suzuki–Miyaura couplings. In 2019, Szostak reported that a Pd/dppb-catalyzed decarbonylative coupling of carboxylic acids **1D** with arylboronic acids **8** in the presence of pivalic anhydride acts as an activating reagent, giving biaryls **6** [28].

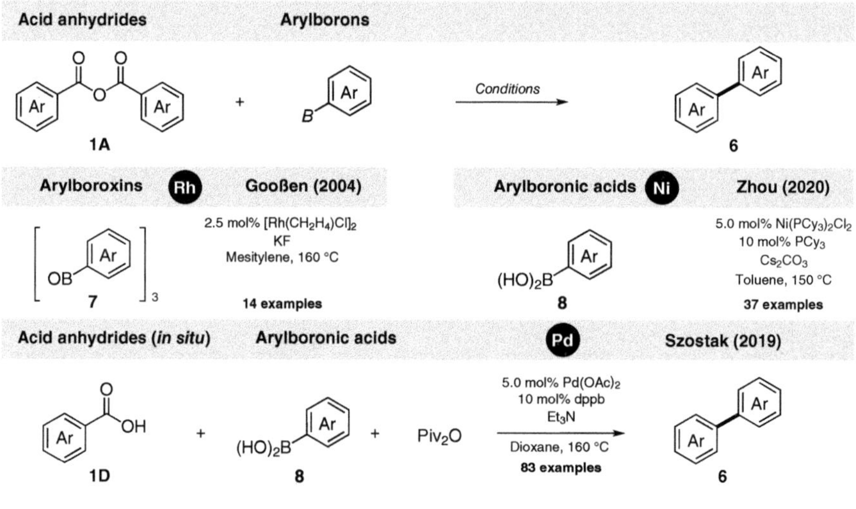

Scheme 15.7 Decarbonylative Suzuki–Miyaura coupling of arenecarboxylic acid anhydrides. Source: Adapted from Gooßen and Paetzold [26].

The ester version of Suzuki–Miyaura coupling was disclosed by Zhao and Wang in 2012 (Scheme 15.8). They reported a decarbonylative cross-coupling of ethyl benzo[*h*]quinoline-10-carboxylates (**1F**) with arylboronic acids **8** in the presence of a rhodium catalyst and CuCl [17]. The key for this reaction was the utilization of chelation-assisted oxidative addition of esters **1F**. The group performed computational studies to gain insight on the reaction mechanism, concluding that a direct C–C oxidative addition of the ester would be the preferred pathway. Although only one example was shown, Chatani showed that an oxazoline can

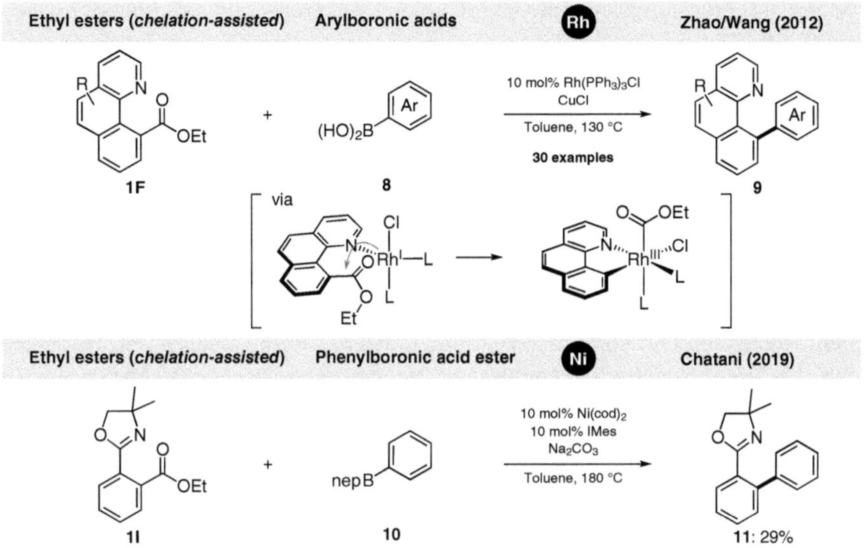

Scheme 15.8 Chelation-assisted decarbonylative Suzuki–Miyaura coupling of ethyl esters.

also function as a directing group. By using Ni/IMes as a catalyst, aromatic ester **1I** can undergo decarbonylative Suzuki–Miyaura coupling with phenylboronic acid ester **10** [29].

A more general decarbonylative coupling of aromatic esters was later reported by the group of Itami/Yamaguchi (Scheme 15.9) [19]. They developed that Ni-catalyzed decarbonylative coupling of aromatic phenyl esters **1G** with arylboronic acids **8**.

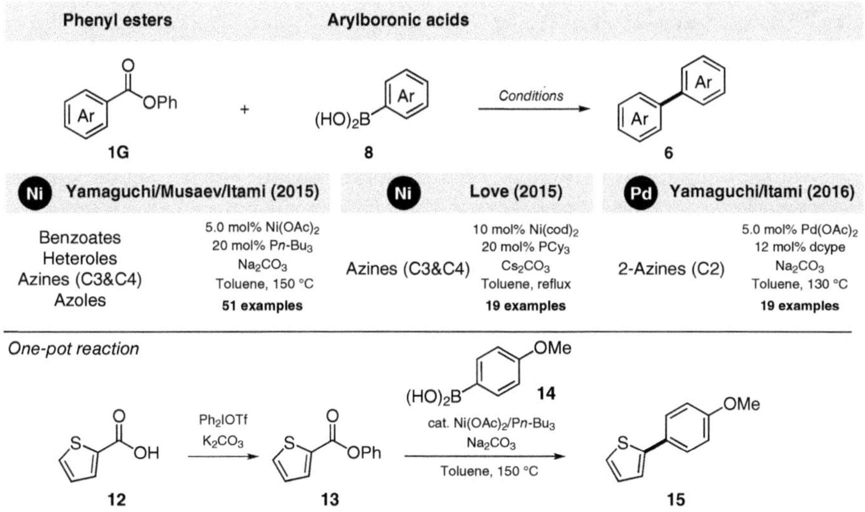

Scheme 15.9 Decarbonylative Suzuki–Miyaura coupling of aromatic phenyl esters. Source: Adapted from Muto et al. [19].

The key of their findings was the use of a highly electron-donating monophosphine ligand, PBu_3, and phenyl esters as substrates. A broad range of aromatic esters including heteroaromatic esters and pharmaceutical compounds can couple with arylboronic acids **8**, furnishing the corresponding biaryls **6** in generally good yields. Starting from carboxylic acid **12**, a one-pot protocol was also established. After the formation of phenyl ester **13** using Ph_2IOTf and K_2CO_3, the nickel catalyst and Na_2CO_3 were added to the same flask, giving heterobiaryl **15**. On the contrary to this broad applicability, this Ni/PBu_3 method did not allow for the use of phenyl 2-azinecarboxylates. A similarly low reactivity of 2-azinecarboxylates under the nickel catalysis was also observed in a report from the Love group in 2015 [30]. Then, Yamaguchi and Itami in 2016 found that a palladium catalyst (Pd/dcype) can solve this issue, enabling the coupling of 2-azinecarboxylates with boronic acids to form biaryls [31].

The Pd/dcype catalysis was effective only for 2-azinecarboxylates. By utilizing this unique character, the Itami/Yamaguchi group achieved a catalyst-enabled site-selective decarbonylative coupling of 2,4-pyridinedicarboxylate **16** (Scheme 15.10) [31]. By using arylboronic acid **14**, the Pd/dcype catalysis arylated only at the C2 position of **16**, giving **17** in 61% yield. Subsequently, **17** was subjected to the Ni/PBu_3 catalyst conditions with arylboroxine **18** to give triaryl **19** in 65% yield.

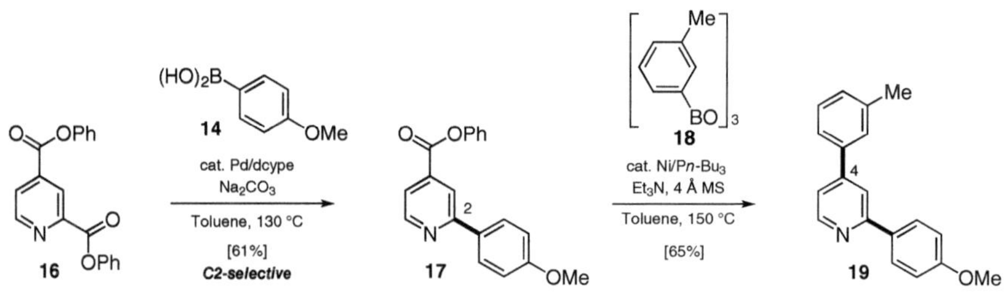

Scheme 15.10 Catalyst-enabled site-selective double decarbonylative coupling of **16**. Source: Adapted from Muto et al. [31].

Alkylative coupling was implemented in decarbonylative reactions by Rueping and coworkers in 2018. They reported an Ni/dcype-catalyzed decarbonylative Suzuki–Miyaura coupling of aromatic phenyl esters **1G** with alkyl–BBN **21** (BBN = 9-borabicyclo[3.3.1]nonane) as an alkylating reagent (Scheme 15.11) [32]. Functional group tolerance including ketones and alkyl esters was noteworthy. They also found that BEt_3 (**22**) is also an applicable nucleophile for ethylation. Soon thereafter, the Newman group reported a Pd-catalyzed decarbonylative coupling between aromatic esters **1G** and **21** [33]. Newman demonstrated that the reaction is applicable to a diversification of pharmaceutical-derived molecule repaglinides **23** with reactive functional groups such as amines and secondary amides. However, unfortunately, these reactions did not allow for the use of secondary and tertiary alkylboranes.

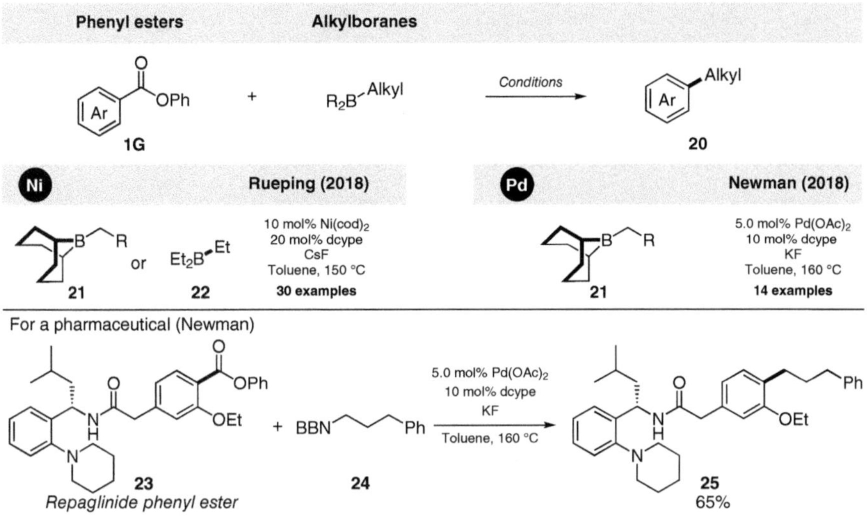

Scheme 15.11 Decarbonylative Suzuki–Miyaura-type alkylation of aromatic esters. Source: Adapted from Chatupheeraphat et al. [32].

15.6 Negishi Coupling

Rueping reported a Negishi-type decarbonylative coupling of aromatic esters **1G** with alkylzinc reagents **26** through Ni/dcype catalysis (Scheme 15.12) [34]. The reaction tolerated sensitive functional groups such as amides, ketones, and heteroarenes. Moreover, the introduction of secondary alkyl group was possible.

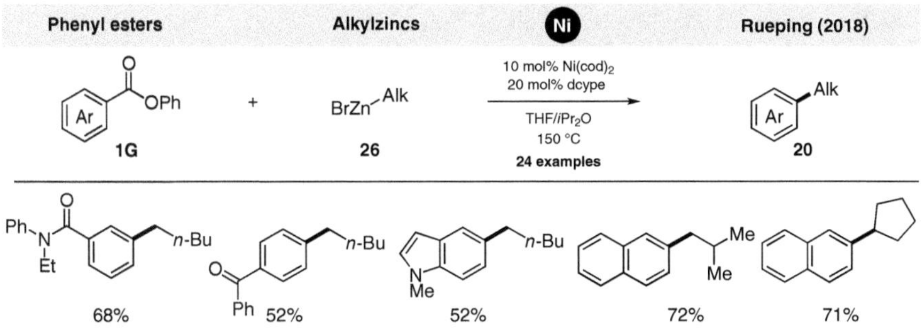

Scheme 15.12 Ni/dcype-catalyzed decarbonylative Negishi coupling of aromatic esters **1G** with alkylzincs **26**. Source: Adapted from Liu et al. [34].

Yamaguchi also developed a different Negishi-type decarbonylative coupling of aromatic esters **1G**, where they used dimethylaluminum chloride (**27**) as a methylating reagent (Scheme 15.13) [35]. The reaction proceeded under Ni/dcypt or Ni/dppp catalyst, giving methylarenes **28** in good to moderate yields. Interestingly,

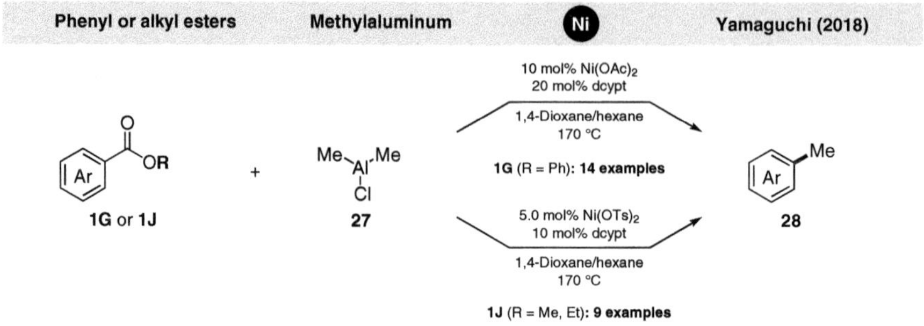

Scheme 15.13 Ni-catalyzed decarbonylative methylation of aromatic esters **1G** or **1J** with dimethylaluminum chloride. Source: Adapted from Okita et al. [35].

the use of **27** was the key for this reaction, as other methyl aluminum reagents such as trimethylaluminum and methylaluminum dichloride resulted in low yields. Moreover, it was revealed that a similar catalytic system allows for less reactive aromatic alkylesters **1J** to react, giving **28** in good yields. It was postulated that the Lewis acidity of the aluminum reagent enhanced the reactivity of the alkylesters.

15.7 Sonogashira Coupling

Pd/Cu-catalyzed Sonogashira coupling of haloarenes with terminal alkynes is a useful catalytic method to construct arylalkynes and is often utilized to synthesize organic materials [36]. Sonogashira type decarbonylative coupling was established by Yamaguchi/Itami in 2017 (Scheme 15.14) [37]. Cooperative Pd/dcypt and copper catalysis was effective for the reaction of phenyl esters **1G** and silylacetylene **29** to give a range of arylalkynes in moderate to good yields. The copper species likely functions to catalytically generate a reactive copper acetylide, which is also

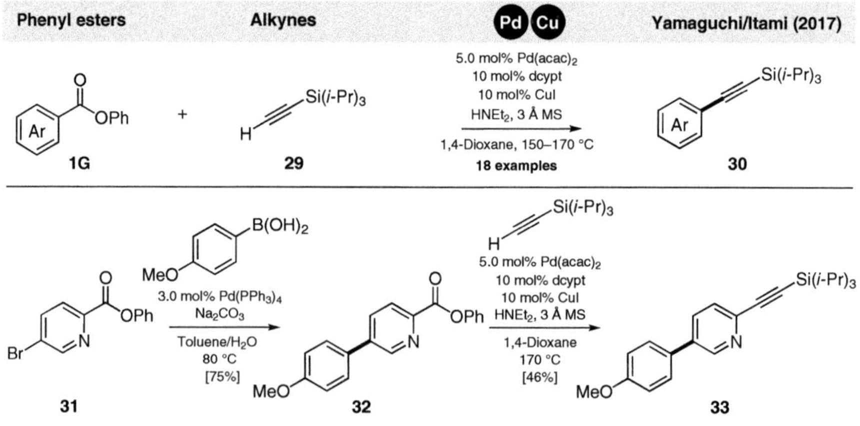

Scheme 15.14 Pd/dcypt-catalyzed decarbonylative Sonogashira coupling of aromatic esters **1G** with terminal alkynes **29**. Source: Adapted from Okita et al. [37].

observed in halogen-based Sonogashira couplings. The utility of this decarbonylative Sonogashira coupling was demonstrated by the sequential coupling of bromo aromatic ester **31**. Suzuki–Miyaura coupling of **31** furnished biaryl **32**, keeping the phenyl ester totally intact. Subsequently, Pd/dcypt-catalyzed decarbonylative alkynylation of **32** successfully gave arylalkyne **33**.

15.8 α-Arylation

α-Aryl carbonyls are a fundamental framework as both bioactive molecules and synthetic intermediates. Metal-catalyzed α-arylation is one of the most efficient methodologies to access to such compounds [38]. In order to make use of aromatic esters in α-arylation, competitive Claisen condensation should be avoided. To date, only one example of decarbonylative α-arylation was achieved (Scheme 15.15). Yamaguchi reported a Pd/dcypt-catalyzed α-arylation of aromatic esters **34** with ketones **35**, albeit limited to only phenyl 4-quinolinecarboxylates [39].

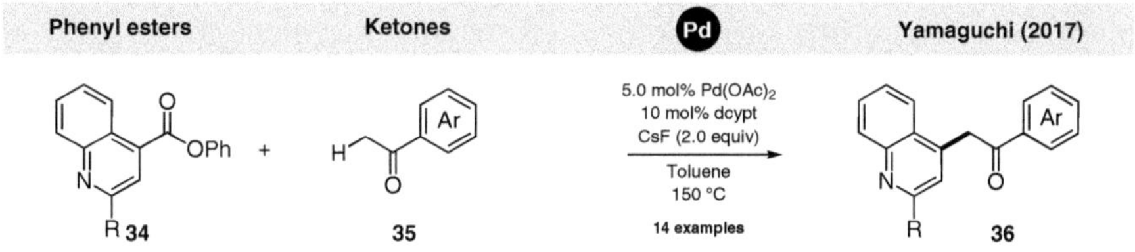

Scheme 15.15 Pd/dcypt-catalyzed decarbonylative α-arylation of aromatic esters with ketones.

15.9 Cyanation

Aryl nitriles are important aromatic molecules that are widely seen in pharmaceutical molecules. Metal-catalyzed cyanations of haloarenes have often been employed to access to aryl nitriles [40]. For decarbonylative cyanation of aromatic esters, Ni/dcype and Ni/dcypt were found to be effective (Scheme 15.16). The first example

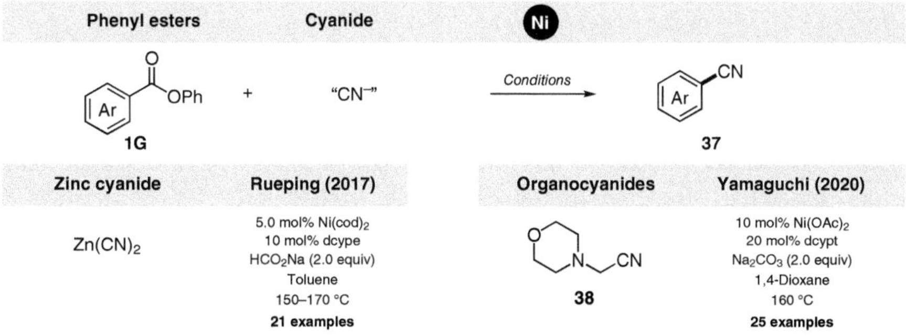

Scheme 15.16 Ni-catalyzed decarbonylative cyanation of aromatic esters **1G**.

was from the group of Rueping, who developed Ni/dcype-catalyzed cyanation of aromatic esters **1G** with $Zn(CN)_2$ [41]. Yamaguchi made this reaction cleaner by using organocyanides (**38**) as a nonmetal cyanating reagent under the influence of Ni/dcypt catalyst, successfully avoiding the production of metal salts during the reaction [42].

15.10 C–H Arylation

C–H functionalization of arenes has been broadly recognized as a powerful method to realize streamlined organic synthesis, which includes a short-step synthesis and late-stage functionalization [43]. In 2009, Yu and coworkers developed an Rh-catalyzed decarbonylative C–H arylation of benzo[*h*]quinolines or pyridin-2-ylbenzenes **39** with acid anhydrides **1A** (Scheme 15.17) [44]. One of the key aspects for this C–H arylation was the utilization of a nitrogen-chelation assisted C–H cleavage. In 2013, Shi and coworkers developed an Rh-catalyzed decarbonylative C–H arylation of arenes **41** using benzoic acids through an *in situ* anhydride formation [45]. In Shi's conditions, 2-pyridyl and imine groups can work as a directing group for the C–H bond cleavage.

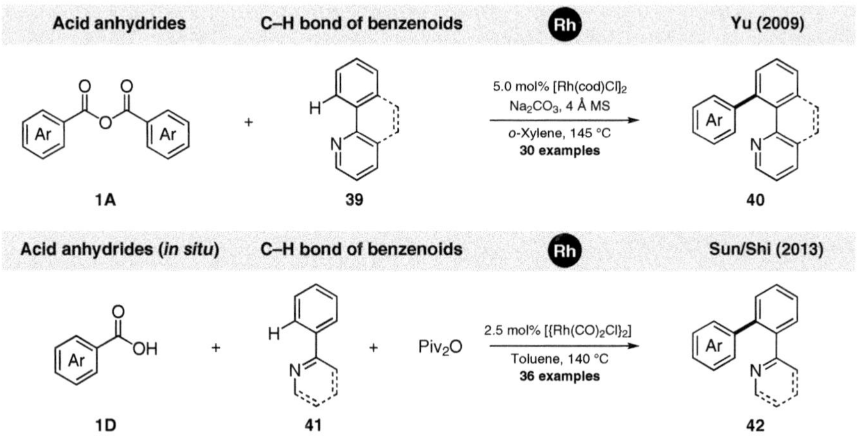

Scheme 15.17 Rh-catalyzed decarbonylative C–H arylation using acid anhydrides. Source: Adapted from Jin et al. [44].

Itami and Yamaguchi in 2012 developed an Ni-catalyzed C–H arylation of aromatic esters **1G** with azoles **43** (Scheme 15.18) [18]. A broad range of aromatic esters, particularly heteroaromatic esters, can couple with 1,3-azoles such as benzoxazoles, oxazoles, and thiazoles, delivering the corresponding heterobiaryls **44** in good yields. A similar transformation was achieved by Gade, who found that their original nickel catalyst **45** with a diphosphine ligand was effective [46]. Later, the Yamaguchi group reported that Pd/dcypt can also catalyze the C–H arylation of azoles with aromatic esters, successfully expanding the substrate generality involving an azole counterpart [47]. This palladium catalysis allowed for benzothiazoles and benzimidazoles to couple with aromatic esters.

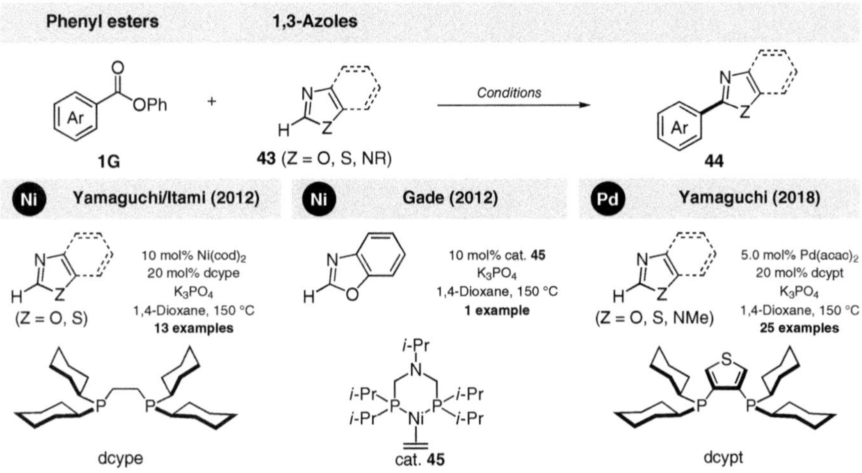

Scheme 15.18 Decarbonylative C–H arylation of aromatic esters **1G** with 1,3-azoles **43**. Source: Adapted from Amaike et al. [18].

It should be noted here that this type of transformation was also translated to the C–H alkenylation of α,β-unsaturated esters by Ni/dcype catalysis [48]. This C–H alkenylation has been applied to a *de novo* synthesis of a naturally occurring molecule.

The utility of the decarbonylative C–H arylation was demonstrated by the formal synthesis of muscoride A (**50**), a natural product with antibacterial activity (Scheme 15.19) [49]. They constructed a bisoxazole framework by an Ni-catalyzed decarbonylative C–H arylation between **46** and **47**. Accessing Wipf's synthetic intermediate **49** [50], they accomplished the formal synthesis of **50** [18].

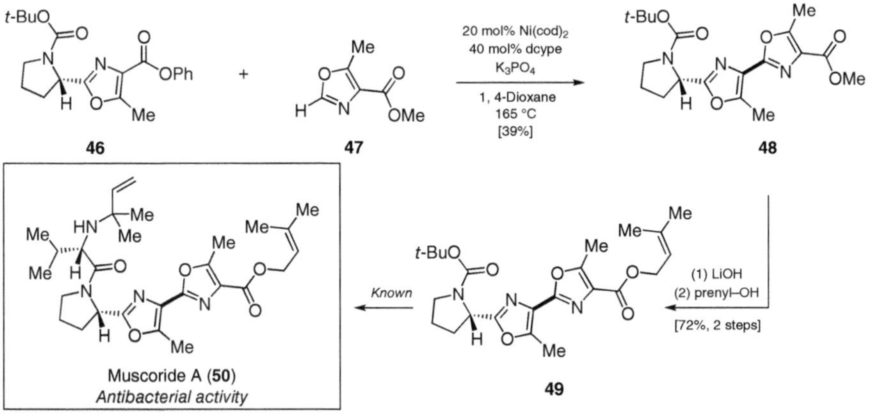

Scheme 15.19 Formal synthesis of muscoride A (**50**) by Ni/dcype-catalyzed decarbonylative C–H arylation. Source: Adapted from Nagatsu et al. [49].

Furthermore, the Yamaguchi group extended their developed methodology to the synthesis of triarylpyridines **55** through double decarbonylative couplings and a ring transformation strategy (Scheme 15.20a) [53]. Starting from 4-oxazole phenyl

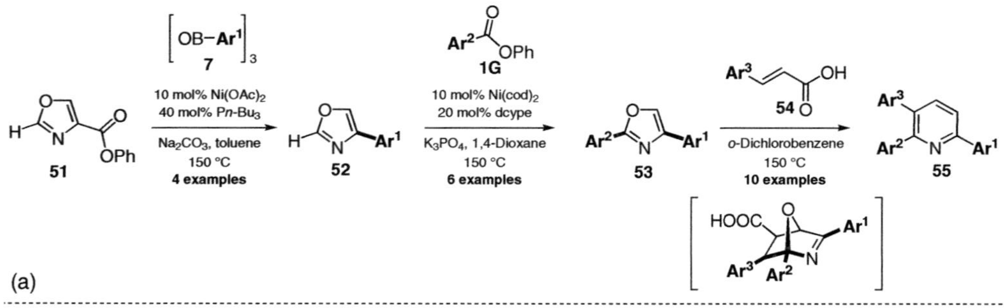

Scheme 15.20 (a) Synthesis of triarylpyridines **55** by a decarbonylative C–H coupling/[4+2] cycloaddition strategy. Source: Adapted from Wipf and Venkatraman [50]. (b) Formal synthesis of GE2270s **63**. Source: Adapted from Selva et al. [51] and Nicolaou et al. [52].

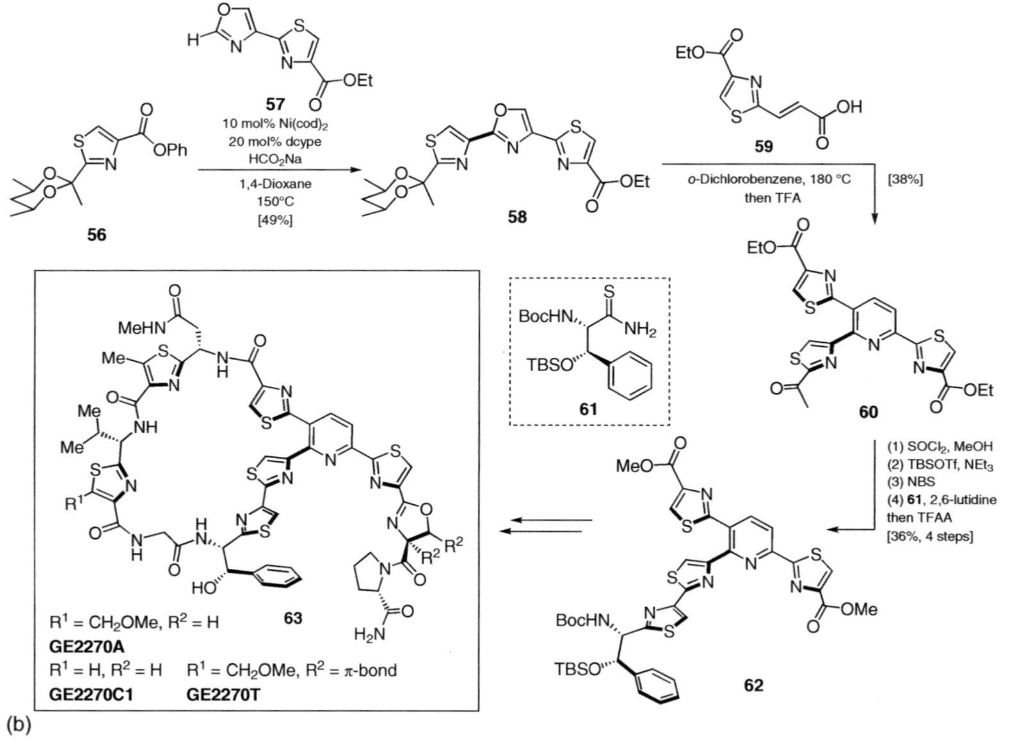

Scheme 15.20 (*Continued*)

ester **51**, sequential decarbonylative Suzuki–Miyaura coupling/C–H arylation gave 2,4-diaryloxazoles **53**. Subsequently, [4+2] cycloaddition of **53** with cinnamic acids **54** led to 2,3,6-triarylpyridines **55**. By applying this synthetic strategy, they accomplished the formal synthesis of GE2270s **63**, which are a family of thiopeptide antibiotics (Scheme 15.20b) [51, 52]. The synthesis of GE2270s **63** was initiated by the decarbonylative C–H arylation of oxazole **57** with thiazole phenyl ester **56**, furnishing dithiazolyl oxazole **58** in a moderate yield. Then, [4 + 2] cycloaddition of **58** with **59** allowed for the rapid access of a key framework of GE2270s, trithiazolylpyridine **60**. By synthesizing Nicolaou's intermediate **62** [52] from **60** through four steps, the formal synthesis of GE2270s **63** was accomplished.

Metal-catalyzed intramolecular C–H arylation of diarylethers has often been utilized to construct dibenzofuran motifs (Scheme 15.21) [54]. Among these, diarylethers bearing a leaving group such as (pseudo)halogen are one of the representative substrates for C–H arylation. Decarbonylative coupling was first applied to these C–H arylations by Ryu and Fukuyama. They developed an Rh-catalyzed decarbonylative intramolecular C–H arylation of a diarylether bearing a carboxylic acid **64** through an *in situ* anhydride formation [55]. In 2018, Yamaguchi and coworkers reported that a Pd-catalyzed intramolecular decarbonylative C–H arylation of phenyl esters **66** leads to dibenzofurans **65** [56].

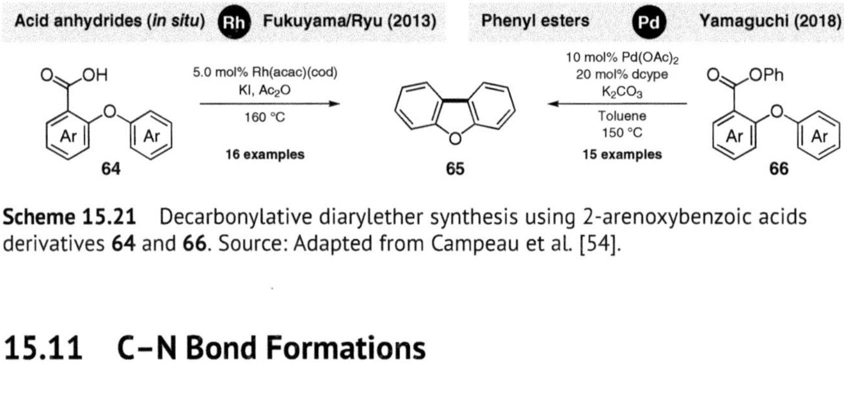

Scheme 15.21 Decarbonylative diarylether synthesis using 2-arenoxybenzoic acids derivatives **64** and **66**. Source: Adapted from Campeau et al. [54].

15.11 C–N Bond Formations

For arylamine synthesis, benzoic acid derivatives are often used as a starting material, and Schmidt or Curtius rearrangement reaction is often employed. Decarbonylative amination of aromatic esters would enable a new arylamine synthesis from benzoic acid derivatives. However, an undesired amide formation is a significant issue to be overcome. By using imine **67** as a nitrogen nucleophile, Rueping developed an Ni/dcypt-catalyzed decarbonylative amination of aromatic esters **1G** (Scheme 15.22) [57]. After this coupling, acidic hydrolysis gave arylamines **68** in generally good yields. In addition to benzoates, heteroaromatic esters such as azines and heteroles were applicable. They emphasized the importance of imine nucleophiles by a control experiment using morpholine instead of imine **67**, resulting in only the corresponding morpholino amide.

As another approach, Sanford and coworkers utilized silylamines as a nitrogen nucleophile both to suppress undesired amide formation and to accelerate

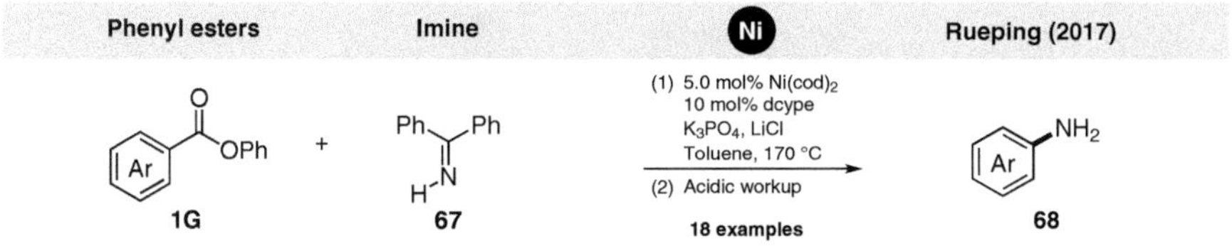

Scheme 15.22 Ni/dcype-catalyzed decarbonylative aniline synthesis of aromatic esters **1G**. Source: Adapted from Yue et al. [57].

transmetalation (Scheme 15.23) [58]. They found that Ni/dcype catalyst enables the decarbonylative coupling between aromatic esters **1G** and TMS-morpholine **69** to give morpholinoarenes **70** in good yields. An alternative protocol was also established, where a silyl transfer reagent (*N*-methyl-*N*-(trimethylsilyl)trifluoroacetamide: MSTFA) was used, introducing various amines **71** ranging from secondary amines to primary amines onto aromatic cores.

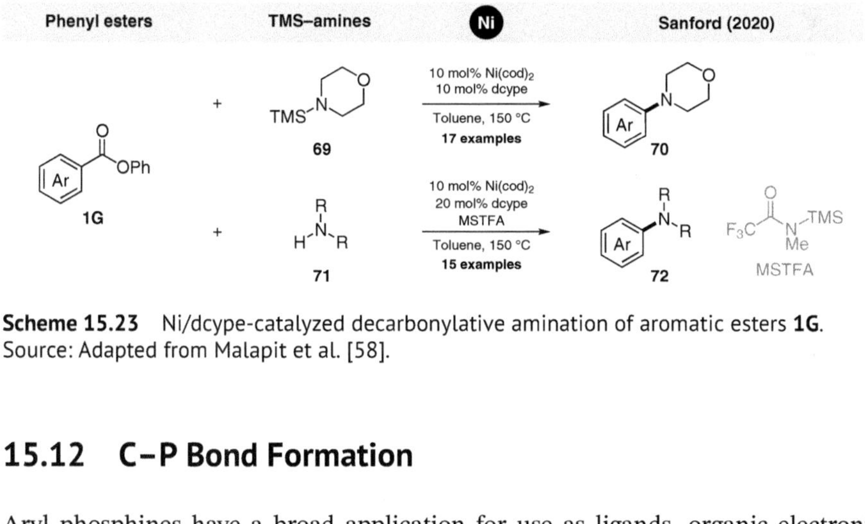

Scheme 15.23 Ni/dcype-catalyzed decarbonylative amination of aromatic esters **1G**. Source: Adapted from Malapit et al. [58].

15.12 C–P Bond Formation

Aryl phosphines have a broad application for use as ligands, organic electronic molecules, and bioactive molecules. Pd-catalyzed Hirao coupling of haloarenes with phosphorus nucleophiles has been recognized as one of the standard synthetic methods of aryl phosphines [59]. Yamaguchi developed an Ni/dcypt-catalyzed decarbonylative phosphination of aromatic esters **1G** with diphenylphosphine oxide (**73**) (Scheme 15.24) [60]. The use of dcypt was dramatically important, since other ligands resulted in low yields of products. Under the slightly modified conditions using NaF, phosphonates **75** participated in this decarbonylative coupling to give arylphosphonates **76**.

An *in situ* anhydride formation strategy was made possible for decarbonylative phosphonation. Szostak and Hong reported a Pd/dppf-catalyzed decarbonylative phosphonation of carboxylic acids **1D** with phosphonates **75** in the presence of pivalic anhydride (Scheme 15.25) [61]. The Han group also established a similar protocol using Boc_2O [62].

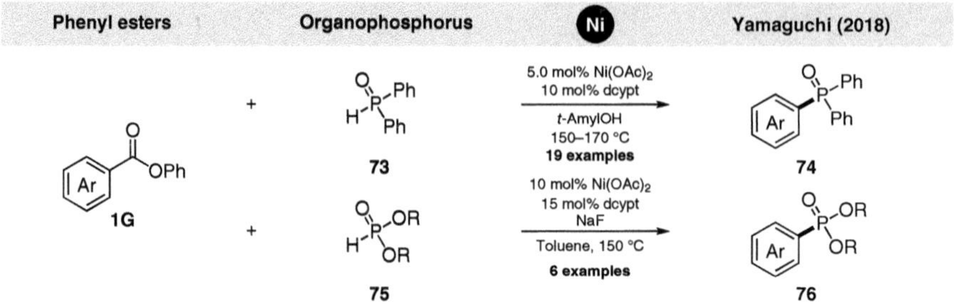

Scheme 15.24 Ni/dcypt-catalyzed decarbonylative C–P bond formation of aromatic esters **1G**. Source: Adapted from Isshiki et al. [60].

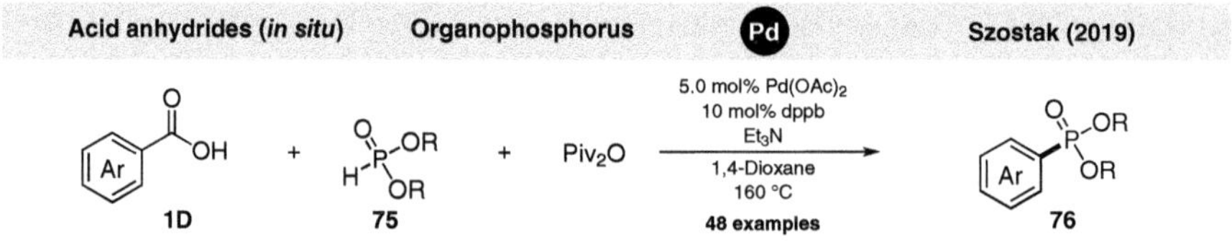

Scheme 15.25 Pd-catalyzed decarbonylative phosphonation of carboxylic acids **1D** through an *in situ* anhydride formation strategy. Source: Adapted from Liu et al. [61].

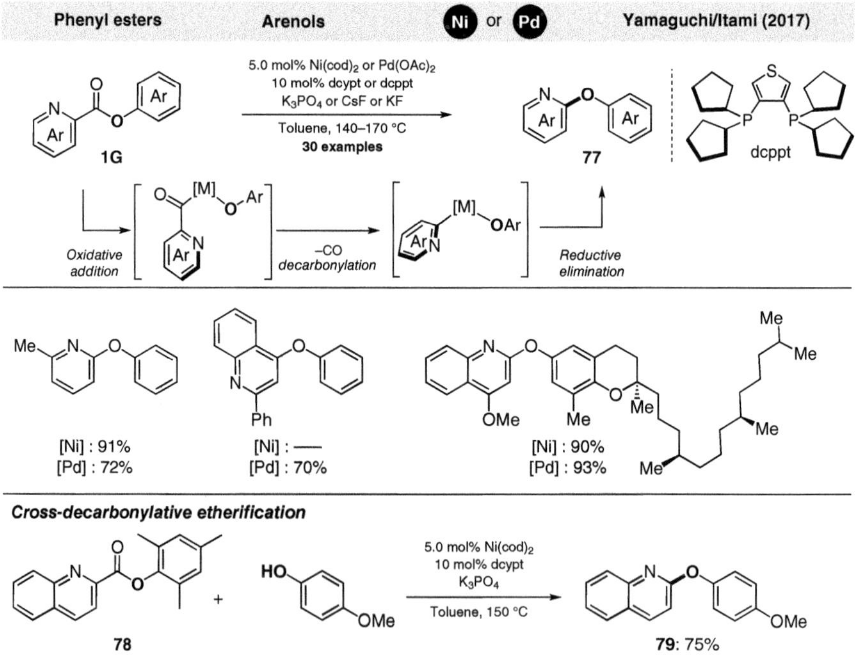

Scheme 15.26 Pd- or Ni-catalyzed decarbonylative diarylether synthesis of aromatic aryl esters. Source: Adapted from Takise et al. [20].

15.13 Etherification

The Yamaguchi/Itami group postulated that decarbonylative diarylether synthesis could be realized by treating aromatic esters **1G** with an appropriate metal catalyst in the absence of external nucleophilic counterparts. To this end, they found that nickel or palladium with dcypt or 3,4-bis(dicyclopentylphosphino)thiophene (dcppt) allows for the decarbonylative synthesis of 2- or 4-arenoxyazines **77** (Scheme 15.26) [20]. By employing bulky aromatic ester **78**, cross-decarbonylative etherification proceeded to give diarylether **79** in a good yield. One caution regarding this reaction is that it occurs only on 2-azinecarboxylates.

15.14 Thioetherification

Aryl sulfides are widespread in bioactive molecules, polymers, and organic electronic molecules. Rueping succeeded in developing decarbonylative thioetherification of aromatic esters **1G** with thiols **80** by an Ni/dppp catalyst and Mg (Scheme 15.27) [63]. The Mg salt was proposed to function as a reducing reagent to generate an active Ni(0) species from the Ni(II) precursor. This reaction allows to the installation of both aryl sulfides and alkyl sulfides on aromatic cores. This reaction likely proceeds through the formation of aromatic thioesters **82** from esters and thiols. Thereafter, an oxidative addition of the C–S bonds of **83**, a decarbonylation, and then a reductive elimination takes place to give aryl sulfide **81** [64].

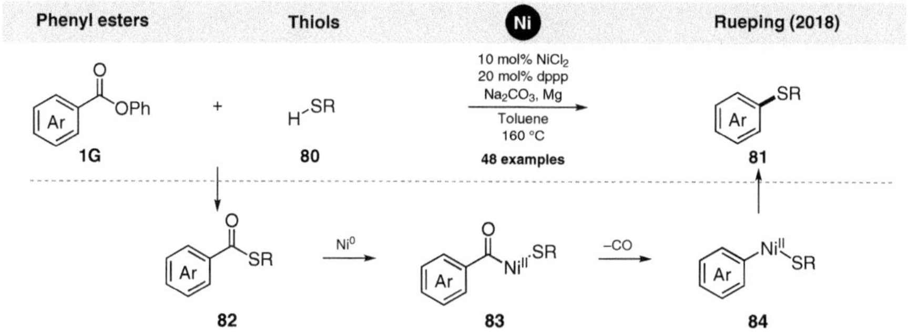

Scheme 15.27 Ni-catalyzed decarbonylative thioetherification of aromatic esters **1G** with thiols **80**. Source: Adapted from Lee and Liao [63].

15.15 Hydrogenation

Hydrogenation of esters through metal hydride catalysis often gives aldehydes and alcohols. In contrast to these reactions, decarbonylative hydrogenation can be a complementary methodology, removing the ester functionality entirely to give arenes. This type of reaction was first realized by the Murai group (Scheme 15.28) [14].

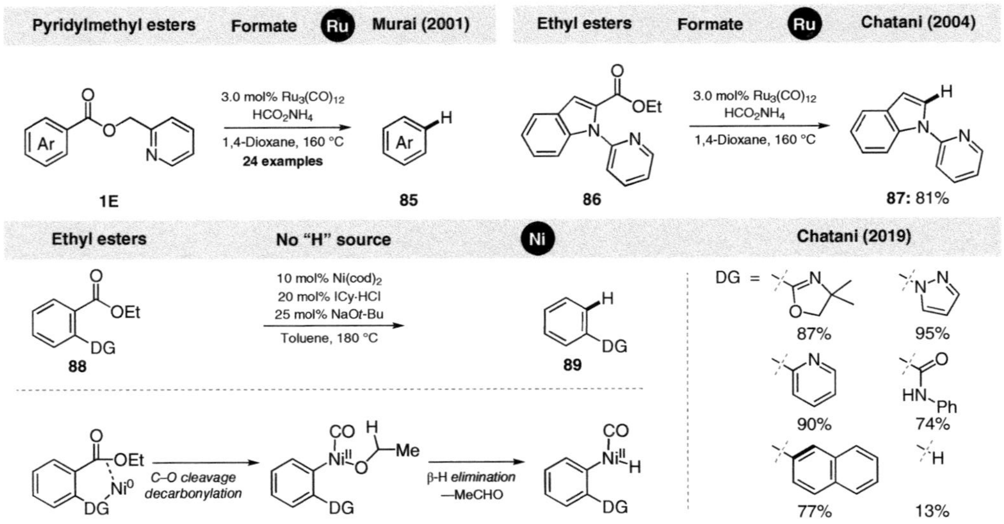

Scheme 15.28 Decarbonylative hydrogenation of aromatic esters through a chelation-assisted strategy. Source: Adapted from Chatani et al. [14].

They achieved an Ru-catalyzed decarbonylative hydrogenation of aromatic pyridylmethyl esters **1E** by using ammonium formate. It was necessary for this reaction to employ pyridylmethyl ester, which helps the C(aroyl)–O bond oxidative addition by chelation of the nitrogen atom to the metal. Later, Chatani reported that triphenylsilane can be used as an alternative hydride source [65].

Applying a similar chelation-assisted system, the hydrogenation of aromatic ethyl esters was achieved by Chatani and coworkers. Under the action of $Ru_3(CO)_{12}$ catalyst and ammonium formate in dioxane at 160 °C, Chatani showed that ethyl 1-(pyridin-2-yl)-1*H*-indole-2-carboxylate (**86**) can be hydrogenated to give 1-*N*-pyridylindole **87** [65]. In 2019, Chatani reported a similar but superior protocol using Ni/ICy catalyst [29]. This nickel catalytic system does not require external hydrogen sources, enabling hydrogen transfer of ethyl esters **88** from the ethoxide moiety through a β-hydrogen elimination. Various directing groups such as nitrogen heterocycles, amides, and naphthyl can assist this decarbonylative hydrogenation.

Hydrogenation of aromatic esters without chelation assistance was developed by Rueping in 2017 (Scheme 15.29). They reported an Ni/dcype-catalyzed decarbonylative hydrogenation of phenyl esters **1G** with polymethylhydrosiloxane (PMHS) to provide arenes **85** [66]. This reaction has high functional group tolerance, since olefins, ketones, and acetals were compatible with the reaction conditions.

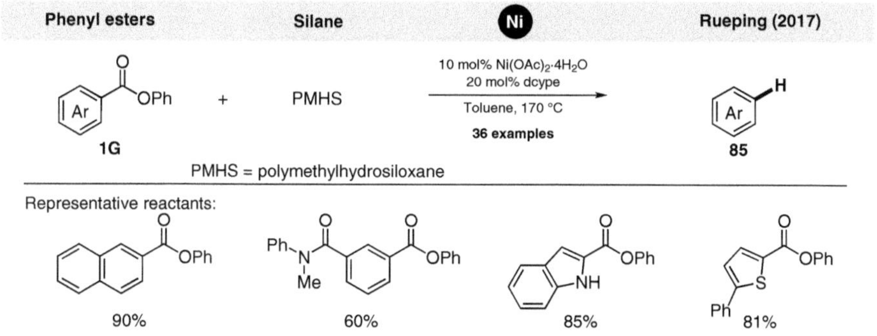

Scheme 15.29 Ni-catalyzed decarbonylative hydrogenation of aromatic esters.

15.16 Borylation, Silylation, and Stannylation

From the viewpoint of the ease of functionalization, boryl-, silyl-, and stannylarenes have been employed as versatile building blocks in organic synthesis. To date, tremendous efforts have been devoted to developing efficient synthetic methods toward these molecules [67]; however, only a few reports of decarbonylative methods of arenecarboxylic acid derivatives were known until recently [68]. In 2016, the Shi group reported the first decarbonylative borylation of aromatic esters by an Ni catalyst (Scheme 15.30) [21]. They found that Ni/ICy catalyst engages aromatic esters **1G** with diboron (B_2nep_2) in the presence of Na_2CO_3 and NaCl,

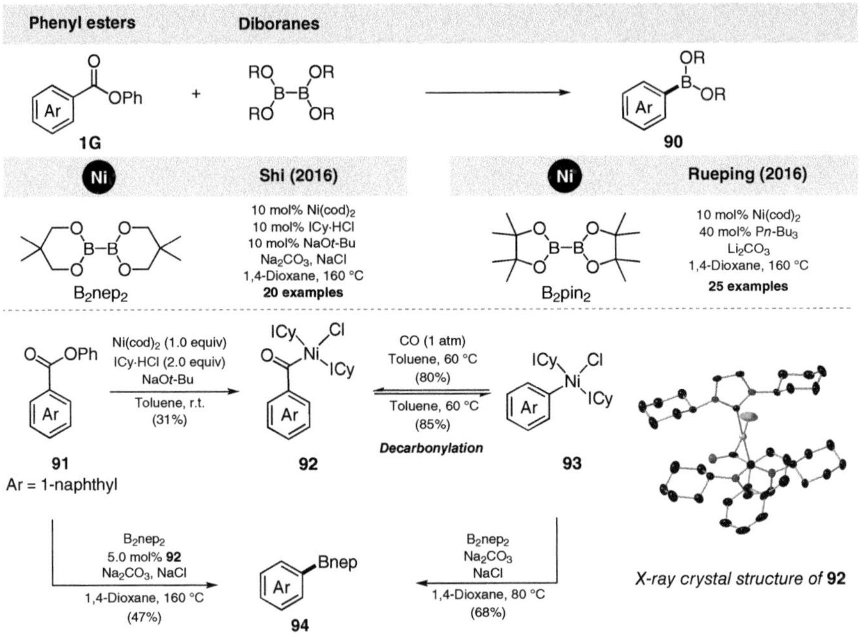

Scheme 15.30 Ni-catalyzed decarbonylative borylation of aromatic esters **1G**. Source: Adapted from Pu et al. [21].

giving aryl neopentylborates **90** in generally good yields. This is the first example involving an NHC ligand in the decarbonylative coupling of aromatic esters. Concurrently, the Rueping group also reported the Ni-catalyzed decarbonylative borylation of aromatic esters **1G** [69]. They used Ni/P*n*-Bu_3 catalyst and B_2pin_2 as a borylating reagent, achieving the transformation of aromatic esters **1G** to aryl–Bpin **90**. Shi experimentally studied the mechanism of this borylation by the isolation and reaction of possible catalytic intermediates. Aromatic ester **91** was reacted with $Ni(cod)_2$/ICy to form aroyl–Ni complex **92**, the structure of which was confirmed by X-ray crystallographic analysis. Upon heating, **92** underwent decarbonylation to give aryl–Ni species **93**. Both nickel complexes **92** and **93** were suggested as catalytic intermediates, as the catalytic reaction of **91** with B_2nep_2 by using **92** as a catalyst, and as a stoichiometric reaction of **93**; both afforded aryl–Bnep **94**.

An *in situ* anhydride formation strategy was also utilized to achieve borylation (Scheme 15.31). Hong and Szostak reported a Pd-catalyzed decarbonylative borylation of arenecarboxylic acids **1D** with B_2pin_2 in the presence of pivalic anhydride [70]. This decarbonylative borylation of **1D** was applied to the late-stage borylation of pharmaceutical molecules such as probenecid, diflufenican, and an estrone derivative.

Although the decarbonylative silylation of aroyl chlorides had been reported in 1990, the reaction of aromatic esters was developed only recently [68]. In 2016, Rueping and Shi independently reported an Ni/Cu-catalyzed decarbonylative

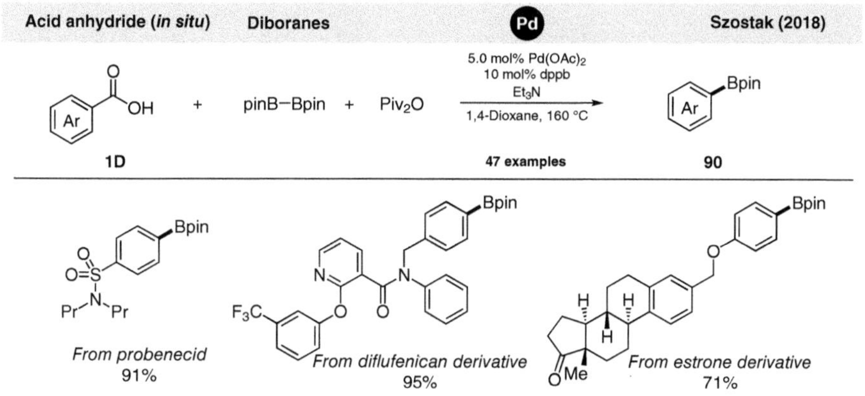

Scheme 15.31 Pd-catalyzed decarbonylative borylation of arylcarboxylic acids through an *in situ* anhydride formation.

silylation of aromatic esters **1G** with silylborane **95** as a silylating reagent (Scheme 15.32) [21, 69]. In this catalysis, the copper cocatalyst works to generate a silyl copper species capable of transmetalation with the nickel species.

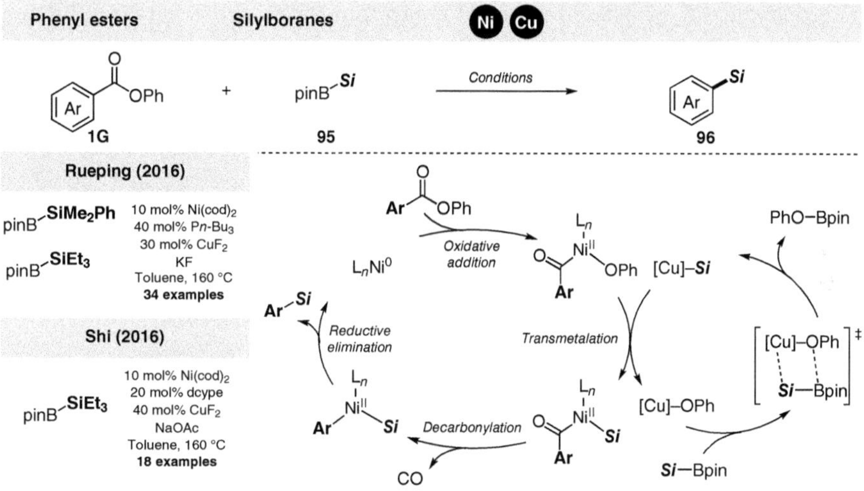

Scheme 15.32 Ni/Cu-catalyzed decarbonylative silylation of aromatic esters **1G**. Source: Adapted from Pu et al. [21] and Guo and Rueping [69].

Despite their toxicity, arylstannanes are still used in academic laboratories, and are often used not only for Migita–Kosugi–Stille coupling to form C–C bonds but also for C–N, C–F, and C–OCF_3 bond formations. A decarbonylative stannylation was reported in 2018 by the group of Rueping (Scheme 15.33) [71]. They developed an Ni/dppp-catalyzed decarbonylative stannylation of aromatic esters **1J** with silylstannanes **97**. Importantly, in addition to phenyl esters (**1G**), this reaction allowed for inactive alkylesters **1J** such as methyl, ethyl, cyclohexyl, and benzyl esters as substrates.

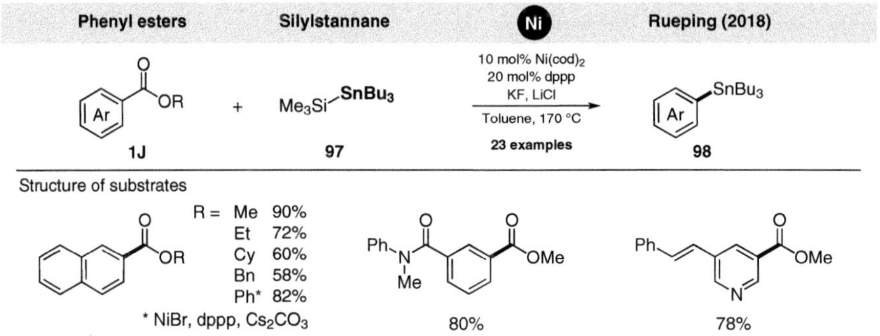

Scheme 15.33 Ni-catalyzed decarbonylative stannylation of aromatic esters. Source: Adapted from Yue et al. [71].

15.17 Miscellaneous

In this section, conceptually distinct metal-catalyzed transformations of esters involving decarbonylation are described.

Recently, metal-catalyzed functional group transfer reactions between different aryl electrophiles have gained attention as an emerging method to synthesize functionalized aromatic molecules without nucleophilic reagents [72]. The Yamaguchi group discovered an ester transfer reaction of aromatic esters **99** with haloarenes **100** by using Ni/dcypt catalyst, Zn^0, Na_2CO_3, and DMAP (Scheme 15.34) [73]. In this reaction, aryl nicotinate **99** works as an ester source. This reaction is thought to proceed through two oxidative additions of **99** and **100** to generate 3-pyridyl–Ni–OAr **102** and aryl–Ni–halogen **103**, respectively. These intermediates can undergo transmetalation to give aryl–Ni–OAr **104**, followed by carbonylation and reductive elimination to give ester transferred product **101**. The use of a metal catalyst allowing the cleavage of aryl–X bonds (in this case, X = halogen and esters) was quite important; otherwise, it resulted in no reaction.

The group also found that this catalytic system is applicable to arenol derivatives instead of haloarenes. For example, pivalate **105** and carbamate **106** can participate in this ester transfer reaction. Of note, simple arenol **108** can also be used in the reaction with aromatic ester **107** to form **109**. In this case, **107** occupies a dual role; one is to act as an ester source, and the other is to serve as an activator of the Ar–OH group to form ester **110**.

Translocation of substituents on arenes is a powerful method to allow for the access of positional isomers. Among such methods, the halogen dance reaction is recognized as a practical reaction [74]. The Yamaguchi group started a new entry in this area, as they reported a Pd/dcypt-catalyzed ester translocation reaction on an aromatic ring, namely, the "ester dance" reaction (Scheme 15.35) [75]. This reaction was suggested to proceed through the following mechanism: (i) an oxidative addition of the C–O bond of esters **112** to give aroyl–Pd species **114**; (ii) *ortho*-C–H cleavage to generate aryne–Pd or η^2-arene–Pd species **115**; (iii) protonation,

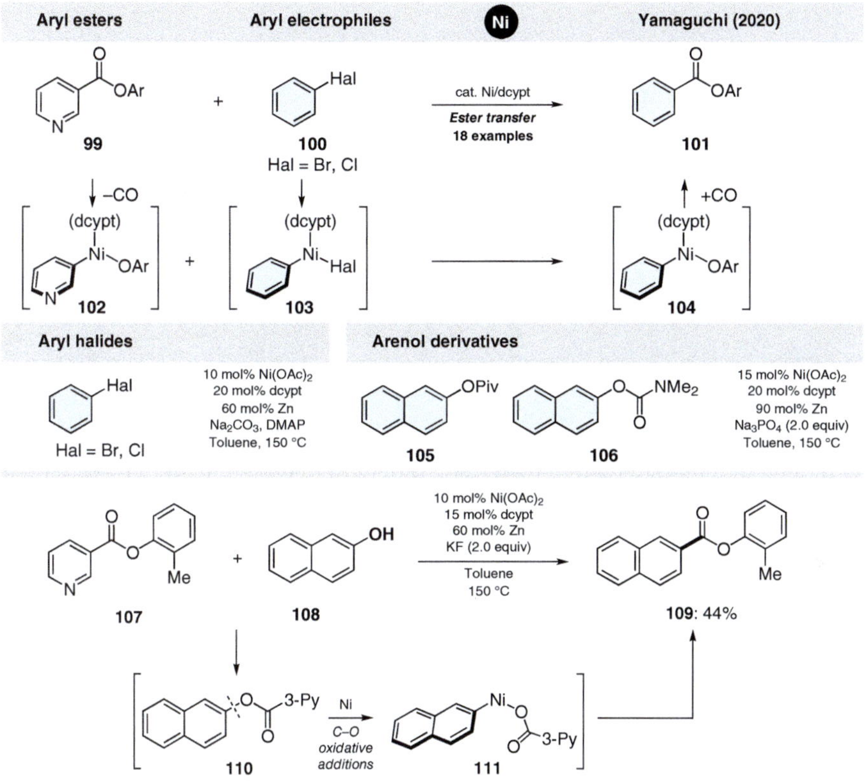

Scheme 15.34 Ni/dcypt-catalyzed ester transfer reaction. Source: Adapted from Isshiki et al. [73].

carbonylation, and reductive elimination to give ester-translocated product **113**. It was postulated that this catalytic cycle is reversible; the regioselectivity of this reaction is being controlled by the thermodynamic stability of the product. For example, phenyl 1-naphthoates underwent this reaction to give 2-naphthoates **116** and **117** in good yields. C4-substituted benzoate was also converted to C3-ester product **118** albeit in moderate yield. 4-Pyridyl ester reacted under the conditions to give nicotinate **119**. Moreover, an ester on C2-substituted nicotinate was translocated to the sterically less hindered C4 position (**120**) in a good yield.

Combining decarbonylative cross-couplings, the Yamaguchi group successfully achieved sequential ester dance/cross-coupling reactions (Scheme 15.36). For instance, the reaction of phenyl 3-thenoate (**121**) with benzothiazole (**43**) under the influence of Pd/dcypt and K_3PO_4 gave 2-benzothiazolylthiophene **122** in 60% yield. Biaryl **122** was likely generated through an ester dance reaction, followed by a decarbonylative C–H arylation. In another example, the combination of ester dance and decarbonylative amination was demonstrated. As mentioned in Scheme 15.35, ester dance of an isonicotinate furnished a nicotinate **119** as a major product (C4 to C3). However, for this amination, the ester dance reaction of **119** occurred in a reverse fashion (C3 to C4), followed by decarbonylative amination

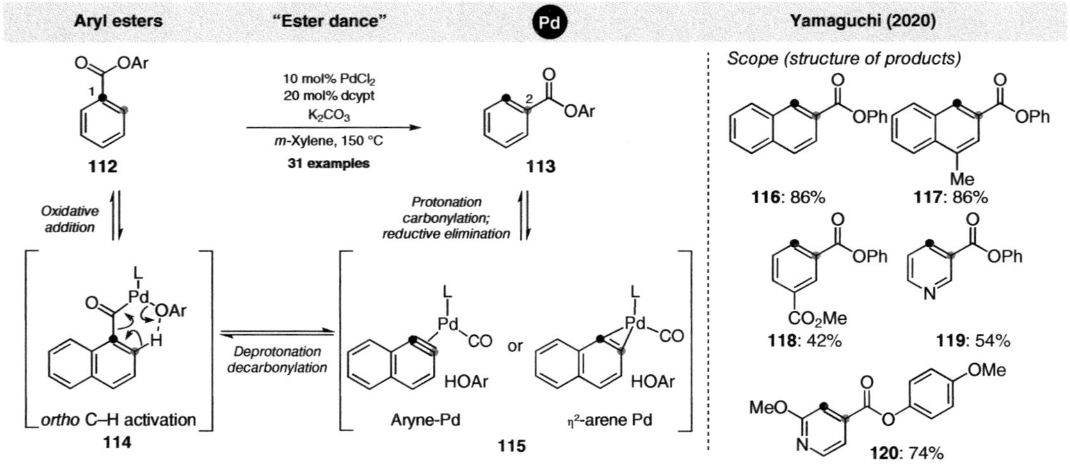

Scheme 15.35 Pd/dcypt-catalyzed "ester dance" reactions of aromatic esters. Source: Modified from Matsushita et al. [75].

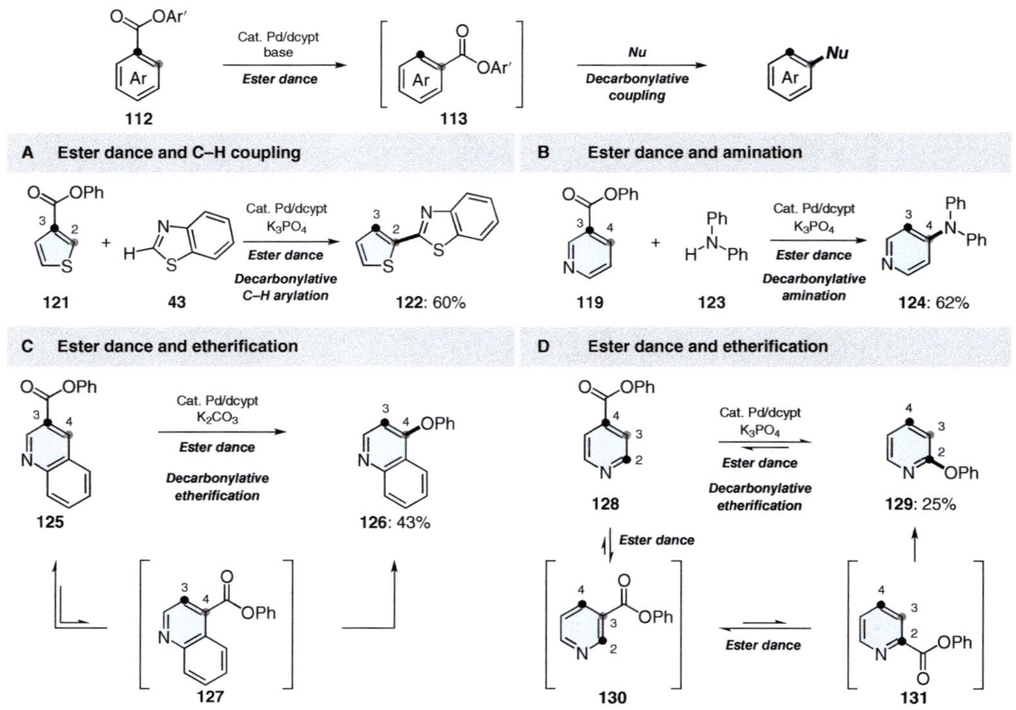

Scheme 15.36 Sequential ester dance and decarbonylative coupling.

with diphenylamine **123** to give 4-aminopyridine **124** in 62% yield. Moreover, when quinoline **125** was treated with Pd/dcypt catalysis, diarylether **126** was obtained by an ester dance and decarbonylative etherification sequence. Although possible intermediate **127** was thermodynamically less stable than **125**, the decarbonylative etherification occurred at only the C4 position, giving **126** as a sole product. As an interesting result, 1,3-translocation of esters was achieved by the reaction of phenyl isonicotinate (**128**) with Pd/dcypt catalyst, eventually furnishing 2-phenoxypyridine (**129**) as a major product albeit in 25% yield. This regioselectivity was also driven by the facile C–O bond reductive elimination at the C2 position of azines.

Strictly speaking, it is not a decarbonylative reaction, but this section also introduces a new type of metal-catalyzed transformation of aromatic esters. In 2020, Yamaguchi and coworkers reported a Pd/dcype-catalyzed deoxygenative coupling between aromatic esters **1G** with organophosphorus compounds **73** in the presence of sodium formate to give benzylphosphines **132** (Scheme 15.37) [76]. In this transformation, aromatic esters **1G** function as benzylating reagents. A variety of aromatic esters including pharmaceutical derivatives were applicable to this reaction. The mechanism of this reaction was proposed as follows: an oxidative addition of **1G** gives aroyl–Pd species **133**, which then reacts with **73** to generate aroyl phosphine species **134**. This intermediate is highly reactive to be directly hydrogenated by palladium and sodium formate to give **135**. Moreover, another process from **134** would be possible. Phosphorous compound **73** reacts with **134**,

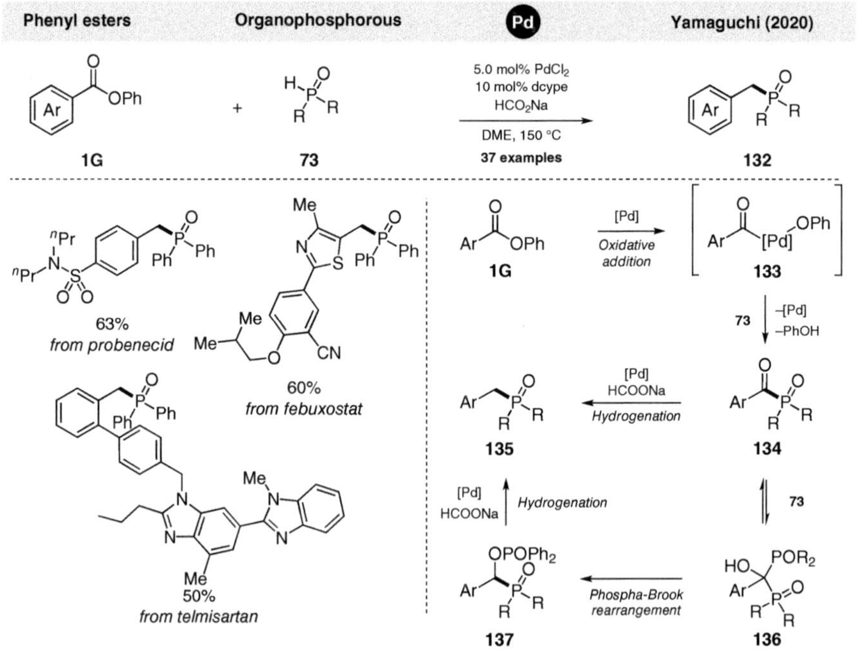

Scheme 15.37 Deoxygenative C–P bond formation of aromatic esters **1G**. Source: Modified from Kurosawa et al. [76].

followed by a phospha-Brook rearrangement to give intermediate **137**. Finally, under the action of a palladium catalyst and sodium formate, **137** is hydrogenated to afford benzylphosphine oxide. This mechanism was supported by several controlled experiments using all plausible intermediates.

15.18 Summary

Starting from the fundamental discovery by Yamamoto, aromatic esters have been utilized in various decarbonylative couplings to give functionalized arenes. These advancements have already allowed aromatic esters to be recognized as practical aryl electrophiles. Additionally, with the emergence of an ester transfer, an ester dance, and a deoxygenative coupling, this field is facing a major turning point. However, there still remain several drawbacks to be overcome. For example, most of these reactions require high temperature (generally over 140 °C) and carefully engineered C–O bond activation. With the development of more effective metal catalysts, it is highly anticipated that these drawbacks are addressed, and decarbonylative ester coupling finds wide use in various synthetic applications.

References

1 Miyaura, N. and Suzuki, A. (1995). *Chem. Rev.* 95: 2457–2483.
2 Rosen, B.M., Quasdorf, K.W., Wilson, D.A. et al. (2011). *Chem. Rev.* 111: 1346–1416.
3 Roglans, A., Pla-Quintana, A., and Moreno-Manãs, M. (2006). *Chem. Rev.* 106: 4622–4643.
4 Modha, S.G., Mehta, V.P., and Van der Eycken, E.V. (2013). *Chem. Soc. Rev.* 42: 5042–5055.
5 Takise, R., Muto, K., and Yamaguchi, J. (2017). *Chem. Soc. Rev.* 46: 5864–5888.
6 Lu, H., Yu, T.-Y., Xu, P.-F., and Wei, H. (2021). *Chem. Rev.* 121: 365–411.
7 Meng, G., Shi, S., and Szostak, M. (2016). *Synlett* 27: 2530–2540.
8 (a) Blum, J., Oppenheimer, E., and Bergmann, E.D. (1967). *J. Am. Chem. Soc.* 89: 2338–2341. (b) Verbicy, J.W., Dellacoletta, B.A., and Williams, L. (1982). *Tetrahedron Lett.* 23: 371–372. (c) Blaser, H.-U. and Spencer, A. (1982). *J. Organomet. Chem.* 233: 267–274.
9 (a) Osakada, K., Yamamoto, T., and Yamamoto, A. (1987). *Tetrahedron Lett.* 28: 6321–6324. (b) Wenkert, E. and Chianelli, D. (1991). *J. Chem. Soc., Chem. Commun.* 627–628. (c) Yamashita, F., Kuniyasu, H., Terao, J., and Kambe, N. (2008). *Org. Lett.* 10: 101–104. (d) Prokopcová, H. and Kappe, C.O. (2009). *Angew. Chem. Int. Ed.* 48: 2276–2286. (e) Ochiai, H., Uetake, Y., Niwa, T., and Hosoya, T. (2017). *Angew. Chem. Int. Ed.* 56: 2482–2486.
10 (a) Murahashi, S.-I., Naota, T., and Nakajima, N. (1986). *J. Org. Chem.* 51: 898–901. (b) Hirata, Y., Yada, A., Morita, E. et al. (2010). *J. Am. Chem. Soc.* 132: 10070–10077.

11 (a) Ishizu, J., Yamamoto, T., and Yamamoto, A. (1976). *Chem. Lett.* 5: 1091–1094. (b) Yamamoto, T., Ishizu, J., Kohara, T. et al. (1980). *J. Am. Chem. Soc.* 102: 3758–3764.
12 Stephan, M., Teunissen, A.J.J.M., Verziji, G.K.M., and de Vries, J.G. (1998). *Angew. Chem. Int. Ed.* 37: 662–664.
13 Gooßen, L.J., Paetzold, J., and Winkel, L. (2002). *Synlett* 10: 1721–1723.
14 Chatani, N., Tatamidani, H., Ie, Y. et al. (2001). *J. Am. Chem. Soc.* 123: 4849–4850.
15 Gooßen, L.J. and Paetzold, J. (2002). *Angew. Chem. Int. Ed.* 41: 1237–1240.
16 Gooßen, L.J. and Paetzold, J. (2004). *Angew. Chem. Int. Ed.* 43: 1095–1098.
17 Wang, J., Liu, B., Zhao, H., and Wang, J. (2012). *Organometallics* 31: 8598–8607.
18 Amaike, K., Muto, K., Yamaguchi, J., and Itami, K. (2012). *J. Am. Chem. Soc.* 134: 13573–13576.
19 Muto, K., Yamaguchi, J., Musaev, D.G., and Itami, K. (2015). *Nat. Commun.* 6: 7508.
20 Takise, R., Isshiki, R., Muto, K. et al. (2017). *J. Am. Chem. Soc.* 139: 3340–3343.
21 Pu, X., Hu, J., Zhao, Y., and Shi, Z. (2016). *ACS Catal.* 6: 6692–6698.
22 (a) Beletskaya, I.P. and Cheprakov, A.V. (2000). *Chem. Rev.* 100: 3009–3066. (b) Oestreich, M. (2009). *The Mizoroki–Heck Reaction*. Wiley.
23 Yu, W., Liu, L., Huang, T. et al. (2020). *Org. Lett.* 22: 7123–7128.
24 Nishii, Y. and Miura, M. (2020). *ACS Catal.* 10: 9747–9757.
25 Kajita, Y., Kurahashi, T., and Matsubara, S. (2008). *J. Am. Chem. Soc.* 130: 17226–17227.
26 Gooßen, L.J. and Paetzold, J. (2004). *Adv. Synth. Catal.* 346: 1665–1668.
27 Zhou, J.-Y., Liu, R.-Q., Wang, C.-Y., and Zhu, Y.-M. (2020). *J. Org. Chem.* 85: 14149–14157.
28 Liu, C., Ji, C.-L., Qin, Z.-X. et al. (2019). *iScience* 19: 749–759.
29 Iyori, Y., Takahashi, K., Yamazaki, K. et al. (2019). *Chem. Commun.* 55: 13610–13613.
30 LaBerge, N.A. and Love, J.A. (2015). *Eur. J. Org. Chem.* 2015: 5546–5553.
31 Muto, K., Hatakeyama, T., Itami, K., and Yamaguchi, J. (2016). *Org. Lett.* 18: 5106–5109.
32 Chatupheeraphat, A., Liao, H.-H., Srimontree, W. et al. (2018). *J. Am. Chem. Soc.* 140: 3724–3735.
33 Masson-Makdissi, J., Vandavasi, J.K., and Newman, S.G. (2018). *Org. Lett.* 20: 4094–4098.
34 Liu, X., Jia, J., and Rueping, M. (2017). *ACS Catal.* 7: 4491–4496.
35 Okita, T., Muto, K., and Yamaguchi, J. (2018). *Org. Lett.* 20: 3132–3135.
36 (a) Sonogashira, K. (2002). *J. Organomet. Chem.* 653: 46–49. (b) Chinchilla, R. and Nájera, C. (2011). *Chem. Soc. Rev.* 40: 5084–5121.
37 Okita, T., Kumazawa, K., Takise, R. et al. (2017). *Chem. Lett.* 46: 218–220.
38 (a) Bellina, F. and Rossi, R. (2010). *Chem. Rev.* 110: 1082–1146. (b) Johansson, C.C.C. and Colacot, T.J. (2010). *Angew. Chem. Int. Ed.* 49: 676–707. (c) Noväk, P. and Martin, R. (2011). *Curr. Org. Chem.* 15: 3233–3262.
39 Isshiki, R., Takise, R., Itami, K. et al. (2017). *Synlett* 28: 2599–2603.

40 Yan, G., Zhang, Y., and Wang, J. (2017). *Adv. Synth. Catal.* 359: 4068–4105.

41 Chatupheeraphat, A., Liao, H.-H., Lee, S.-C., and Rueping, M. (2017). *Org. Lett.* 19: 4255–4258.

42 Iizumi, K., Kurosawa, M.B., Isshiki, R. et al. (2021). *Synlett* 32: 1555–1559.

43 (a) Ackermann, L., Vicente, R.N., and Kapdi, A.R. (2009). *Angew. Chem. Int. Ed.* 48: 9792–9826. (b) Chen, X., Engle, K.M., Wang, D.-H., and Yu, J.-Q. (2009). *Angew. Chem. Int. Ed.* 48: 5094–5115. (c) Alberico, D., Scott, M.E., and Lautens, M. (2007). *Chem. Rev.* 107: 174–238. (d) Yamaguchi, J., Yamaguchi, A.D., and Itami, K. (2012). *Angew. Chem. Int. Ed.* 51: 8960–9009. (e) Wencel-Delord, J. and Glorius, F. (2013). *Nat. Chem.* 5: 369–375. (f) Meng, G., Lam, N.Y.S., Lucas, E.L. et al. (2020). *J. Am. Chem. Soc.* 142: 10571–10591.

44 Jin, W., Yu, Z., He, W. et al. (2009). *Org. Lett.* 11: 1317–1320.

45 Pan, F., Lei, Z.-Q., Wang, H. et al. (2013). *Angew. Chem. Int. Ed.* 52: 2063–2067.

46 Kruckenberg, A., Wadepohl, H., and Gade, L.H. (2013). *Organometallics* 32: 5153–5170.

47 Matsushita, K., Takise, R., Hisada, T. et al. (2018). *Chem. Asian J.* 13: 2393–2396.

48 Meng, L., Kamada, Y., Muto, K. et al. (2013). *Angew. Chem. Int. Ed.* 52: 10048–10051.

49 Nagatsu, A., Kajitani, H., and Sakakibara, J. (1995). *Tetrahedron Lett.* 36: 4097–4100.

50 Wipf, P. and Venkatraman, S. (1996). *J. Org. Chem.* 61: 6517–6522.

51 (a) Selva, E., Beretta, G., Montanini, N. et al. (1991). *J. Antibiot.* 44: 693–701. (b) Kettenring, J., Colombo, L., Ferrari, P. et al. (1991). *J. Antibiot.* 44: 702–715. (c) Selva, E., Ferrari, P., Kurz, M. et al. (1995). *J. Antibiot.* 48: 1039–1042.

52 (a) Nicolaou, K.C., Zou, B., Dethe, D.H. et al. (2006). *Angew. Chem. Int. Ed.* 45: 7786–7792. (b) Nicolaou, K.C., Dethe, D.H., Leung, G.Y. et al. (2008). *Chem. Asian J.* 3: 413–429.

53 Amaike, K., Itami, K., and Yamaguchi, J. (2016). *Chem. Eur. J.* 22: 4384–4388.

54 (a) Campeau, L.-C., Parisien, M., Jean, A., and Fagnou, K. (2006). *J. Am. Chem. Soc.* 128: 581–590. (b) Wang, C., Piel, I., and Glorius, F. (2009). *J. Am. Chem. Soc.* 131: 4194–4195. (c) Liégault, B., Lee, D., Huestis, M.P. et al. (2008). *J. Org. Chem.* 73: 5022–5028.

55 Maetani, S., Fukuyama, T., and Ryu, I. (2013). *Org. Lett.* 15: 2754–2757.

56 Okita, T., Komatsuda, M., Saito, A.N. et al. (2018). *Asian J. Org. Chem.* 7: 1358–1361.

57 Yue, H., Guo, L., Liao, H.-H. et al. (2017). *Angew. Chem. Int. Ed.* 56: 4282–4285.

58 Malapit, C.A., Borrell, M., Milbauer, M.W. et al. (2020). *J. Am. Chem. Soc.* 142: 5918–5923.

59 Hirao, T., Masunaga, T., Ohshiro, Y., and Agawa, T. (1980). *Tetrahedron Lett.* 21: 3595–3598.

60 Isshiki, R., Muto, K., and Yamaguchi, J. (2018). *Org. Lett.* 20: 1150–1153.

61 Liu, C., Ji, C.-L., Zhou, T. et al. (2019). *Org. Lett.* 21: 9256–9261.

62 Zhang, J.-S., Chen, T., and Han, L.-B. (2020). *Eur. J. Org. Chem.* 2020: 1148–1153.

63 Lee, S.-C., Liao, H.-H., Chatupheeraphat, A., and Rueping, M. (2018). *Chem. Eur. J.* 24: 3608–3612.

64 (a) Ichiishi, N., Malapit, C.A., Woźniak, Ł., and Sanford, M.S. (2018). *Org. Lett.* 20: 44–47. (b) Liu, C. and Szostak, M. (2018). *Chem. Commun.* 54: 2130–2133. (c) Ishitobi, K., Isshiki, R., Asahara, K.K. et al. (2018). *Chem. Lett.* 47: 756–759. (d) Brigham, C., Malapit, C., Lalloo, N., and Sanford, M.S. (2020). *ACS Catal.* 10: 8315–8320.

65 Tatamidani, H., Yokota, K., Kakiuchi, F., and Chatani, N. (2004). *J. Org. Chem.* 69: 5615–5621.

66 Yue, H., Guo, L., Lee, S.-C. et al. (2017). *Angew. Chem. Int. Ed.* 56: 3972–3976.

67 (a) Ishiyama, T. and Miyaura, N. (2004). *Chem. Rec.* 3: 271–280. (b) Hartwig, J.F. (2012). *Acc. Chem. Res.* 45: 864–873.

68 Krafft, T.E., Rich, J.D., and McDermott, R.J. (1990). *J. Org. Chem.* 55: 5430–5432.

69 Guo, L. and Rueping, M. (2016). *Chem. Eur. J.* 22: 16787–16790.

70 Liu, C., Ji, C.-L., Hong, X., and Szostak, M. (2018). *Angew. Chem. Int. Ed.* 57: 16721–16726.

71 Yue, H., Zhu, C., and Rueping, M. (2018). *Org. Lett.* 20: 385–388.

72 (a) Lee, Y.H. and Morandi, B. (2018). *Nat. Chem.* 10: 1016–1022. (b) De La Higuera Macias, M. and Arndtsen, B.A. (2018). *J. Am. Chem. Soc.* 140: 10140–10144.

73 Isshiki, R., Inayama, N., Muto, K., and Yamaguchi, J. (2020). *ACS Catal.* 10: 3490–3494.

74 Schnürch, M., Spina, M., Khan, A.F. et al. (2007). *Chem. Soc. Rev.* 36: 1046–1057.

75 Matsushita, K., Takise, R., Muto, K., and Yamaguchi, J. (2020). *Sci. Adv.* 6: eaba7614.

76 Kurosawa, M.B., Isshiki, R., Muto, K., and Yamaguchi, J. (2020). *J. Am. Chem. Soc.* 142: 7386–7392.

Index

Amide Bond Activation: Concepts and Reactions, First Edition. Edited by Michal Szostak.

d

o

p

t